The Collected Works of George E. P. Box

Volume I

The Wadsworth Statistics/Probability Series

Series Editors

Richard A. Becker and John M. Chambers
S: An Interactive Environment for Data Analysis and Graphics

Peter J. Bickel, Kjell Doksum, and John L. Hodges, Jr.
Festschrift for Erich L. Lehmann

George E. P. Box
The Collected Works of George E. P. Box
Volumes I and II, edited by George C. Tiao

Leo Breiman, Jerome H. Friedman, Richard A. Olshen, and Charles J. Stone
Classification and Regression Trees

John M. Chambers, William S. Cleveland, Beat Kleiner, and Paul A. Tukey
Graphical Methods for Data Analysis

Franklin A. Graybill
Matrices with Applications in Statistics, Second Edition

John W. Tukey
The Collected Works of John W. Tukey
Volume I: *Time Series, 1949–1964,* edited by David R. Brillinger
Volume II: *Time Series, 1965–1984,* edited by David R. Brillinger

The Collected Works of George E. P. Box

Volume I

Editor in Chief

George C. Tiao
University of Chicago

Editors

C. W. J. Granger
University of California, San Diego

Irwin Guttman
University of Toronto

Barry H. Margolin
National Institute of Environmental Health Sciences

Ronald D. Snee
E. I. DuPont de Nemours & Co.

Stephen M. Stigler
University of Chicago

Wadsworth Advanced Books & Software
Belmont, California
A Division of Wadsworth, Inc.

Acquisitions editor: John Kimmel
Production editor: Mary Roybal
Cover and jacket: Lois Stanfield

Printed in the United States of America

1 2 3 4 5 6 7 8 9 10 — 89 88 87 86 85

Acknowledgments

Articles 1.12, 1.13, 2.7 from the *Journal of the American Statistical Association* and articles 1.8, 1.9, 1.10, 2.9, 2.10, 2.11 from *Technometrics* are reprinted by permission from the AMERICAN STATISTICAL ASSOCIATION. Articles 1.4, 2.3, 2.4 are reprinted with permission from the BIOMETRIC SOCIETY. Articles 1.1, 1.2, 1.5, 1.6, 1.7, 1.11, 1.14, 2.2, 2.6, 2.12 are reprinted with permission of the BIOMETRIKA TRUST. Articles 2.5, 2.8 are reprinted with permission of the INSTITUTE OF MATHEMATICAL STATISTICS. Article 2.13 is reprinted with permission of INTERNATIONAL BUSINESS MACHINES CORPORATION. Article 1.16 from *Applied Statistics* and articles 1.3, 1.15, 2.1, 2.14 from the *Journal of Royal Statistical Society* are reprinted with permission from the ROYAL STATISTICAL SOCIETY.

Library of Congress Cataloging in Publication Data

Box, George E. P.
The collected works of George E. P. Box.

(The Wadsworth statistics/probability series)
Includes bibliographies.
1. Mathematical statistics—Collected works. I. Tiao, George C., 1933- . II. Stigler, Stephen M. III. Title. IV. Series.
QA276.A12B6825 1984 519.5 84-17310
ISBN 0-534-03307-5 (v. 1)
ISBN 0-534-03308-3 (v. 2)

ISBN 0-534-03307-5

Contents

Volume II

Preface

The writings of George E. P. Box over the last thirty-five years have had a major impact on statistical theory and practice. His contributions cover a wide spectrum of topics, including statistical inference, robustness, time series analysis, design of experiments, response surface methodology, and evolutionary operation. He is a true believer in statistical methods as necessary tools in scientific investigation. To him, development of sound statistical theory and methods must go hand in hand with practice, and a statistician should be a genuine partner in scientific investigation, in the course of which his tools are judiciously applied and are modified and expanded as needs arise.

George Box's philosophy on statistics is amply reflected in the varied contributions he has made to the profession. These contributions include his books and articles on several major areas of statistics, some of which he himself initiated; the founding of a statistics department noted for its healthy blending of theory and practice in teaching and research; and his work as a consultant to industry, which provides a source, as well as an outlet, for his tools.

The collected works of Box presented in these volumes cover a major portion of his articles. The papers are divided into the following five main parts:

1. Statistical Inference, Robustness, and Modeling Strategy
2. Experimental Design and Response Surface Methodology
3. Time Series Analysis and Forecasting
4. Distribution Theory, Transformation of Variables, and Nonlinear Estimation
5. Application of Statistics

For each area, a member of the board of editors (Granger, Guttman, Margolin, Snee, Stigler, and Tiao) has selected articles as well as written an introduction to the articles highlighting their basic ideas, major methodological developments, and impact. A complete bibliography of the books and articles written by Box as of 1983 is given at the end of each volume.

On a personal note, as one of Box's students some twenty years ago and a long-time colleague, I am more pleased than I can say to see the publication of this collection. It represents a token tribute to his accomplishments, but, more important, it provides a perspective on the origins and development of his ideas. I wish to thank the other five editors for the time they have generously contributed to make publication possible, and David M. Steinberg and Chung Chen for their assistance in the preparation of the work. Finally, thanks are owed to many journals for allowing these articles to be reprinted.

George C. Tiao
University of Chicago

January 1984

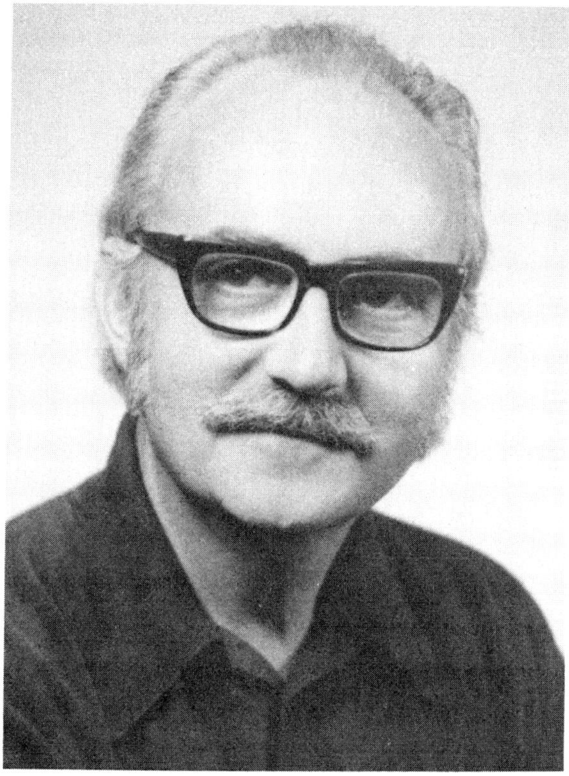

George E. P. Box

Biography

George E.P. Box was born October 18, 1919, in Gravesend, England, and did undergraduate work in chemistry at the University of London. While serving in the British Army during World War II, he was involved in experimental work to combat the effects of chemical weapons that required the use of statistics and the design of experiments. His formal education in statistics resulted in a Bachelor of Science degree in Mathematical Statistics in 1947 and a Doctor of Philosophy in Mathematical Statistics in 1952, both from the University of London. He also received a Doctor of Science degree from the University of London in 1961 and a honorary Doctor of Science degree from the University of Rochester in 1975.

Box began his career in statistics as Statistician and Head of the Statistical Techniques Research Section at Imperial Chemical Industries in England in the early- and mid-1950s. He took a leave of absence from Imperial Chemical Industries from 1953 to 1954 to accept a position as visiting Research Professor at the University of North Carolina and later returned to the United States as Director of the Statistical Techniques Research Group at Princeton University in 1957. He moved to the University of Wisconsin, Madison, in 1960, where he was founder, Professor, and Chairman of the Statistics Department until 1965. In 1965–1966, Box was a visiting Ford Foundation Professor at Harvard Business School. He returned to the University of Wisconsin as Professor and Chairman of the Statistics Department in 1966, leaving again in 1970 to become visiting Professor at the University of Essex in Colchester, England. In 1971, he was appointed to the newly created Ronald Aylmer Fisher Chair of Statistics at the University of Wisconsin, Madison, where he was appointed Vilas Research Professor of Mathematics and Statistics in 1980.

Dr. Box has published more than 110 research papers and is a coauthor of six books. The importance of his work has been recognized by the bestowal of many major awards, among them the British Empire Medal, the Royal Statistical Society Guy Medal in Silver, the Smith Reynolds Teaching Award from the University of Wisconsin, the Wilks Memorial Medal (ASA and U.S. Army), the American Society for Quality Control Shewhart Medal, and the American Institute of Chemical Engineers Professional Progress Award. He is a Fellow of the Institute of Mathematical Statistics, the American Academy of Arts and Sciences, the American Statistical Association, the Royal Statistical Society, and the American Society for Quality Control, as well as a member of the International Statistical Institute and the Biometrics Society. He has also served as president of the American Statistical Association and the Institute of Mathematical Statistics.

Few persons have had a greater impact on the use of statistics in science and engineering. Dr. Box is the principal architect of response surface methods and evolutionary operation and has made major contributions to experimental design, time series analysis, inference, nonlinear estimation, and robustness. In each instance, his contributions were stimulated by practical problems. He has served as a consultant to American Cyanamid, Monsanto Company, the World Bank, Pillsbury Company, and the Federal Reserve Board.

Dr. Box is also known worldwide for his pioneering work with the late Dr. Gwilym Jenkins on time series analysis. The Box-Jenkins approach, as it is fondly referred to, has been found useful in the modeling and control of industrial processes, economic forecasting, inventory management, and the analysis of environmental data. In addition to his professional activities, Dr. Box has many other interests. He is also known for his story telling, song writing, acting, poetry reading, and his interest in politics.

1

Statistical Inference, Robustness, and Modeling Strategy

Contents

1.11 "A Bayesian Approach to Some Outlier Problems" (with G. C. Tiao). *Biometrika,* vol. 55, no. 1 (1968), pp. 119–129 179

1.12 "Science and Statistics." *J. Amer. Stat. Assoc.,* vol. 71, no. 356 (1976), pp. 791–799 191

1.13 "Some Problems of Statistics and Everyday Life." *J. Amer. Stat. Assoc.,* vol. 74, no. 365 (1979), pp. 1–4 201

1.14 "Bayesian Analysis of Some Outlier Problems in Time Series" (with B. Abraham). *Biometrika,* vol. 66, no. 2 (1979), pp. 229–236 205

1.15 "Sampling and Bayes' Inference in Scientific Modelling and Robustness." *J. Roy. Stat. Soc.,* Series A, vol. 143, part 4 (1980), pp. 383–430 213

1.16 "Linear Models and Spurious Observations" (with B. Abraham). *Applied Statistics,* vol. 27, no. 2 (1978), pp. 131–138 261

1.0
Introduction

STEPHEN M. STIGLER
University of Chicago

Philosophy plays a more conscious role in modern statistics than in any other science: A chemist or a physicist can generally fulfill a routine research program without worrying about the fundamental nature of matter, but a statistician cannot even calculate an interval (confidence, fiducial, or posterior) without coming face to face with disputed questions about the nature of inference. This consciousness has been slow in developing. Before this century, few scientists thought seriously about the foundations of inference and its interaction with statistical methodology — only Venn and Edgeworth come quickly to mind among nineteenth-century Englishmen; however, since Fisher all that has changed. Today virtually all practicing research statisticians are acutely aware of the major philosophical issues confronting their area of specialization and have thought about the methodological principles underlying their approach. What other science can make such a claim?

In the years since he first acquired the title "statistician" during World War II, no one has done more than George Box to influence practical statistical philosophy. That part of Box's work presented here is rich and varied. It encompasses his strikingly influential early work on the nature and importance of assumptions in classical analyses, several penetrating essays on the principles of experimentation and model building, the germs of his contributions to a practical Bayesian methodology (with particular emphasis on its use for checking assumptions), and several major statements about his scientific perspective. And yet, despite this variety, there are several common threads to be found. The most prevalent one is the insistence that a statistical investigation is a scientific investigation, with two corollaries. First, we should be certain our statistical technique addresses the question of real scientific interest (and not an irrelevant one imposed by some aspect of the mathematical model that is tangential to the main issue). Second, our conclusions must be expressed in terms that comment on the question posed. Put this way, Box's criteria sound like platitudes. Who could possibly disagree with the mandate that we should address a real question and try to answer it? Yet Box's work teaches us again that what can be accepted as near tautology when expressed in general terms may be far more difficult to grasp at a more specific level.

The earliest work represented here (1.1) was done in a classical mode, building on investigations of E. S. Pearson and Geary. Those earlier workers had noted the influence of non-normality on tests of comparing variances, but in Box's hands the topic was transformed. As early as in his 1953 paper (1.1), he introduced the felicitous term "robustness" to statistics, and it has stuck with us ever since, because Box saw robustness as a general problem and diagnosed its causes and prescribed solutions, where his predecessors had mostly been content with observing the phenomenon in isolated cases. Starting by evaluating Bartlett's test for heterogeneity of variance in the presence of non-normal errors, Box arrived at the then startling idea that Bartlett's test was far more effective as a test *of* normality. One casualty of his work was the common statistical procedure of applying Bartlett's test as a preliminary step to performing the much more robust analysis-of-variance test. As Box put it, "To make the preliminary test on variances is rather like putting to sea in a rowing boat to find out whether conditions are sufficiently calm for an ocean liner to leave port!"

By 1955, Box (with S. L. Andersen [1.3]), in an important investigation of permutation tests, provided us with a working definition of "robust" as fulfilling the second of these two requirements: "Statistical criteria should (1) be sensitive to change in the specific factors tested, (2) be insensitive to changes, of a magnitude likely to occur in practice, in extraneous factors." This notion was singled out for special comment by Barnard in his contribution to th discussion, where he predicted "the idea will have a long and vigorous life." The idea has indeed proved so fruitful that subsequent work has almost amounted to an industry! Box himself contributed one more work in the classical vein (with G. S. Watson [1.5]) on the robustness of regression tests and the influence of design upon robustness, but most of his own subsequent work has taken a somewhat different slant. Before moving on, however, it should be noted that one major reason for rereading these early papers has nothing whatsoever to do with philosophy: In all cases they display exquisite mathematical technique, particularly through the use of expansions for the study of the distributions of sums of squares. For technical excellence alone they reward attention.

Starting in the early 1960s, in joint work with G. C. Tiao, Box moved to a consideration of the practical implications of Bayesian analysis. This work was to occupy a major portion of Box's attention for much of the following decade, culminating in the full-length 1973 monograph *Bayesian Inference in Statistical Analysis,* with Tiao. Because several of the papers of this period are subsumed by that book, they are not included; however, the earliest papers are reprinted to show the continuity in the development of Box's thought. For it is clear from even the first paper with Tiao (1.6) that Bayesian inference was valued primarily as a liberating device, a way of achieving robustness and model flexibility without sacrificing power. This aspect had not been stressed by Box's predecessors, but it was to become a key focus in the work that followed. For example, it permitted economical ways of dealing with outliers (with Tiao [1.11] and with Abraham [1.14]), treating them as intrinsic to the model as opposed to the then conventional (and much less satisfactory) view that outliers were wholly extrinsic. (One tested for "one outlier" as one might try to determine on the basis of examination scores alone whether a class of students contained "one football player," rather than conceiving of the possibility that athletic scholarships were a standard part of school policy and taking the whole student body as the object of interest.) The simple fruitful idea of Box and Tiao of enlarging the family of normal densities

$$p(y \mid \theta, \sigma) = k \exp\left\{-\frac{1}{2}\left[\frac{y-\theta}{\sigma}\right]^2\right\}$$

to the exponential power family

$$p(y \mid \theta, \sigma, \beta) = k \exp\left\{-\frac{1}{2}\left[\frac{y-\theta}{\sigma}\right]^{2/(1+\beta)}\right\}$$

is already found in the 1962 paper (and foreshadowed in Box's 1953 note on kurtosis [1.2]). The further step to Bayesian analyses in binomial mixture models came in 1968 and is still shedding considerable light in unexpected ways upon Tukey's biweight estimate (see Box's 1980 J.R.S.S. discussion paper [1.15]).

Two minor gems in this collection are the note (with Tiao [1.7]) on "criterion" and "inference" robustness, and the note (1.9) on the "use and abuse of regression." The first of these makes the important and neglected point that as we entertain different distributional hypothesis in a study of robustness, we should not maintain a static criterion, but should let the criterion (e.g., a loss function) vary so as to be appropriate to the distribution in question. For example, if the errors are really Cauchy, squared error loss is an absurd criterion. I think the terminology used in this note is misleading and may have contributed to its not receiving its due attention ("criterion robustness" is used for robustness with respect to a static criterion when we would, I think, expect it to mean robustness with respect to changing criteria). But the *concept* is extremely novel and important, and this republication may help call renewed attention to it. The second note is distinguished in particular for the way it says what it says. Books have been written on the pitfalls of latent variables in regression, but Box captured the essence in one word: They are "lurking" variables! This paper also contains Box's oft-quoted admonition, "To find out what happens to a system when you interfere with it you have to interfere with it (not just passively observe it)." As a guide to practice (particularly in social science), this directive may be too stringent, but it does express an important truth.

Another of Box's areas represented here is his approach to experimentation, with his emphasis upon its iterative nature and with special attention to the study of mechanistic models. The fundamental ideas are already present and well expressed in the 1955 paper with Youle (1.4); section 9 of that paper contains as succinct and wise a discussion of experimental strategy as the statistical literature can offer. Box's later papers enlarge upon different aspects of his approach. In 1965 (with W. G. Hunter [1.8]) we find the importance of diagnostics being forcefully argued; a good presentation of the insights available from Bayesian analysis into the choice of a parameterization is also contained in this paper. In 1967 (with W. J. Hill [1.10]) the topic is the use of information metrics and sequential experimentation for discrimination between different mechanistic models. But always the main point is the use of sound statistical practice to understand, even to discover, the basic physical laws of a system.

The 1980 J.R.S.S. paper (1.15) is a landmark as an attempt to synthesize Bayesian model fitting with non-Bayesian model criticism — a formal combination of the two elements Box has stressed almost from the beginning. The paper brings together many elements of vintage Box, and the discussion shows that his ideas have lost none of their power to stimulate and excite.

One minor thread running through these papers is Box's use of schematic diagrams as tools for developing and expressing his ideas. Box's own mode of mathematical thought is strongly geometric — the 1953 note (1.2) on tests of kurtosis illustrates this neatly. And so it is natural for him to appeal to geometric diagrams to express his experimental philosophy as well. The best known of these diagrams, showing iteration between theory and experiment, between criticism and estimation, between deduction and induction (a sort of scientific version of the game of Ping-Pong), dates from the 1955 paper with Youle (1.4); the most ornate elaboration of this idea, a Rube Goldbergesque rendition of the entire process of scientific investigation, is in the 1974 Fisher lecture (published in J.A.S.A. in 1976 [1.12]). My own view is that these schematics were more helpful to Box in framing and developing his ideas than they are in expressing the ideas to others, though there can be no overestimating the force of the ideas. His use of graphical devices to capture messages in data is far more telling. Box's graphs are simple in execution, usually nothing more than dot diagrams or sketches of densities or histograms or contour plots, but they can be devastating in consequence. My personal favorite is the joint posterior of a regression coefficient and an autoregressive parameter that he gave as Figure 3 in the 1980 J.R.S.S. paper. That one picture tells more about the effects of correlation on analysis than any combination of regression textbooks on the market. As we enter an age of computer graphics, where cheap graphics often seem to be intended primarily for mere decoration and all too frequently a thousand pictures are worth one word, we should pay renewed attention to Box's statistical drawings as magnificent examples of clear thought expressed clearly.

Finally, as I reread and reconsider this segment of Box's work, I am struck by the elegance of his writing style and its aptness to the subject. His later major addresses — the 1974 Fisher lecture (reprinted in 1.12), the 1978 A.S.A. presidential address published in J.A.S.A., in 1979 [1.13]), and the more technical 1980 J.R.S.S. discussion paper (1.15) — are prime exemplars. They are direct, clear, provocative, learned, contentious, and irreverent (except where Fisher is concerned). They are the words of a scientist who profoundly believes in the worth of his science, who cares deeply about what he is doing and knows he does it well, and who has fun learning and teaching others the fruits of this learning.

1.1
Non-Normality and Tests on Variances

G. E. P. BOX
Imperial Chemical Industries Ltd. and University of North Carolina

1. Introduction

For many experimenters the most commonly used statistical tests are those for *comparing* sample means and sample variances. The test used for the equality of the means of k groups of observations is usually the analysis of variance test (or equivalently the t test when $k = 2$), whilst the test used for the equality of variances of k groups of observations is Bartlett's (1937) modification of the Neyman-Pearson (1931) L_1 test (or the equivalent F test when $k = 2$).

The tests mentioned are derived on a number of assumptions, in particular, that the observations are normally distributed. Usually, however, since little is known of the populations from which the samples are drawn, these tests are used, of necessity, as if the assumption of normality could be ignored. So far as comparative tests on means are concerned it appears (perhaps rather surprisingly) that this practice is largely justifiable, for thanks to the work of Pearson (1931), Bartlett (1935), Geary (1947), Gayen (1950 a, b), David & Johnson (1951 a, b) there is abundant evidence that these *comparative* tests on means are remarkably insensitive to general* non-normality of the parent population.

It would appear, however, that this remarkable property of 'robustness' to non-normality which these tests for comparing means possess, and without which they would be much less appropriate to the needs of the experimenter, is not necessarily shared by other statistical tests, and in particular is not shared by the tests for equality of variances, mentioned above.

The sensitivity to non-normality of the tests for comparing *two* variances was first pointed out by E. S. Pearson (1931) whose findings were confirmed by Geary (1947), Finch (1950) and Gayen (1950a). These authors showed that this test is particularly sensitive to changes in γ_2 from the normal theory value of zero. (The notation γ_1, γ_2, etc., will be used for the standardized parent cumulants. In Karl Pearson's notation $\gamma_1 = \surd\beta_1$, $\gamma_2 = \beta_2 - 3$.) In the present paper it is shown that the sensitivity is even greater when the number of variances to be compared exceeds two, and, indeed, that the sensitivity to non-normality of the L_1 criterion or the equivalent Bartlett test can be of the same order of magnitude as the sensitivity of criteria such as b_2 specifically designed to *test* normality.

It is further shown that the difference in sensitivity of the test on means on the one hand and the test on variances on the other arises from an essential difference in the nature of these two types of tests and this suggests how the extreme sensitivity to non-normality of the variance tests can be remedied.

* By 'general' parent non-normality is meant that the departure from normality, in particular skewness, is the same in the different groups, as could usually be assumed when the data were from an experiment in which the groups corresponded with different applied treatments to be compared. In tests in which sample means are *compared*, general skewness tends to be cancelled out; larger effects are found, however, if the skewness is in different directions in the different groups.

The Collected Works of George E. P. Box, 1984, Wadsworth, Inc., Belmont, CA 94002.
Originally published in *Biometrika,* vol. 40, parts 3 and 4 (1953), pp. 318–335.

2. Distribution of variance test for large samples

Suppose there are k groups of independently distributed observations with n_t in the tth group and $\sum_t n_t = N$. The ith observation of the tth group is denoted by y_{ti} and the usual estimate of variance in the tth group, having $\phi_t = n_t - 1$ degrees of freedom, by s_t^2. The average of the estimated variances is denoted by s^2, where $s^2 = \sum_t \phi_t s_t^2/\Phi$ and Φ is the total number of degrees of freedom $\Sigma\phi_t = \Phi$.

The L_1 criterion due to Neyman & Pearson (1931) tests the hypothesis that the *variances* in the groups are all equal, given that the observations are normally distributed. Alternatively, M_1, the logarithmic modification due to Bartlett (1937), may be employed:

$$M_1 = \Phi \ln s^2 - \sum_t \phi_t \ln s_t^2. \tag{1}$$

Following Neyman & Pearson (1931) it may be shown that when the appropriate null hypothesis is true and on the assumption of parent normality, M_1 is distributed in large samples as χ^2_{k-1} (where χ^2_{k-1} means χ^2 with $k-1$ degrees of freedom). This is the basis of the test proposed by Bartlett (1937). He shows that, even for fairly small samples, to a close approximation M_1 is distributed as $(1+A)\chi^2_{k-1}$, where A, an adjustable constant which tends to zero for large values of ϕ_t, is given by

$$A = \frac{1}{3(k-1)}\left\{\sum_t \frac{1}{\phi_t} - \frac{1}{\Phi}\right\}. \tag{2}$$

It is convenient to define the quantity

$$d = \left(1 + \frac{1}{2}\frac{\phi}{\phi+1}\gamma_2\right)^{-1}. \tag{3}$$

When ϕ tends to infinity this becomes

$$\delta = (1 + \tfrac{1}{2}\gamma_2)^{-1}. \tag{4}$$

Thus δ is an alternative measure of kurtosis and the quantities β_2, γ_2 and δ are related as follows:

β_2	1	1·8	2	3	6	∞
γ_2	-2	$-1·2$	-1	0	3	∞
δ	∞	2·5	2	1	0·4	0

Denote by $s_\phi^2(\gamma_2)$ an estimate of variance having ϕ degrees of freedom and based on observations drawn from a population with finite cumulants whose measure of kurtosis is γ_2. Then $s_\phi^2(0)$ is an estimate from a normal population. The estimate $s_\phi^2(\gamma_2)$ has mean σ^2 and variance $2\sigma^4/(d\phi)$. Also from the general expressions for the higher cumulants of such an estimate it is seen that as ϕ tends to infinity the distribution of $s_\phi^2(\gamma_2)$ tends to normality, and for sufficiently large values of ϕ, $s_\phi^2(\gamma_2)$ is distributed like $s_{d\phi}^2(0)$.

Denote by $M_1(0;\, \phi_1\phi_2, \ldots, \phi_k)$ the criterion calculated from normally distributed observations and by $M_1(\gamma_{21}\gamma_{22}\ldots\gamma_{2k};\, \phi_1\phi_2\ldots\phi_k)$ the criterion calculated from k groups of observations drawn from populations with measures of kurtosis $\gamma_{21}, \gamma_{22}, \ldots, \gamma_{2t}\ldots\gamma_{2k}$. Then for large samples

$$M_1(\gamma_{21}\gamma_{22}\ldots\gamma_{2k};\, \phi_1\phi_2\ldots\phi_k)$$

is distributed as

$$\Phi \ln\{\Sigma\, \phi_t s^2_{d_t\phi_t}(0)/\Phi\} - \sum_t \phi_t \ln s^2_{d_t\phi_t}(0). \tag{5}$$

When the samples are sufficiently large we may write δ_t for d_t, and in the important case where the kurtosis in each of the sampled populations is the same, so that $\delta_t = \delta$ for all t, $M_1(\gamma_2; \phi_1\phi_2 \dots \phi_k)$ is distributed as

$$\delta^{-1}\{\delta\Phi \ln s^2_{\delta\Phi}(0) - \sum_t \delta\phi_t \ln s^2_{\delta\phi_t}(0)\} = \delta^{-1}M_1(0; \delta\phi_1, \delta\phi_2, \dots, \delta\phi_k). \tag{6}$$

Consequently $M_1(\gamma_2)$ is distributed asymptotically not as χ^2_{k-1} but as $(1+\frac{1}{2}\gamma_2)\chi^2_{k-1}$ for any parent distribution having finite cumulants. We see therefore that M_1 is asymptotically biased if γ_2 is not zero. In particular, in large samples the mean of the distribution curve of M_1 would be $(1+\frac{1}{2}\gamma_2)(k-1)$ instead of $k-1$ and the standard deviation $(1+\frac{1}{2}\gamma_2)\{2(k-1)\}^{\frac{1}{2}}$ instead of $\{2(k-1)\}^{\frac{1}{2}}$. The discrepancy in means relative to the standard deviation would thus become larger as k was increased. Consequently divergences from normal theory probabilities, known to be large when $k = 2$, would become progressively larger for $k > 2$. Table 1 shows the true probability of exceeding the normal-theory 5 % significance levels in large samples for various values of k and γ_2 when the null-hypothesis is true.

Table 1. *True percentage chance of exceeding* 5 % *normal-theory significance level of* M_1 *in large samples from non-normal populations for various values of* γ_2

γ_2	No. of groups = k					
	2	3	5	10	20	30
2	16·6	22·4	31·5	48·9	71·8	84·9
1	11·0	13·6	17·6	25·7	38·9	49·8
0	5·0	5·0	5·0	5·0	5·0	5·0
−1	0·56	0·25	0·08	0·010	0·0004	0·00001

In the case when the null-hypothesis is not true it will be seen that changes in the level of the criterion due to non-normality and due to real differences in the group variances will tend to be 'confounded'. With leptokurtic populations the criterion will tend to show differences when none exist, while with platykurtic populations real differences will tend to be masked.

The behaviour of the test for means is in sharp contrast to that for variances discussed above. The appropriate test for the hypothesis—that the *means* in the groups are all equal, given that the variances are all equal, and that the observations are normally distributed—is the 'within and between groups' analysis of variance test. In its commonest form the criterion used is

$$F = (\text{between groups mean square})/(\text{within groups mean square})$$
$$= \left\{\sum_t \frac{n_t(\bar{y}_t - \bar{y})^2}{k-1}\right\} \Big/ s^2.$$

Alternatively, to correspond with M_1, we could employ a form obtained from the logarithm of the likelihood ratio

$$M_2 = (N-1)\ln\{1 + (k-1)F/(N-k)\}.$$

For large samples Neyman & Pearson (1931) showed that assuming normality a criterion equivalent to M_2 would be distributed as χ^2_{k-1}. With certain not very severe restrictions on the parent population it is possible to show that this result is true *whether the population is normal or not.* Thus for large samples the criterion M_2 for means tends to follow the normal-theory distribution under conditions of non-normality. This explains to some extent the insensitivity to non-normality found by Pearson and other workers mentioned above. In contrast, the criterion M_1 for variances follows a distribution directly dependent on γ_2 however large are the samples.

It is perhaps unfortunate that the test for homogeneity of means, and the test for homogeneity of variances in the particular case of two groups, can each be brought to the form of a variance ratio test. This has sometimes led to apparently contradictory statements about the sensitivity of the 'variance ratio' to non-normality. As originally pointed out by Pearson in 1931, the two criteria are really essentially different in character but follow the same distribution when the parent is normal. The marked difference in sensitivity of the criteria is well brought out by an example taken from Gayen's work (1950*a*). A test for homogeneity of *means* of five groups of five observations leads to an F test on 4 and 20 degrees of freedom; on the normal assumption this will be distributed in the same form as the F criterion used to compare the *variances* in two independent groups of 5 and 21 observations. Gayen shows, for example, that for a certain type of non-normal distribution in which $\gamma_1 = 0$ and $\gamma_2 = 2$ the true probability of exceeding the 5 % point for the test on means would be 4·5 % (a quite trivial discrepancy), whilst that for the test on variances would be 10·2 %.

3. Distribution of M_1 in small samples for certain non-normal parent distributions

It will now be shown that parent distributions exist for which relations similar to (6) are obtained even for small samples.

(3·1) *Population means known*

For the moment it will be assumed that the means of the k sampled populations are known. Denote by $s^2_{.n}$ an estimate of variance having n degrees of freedom calculated from the n deviations from the known group mean η. The criterion calculated from such estimates is denoted by $M_{.1}$. If the distribution of the observations is such that $s^2_{.n}(\gamma_2)$ is distributed like $s^2_{.\delta n}(0)$ for any sample size then $ns^2_{.n}(\gamma_2)$ is distributed like $\delta^{-1}\chi^2_{\delta n}\sigma^2$, which implies that $(y-\eta)^2$ is distributed like $\delta^{-1}\chi^2_{\delta}\sigma^2$. This is true if

$$p(y) = (\delta/2\sigma^2)^{\frac{1}{2}\delta}\{\Gamma(\tfrac{1}{2}\delta)\}^{-1}\,|\,y-\eta\,|^{\delta-1}\exp\{-\delta(y-\eta)^2/2\sigma^2\} \quad (-\infty<y<+\infty;\ 0<\delta<\infty). \quad (7)$$

This is a double χ distribution, that is to say, it is a symmetrical distribution of the form of two χ distributions having δ degrees of freedom 'back to back'. It has mean η, variance σ^2, $\gamma_1 = 0$ and $\gamma_2 = 2(1-\delta)/\delta$ (that is, $\delta^{-1} = 1+\frac{1}{2}\gamma_2$ as previously defined). When $\delta = 1$ the term in $|\,y-\eta\,|$ vanishes and the distribution is normal. When δ is less than 1, γ_2 is greater than zero, and a leptokurtic distribution results. When δ is greater than 1, γ_2 is less than zero and a bimodal platykurtic distribution is obtained. For such parent distributions it is easily seen by the previous argument that

$$M_{.1}(\gamma_2; n_1, n_2, \ldots, n_k) \text{ is distributed } \textit{exactly} \text{ as } \delta^{-1}M_{.1}(0; \delta n_1, \delta n_2, \ldots, \delta n_k), \quad (8)$$

where $n_1, n_2, \ldots, n_k$ refer to the sample sizes.

(3·2) *Population means unknown*

In 1931 Le Roux carried out a very painstaking research into the distribution of the variance calculated from small samples drawn from non-normal distributions of Pearson type. He studied changes in the distribution of s^2 as the parent distribution changed, by calculating its first four moments from the moments of the parent distribution, and confirmed his findings by extensive sampling experiments. One of his discoveries was that in addition to the normal distribution a series of non-normal distributions existed (which he called $D(s^2)$ III distributions) for which the first four moments of s^2 were almost exactly those of a Pearson type III curve. These $D(s^2)$ III distributions show kurtosis accompanied by marked skewness. A selection of these adapted from Le Roux's table are shown below.

$D(s^2)$ III *distributions*

γ_1^2	0·00	0·10	0·20	0·40	0·60	0·80	1·00	1·20	1·40
γ_2	0·00	−0·04	0·03	0·24	0·48	0·73	0·99	1·25	1·51

For such curves Le Roux deduced that to a close approximation $s_\phi^2(\gamma_2)$ was distributed as $s_{d\phi}^2(0)$ for any sample size $\left(\text{where as before } d^{-1} = 1+\frac{1}{2}\frac{\phi}{\phi+1}\gamma_2\right)$, a conclusion supported by his sampling experiments. We may conclude therefore that the relation (5) is approximately true for samples of any size chosen from such populations.

In particular, if the degree of kurtosis is the same in each of the populations sampled and each variance is based on the same number ϕ of degrees of freedom, then for $D(s^2)$ III populations

$$M_1(\gamma_2;\ \phi, \phi, \ldots, \phi) \text{ is distributed approximately as } d^{-1}M_1(0;\ d\phi, d\phi, \ldots, d\phi).$$

Using Bartlett's method we see that for $D(s^2)$ III populations therefore $M_1(\gamma_2; \phi_1, \phi_2, \ldots, \phi_k)$ is distributed approximately not as $(1+A)\,\chi_{k-1}^2$ but as

$$d^{-1}(1+Ad^{-1})\,\chi_{k-1}^2, \tag{9}$$

where A is the correcting factor of order ϕ^{-1} defined in (2). We note in passing that departure from normality of this form would not need to be very great or the sample size very big before the effect of the constant d^{-1} completely swamped the effect of the correction term A.

4. Two groups of observations

When $k = 2$ we may use the variance ratio F as an alternative form of the M_1 criterion. Using the same argument as before and a corresponding notation it is apparent that for samples of any size drawn from double χ distributions with measures of kurtosis γ_{21} and γ_{22},

$$F(\gamma_{21}, \gamma_{22};\ n_1, n_2) \text{ is distributed exactly as } F(0;\ \delta_1 n_1, \delta_2 n_2), \tag{10}$$

whilst for samples of any size from $D(s^2)$ III populations

$$F(\gamma_{21}, \gamma_{22};\ \phi_1, \phi_2) \text{ is distributed approximately as } F(0;\ d_1\phi_1, d_2\phi_2). \tag{11}$$

These results, which are, of course, equivalent asymptotically, would be expected to apply for *any population* when the sample sizes were sufficiently large. To provide some indication of how large the samples would have to be, comparison was made with the results given by (11) and the more exact values calculated by other authors. Gayen (1950*b*) and Finch (1950) have considered the distribution of the variance ratio or the equivalent z criterion for populations defined by the first few terms of Edgeworth's series. Both these authors note that for these populations the effect of changes in γ_1 is small but large discrepancies are produced by changes in γ_2. In Fig. 1, γ_1 is assumed zero and the continuous lines represent the probability as determined by these authors, of exceeding the normal theory 5 % point for various values of γ_2. The values of γ_2 are assumed to be the same in each of the two populations. The first graph illustrates the case $\phi_1 = 24$, $\phi_2 = 60$ considered by Finch, and the second the case $\phi_1 = 4$, $\phi_2 = 20$ considered by Gayen. The dotted lines are the values given by equation (11) above. A further comparison is supplied by Pearson's (1931) sampling experiments. Pearson drew 500 pairs of samples of 5 and 20 from six experimental populations and calculated a criterion equivalent to F; this corresponds very nearly to the case $\phi_1 = 4, \phi_2 = 20$ above (ϕ_2 is actually 19 instead of 20). In his table Pearson shows the number of pairs of samples for which the criterion exceeds the 4·86 % point (which again is very close to the 5 % point considered above). These numbers reduced to percentages are plotted as circles about the second line.

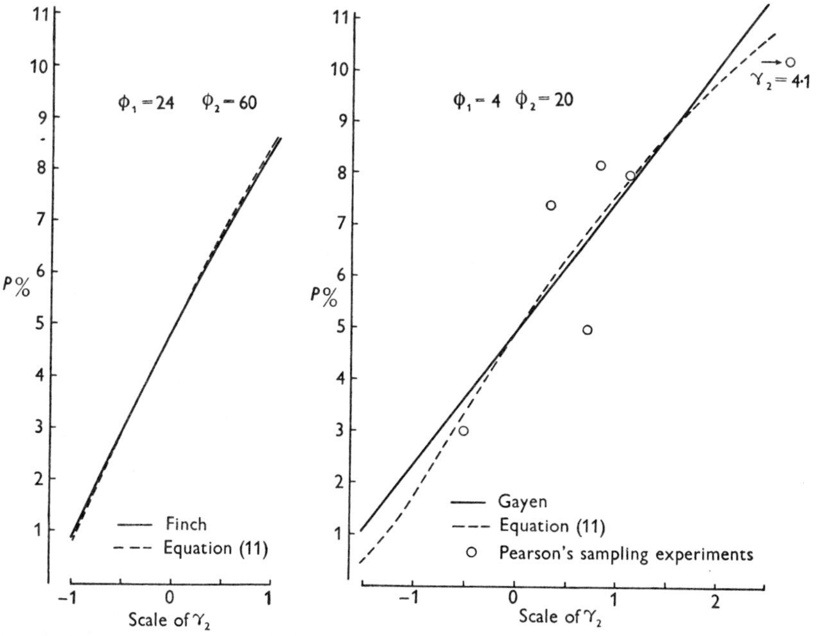

Fig. 1. Percentage probability P of exceeding normal-theory 5 % level plotted against value of γ_2.

The agreement between the simple approximation (11) and the values found by Finch is very close, as would be expected for the comparatively large sample sizes used. The extent of agreement with Gayen's results is rather surprising when the small size of one of the samples is remembered. In both approaches it is assumed, however, that the higher cumulants are finite. The marked departure from both lines which occurs when this is not so is shown by Pearson's sampling experiment from the type VII (i.e. the Student t) distribution for which $\gamma_2 = 4 \cdot 1$ and higher cumulants are infinite.

5. Sampling experiments with a $D(s^2)$ III distribution

Confirmation of (9) in the case $k = 20$ was obtained from an experiment in which samples were drawn from a population nearly of type $D(s^2)$ III form. Le Roux drew 1000 samples of $n = 20$ observations from such a distribution in which $\gamma_1^2 = 1 \cdot 0$ and $\gamma_2 = 0 \cdot 8$. (The distribution was actually a Pearson type I.) He calculated the quantity $(n-1)s^2/n$ for each of the 1000 samples and published the resulting frequency distribution. The samples were grouped but the grouping interval was fairly fine. In the present work Le Roux's observed distribution of s^2 has been resampled, the sampling being without replacement. In each experiment twenty-five groups of twenty variance estimates were reconstructed and from these, twenty-five values of M_1 were calculated. The experiment was then repeated using Le Roux's observed distribution of variance estimates for samples of $n = 5$ from the same parent distribution.

Table 2. *Characteristics of observed distributions of twenty-five values of M_1 for $k = 20$ groups, with (a) $n = 20$ observations and (b) $n = 5$ observations, drawn from a Pearson type I parent distribution for which $\gamma_1^2 = 1{\cdot}0$, $\gamma_2 = 0{\cdot}8$*

			Expected on normal theory (χ^2 approximation)	Found in samples of 25 values of M_1	Expected from equation (9) (χ^2 approximation)
$n = 20$	M_1 distributed as:		$1{\cdot}018\chi^2_{19}$		$1{\cdot}415\chi^2_{19}$
	Mean		19·4	28·6 ± 1·9*	26·9
	Variance		39·4	93·3 ± 30·3†	76·1
	No. out of 25 significant at P % point	P = 10	2·5	13 (52 %)	10·3 (41·2 %)
		5	1·25	9 (36 %)	7·5 (30·0 %)
		1	0·25	4 (16 %)	3·2 (12·8 %)
		0·1	0·025	1 (4 %)	0·9 (3·6 %)
$n = 5$	M_1 distributed as:		$1{\cdot}088\chi^2_{19}$		$1{\cdot}473\chi^2_{19}$
	Mean		20·7	28·0 ± 1·8*	28·0
	Variance		44·9	79·1 ± 25·6†	82·4
	No. out of 25 significant at P % point	P = 10	2·5	13 (52 %)	9·6 (38·2 %)
		5	1·25	9 (36 %)	6·8 (27·1 %)
		1	0·25	2 (8 %)	2·8 (11·0 %)
		0·1	0·025	0 (0 %)	0·7 (2·9 %)

* The quantity following the ± signs is the approximate standard error of the mean obtained from S.E.(Mean M_1) $\simeq \{V(M_1)/25\}^{\frac{1}{2}}$.

† The quantity following the ± signs is a rough estimate of the standard error of the variance of M_1 obtained by assuming that M_1 is approximately distributed in Pearson type III form and substituting $\gamma_2 = 12/(k-1)$ in the approximate formula, S.E.$\{V(M_1)\} \simeq V(M_1)\{(\gamma_2+2)/25\}^{\frac{1}{2}}$.

The two observed distributions of M_1 are shown in the sixth and seventh columns of the table in the Appendix. The main features of the observed distributions are shown in Table 2. It will be seen that the means and variances of the values of M_1 are in close agreement with the values anticipated from equation (9) and differ very markedly from those expected on normal theory. The excess of 'significant' values is also verified. For example, both for samples of $n = 20$ and $n = 5$ observations, nine out of twenty-five (or 36 %) of the values were significant at the 5 % point.

6. Sampling experiments with other distributions

Besides the population of the $D(s^2)$ III type, Le Roux considered a variety of other populations, mostly using sampling results used by E. S. Pearson in earlier investigations, to determine empirically the distribution of the sample variance. In the present investigation Le Roux's distributions derived from the following parent distributions have been resampled in the manner already described:

Pearson type	II	III	I	IV	VII
(γ_1^2, γ_2)*	(0·0, −0·5)	(0·5, 0·75)	(1·0, 0·8)	(1·2, 2·8)	(0·0, 4·1)

From each distribution, twenty-five sets of twenty samples of s^2 have been obtained, and for these twenty-five values of M_1 calculated. These observed values of M_1 are set out in the Appendix. Table 3 shows the values of the means and variances of M_1 observed and those expected on normal theory, together with the number of values exceeding the normal theory significance points.

It is seen that large discrepancies from the normal theory distributions occur, and the discrepancies are in the directions expected. The discrepancies are larger when $n = 20$ than when $n = 5$, and for the populations studied the asymptotic result seems to give an upper limit to the discrepancy which is approached as the sample size is increased. The asymptotic result appears to be more rapidly approached when kurtosis is accompanied by skewness till for the skew $D(s^2)$ III populations, the asymptotic value is attained even for small sample sizes. The asymptotic result is approached more slowly when kurtosis is large and for the very leptokurtic type IV and type VII distributions, sample sizes greater than twenty appear to be needed before the asymptotic values are closely approached.

7. M_1 as a test for normality

The seriousness of the discrepancies to be expected, even in small samples, may be appreciated by considering M_1 as a *test* for normality.

Suppose a sample of N observations is drawn from a population distributed in the double χ form of equation (7). We shall assume that the mean η of the population is known, so that without loss of generality we may take it to be zero. We wish to test the hypothesis H_0 that the distribution is normal, i.e. that $\delta = 1$, $\sigma > 0$, against the class of alternatives H_1 which specify that the distribution is not normal, i.e. that $\delta \neq 1$, $\sigma > 0$. Then, following through the method set out by Neyman & Pearson (1933) for the selection of best critical regions, it is easy to show that the appropriate criterion is $M_{.1} = N \ln(\Sigma y^2/N) - \Sigma \ln y^2$, and that if $M_{.1}(\alpha)$ and $M_{.1}(1-\alpha)$ are the α and $1-\alpha$ significance points for $M_{.1}$, then the inequalities

$$M_{.1} > M_{.1}(\alpha) \quad \text{and} \quad M_{.1} < M_{.1}(1-\alpha) \tag{12}$$

define a pair of common best critical regions for testing H_0 against the alternative hypotheses $\delta < 1$ $(\gamma_2 > 0)$ and $\delta > 1$ $(\gamma_2 < 0)$, respectively, where in each case σ is unknown. Thus in the particular case when there is only one observation per group, $M_{.1}$ is itself the most sensitive criterion possible for testing for a certain type of non-normality. Non-normal distributions of the class considered, which are defined by equation (7), would rarely be expected to approximate to the distribution of actual data, but the above investigation suggests that we may form some idea of the sensitivity to non-normality of $M_{.1}$ for small samples by seeing how good it is as a *test* for non-normality.

* The actual standardized cumulants given above differ slightly in some cases from the theoretical values due to the finite size (10,000) of the populations which Le Roux sampled.

Table 3. *Characteristics of observed distributions of twenty-five values of M_1 for $k = 20$ groups of (a) $n = 20$*, (b) $n = 5$ observations, drawn from a variety of Pearson-type parent distributions*

			Mean			Variance			No. out of 25 exceeding normal theory significance point, P			
n	γ_1^2	γ_2	Expected, normal theory	Found in samples of 25 values of M_1	Expected for $D(s^2)$ III population with same γ_2	Expected, normal theory	Found in samples of 25 values of M_1	Expected $D(s^2)$ III population with same γ_2	$P = 10\%$	$P = 5\%$	$P = 1\%$	$P = 0{\cdot}1\%$
20	0·0	−0·5	19·4	15·0	14·7	39·4	16·5	22·7	0 (0%)	0 (0%)	0 (0%)	0 (0%)
20	0·5	0·7	19·4	25·8	25·9	39·4	92·7	70·8	10 (40%)	5 (20%)	3 (12%)	2 (8%)
20	1·0	0·8	19·4	28·6	26·9	39·4	93·3	76·1	13 (52%)	9 (36%)	4 (16%)	1 (4%)
25	1·2	2·8	19·4	37·4	46·5	39·4	159·0	227·2	21 (84%)	16 (64%)	10 (40%)	5 (20%)
20	0·0	4·1	19·4	39·4	59·1	39·4	329·4	367·1	19 (76%)	17 (68%)	11 (44%)	7 (28%)
5	0·0	−0·5	20·7	19·3	16·2	44·9	36·8	27·8	2 (8%)	1 (4%)	0 (0%)	0 (0%)
5	0·5	0·7	20·7	23·0	27·0	44·9	29·4	76·9	3 (12%)	2 (8%)	0 (0%)	0 (0%)
5	1·0	0·8	20·7	28·0	28·0	44·9	79·0	82·4	13 (52%)	9 (36%)	2 (8%)	0 (0%)
5	1·2	2·8	20·7	26·8	47·8	44·9	67·5	240·0	11 (44%)	8 (32%)	1 (4%)	0 (0%)
5	0·0	4·1	20·7	28·5	61·8	44·9	192·1	401·4	9 (36%)	5 (20%)	1 (4%)	1 (4%)

* In the single case $\gamma_1^2 = 1{\cdot}2$, $\gamma_2 = 2{\cdot}8$ the sample n number taken by Le Roux was 25 instead of 20.

(7·1) *Comparison of $M_{.1}$ with criteria for testing normality*

The usual test criterion for kurtosis is b_2, the sample estimate of the moment coefficient β_2. Geary (1935*a*) proposed as a new test for kurtosis the criterion

$$a = (\text{sample mean deviation})/(\text{sample standard deviation}).$$

This criterion had the advantage that unlike b_2, the distribution when the null-hypothesis of normality was true could be found fairly readily. Further investigation was carried out by Geary (1935*b*, 1947) to discover whether the power of this test to pick out a departure from normality was comparable to that of b_2. Pearson (1935) studied this question by means of sampling experiments. One such experiment consisted of drawing ten samples each of seventy-six observations from each of three symmetrical populations:

Description	γ_1^2	γ_2^*
Rectangular	0·0	−1·2
Pearson type VII	0·0	4·1
Double exponential	0·0	2·9

* As before the actual standardized cumulants given above differ slightly from the theoretical values for the last two populations due to the finite size (10,000) of the population sampled.

Pearson used his results to show the relative frequency with which the test criteria b_2 and a detected departures from normality and also to show the correlation between the results of the two tests. He published not only his conclusions but also the actual samples which he used and I have used these again to calculate the corresponding values for $M_{.1}$. In my calculation of $M_{.1}$ I have used the population mean (zero in each case), whereas in Pearson's calculations of b_2 and a the sample mean was used. In samples of seventy-six, however, the discrepancy this introduces should not be large. A further difficulty arises due to the fact that the samples are grouped to the nearest unit, resulting in a frequency class having zero deviation from the mean which would give rise to an infinite value of $M_{.1}$. To overcome this I have spaced the values evenly in the interval 0·0–0·5. Thus if there were n values in the zero frequency class these have been assumed to be at $1/(4n)$, $3/(4n)$, ..., $(2n-1)/(4n)$. Fig. 2*a* shows the thirty values of $M_{.1}$ plotted against the corresponding values of b_2 and Fig. 2*b* shows the same values plotted against the values of a. It is seen that in each case marked correlation occurs, and if these graphs are compared with Pearson's Fig. 4 in which he plots a against b_2 it will be seen that the extent of correlation between $M_{.1}$ and b_2 and between $M_{.1}$ and a is similar to that between a and b_2. So far as is known, no tables are available for the significance points of M_1 (or for the equivalent criterion L_1) for the case of seventy-six groups with variance estimates each based on 1 degree of freedom. Also the asymptotic approximations of Bartlett (1937), Hartley (1940) and Box (1949) break down for large k and small numbers of degrees of freedom. Approximate significance points for $M_{.1}$ were therefore calculated by considering the distribution of $L_1^{-1} = \exp\{M_1/k\}$ which ranges from 1 to ∞. General expressions for the moments of L_1 were given by Neyman & Pearson (1931). From these the first four moments of L_1^{-1} were calculated and the appropriate Pearson type curve to approximate the distribution was found to be of type VI. It was assumed therefore that L_1^{-1} was approximately distributed as $cF_{\phi_1\phi_2}$, the curve starting at 1. By equating the first three moments of the two curves the values

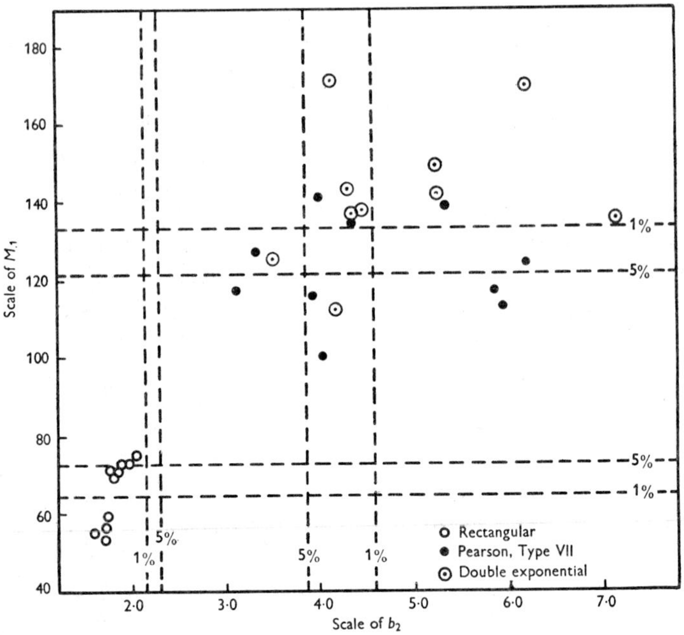

Fig. 2*a*. Comparison of $M_{.1}$ and b_2 in samples of seventy-six.

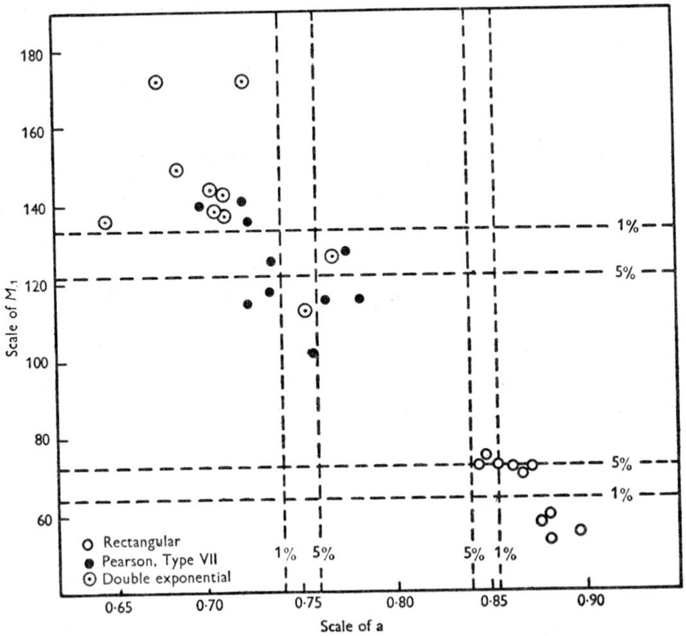

Fig. 2*b*. Comparison of $M_{.1}$ and a in samples of seventy-six.

$c = 2{\cdot}4705$, $\phi_1 = 67{\cdot}1998$, $\phi_2 = 45{\cdot}2904$ were obtained, and the approximate significance points calculated by two-way, four-point harmonic interpolation in the F tables of Merrington & Thompson (1943). The approximate significance points thus found were:

Lower		Upper	
1 %	5 %	5 %	1 %
64·1	72·3	121·2	133·3

Table 4 is an extension of Pearson's Table VII showing the number of values of b_2, a and $M_{.1}$ exceeding their 5 and 1 % significance levels.

Table 4. *Samples of seventy-six; comparison of criteria*

Distribution	Rectangular $\gamma_1^2 = 0{\cdot}0,\ \gamma_2 = -1{\cdot}2$			Pearson type VII $\gamma_1^2 = 0{\cdot}0,\ \gamma_2 = 4{\cdot}1$			Double exponential $\gamma_1^2 = 0{\cdot}0,\ \gamma_2 = 2{\cdot}9$		
	b_2	a	$M_{.1}$	b_2	a	$M_{.1}$	b_2	a	$M_{.1}$
No. of samples:									
within 5 % limits	0	0	3	2	3	5	1	1	1
between 5 and 1 % limits	0	2	3	4	1	2	5	1	1
beyond 1 % limits	10	8	4	4	6	3	4	8	8

For the rectangular and type VII distributions the criterion appears to be less sensitive to kurtosis than b_2 or a; for the double exponential distribution, however, $M_{.1}$ appears to be as sensitive as a and more sensitive than b_2. It is not of course contended that $M_{.1}$ is necessarily a practical test for kurtosis; we are concerned only to show that since b_2 and a are presumably very sensitive criteria for detecting kurtosis, $M_{.1}$ is much more sensitive *even when the sample size is small* than we should desire a test *for homogeneity of variances* to be.

In this connexion it is of some interest to reconsider some data discussed by Bartlett & Kendall (1946). The results appear as a two-way table with fifteen columns and three rows. The entries in the table are the logarithms of variance estimates s^2, each based on about 48 degrees of freedom. The authors show that by taking logarithms the data are brought to a suitable form for the application of analysis of variance. Their analysis is as follows:

	D.F.	S.S.	M.S.
Between rows	2	0·2667	0·1333
Between columns	14	0·1047	0·0075
Residual	28	0·1005	0·0036
Theoretical variance	—	—	0·0020

The first three items are calculated in the usual way, and the theoretical variance is based on the normal theory value $\kappa_2(\ln s^2) \sim 2/(\phi - 1)$. Compared with this theoretical variance the residual mean square is significant ($P = 0{\cdot}01$). It is deduced, therefore, that there is heterogeneity in the residual variance, and methods for further analysing this residual heterogeneity are discussed. Although familiarity with the distribution of their data (of which clearly a large quantity was available) no doubt rendered quite legitimate in this example the assumption by these authors of approximate normality, it is perhaps worth while emphasizing the dependence of this test on that assumption. For in general

$$\kappa_2(\ln s^2) \sim 2/(d\phi - 1),$$

consequently a value for γ_2 of 1·57 would have made the residual and theoretical variances exactly equal. Furthermore, since any value for the theoretical variance greater than 0·0024 would fail to show the residual variance significant at the 5 % level, any hypothesis that the residual variance is homogeneous and γ_2 greater than 0·44 is not contradicted by the data at this level of significance.

8. Other tests of variance homogeneity

The reason for the difference in sensitivity to non-normality between the criterion M_2 for testing homogeneity of means and M_1 for testing homogeneity of variances can be seen as follows. In the analysis of variance (F-form) the criterion for means may be written as P/Q, where $P = \{\Sigma n_t(\bar{y}_t - \bar{y})^2\}/(k-1)$ is the between-groups mean square and $Q = s^2$ is the within-groups mean square. When the null hypothesis is true, P and Q each provides an unbiased estimate of σ^2 whether the observations are drawn from normal populations or not, so that Q is always a standard with which P may be usefully compared. But if in M_1, the criterion for variances, we write $s_t^2 = \sigma^2(1+x_t)$ and formally expand the logarithms, we obtain $M_1 = \frac{1}{2}\sum_t \phi_t(x_t - \bar{x})^2$ plus terms in higher powers of x_t. For sufficiently large values of ϕ_t we can ignore the higher powers of x_t and we have $M_1/(k-1) \sim p/q$, where

$$p = \{\Sigma\phi_t(s_t^2 - s^2)^2\}/(k-1) \quad \text{and} \quad q = 2\sigma^4.$$

Comparing these with the expressions above we see that asymptotically the M_1 test is like an analysis of variance on the sample variances instead of sample means, *but the quantity p corresponding to the between-groups mean square is compared not with an estimate from the internal evidence of the samples but with a theoretical value of the variance which is appropriate only when the parent distribution is normal.* In general, the variance of a variance estimate s_t^2 is $d^{-1} \times 2\sigma^4/\phi_t$; consequently the appropriate value with which to compare p would be $d^{-1} \times 2\sigma^4$ and not $2\sigma^4$.

From the above discussion it will be seen that to obtain a criterion less dependent on γ_2, information on the variation to be expected in the sample variances gathered from internal evidence in the samples should be utilized. A test of variances should be 'studentized' for the fourth moment just as a test of means is studentized for the second moment.

(8·1) *Other tests not utilizing information within groups*

Other tests of variance homogeneity which do not utilize evidence on variance variability within the samples are equally sensitive to non-normality. In fact for $D(s^2)$ III parent distributions and, asymptotically, for all distributions with finite cumulants any criterion

$f\{s^2_{\phi_1}(\gamma_{21}), s^2_{\phi_2}(\gamma_{22}), \ldots, s^2_{\phi_k}(\gamma_{2k})\}$ will be distributed like $f\{s^2_{d_1\phi_1}(0), s^2_{d_2\phi_2}(0), \ldots, s^2_{d_k\phi_k}(0)\}$. Examples of such tests are those proposed by Cochran (1941) and by Hartley (1950). Cochran's criterion is $g = s^2_{\text{max.}}/\Sigma s^2_t$, whilst that proposed by Hartley is $F_{\text{max.}} = s^2_{\text{max.}}/s^2_{\text{min.}}$, where $s^2_{\text{max.}}$ and $s^2_{\text{min.}}$ are the largest and smallest of the group variances $s^2_1, s^2_2, \ldots, s^2_t, \ldots, s^2_k$. Using the same approximation as that adopted by Hartley it is a simple matter to calculate the true chance of exceeding the normal theory significance points for a $D(s^2)$ III parent population. For instance with k groups of 21 observations drawn at random from such a population in which $\gamma_2 = 1$ the approximate percentage chance of exceeding the normal theory 5 % significance point would be:

	$k = 5$	$k = 10$	$k = 20$
M_1	17·5	25·2	37·3
$F_{\text{max.}}$	17·3	23·3	30·3

The discrepancies using $F_{\text{max.}}$ are seen to be of similar magnitude to those found with M_1, and similar results are to be expected using Cochran's g. The multivariate tests for the constancy of the variance covariance matrix from one group of observations to another (Wilks, 1932; Bishop, 1939; Box, 1949, 1950) would be expected to be equally dependent on the assumption of multivariate normality.

Sensitivity to non-normality is found also with sequential tests. For example, Wald (1947) discussed the use of the sequential likelihood ratio test of the hypothesis H_0 that $\sigma^2 = \sigma^2_0$ when the alternative H_1 is that $\sigma^2 = \sigma^2_1$ and the observations $y_1, y_2, \ldots$ were drawn from normal populations with known mean η. Suppose the observations were drawn, not from a normal universe, but from the double χ population whose distribution is given by equation (7). Then it will be found that the logarithm of the likelihood ratio $L(\gamma_2)$ is equal to $\delta L(0)$ (where $L(0)$ is the logarithm of the likelihood ratio when the parent distribution is normal). Consequently if the quantity $L(0)$ is calculated when γ_2 is not zero and referred to limits $\ln(\beta/(1-\alpha))$ and $\ln((1-\beta)/\alpha)$, this is equivalent to referring the actual likelihood $\delta L(0)$ to the limits $\delta \ln\{\beta/(1-\alpha)\}$ and $\delta \ln\{(1-\beta)/\alpha\}$. The actual risks of error of the two kinds will therefore be α' and β' chosen so that $\beta'/(1-\alpha') = \{\beta/(1-\alpha)\}^\delta$ and $(1-\beta')/\alpha' = \{(1-\beta)/\alpha\}^\delta$. For $100\alpha = 100\beta = 5$ the percentage risk of errors of each of the two kinds is given below for various values of γ_2:

γ_2	-1	0	1	2
$100\alpha' = 100\beta'$	0·28	5·00	12·32	18·66

As with the tests with fixed sample sizes the result will be approximately true for any population with finite cumulants when the average sample sizes are not small. Roughly $\alpha' = \alpha^\delta$, $\beta' = \beta^\delta$ for small values of α and β. Similar difficulties would be expected with other sequential tests relating to the true value of and equality of variances; for example, with the tests proposed by Girshick (1946) and the tests based on ranges suggested by Cox (1949).

(8·2) *A test utilizing within-group information*

A test on variances less sensitive to parent non-normality must clearly utilize the within-group information in some way. Although the author believes that a better approach is available, one immediately practical method is to split up the groups of observations into sub-groups, and carry out an analysis of variance between and within groups on the logarithms of the sub-group variances, following Bartlett & Kendall (1946), who have shown the value of the logarithmic transformation in bringing variance data to a form suitable for the application of analysis of variance. The question of what sizes should best be taken for the subgroups requires further investigation.

Table 5. *Criteria calculated for ten samples of twenty groups of twenty observations drawn from a rectangular population*

Source of sample*		Test of means	Tests of variances			
				M_2 on logarithms of subgroup variances		
Page	Rows	M_2	M_1	2 sub-groups of 10	5 sub-groups of 4	10 sub-groups of 2
I	1–20	14·5	6·7	26·2	10·3	34·0
I	21–40	22·7	8·6	18·6	16·1	18·1
I and II	41–50 1–10	13·8	10·8	29·6	29·3	19·3
II	11–30	12·6	16·7	30·7	35·0	18·1
II	31–50	19·6	9·9	40·6	20·3	10·3
III	1–20	26·9	5·2	17·9	10·8	11·6
III	21–40	12·5	16·1	40·8	29·6	19·8
III and IV	41–50 1–10	18·0	7·4	34·1	21·6	34·1
IV	11–30	23·8	10·3	37·9	28·2	19·5
IV	31–50	11·8	10·3	30·6	32·4	29·6
Mean:	Found	17·6 ± 1·7	10·2 ± 1·2	30·7 ± 2·6	23·4 ± 2·8	21·4 ± 2·7
	Expected on normal theory	19·5	19·4	27·0	22·1	21·3
Variance:	Found	29·3 ± 15·6	13·9 ± 7·2	66·1 ± 35·3	78·7 ± 42·0	71·0 ± 37·9
	Expected on normal theory	40·1	39·4	79·0	51·5	48·1

* The groups of observations consisted in each case of the first twenty columns of numbers from pages of Fisher & Yates's tables. The page numbers and the numbers of the rows are shown above.

The results of a small sampling experiment of some interest in this connexion are set out in Table 5. Ten samples of twenty groups of twenty observations were drawn from the table of random numbers prepared by Fisher & Yates (1938). The parent distribution was thus effectively rectangular. As a test of group to group homogeneity of *means* a 'between and within groups' analysis of variance was performed for each of the ten samples of twenty groups and the ten resulting values of M_2 calculated. These are shown in the third column of the table. Four different tests for homogeneity of the variances were applied to each sample. The first was the M_1 test, the results for which are shown in the fourth column of the table. In the remaining three columns are shown the values of M_2 for analysis of variance performed on the logarithms of subgroup variances. The groups of twenty observations were divided into subgroups in three different ways: (i) two subgroups of ten observations, (ii) five subgroups of four, (iii) ten subgroups of two. The means and variances of the observed values together with their approximate standard errors and the values expected on normal theory are shown at the bottom of the table.

With so few samples, only large discrepancies of course would be detectable. We see that the test on means shows no evidence of departure from the values expected on normal theory even though the parent population is rectangular. The M_1 test for homogeneity of variance on the other hand shows extremely large departures, the mean being only about half the value expected on normal theory (for the rectangular parent distribution the asymptotic mean value is two-fifths of the normal-theory value). In contrast, it is seen that all the tests for homogeneity of variance based on M_2 give values agreeing fairly well with what would be expected if it could be assumed that the distribution of the logarithm of the variance was exactly normal, an assumption far from true, particularly for subgroups with only two observations.

9. Discussion

It has frequently been suggested that a test of homogeneity of variances should be applied before making an analysis of variance test for homogeneity of means in which homogeneity of variance is assumed. The present research suggests than when, as is usual, little is known of the parent distribution, this practice may well lead to more wrong conclusions than if the preliminary test was omitted. It has been shown (Welch, 1937; David & Johnson, 1951*b*; Box, 1952; and Horsnell, 1953) that in the commonly occurring case in which the group sizes are equal, or not very different, the analysis of variance test is affected surprisingly little by variance inequalities. Since this test is also known to be very insensitive to non-normality it would be best to accept the fact that it can be used safely under most practical conditions. To make the preliminary test on variances is rather like putting to sea in a rowing boat to find out whether conditions are sufficiently calm for an ocean liner to leave port!

When the groups of observations were of unequal size and differences in variances might occur it would seem logical to replace the usual analysis of variance criterion which uses a pooled estimate of within-groups variance by the alternative criterion proposed by Welch (1951) and by James (1951), i.e. $\sum_t w_t(\bar{x}_t - \bar{x})^2$, where $w_t = n_t/s_t^2$. This criterion is robust to inequality of variance and almost certainly to non-normality also. (In fact, where inequality of variance might occur it would seem most logical to use Welch's criterion even if the groups were equal.)

When a criterion for testing a statistical hypothesis is derived (for example, by the likelihood ratio method), it is usually necessary for purposes of mathematical convenience to over-simplify the specification of the problem. We should not be surprised therefore if an examination of the resulting criterion shows that the assumptions have sometimes, so to speak, been interpreted rather too literally. For this reason it is most important that derived criteria should be studied for robustness.

The property of robustness I believe to be even more important in practice than that the test should have maximum power and that the statistics employed should be fully efficient.

Where necessary I believe that the latter qualities should be sacrificed to ensure the former.* On the other hand, I do not think that we need necessarily go to the extreme of using non-parametric tests when it may well be that more powerful robust parametric tests can be found.

I am greatly indebted to Prof. E. S. Pearson for his interest and his many valuable suggestions for the improvement of the presentation of this paper. In conclusion, I wish to thank Mrs Margaret Edmondson for valuable assistance with the computations.

REFERENCES

Bartlett, M. S. (1935). *Proc. Camb. Phil. Soc.* **31**, 223.
Bartlett, M. S. (1936*a*). *Proc. Camb. Phil. Soc.* **32**, 560.
Bartlett, M. S. (1936*b*). *Proc. Roy. Soc.* A, **154**, 124.
Bartlett, M. S. (1937). *Proc. Roy. Soc.* A, **160**, 268.
Bartlett, M. S. (1938). *Proc. Camb. Phil. Soc.* **34**, 33.
Bartlett, M. S. & Kendall, D. G. (1946). *J. R. Statist. Soc. Suppl.* **8**, 128.
Bishop, D. J. (1939). *Biometrika*, **31**, 31.
Box, G. E. P. (1949). *Biometrika*, **36**, 317.
Box, G. E. P. (1950). *Biometrics*, **6**, 362.
Box, G. E. P. (1952). Unpublished thesis, Ph.D., London University.
Cochran, W. G. (1941). *Ann. Eugen., Lond.*, **11**, 47.
Cox, D. R. (1949). *J. R. Statist. Soc.* B, **11**, 101.
David, F. N. & Johnson, N. L. (1951*a*). *Biometrika*, **38**, 43.
David, F. N. & Johnson, N. L. (1951*b*). *Trabajos de estadistica*, **2**, 179.
Finch, D. J. (1950). *Biometrika*, **37**, 187.
Fisher, R. A. & Yates, F. (1938). *Statistical Tables for Biological, Agricultural and Medical Research*, 1st ed. Edinburgh and London: Oliver and Boyd.
Gayen, A. K. (1949). *Biometrika*, **36**, 353.
Gayen, A. K. (1950*a*). *Biometrika*, **37**, 236.
Gayen, A. K. (1950*b*). *Biometrika*, **37**, 399.
Geary, R. C. (1935*a*). *Biometrika*, **27**, 310.
Geary, R. C. (1935*b*). *Biometrika*, **27**, 353.
Geary, R. C. (1947). *Biometrika*, **34**, 209.
Geary, R. C. & Pearson, E. S. (1938). *Tests of Normality*. Biometrika Office.
Girshick, M. A. (1946). *Ann. Math. Statist.* **17**, 123.
Hartley, H. O. (1940). *Biometrika*, **31**, 249.
Hartley, H. O. (1950). *Biometrika*, **37**, 308.
Horsnell, G. (1953). *Biometrika*, **40**, 128.
James, G. S. (1951). *Biometrika*, **38**, 324.
Kendall, M. G. (1943). *The Advanced Theory of Statistics*, **1**. London: Charles Griffin and Co.
Le Roux, J. M. (1931). *Biometrika*, **23**, 134.
Merrington, M. & Thompson, C. M. (1943). *Biometrika*, **33**, 73.
Neyman, J. & Pearson, E. S. (1928). *Biometrika*, **20** A, 175 and 263.
Neyman, J. & Pearson, E. S. (1930). *Bull. int. Acad. Cracovie*, A, p. 73.
Neyman, J. & Pearson, E. S. (1931). *Bull. int. Acad. Cracovie*, A, p. 460.
Neyman, J. & Pearson, E. S. (1933). *Phil. Trans.* A, **231**, 289.
Pearson, E. S. (1931). *Biometrika*, **23**, 114.
Pearson, E. S. (1935). *Biometrika*, **27**, 333.
Tukey, J. W. (1948). *Human Biology*, **20**, 205.
Wald, A. (1947). *Sequential Analysis*. New York: John Wiley and Sons.
Welch, B. L. (1937). *Biometrika*, **29**, 350.
Welch, B. L. (1947). *Biometrika*, **34**, 28.
Welch, B. L. (1951). *Biometrika*, **38**, 330.
Wilks, S. S. (1932). *Biometrika*, **24**, 471.

* Since writing the above a very interesting paper by J. W. Tukey (1948) has come to my notice which has many points of contact with the present paper and which expresses similar views to the above.

APPENDIX

Details of sampling experiments. Values of M_1 for $k = 20$ variance estimates, based on sets of $n = 20$, 25 and 5 observations randomly drawn from various parent populations

Parent population: Pearson type and values of (γ_1^2, γ_2)	II (0·0, − 0·5)		III (0·5, 0·75)		I (1·0, 0·8)		IV (1·2, 2·8)		VII (0·0, 4·1)	
Sample	$n = 20$	$n = 5$	$n = 20$	$n = 5$	$n = 20$	$n = 5$	$n = 25$	$n = 5$	$n = 20$	$n = 5$
1	21·26	13·18	24·70	22·76	21·06	33·06*	50·74***	33·69*	94·48***	25·49
2	18·22	20·75	20·59	33·14*	42·46**	21·75	26·92	36·16*	44·32***	16·47
3	15·80	16·06	30·39?	35·76*	22·85	21·24	42·64**	33·95*	22·18	25·25
4	18·26	13·49	29·23?	18·00	20·70	13·28	80·00***	25·83	47·21***	32·19?
5	10·49	13·31	37·26**	19·34	33·93*	27·84	21·96	25·08	36·78*	14·58
6	12·66	22·95	23·84	18·29	57·62***	32·44?	18·70	32·27?	32·58*	17·41
7	10·87	19·16	16·64	28·49	42·16**	39·57**	35·44*	30·66?	40·81**	34·02*
8	19·83	32·19?	21·68	16·70	19·64	33·59*	42·77**	9·27	29·91?	36·64*
9	14·65	16·76	18·84	17·79	22·57	17·88	29·44?	19·40	24·42	28·46
10	13·63	14·90	19·42	30·27?	33·70*	38·44*	36·11*	15·59	36·09*	32·77*
11	18·43	20·12	16·94	25·75	30·41?	39·10*	35·14*	32·87*	37·86**	32·44?
12	20·27	17·61	27·26?	24·33	36·27*	33·83*	34·18*	18·44	15·45	24·58
13	10·84	14·88	28·72?	23·34	21·43	20·79	42·70**	33·08*	49·13***	25·01
14	15·98	27·92	18·77	17·01	37·74**	26·30	36·25*	28·50	37·89**	32·31?
15	21·22	20·55	31·54*	21·82	17·31	35·89*	52·25***	20·86	37·84**	21·01
16	12·53	14·30	19·39	22·08	31·55*	20·83	50·98***	27·67	24·68	17·64
17	8·35	17·80	30·62?	12·40	16·00	14·03	42·23**	35·38*	24·92	19·03
18	9·03	18·68	24·97	23·11	20·36	16·03	28·58?	42·13**	36·73*	27·42
19	16·55	17·26	19·97	29·34	20·30	30·89?	34·33*	26·62	72·17***	31·93?
20	10·07	18·12	13·09	25·70	21·16	46·49**	37·14**	28·14	57·89***	23·77
21	16·29	14·66	44·74***	24·12	28·55?	33·38*	45·22***	32·37?	31·22*	27·95
22	10·72	14·26	55·37***	25·27	31·44*	32·70?	30·32?	17·93	28·72?	87·95***
23	11·69	36·22*	32·86*	21·66	30·69?	30·22?	24·36	16·63	12·91	25·87
24	15·39	28·80	15·32	20·73	27·57	21·75	27·81?	15·16	71·40***	34·10*
25	21·61	21·05	22·30	19·05	28·70	19·75	29·25?	34·18*	36·56*	19·96

		10%	5%	1%	0·1%
Significance points for M_1 (Bartlett's approx.):	$n = 25$	27·60	30·58	36·72	44·46
	$n = 20$	27·70	30·70	36·86	44·63
	$n = 5$	29·58	32·78	39·36	47·65
Significance at these levels denoted by		?	*	**	***

1.2

A Note on Regions for Tests of Kurtosis

G. E. P. BOX
Imperial Chemical Industries Ltd.

The most common test criterion for kurtosis is b_2, the sample fourth moment divided by the fourth power of the standard deviation; an alternative criterion proposed by Geary (1935) and denoted by a is the ratio of the sample mean deviation to the sample standard deviation. The present author has shown in a paper appearing on p. 318 of this volume that Bartlett's (1937) modification of the Neyman-Pearson L_1 criterion (1931) for testing for the equality of variances is so sensitive to kurtosis that when there is only one observation per group (group means assumed known) this criterion denoted by $M_{.1}$ is of the same order of sensitivity to kurtosis as is b_2 or a.

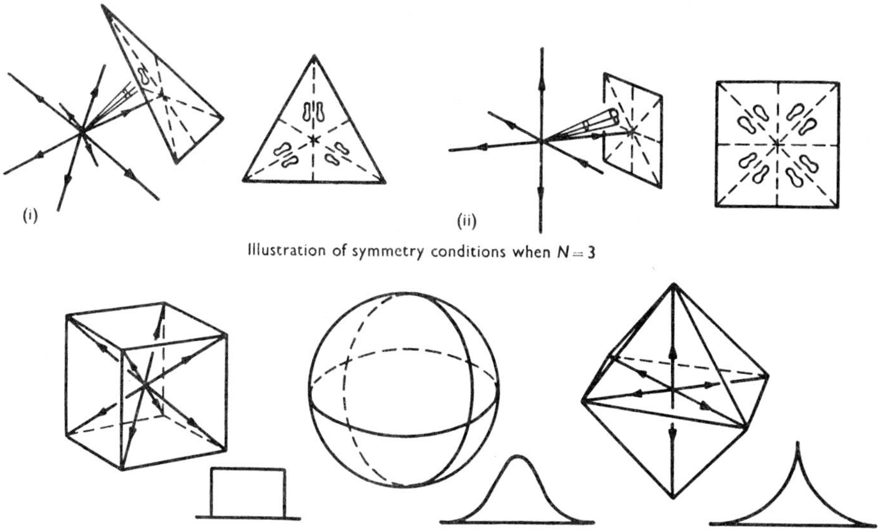

(i) (ii)
Illustration of symmetry conditions when $N = 3$

(iii) Rectangular (iv) Normal (v) Double exponential
Distributions and probability contours showing directions of accumulation of probability density

Fig. 1. Geometrical properties of tests of kurtosis.

A better understanding of the relations which the various tests for kurtosis have to one another and to $M_{.1}$ is gained by considering the general nature of critical regions for such tests. For simplicity we shall need to assume that the mean is known and equal to zero. We notice first that any criterion to detect kurtosis must be independent of scale; if we reach a certain conclusion from a sample $y_1, y_2, \ldots, y_N$ we must reach the same conclusion from the sample $ky_1, ky_2, \ldots, ky_N$. It follows that the boundary of the critical region in the sample space consists of one or more cones with apices at the origin. Also each observation has equal weight so that if the point $y_1, y_2, \ldots, y_N$ is on the boundary of a critical region then so must the remaining $N! - 1$ points obtained by permuting the order of the values. Finally, since a test of kurtosis should detect departures from normality other than those due to asymmetry of the distribution, the test criterion must be independent of the signs of the observations. Thus the pattern of critical subregions in the region of the space in which the observations are all positive will be repeated in all the remaining $2^N - 1$ regions of the space generated by possible differences in sign. The conditions of symmetry thus imposed are illustrated for $N = 3$ in Fig. 1 (i) and (ii) which shows cross-sections of subsets of cones satisfying these conditions. When the null-hypothesis of normality is true, the probability density in the sample space is constant over the surfaces of N-dimensional hyperspheres with centre at the origin. Consequently if the hypercones of the critical region are such that they include a proportion α of the total surface of such hyperspheres the error of the first kind is controlled at the level α irrespective of the value of the scale parameter.

The Collected Works of George E. P. Box, 1984, Wadsworth, Inc., Belmont, CA 94002.
Originally published in *Biometrika,* vol. 40, parts 3 and 4 (1953), pp. 465–468.

In addition to controlling this error, we require the test to pick out as often as possible a departure from normal-kurtosis when it occurs. Consider the probability contours corresponding to symmetrical distributions having varying amounts of kurtosis in the case $N = 3$. For the normal distribution the contours are spheres (Fig. 1 (iv)). As γ_2 falls below zero, the contours tend to belly out and for the rectangular distribution ($\gamma_2 = -1{\cdot}2$) the single probability contour is the surface of a cube (Fig. 1 (iii)), the density within the cube being uniform. As γ_2 is made larger than zero the contours tend to sag

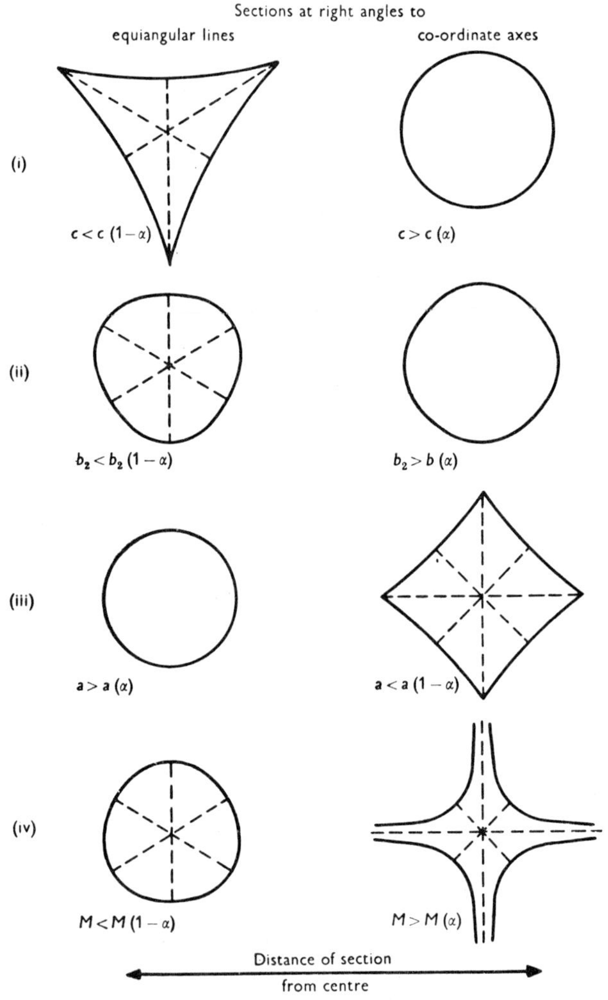

Fig. 2. Sections of critical regions.

inwards till for the double exponential distribution ($\gamma_2 = 3$) the contours are surfaces of regular octahedra (Fig. 1 (v)). In general for platykurtic distributions ($\gamma_2 < 0$) there is an accumulation of probability density along the equiangular lines (and a depletion along the co-ordinate axes), whilst for leptokurtic distributions ($\gamma > 0$) there is an accumulation of probability density along the co-ordinate axes (and a depletion in the direction of the equiangular lines). It follows that (i) a good critical region for testing for platykurtosis would be the 2^N hypercones having apices at the centre and axes along the 2^N equiangular lines, whilst (ii) a good critical region for testing for leptokurtosis would consist of the $2N$ hypercones having their apices along the $2N$ branches of the co-ordinate axes.

The directions of the axes of cones forming such regions when $N = 3$ are shown by arrows in Fig. 1 (iii) and (v) respectively. In particular b_2 and a have critical regions of these types and so has $M_{.1}$. In each case, the particular type which occurs depends on whether we are using the criterion to test for leptokurtosis or for platykurtosis. The regions differ only in the shapes of their cross-section and even these differences are limited to some extent by the symmetry conditions listed above. For the case $N = 3$, sections of cones corresponding to the approximate upper and lower 5 % levels for the test criteria b_2, a and $M_{.1}$ are shown in Fig. 2 (ii), (iii) and (iv). (In order to make clear that the region in Fig. 2 (ii) defined by $b_2 > b_2(\alpha)$ has a non-circular section, it has been necessary slightly to exaggerate the departure from circularity.)

2. It is of some interest to consider kurtosis tests from the point of view of the Neyman-Pearson theory of best critical regions. Suppose that the distribution specified by the alternative hypothesis is a member of the family

$$p(y) = \left\{\frac{2\rho}{q}\Gamma\left(\frac{1}{q}\right)\right\}^{-1} \exp-\left\{\frac{|y|}{\rho}\right\}^q \quad (-\infty<y<+\infty, 0<q<\infty, 0<\rho<\infty), \tag{1}$$

and *q is given*. This is a symmetrical distribution with mean zero and

$$\kappa_2 = \rho^2\frac{\Gamma(3/q)}{\Gamma(1/q)}, \quad \gamma_2 = \frac{\Gamma(5/q)\,\Gamma(1/q)}{\{\Gamma(3/q)\}^2} - 3. \tag{2}$$

When $q = 2$, γ_2 is zero and the distribution is normal. When $q > 2$ the distribution is platykurtic and, in particular, when q tends to infinity the distribution tends to the rectangular, for (1) tends to

$$p(y)\begin{cases} = 1/(2\rho) & (y<\rho) \\ = 0 & (y>\rho). \end{cases} \tag{3}$$

When $q < 2$ the distributions are leptokurtic in particular when $q = 1$ the distribution is the double exponential.

The likelihood ratio test criterion (Neyman & Pearson, 1928) for testing the null hypothesis of normality ($q = 2$) against the *single alternative* that the distribution is of the form of the above equation with q equal to some other *specific* value q_0 is readily found to be proportional to

$$\Lambda_{q_0} = (m_{q_0})^{1/q_0}/s, \tag{4}$$

where m_q is the qth absolute moment,

$$m_q = \Sigma\,|\,y\,|^q/N, \quad \text{and in particular } s = m_2^{\frac{1}{2}}. \tag{5}$$

The inequality

$$\Lambda_{q_0} < \Lambda_{q_0}(1-\alpha), \tag{6}$$

where $\Lambda(1-\alpha)$ and $\Lambda(\alpha)$ refer respectively to the lower and upper significance points of the criterion, defines a set of cones lying along the co-ordinate axes when $q_0 < 2$ and a set of cones lying along the equiangular lines when $q_0 > 2$. That every such set defines a *best* critical region for the appropriate value of q_0 can be seen from the fact that on every* hyperspherical shell (4) and (6) define the region, independent of scale factors in the null distribution or in the alternative distribution, the boundary of which is given by the fundamental inequality $p_t \geq kp_0$ of Neyman & Pearson (1933). This region on every such shell thus contains for its size the greatest possible concentration of probability density when the appropriate alternative hypothesis is true.

We see that, $a < a(1-\alpha)$ defines the most powerful test when the alternative distribution is the double exponential, for which $\gamma_2 = 3$, also $b_2 < b_2(1-\alpha)$ defines the most powerful test when the alternative distribution is $p(y) = \text{const.}\exp-(y/\rho)^4$ for which $\gamma_2 = -0{\cdot}812$. When the alternative distribution is rectangular ($\gamma_2 = -1{\cdot}2, q = \infty$) the test is defined by $c < c(1-\alpha)$, where

$$c = |\,y_L\,|/s \tag{7}$$

and y_L is the observation largest in absolute magnitude. It should be noted that with this criterion we should be testing whether the largest deviate in absolute magnitude were *too small* and the test would be based on the *lower* tail area of the distribution of c. The same criterion, but using the upper tail area of

* This is literally true if it is assumed as in (1) that $q_0 < \infty$. In the limiting case when $q_0 = \infty$, if we suppose a *particular* rectangular distribution of semi-range ρ to be the alternative and if r is the radius of a hemispherical shell, then the region defined on the shell will contain an amount of probability density *greater* than any other region of similar size if $\rho < r < \sqrt{N}\rho$. For all other values of r it will contain an amount of probability density *equal* to that of other regions of the same size. The test is thus still most powerful in this limiting case.

c, has been proposed by Thompson (1935) as a test for a single outlier (see also Pearson & Chandra Sekar, 1936). Sections of critical regions for this criterion are shown in Fig. 2 (i). It will be noted that only the lower tail areas of c, b_2 and a define most powerful tests for the alternatives considered. In Fig. 2 the regions are differentiated by drawing in the axes of symmetry only for those sections which correspond to most powerful tests. We should expect c and b_2 to be particularly good criteria, therefore, when the alternative hypothesis was that the distribution showed marked platykurtosis; on the other hand, in testing for marked leptokurtosis a would be expected to be better. Pearson's (1935) sampling experiments (his Table 7) which are shown again in table 4 of the article appearing on p. 318 of this journal support this suggestion (although with so few results no firm conclusions can be drawn). For the rectangular distribution a larger number of significant results is obtained with b_2, whilst for the double exponential a larger number is obtained with a. It does not of course follow that a would become better than b_2 as soon as γ_2 was greater than zero; the change-over might well be at some other point.

In a study of possible tests of kurtosis of this type, and assuming the alternative distribution to be the symmetrical Gram-Charlier, Geary (1947) reached the conclusion that b_2 was the most 'efficient' test for *infinitely large samples* (and correspondingly small departures from normality). He also showed that, over a wide range of values of q, the power of the various possible tests would not be expected to be greatly different. The present research emphasizes that the choice of criterion for samples of finite size must depend on the type of alternative hypothesis which is in mind. The similarity of the critical regions for the various criteria confirm Geary's second conclusion.

REFERENCES

Bartlett, M. S. (1937). *Proc. Roy. Soc.* A, **160**, 268.
Geary, R. C. (1935). *Biometrika*, **27**, 310.
Geary, R. C. (1947). *Biometrika*, **34**, 209.
Neyman, J. & Pearson, E. S. (1928). *Biometrika*, **20** A, 175.
Neyman, J. & Pearson, E. S. (1933). *Phil. Trans.* A, **231**, 289.
Pearson, E. S. (1935). *Biometrika*, **27**, 333.
Pearson, E. S. & Chandra Sekar, C. (1936). *Biometrika*, **28**, 308.
Thompson, W. R. (1935). *Ann. Math. Statist.* **6**, 214.

1.3

Permutation Theory in the Derivation of Robust Criteria and the Study of Departures from Assumption*

G. E. P. BOX and S. L. ANDERSEN
Imperial Chemical Industries Ltd. and Institute of Statistics, University of North Carolina

[Read before the RESEARCH SECTION of the ROYAL STATISTICAL SOCIETY, November 3rd, 1954, Mr. E. C. FIELLER in the Chair]

SYNOPSIS

IN the practical circumstances in which statistical procedures are applied, little is usually known of the validity of assumptions such as the normality of the error distribution. Procedures are required which are "robust" (insensitive to changes in extraneous factors not under test) as well as powerful (sensitive to specific factors under test). Permutation theory, which provides one method for deriving robust criteria, is discussed and applied to the problem of comparing variances.

1. INTRODUCTION

IN this paper attention is confined to problems in hypothesis testing although we believe that much of this discussion applies in other fields.

To fulfil the needs of the experimenter, statistical criteria should

(1) be sensitive to change in the specific factors tested,
(2) be insensitive to changes, of a magnitude likely to occur in practice, in extraneous factors.

A test which satisfies the first requirement is said to be powerful and we shall typify a test which satisfies the second by calling it "robust".

In the derivation of parametric tests (for example by the methods of Neyman and Pearson) it is usual to assume a form of mathematical model involving some specific probability distribution and then carefully to select the form of criterion so as to satisfy the first requirement listed above. Because this procedure does not necessarily result in tests which satisfy the second requirement "non-parametric" tests have been devised in which the form of criterion is selected, usually somewhat arbitrarily, but in such a way that the assumptions which need to be made are of a less specific character. Thus parametric tests tend to satisfy the first requirement listed above (at least when the assumptions are true) but not necessarily the second, whilst non-parametric tests tend to satisfy the second requirement but not necessarily the first. For this reason much research has been conducted on (1) the robustness of various parametric tests and (2) the power of various non-parametric tests.

It is a disconcerting fact that while for one problem a set of apparently restrictive assumptions may lead to a remarkably robust criterion, yet for some other problem the same set of assumptions will give a criterion which is of little value unless the assumptions are very nearly justified. For example, both the analysis of variance test to compare k means and the L_1 test or Bartlett (M_1) test to compare k variances may be derived assuming normally distributed error terms. The very different effects on the null probability that accompany departures from normality are illustrated in Table 1.

* Sponsored by the Office of Ordnance Research, United States Army under contract DA-36-034-ORD-1177.

The Collected Works of George E. P. Box, 1984, Wadsworth, Inc., Belmont, CA 94002. Originally published in *J. Roy. Stat. Soc.*, Series B, vol. XVII, no. 1 (1955), pp. 1-34.

TABLE 1

Comparison of Effect of Kurtosis in Tests to Compare k *Means and Tests to Compare* k *Variances*

Departure from Normal Kurtosis	*Percentage Chance of Exceeding Nominal 5 per cent. point when Null Hypothesis True*			
	Comparison of k *means of 10 Observations (Analysis of Variance)**		*Comparison of* k *Variances (Bartlett Test* M_1*)*†	
$\gamma_2 = \beta_2 - 3$	$k = 2$	$k = 20$	$k = 2$	$k = 20$
2	4·74	4·90	16·6	71·8
0	5·00	5·00	5·0	5·0
−1	5·13	5·05	0·56	0·0004

* Taken from Gayen's (1950) values assuming parent Edgeworth population. Similar values are obtained from the method of Section 4.1 of this paper.

† Asymptotic values. Values of similar magnitude are obtained for any sample size for specific populations discussed by Box (1953*a*).

It is clear that the test on variances can be so misleading as to be almost valueless unless we can assert that in most situations met in practice the distribution is very close to the normal. The authors' belief is that such an assertion would certainly not be justified. Published data are comparatively meagre but the frequency distributions given in the older issues of *Biometrika* give little ground for supposing that distributions usually follow the normal law. For example, many of the curves in the Monier-Williams data on percentage butter-fat in milk quoted by Tocher (1928) while looking "reasonably normal" in fact show values of γ_2 between 1 and 2. Similarly of the eight sets of data supplied by Shewart to E. S. Pearson (1931) as typical of observations collected at Bell Telephone laboratories three showed values of γ_2 between 1 and 2.

Many writers including Pearson (1931), Geary (1947), Gayen (1950) and David and Johnson (1951*a*, *c*, 1952) have studied the analysis of variance criterion when the distribution is non-normal. It has been found to be remarkably insensitive to general non-normality.* It has also been shown (Welch, 1937*b*; David and Johnson, 1951*b*; Horsnell, 1953; Box 1954*a* and *b*) that in the commonly occurring case where the group sizes are equal this test is not very sensitive to variance inequalities from group to group. The analysis of variance test for equal group sizes can probably therefore be used with confidence in most practical situations and since we could not expect to obtain a criterion of greater power, unless the nature of the population sampled were specifically known, it may be regarded as fulfilling remarkably well both the requirements of power and robustness. This is perhaps a reason why this test has proved of such great practical utility.

It is clearly desirable that other tests should have properties as satisfactory as those of the analysis of variance test and the problem arises of what is to be done when a standard criterion is found to be unsatisfactorily sensitive to departures from assumption. It has been suggested that before using such a test we should employ one or more preliminary tests to "determine whether the assumptions are justified". This idea seems to us to be a mistaken one, for whether or not a departure from assumption is detected will depend upon the power of the preliminary test which will in turn depend upon the number of observations available. On the other hand, whether or not such a departure is of *importance* depends upon something quite different, namely the robustness of the main test. Thus in some circumstances (for example a preliminary test for equality of variances made before a test to compare means from equally sized groups) a detectable discrepancy might not be large enough to upset the main test, while in other circumstances (for example, a test of normality applied before a Bartlett test on variances) a discrepancy too small to be detected could very seriously upset the behaviour of the main test. It would seem that if this idea of preliminary testing were taken to its logical conclusion we ought to perform another test to check the assumptions made in the preliminary test and so on. We would thus be faced with an endless, and possibly circular, series of tests.

* "General" non-normality is meant to imply that the observations all have the same non-normal parent distribution with possibly different means. This would seem to provide a likely approximation to many experimental situations. Somewhat larger effects have been demonstrated (for example Gayen, 1950) when the distribution is different for observations in different groups.

What are really required are test criteria which 'can stand on their own feet', so that no preliminary testing is necessary. In situations where the standard criterion does not satisfy this requirement some alternative or modified criterion should be sought. One instance of such a modified criterion is that proposed by Cochran (1937) and by Welch (1937*b*, 1951) and James (1951) for the comparison of means in a one-way classification when the variances may differ. As we have noticed the divergencies occurring in the analysis of variance test due to inequality of variance are usually not very serious if the group sizes are equal. However, with unequal groups much larger effects can occur. The standard analysis of variance criterion for the comparison of k means $\bar{x}_1, \bar{x}_2, \ldots, \bar{x}_t, \ldots, \bar{x}_k$ is of the form $\Sigma\, w_t(\bar{x}_t - \bar{x})^2$ where the weighting coefficient $w_t = n_t/s^2$ and $\bar{s}^2$ is the pooled estimate of variance. With this criterion the expected values of treatment and error mean squares have the ratio

$$b = 1 + \frac{k\,(N-1)}{(k-1)\,N}\left(\frac{\bar{\sigma}^2}{\dot{\sigma}^2} - 1\right)$$

where N is the total number of observations, $\bar{\sigma}^2$ is the unweighted mean variance and $\dot{\sigma}^2$ is the weighted mean variance, the weights being the numbers of degrees of freedom in the groups. When the group sizes are unequal, $\bar{\sigma}^2$ and $\dot{\sigma}^2$ will in general be different, b will differ from unity, and serious bias may be introduced into the comparison. To cure this deficiency weights $w_t = n_t/s_t^2$ are used in the modified criterion which is clearly more appropriate where the variances differ.

The exact null distribution of the modified criterion on the normal assumption has not been found but the approximation supplied by Welch (1951) is probably quite adequate in practice. Since the assumptions on which "exact" distributions are determined are seldom justified in practice, and since in any case the mind cannot appreciate small differences in probability, reasonable approximations to probability distributions are all that are really required.

The modified criterion would be expected to be insensitive to differences in group variances (and by analogy with the standard test) to departures from normality also. Its power has not been investigated but assuming that this is satisfactory it would seem to fill the need for a reliable test to compare means when the variances and sample sizes are different.

Modified tests of this kind are required for other situations. For example:

(1) It has been shown by Geary (1936) that the single-sided t test for the comparison of a sample mean with some hypothetical value is particularly sensitive to skewness in the parent population. Remedial measures have been proposed by Tukey (1948).

(2) It is almost certainly true that certain of the multivariate tests can be grossly misleading under practical conditions, but little has been done to determine the extent of these deficiencies or to remedy them.

(3) Some of the tests proposed for "Poisson variates" will almost certainly be upset in the commonly occurring case where the distribution deviates from the Poisson form.

(4) The tests to compare variances are extremely sensitive to non-normal kurtosis and alternative robust criteria are required.

2. Permutation Tests

A remarkable new class of tests which have since been called permutation tests (or randomisation tests) were introduced by Fisher (1935). After discussing the application of the "paired" t test to some experimental data of Darwin's, Fisher remarked:

> "It seems to have escaped recognition that the physical act of randomisation, which, as has been shown, is necessary for the validity of any test of significance, affords the means, in respect of any particular body of data, of examining the wider hypothesis in which no normality of distribution is implied."

He went on to show how the null hypothesis could be tested simply by counting how many of the mean differences obtained by rearranging the pairs exceeded the actual mean difference observed. He showed that for the particular set of data he examined, the null probability given by the permutation test and that given by the t test were almost identical.

In connection with a later application of the permutation principle to comparing means of unpaired data, Fisher (1936) said:

> "Actually, the statistician does not carry out this very simple and very tedious process but his conclusions have no justification beyond the fact that they agree with those which could have been arrived at by this elementary method."

In a discussion of this type of test, E. S. Pearson (1937) emphasized that in a permutation test, as in any other, a choice of the criterion to be used still had to be made. For example, a two-sample permutation test of the type discussed by Fisher could be based on the differences in means, medians, mid-points or any other "position" statistics of the samples. He carried out a sampling experiment performed with a rectangular parent population in which the permutation test based on the differences of the mid-points detected departures from the null hypothesis more often than did the test using the sample means. He thus emphasized that if the permutation test was to be powerful, the choice of criterion would have to depend on the type of alternative hypothesis which the experimenter had in mind.

The points of view expressed by Fisher and Pearson are of course in no way contradictory. Fisher is concerned with the validity of the test of the null hypothesis, while Pearson is concerned with the power of the test when some alternative hypothesis is true.

Two alternative views of the nature of the inference in the permutation test can be taken. These differ in the conception of the population of samples from which the observed sample is supposed to have been drawn. On the first view our attention is confined only to that finite population of samples produced by rearrangement of observations of the experiment. We prefer to adopt the second view which is that the samples are regarded as being drawn from some hypothetical infinite population in the usual way.

It is of some interest to consider the permutation test from the point of view of Neyman-Pearson theory.

2.1. *Permutation Tests from the Point of View of Neyman-Pearson Theory*

In the Neyman-Pearson theory it is supposed that we wish to test some hypothesis, H_0, concerning the nature of the probability law governing N observations, $x_1, x_2, \ldots, x_N$. We have in mind some alternative hypothesis H_1 concerning the nature of this law. To make the test we select a region in the sample space, w, called the "critical region" which is such that, if the sample point is contained in w the null hypothesis will be rejected, and otherwise accepted. The criterion which defines the critical region w is chosen such that

(i) When H_0 is true the chance of rejecting H_0 will always be controlled at some level, α, called the risk of error of the first kind.

(ii) When H_1 is true the chance of rejecting H_0, called the power of the test, will be as large as possible.

Thus, if X_r is a vector of observations $(x_1, x_2, \ldots, x_N)$ and $p_0(X_r)$ and $p_1(X_r)$ are the probability laws for the null hypothesis and alternative hypothesis, we require that w is such that

$$\int_w p_0(X_r)\, dX_r = \alpha \qquad (1)$$

$$\int_w p_1(X_r)\, dX_r \text{ is a maximum} \qquad (2)$$

Now usually we do not know what form is appropriate for $p_0(X_r)$, and if we assume some specific function, such as the normal law, we run the risk of the test being valueless under practical circumstances when the real distribution is unknown. The object of the permutation procedure is to satisfy equation (1) with the minimum of assumption.

We can regard the sample X_r as a member of a set X containing the $N!$ samples $X_1, X_2, \ldots, X_r, \ldots, X_{N!}$ which contain the same observations as X_r in all possible arrangements. The chance that we shall draw the sample X_r given that it belongs to the set X is

$$p_0(X_r/X) = \frac{p_0(X_r)}{\sum_{r=1}^{N!} p_0(X_r)} = \frac{p_0(X_r)}{p_0(X)} \qquad (3)$$

Thus $p_0(X_r)$ can always be written

$$p_0(X_r) = p_0(X)\, p_0(X_r/X) \qquad (4)$$

and equation (1) becomes

$$\int_w p_0(X)\, p_0(X_r/X)\, dX_r = \alpha \qquad (5)$$

To satisfy this equation we now need only assume that $p_0(X_r)$ is some *symmetric* function of the observations $x_1, x_2, \ldots, x_N$. (This would be so if the observations were *independently* and *identically* distributed in any form whatever, or, if each observation were equally dependent on all the others, but not, for example, if the observations were not identically distributed or were serially correlated).

If then $p_0(X_r)$ is a symmetric function, we have $p_0(X_r) = p_0(X_q)$, $(r, q = 1, 2, \ldots N!)$ and

$$p_0(X_r/X) = (N!)^{-1} \quad . \quad . \quad . \quad . \quad . \quad . \quad . \quad (6)$$

Provided, therefore, that α is such that an integer I exists* for which $I = \alpha N!$ we can ensure that equation (1) is exactly satisfied by arranging that I out of the $N!$ permutations of each set X are contained in w and $N! - I$ are outside w, when we have for the integral in (5)

$$\int_\Delta p_0(X) \sum_{r=1}^{I} p_0(X_r/X)\, dX = \frac{I}{N!} \int_\Delta p_0(X)\, dX = \alpha \quad . \quad . \quad . \quad . \quad (7)$$

where Δ is the entire region of the sample space.

2.2. *More than One Population*

When two or more treatments are compared within n blocks, each containing s observations, we need only assume that within any particular block the probability density function is symmetric. The nature of the functions can differ from block to block. If the vector of observations X_t^j in the jth block is regarded as a member of a set X^j containing the $s!$ samples $X_1^j, X_2^j, \ldots X_t^j \ldots X_{s!}^j$ in all possible arrangements, then as before we can write the null density function in the form

$$p_{0j}(X_t^j) = p_{0j}(X^j)\, p_{0j}(X_t^j/X_j) \quad . \quad . \quad . \quad . \quad . \quad . \quad (8)$$

and we have to choose w to satisfy the equation

$$\int_w \prod_{j=1}^{n} p_{0j}\,(X^j) \prod_{j=1}^{n} p_{0j}\,(X_t^j/X^j) \prod_{j=1}^{n} dX_t^j = \alpha \quad . \quad . \quad . \quad . \quad . \quad (9)$$

If we take a vector from each of the n blocks to form a new vector X_r of ns elements

$$X_r = (X_t^1.\ X_u^2, \ \ldots\ , X_v^n)$$

where $r = 1, 2, \ldots (s!)^n$; $t, u, v, = 1, 2, \ldots s!$. Then as before we can write the integral in (9) as

$$\int_w p_0(X)\, p_0(X_r/X)\, d\, X_r \quad . \quad . \quad . \quad . \quad . \quad . \quad . \quad . \quad (10)$$

If the within-block distributions are symmetric functions then $p_0(X_r/X) = 1/(s!)^n$ and we can again construct a region w of size α by arranging that $I = \alpha(s!)^n$ out of the $(s!)^n$ within-block permutations of the $ns = N$ observations are contained in w. We have therefore

$$\int_\Delta p_0(X) \sum_{r=1}^{I} p_0(X_r/X)\, dX = \frac{I}{(s!)^n} \int_\Delta p_0(X)\, dX = \alpha \quad . \quad . \quad . \quad . \quad (11)$$

where Δ is the entire region of the sample space.

2.3. *Possible Procedures for Controlling the Value of* α

To maintain α at its nominal value when the null hypothesis is true we note three possible procedures which are, in order of reliability:

* When no such integer exists we cannot maintain the first kind of error exactly at the level α but may use the nearest value α' such that the integer I equals $\alpha' N!$. In practice this will usually be quite satisfactory.

(1) Construct a critical region of size α on the supposition that $p_0(X_r)$ follows some specific distribution function. If then the true distribution is not that assumed we shall be in error by a lesser or greater amount depending on the "robustness" of the criterion which we evolve. We might of course postulate a population sufficiently flexible to cover all the circumstances likely to be met in practice. However, such populations are usually difficult to define and to deal with mathematically and some more specific assumption such that the distribution follows the normal law, is usually made. The value of a test so derived will then depend on factors which are often unforeseeable and certainly its practical utility cannot be assumed.

(2) Assume only that $p_0(X_r)$ belongs to the class of symmetric distribution functions and construct a critical region based on permutation theory. We shall then be in error only if $p_0(X_r)$ does not belong to this very wide *class* of distributions (for example if the observations are serially correlated).

(3) Carry out a process of randomization (in cases where this is possible) so guaranteeing that the effective distribution $p_0(X_r)$ is symmetric and use a critical region based on permutation theory. We cannot now be in error so far as α is concerned unless further disturbances are introduced after the randomization has been performed.

2.4. *Power of the Permutation Test*

Now for each set of vectors X there are of course a very large number of ways in which we can choose the subset of I vectors which are to be included in the critical region. For the single-population test of Section 2.1, for instance, there are $(N!)/\{(N-I)!I!\}$ different ways of doing this.

To obtain a powerful test we should choose the particular subset of I samples so that

$$\int_A p_1(X) \sum_{r=1}^{I} p_1(X_r/X)\, dX \text{ is a maximum} \quad . \quad . \quad . \quad . \quad (12)$$

Now if, in fact, some region is better than another $p_1(X_r)$ cannot be a symmetric distribution. So that in all cases in which we are interested the value of this integral depends upon the form of $p_1(X_r)$. Thus, although we need not specify the particular form of the null distribution, we can only satisfy Neyman and Pearson's second condition if we are prepared to be specific about the class of probability density functions which we have in mind in our alternative hypothesis. If the alternative H_1 is so specified then as has been shown by Lehman and Stein (1949) it is a fairly simple matter to select a best critical region.

For example, suppose that the object of the permutation procedure was to test the hypothesis that each of two samples came from the same distribution against the alternative that they came from two different distributions, one of which had a larger location parameter than did the other.

If we assumed as the specific alternative hypothesis that the observations were drawn from normal populations having the same variance but different means $\mu_2 > \mu_1$ then

$$p_1(X_r) = \text{constant} \times \exp.\left\{-\frac{1}{2\sigma^2}\left(\sum_{\alpha=1}^{n_1}(x_\alpha-\mu_1)^2 + \sum_{\beta=n_1+1}^{n_1+n_2}(x_\beta-\mu_2)^2\right)\right\} \quad . \quad (13)$$

Clearly (12) will be satisfied only if $\sum_{r=1}^{I} p_1(X_r/X)$ is made a maximum for each set X of observations $x_1 \;\ldots\; x_{n_1}, x_{n_1+1} \;\ldots\; x_{n_1+n_2}$ and since $p_1(X) = \sum_{r=1}^{N!} p_1(X_r)$ is constant for all the samples in X, we have in fact to maximize $\sum_{r=1}^{I} p_1(X_r)$ for each set X.

Now if we consider the value of $p_1(X_r)$ for each of the $N!$ possible rearrangements of the observations we see that it takes its largest value for the $n_1!n_2!$ arrangements which are such that the smallest n_1 observations fall in the first sample and the largest n_2 in the second sample. Equivalently, these are the samples which have the largest difference in means. $p_1(X_r)$ takes its next largest value for the $n_1!n_2!$ samples with the next largest difference in means and so on.

Thus the assumption that the alternative distribution is normal leads to a test in which the I samples included in the critical region are those having the largest differences in group means.* If the distribution had been assumed to be rectangular, a more powerful test would be based on the comparison of mid-points as was found in Pearson's sampling experiment.

It will be noted that if the distribution were really normal then the permutation test would necessarily be less powerful than the normal theory test (i.e., the single-sided t test) since the latter is uniformly most powerful for the normal distribution. This difference in power represents the price we must pay to ensure greater robustness of the criterion in the practical situation where we do not know whether the distribution is normal or not. Some illustration of the amount of power lost (which seems to be remarkably small in the cases considered) is given later in the discussion of tests to compare variances.

2.5. *Rationale for Choice of Alternative Hypothesis*

On permutation theory then we need make no very restrictive assumption concerning the null distribution, but to obtain a "most powerful" criterion for some particular hypothesis we must define the precise alternative we have in mind. The actual test we make is a test of the null hypothesis so that lack of restrictive assumptions when the null hypothesis is true is all that we require to ensure the validity of the test procedure. So far as the practical choice of criterion is concerned, it seems we could argue as follows. Our feeling about the type of distribution likely to be encountered could usually best be expressed in terms of a distribution of possibilities rather than any one possibility. This mental "prior distribution of distributions" might be imagined to have some central value. We should be reluctant to treat this central value as if it were the only one that could occur so far as the null distribution was concerned since this might limit the validity of the test of the null hypothesis to cases where this central distribution really applied. On the other hand, once the validity of the test of the null hypothesis had been safeguarded by the use of permutation theory, it would seem natural to base our criterion on a statistic appropriate for what we supposed to be the central alternative distribution. Even though we might expect this distribution seldom, if ever, to be exactly realized, we would expect that the loss of power suffered in the long run for a series of tests on a series of varying distributions would be smallest for such a statistic.

Using this rationale, the choice in the above example between the difference in means and the difference in midpoints as the appropriate criterion would be based on whether the statistician's mental picture of the distribution of distributions likely to be met in practice in the particular circumstances was centred about the normal or about the rectangular distribution. In most cases the normal distribution would be chosen, though there could be experimental circumstances which would lead the statistician to choose some other distribution as the central one for experiments of a particular type. A test derived in this way would seem to satisfy as nearly as possible the two requirements of Section 1.

We have seen how we are led to a permutation test based on some particular function of the observations. We now consider the properties of such functions when the observations are permuted.

2.6. *Permutation Distribution*

Suppose $g(X_r)$ is a function of the N ordered observations, $x_1, x_2 \; . \; . \; . \; x_N$. Then the probability $Pr\{(g(X_r) = g)/X\} = \sum_{g(X_r)=g} p(X_r/X)$ tabulated as a function of g is the "permutation distribution" of $g(X_r)$. Evaluation of the permutation distribution, or of such part of it as is necessary to determine the critical value of the statistic, is laborious. To make the permutation theory of practical value Pitman (1937*a*, *b*) and Welch (1937*a*, 1938) used an approximation to the permutation distribution based on the values of its moments.

* Although there are $N!$ possible arrangements of the sample, there are only $N!/n_1!\,n_2!$ arrangements which result in possibly different mean differences. Thus in practice α would have to be taken so that $I/n_1!\,n_2! = \alpha N!/n_1!\,n_2!$ was an integer.

2.7. *Permutation Moments*

The hth permutation moment of $g(X_r)$ denoted by $\underset{P}{E}\{g(X_r)\}^h$ is the hth moment of the permutation distribution of $g(X_r)$ and is defined by

$$\underset{P}{E}\{g(X_r)\}^h = \sum_{r=1}^{\Omega} \{g(X_r)\}^h \, p(X_r/X) \quad . \quad . \quad . \quad . \quad . \quad (14)$$

Where the summation is over all permissible arrangements Ω in number.

2.8. *The Parent Probability Density Function and the Permutation Distribution*

The probability density function of any statistic $g(X_r)$ can be regarded as a weighted aggregate of permutation distributions of $g(X_r)$; for

$$Pr\{g < g(X_r) < g + \delta g\} = \int_{g(X_r)=g}^{g(X_r)=g+\delta g} p(X_r)\, d\, X_r \quad . \quad . \quad . \quad . \quad . \quad (15)$$

$$= \int_{g(X_r)=g}^{g(X_r)=g+\delta g} p(X)\, p(X_r/X)\, dX = \int_{\Delta} p(X) \sum_{g(X_r)=g}^{g(X_r)=g+\delta g} p(X_r/X)\, d\, X \quad . \quad . \quad (16)$$

2.9. *Ordinary Moments and Permutation Moments*

As originally indicated by Welch (1937), the hth overall moment of $g(X_r)$ denoted by $E\{g(X_r)\}^h$ may be evaluated by taking the expectation of the hth permutation moment over all values of X; for

$$E\{g(X_r)\}^h = \int_{\Delta} \{g(X_r)\}^h \, p(X_r)\, d\, X_r \quad . \quad . \quad . \quad . \quad . \quad . \quad . \quad (17)$$

$$= \int_{\Delta} p(X) \sum_{r=1}^{\Omega} \{g(X_r)\}^h \, p(X_r/X)\, d\, X \quad . \quad . \quad . \quad . \quad . \quad (18)$$

$$= \int_{\Delta} p(X)\, \underset{P}{E}\{g(X_r)\}^h d\, X \quad . \quad . \quad . \quad . \quad . \quad . \quad . \quad (19)$$

The above theory may be employed to provide two useful results:

(1) Robust tests may be formulated by approximating to the permutation tests.

(2) The effect on standard test procedures of non-normality and certain other departures from assumption may be evaluated.

3. AN EXAMPLE

It is helpful to study in some detail the following simple example. Suppose that an experiment has been carried out in which n pairs of observations $x_{ti}(t = 1, 2;\ i = 1, 2, \ldots n)$ have been made. One observation within each pair has treatment A applied and the other has treatment B, as in Fisher's first example. This is the familiar situation encountered in the paired t test. It can equivalently be regarded as an example of a randomized block design having n blocks and $s = 2$ treatments with a total of $sn = N$ observations.

The null hypothesis is that within each pair the probability density is unchanged by interchanging the observations. Suppose the alternative hypothesis was that

$$x_{ti} = \alpha_t + \beta_i + z_{ti} \quad . \quad . \quad . \quad . \quad . \quad . \quad . \quad (20)$$

where α_t and β_i are treatment and block constants respectively and z_{ti} was a normally distributed random variable with mean zero and constant but unknown variance σ^2. Then using the type of argument indicated in Section 2.4, Lehman and Stein (1949) show that the best permutation critical region is that based on the difference between the sample means, $\bar{x}_1 - \bar{x}_2 = \bar{y}$. The observed difference in means is referred to the permutation distribution of mean differences generated by all rearrangements within pairs. This is equivalent to the distribution obtained by associating all possible plus and minus signs with the individual differences $y_i = x_{1i} - x_{2i}$.

3.1. *Approximation to the Test*

The same critical region is obtained if we calculate the permutation distribution for the t statistic itself because $t = \{n(n-1)\}^{\frac{1}{2}}\,\bar{y}(\Sigma\, y^2 - n\bar{y}^2)^{-\frac{1}{2}}$ is a monotonic increasing function of $\bar{y}$. Equivalently we can use the analysis of variance criterion $F = t^2 = \{S_T/1\}/\{S_E/(n-1)\}$ where S_T and S_E are the treatment and error sums of squares respectively. Following Pitman and Welch, it is best to consider the form $W = S_E/(S_E + S_T)$ which in the present example is

$$W = \Sigma\,(y-\bar{y})^2/\Sigma\, y^2 = \{1 + t^2/(n-1)\}^{-1}$$

a monotonic decreasing function of t^2 in which the denominator $\Sigma\, y^2$ remains constant in all permutations. The permutation moments and normal theory moments for W are as follows:

$$\underset{P}{E}(W) = \frac{n-1}{n} \qquad \underset{N}{E}(W) = \frac{n-1}{n} \quad . \quad . \quad . \quad (21)$$

$$\underset{P}{V}(W) = \frac{2(n-1)}{n^2(n+2)}\left(1 - \frac{b_2 - 3}{n-1}\right) \qquad \underset{N}{V}(W) = \frac{2(n-1)}{n^2(n+2)} \quad . \quad . \quad . \quad (22)$$

The mean of the permutation distribution of W is the same as that for normal theory. The variance differs from that of normal theory by the inclusion of a term of order n^{-1}, involving the sample value of the fourth moment ratio which is defined as

$$b_2 = (n+2)\,\Sigma\, y^4/(\Sigma\, y^2)^2 \quad . \quad . \quad . \quad . \quad . \quad (23)$$

so that $E(b_2) = 3$ when the y's are normal. It will be recalled that for normal theory, W follows a Beta distribution $Pr(W < W_0) = I_{w_0}(\frac{1}{2}\nu_2, \frac{1}{2}\nu_1)$ where ν_1 and ν_2, the degrees of freedom of the distribution, are in this case $\nu_1 = 1$ and $\nu_2 = n-1$.

The permutation distribution of W is of course discontinuous. However, its value lies between zero and one and Pitman (1937) has shown that its third and fourth moments agree reasonably closely with those of the Beta distribution. It is therefore reasonable to approximate the permutation distribution by a Beta distribution, equating the first two moments of the two distributions.

For a Beta distribution with mean and variance μ_1 and μ_2 and degrees of freedom ν_1 and ν_2

$$\nu_1 = \frac{(1-\mu_1)}{\mu_1}\,\nu_2 \text{ and } \nu_2 = \frac{2\mu_1(\mu_1 - \mu_1{}^2 - \mu_2)}{\mu_2} \quad . \quad . \quad . \quad (24)$$

By substituting the values of the permutation moments for μ_1 and μ_2 a Beta distribution is obtained with modified degrees of freedom which approximates the permutation distribution Since ν_1/ν_2 involves μ_1 only which in this example, and in all the other examples we consider, is the same for permutation theory as for normal theory, both degrees of freedom in the approximation are multiplied by the same factor d.

In the present case

$$d = 1 + \frac{b_2 - 3}{n\{1 - b_2/(n+2)\}} \text{ or to order } n^{-1}\ d = 1 + \frac{b_2 - 3}{n} \quad . \quad . \quad (25)$$

We can now transform back to the t or F form and finally we have that, as an approximation for the permutation test, we should perform the usual t test or F test but, instead of employing 1 and $n-1$ degrees of freedom, we should use d and $d(n-1)$ degrees of freedom, where d is given by equation (25).* Thus a test is provided which, unlike the full permutation test, is readily carried out in practice.

We see that b_2 occurs in (25) only to order n^{-1} and that consequently for moderate or large values of n the effect of the modifying factor is negligible.

* The large number of approximate procedures which employ an F statistic with modified degrees of freedom would seem to justify a new table in which the significance points were tabulated at fractional values of ν_1 and ν_2 paying particular attention to the lower values.

3.2. *Comparison of Critical Regions*

Although in many practical cases the modified test would differ only slightly from the normal theory test, it is nevertheless instructive to consider in what ways the two tests differ. This can best be done by studying the relative shapes of the critical regions. It is only possible to compare the critical regions geometrically for n, the number of pairs of observations, as large as three. The permutation test would, of course, not be of any real value for so small a number of observations since the permutation distribution contains only eight distinct values and certainly if the modified test had to be justified only on the grounds of an approximation to the permutation test, it too would be of little value. However, we shall see later that the modified criterion can to some extent be justified independently of its approximation to the permutation test. In any case, the general tendencies shown for this case are preserved in a somewhat less extreme form when the sample size is larger. In Fig. 1 the regions are compared. It is supposed that the tests are single-sided and are for an increase in mean at the 5 per cent. level of significance. Both regions then lie entirely in the octant of the sample space shown in the diagram in which all the signs are positive. Since both tests are independent of scale the critical regions are necessarily conical. Sections of the cones are shown on the plane $y_1 + y_2 + y_3 = 30$ in which the mean $\bar{y}$ is equal to ten units.

The critical region for the normal theory t test lies within the cone having its apex at the origin and the circular section shown on the plane $\Sigma\, y = 30$. The critical region for the modified test (the approximate permutation test) lies within the cone having its apex at the origin and the propeller-like section shown on the plane $\Sigma\, y = 30$. The types of differences which are found between the tests are illustrated by the samples P and Q corresponding to the points (0, 15, 15) and (5, 5, 20). These samples are shown diagrammatically at the base of the figure. Both samples are of equal significance using the t test and fall outside the 5 per cent. critical region. With the modified test Q is significant and P not significant at the 5 per cent. point.

	Sample			t *Test*	*Approximate Permutation Test*
	y_1	y_2	y_3	(*per cent.*)	(*per cent.*)
P	0	15	15	9·2	12·0
Q	5	5	20	9·2	4·3

We see that the permutation test tends to select as "significant" samples like Q at the expense of those like P.

3.3. *An Alternative Derivation of the Modified Test*

The full permutation test ensures that a proportion is selected from each set of samples for which the observations $y_1, y_2, \ldots y_n$ are numerically the same but with possibly different signs attached. Those selected have the largest means. Now the form of modified test tells us that, as an approximation, we can ignore all properties of the sample except $\Sigma\, y$, $\Sigma\, y^2$ and $\Sigma\, y^4$. Thus approximately the modified test selects a proportion α from each set of samples for which $\Sigma\, y^2$ and $\Sigma\, y^4$ are constant again choosing those samples with the largest means.

Now a test which is exactly of this type can be derived independently of permutation theory. First, following Neyman and Pearson (1933) let us sketch a rather more general derivation than that which is usually given for the ordinary t test. If the null distribution of the y's follows any spherical distribution, that is to say if*

$$p_0(y) = p_0(y_1, y_2, \ldots y_n) = f(\Sigma\, y^2) \qquad 0 < \Sigma\, y^2 < L \quad . \quad . \quad . \quad (26)$$

(where L may be infinite) then we can choose a region w for which, whatever be $f(\Sigma\, y^2)$

$$\int_w p_0(y) = \alpha \quad . \quad . \quad . \quad . \quad . \quad . \quad . \quad (27)$$

* This is the only assumption (Box, 1952, 1953*b*) that need be made in the derivation of those normal theory criteria, which are independent of scale such as the t and F tests, the tests of normality, and the Bartlett test.

by combining together sub-regions of size α on every region of the sample space for which $\Sigma\, y^2$ is constant (that is to say on spherical shells centred at the origin). A best critical region may now be obtained by choosing each sub-region of size α to contain the maximum probability density when $E(y_i) = \eta > 0$. If the distribution of the y's about η is assumed to be normal or more

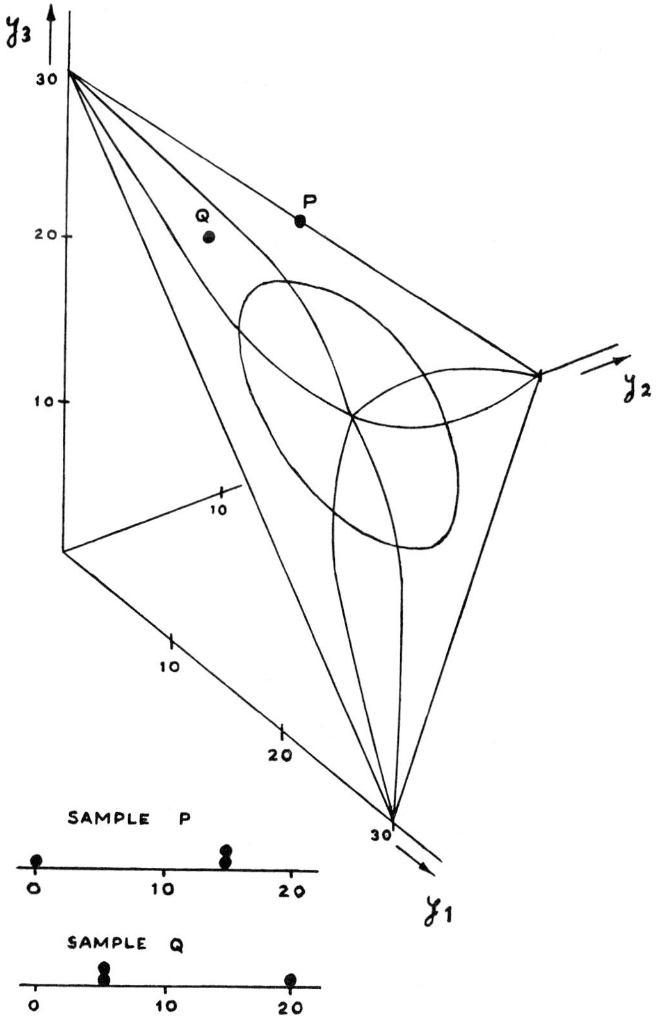

FIG. 1.—Comparison of critical regions for approximate permutation test and normal theory test.

generally spherical with $f(\Sigma\, y^2)$ a decreasing function of $\Sigma\, y^2$ then (as can be seen geometrically and may be proved following the lines given by Neyman and Pearson for the normal distribution) the best critical region is built up by those sub-regions of the spherical shells for which $\bar{y}$ is largest (i.e., for which $\bar{y} >$ constant where the constant is suitably chosen so that the size of each sub-region is α).

If now, instead of supposing that the null distribution $p_0(y)$ is a function of Σy^2 only, we suppose that it is a function of Σy^2 and Σy^4, so that

$$p_0(y) = f(\Sigma y^2, \Sigma y^4) \quad . \quad . \quad . \quad . \quad . \quad . \quad . \quad (28)$$

then we can choose a critical region for which (27) is true by compounding sub-regions of size α on every region of the sample space for which Σy^2 is constant and Σy^4 is constant also. If the alternative hypothesis was the same as before, namely, that $E(y_i) = \eta > 0$ and the distribution of the y's about η was normal or followed a decreasing spherical distribution function, then we should obtain a best critical region by including in it those parts of the regions for which Σy^2 and Σy^4 were constant which gave the largest values of $\bar{y}$. (That is to say those parts for which $\bar{y} >$ constant, again choosing the constant so that a part of size α was taken on each sub-region.)

To picture the form of critical region more clearly imagine the case of three observations. Σy^2 is constant on a sphere. Suppose we draw on this sphere the lines of Σy^4 = constant. We now take a fraction α of each of these lines choosing the parts so that we include the largest possible values of $\bar{y}$. The region so obtained will be similar to that obtained for the approximate permutation test illustrated in Fig. 1.

3.4. *Use of the Theory to Determine the Effect of Departures from Assumptions for Standard Statistical Tests*

In equations (21) and (22) for the particular example of the "paired t test" the mean and variance of the permutation distribution of W and of the normal theory distribution of W were compared. As already noted we may obtain the overall (i.e., the ordinary) moments of any function of the observations by taking the expectation of the permutation moments for that function over all samples. In the present case we have, therefore, that the ordinary moments of the W criterion for the paired t test when the distribution is not necessarily normal are

$$E(W) = \frac{n-1}{n} \quad . \quad . \quad . \quad . \quad . \quad . \quad . \quad . \quad . \quad (29)$$

$$V(W) = \frac{2(n-1)}{n^2(n+2)}\left\{1 - \frac{E(b_2-3)}{n-1}\right\} \quad . \quad . \quad . \quad . \quad . \quad (30)$$

Using the Beta approximation as before we see that *for any parent population* which is such that the joint density function of the sample is symmetric in the observations the approximate probability in the paired t test will be given by referring the usual t or F criteria to tables with δ and $\delta(n-1)$ degrees of freedom where

$$\delta = 1 + \frac{E(b_2-3)}{n\{1 - E(b_2)/(n+2)\}} \text{ or to order } n^{-1}\ \delta = 1 + \frac{E(b_2-3)}{n} \quad . \quad . \quad (31)$$

We are thus provided with an extremely simple and readily appreciated method for assessing the approximate effect of non-normality.

We are now approximating to the ordinary (continuous) distribution of W when the parent population may not be normal. We should expect the Beta approximation to provide an even better representation of the curve in this case than it does for the (discrete) permutation distribution. We shall see that the approximation does represent the permutation distribution remarkably well even under very extreme conditions and we shall therefore expect this approximation to be adequate in the present case. It may be noted that when the distribution is normal $E(b_2) = 3$ and the approximation is exact.

To obtain numerical results we need to calculate the value of $E(b_2)$ for non-normal parent distributions. Except when n is very large or the distribution is normal, $E(b_2)$ differs from the population value β_2. The sort of values to be expected can be seen from some sampling results of E. S. Pearson (1935). He found that in sampling from curves of the Pearson type the sample values of b_2 tended to be heavily biased. His definition of b_2 differs slightly from ours but the results indicate the general trend. Thus for a population with $\sqrt{\beta_1} = 0 \cdot 7$ and $\beta_2 = 3 \cdot 75$ the mean of six values for b_2 for samples of 50 was $2 \cdot 81$ while for a symmetric population with $\beta_2 = 7 \cdot 1$ the mean of ten values of b_2 for samples of 76 was $4 \cdot 61$.

We can proceed following Fisher (1928) by expanding the denominator of b_2 and taking expectations when we obtain to order n^{-2}

$$E(b_2 - 3) = \gamma_2' - n^{-1}(2\gamma_4' - 3\gamma_2'^2 + 11\gamma_2') + n^{-2}(3\gamma_6' - 16\gamma_4'\gamma_2' + 15\gamma_2'^3 + 38\gamma_4' - 3\gamma_2'^2 + 86\gamma_2') \quad (32)$$

where $\gamma'_{r-2} = \kappa_r(y)/\{\kappa_2(y)\}^{\frac{1}{2}r}$ are the standardized cumulants of the y's (*the differences* of the original observations). In terms of the standardized cumulants of the original observations x_{ti} we find

$$E(b_2 - 3) = \tfrac{1}{2}\gamma_2 - \tfrac{1}{4}n^{-1}(2\gamma_4 - 3\gamma_2^2 + 22\gamma_2) + \tfrac{1}{8}n^{-2}(3\gamma_6 - 16\gamma_4\gamma_2 + 15\gamma_2^3 + 76\gamma_4 - 6\gamma_2^2 + 344\gamma_2) \quad . \quad (33)$$

Calculations of this sort are greatly facilitated by the use of the tables of symmetric functions provided by David and Kendall (1949).

4. Tests to Compare Means

We have seen how approximate permutation theory as developed by Welch and Pitman may be employed in two distinct ways:

(1) to provide a test which is "robust" in the sense that the null probability is approximately correct provided only that the null distribution is a symmetric function of the observations (a requirement which may often be guaranteed by randomization);

(2) to provide an additional method for assessing the consequences of departure from assumptions in which the effects are shown in the convenient and readily comprehended form of a modification in degrees of freedom in the standard tests.

It is instructive to consider these two applications together and thus to study the nature of the "correction factors" which the permutation tests apply to the standard procedures. We do this first for the comparison of means using the results of Pitman and Welch and then, in the next section, apply the theory to tests on variances.

The correction factors d, by which the degrees of freedom must be multiplied in the one way classification analysis of variance test and in the randomized block tests, are set out below.

4.1. *One-way Classification Analysis of Variance*

4.1.1. *Approximate Permutation Test*

(a) *Equal Groups*

Suppose there are s groups of n observations and $sn = N$, then

$$d = 1 + \frac{N+1}{N-1}\,\frac{c_2}{N - c_2} \text{ or to order } N^{-1}\; d = 1 + \frac{c_2}{N} \quad . \quad . \quad . \quad (34)$$

(b) *Unequal Groups*

Suppose there are s groups with n_t observations in the tth group and $\Sigma n_t = N$, also x_{ti} is the ith observation in the tth group ($i = 1, 2, \ldots n_t$; $t = 1, 2, \ldots s$). Then

$$d = 1 + \frac{N+1}{N-1}\,\frac{c_2}{(N^{-1}+A)^{-1} - c_2} \qquad A = \frac{N+1}{2(s-1)(N-s)}\left(\frac{s^2}{N} - \Sigma\frac{1}{n_t}\right) \quad . \quad . \quad (35)$$

where

$$c_2 = k_4/k_2^2$$

and k_4 and k_2 are k statistics for the whole sample

$$k_4 = \{N(N+1)\,S_4 - 3\,(N-1)\,S_2^2\}/(N-1)(N-2)(N-3),$$

$$k_2 = S_2/(N-1) \text{ and } S_r = \sum_{t=1}^{s}\sum_{i=1}^{n_t}(x_{ti} - \bar{x})^r \quad . \quad . \quad . \quad . \quad . \quad (36)$$

4.1.2. *Effect of Non-normality on the Null Distribution*

The effect of general non-normality is represented by a modification of the degrees of freedom by the factor δ obtained by substituting $E(c_2)$ for c_2 in the formula for d. To order N^{-2}

$$E(c_2) = \gamma_2 - N^{-1}\{2\gamma_4 - 3\gamma_2^2 + 10\gamma_2 + 12\gamma_1^2\} + N^{-2}\{3\gamma_6 - 16\gamma_4\gamma_2 + 15\gamma_2^3 + 36\gamma_4 + 120\gamma_3\gamma_1 - 88\gamma_1^2\gamma_2 + 66\gamma_2 + 204\gamma_1^2\} \quad (37)$$

Since the modifying factor in the permutation test is of the approximate form $d = 1 + c_2/N$ we see that, for this case of comparison of means the normal theory test provides a close approximation to the permutation test for all but very small values of N. This fact justifies its use, as Fisher pointed out. It follows that the effect of non-normality, obtained by replacing c_2 by $E(c_2)$, is also small. The following table shows a number of calculated values for various levels of γ_1^2 and γ_2 (*a*) assuming the parent population is a Pearson curve and (b) assuming as does Gayen (1950) that the parent population follows the Edgeworth series

$$p(x) = \varphi(x) - \varphi^{(3)}(x)\,\gamma_1/6 + \varphi^{(4)}(x)\,\gamma_2/24 + \varphi^{(6)}(x)\,\gamma_1^2/72$$

where $\varphi(x)$ is the normal function and $\varphi^{(r)}(x)$ its rth derivative. For the former distributions we can find the values of the higher moments and hence the higher cumulant ratios using Pearson's recurrence formula. For the Edgeworth series we have $\gamma_3 = 0$, $\gamma_4 = 0$, $\gamma_5 = -35\,\gamma_1\gamma_2$, $\gamma_6 = -35\,\gamma_2^2$.

TABLE 2

Effect of Departures from Normality on the Null Probability in the Analysis of Variance Test for 5 Groups of 5 Observations

Percentage Chance of Exceeding Nominal 5 per cent. point when Null Hypothesis is True

Measure of Skewness $\gamma_1^2 = \beta_1$		Measure of Kurtosis $\gamma_2 = \beta_2 - 3$: −1	−0·5	0	0·5	1
0·0	Pearson curve .	5·29	5·13	5·00	4·85	*
	Edgeworth series	5·26	5·12	5·00	4·88	4·77
	Gayen's result .	5·24	5·12	5·00	4·88	4·76
0·5	Pearson curve .	5·24	5·10	5·00	4·90	*
	Edgeworth series	5·27	5·14	5·02	4·92	4·82
	Gayen's result .	5·29	5·17	5·05	4·93	4·81
1·0	Pearson curve .	5·16	4·99	4·89	4·84	4·79
	Edgeworth series	5·31	5·17	5·05	4·96	4·87
	Gayen's result .	5·34	5·22	5·10	4·98	4·86

* More terms would be needed in the asymptotic series to give reliable values for $E(c_2)$ in this region.

We note the good agreement obtained between the results of the present technique and the results of Gayen (who used an entirely different method) when the same type of parent distribution is assumed. In view of the known shortcomings of the Edgeworth series in regions not close to the normal curve it is also of some interest to compare the Edgeworth results with those obtained assuming a Pearson type curve.

4.2. *Randomized Blocks Test*

4.2.1. *Approximate Permutation Test*

Assuming s observations per block with n blocks and $sn = N$, with x_{ti}, the tth observation in the ith block ($i = 1, 2, \ldots n$; $t = 1, 2, \ldots s$) then

$$d = 1 + \frac{(ns - n + 2)\,V_2 - 2n}{n(s-1)(n - V_2)} \quad \text{or to order } n^{-1} \quad d = 1 + \frac{V_2}{n} \quad (38)$$

where

$$V_2 = (n-1)^{-1} \sum_{i=1}^{n} (s_i^2 - \bar{s^2})^2/(\bar{s^2})^2$$

is the square of the coefficient of variation of the sample block variances,

$$s_i^2 = \sum_{t=1}^{s} (x_{ti} - \bar{x}_i)^2/(s-1)$$

is the sample variance in the ith block and

$$\bar{s^2} = \sum_{i=1}^{n} s_i^2/n.$$

4.2.2. *Effect of Departures from Assumption on the Null Distribution*

To determine the effects of departures from assumptions the factor δ modifying the degrees of freedom is obtained by substituting $E(V_2)$ for V_2 in the factor for d. If we assume normality and equal variances it is readily shown that $E(V_2) = 2n/(ns - n + 2)$ whence for this case δ is equal to 1 as it should be. We have not had to assume that the distribution within each block is the same so it is possible to determine the effects due to possibly different populations in the blocks We here consider two particular cases, (a) the effect of unequal block variances assuming normality and (b) the effect of non-normality assuming equal block variances, by evaluating approximately the values of $E(V_2)$ appropriate to these two assumptions:

(a) *Inequality of Block Variances Assuming Normality*

If $E(s_i^2) = \sigma_i^2$, $\sum_{i=1}^{n} \sigma_i^2/n = \bar{\sigma}^2$ and $C_r = n^{-1} \sum_{i=1}^{n} (\sigma_i^2)^r/(\bar{\sigma^2})^r$ then by expanding the denominator of V_2 and taking expectations we have the following expression which is taken to terms as high as $\{n(s-1)\}^{-4}$ so that the examples in Table 3, in which n and s are small, may be studied.

$$E_1(V_2) = \frac{2n}{(ns - n + 2)} + \frac{n(s+1)}{(n-1)(s-1)} \times \left(a_1 - \frac{2}{n(s-1)} a_2 + \frac{4}{n^2(s-1)^2} a_3 - \frac{8}{n^3(s-1)^3} a_4 + \frac{16}{n^4(s-1)^4} a_5 \cdots \right). \quad (39)$$

where

$$a_1 = C_2 - 1,$$
$$a_2 = 4C_3 - 3C_2^2 - 1$$
$$a_3 = 18C_4 - 32C_3C_2 + 15C_2^3 - 1$$
$$a_4 = 96C_5 - 210C_4C_2 - 80C_3^2 + 300C_2^2C_3 - 105C_2^4 - 1$$
$$a_5 = 600C_6 - 1584C_5C_2 - 1080C_4C_3 + 2520C_4C_2^2 - 3360C_3C_2^3 + 1960C_3^2C_2 + 945C_2^5 - 1.$$

(b) *Effect of Non-Normality Assuming Block Variances Constant*

With this assumption we find

$$E_2(V_2) = \frac{2n}{ns - n + 2} + \frac{1}{s}\gamma_2 - \frac{1}{n}[12\gamma_2/s(s-1) - 3\gamma_2^2/s^2 + 2\gamma_4/s^2 + 8(s-2)\gamma_1^2/s(s-1)^2]$$
$$+ \frac{1}{n^2}[100\gamma_2/s(s-1)^2 - 6\gamma_2^2/s^2(s-1) + 40\gamma_4/s^2(s-1)$$
$$+ 160(s-2)\gamma_1^2/s(s-1)^3 + 3\gamma_6/s^3 + 96(s-2)\gamma_3\gamma_1/s^2(s-1)^2 - 16\gamma_4\gamma_2/s^3$$
$$- 64(s-2)\gamma_2\gamma_1^2/s^2(s-1)^2 + 15\gamma_2^3/s^3]. \quad (40)$$

The modifying factor δ contains $E(V_2)$ to order n^{-1} thus if the number of blocks was small the effect of unequal block variances could be appreciable. On the other hand, assuming non-normality but equality of variance, the factor δ contains γ_2 only to order $(ns)^{-1}$. Thus the effect of general non-normality would be small unless both n and s were small.

As an example we may compare the results for inequality of variances in randomized blocks with exact values obtained using the theory of quadratic forms in multi-normally distributed variates (Box, 1954*b*).

TABLE 3

Approximate and Exact Probabilities of Exceeding Normal Theory 5 per cent. Point when Block Variances are Unequal, Assuming a Normal Population

Number of Treatments (s)	*Number of Blocks* (n)	*Block Variances*	*Percentage Chance of Exceeding 5 per cent. Point when Null Hypothesis is True*	
			Approx.	*Exact*
11	3	1, 2, 3	4·4	4·3
5	3	1, 2, 3	4·3	4·3
11	3	1, 1, 3	3·8	3·8
5	3	1, 1, 3	3·7	3·9
3	11	1, 1, . . . 1, 3	4·8	4·9

5. ROBUST TESTS FOR VARIANCES

In the tests to compare means studied in the last section the corrective factors were of order N^{-1} so that the normal theory tests were for these examples "non-parametric to order N^0". As the sample size was increased the sampling distribution of the criterion considered would thus ordinarily tend to its normal theory form whatever the parent distribution. It has been shown in an earlier paper (Box, 1953) that for tests on variances the corrective factors are of order N^0, and these tests depend directly upon the assumption of normality, therefore, for all sample sizes. It was also shown that this difference in behaviour arose because, whereas in tests to compare means we compare the variation among the means with an estimate of the variation obtained from internal evidence within the groups, in current tests to compare variances (F test on two independent samples, L_1 test, Bartlett test, Cochran's test, F(max) test, Wald's sequential test) we tacitly compare some measure of variation among the variances with a theoretical value which is correct only for the normal distribution. The variation among a set of variances depends upon the fourth moment just as the variation in a set of means depends upon the second moment so that what is needed is to "studentize" the variance tests for the fourth moment just as the tests on means are "studentized" for the second moment.

A simple way of doing this, which was discussed, involved the conversion of the test on variances to a test on means. The manner in which this had to be done was, however, somewhat arbitrary and we shall here investigate an alternative procedure based on the permutation theory discussed above.

5.1. *Tests to Compare Two Variances. Means Assumed Known*

The simplest case in which a test may be made for equality of variances is that of two groups with the mean of each group known. If a typical observation is denoted by x_{ti} ($i = 1, 2, \ldots n_t$; $t = 1, 2$) we can assume without loss of generality that $E(x_{ti}) = 0$. The normal theory criterion is then

$$F = n_2 \sum_{i=1}^{n_1} x_{1i}^2 \Big/ n_1 \sum_{j=1}^{n_2} x_{2j}^2 \qquad (41)$$

An equivalent statistic for testing the same hypothesis is

$$W = \sum_{i=1}^{n_1} x_{1i}^2 \Big/ \left(\sum_{i=1}^{n_1} x_{1i}^2 + \sum_{j=1}^{n_2} x_{2j}^2 \right) \qquad (42)$$

When the distribution is normal W follows the Beta distribution with degrees of freedom n_1 and

n_2. The first two moments of the permutation theory distribution are readily calculated and may be compared with those for the normal theory distribution

$$\underset{P}{E} = n_1/N \qquad \underset{N}{E} = n_1/N \quad . \quad . \quad . \quad . \qquad (43)$$

$$\underset{P}{V} = \frac{2n_1n_2}{N^2(N+2)}\left\{1 + \frac{1}{2}\frac{N}{N-1}(b_2 - 3)\right\} \quad \underset{N}{V} = \frac{2n_1n_2}{N^2(N+2)} \quad . \quad . \quad . \qquad (44)$$

where

$$b_2 = (N+2)\left(\sum_{i=1}^{n_1} x_{1i}^4 + \sum_{j=1}^{n_2} x_{2j}^4\right)\Bigg/\left(\sum_{i=1}^{n_1} x_{1i}^2 + \sum_{j=1}^{n_2} x_{2j}^2\right)^2 \quad . \quad . \quad . \qquad (45)$$

Using the same argument as before we find that to carry out the approximate permutation test we should enter the F criterion in the usual tables but with dn_1 and dn_2 degrees of freedom where

$$d = \left[1 + \tfrac{1}{2}\left\{\frac{N+2}{N-1-(b_2-3)}\right\}(b_2 - 3)\right]^{-1} \text{ or to order } N^0,\ d = [1 + \tfrac{1}{2}(b_2 - 3)]^{-1} \qquad (46)$$

As before, the approximate effect of non-normality is obtained by replacing b_2 by $E(b_2)$. We find that to order N^{-2}

$$\begin{aligned} E(b_2 - 3) = \gamma_2 &- N^{-1}(2\gamma_4 + 11\gamma_2 + 20\gamma_1^2 - 3\gamma_2^2) \\ &+ N^{-2}(3\gamma_6 - 16\gamma_4\gamma_2 + 15\gamma_2^3 + 38\gamma_4 + 168\gamma_3\gamma_1 - 3\gamma_2^2 \\ &\qquad - 160\gamma_2\gamma_1^2 + 86\gamma_2 + 380\gamma_1^2) \quad . \qquad (47) \end{aligned}$$

The result agrees to order N^0 with that previously given when it was shown by a different argument that the effect of non-normality on this test could be represented approximately by a modification in degrees of freedom by a factor $\delta = (1 + \frac{1}{2}\gamma_2)^{-1}$.

If we compare this result with that obtained from the test to compare means we have approximately for the modifying factors

comparison of means	comparison of 2 variances	
$1 + \gamma_2/N$	$(1 + \frac{1}{2}\gamma_2)^{-1}$	. . . (48)

We note that in the modifying factor for a test on means γ_2 appears only to order N^{-1} whereas for the test on variances it appears to order N^0. Thus the sample size would not have to be very large before a discrepancy which seriously upset the variance test would be negligible for the test on means. We note also that the effects will be opposite in direction in the two tests; for example in sampling from a leptokurtic population the significance of the test on means would be slightly increased but that for the test on variances reduced.

Before the modified test on variances could be recommended as a useful procedure, two questions needed to be considered.

(1) How good is the moment approximation to the permutation test?

(2) How much power is lost by using the modified test when the distribution happens to be normal?

It may be remarked that since the divergencies of the permutation moments from those of normal theory are far larger for the tests on variances than for the tests on means the Beta function approximation would be expected to be much more heavily strained.

To shed light on these questions an extensive sampling experiment was conducted with the object of comparing the behaviour of the standard F test and the modified F test in regard to their robustness and power.

5.2. *Sampling Experiment to Compare Standard and Modified Procedures for Comparing Two Variances*

The power and robustness of the standard F test and the modified F test were investigated for the rectangular, normal and double-exponential parent distributions.

The empirical sampling procedure involved drawing 2000 samples of size 20 from each of these

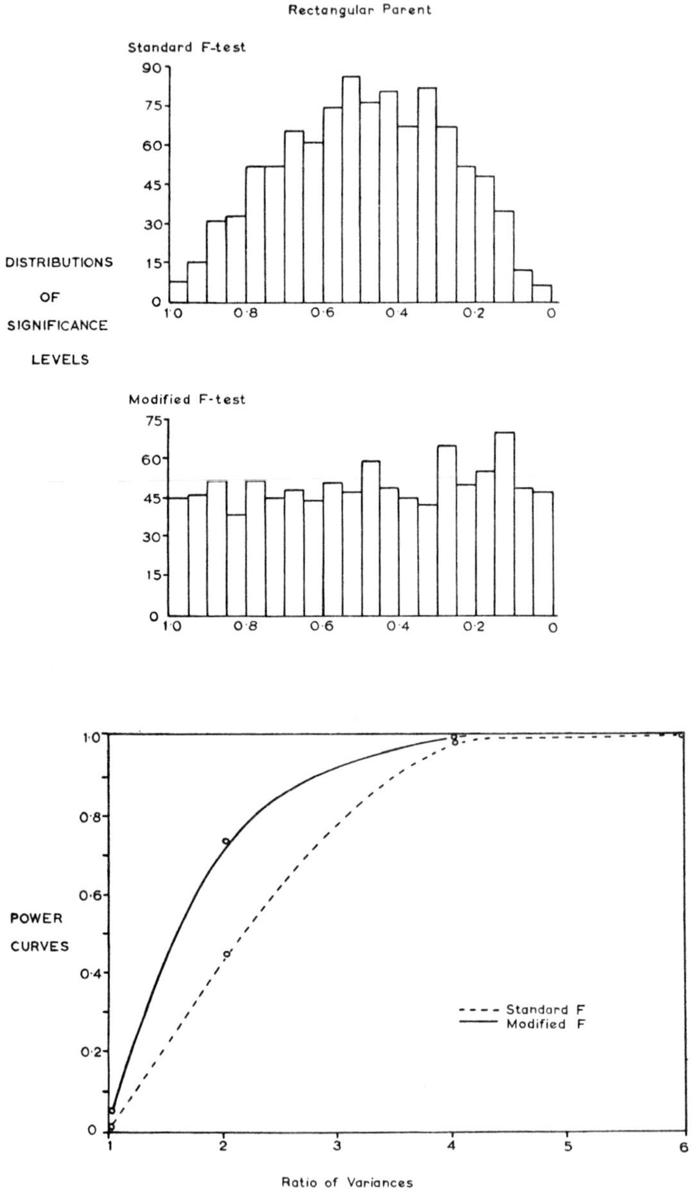

Fig. 2(a).—Behaviour of Standard F-test and Modified F-test for a 'Rectangular' Parent distribution.

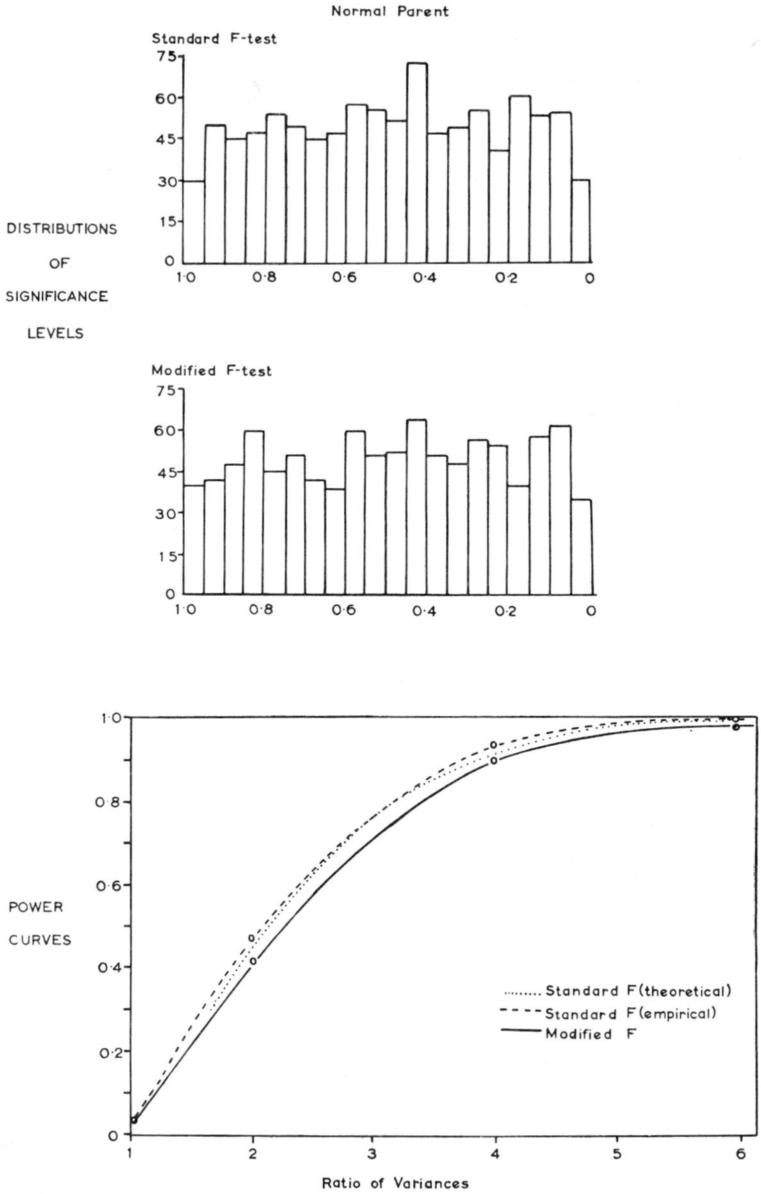

FIG. 2(b).—Behaviour of Standard F-test and Modified F-test for a 'Normal' Parent distribution.

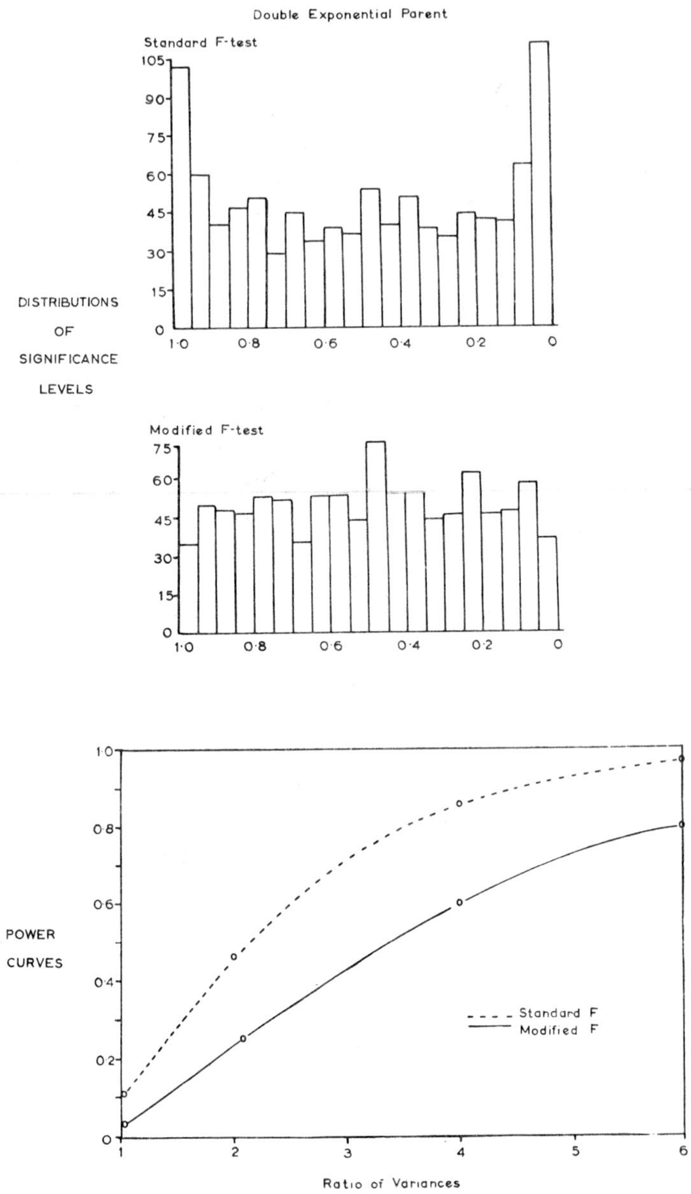

FIG. 2(c).—Behaviour of Standard F-test and Modified F-test for a 'Double Exponential' Parent distribution.

populations. The samples were paired to give 1000 values of the standard F statistic for each population. The appropriate probability associated with each of these F's was estimated by the standard and modified methods from a set of graphs prepared from Pearson's table of the incomplete Beta function.

In addition to the calculations above, which evaluate the behaviour of the criteria when the null hypothesis is true, the distributions were estimated for three alternative hypotheses:

$$H_1: 2\sigma_1^2 = \sigma_2^2 \qquad H_2: 4\sigma_1^2 = \sigma_2^2 \qquad H_3: 6\sigma_1^2 = \sigma_2^2$$

In drawing the original 40,000 observations, 100 values of the deviates of the density functions were calculated at the 0·5, 1·5, 2·5 . . . 99·5 percentile points for each of the three populations. The deviate together with its square and fourth power was punched on an IBM card for each of the 100 percentile points and for the three distributions, making a master deck of 100 cards for each distribution.

The sampling deck of 40,000 cards was then made up by putting on blank cards a two digit random number and a sequence number running from 1 to 40,000. These were then sorted into 100 groups according to the random number appearing on the card. On each card containing random number 00 was punched the 0·5 percentile value of the variate with its square, cube and fourth power; on the card with random number 01 was punched the 1·5 percentile value, etc. The 40,000 cards were then sorted into their original sequence by the sequence number to give the sampling deck from which 2000 samples of 20 were drawn.

The imperfection in the three populations caused by selecting only 100 values to approximate the continuous distributions may be noted. Comparisons of the theoretical and actual measures of kurtosis γ_2 are shown below.

Values of γ_2 in Parent Populations for Sampling Experiment:

	Rectangular	*Normal*	*Double Exponential*
"Theoretical" . .	−1·2000	0·0000	3·0000
Actual	−1·2002	−0·1660	1·7301

The deviations of these populations from their theoretical counterparts is of no serious consequence as we are concerned only to obtain populations representing three degrees of kurtosis centred about the normal.

If the tests behaved as we should wish, then when the null hypothesis was true, 5 per cent. of the samples would show a "level of significance" between 0 and 5 per cent., a further 5 per cent. between 5 and 10 per cent. and so on. Thus in making 1000 tests when the null hypothesis was true the expected number showing a "level of significance" within each 5 per cent. range would be 50. Fig. 2 shows the distributions of significance levels actually found. The observed distributions illustrate the failure of the standard F test and the improvement to be obtained by using the modified test. The χ^2 goodness of fit test may be used to compare the actual frequencies in the 20 groups with the theoretical expectation of 50 per group which would be appropriate if the test were unaffected by the non-normality of the parent universe. For the rectangular population the value of χ^2 is 254·2 ($P < 10^{-7}$ per cent.) for the standard test and 21·8 ($P = 29{\cdot}0$ per cent.) for the modified test. The standard F test gives only 0·7 per cent. of the values below the 95 per cent. point and 0·6 per cent. of the values above the 5 per cent. point. The modified test corrects almost perfectly for the non-normality giving 4·5 per cent. of the values below the 95 per cent. point and 4·7 per cent. of the values above the 5 per cent. point.

For the "normal" parent distribution it is of some interest to note that the slight imperfection of the parent population and in particular its truncation is apparently enough to upset the standard test. The value of χ^2 for this population is 36·1 ($P = 1{\cdot}1$ per cent.), for the standard test and 27·8 ($P = 8{\cdot}6$ per cent.) for the modified test. In particular, the percentages below the 95 per cent. and above the 5 per cent. point are increased from 3·0 to 4·0 per cent. and from 3·0 to 4·5 per cent. respectively as a result of using the modified test.

For the double-exponential distribution the value of χ^2 is reduced from 166·2 ($P < 10^{-7}$ per cent.) to 33·0 ($P = 4{\cdot}13$ per cent.). In this case the modified test appears to slightly over-correct, the percentages below the 95 per cent. point and above the 5 per cent. point being reduced from 10·2 to 3·6 and 11·0 to 3·6 per cent. respectively.

Having demonstrated the relative insensitivity of the modified test to departures from normality the question arises as to how much power is lost by the modification of the standard test. Smooth curves drawn through the points obtained from the sampling experiments with populations corresponding to the alternatives H_1, H_2 and H_3 are shown in the lower part of Fig. 2.

It will be seen that, when the distribution was normal, close agreement was found between the empirical and theoretical curves for the standard F test and there appeared to be little loss of power with the modified test. For example, the probability of detecting a variance ratio of 3 : 1 was reduced from about 75 per cent. to about 71 per cent. by the use of the modified test. Since this difference in power is subject to a sampling error of 2 per cent. the 95 per cent. confidence range for the loss of power is 0–8 per cent.

The power curve for the standard F test with 18 and 18 degrees of freedom and $\alpha = 5$ per cent. coincides very closely with the modified F test curve for 20 and 20 degrees of freedom. It appears therefore that, in the case studied, a loss of power of about 10 per cent. occurs in the sense that 10 per cent. more observations are required with the modified test than with the standard test to obtain the same power.

The standard F test does not have a 5 per cent. intercept for the non-normal populations and consequently the comparison of power curves shown in Fig. 2 cannot be made directly. What can be noted is that the modified test is more powerful for the rectangular and less powerful for the double-exponential distribution.

5.3. *Tests to Compare* k *Variances*

The statistic most commonly used to test the equality of k variances when k is greater than 2 is the Bartlett statistic

$$M_1 = \nu \ln \bar{s}^2 - \sum_{t=1}^{k} \nu_t \ln s_t^2 \quad . \quad . \quad . \quad . \quad . \quad (49)$$

where s_t^2 $(t = 1, \ldots, k)$ is an estimate of variance based on ν_t degrees of freedom, $\bar{s}^2$ is the average variance $\nu \bar{s}^2 = \Sigma \nu_t s_t^2$ and $\nu = \Sigma \nu_t$.

Bartlett showed that for a normal parent population M_1 is distributed very nearly as $(1 + A)$ χ^2_{k-1}, where χ^2_{k-1} follows the chi-square distribution with $k - 1$ degrees of freedom. The modifying factor $A = \{3(k-1)\}^{-1} \{\Sigma \nu_t^{-1} - \nu^{-1}\}$ is of order ν^{-1} and is small for moderate and large samples.

5.3.1. *Means Assumed Known*

For two groups the Bartlett statistic and the "double-sided" F test are equivalent. Thus, if we assume the means are known (so that $n_t s_t^2 = \sum_{t=1}^{n_t} x^2_{ti}$ and $\nu_t = n_t$), the approximate permutation test could be alternatively carried out in the Bartlett form instead of in the F form. It is readily seen that to the same degree of approximation as before, M_1 would be distributed for this case $k = 2$, on permutation theory, as $d^{-1}(1 + Ad^{-1})\chi^2_{k-1}$. The constant A is of order ν^{-1} and since the test depends on d to order ν^0 there seems little point in using this refinement. We should therefore obtain a test (which, like the test on means, is "non-parametric" to order ν^0) by referring $\{1 + \frac{1}{2}(b_2 - 3)\}^{-1} M_1$ to the chi-squared distribution with $k - 1$ degrees of freedom.

One might expect that such a procedure could be generalized to more than two groups. As a confirmation of this we have evaluated the first two permutation moments of M_1 to order N^{-1}. The derivation is outlined below:

In this case where the means are assumed known and equal to zero,

$$n_t s_t^2 = \sum_{i=1}^{n_t} x^2_{ti}, \quad N\bar{s}^2 = \sum_{t=1}^{k} \sum_{i=1}^{n_t} x^2_{ti}, \quad \nu_t = n_t \text{ and } \nu = N.$$

If we write $z_t = (s_t^2 - \bar{s}^2)/\bar{s}^2$ the statistic M_1 may then be written

$$M_1 = -\sum_{t=1}^{k} n_t \ln(1 + z_t) \quad . \quad . \quad . \quad . \quad . \quad (50)$$

which we expand to give

$$\sum_{t=1}^{k} n_t(z_t - \tfrac{1}{2} z_t^2 + \tfrac{1}{3} z_t^3 - \ldots) \qquad (51)$$

Now it will be noted that the denominator of z_t is a constant for all permutations of the observations. By straightforward but tedious algebra it is now possible to evaluate the permutation moments of each term in the expression and so to obtain the expected value of M_1. In a similar way $E(M_1^2)$ may be found and hence the variance of M_1. Proceeding in this way we find to order N^{-1}

$$\underset{P}{E}(M_1) = (k-1)\left[\tfrac{1}{2}(b_2' - 1) + N^{-1}\{\tfrac{1}{2}(b_2' - 1) + (NA - 3)(3b_2' - b_4' - 2) + \frac{9}{4}(3NA - 2)(b_2' - 1)^2\}\right] \qquad (52)$$

where as before

$$A = \frac{1}{3(k-1)}\left(\sum \frac{1}{n_t} - \frac{1}{N}\right)$$

and is of order N^{-1}.

Also

$$\underset{P}{V}(M_1) = \tfrac{1}{2}(k-1)(b_2' - 1)^2$$

plus terms of order N^{-1} as follows:

$\frac{1}{36N}\times$

	$N\sum \frac{1}{n_t}$	k^2	k	(1)
b_6	+9	−9	−18	+18
b_4	+72		−144	+72
b_4b_2	−108	+36	+216	−144
b_2^3	+168	−36	−288	+156
b_2^2	−207	+9	+306	−108
b_2	+72		−72	
(1)	−6			+6

(53)

where

$$b_r' = N^{\frac{1}{2}r} \sum_{t=1}^{k}\sum_{i=1}^{n_t} x_{ti}^{r+2} \Big/ \{\sum_{t=1}^{k}\sum_{i=1}^{n_t} x_{ti}^2\}^{\frac{r+2}{2}}.$$

The terms in N^{-1} involve the higher moment ratios b_4 and b_6. However, to order N^0 we have

$$\underset{P}{E}(M_1) = \{1 + \tfrac{1}{2}(b_2' - 3)\}(k-1) \qquad \underset{P}{V}(M_1) = 2\{1 + \tfrac{1}{2}(b_2' - 3)\}^2(k-1) \qquad (54)$$

which are the first two moments of $\{1 + \frac{1}{2}(b_2' - 3)\}\chi^2$. This, together with the fact that it has been shown elsewhere that for a non-normal parent M_1 is distributed asymptotically as $(1 + \frac{1}{2}\gamma_2)\chi^2$, confirms that the statistic $M_1/\{1 + \frac{1}{2}(b_2' - 3)\}$ should supply a criterion of greater robustness for comparison of variances.

5.3.2. *Means Not Assumed Known*

If we assume group means known, we can derive the permutation moments of the W form of the F statistic and an expansion of the permutation moments of M_1. We are able to do this because the denominators in the expansions to be considered contain the sum of squares of the N observations and are fixed for all permutations of the sample. If the means are not known then this quantity is replaced by the sum of the squares of deviations from the group means, which changes in different permutations, and we are faced with the problem of determining permutation moments for ratios, which greatly complicates the problem.

We have proceeded therefore by assuming a criterion which has a form analogous to that found for the case in which the means were assumed known.

$$M_1' = M_1/(1 + \tfrac{1}{2}c_2) \qquad (55)$$

where in the case of equal groups

$$c_2 = k \, \Sigma \, k_{4t}/(\Sigma \, k_{2t})^2 \qquad (56)$$

and k_{4t}, k_{2t} are k statistics for the tth group.

It is possible to confirm by series expansion that, to order N^0, M_1' does in fact have mean and variance $k - 1$ and $2(k - 1)$.

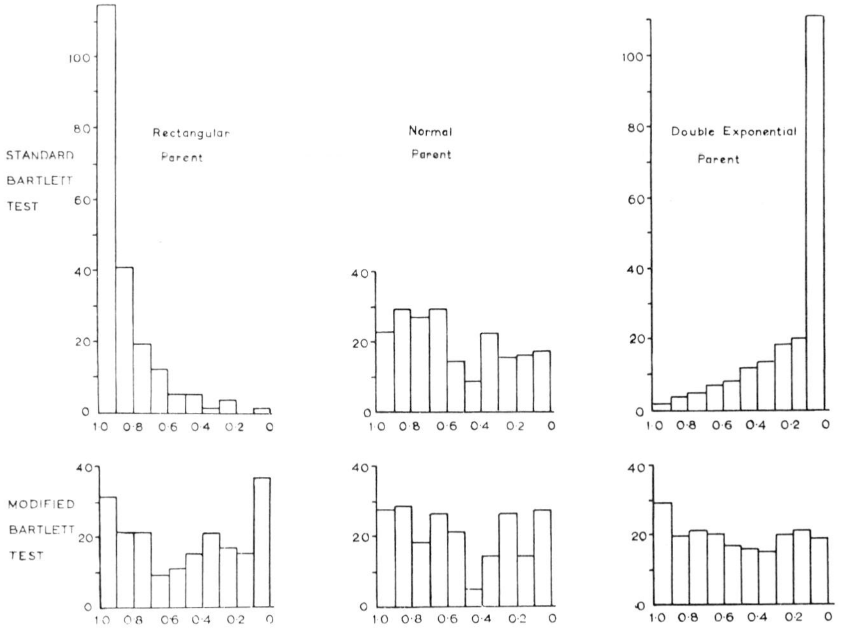

Fig. 3.—Behaviour of Standard and Modified Bartlett test for various Parent distributions.

5.4. *Sampling Experiment for* M_1'

2000 samples of 20 observations were drawn from each of the three populations described in Section 5.2. These were subdivided into 200 sets of 10 samples each and for each set of 10 samples both the standard M_1 and the modified M_1' were computed to obtain three sampling distributions of each statistic. The results of the sampling when the null hypothesis of equal variances was true are shown in Fig. 3. These verify the extreme sensitivity, already noted by Box (1953) and by Kempthorne and Barclay (1953), of the normal theory test under conditions of non-normality and the considerable improvement achieved by using the modified statistic M_1'.

The most serious aberration from ideal behaviour in the modified test is the occurrence of 37 cases as against the expected 20 in the upper decile when sampling from a rectangular population. However, some solace can be found in the fact that this population represents a more severe deviation from normality than is commonly encountered.

As before we may make a chi-squared test of the hypothesis that the underlying distribution of probabilities is rectangular and contains 20 samples in each decile grouping. The results of these computations show

Parent Distribution	*Test Criterion*	$\chi^2\,(9)$	*Probability (per cent.)*
Rectangular . . .	M_1	550·8	0·00001
	M_1'	36·1	0·004
Normal	M_1	21·2	1·170
	M_1'	22·9	0·738
Double-exponential . .	M_1	475·8	0·00001
	M_1'	6·7	66·9

While inspection of the diagrams for the rectangular parent population and the marked reduction in χ^2 indicates that the modified criterion is much less sensitive to non-normality than the unmodified criterion, yet the significance of 0·004 per cent. for the chi-square test indicates that M_1' is not behaving entirely as we should wish.

Again in the case of the normal population the results are not ideal for either test. In addition to the investigation of the properties of the test criteria in the null case, a further 2000 samples of 20 observations was drawn from normal populations with different variances to estimate the power with respect to the alternative hypothesis that

$$\sigma_1^2 = \sigma_2^2 = \sigma_3^2 = \sigma_4^2 = 1; \quad \sigma_5^2 = \sigma_6^2 = \sigma_7^2 = 1{\cdot}7; \quad \sigma_8^2 = \sigma_9^2 = \sigma_{10}^2 = 2{\cdot}6$$

For a significance level of 5 per cent. the probabilities of detecting heterogeneity of variance of the magnitude of H_1 for both the standard and modified statistics when sampling from the normal parent population are found to be for the standard test 81·5 per cent. and for the modified test 81·0 per cent., each value being subject to a sampling error of about 2·8 per cent.

6. Discussion and Summary

One of the simplest statistical procedures is the test of the hypothesis that the mean of a population is equal to μ when the standard deviation σ is known. If a sample of N observations $x_1, x_2, \ldots x_N$ is available the criterion usually chosen is $\sqrt{N}\,\bar{x}/\sigma$ which is referred to tables of the unit normal curve.

The validity of this test of the null hypothesis does not rest on the supposition that the observations are exactly normally distributed. We can appeal to the central limit theorem which tells us that for almost all parent distributions the criterion is distributed asymptotically in the form assumed; moreover that the convergence to this form is *rapid*. In particular, since $\gamma_r(\bar{x}) = \gamma_r(x)/N^{r/2}$, the coefficients of skewness and kurtosis for the mean are $\gamma_1(\bar{x}) = \gamma_1(x)/N^{\frac{1}{2}}$ and $\gamma_2(\bar{x}) = \gamma_2(x)/N$. Thus for all but extremely small sample sizes and "pathological" parent distributions the null test is approximately valid.

A very similar argument may be employed to establish the practical validity of the analysis of variance tests. It appears from permutation theory that the criterion tends to the normal theory form whatever the parent distribution and since the modifying factor in the degrees of freedom is approximately $\delta = 1 + E(\gamma_2)/N$ the convergence is rapid.

When this 'central limit' property is lacking the criterion is of much less practical utility and it seems necessary to seek alternative tests which have greater robustness. One way of doing this is by approximating to the appropriate permutation test. The form of permutation test statistic can be made to depend on the "most likely" alternative distribution.

In the test for equality of variances it is shown how a more robust test based on Bartlett's criterion, but employing the sample estimate of b_2, may be obtained. This form of criterion is however somewhat complicated and other simpler criteria, for example a "studentized" form of Hartley's F(max) criterion, might be developed on similar lines.

We are greatly indebted to Professor Gertrude Cox who provided the opportunity for us to carry out this research, and to Dr. R. J. Hader who was responsible for the administration of the ordnance contract. Our thanks are also due to F. J. Verlinden, the computing staff at North Carolina State College and to Mr. E. Broughton for valuable assistance with the computations.

References

BARTLETT, M. S. (1934), *Proc. Camb. Phil. Soc.*, **30**, 164–169.
—— (1937), *Proc. Roy. Soc., A*, **160**, 268–282.
BOX, G. E. P. (1952), *Biometrika*, **39**, 49–57.
—— (1953*a*), *Biometrika*, **40**, 318–335.
—— (1953*b*), *Ann. Math. Statist.*, **24**, 687.
—— (1954*a*), *Ann. Math. Statist.*, **25**, 290–302.
—— (1954*b*), *Ann. Math. Statist.*, **25**, 484–498,
COCHRAN, W. G. (1937), *J.R. Statist. Soc. Suppl.*, **4**, 102–118.
DAVID, F. N., and JOHNSON, N. L. (1951*a*), *Biometrika*, **38**, 43–57.
—— —— (1951*b*), *Trobajos de Estadistica*, **2**, 179–188.
—— —— (1951*c*), *Ann. Math. Statist.*, **22**, 382–392.
DAVID, F. N., and KENDALL, M. G. (1949) *Biometrika*, **36**, 431–449.
FISHER, R. A. (1928), *Proc. London Math. Soc.*, Series 2, **30**, 199–238.
—— (1935), *The Design of Experiments.* Edinburgh: Oliver & Boyd.
—— (1936), *J.R. Anthrop. Soc.*, **66**, 57–63.
GAYEN, A. K. (1950), *Biometrika*, **37**, 236–255.
GEARY, R. C. (1936), *J.R. Statist. Soc. Suppl.*, **3**, 178–184.
—— (1947), *Biometrika*, **34**, 209–242.
HORSNELL, G. (1953), *Biometrika*, **38**, 128–186.
JAMES, G. S. (1951) *Biometrika*, **38**, 324–329.
KEMPTHORNE, O., and BARCLAY, W. D. (1953), *J. Amer. Stat. Soc.*, **48**, 610–614.
LEHMAN, E. L., and STEIN, C. (1949), *Ann. Math. Statist.*, **20**, 28–45.
NEYMAN, J., and PEARSON, E. S. (1933), *Phil. Trans. A.*, **231**, 289–337.
PEARSON, E. S. (1931), *Biometrika*, **23**, 114–133.
—— (1935), *Biometrika*, **27**, 333–352.
—— (1937), *Biometrika*, **29**, 53–64.
PITMAN, E. J. G. (1937*a*), *J.R. Statist. Soc. Suppl.*, **4**, 119–130.
—— (1937*b*), *Biometrika*, **29**, 322–335.
TOCHER, J. F. (1928), *Biometrika*, **20**, 105–244.
TUKEY, J. W. (1928), *Human Biology*, **20**, 205–214.
WELCH, B. L. (1937*a*), *Biometrika*, **29**, 21–52.
—— (1937*b*), *Biometrika*, **29**, 350–362.
—— (1938), *Biometrika*, **30**, 149–158.
—— (1951), *Biometrika*, **38**, 330–336.

DISCUSSION ON THE PAPER BY DR. BOX AND DR. ANDERSEN

Dr. B. L. WELCH: Dr. Box and Dr. Andersen have presented a paper concerned with topics which have often been discussed but which, nevertheless, have not recently received much consideration in the Research Section. With the paper as a whole I am in agreement. For instance, when comparing several variances in samples from non-normal populations, there is no doubt that we shall be wrong in assuming normality, because the standard error of any criterion used will not then be right even in the large sample sense. In the same way, when comparing means of samples of unequal sizes from populations with unequal variances, if we inject into the situation the assumption that the variances of the populations are equal, the standard error of our criterion will again be wrong to order n^0. The position is, however, somewhat different with the standard analysis of variance tests, at least in as far as they are employed to test null hypotheses in experiments where there has been randomization. There is more scope for expressions of personal opinion about the effects of departures from assumption in these cases and I shall accordingly confine my remarks to such tests.

Consider the simplest example—the E^2-test of homogeneity as it is usually carried out assuming normal theory. Now the normal theory variance of E^2 may be compared with the variance of E^2 calculated from permutation theory. The ratio of these variances differs from unity only by a quantity of order N^{-1} where N is the number of individuals in all the groups combined, a number which will usually be large even if the separate groups are small. There seems little likelihood, therefore, that the normal theory E^2-test will prove misleading even when applied to non-normal data. In fact, by making use of a characteristic (0, 1) variable, the normal theory E^2-test can even be used to test the significance of the Lexis measure of dispersion when the numbers in the groups are as small as two—a situation with which the more usual χ^2-test does not cope adequately.

Consider next the usual normal-theory criterion used to assess the significance of treatments in a randomized block experiment. Here again the normal theory variance of the criterion may be compared with that deduced from permutation theory. The possibility that the ratio of these variances will differ appreciably from unity is again seen to be rather remote, although this is more likely to occur than in the case of the E^2-criterion which I have just mentioned. It can happen if the variances between plots within blocks differ very much from block to block, and where this is so one can improve upon the usual test of the null hypothesis by modifying the degrees of freedom in the very convenient way Dr. Box and Dr. Andersen suggest. But if there are one or two blocks obviously very much more variable than the others one might prefer to proceed more simply by ignoring altogether the more variable blocks and carrying out the usual analysis on the remainder.

As one proceeds to more and more complex experimental situations, consideration of the corresponding permutation theory indicates more and more possibilities of departure from the usual normal theory assumptions, and indeed it has been observed in uniformity trials that the permutation variance of the standard analysis of variance criterion in Latin Square experiments can on *individual* fields differ considerably from the normal theory variance of the same criterion. This has never seemed to me in itself conclusive, since the totality of plot yields in an individual experiment is usually to be regarded as in some sense a random sample from a larger population of sets of plots—however difficult it may be to give precision to the delineation of this population. The viewpoint of the present authors is, I gather, here much the same as mine. Empirical investigation on a large scale would be necessary to determine whether permutation theory and normal theory will often lead to differing conclusions when the wider view is taken. But the work which has been done so far on this problem does at least have the merit of emphasizing that the more complicated the experimental situation becomes, the less heavily one can lean on permutation theory to validate the usual normal theory tests and the more heavily one must lean on the mathematical assumptions made.

It is with very great pleasure that I propose the vote of thanks to the authors.

Dr. H. E. Daniels: The authors are to be congratulated on their worthwhile study of an important practical problem. Sooner or later every statistician is confronted with data to which he hesitates to apply the standard tests, and in this paper an honest attempt is made to meet the difficulty by providing tests which avoid it at not too great a cost. I have only two observations to make, the first of a cautionary nature, the second perhaps more constructive.

It is pointed out in the paper that subsidiary tests of the assumptions underlying the main test may lead to quite a wrong decision as to whether or not the latter may be safely used. We are urged to avoid such preliminary explorations and to use whenever possible a robust test. Where the assumptions concern "extraneous" factors I would certainly agree, but it must be remembered that situations exist where a failure in the assumptions not merely affects the significance level and power of the test but renders pointless the very hypothesis being tested. Robustness is then irrelevant; the experimenter is obliged to do preliminary tests unless he has strong grounds for believing them unnecessary.

Consider, for example, the typical analysis of covariance situation. In comparing the adjusted treatment means it is assumed that neither the concomitant variable x nor the regression on it is affected by treatments. If the latter assumption is violated I suppose one can still regard the test as comparing treatment means adjusted to the average x for the whole experiment. With this more restricted point of view it would still be feasible to examine the test for robustness. But if x is also affected by the treatments, comparison of the adjusted means is an arbitrary procedure and a multivariate analysis becomes appropriate. In cases like this the question being asked is in fact conditional on the assumptions made.

My other point concerns permutation distributions. The authors have followed Pitman and Welch in approximating to the permutation distribution of a statistic lying between 0 and 1 by the corresponding beta distribution having the same mean and variance. Pitman showed that except in very unlikely samples the third and fourth moments also matched well. Nevertheless one is usually interested in the "tail" region where the relative error of the approximation might still be appreciable, and, as Dr. Box has emphasized, the matter requires further study, particularly in regard to the test for variances.

There does exist another type of approximation obtained by the method of steepest descents familiar to students of statistical mechanics. This so-called saddlepoint approximation, though a good deal more elaborate to compute than the beta type, usually involves a relative error of $O\left(\frac{1}{n}\right)$ in the tails and could perhaps serve as a standard of comparison in place of the exact permutation distribution.

As a simple illustration let us consider the example of section 3, though the beta type approximation is known to be satisfactory in that case. The sample consists of n differences $y_i = x_{1i} - x_{2i}$. Let z_i take the values $\pm y_i$ with probability $\frac{1}{2}$ and consider the distribution of $Z = \Sigma z_i$, from which that of t or w can be derived. Its cumulant generating function is $K(T) = \Sigma \log \cosh Ty_i$ and the saddlepoint approximation to the distribution of Z turns out to have probability density

$$g(Z) = \frac{1}{\sqrt{2\pi K''(T)}} e^{K(T)-TZ}$$

where T is the unique real root of the equation

$$Z = K'(T) = \Sigma y_i \tanh Ty_i$$

(In the extreme case where $y_i = \pm 1$ the permutation distribution is binomial and

$$g(Z) = \frac{1}{\sqrt{2\pi}} \frac{n^{n+\frac{1}{2}}}{(n-Z)^{\frac{1}{2}(n-z+1)}(n+Z)^{\frac{1}{2}(n+z+1)}}$$

which is its Stirling approximation.) Numerical integration then yields the distribution function after renormalizing to make the total probability unity. In practice it is more convenient to work in terms of T which has probability density

$$h(T) = \sqrt{\frac{K''(T)}{2\pi}}\, e^{K(T)-TK'(T)}$$

and to interpolate at the required values of $Z = K'(T)$.

I tried this out on the following eight values for y_i extracted from Fisher and Yates's random numbers: 81, 91, 57, 7, 15, 35, 60, 9. Taking a series of probability levels for the normal t test the corresponding probabilities for the permutation distribution and the two approximations were computed as shown in the table:

Normal t	·90	·70	·50	·30	·10	·05	·02	·01	·001
Exact	·891	·672	·492	·259	·117	·055	·031	·008	—
Beta	·884	·680	·487	·271	·105	·055	·023	·012	·0015
Saddlepoint	·899	·697	·499	·278	·114	·063	·027	·010	—

For a sample size as small as eight the approximations fit reasonably well, but an interesting feature is that in both the t and the fitted beta cases non-zero probabilities occur beyond the range of the exact test which is determined by $|Z| \leqslant \Sigma |y_i|$. The beta approximation depends on the inequality $Z^2 < \Sigma y_i^2$, which is not attained except when every $|y_i|$ is the same. Possibly a beta approximation covering the correct range might be an improvement.

In the more interesting case of variances the saddlepoint approximation is unfortunately too elaborate to give here but I hope to continue the discussion elsewhere.

I have much pleasure in seconding the vote of thanks to Dr. Box and Dr. Andersen for this excellent paper.

The vote of thanks was put to the meeting and carried unanimously.

Dr. D. R. Cox: On a point of presentation, it might help to plot the power curves in Fig. 2 on log-probability paper. The curves for one test at different significance levels would then be, approximately, a set of parallel straight lines so that the curves for different tests at unequal significance levels can be compared easily in a rough and ready way.

I should like to say something about tests for dispersion based on a number of small samples, a matter discussed by Dr. Box in his earlier paper on robust tests. Suppose, for example, that we have a number k of small samples each of size n from each of two populations whose variations are to be compared, and that k is so small that the observed variation in dispersion within populations cannot be used. Then the question arises as to whether tests based on the sample range are more robust than tests based on the sample r.m.s.; for normal populations and for $n \leqslant 5$ the difference in power between the tests is very small.

For general non-normality bias in the estimate derived from the range is irrelevant and what we have to consider is the coefficient of variation, as a function of β_2, of estimates derived from (*a*) the mean range and (*b*) the r.m.s. within samples. The coefficient of variation of range is, of course, not a function of β_2, but it is possible to estimate the general form of the relation and then to get results like the following.

Samples of four. Coefficient of variation of estimates from (*a*) range, (*b*) r.m.s. as a ratio of the coefficient of variation at $\beta_2 = 3$.

β_2	1	3	5	7	9
(*a*)	0·83	1·00	1·23	1·34	1·41
(*b*)	0·50	1·00	1·32	1·58	1·80

The tests based on range are therefore more robust than those based on r.m.s.

This is for a number of small samples. If the method is to be applied to larger samples these have first to be divided into subsamples. The best size of subsample is known to be seven or eight, although a moderate gain in efficiency can be achieved by using samples of about four or five and recovering some of the information in the variation of the subsample totals.

Mr. F. J. Anscombe: Dr. Box and Dr. Andersen have given a most helpful account of what may be done in situations where we should like to apply the customary normal-law methods of analysis to a set of data but where we have no confidence that the assumptions underlying those methods are satisfied. The authors have shown how the customary methods may be modified so as to remain valid under very weak conditions, and have pointed out how great are the departures from the usual assumptions that can be tolerated if the customary methods are not modified in this way.

I should like to outline briefly some work done this last summer by Professor J. W. Tukey and me, which is complementary to the work of the present authors. Tukey and I have been considering the transformation of observations to improve the appropriateness of the usual normal-law methods. Ideally, if one could so arrange it, one would like the distribution of the observations to be normal, with constant variance, so that the usual procedures would be strictly valid. One would also like all regressions to be linear and all factors (in a factorial experiment) to have additive effects, i.e. no interactions, so that the interpretation of the observations would be as clear and easy to grasp as possible. An approach can sometimes be made towards this ideal by transforming the observations before analysis. But since we generally have no reliable theoretical information about the distribution of the observations or the structure of responses, we need some practicable method of investigating a set of observations, either in their original state or after a tentative transformation has been applied, to see how far these various requirements are met. It is convenient to begin by a graphical treatment, as follows. Corresponding to each observation y, calculate the fitted value Y and the residual $z = y - Y$. For example, in a two-way classification, with observations y_{ij}, we have, in the usual notation,

$$Y_{ij} = \bar{y}_{i\cdot} + \bar{y}_{\cdot j} - \bar{y}_{\cdot\cdot},$$

and then

$$z_{ij} = y_{ij} - Y_{ij}.$$

Now plot the z's against the Y's on a scatter diagram. Since

$$\Sigma\, z = \Sigma\, zY = 0,$$

the ordinates have zero mean and zero linear component of regression on Y. If there is non-additivity present of the sort that could be removed by a transformation of the observations, there will be a curved (quadratic) regression of the z's on the Y's. If the variance of the y's changes progressively with the mean, the variance of the z's will change progressively with Y and the points may have a wedge-shaped outline. If the distribution of the y's is not normal, the marginal distribution of the z's will be not normal (in general). So non-additivity, heteroscedasticity and non-normality can be investigated at once on the same diagram. It is possible to calculate criteria to test these effects, or estimate their magnitude; the criteria are based on the statistics

$$\text{(i) } \Sigma\, zY^2, \quad \text{(ii) } \Sigma\, z^2 Y, \quad \text{(iii) } \Sigma\, z^3, \quad \text{(iv) } \Sigma\, z^4,$$

(i) being for non-additivity, (ii) for heteroscedasticity, and (iii) and (iv) for non-normality. The complete formulae depend on the design of the experiment and are liable to be tedious to obtain, owing to the correlations between the z's.

The relevance of the present paper to these studies is that we require investigations of the sort that Dr. Box has carried out if we are to decide how far we can afford to sacrifice constancy of variance and normality for the sake of additivity.

Professor CHAPMAN: This study by Dr. Box and Dr. Andersen is indeed valuable from the point of view of evaluating some of the *ad hoc* procedures which have been adopted in statistics, such as the procedure of piling one test on another in order to try to improve the validity or efficiency of the final test; the present and previous papers by Dr. Box show that one does not necessarily gain anything by so doing. We should, however, not necessarily condemn all preliminary tests because in some cases they were unnecessary or have been used without adequate study. A study made by Dr. Paull (*Annals of Mathematical Statistics*, **21** (1950), 539), showed, for example, that a preliminary test could be used and a procedure derived which was better than standard procedures which did not involve a preliminary test. Therefore, in the long run, particularly in complicated situations, it may be necessary to use preliminary tests, but they must be used with care and only after adequate study has shown what is the overall behaviour of the whole series of tests.

Of course the applied statistician will frequently prefer to use a single robust test, in which connection the criteria that Dr. Box has introduced may well be extended into four criteria, using the familiar ideas of type 1 and type 2 error. First, the statistician looks for a test which satisfies the prescribed limits of type 1 error under the assumptions. He will also want to minimize the type 2 error under the basic assumption, i.e. to have a powerful test (this is requirement 1 in Dr. Box's paper). The statistician would also like a test which would keep the type 1 error fairly constant if the assumptions are not fulfilled. This is the idea of robustness. Still further, the statistician would like to have a test which keeps small the type 2 error if the assumptions fail; thus robustness has two aspects, and this particular point is of some interest in connection with analysis of variance tests.

If we consider the test for equality of means of two normal populations with common variances, then the analysis of variance test (which specializes in the case of two populations to the two sample t-test) is known to be most powerful among all unbiased tests. Recent papers by Dr. Mood and Professor Dixon (in the *Annals of Mathematical Statistics*, **25** (1954), 514, 610) on the power of distribution free tests show that the Wilcoxon test compares quite favourably with this test in this situation, i.e. where the assumptions underlying the t-test are valid. Not much is yet known of the comparison of the power of these tests when the assumptions are not fulfilled. Results of a study announced by Professor Lehmann at Stanford, California, in June 1953, comparing the asymptotic relative efficiency of these tests suggest however that for some non-normal distributions the Wilcoxon test may be much more powerful, for sufficiently large samples, than the two sample t-test.

This second aspect of robustness is in fact considered by Dr. Box and Dr. Andersen in the sampling study of the power of their proposed permutation test to compare variances. Moreover the sampling study suggests again that the permutation test may show to much greater advantage from the power point of view when the basic assumptions are not valid.

Professor BARTLETT: I am in sympathy with Dr. Box and Dr. Welch in their endeavour to ascertain how the results reached would affect practical work. Very many applied statisticians naturally want fairly simple rules, but there is the possible conflict between such simple rules and the assumptions on which these are based not always holding. As has already been mentioned, the analysis of variance technique in particular is robust in this way, in fact the actual technique of analysing variance and getting the various mean squares is, in itself, independent of any further testing, and the randomization in experimental work moreover makes these tests very fool-proof, at least in the case of equal numbers of observations for the different treatments. So that while it may be advisable occasionally to look at one's assumptions to see when they will break down, Dr. Box has confirmed in the case of analysis of variance that one does get this robustness. In other situations there may be the difficulty of not knowing quite where to stop in the search for robust tests. For example, when there is no randomization there will be the question what to do about independence. There is a whole field of problems in regard to which one is aware that the question of dependence has to be considered, and many previous analyses in which independence had been assumed are incorrect. One can consider, if necessary, making robust tests which allow for dependence, but in this broad field one can get into rather deep water if one attempts to make robustness cover everything, because there will then be a danger of the tests becoming not powerful enough to be very useful. So that in some cases at least one may require to specify rather narrower assumptions than one would ideally wish from Dr. Box's standpoint.

To mention a small point in connection with the homogeneity of variances test, Dr. Box has not looked at this so much in the context of analysis of variance, in the sense of disentangling variances to see whether they are homogeneous. One may there have to be a little careful. It is well to bear in mind that if one takes linear combinations of observations the variances of such linear combinations can be the same as before (for an orthogonal transformation), but if the original observations are non-normal the linear combinations of them will be nearer normality. There is, of course, no paradox here; the fallacy would be to suppose that the resulting new observations were independent if the original observations were. This emphasizes the need to take care that the observations being studied are the original ones for which the assumption of independence may be reasonable.

Dr. F. YATES: I should like to give a word of warning concerning the approach to tests of significance adopted in this paper. It is very easy to devise different tests which, on the average, have similar properties, i.e. they behave satisfactorily when the null hypothesis is true and have approximately the same power of detecting departures from that hypothesis. Two such tests may, however, give very different results when applied to a given set of data. This situation leads to a good deal of contention amongst statisticians and much discredit of the science of statistics. The appalling position can easily arise in which one can get any answer one wants if only one goes around to a large enough number of statisticians.

This is itself an argument against lightheartedly elaborating a multiplicity of tests of significance to deal with data of the same type. But the tests given in the paper seem open to objection on other grounds also. It is stated in section 3.2 that the points P and Q shown in Fig. 1, which give the same significance level of 9·2 per cent. on the t-test, give significance levels of 12·0 per cent. and 4·3 per cent. respectively on the approximate permutation test. These differences are perhaps not very alarming, but the same argument will show that three observations having the values 9·8, 10·1 and 10·1 which, on the normal theory test, or indeed on any ordinary test of significance, will give a high level of significance ($t = 100$, giving a probability of about ·01 per cent.) will be judged non-significant on the approximate permutation test, while the values 9·9, 9·9 and 10·2 will be judged significant on the approximate permutation test. I, personally, would certainly find it difficult to convince an intelligent man that there is any profound difference between these two sets of observations.

I will not elaborate on these questions here, but should like to suggest that a good deal of the trouble has arisen through adopting the Neyman-Pearson attitude to tests of significance. This attitude, if I understand it aright, is a consequence of regarding tests of significance as criteria for definite action, i.e. essentially the decision function approach, whereas, in fact, tests of significance are used in scientific research as an aid to assessing the weight of evidence for or against a given hypothesis. Scientific research workers do not reject a hypothesis once and for all if a certain set of observations show a significant departure at the 5 per cent. point, any more than they accept it if the observations do not show any significant departure from the hypothesis. If scientific research proceeded on these lines it would in a very short time be in a state of inextricable confusion.

Incidentally, I should like to question the statement in section 2 that the points of view expressed by Fisher and Pearson are in no way contradictory. To suggest that Fisher, or scientists who use Fisherian tests, are not concerned with the power of tests of significance to detect relevant departures from the hypothesis being tested is a gross misrepresentation.

Mr. A. J. MAYNE: A possible new method of forming robust statistics is as follows. Consider a statistic S, which is a function of a sample drawn from a population with any distribution; S can be a univariate or multivariate statistic. Then polynomial transformations of S, with coefficients formed from the sample data, can usually be found, whose distributions can be made successive approximations to a standard normal distribution, or the distribution of S for a normal population. The r'^{th} approximation is usually a polynomial of degree r in S, with dominant term S for all except very small sample sizes. As r increases, the r'^{th} approximation in general yields a progressively more robust statistic.

Although little work has yet been done on this new method, and more investigations on its efficiency are needed, it has been applied to the t-statistic of §3.1 of Box and Andersen's paper. The null hypothesis considered for this example of application of the new method is that the parent distributions of both populations are the same, thus that the distribution of $y \equiv x_1 - x_2$ is a general *symmetric* univariate distribution. Let k_2 be the r'^{th} order *sample* k-statistic, formed from the sample items which are the differences between the values of the observations in each pair, i.e. formed from $y_i \equiv x_{1i} - x_{2i}$ ($i = 1$ to n). Let

$$l_4 \equiv k_4 k_2^{-2}$$

Then

$$[1 - \tfrac{1}{4} l_4 n^{-1}]\, t + \frac{1}{12} l_4 n^{-1}\, t^3$$

is a statistic with improved fit to the distribution of t for normal x_1 and x_2 populations, and

$$[1 - \tfrac{1}{4}(1 + l_4)\, n^{-1}]\, t - \tfrac{1}{4}(1 - \tfrac{1}{3}\, l_4)\, n^{-1}\, t^3$$

gives an improved fit to the standard normal distribution.

The powers of the tests based on these transformed statistics have not yet been investigated, but it seems reasonable to suppose that, if the departure of the population from normality is not too large, they are comparable to the power for the original statistic when the population is normal, if the transformed statistics are fitted to the distribution of the original statistic for normal population.

Professor BARNARD: The strategic as opposed to detailed tactical advances in statistics are usually associated with the birth and christening of new ideas. "Efficiency", "sufficiency", "likelihood", "power", are some examples. "Robustness" is a new addition to the family, and Dr. Box is to be congratulated not only for giving birth to it, but also for such aptness in its christening. It can confidently be predicted that the idea will have a long and vigorous life.

However, the ideas we use in statistics form a somewhat unruly family, and the task of keeping order will not be made easier by the new arrival. I share Dr. Yates's concern lest we find ourselves in the position where two statistical tests, while applying with apparently equal appropriateness to answering apparently the same question on the basis of the same data, yet give different answers. It must not be forgotten that statistical inference owes its existence as a science to the recognition of the fact that there are good and bad ways of analysing data; and the existence of cases where the "good" answer is not determined tends to blur this recognition. While we must expect some difficult cases to occur, they should be rare.

From this point of view it is comforting to learn that the t-test and the F-test applied to the analysis of variance are robust tests. Thus we are able to say, if asked why we test the hypothesis that the observations are *normally* distributed with mean μ, that it would make little difference if our observations were not normally distributed, provided they are nearly so, and this latter we can usually justify by an appeal to the nature of the data (suitably transformed, perhaps). And these two tests seem to account for practically all the tests we make on measured data.

It is curious how difficult it is to get away from situations where the t- and F-tests apply. Just as attempts to generalize to multivariate tests leave me rather cool—I have yet to see a multivariate test not reducible to a t- or F-test which really seemed to mean something—so attempts to extend in the direction of other moments than the first seem more difficult than one might think. For example, we might be tempted, if challenged to give an instance where two wholly independent estimates of variance are to be compared, to quote the comparison of the accuracy of two methods of measurement, or of two guns; but here we can argue that what we are really interested in is not the mean square deviation from the sample mean, but the mean square deviation from the true value; and if we measure deviations from the true value (and perhaps make a transformation), the comparison reduces to a t-test.

Mr. EHRENBERG: I think that robust tests of significance are obviously desirable, but at the same time I should like to make a critical remark upon their use. There is, in these days, a tendency to treat variance heterogeneity and departures from normality as being of importance only in as far as they may affect tests of significance. In practice, however, it is surely necessary, before considering how to test the significance of the difference of means, say, to see whether such means are useful descriptive parameters; the data may, for example, be skew. This putting of the cart before the horse is suggested on p. 2, para. 3, for example, in which the authors seem to warn us against even bothering to look for such things as skewness, kurtosis, variance heterogeneity, and so on; they seem to warn us, that is, against seeing what the data are really like.

It may be that the authors' attitude can hardly be considered at fault within the framework adopted in their initial sentence, namely the restriction to the testing of *a priori* hypotheses. But that framework is extraordinarily narrow. To cling to a hypothesis which turns out to be technically inappropriate—for example one formulated in terms of mean values when some of the data turn out to be skew—will, of course, prevent an understanding of the data. And practically any experiment, however well-considered in the first instance, will throw up many facts or implications which have not been considered beforehand. Fisher's famous dictum that every experiment may be said to exist to give the facts a chance to disprove the null hypothesis is stimulating, though hardly true. Every experiment may be said to exist, not to give the facts a chance to disprove the null hypothesis, but just to give the facts a chance. In view of Dr. Yates's remarks, I feel that an experimental scientist may say (if no statistician is involved) that every experiment exists merely to give the facts. The methodological notion of explicit hypothesis-making is nowadays rather allowed to master us instead of being used merely in as far as it can be helpful.

In using robust tests, as in any other application of random sampling theory, it is as well to ask oneself first how one would analyse the data if they were not random data at all. After all, tests of significance only become necessary and valid, as Fisher stressed in the quotation given at the beginning of section 2 of the paper, when there has been a physical act of randomization.

Professor BARTLETT: I think that Mr. Ehrenberg is somewhat exaggerating the possibility of allowing facts to speak for themselves. Always there is a compromise between theory and facts; facts of themselves are meaningless unless one is asking some kind of question. Certainly in some cases one allows the facts to suggest generalities which one then investigates further, but merely to search for apparent results from the facts can be taken too far.

Mr. EHRENBERG: I cannot help feeling that the test-book methods over-emphasize looking at only those facts about which one has made a hypohesis beforehand. Often one does not make *a priori* hypotheses, but has to see how existing facts fit in with other facts or theories.

The authors subsequently replied in writing as follows:—

We are happy that our paper provoked what to us at least was a most stimulating discussion and are grateful to all who took part. With most of the speakers we are in agreement, but a few points call for comment. As we have tried to emphasize, most of the standard normal theory tests to *compare means* are remarkably robust and admirably fulfil the needs of the experimenter. Our object in discussing permutation theory for *these* tests is to demonstrate this, and to consider more closely the behaviour of the permutation tests in those cases with which we are most familiar. Our object is not, as Dr. Yates supposes, to suggest alternative tests for these situations.

If the state of affairs which exists for tests to compare means was found for all other tests, then to introduce alternative procedures with the same general properties might be as pointless and harmful as Dr. Yates believes. Unfortunately the happy state of affairs found for tests on means is *not* a general one. For example, it has been shown that the current tests to compare several variances, which have for many years been recommended in the text-books, are grossly misleading for populations showing a degree of non-normality frequently encountered in practice but not readily detectable for sample sizes usually considered. There is reason to believe that a similar state of affairs may be found in other instances including those we mention. To ignore this situation for the sake of uniformity would produce the result that, presented with the same data, all statisticians arrived at the same *wrong* conclusions. This would lead to more discredit of the science of statistics than would an honest attempt to face the situation and remedy it. Incidentally, we see no particular reason why statisticians should be expected to give the same answer if they are asked different questions.

Concerning the behaviour of the approximations to Fisher's permutation test, our object in preparing Fig. 1 was to make the point that Dr. Yates has restated namely that for individual samples the normal theory and permutation test can give dissimilar results. The permutation test has real relevance only for significance levels corresponding with a probability level $\alpha > 1/2^n$. For such levels the difficulties mentioned by Dr. Yates do not of course arise, although quite large differences between the two sorts of test can be observed for particular samples. Unfortunately, it is not easy to illustrate diagrammatically the situation for $\alpha > 1/2^n$ and $n > 3$ and we feel that Fig. 1 is of some interest provided the reservations we made are taken into account. Incidentally, we find it difficult to detect the connection between the behaviour of this test and the Neyman-Pearson theory which plays no essential part in the test's derivation.

The basic difficulty is that whatever we assume may be "taken too literally". For example, on the one hand, if we assume normality, a test based on that assumption may fail unless that assumption is almost exactly true. On the other hand, a permutation test is derived on the supposition that we are prepared to believe almost anything about the population however unlikely and may not therefore give sensible results for certain extreme samples. In most practical situations the user of the test believes neither in exact normality nor in the possibility of extreme departures from it. The mental concept of a prior distribution of distributions is, however, difficult to use as a basis for deriving a real test statistic, and it seems that in practice we must consider each on its merits and proceed, as in other sciences, by making the simplest assumptions first and elaborating these only if it transpires that to obtain a test of practical value it is necessary to do so. In this connection, it would be safest to behave as though the robustness of tests on means was a lucky accident. When new methods are suggested their authors should be required to show not only that their procedures have desirable properties when the assumptions are true, but also that they can be expected to work reasonably well with the degree of departure from assumption likely to be encountered with real data. This might have the secondary but desirable result of alleviating to some extent the present pressure on space in some of the statistical journals.

The difficulties that face us in attempting to reply to Dr. Yates's final point are similar to those which must be experienced by neutral countries who try to keep out of the "cold war". We suspected that if we included statements by Fisher and Pearson in the same paragraph, even though these were about different things, this could lead to a discussion generating more heat than light. We therefore were at some pains to explain that the points of view we quoted were complementary rather than contradictory. Alas, we are now not only accused of saying that Pearson and Fisher agree but in the very next sentence are charged with "gross misrepresentation" for saying (which we did not) that Fisher and Pearson do not agree about the importance of powerful tests. Against this double-edged weapon no defence is possible and we admit defeat.

We have confined consideration to hypothesis testing only for purposes of simplicity. We do not regard such tests as the be-all and end-all of statistics and we point out in the first sentence of our paper that similar considerations of robustness should be applied for other statistical procedures. We conceive scientific research as developing by dual processes which may be called synthesis and analysis. In the process of synthesis the hypothesis is built up. The appropriate data are then collected. The process of analysis now begins and out of it a new synthesis is generated and so on. The two processes used in alternation are applied not once but many times during the course of most real investigations. Of the two, we feel that process of synthesis (almost completely neglected by statisticians) is the most important to the ultimate success of real investigations. One of us has discussed these ideas in more detail in a paper shortly to appear in *Biometrics*. What we are concerned with in the present paper is that tests which may be applied at each stage of this process *should not mislead* the experimenter. On the question of the usefulness of tests other than those to compare means we agree with Mr. Ehrenberg rather than with Professor Barnard. However, our intention seems to have been misunderstood by the former speaker for, far from warning against the use of a test such as that for variance heterogeneity, we go to some trouble to try to formulate a procedure which will be useful for this purpose. We also point out that we consider that such tests should be made because we are interested in variance heterogeneity, normality, etc., in their own right and not as auxiliary features only of importance in relation to their effect on other tests. We believe with Mr. Ehrenberg that one must examine the data in various ways, particularly to decide what has to be done next. All that we would plead is that the tests used, whether they be few or many, shall be robust so that having applied them, we know what it is we have really tested.

1.4

The Exploration and Exploitation of Response Surfaces: An Example of the Link between the Fitted Surface and the Basic Mechanism of the System

G. E. P. BOX and P. V. YOULE
Imperial Chemical Industries Ltd.

This is a sequel to an article which recently appeared in this journal [1] and had the same general title. The previous article described a number of applications of newly developed techniques [2] for the study of response surfaces. The present article shows how study of the form of the empirical surface can throw important light on the basic mechanism operating and can thus make possible developments in the *fundamental theory* of a process. This idea is illustrated in some detail with an example previously discussed only from the empirical standpoint. A theoretical surface, based on reaction kinetics is now derived, rate constants are estimated from the data and the theoretical surface is compared with the empirical surface previously obtained. It is then shown how the canonical variables of the empirical surface can relate to the basic physical laws controlling the system. In this connection the problem of suitable choice of metrics for the variables is discussed. In a final section some general remarks on the process of scientific investigation are appended.

I. INTRODUCTION

A response surface is a graphical representation of a relationship

$$\eta = \phi(x_1, x_2, \cdots, x_k)$$

between some response such as yield, whose level is denoted by η, and a number of quantitative variables (or factors), such as temperature, time and concentration, whose levels are denoted by $x_1, x_2, \cdots, x_k$.

The feature of the surface of greatest interest is often the value or values of the variables $x_1, x_2, \cdots x_k$ for which η is a maximum.

In the previous paper it was emphasised that the study of numerous examples had indicated that sharply defined point maxima appeared to be something of a rarity. The typical situation was that in which the response was found to be insensitive in the region of the maximum

The Collected Works of George E. P. Box, 1984, Wadsworth, Inc., Belmont, CA 94002.
Originally published in *Biometrics,* vol. 11, no. 3 (1955), pp. 287–323.

to certain joint changes in the levels of variables, indicating the existence of 'factor dependence'. An extreme form of this phenomenon occurred where there was a line, plane, or space of near-maxima rather than a single point maximum. Such a response surface was said to contain a 'stationary ridge system'. A second type of surface of common occurrence contained a rising ridge. It was suggested that the nature of a ridge system could indicate the physical laws which underlay the process studied.

The method recommended for exploring a response surface consisted first of performing a simple pattern of experiments designed to detect, in the initial region explored, any general sloping tendency of the surface. If such a tendency was found, further experiments were performed in the indicated direction of increasing response. Either initially, or after one or two cycles of this 'steepest ascent' procedure had brought the experimenter to a region of higher response, it was usually found that no sloping tendency could be detected and exploited. The region so attained was then examined by performing a slightly more elaborate pattern of experiments and fitting a suitable function which enabled curvature in the surface and dependence between the variables to be taken account of.

In the absence of prior knowledge concerning the form of the response function, a local representation could be obtained by fitting a polynomial in $x_1, x_2, \cdots, x_k$, in which all terms up to a given order d were included. This was of course equivalent to supposing that the true function could be locally represented to a sufficient approximation by its Taylor series ignoring terms of order higher than d.

In the majority of applications, where the object was not so much to graduate the response surface accurately but rather to determine approximately its general characteristics in the optimum region, an equation of only second degree has usually been adequate.* Reduction of this fitted second degree equation to canonical form has allowed the nature of the fitted surface to be readily appreciated and has indicated in what regions further experiments were necessary.

It has been found that:

(1) This approach has made it possible to comprehend features of the surface which could be exploited to attain further gain when possibility of improvement by simpler means had been exhausted.

(2) By considering the features of the surface for the principal response such as yield, or cost, in relation to the features of the surface for 'auxiliary' responses such as purity, it has been possible to discover

*Where more accurate *graduation* was required (as for example in the work on pulse columns performed for the Atomic Energy Commission) an equation of third degree has been used [4] [5].

conditions which were 'best' in the practical sense of bringing all the responses to 'most satisfactory' compromise levels.

(3) Consideration of the shape of a fitted response surface has suggested new theories of behaviour of the system.

It is this last aspect which we shall here discuss further.

In the analysis of the fitted second-degree equation the existence of a ridge is indicated by one or more of the coefficients in the canonical form of the equation being small in comparison with the others. Where these small coefficients are negligible it is implied that the system in k variables can be more economically described in terms of less than k canonical or 'compound' variables. It appears that these compound variables can have greater significance than a purely representational one. In fact they can indicate the fundamental mechanism of the system. To make this clear we first consider a simple hypothetical example.

Suppose that the effect on yield of the concentrations c_1 and c_2 of two reactants were being studied and that previous experimentation had suggested that we should now explore the ranges of concentration: $c_1 = 50$–60 grams per litre and $c_2 = 30$–40 grams per litre which were expected to be near their optimum values. It is usually simplest in such examples to work with coded values of the variables and we will suppose that 'standardised variables' were chosen as follows:

$$x_1 = (c_1 - 55)/5 \qquad x_2 = (c_2 - 35)/5$$

so that the region explored with suitably placed experiments was defined by

$$-1 < x_1 < +1, \qquad -1 < x_2 < +1$$

Suppose finally that the fitted second degree equation was

$$Y = 78.56 + 0.50x_1 - 0.21x_2 - 2.31x_1^2 - 2.15x_2^2 + 4.08x_1x_2$$

and that the errors of estimate of the coefficients were sufficiently small so that the equation was as a whole meaningful.

Now this, like any other second degree equation, can be written in canonical form. (That is by changing the origin and rotating the co-ordinate axes we can write it in form containing only quadratic terms). In the present case the equation, written in this way becomes

$$\left.\begin{aligned} Y &= 78.63 - 4.27X_1^2 - 0.19X_2^2 \\ \text{where} \quad X_1 &= 0.72x_1 - 0.69x_2 - 0.06 \\ X_2 &= 0.69x_1 + 0.72x_2 - 0.53 \end{aligned}\right\} \qquad (1)$$

and these last two equations define the positions and directions of the new coordinate axes.

The centre of the system (that is the point $X_1 = 0$, $X_2 = 0$) has coordinates $x_1 = 0.41$, $x_2 = 0.34$. Thus the axes of the system defined by the lines $X_1 = 0$ and $X_2 = 0$ pass close to the original origin and through the region in which the experiments have been performed. A discussion of the surface in terms of these axes is relevant therefore.

We notice that very nearly the equation is:

$$Y = 78.6 - 4.27(0.7)^2(x_1 - x_2 - 0.1)^2 - 0.19(0.7)^2(x_1 + x_2 - 0.8)^2$$

or in terms of the original units

$$Y = 78.6 - 0.084(c_1 - c_2 - 20.5)^2 - 0.004(c_1 + c_2 - 94.0)^2$$

Thus the canonical variables correspond very nearly, to the *difference* of the concentrations and the *sum* of the concentrations, the coefficient of the difference being much larger than that of the sum. Now remembering that our estimates of the coefficients are subject to error and also that the form of equation is probably not entirely adequate, it would seem that the data might be explained on the hypothesis that yield depended *only* on the difference between the concentrations and not at all on their 'overall' level, the best yield being attained whenever c_1 was about 20 grams per litre greater than c_2. This hypothesis could be readily checked over wider ranges of the variables by further experiment.

Assuming this hypothesis was shown to be substantially correct, attention which had so far been focussed on the mathematical analysis would be shifted to physico-chemical theory. The experimenter would ask himself "What mechanism could produce the phenomenon of yield being dependent on this concentration difference?" If he could answer that question further experiments would be devised to test his theory. Such a theory by contributing to a basic understanding of the mechanism of the reaction could, for example, lead to new methods of overcoming yield-limiting factors either by modification of the physical conditions or by the introduction of other reactants into the system. The fitted equation may thus provide not merely an empirical representation of the surface near the maximum (useful though this is) but also a valuable indication of how the system works.

Now as a consequence of the form of our fitted equation the canonical variables X_1 and X_2 are necessarily expressed as *linear functions* of the quantities x_1 and x_2 but it is obvious that usually an underlying 'compound variable' will be some less simple function. For example it will frequently happen that the level of yield will depend on the *ratio* of two concentrations rather than their difference and we shall see later that in real examples more complicated relationships occur. The difficulties which this presents are not as serious as they first appear.

Let us consider the particular example of the yield depending on the ratio of the concentrations of two reactants, there being a certain optimum ratio. A second degree equation fitted to the *logarithms* of c_1 and c_2 would give an equation similar to that obtained before but with a dominant canonical variable ($\log c_1 - \log c_2$) instead of $(c_1 - c_2)$ and the yield surface plotted in terms of $\log c_1$ and $\log c_2$ would contain a stationary ridge system.

Now in practice we should usually be attempting to represent the relationship over ranges of concentration which were fairly small compared with the overall magnitudes of the concentrations. The appearance of the surface would then not be very different whether it was plotted in terms of c_1 and c_2 or in terms of $\log c_1$ and $\log c_2$ and a ridge system which was represented by equations like (1) would still be found even though the second degree equation were fitted to c_1 and c_2 rather than to $\log c_1$ and $\log c_2$. That this is generally true for other types of functions can readily be seen.

Suppose Y depends only on $X = f(c_1, c_2)$ and $f(c_1, c_2)$ can be represented locally reasonably well by the first order terms of its Taylor series

$$X = f_0(c_1, c_2) + (\partial f/\partial c_1)_0 (c_1 - c_{10}) + (\partial f/\partial c_2)_0 (c_2 - c_{20}) \qquad (2)$$

where the subscript zero denotes that the value at the point c_{10}, c_{20} is taken. Now this equation is of the linear form

$$X = ac_1 + bc_2 + d \qquad (3)$$

so that if, in the region of the optimum, Y, could be approximated by a quadratic function of X, we should have

$$Y = Y_0 + B(X - X_0)^2$$

We see therefore that while we should expect that our procedure for *detecting* local factor dependence would have fairly wide applicability, the question of what was the relevent form for the compound variable would usually be a matter for further speculation and experiment.

Returning to the case where yield is dependent on the level of c_1/c_2 equation (2) gives for points in the neighbourhood of c_{10}, c_{20}

$$\frac{c_1}{c_2} \simeq \frac{c_{10}}{c_{20}} \left[1 + \frac{c_1}{c_{10}} - \frac{c_2}{c_{20}}\right] \qquad (4)$$

Here therefore the experimenter working with the untransformed variables c_1 and c_2 would find a dominant canonical variable $X = ac_1 + bc_2 + d$ in which a/b was roughly equal to c_{20}/c_{10} the ratio of *the average concentrations.* This would indicate that proportional changes in the concentrations would leave X and hence the yield Y unchanged and would lead to recalculation of the equation in terms of log c_1 and log c_2 when a closer fit would probably be obtained.

Usually where the first analysis indicates some ridge system we must rely on possible theories of the system to indicate the correct function to employ. These theories must of course be checked by further experimentation.

The simple numerical illustration quoted above is hypothetical but serves to introduce the following genuine but more complicated example.

2. THE EMPIRICAL STUDY

An experimental study of the system concerned has been described in reference [1] (first example) and some of the detailed calculations will be found in [3].* It was desired to maximise the yield of one of the products of a chemical reaction. To do this the yields of this and other products of the reaction were experimentally determined under various conditions of temperature (T), concentration of one of the reactants (c), and time of reaction (t). The conditions T, c and t were measured as degrees centigrade, % concentration and 'hours on temperature' respectively.

The results are shown in columns 1, 2, 3, 4, 11, 12, 13 and 14 of Table 1. The quantity $\hat{\eta}_2$ is the estimated fraction of unchanged starting material, $\hat{\eta}_3$ the estimated fraction converted to the desired product and $\hat{\eta}_5$ the estimated fraction occurring as an unwanted by-product. The fractions are called the 'yields' and are sometimes quoted as percentages. The circumflex accent will be used throughout this paper to indicate *observed* or *estimated* quantities, the 'true' values will be unaccented.

For convenience the levels of the variables are coded in columns (5), (6) and (7) of Table 1 as follows:

$$x_1 = (T - 167)/5, \qquad x_2 = (c - 27.5)/2.5, \qquad x_3 = (t - 6.5)/1.5$$

The coded values for the first eight experiments are then all at the levels $+1$ and -1 forming a 2^3 factorial design. When this is augmented with experiments 9–15 a 'central composite' experimental design [2]

*The 'natural units' given in [3] differ slightly from those quoted in reference [1]. The yields given in Table 1 of this paper differ slightly from those in [3] due to refinements previously ignored.

TABLE I. EXPERIMENTAL DATA

(1)	(2)	(3)	(4)	(5)	(6)	(7)	(8)	(9)	(10)	(11)	(12)	(13)	(14)	(15)	(16)	(17)
Expt.	Levels of Variables			First Coding			Second Coding			Observed Yields				$\hat{z}_t$	$\hat{u}$	$x = (T + 273)^{-1}$
	$T(°C)$	$c(\%)$	t(hr.)	x_1	x_2	x_3	$\dot{x}_1$	$\dot{x}_2$	$\dot{x}_3$	$\hat{\eta}_2$	$\hat{\eta}_3$	$\hat{\eta}_5$	Total			
1	162	25	5	−1	−1	−1	−1	−1	−1	0.415	0.459	0.112	0.986	0.739	−6.29	10^{-7} x 22989
2	162	25	8	−1	−1	1	−1	−1	1	0.338	0.533	0.112	0.983	0.714	−6.64	22989
3	162	30	5	−1	1	−1	−1	1	−1	0.277	0.575	0.127	0.979	0.683	−6.23	22989
4	162	30	8	−1	1	1	−1	1	1	0.217	0.588	0.160	0.965	0.632	−6.49	22989
5	172	25	5	1	−1	−1	1	−1	−1	0.199	0.606	0.162	0.967	0.627	−5.80	22472
6	172	25	8	1	−1	1	1	−1	1	0.150	0.580	0.226	0.956	0.559	−6.04	22472
7	172	30	5	1	1	−1	1	1	−1	0.122	0.586	0.245	0.953	0.536	−5.70	22472
8	172	30	8	1	1	1	1	1	1	0.043	0.524	0.380	0.947	0.413	−5.79	22472
9	167	27.5	6.5	0	0	0	0.01	0.05	0.12	0.193	0.569	0.213	0.975	0.595	−6.05	22727
10	177	27.5	6.5	2	0	0	1.97	0.05	0.12	0.064	0.554	0.308	0.926	0.391	−5.42	22222
11	157	27.5	6.5	−2	0	0	−2.03	0.05	0.12	0.376	0.469	0.147	0.992	0.707	−6.51	23256
12	167	32.5	6.5	0	2	0	0.01	1.92	0.12	0.180	0.575	0.222	0.977	0.586	−6.21	22727
13	167	22.5	6.5	0	−2	0	0.01	−2.19	0.12	0.263	0.550	0.183	0.996	0.651	−6.05	22727
14	167	27.5	9.5	0	0	2	0.01	0.05	1.73	0.099	0.589	0.280	0.968	0.515	−6.18	22727
15	167	27.5	3.5	0	0	−2	0.01	0.05	−2.52	0.250	0.503	0.221	0.974	0.605	−5.47	22727
16	177	20	6.5	2	−3	0	1.97	−3.54	0.12	0.141	0.611	0.230	0.982	0.560	−5.60	22222
17	177	20	6.5	2	−3	0	1.97	−3.54	0.12	0.152	0.629	0.207	0.988	0.596	−5.72	22222
18	160	34	7.5	−1.4	2.6	0.7	−1.41	2.36	0.72	0.159	0.600	0.221	0.980	0.583	−6.39	23095
19	160	34	7.5	−1.4	2.6	0.7	−1.41	2.36	0 72	0.196	0.606	0.193	0.995	0.624	−6.54	23095

[1] is formed suitable for fitting an equation of second degree to any observed response.

A second degree equation fitted to the yields of desired product $\hat{\eta}_3$ for experiments 1–15 indicated a possible planar ridge system. Further experiments 16–19 were carried out therefore on the estimated maximum plane. In spite of the great differences in the actual conditions employed (see table 1) these gave yields close to the maximum value of about 60% in accordance with prediction. These observations were now included in the calculation and the best fitting second degree equation using all 19 observations was then:

$$Y = 58.78 + 1.90x_1 + 0.97x_2 + 1.06x_3 - 1.88x_1^2 - 0.69x_3^2 - 0.95x_2^2 - 2.71x_1x_2 - 2.17x_1x_3 - 1.24x_2x_3 \quad (5)$$

where Y denotes the % yield predicted by the equation.

From Table 2 it will be seen that the sum of squares due to regression accounted for 92.6% of the total variation after elimination of the mean. From the residual sum of squares an estimate $\sigma = 1.81$ of the experimental error standard deviation was obtained (this may be biased upwards due to some inadequacy of the assumed form of the equation). Using this estimate for σ the standard errors of the coefficients in equation (5) could be calculated. For linear, quadratic and interaction terms these were all between 0.3 and 0.5. It appeared therefore that equation (5) was reasonably well determined. It was found to have the canonical form

$$Y - 59.15 = -3.40X_1^2 - 0.32X_2^2 + 0.20X_3^2 \quad (6)$$

where

$$X_1 = 0.751x_1 + 0.479x_2 + 0.455x_3 - 0.349 \quad (7)$$

$$X_2 = 0.308x_1 + 0.356x_2 - 0.882x_3 + 0.013 \quad (7a)$$

$$X_3 = 0.584x_1 - 0.803x_2 - 0.120x_3 + 0.485 \quad (7b)$$

The centre of the system (that is the point $X_1 = 0, X_2 = 0, X_3 = 0$) had coordination $x_1 = -0.03$, $x_2 = 0.55$, $x_3 = 0.23$. Thus the axes of the system passed through the region in which the experiments had been performed and a discussion in terms of these axes was therefore immediately relevant.*

*In the case of ridges, especially rising ridges, the 'centre' of the canonical system may be found almost anywhere on the line or plane of the crest of the ridge. When this centre is remote from the region of experiments, we cannot, of course, draw any conclusions about the nature of the surface at this remote point. However the preliminary use of the steepest ascent procedure will have ensured that the experimenter has already been brought to a point which is close to the ridge. The canonical equation can therefore be rewritten in terms of a new origin close to the original origin and on the line or plane of the ridge. From this form of the equation the principal features of the response surface in the *immediate region* of the experiments may be readily comprehended (see for example reference [1] pp. 37 and 53 and reference [3] p. 531.

Now the canonical variables have the same scale as the original variables. That is to say in solid models (like those in Figures 3 and 4) in which unit change in x_1, x_2, or x_3 is represented by the same distance, unit change in X_1, X_2 and X_3 is also represented by this same distance. Consequently the relative magnitudes of the coefficients in the canonical equation indicate the relative *importance* of the canonical variables in describing the function over the experimental region which has been chosen as appropriate for study. The coefficient (-3.40) of X_1^2 in equation (6) is over 10 times larger than either of the other two coefficients (-0.32 and 0.20). Furthermore the latter are somewhat less in magnitude than their errors of estimation. (Appropriate standard errors for these constants can be shown to be of the same order of magnitude as those of the original quadratic and interaction terms, namely about 0.3 to 0.5.) To express the matter a little differently within the region in which the fitted equation has some relevance which may be roughly defined as $-2 < X_1 < +2$, $-2 < X_2 < +2$, $-2 < X_3 < +2$, the maximum contribution to the % yield Y of the terms in X_1 is about 14% whilst that of each of the terms in X_2 and X_3 is only about 1% which is of the same order of magnitude as the experimental standard deviation. These facts suggest that the system may be described by an equation containing a single canonical variable only.

The refitting *ab initio* of an equation of the form

$$Y - Y(\text{max}) = BX^2$$

containing only a single variable X which is itself linear in x_1, x_2 and x_3 is possible but laborious. An approximation used in [1], [2] and [3] may be obtained simply by ignoring the smaller coefficients in equation (6) when we have

$$Y - 59.15 = -3.40X_1^2$$

with X_1 defined as before (equation (7)).

A closer approximation is obtained by fitting by least squares the expression

$$Y = A + BZ + CZ^2 \tag{8}$$

where Z is the linear aggregate $ax_1 + bx_2 + cx_3$ obtained by omitting the constant term in X_1. In the present example

$$Y = 59.50 + 2.65Z - 3.80Z^2 \tag{9}$$

where

$$Z = 0.751x_1 + 0.479x_2 + 0.455x_3 \tag{9a}$$

That is $$Y - 59.96 = -3.80X_2'^2 \tag{10}$$

where $$X_1' = 0.751x_1 + 0.479x_2 + 0.455x_3 - 0.348 \tag{11}$$

or in terms of the original variables

$$X_1' = 0.150T + 0.191c + 0.303t - 31.969 \tag{11a}$$

An analysis of variance is shown in table 2.

TABLE 2

Analysis of variance table for fitted equations

Degrees of Freedom	Source		Sum of squares	
9	Full 2nd degree equation (equation 5)	(equation 9)	371.4	357.4
		Remainder		14.0
9	Residual			29.5
18	Total after eliminating mean			400.9

The sum of squares due to the 'full second degree equation' (equation 5) has nine degrees of freedom associated with the nine independent constants fitted in addition to the mean. The 'canonical equation' (equation 9) contains only four independent constants apart from the mean. These are two constants of equation (9) and two of the coefficients in (9a) (the third is fixed by the requirement that the squares of these coefficients sum to unity). We see that the simpler expression is associated with a sum of squares of 357.4 and that the fitting of an equation containing five more constants accounts only for a further 14.0 of the sum of squares.

The estimates in equations (9) and (9a) are not linear functions of the observations and consequently the number of constants in this equation and the number of extra constants associated with the remainder sum of squares cannot be directly associated with degrees of freedom in the usual sense. However the analysis serves to show that the simpler expression probably accounts for the data as well as does the more complicated one.

If we put Y equal to its maximum value in (10) we get $X_1 = 0$ which, when substituted in (11) or (11a), gives a plane of alternative conditions on which Y attains its maximum value. In general on substituting a lower value for Y in equation (10) we obtain two values for X_1 equal in magnitude but opposite in sign which when entered in (11) give the equations of parallel planes of lower yield 'sandwiching' the maximum plane as illustrated in figure 3. Sections of this system for three levels of the concentration variable are shown on the right hand side of this figure.

At the time when these experiments were carried out (some five years ago) the number of chemical yield surfaces which had been approximately determined was small and this surface, showing such marked dependence between all the variables, was somewhat unexpected. Further experiments having confirmed the reality of the system it was realised that this was a case where the reaction was sufficiently simple to allow a theoretical study which, as it turned out, explained the type of surface found.

Although most chemical systems are more complicated, the study serves to show that, as a result of the laws which govern chemical systems, ridge surfaces of one sort or another are to be generally expected (as experience has in fact confirmed).

3. THE THEORETICAL STUDY

The chemical system could be represented by the following sequence of competitive reactions

$$2a + bNb \rightarrow aNb + ab \qquad \text{(reaction 1)}$$

$$2a + aNb \rightarrow aNa + ab \qquad \text{(reaction 2)}$$

The substance bNb contained a large molecular nucleus N to which were attached the two groupings b. In the part of the sequence denoted by 'reaction 1', one of the b groupings was replaced by an a to form aNb which was the required product. However, as is shown in 'reaction 2', under the conditions in which the first reaction could take place aNb could destroy itself by combining with more a to form the unwanted product aNa.

The concentration of the starting material bNb was kept constant throughout the experiments, the concentration which was varied being that of the substance a. The substance ab was chemically inert and no reverse reaction took place.

If the reactions were allowed to continue for a time t a mixture of unchanged a and bNb and of the products aNb, aNa and ab was produced. The required product aNb had to be separated from the other products in subsequent purification stages.

To proceed we need to use some simple chemical ideas. As they may be unfamiliar to those readers who are not chemists we shall briefly explain them as they are needed.

Concentrations of reactants.

The chemical equations above indicate the proportions in which the actual molecules combine. It is convenient to measure the amounts of the various substances in 'gram moles', one gram mole being the molecular weight in grams of the substance concerned. Thus the equation for reaction (1) implies that two gram moles of a combine with one gram mole of bNb to form one gram mole of aNb and one gram mole of ab. We shall be concerned with the *concentrations* of the reactants in the solvent and these will be expressed as 'gram moles per litre'. The following symbols are used to denote the concentrations and fractional yields of the various reactants in the system.

	Reactant	Conc. at time t	Conc. Initially	"Fractional" yield at time t = (concentration)/c_{20}
Starting materials	a	c_1	c_{10}	η_1
	bNb	c_2	c_{20}	η_2
Products	aNb	c_3	0	η_3
	ab	c_4	0	η_4
	aNa	c_5	0	η_5

In addition the symbol C denotes the ratio c_{10}/c_{20} of the initial concentrations of the starting materials. In the discussion of section 2 and in [1] and [3] the concentration of a was measured as a 'percentage' which was denoted by c. Since the concentration c_{20} was kept constant in our experiment c and C are alternative measures of the concentration of a. In fact $C = 0.4555c$.

We can now derive a theoretical expression for the yield η_3 of aNb in terms of T the temperature, C the relative concentration of a, and t the time, which can be directly compared with the empirical equation.

Since the number of gram moles of a present in the system *in some form or other* at any time t must be constant and equal to c_{10} it follows that

$$c_1 + c_3 + c_4 + 2c_5 = c_{10} \qquad (12)$$

Applying the same reasoning to the nucleus N we have

$$c_2 + c_3 + c_5 = c_{20} \tag{13}$$

whence

$$\eta_2 + \eta_3 + \eta_5 = 1 \tag{14}$$

Finally

$$c_3 + 2c_5 = c_4 \tag{15}$$

By subtracting four times equation (13) from equation (12) and adding equation (15) we obtain

$$c_1 = c_{10} - 4c_{20} + 2c_3 + 4c_2$$

or

$$\eta_1 = C - 4 + 2\eta_3 + 4\eta_2 \tag{16}$$

Kinetic Theory.

We now use two simple concepts from the kinetic theory of chemical reactions. The law of 'mass action' states that, in dilute solution, the speed of a chemical reaction is proportional to the molecular concentrations of the substances reacting. Thus in particular if, at time t, p and q are the molecular concentrations of two substances P and Q which are taking part in a non-reversible reaction to form a substance R, such that one molecule of P reacts with one molecule of Q to form one molecule of R, then the rate of reaction (that is the rate of decrease of the molecular concentrations of P or Q or the rate of increase of the molecular concentration of R) is given by

$$-\frac{dp}{dt} = -\frac{dq}{dt} = \frac{dr}{dt} = kpq \tag{17}$$

where r is the molecular concentration of R at time t. Experimental results have shown that the law often holds *approximately* in moderately concentrated solutions such as we consider.

The quantity k occuring in equation (17) is called the *rate constant.* Its magnitude depends on the temperature (T). Study of a variety of chemical reactions has shown that the relationship between k and T is represented fairly satisfactorily by the empirical equation due to Arrhenius

$$k = \alpha \exp \{-\beta/(T + 273)\} \tag{18}$$

where α and β are constants depending on the reaction studied.

In the sequence with which we are concerned we denote the rate constant for the first reaction by k' and the rate constant for the second reaction by k and define constants α', β', α and β so that

$$k' = \alpha' \exp\{-\beta'/(T + 273)\} \qquad k = \alpha \exp\{-\beta/(T + 273)\} \tag{19}$$

Then the ratio $\rho = k'/k$ of the rate constants is given by

$$\rho = \gamma \exp\{-\delta/(T + 273)\} \tag{20}$$

where

$$\gamma = \alpha'/\alpha \qquad \text{and} \qquad \delta = \beta' - \beta \tag{21}$$

The equation of the 'theoretical' surface:

Now reaction (1) occurred as follows:

$$a + bNb \xrightarrow{k'} aNb + b$$

$$a + b \xrightarrow{\infty} ab$$

the second part of the reaction being instantaneous.

Similarly for reaction (2)

$$a + aNb \xrightarrow{k} aNa + b$$

$$a + b \xrightarrow{\infty} ab$$

At some particular temperature T then, the rate of disappearance of bNb in reaction (1) is $\rho k c_1 c_2$. Also the rate of formation of aNb from reaction (1) is $\rho k c_1 c_2$ while the rate at which it is destroyed in reaction (2) is $k c_1 c_3$.

We have therefore the pair of differential equations

$$-dc_2/dt = \rho k c_1 c_2 \tag{22}$$

$$dc_3/dt = \rho k c_1 c_2 - k c_1 c_3 \tag{23}$$

These together with equations (14), (15), (16) and (19) allow us to obtain expressions for η_1, η_2, η_3, η_4 and η_5, the yields of the products at time t. The derivation is given in section 1 of the appendix. In the particular case of the desired product aNb the yield at time t is

$$\eta_3 = \frac{\rho}{\rho - 1} z_t(1 - z_t^{\rho-1}) \tag{24}$$

where z_t is a function of T, C, t and c_{20} depending on the constants α, β, γ and δ and defined by

$$c_{20} t\alpha \exp\{-\beta/(T + 273)\}$$

$$= (\rho - 1) \int_{z_t}^{1} z^{-1}\{2(\rho - 2)z^{\rho} + 2\rho z + (\rho - 1)(C - 4)\}^{-1}\, dz \quad (25)$$

where
$$\rho = \gamma \exp \{-\delta/(T + 273)\} \quad (25a)$$

To the extent that the various assumptions are justified therefore we have an equation for the theoretical yield surface for η_3 and (see equations 74, 76, 77 and 78 of the appendix) incidentally for η_1, η_2, η_4 and η_5 also. In the form in which it is expressed by equations (24) and (25), the characteristics of this surface cannot be readily appreciated so we shall proceed by actually fitting this form of expression to our data and comparing the resulting surface with that obtained empirically. To do this we need to estimate the values of the unknown constants from the data.

The ratio ρ of the rate constants.

Considering again the equations describing the reactions we see that the ratio ρ of the rate constants is the ratio of the probability that an a will replace a b from bNb to the probability that an a will replace a b from aNb.

Now if the chance of an a replacing a b at a particular position on the nucleus N is independent of the type of grouping (a or b) which occupies the other position, then the chance of replacement will be twice as great with bNb, where there are two positions at which replacement can occur, as with aNl, where there is only one. Thus ρ will equal 2 independently of the temperature T.

Now it is readily shown (see section 2 of the appendix) that the maximum yield for aNb is given by

$$\eta_3(\max) = \rho^{-1/(\rho-1)} \quad (26)$$

at which value

$$\eta_2 = \rho^{-\rho/(\rho-1)} \quad \text{and} \quad \eta_5 = 1 - \rho^{-\rho/(\rho-1)} - \rho^{-1/(\rho-1)} \quad (27)$$

Consequently if $\rho = 2$ then η_3 (max) $= 0.5$, and at this maximum value of η_3, $\eta_2 = 0.25$ and $\eta_5 = 0.25$.

The maximum yield actually found was not 50% but about 60%. We must conclude therefore that the probability of replacement of a particular grouping on a half-substituted molecule is not the same as the probability of replacement on an unsubstituted molecule (a conclusion we should in any case expect from chemical considerations). We might however still expect the ratio of these probabilities to be largely independent of temperature, at least over the range we have considered. This implies that the temperature constants β' and β in equations (19) would be equal and that in equation (20) $\rho = \gamma$ and $\delta = 0$. If we substitute the value $\eta_3(\max) = 0.60$ in (26) we obtain $\rho = 3.4$ and the theoretical distribution of yield between the products bNb, aNb and aNa when η_3 was at its maximum value would then be

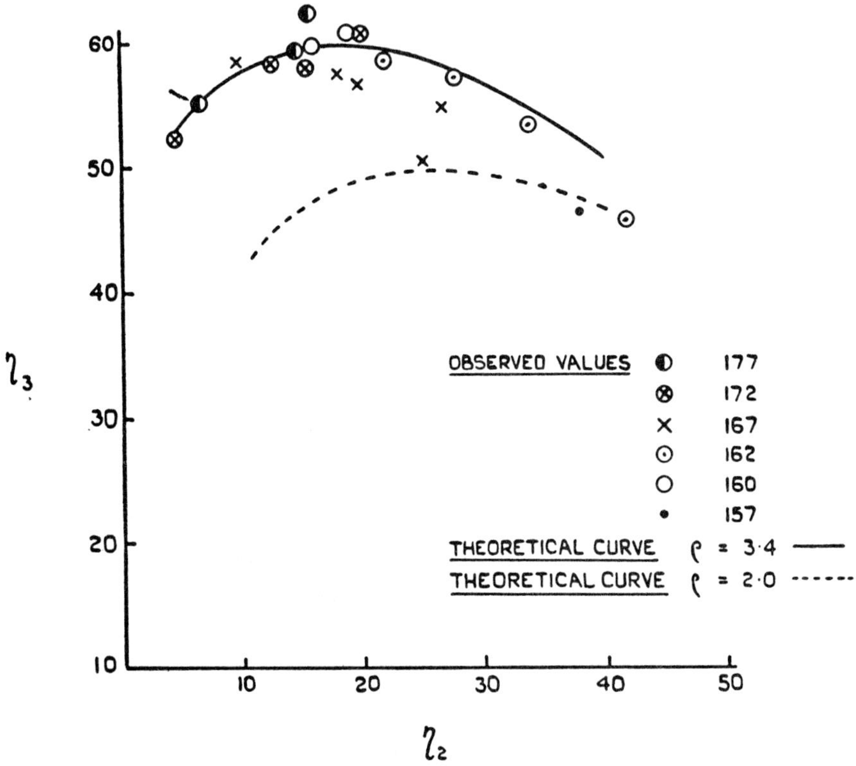

FIGURE 1. YIELD OF aNb (η_3) PLOTTED AGAINST YIELD OF bNb (η_2) WITH THEORETICAL CURVES FOR $\rho = 3.4$ AND $\rho = 2$.

$\eta_2 = 0.176$, $\eta_3 = 0.600$, $\eta_5 = 0.224$. These values agree quite well with those found in the final four experiments 'on the maximum plane' of the empirical surface for which the averages were

$$\eta_2 = 0.162 \qquad \eta_3 = 0.611 \qquad \eta_5 = 0.213$$

That the value $\rho = 3.4$, is reasonably consistent with the data in other respects can be seen from figure 1 where the observed value $\hat{\eta}_3$ is plotted against $\hat{\eta}_2$. The theoretical relationship between η_3 and η_2 is

$$\eta_3 = \frac{\rho}{\rho - 1} \{\eta_2^{1/\rho} - \eta_2\} \tag{28}$$

Most of the experimental observations are in fairly close agreement with the theoretical curve for $\rho = 3.4$ although there seems to be some departure for high values of η_2. Also there seems to be no evidence that the points corresponding to different temperatures follow different lines whose general form is changing steadily with temperature.

The characteristics of our solution are not very sensitive to the value of ρ chosen and we proceed by supposing that ρ is equal to 3.4 and

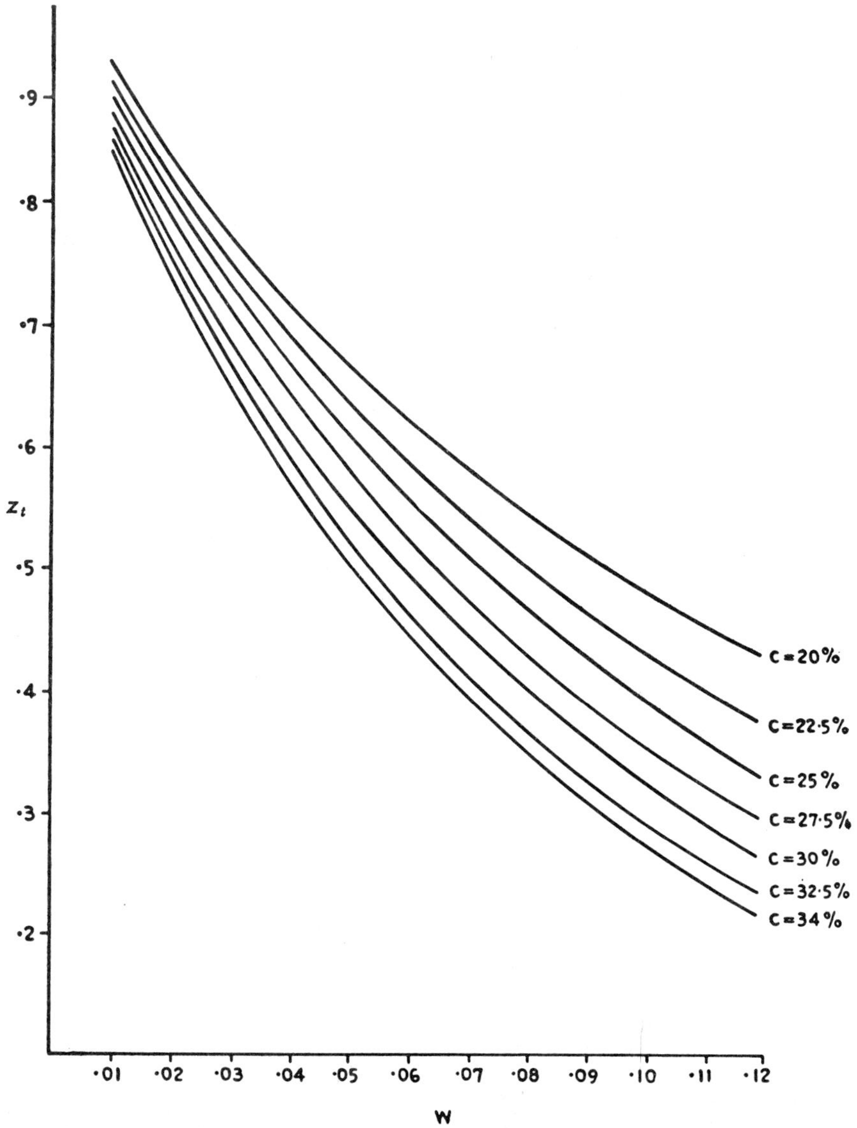

FIGURE 2. GRAPHS SHOWING VALUES OF THE INTEGRAL w FOR VARIOUS VALUES OF z_t AND c.

is independent of temperature. This implies that, the first substitution having occurred, the probability of the second substitution is reduced by a factor of 1.7.

Putting the value $\rho = 3.4$ in (25) we have for our theoretical equation

$$c_{20}t\alpha \exp\{-\beta/(T + 273)\}$$
$$= 2.4 \int_{z_t}^{1} z^{-1}\{2.8z^{3.4} + 6.8z + 2.4(C - 4)\}^{-1}\, dz \qquad (29)$$

Estimates of α and β

The value of the expression on the right-hand side of the equation (29) which we shall denote by $w(z_t, C)$ or more simply by w depends on z_t and C alone or (remembering that the percentage concentration c which we have considered in our experiments is equal to $C/0.4555$) on z_t and c alone. The integral cannot be expressed in terms of elementary functions but its value, for any level of z_t and for the values of c we have employed in our experiments, can be read off from the graphs in figure 2. Each of these graphs was obtained by setting c equal to the appropriate value, calculating ordinates of the curve

$$y = z^{-1}\{2.8z^{3.4} + 6.8z + 2.4(C - 4)\}^{-1} \qquad (30)$$

at the 7 equally spaced values $z = 1.000, 0.875, 0.750, 0.625, 0.500, 0.375, 0.250$ and then calculating the area between $z = 1$ and $z = z_t$ under the curve by numerical quadrature.

The values of the constants α and β can now be estimated from the data as follows. Taking natural logarithms equation (29) may be written in the form

$$u = \ln \alpha - \beta/(T + 273) \qquad (31)$$

where

$$u = \ln w - \ln c_{20} - \ln t \qquad (32)$$

For each experiment the value $\hat{z}_t$ (shown in column 15 of table 1) can be estimated from the formula*

$$\hat{z}_t = \hat{\eta}_2 + \hat{\eta}_3(\rho - 1)/\rho \qquad (33)$$

*We notice that in every case the total of $\hat{\eta}_2$, $\hat{\eta}_3$ and $\hat{\eta}_5$ in table 1 is less than the theoretical value of 100%. This is due partly to difficulties of accurate determination of aNa in the presence of other substituents and partly due to some degradation of this product. Because of uncertainty concerning the estimate $\hat{\eta}_5$, z_t was calculated from $\hat{\eta}_2$ and $\hat{\eta}_3$ alone. In references [1] and [3] an empirical surface for aNa was fitted and a region was shown in the maximal plane of aNb where less than 20% of aNa was obtained. In the theoretical equations the yields of both products depend only on z_t and consequently for any surface for which the yield η_3 of aNb was constant the yield η_5 of by-product aNa would be constant also. The region found in the empirical study where aNa was less than 20% probably occurred because degradation of this product was favoured by reaction conditions in this neighbourhood. The effect of this degradation is not allowed for in the theoretical study.

or putting $\rho = 3.4$

$$\hat{z}_t = \hat{\eta}_2 + 0.706\hat{\eta}_3 \tag{34}$$

The corresponding values of $\hat{w}$ for each experiment may now be obtained from the values of z_t by reading from the appropriate graph in figure 2. Finally the values of $\hat{u}$ shown in column 16 of table 1 may be obtained by substituting values of t and $\hat{w}$ for each experiment in (32) remembering that the value for c_{20} was kept constant at 3.10.

From the form of equation (31) we see that we may now obtain estimates of $\ln \alpha$ and β by fitting a regression line of $\hat{u}$ on $x = (T + 273)^{-1}$ by the method of least squares.

We find

$$\ln \hat{\alpha} = 16.86 \pm 3.62 \tag{35}$$

$$\hat{\beta} = 10{,}091 \pm 1{,}595 \tag{36}$$

The quantities following the plus and minus signs in (35) and (36) are the formal 'standard errors' calculated in the usual way from the residual sum of squares. It is clear from inspection of the table that the deviations from the regression line contributing to this sum of squares still contain components due to c and t indicating that the theoretical expression does not give a perfect fit to the data. This is not surprising first because of the assumptions we have had to make in the derivation and second because the levels assumed for time t and concentration c are not entirely appropriate. Doubt concerning the level of t exists because, in addition to the reaction occuring during the 'time on temperature', some reaction will also occur while the reaction vessel is being heated up and this is difficult to allow for. The value of c may not be entirely appropriate because owing to solubility factors the effective concentrations in the solvent may be slightly different at different temperatures and at different stages of the reaction. In spite of these limitations a reasonably close representation of the experimental data is achieved by the theoretical expression (29) which contains only three adjustable parameters (ρ, α and β) as compared with the ten adjustable parameters (β_0, β_1, $\cdots$, β_{23}) of the empirical expression.

Comparison of the theoretical and empirical surfaces.

Using our estimates, $\hat{\alpha}$ and $\hat{\beta}$, we may now calculate the value of z_t and hence the values of η_2, η_3 and η_5 predicted by our 'theoretical' equation for any desired level of T, c and t. Figure 4 shows the contours of the resulting 'theoretical surface' for comparison with Fig. 3.

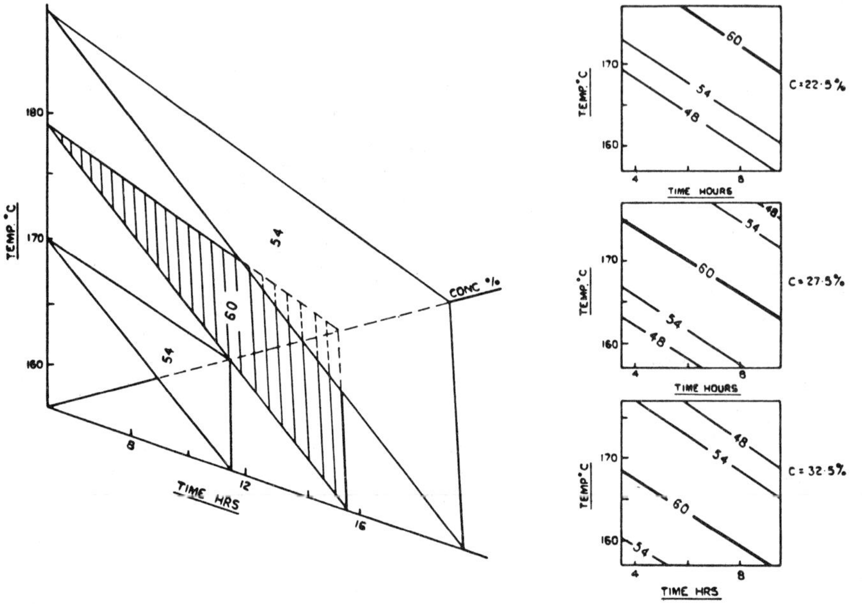

FIGURE 3. CONTOURS OF EMPIRICAL YIELD SURFACE WITH SECTIONS AT THREE LEVELS OF CONCENTRATION

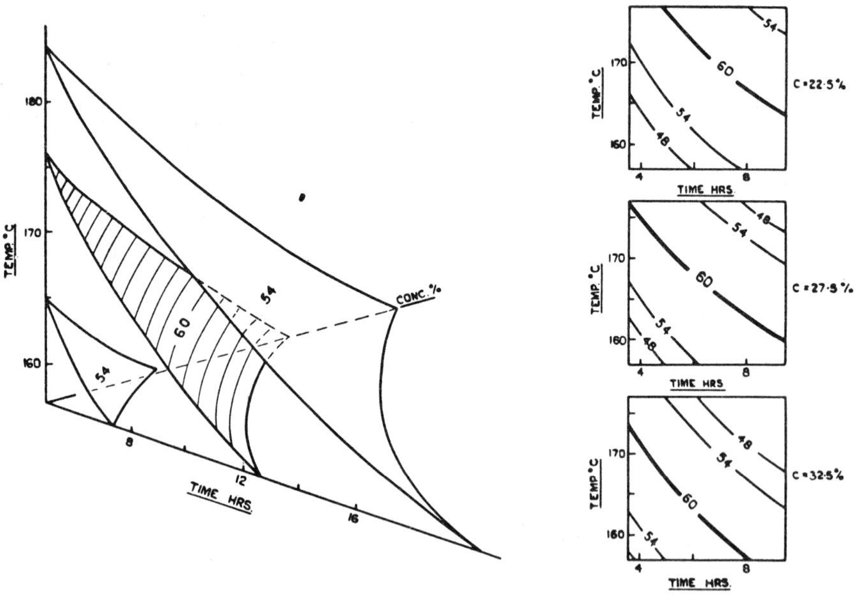

FIGURE 4. CONTOURS OF THEORETICAL YIELD SURFACE WITH SECTIONS AT THREE LEVELS OF CONCENTRATION. ρ INDEPENDENT OF TEMPERATURE

It is seen that there is remarkably close agreement between the general characteristics of the two surfaces one of which has been obtained entirely empirically and the other derived on a particular theory of the mechanism of the reaction. In each case there is a whole surface rather than a unique point on which the maximum yield is obtained surrounded on either side by surfaces of lower yield.

A final point of some interest is that z_t and hence the yield η_3 is in fact a function of *four* quantities T, C, t *and* c_{20} which determines the 'overall concentration' of the reactants. Of these only the first three were varied in the experiments, and c_{20} was kept constant. Had we included c_{20} as a variable which would have been a perfectly sensible thing to do, then the yield surface would still have been completely described in terms of a *single* canonical variable and there would have been a redundancy of three variables instead of two.

An Analogy.

A simple picture* of what is occurring can perhaps be gained from the following analogy.

Imagine a billiard table on which a number of black and white balls are in continuous motion. Suppose that when a black ball collides with a white ball a blue ball is produced and that when a black ball collides with a newly formed blue ball a red ball results.

In this analogy black and white balls correspond to the molecules of the starting materials a and bNb, blue balls to molecules of the required product aNb, and red balls to the unwanted aNa. Starting off with a given number of black and white balls we can see that as soon as the system is set in motion blue balls will begin to appear and these in turn give rise to an increasing number of red balls. Provided that initially there is a sufficient excess of black balls the number of blue balls will increase to a maximum then fall off until finally only red balls and excess black balls remain.

Clearly the proportions of the various sorts of balls on the table at a given instant will depend on the following variables:

(1) The time which has elasped since the start.
(2) The speed with which the balls move.
(3) The relative numbers of black to white balls at the start.
(4) The total number of white balls at the start.

---

*This is intended to provide only a very rough parallel which may be of assistance to non-chemist readers. The real mechanism of chemical reactions is known to be much more complicated and cannot be accounted for by simple collision. For example only a small proportion of molecular 'collisions' actually result in reaction and this proportion is dependent on temperature.

These correspond to the variables t, T, C and the overall concentration c_{20} (temperature T being linked to speed by the Ahhrenius equation). Suppose for fixed conditions of (2) (3) and (4) the time is noted for the maximum proportion of blue balls to be produced. If now conditions (3) and (4) are kept the same but the speed with which each ball moves is doubled the effect will be like that of showing a cinematograph film at twice the rate, an identical sequence of events will be gone through twice as fast and in particular the *same* maximum proportion of blue balls to the initial number of white balls will be produced but in half the time. Similarly if conditions (2) and (3) are kept constant but the initial number of black and white balls on the table is doubled and if we ignore the effect of interference then again a similar sequence of events will occur but at twice the speed and again the *same* maximum proportion of blue balls to the initial number of white balls will be produced but in half the time.

The effect of change in factor (3), the relative number of black and white balls is less easily appreciated intuitively. However we can see that since the *relative* rates at which white balls are disappearing and red and blue balls appearing is at any stage completely independent of the number of black balls (since any change in the number of black balls effects both these rates proportionally) it follows that the proportion of blue balls relative to white balls and the proportion of red balls relative to white balls follows precisely the same course whatever the number of black balls. Consequently once again the same maximum proportion of blue balls to white is produced whatever the proportion of black to white balls. It is evident that for such a system a maximum can be obtained for almost any level of a particular variable provided the other three variables are suitably adjusted.

4. THE CANONICAL VARIABLE

The part played by z_t in the theoretical equation is exactly parallel to that played by the canonical variable X_1 in the empirical equation. This is seen most clearly if we consider a case where z_t may be obtained as an explicit function of T, c, and t. That is to say a case where the integral w in equation (25) may be explicitly evaluated. We have noted already that on the simplest view of the reaction we would expect the ratio ρ of the rate constants to equal 2, and it is readily seen from the form of equations (24) and (25) that, apart from the maximum yield of η_3 being 50% rather than 60%, the general characteristics of the resulting surface will be the same with this value as they are with the value $\rho = 3.4$. Taking $\rho = 2$ we have

$$\eta_3 = 2z_t(1 - z_t) \tag{37}$$

where (see section 3 of the appendix) z_t is now explicitly defined in terms of T, c and t by the equation

$$z_t = (C - 4)/\{C \exp [c_{20}(C - 4)t\alpha \exp \{-\beta/(T + 273)\}] - 4\} \quad (38)$$

Now subtracting η_3 (max) $= 0.50$ from both sides of (37), and writing Y for 100 η_3 (to agree with the notation of the empirical surface) we have

$$Y - Y(\max) = 200z_t(1 - z_t) - 50 \quad (39)$$

$$= -\{7.07(2z_t - 1)\}^2 \quad (40)$$

Writing $\dot{W} = 7.07\ (2z_t - 1)$ we see that the theoretical surface is completely described by the pair of equations

$$\begin{cases} Y - Y(\max) = -\dot{W}^2 & (41) \\ \dot{W} = 7.07\left\{\dfrac{2(C - 4)}{\{C \exp [c_{20}(C - 4)t\alpha \exp \{-\beta/(T + 273)\}] - 4\}} - 1\right\} & (42) \end{cases}$$

Now (equation 10) the empirical surface is closely approximated by

$$Y - Y(\max) = -3.80X_1'^2 \quad (43)$$

where substituting $C = 0.4555c$ in (11a) we have

$$X_1' = 0.150T + 0.419C + 0.303t - 31.969 \quad (44)$$

If we write $W = (3.80)^{\frac{1}{2}}X_1'$ we see that the empirical surface is approximately described by the equations

$$\begin{cases} Y - Y(\max) = -W^2 & (45) \\ W = 0.292T + 0.817C + 0.591t - 62.320 & (46) \end{cases}$$

which are directly comparable with (41) and (42).

The 'theoretical canonical variable' $\dot{W}$ is a more complicated function of T, C and t than is the empirical canonical variable W. The latter is necessarily a simple *linear* function of T, C and t, and consequently contour surfaces of constant yield in figure 3 are necessarily planes. However over the regions considered these planes do provide a reasonable approximation to the curved contour surfaces of Figure 4 as (for the reason given in Section 1) we might expect them to.

In the discussion above we have compared the canonical variable of the empirical surface with the 'theoretical canonical variable' arising in the particular case when $\rho = 2$. When $\rho = 3.4$ a similar situation will exist and although η_3 will not be a quadratic function of z_t yet a quadratic function will still closely approximate the true curve near the maximum.

5. TEMPERATURE DEPENDANCE OF ρ

In our derivation we have supposed that the ratio $\rho = k'/k$ of the rate constants, was itself independent of temperature. To put it another way we have supposed that the temperature constants β' and β of equation (19) were equal. This supposition is supported by the data over the range of values studied. It is interesting however to consider how the surface would be affected if this were not true and this is perhaps best done by considering an example. Let us suppose that, at the temperature 157°C, ρ was equal to 2 and that, at the temperature 177°C, ρ had increased to 3.4. Substituting these values in equation (20) we find that this implies that $\gamma = 12.63$ and $\delta = 5.132$. If we suppose that the values for the constants α and β were the same as before we then have

$$\ln k' = 29.49 - 15{,}223/(T + 273) \tag{47}$$

$$\ln k = 16.86 - 10{,}091/(T + 273) \tag{48}$$

$$\ln \rho = \ln k' - \ln k = 12.63 - 5{,}132/(T + 273) \tag{49}$$

The solid contour model for the (yield: temperature, concentration, time) surface which would then be found is shown in Figure 5 together with sections taken at various levels of concentration. The diagrams

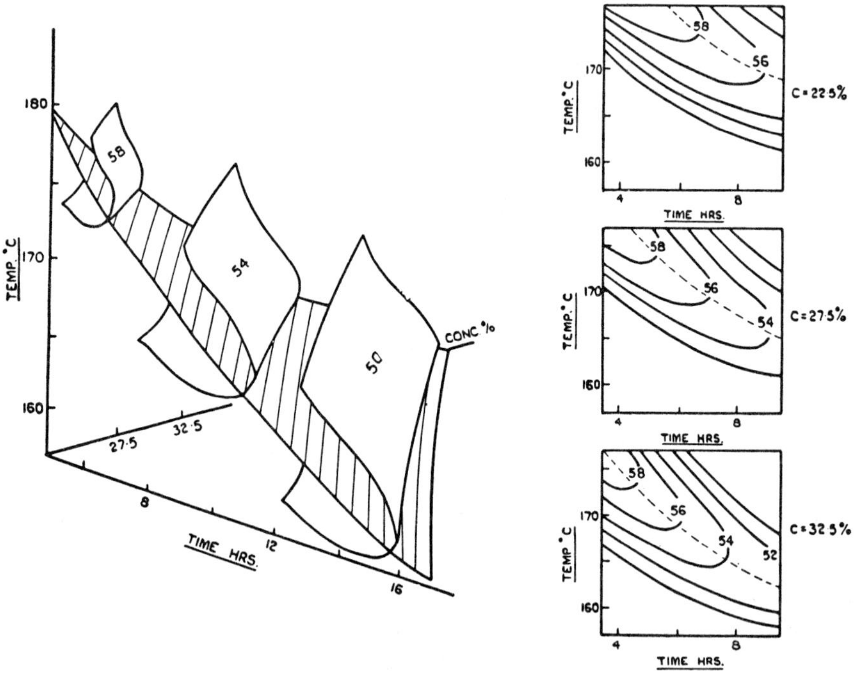

FIGURE 5. CONTOURS OF THEORETICAL YIELD SURFACE WHEN ρ DEPENDS ON TEMPERATURE.

were prepared by carrying through the numerical integration and subsequent calculations as before for each of the concentrations 22.5%, 27.5%, and 32.5% and for five values of ρ corresponding to five values of the temperature calculated from equation (49) as follows:

$T =$	157	162	167	172.5	177
$\rho =$	2.00	2.34	2.62	3.00	3.40

The three concentration sections were then prepared by drawing smooth contour lines through the points calculated at these temperatures and finally from these sections the representation of the solid model was obtained.

Considering this solid model we see that we now have a situation where there is a 'rising' ridge instead of a stationary one. The value of the yield steadily increases on the ridge as the temperature is increased. A section of the solid model for a particular value of time or concentration gives a two-dimensional rising ridge system running diagonally to the axes of the variables like the concentration sections shown. A section of the solid model for a particular value of temperature on the other hand gives a two dimensional *stationary* ridge system of the type considered before.

In general we must expect a surface of this sort to occur in the common case where a competitive system is influenced by a highly dependent set of variables and one competitor is favoured by a certain direction of movement in the variables.

In the example we have considered the rising ridge results from the first reaction in the sequence being favoured by high temperatures. It is easy to imagine other examples of this sort of phenomenon. For instance in some systems the rates of competing reactions depend on different powers of the concentration terms (see for example reference [5]). In these circumstances a rising ridge associated with concentration would be expected.

It is of some interest to consider the behaviour of the empirical method when a rising ridge of this sort occurs. The typical situation encountered is as follows. Analysis of the fitted second degree equation yields a canonical equation

$$Y - Y_S = B_{11}X_1^2 + B_{22}X_2^2 + B_{33}X_3^2 \qquad (50)$$

in which (as with the stationary ridge system discussed in section 2) one of the coefficients (say B_{11}) is negative and comparatively large and the other two (B_{22} and B_{33}) are small. The centre S of the fitted system is remote from the design. To determine the nature of the system in the region where it applies (in the neighborhood of the design) a new origin S', situated as closely as possible to the design centre and in the plane $X_1 = 0$ is taken. Suppose this new origin is at the point $X_2 = X_{2S'}$ and $X_3 = X_{3S'}$ in the plane $X_1 = 0$. Then writing $X_2' = X_2 - X_{2S'}$ and $X_3' = X_3 - X_{3S'}$ for the new coordinates and substituting these in equation (50) we have

$$Y - Y_{S'} = B_{11}X_1^2 + B_2X_2' + B_3X_3'(+ B_{22}X_2'^2 + B_{33}X_3'^2) \tag{51}$$

Where B_2 and B_3 measure the slopes of the yield surface *on the plane of the ridge* and will not be negligible if the ridge is non-stationary. If we can ignore B_{22} and B_{33} as negligible in comparison with B_{11}, equation (51) is that of a system having contour surfaces which are parabolic cylinders like that in Figure 11(f) of [1] or Figure 11.8(E) of [3].

We see from equation (51) that, if we wish to move in a direction so that $Y - Y_{S'}$ is made as large as possible, we should, (since B_{11} is negative) make the contribution of the first term equal to zero by keeping $X_1 = 0$ (by remaining on the plane of the ridge). Also we should proceed so that the contribution of the terms B_2X_2' and B_3X_3' is as large as possible. For movement through a given distance r on the plane this will be achieved by following the direction of steepest ascent. Thus X_2' and X_3' should be varied in proportion to B_2 and B_3. We see that this movement which would be at right angles to the yield contours on the plane of the ridge would in the present example lead in the *direction of rising temperature.*

In general where a competitive system is affected by a single factor like temperature or concentration this type of analysis will be helpful in identifying the factor responsible. At the same time it should be borne in mind that in the presence of a ridge system unequivocal identification by this means is not possible. For instance in the present example we see from figure 5 that we could attribute the effect found to the joint influence of time and concentration instead of to temperature. As always it is necessary to consider evidence of this kind in the light of possible theoretical explanations for the phenomenon observed.

6. CHOICE OF METRICS FOR VARIABLES

In the experiments described the standardised variables x_1, x_2, x_3 of the design were linearly related to the natural variables T, c and t as follows $x_1 = (T - 167)/5$, $x_2 = (c - 27.5)/2.5$, $x_3 = (t - 6.5)/1.5$. The unit change of a given variable in the design, taken as a percentage of the average departure of the variable from its natural origin we may call 'the coefficient of unit change', U. Remembering that the natural origin for temperature is $-273°$C, in the present case we have

$$U \text{ (Temp.)} = 100 \times 5/440 = 1.1\%$$

$$U \text{ (conc.)} = 100 \times 2.5/27.5 = 9.1\%$$

$$U \text{ (time)} = 100 \times 1.5/6.5 = 23.1\%$$

When experiments are to be conducted with the intention of fitting an empirical response surface doubt may exist as to whether we should relate the standardised variables of the design to the natural variables by a linear scale, as was done in this experiment, or by a log scale, or a reciprocal scale, or in some other manner.

When the coefficients of unit change are small the surface plotted in terms of transformed variables, like those above, will usually be almost the same in appearance as when plotted in terms of the untransformed variables, since the relationships over the ranges studied between transformed and untransformed variables will be almost linear in this circumstance. Even so by appropriate choice of metrics the interpretation of the fitted equation may be greatly simplified as will be illustrated in section 7.

Theoretical Surface in terms of the New Metrics.

In the present example we have seen (equation 25) the important part which is played by the function

$$c_{20}t\ \alpha\ \{\exp - \beta/(T + 273)\} \tag{52}$$

in describing the yield surfaces. In fact the time t, the *overall* concentration c_{20} , and (if ρ is independent of temperature) the temperature T enter the theoretical equation only through this expression. Its logarithm is

$$\ln c_{20} + \ln t + \ln \alpha - \beta/(T + 273) \tag{52a}$$

which is a linear expression in functions of T, c_{20} and t. If therefore we use a reciprocal scale for absolute temperature and logarithmic scales for the time and the overall concentration the contours of the ridge system will appear as planes in the space of these variables. In figure 6(a) a section of the theoretical yield surface already given in Figure 4 is shown with time plotted on a log scale and temperature on a reciprocal scale.

When ρ is assumed to be temperature dependent, T enters the expression (25) on the right hand side as well as on the left. However as will be seen from figure 6(b), over the ranges considered, the ridge is again rendered almost straight by plotting on the basis of reciprocal absolute temperature and log time. From these diagrams it will be seen that, when the variables are scaled in terms of these new units, a

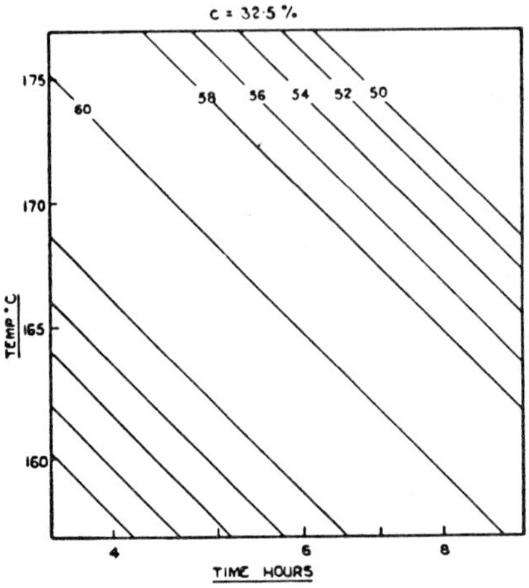

FIGURE 6(a)

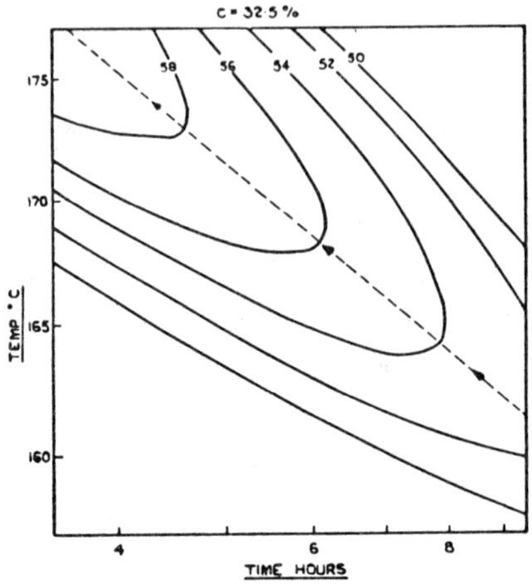

FIGURE 6(b)

CONTOURS OF THEORETICAL YIELD SURFACES WITH TIME PLOTTED ON A LOG SCALE AND TEMPERATURE ON A RECIPROCAL SCALE.

considerably closer fit might be expected from a second degree equation.

For many other chemical reactions the compound variable in (52) will play an equally important part in the response function and in the absence of other evidence the choice in empirical investigations of reciprocal scales for absolute temperature and log scales for time and overall concentration would seem to be indicated. One would expect that, on these scales of measurement, the system could be more precisely represented by a simple equation.

The choice, in the present example, of an appropriate metric for C, the concentration of reactant a relative to that of reactant bNb, is less easily decided. However, if we consider the particular case where $\rho = 2$ the equation of the surface for z_t may be written

$$\ln w = \ln \alpha - \beta(T + 273)^{-1} + \ln c_{20} + \ln t$$

$$= \ln \ln \left[\frac{(4z_t + C - 4)}{Cz_t}\right] - \ln (C - 4) \qquad (53)$$

We are particularly interested in the region of the surface where η_3 takes its maximum value. Here $z_t = \frac{1}{2}$. Substituting this value in equation (53) we have for the equation of the surface on which η_3 is a maximum

$$-\beta(T + 273)^{-1} + f(C) + \ln t + \ln \alpha + \ln c_{20} = 0 \qquad (54)$$

where

$$f(C) = \ln \left\{ \frac{C - 4}{\ln (2C - 4) - \ln C} \right\} \tag{55}$$

would seem to be an appropriate metric for C.

Empirical Surface in Terms of the New Metrics.

The expectation that an equation of second degree might fit the data more closely when the variables were expressed in terms of the new metrics $(T + 273)^{-1}$, $f(C)$, and $\ln t$, is borne out, as is seen from the analysis of variance in Table 3. Since, in this example, the coefficients of unit change are not large no dramatic reduction in the residual sum of squares is to be expected however.

TABLE 3

Analysis of Variance before and after transformation of the variables

Source	DF	Sums of squares	
		Original Metrics	New Metrics
Due to Regression (after elimination of the mean)	9	371.4	380.2
Residual	9	29.5	20.7
Total (after elimination of the mean)	18	400.9	400.9

In refitting the equation after changes in the metrics use may be made of the estimates already obtained in the following way. When recoding the data for the new metrics we need only ensure that the coded data are *linearly related* to the chosen functions. We can therefore arrange matters so that the coded values of the independent variables $\dot{x}_1$, $\dot{x}_2$, $\dot{x}_3$ are "close" to those of the original independent variables x_1, x_2, x_3. In the present example this was done by arranging that the re-coded levels of the variables for the first eight experiments (the 2^3 factorial part) were -1 and $+1$ as before. For example, for temperature we require a coding $\dot{x}_1 = a + b\,(T + 273)^{-1}$ such that $\dot{x}_1 = -1$ when $T = 162°\text{C}$ and $\dot{x}_1 = 1$ when $T = 172°\text{C}$. Substituting these values in the equation and solving we obtain $a = 88$, $b = 38{,}715$ whence the coding used for temperature was

$$\dot{x}_1 = 88 - 38{,}715(T + 273)^{-1} \tag{56}$$

Proceeding in a similar way with the remaining variables we have

$$\dot{x}_2 = -27.9299 + 9.9965f(C) \tag{57}$$

$$\dot{x}_3 = -7.84868 + 4.25532 \ln t \tag{58}$$

These recoded values are shown in columns (8), (9) and (10) of table 1. The differences from the original coding are not very large and we can therefore regard the coefficients b_0, b_1, b_2 etc. already obtained as *first approximations* to the new coefficients $\dot{b}_0$, $\dot{b}_1$, $\dot{b}_2$ etc. Accurate values of $\dot{b}_0$, $\dot{b}_1$, $\dot{b}_2$ etc. were obtained by writing down the normal equations (after elimination of the mean*) for the new recoded variables, inserting b_1, b_2, b_3 etc. as first approximations and then obtaining successively closer and closer approximations by "one at a time" and "steepest ascent" relaxation (see for example [7] and [8]). If the elements of the new inverse matrix are required these can be obtained, (for example, by Hotelling's method [9]), using the elements of the known inverse from the original coding as the first approximation.

7. BASIC CONSTANTS AND CANONICAL VARIABLES

We find for the newly fitted equation

$$\dot{Y} = 59.15 + 2.00\dot{x}_1 + 1.01\dot{x}_2 + 0.67\dot{x}_3 - 2.00\dot{x}_1^2 - 0.72\dot{x}_2^2 - 1.00\dot{x}_3^2 - 2.78\dot{x}_1\dot{x}_2 - 2.18\dot{x}_1\dot{x}_3 - 1.16\dot{x}_2\dot{x}_3 \tag{59}$$

On comparison with equation (5) it will be seen that, as would be expected, the coefficients are quite close to those obtained before.

The canonical form of the equation is also similar. We have

$$\dot{Y} - 59.51 = -3.50\dot{X}_1^2 - 0.41\dot{X}_2^2 + 0.19\dot{X}_3^2 \tag{60}$$

where $$\dot{X}_1 = 0.760\dot{x}_1 + 0.473\dot{x}_2 + 0.446\dot{x}_3 - 0.329 \tag{61}$$

On decoding we now have an expression for the canonical variable $\dot{X}_1$ in more appropriate functions of the natural variables.

$$\dot{X}_1 = -29{,}423(T + 273)^{-1} + 4.728f(C) + 1.897 \ln t + 49.839 \tag{62}$$

Putting $\dot{X}_1 = 0$ we see that on the new scales the empirically fitted equation gives for the plane of maxima

$$-29{,}423(T + 273)^{-1} + 4.728f(C) + 1.897 \ln t + 49.839 = 0 \tag{63}$$

Now we have seen (54) that for $\rho = 2$ the theoretical equation of the

*By fitting the equation in the form of $y - \bar{y} = b_1(x_1 - \bar{x}_1) + b_2(x_2 - \bar{x}_2) + \cdots + b_{11}(x_1^2 - \overline{x_1^2}) + \cdots$, etc. the convergence of the iteration is speeded up.

maximum plane is

$$-\beta(T + 273)^{-1} + f(C) + \ln t + \ln \alpha + \ln c_{20} = 0 \qquad (64)$$

with which (63) may be compared.

If the theoretical system exactly fitted, the coefficients of $f(C)$ and $\ln t$ would be equal and on dividing equation (63) by this common coefficient (63) and (64) would be exactly comparable. Here to enable comparison to be made, we divide (63) through by the geometric mean 2.995 of the coefficients 4.728 and 1.897 of $f(C)$ and $\ln t$ to obtain

$$-9{,}824(T + 273)^{-1} + 1.579f(C) + 0.633 \ln t + 16.641 = 0 \qquad (65)$$

Comparison of (65) and (64) shows that the canonical variable X_1 is *carrying as coefficients the constants of the reaction.* The value 9,824 is an estimate of β and (since $\ln c_{20} = 3.10$) $16.64 - 3.10 = 13.54$ is an estimate of α. (Both are in reasonable agreement with the estimates of equations 35 and 36).

The lack of equality of the coefficients of $f(C)$ and $\ln t$ does not support theoretical expectation. However, this is probably because $\rho \neq 2$ and also because for reasons given already, the 'effective reaction times' are greater than the values assumed and the 'effective relative concentrations' are less. It would be possible to calculate appropriate 'correction factors' which could indicate how our basic theory should be modified. We shall not pursue this topic here however.

This example served to point out to us two interesting possibilities which have been borne in mind and developed in later investigations. These are:

1) Where sufficient is known of the nature of the basic mechanism (i.e. the kinetics in chemical examples) we may proceed to fit a surface based on this mechanism rather than on the empirical Taylor series.
2) When we start off with little knowledge of a system careful study of the characteristics of a fitted empirical surface, particularly as elucidated by canonical analysis, can lead to a *conception* of the probable basic mechanism. A first guess can then be tested and improved upon by a process of 'experimental iteration'.

Of these (2) is possibly the more important and may have applications outside chemistry, for example, the characteristics of the surface for a fertilizer trial in agriculture or a nutrition experiment in biology might supply important information on the metabolism of the plant or animal cell.

8. BASIC CONSTANTS AND THE DIRECTION OF STEEPEST ASCENT

We have seen above that, in the example we have studied, essential information concerning the reaction constants was contained in the coefficient of the canonical variable X_1 .

It is perhaps worth noting also that the direction of steepest ascent would also contain much of this information. For we see that if the true surface could be represented in terms of a single canonical variable so that

$$\eta - \eta(\max) = \lambda(px_1 + qx_2 + rx_3 + s)^2 \tag{66}$$

then multiplying out this expression and equating the coefficients to the constants β_0 , β_1 , $\cdots$, β_{23} we have

$$\left.\begin{array}{llll} \beta_0 = \eta \max + \lambda s^2 & \beta_1 = 2\lambda sp & \beta_2 = 2\lambda sq & \beta_3 = 2\lambda sr \\ \beta_{11} = \lambda p^2 & \beta_{22} = \lambda q^2 & \beta_{33} = \lambda r^2 & \\ \beta_{12} = 2\lambda pq & \beta_{13} = 2\lambda pr & \beta_{23} = 2\lambda qr & \end{array}\right\} \tag{67}$$

Thus for this type of example the constants β_1 , β_2 , and β_3 which define the direction of steepest ascent are in fact proportional to the coefficients p, q, r in the canonical variable, as is at once obvious from the consideration that the direction of steepest ascent is at right angles to the contour plane of maxima in the space of the factors.

In examples like the present one estimates of β_1 , β_2 , β_3 , and β_{12} , β_{13} and β_{23} are available after the first eight experiments. If these are such as would support the hypothesis that

$$\frac{\beta_{12}}{\beta_1\beta_2} = \frac{\beta_{13}}{\beta_1\beta_3} = \frac{\beta_{23}}{\beta_2\beta_3} \tag{68}$$

we may begin to suspect (although we are entitled to do no more) that we may be dealing with a system having a single dominant canonical variable.

9. SOME REMARKS ON THE PROCESS OF SCIENTIFIC INVESTIGATION

The technique of scientific investigation contains two essential processes

a) the devising of experiments suggested by the investigator's appreciation of the situation to date and designed to elucidate it further;

b) the examination of results of experiments performed to date in the light of all background knowledge available, with the object of postulating theories susceptible of test in future experimentation.

The first is essentially a movement from 'theory' to 'experiment' indicated in Figure 7 by an arrow pointing upwards, the second is a movement from 'experiment' to 'theory' indicated by an arrow pointing downwards.

During a complete investigation these processes of synthesis and analysis used in alternation will normally be employed many times and, by what we may call 'experimental iteration', the investigator should be led closer and closer to the truth.

Most investigations first pass through a 'speculative' stage. Here statistical methods can rarely be of help but it is nevertheless vital that this early work should be done fully and with imagination, otherwise later effort may be wasted in detailed investigation of the wrong

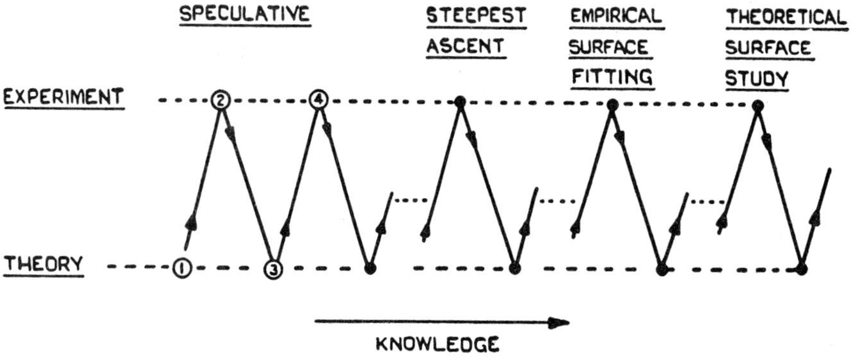

FIGURE 7. DIAGRAMATIC REPRESENTATION OF PROCESS OF EXPERIMENTAL ITERATION.

basic system. Statistical methods provide efficient tools for investigating a system whose general nature has been broadly decided. They provide no substitute for basic scientific thinking about what the system to be investigated should be. It is the duty of a statistician to dissuade the experimenter from employing these methods until he has done sufficient preliminary work to decide what basic system he should explore more fully and incidentally until he has acquired reasonable skill in carrying out experiments with the system.

To appreciate the interplay of processes (a) and (b) let us imagine the beginning of a chemical investigation. At stage 1 in Figure 7, the experimenter would have some, perhaps not very precise, idea as to the general way in which some chemical might be manufactured. Process (a) would begin in his mind something like this—"I believe that in suitable circumstances reactant A would combine with reactant B to form C. From theoretical knowledge, my own experience, and other people's experience of similar reactions cited in the literature, I should think that conditions X might be worth trying".

The appropriate experiment would then be performed (stage 2 in the diagram). As soon as the results were seen the second type of mental process, denoted by (b) above would start—"The reaction did produce a little of the desired product C but there was a very large amount of unwanted product D also present. This could be due to the large amount of water which had to be used to dissolve the reactants and which would favour formation of D".

He has now reached stage (3) at which point the first kind of mental process (a) begins again—"If I carry out the reaction using a non-aqueous solvent I may avoid the large production of by-product D". He is thus led to perform a further experiment at stage (4) using a non-aqueous solvent, and so on.

When the speculative experiments have led to some reasonably well defined system which is sufficiently promising to justify development much will be gained by using the powerful tools provided by applied mathematics such as "steepest ascent", empirical surface fitting and "theoretical surface study".

It should be noted that these techniques still employ the basic processes (a) and (b), and that our applied mathematics helps as much with (b) as it does with (a). Thus we are not only concerned with designing experiments which will estimate the 'effects of the factors' (process *a*) but also with making calculations (for example of the direction of steepest ascent, and of the canonical form of a fitted equation) which suggest what further experimentation should be performed (process *b*).

There has been a tendency for some statisticians to concentrate on the, perhaps rather rare, experimental situation where a single group of experiments is planned and from the result some irrevocable decision is to be made. Such an investigation is concerned exclusively with a single application of procedure (a). It is customary to emphasise in such a situation the danger of taking action on a hypothesis suggested by inspection of the data, but not in mind when the experiment was planned.

This point has sometimes been misunderstood and interpreted to mean that process (b) was in some way suspect and should not be indulged in. In fact of course it is fundamental to investigation which would be quite barren without it. In researches of the type we have discussed the experimenter and the statistician should examine the data most carefully and any hypothesis which appeared possible and important should be submitted to the test of later experimentation.

The particular sequence of techniques shown in Figure (7) is not to be regarded as providing a set pattern which should always be followed. The use of these and other devices will be decided by such circumstances as the degree of basic knowledge concerning the mechanism of the system and the object and importance of the study. If, for example, the experimenter were required merely to make in the laboratory a few pounds of some rare organic chemical for some special research purpose, he would be quite content to do a few speculative experiments sufficient to allow him to prepare this small amount of material in reasonable quality, the finding of an economic process would not be worth the trouble. At the other end of the scale, for a large and expensive manufacture an elaborate and costly study would be justified.

10. ACKNOWLEDGEMENT

We wish to express our thanks to Mr. K. A. Burhouse and Mrs. Margaret Edmondson for valuable assistance in the preparation of this paper.

APPENDIX A—DERIVATION OF CERTAIN RESULTS QUOTED IN THE PAPER

1) *General solution of the differential equations.*

From equations (22) and (23) we have

$$dc_3/dc_2 = \rho^{-1}c_3/c_2 - 1 \tag{69}$$

Write $c_3 = c_2 s$, then

$$dc_3/dc_2 = s + c_2(ds/dc_2) \tag{70}$$

Substituting (70) in (69) we have

$$\frac{-dc_2}{c_2} = \frac{ds}{1 + s(\rho - 1)/\rho} \tag{71}$$

$$-\ln c_2 + \text{constant} = \frac{\rho}{\rho - 1} \ln \{1 + s(\rho - 1)/\rho\} \tag{72}$$

Now when $t = 0$, $c_2 = c_{20}$, and $s = 0$ hence the constant in $\ln c_{20}$.
Now $\eta_2 = c_2/c_{20}$ and $s = c_3/c_2 = \eta_3/\eta_2$
Thus

$$\eta_2 = \left(\frac{\rho - 1}{\rho}\frac{\eta_3}{\eta_2} + 1\right)^{-\rho/(\rho-1)} \tag{73}$$

Write

$$\eta_2 = z^{\rho} \tag{74}$$

After substituting this expression in (73) and rearranging we have

$$\eta_3 = \frac{\rho}{\rho - 1} z(1 - z^{\rho-1}) \tag{75}$$

and using equation (14)

$$\eta_5 = 1 - \frac{\rho}{\rho - 1} z + \frac{1}{\rho - 1} z^\rho \tag{76}$$

From equation (15)

$$\eta_4 = 2 - \frac{\rho}{\rho - 1} z - \frac{\rho - 2}{\rho - 1} z^\rho \tag{77}$$

Finally from equation (16)

$$\eta_1 = C - 4 + \frac{2\rho}{\rho - 1} z + 2 \frac{(\rho - 2)}{\rho - 1} z^\rho \tag{78}$$

Now from equation (22)

$$\frac{-d\eta_2}{dt} = c_{20}\rho k \eta_1 \eta_2 \tag{79}$$

Substituting (78) in (79) and noting that $d\eta_2/dt = \rho z^{\rho-1} dz/dt$ we have after rearrangement

$$-kc_{20} \frac{dt}{dz} = \frac{\rho - 1}{z\{2(\rho - 2)z^\rho + 2\rho z + (\rho - 1)(C - 4)\}} \tag{80}$$

whence

$$kc_{20}t = (\rho - 1) \int_{z_t}^{1} z^{-1}\{2(\rho - 2)z^\rho + 2\rho z + (\rho - 1)(C - 4)\}^{-1} \, dz \tag{81}$$

and using the Arrhenius equation (18) we obtain equation (25). Equations (74), (75), (76), (77) and (78) together with (81) yield the complete solution allowing the yields η_1, η_2, η_3, η_4 and η_5 to be evaluated at any time t.

2) *Maximum yield of aNb.*

If we put $dc_3/dt = 0$ in equation (23) we have

$$\eta_3/\eta_2 = \rho \tag{82}$$

substituting in (73) $$\eta_2 = z^\rho = \rho^{-\rho/(\rho-1)} \tag{83}$$

Whence

$$\eta_3 = \rho^{-1/(\rho-1)} \tag{84}$$

which is readily shown to be a maximum.

At this point $$\eta_5 = 1 - \rho^{-1/(\rho-1)} - \rho^{-\rho/(\rho-1)} \tag{85}$$

3) *Solution when* $\rho = 2$.

In the particular case when $\rho = 2$ equation (81) becomes

$$kc_{20}t = \int_{z_t}^{1} z^{-1}(4z + C - 4)^{-1}\, dz \tag{86}$$

giving
$$kc_{20}t = \{\ln (4z_t + C - 4) - \ln (Cz_t)\}/(C - 4) \tag{87}$$

After rearrangement this yields the explicit function for z_t

$$z_t = (C - 4)/[C \exp \{c_{20}(C - 4)tk\} - 4] \tag{88}$$

whence using the Arrhenius equation (18) we obtain equation (38).

REFERENCES

[1] Box, G. E. P. The Exploration and Exploitation of Response Surfaces: Some general considerations and examples. *Biometrics, 10:* 16, 1954.
[2] Box, G. E. P. and Wilson, K. B. On the Experimental Attainment of Optimum Conditions. *J. R. Stat. Soc., B, 13:* 1, 1951.
[3] *The Design and Analysis of Industrial Experiments*, Box, G. E. P., Connor, L. R., Cousins, W. R., Davies, O. L. (Editor), Himsworth, F. R., and Sillito, G. P., London: Oliver and Boyd, 1954.
[4] Arnold, D. S., Box, G. E. P., Erickson, E. E., Hunter, J. S., Nelli, J. R., Pike, F. P. The application of Statistical Procedures to the Flooding Capacity of a Pulse column, Progress Report No. 3 Contract No. AT-(40-1)-1320.
[5] Erickson, E. E., Hunter, J. S., Nelli, J. R., Pike, F. P. The flooding capacity of a Pulse Column on the Benzene-Water System. Progress Report No. 4. Contract No. AT-(40-1)-1320.
[6] Hammett, L. P. *Physical Organic Chemistry, 1st Ed.*, 1940. New York: McGraw-Hill.
[7] Fox, L. A short account of Relaxation Methods. *Quart J. Mechanics and Applied Mathematics, 1:* 253, 1948.
[8] Booth, A. D. An application of the Method of Steepest Descents to the Solution of systems of Non-Linear Simultaneous equations. *Quart. J. Mechanics and Applied Mathematics, 2:* 460, 1949.
[9] Hotelling, H. Some new methods of Matrix Calculation, *Ann. Math. Stat., 14:* 1, 1943.

1.5
Robustness to Non-Normality of Regression Tests†

G. E. P. BOX and G. S. WATSON
University of Wisconsin and University of Toronto

1. Summary

A number of statistical procedures involve the comparison of a 'regression' mean square with a 'residual' mean square using the normal-theory F distribution for reference. The use of the procedure for the analysis of actual data implies that the distribution of the mean-square ratio is insensitive to moderate non-normality. Many investigators, in particular Pearson (1931), Geary (1947), Gayen (1950), have considered the sensitivity of this distribution to parent non-normality for important special cases and a very general investigation was carried out by David & Johnson (1951 *a*, *b*).

The principal object of this paper is to demonstrate the overriding influence which the *numerical values of the regression variables* have in deciding sensitivity to non-normality and to demonstrate the essential nature of this dependency.

We first obtain a simple approximation to the distribution of the regression F statistic in the non-normal case. This shows that it is 'the extent of non-normality' in the regression variables (the x's), which determines sensitivity to non-normality in the observations (the y's). Our results are illustrated for certain familiar special cases. In particular the well-known robustness of the analysis of variance test to compare means of equal-sized groups and the notorious lack of robustness of the test to compare two estimates of variance from independent samples are discussed in this context. We finally show that it is possible to choose the regression variables so that, to the order of approximation we employ, non-normality in the y's is without effect on the distribution of the test statistic. Our results demonstrate the effect which the choice of *experimental design* has in deciding robustness to non-normality.

2. The mean square ratio R

Suppose that the *response* y_u, observed at the uth set of levels $x_{1u}, x_{2u}, \ldots, x_{pu}$ of p *regression variables*, may be represented by the response function

$$y_u = \beta_0 + \beta_1 x_{1u} + \ldots + \beta_i x_{iu} + \ldots + \beta_p x_{pu} + \epsilon_u \quad (u = 1, 2, \ldots, N), \tag{1}$$

where the *error* $\epsilon_u = y_u - E(y_u)$ is a random variable. The regression variables may be quantitative or merely indicator variables denoting presence or absence of a certain quality. We speak of the N values x_{iu} as the elements of the ith *regression vector* x_i and suppose that the model is set up so that the regression vectors are linearly independent, with $\sum_u x_{iu} = 0$. We denote the usual least squares estimates of $\beta_1, \ldots, \beta_p$ by $b_1, \ldots, b_p$ and write

$$c_{ij} = \sum_{n=1}^{N} x_{iu} x_{ju} \quad (i, j = 1, \ldots, p).$$

To shed light on the plausibility of $\beta_1^*, \beta_2^*, \ldots, \beta_p^*$ as possible values of the coefficients, the ratio of mean squares

$$R_\beta^* = \{S_{\beta *}/p\}/\{S_E/(N-p-1)\} \tag{2}$$

† This research was supported in part by the United States Navy, through the Office of Naval Research, and by the United States Army through the Office of Ordnance Research.

The Collected Works of George E. P. Box, 1984, Wadsworth, Inc., Belmont, CA 94002.
Originally published in *Biometrika,* vol. 49, parts 1 and 2 (1962), pp. 93–106.

may be calculated. The appropriateness of this ratio can be seen from the fact that the *regression* sum of squares

$$S_{\beta *} = \sum_{i=1}^{p} \sum_{j=1}^{p} c_{ij}(b_i - \beta_i^*)(b_j - \beta_j^*) \tag{3}$$

is a measure of overall discrepancy between the least-squares estimates b_i and the contemplated values β_i^*, while the *residual* sum of squares

$$S_E = \sum_{u=1}^{N} (y_u - \hat{y}_u)^2 \quad \text{with} \quad \hat{y}_u = b_0 + \sum_{i=1}^{p} b_i x_{iu} \tag{4}$$

measures the internal consistency of the data independently of the choice of β_i^*. Furthermore, provided that the assumed form of response function is adequate, and that the errors have equal variances and are uncorrelated one with another, the expected values of the component sums of squares are

$$E(S_{\beta *}) = \sum_{i=1}^{p} \sum_{j=1}^{p} c_{ij}(\beta_i - \beta_i^*)(\beta_j - \beta_j^*) + p\sigma^2, \tag{5}$$

$$E(S_E) = (N-p-1)\sigma^2. \tag{6}$$

If in addition the errors could be supposed to follow normal distributions, that is, if the joint distribution of the ϵ_u was a spherical normal distribution, then when $\beta_i^* = \beta_i$ $(i = 1, 2, \ldots, p)$ $R_{\beta *}$ would have an F distribution with $\nu_1 = p$ and $\nu_2 = N - p - 1$ degrees of freedom. Since in the analysis of real data we do not know the precise distribution of the ϵ_u, one would like to know under what circumstances the F distribution still supplied an adequate approximation to the distribution of $R_{\beta *}$. In particular, as has been shown by Fisher (1947) and Scheffé (1953), when the approximation was adequate the statistic $R_{\beta *}$ could be used not only to test hypotheses concerning particular values β^* of the coefficients but also to supply interval estimates for any linear combinations whatever of the β's.

For the 'special' case $\beta_i^* = 0$ $(i = 1, \ldots, p)$ we denote the mean-square ratio by R. Thus

$$R = \frac{S_0/p}{S_E/(N-p-1)}, \tag{7}$$

where

$$S_0 = \sum_{i=1}^{p} \sum_{j=1}^{p} c_{ij} b_i b_j, \quad S_E = \Sigma \hat{y}^2 - N\bar{y}^2 - S_0. \tag{8}$$

In what follows we consider only this case since the more general situation always can be reduced to it. For example, if we calculate $\dot{R}$ from the constructed observations

$$\dot{y}_u = y_u - \beta_1^* x_{1u} - \beta_2^* x_{2u} - \ldots - \beta_p^* x_{pu},$$

then $\dot{R}$ is identical with $R_{\beta *}$. In what follows the reference of R to the normal-theory F distribution is referred to as the *general regression test*.

3. A familiar comparison

That the answer to the question 'What effect has non-normality on the general regression test?' is profoundly influenced by the nature of the x vectors may be demonstrated by a familiar comparison. By different choices of the x vectors, almost the same regression model can be made to reproduce on the one hand a test to compare means which is little affected by non-normality and on the other a comparison of variances test which is notoriously sensitive to non-normality.

3·1. *Comparison of means*

If in the general regression model of equation (1) we write $N = (p+1)n$, $x_{iu} = p/(p+1)$ for $u = n(i-1)+t$ $(i=1,\ldots,p;\ t=1,2,\ldots,n)$ and $x_{iu} = -1/(p+1)$ otherwise, then we obtain the regression model for the n observations in each of $p+1$ groups, with the elements in each regression vector adjusted to add to zero, so that the appropriate variance mean-square ratio is

$$R_m = \frac{n \sum_{i=1}^{p+1} (\bar{y}_i - \bar{y})^2/p}{\sum_{i=1}^{p+1} \sum_{t=1}^{n} (y_{it} - \bar{y}_i)^2/(N-p-1)}, \tag{9}$$

where $\bar{y}_i$ is the mean for the ith group of n observations and $\bar{y}$ the overall mean.

3·2. *Comparison of variances*

Suppose there are $N-1$ observations and in the general regression model let us temporarily relax the provision that $\sum_{u=1}^{N-1} x_{iu} = 0$ and suppose that the constant term β_0 is known. Now put $x_{iu} = 1$ when $u = i$ $(i=1,2,\ldots,p)$ and $x_{iu} = 0$ otherwise. Then the mean-square ratio is

$$R_v = \frac{\sum_{u=1}^{p} (y_u - \beta_0)^2/p}{\sum_{s=p+1}^{N-1} (y_s - \beta_0)^2/(N-p-1)}. \tag{10}$$

This has the form of the standard test for the comparison of variances (the population means being known to equal β_0) for two independent samples of size $n_1 = p$ and $n_2 = N-p-1$.

The criteria R_m and R_v can each be obtained therefore as particular cases of the general regression criterion. Furthermore, on the usual assumptions, the null distribution of each criterion is the normal-theory F with $\nu_1 = p$ and $\nu_2 = N-p-1$ degrees of freedom. It is well known, however, that whereas R_m has a distribution which is remarkably insensitive to departures from normality in the parent population this is not the case for R_v.

Specifically, using a method which we discuss later in more detail, Box & Andersen (1955) have shown that in the non-normal situation R_m and R_v have distributions which may be approximated by F distributions with modified degrees of freedom. They show that R_m is approximately distributed as F with $\nu_1 = \delta_m p$ and $\nu_2 = \delta_m(N-p-1)$ degrees of freedom and R_v is approximately distributed as F with $\nu_1 = \delta_v p$ and $\nu_2 = \delta_v(N-p-1)$ degrees of freedom, where for moderate non-normality and moderate numbers of observations the δ's are approximately

$$\delta_m^{-1} = 1-(1/N)\Gamma_y, \quad \delta_v^{-1} = 1+\tfrac{1}{2}\Gamma'_y. \tag{11}$$

The measures of kurtosis Γ_y and Γ'_y in these expressions are basically similar one to the other and take zero values when the distribution is normal. Explicitly†

$$\Gamma_y = E\{C_y\} = E\left\{\frac{k_4}{k_2^2}\right\}, \quad \Gamma'_y = E\{C'_y\} = E\left\{\frac{N+1}{N-1}\frac{m_4}{m_2^2}\right\} - 3, \tag{12}$$

† Asymptotic expansions for these constants in terms of the standardized cumulants of the parent distributions are given by Box & Andersen (1955).

where k_2 and k_4 are the usual k statistics for the whole sample of N observations and m_q is the qth moment about the mean

$$m_q = \left\{\sum_{u=1}^{n_1} (y_u - \beta_0)^q + \sum_{s=n_1+1}^{n_1+n_2} (y_s - \beta_0)^q\right\} \Big/ (N-1)$$

for the sample of $N-1$ observations.

The insensitivity to non-normality shown by R_m arises because the corrective factor is of order N^{-1}, whereas that for R_v is of order 1. From our present point of view the example serves to show that different choices of the x's in the general regression model can change sensitivity to non-normality by a factor as large as $\frac{1}{2}N$.

We shall now investigate the specific nature of this dependence of sensitivity on the nature of the x vectors. We shall be able to show that the corrective factor for the general regression test is, to order N^{-1}

$$\delta^{-1} = 1 + \Gamma_y C_X/2N, \tag{13}$$

where C_X is a measure analogous to Γ_y of 'non-normality' in the x's. In the two special examples considered above C_X approaches respectively its smallest possible value of -2 and its largest possible value of N.

4. PERMUTATION MOMENTS OF R

In matrix notation our model becomes

$$y = \mathbf{1}'\boldsymbol{\beta}_0 + \mathbf{X}\boldsymbol{\beta} + \boldsymbol{\epsilon}, \tag{14}$$

where $\mathbf{1}$ is a column of N unities, $\mathbf{X}$ the $N+p$ matrix of the levels of the p regression variables, $\mathbf{y}$ the column of the N observed responses, and $\boldsymbol{\beta}$ the column of coefficients $\beta_1, \ldots, \beta_p$. Then, since we suppose that $\mathbf{X}'\mathbf{1} = \mathbf{0}$, the least-squares estimators b_0 and $\mathbf{b}$ of β_0 and $\boldsymbol{\beta}$ are given by

$$b_0 = \bar{y}, \quad \mathbf{b} = (\mathbf{X}'\mathbf{X})^{-1}\mathbf{X}'\mathbf{y}. \tag{15}$$

Writing $\mathbf{z}$ for the column of N deviations $z_u = y_u - \bar{y}$ $(u = 1, \ldots, N)$ and $\mathbf{M} = \{M_{uv}\}$ for the symmetric idempotent matrix $\mathbf{M} = \mathbf{X}(\mathbf{X}'\mathbf{X})^{-1}\mathbf{X}'$, we have $\mathbf{b} = (\mathbf{X}'\mathbf{X})^{-1}\mathbf{X}'\mathbf{z}$, whence the regression and residual sums of squares defined in (8) and (4) are given by

$$S_0 = \mathbf{z}'\mathbf{M}\mathbf{z}, \quad S_E = \mathbf{z}'(\mathbf{I}-\mathbf{M})\,\mathbf{z}. \tag{16}$$

We are concerned with the distribution of $R = \{(N-p-1)\,S_0\}/\{pS_E\}$ when $\boldsymbol{\beta} = \mathbf{0}$, that is when

$$y = \mathbf{1}'\beta_0 + \boldsymbol{\epsilon}.$$

Now if the error vector $\boldsymbol{\epsilon}$ is spherically normally distributed, then R has an F distribution with $\nu_1 = p$ and $\nu_2 = N-p-1$ degrees of freedom. Equivalently

$$W = \frac{S_0}{S_0 + S_E} = \frac{S_0}{\mathbf{z}'\mathbf{z}} \tag{17}$$

has a beta distribution with $\nu_1 = p$ and $\nu_2 = N-p-1$ degrees of freedom.

To study the distribution of R on less specific assumptions we suppose only that the errors have a symmetric distribution, that is to say we suppose that the probability density function $p(\epsilon)$ of the vector $\boldsymbol{\epsilon}$ is a symmetric function of the elements of $\boldsymbol{\epsilon}$. As a special case this includes the chief possibility of interest to us here, that $p(\boldsymbol{\epsilon}) = \prod_{u=1}^{N} p(\epsilon_u)$, where $p(\epsilon_u)$ is any distribution whatever. Now suppose a probability density to be associated with a particular $\boldsymbol{\epsilon}$, then this same probability density is associated with every rearrangement of the elements of $\boldsymbol{\epsilon}$. Also, since $z_u = y_u - \bar{y} = \epsilon_u - \bar{\epsilon}$, it follows that whatever be the probability density associated with a particular $\mathbf{z}$ this same probability density is associated with every rearrangement of the elements of $\mathbf{z}$. The distribution obtained by associating a probability $1/N!$ with each of the possible $N!$ values of any function $f(\mathbf{z})$ obtained by rearranging the elements $\boldsymbol{\epsilon}$ is called the permutation distribution of $f(\mathbf{z})$. The mean and variance of this permutation distribution we denote by $E_P f(z)$ and $E_P f(z)$, respectively.

Now suppose we define *different samples* to mean vectors $\mathbf{z}$ which cannot be made identical by rearrangement of the elements, then if E_S denotes the expected value taken over all different samples $\mathbf{z}$, the overall moments $Ef(\mathbf{z})$ and $Vf(\mathbf{z})$ of $f(\mathbf{z})$ are given by

$$\left.\begin{aligned} Ef(\mathbf{z}) &= \underset{s}{E}\{E_P f(\mathbf{z})\}, \\ Vf(\mathbf{z}) &= \underset{s}{E}\{V_P f(\mathbf{z})\}. \end{aligned}\right\} \tag{18}$$

To approximate the distribution of W for any error distribution of form $p(\epsilon) = \prod_{u=1}^{N} p(\epsilon_u)$ it is convenient for our purpose to find $E(W)$ and $V(W)$ by first finding the appropriate permutation moments and then taking their expected values over all samples.

In what follows it is helpful to express our results in terms of the power sums

$$\sum_{u=1}^{N} z_u^r = S_r$$

which, of course, remain constant under permutations of the ϵ_u. We have in particular

$$S_1 = 0, \quad S_2 = \mathbf{z}'\mathbf{z} = S_0 + S_E, \quad W = S_0/S_2.$$

4·1. $E_P(W)$

We have
$$E_P(z_u^2) = S_2/N, \quad E_P(z_u z_v) = -S_2/\{N(N-1)\}$$
whence
$$E_P(\mathbf{z}\mathbf{z}') = S_2/\{N(N-1)\}\{N\mathbf{I} - \mathbf{1}\mathbf{1}'\}. \tag{19}$$

Now $\mathbf{z}'\mathbf{M}\mathbf{z} = \operatorname{tr}(\mathbf{M}\mathbf{z}\mathbf{z}')$ where $\operatorname{tr}(\mathbf{A})$ denotes the *trace* of the matrix $\mathbf{A}$. Using the linearity of the trace and expectation operators we have

$$\begin{aligned} S_2 E_P(W) = E_P(S_0) = E_P(\mathbf{z}'\mathbf{M}\mathbf{z}) &= E_P \operatorname{tr}(\mathbf{M}\mathbf{z}\mathbf{z}') \\ &= \operatorname{tr}\{\mathbf{M}E_P(\mathbf{z}\mathbf{z}')\}. \end{aligned} \tag{20}$$

But $\mathbf{M}\mathbf{1} = \mathbf{0}$ and $\operatorname{tr}(\mathbf{M}) = \operatorname{tr}\{\mathbf{X}(\mathbf{X}'\mathbf{X})^{-1}\mathbf{X}'\} = \operatorname{tr}\{\mathbf{X}'\mathbf{X}(\mathbf{X}'\mathbf{X})^{-1}\} = p$. Hence on substituting (19) in (20) we obtain finally

$$S_2 E_P(W) = pS_2/(N-1),$$
that is
$$E_P(W) = p/(N-1), \tag{21}$$

as is obtained on classical regression assumptions. In particular we see that $E_P(W) = E_N(W)$, where $E_N(W)$ is the expected value of W on the usual normal theory.

4·2. $V_P(W)$

In what follows summations and permutation expectations are taken over all combinations for which the subscripts are unequal. Thus $E_P\{z_1^\alpha z_2^\beta z_3^\gamma z_4^\delta\}$ means the average value of $z_t^\alpha z_u^\beta z_v^\gamma z_w^\delta$ taken over all permutations for which t, u, v and w are unequal. Similarly $\Sigma M_{13} M_{23}$

means $\sum_{t} \sum_{u \neq t} \sum_{v \neq u \neq t} M_{tv} M_{uv}$. Also, since the x's and hence the M's can be regarded as remaining fixed whilst the combinations of the z's pass through all possible permutations, we have, for example,

$$E_P \Sigma z_1^\alpha z_2^\beta z_3^\gamma z_4^\delta M_{13} M_{23} = E_P(z_1^\alpha z_2^\beta z_3^\gamma z_4^\delta) \Sigma M_{13} M_{23}.$$

On squaring the expression

$$S_0 = \Sigma z_1^2 M_{11} + \Sigma z_1 z_2 M_{12}$$

and taking expectations, we then have

$$\begin{aligned} E_P(S_0^2) = {} & E_P(z_1^4) \Sigma M_{11}^2 + E_P(z_1^2 z_2^2)(2\Sigma M_{12}^2 + \Sigma M_{11} M_{22}) \\ & + E_P(z_1 z_2^2 z_3)(4\Sigma M_{12} M_{23} + 2\Sigma M_{11} M_{23}) \\ & + E_P(z_1^3 z_2)\, 4\Sigma M_{11} M_{12} \\ & + E_P(z_1 z_2 z_3 z_4) \Sigma M_{12} M_{34}. \end{aligned} \tag{22}$$

Using David & Kendall's tables (1949)

$$\left.\begin{aligned} N E_P(z_1^4) &= S_4, & N^{(2)} E_P(z_1^2 z_2^2) &= S_2^2 - S_4, \\ N^{(3)} E_P(z_1 z_2^2 z_3) &= 2S_4 - S_2^2, & N^{(2)} E_P(z_1^3 z_2) &= -S_4, \\ N^{(4)} E_P(z_1 z_2 z_3 z_4) &= 3S_2^2 - 6S_4, & & \end{aligned}\right\} \tag{23}$$

where

$$N^{(r)} = \prod_{i=0}^{r-1} \{N - i\}.$$

Also, using the relations $\mathbf{M1} = \mathbf{0}$, $\mathbf{M}^2 = \mathbf{M}$, $\operatorname{tr} \mathbf{M} = p$, we find rather remarkably that each of the sums involving the elements of M can be expressed in terms of $m = \sum_{u=1}^{N} M_{uu}^2$. In fact

$$\left.\begin{aligned} \Sigma M_{12} &= p - m, & \Sigma M_{11} M_{32} &= p^2 - m, & \Sigma M_{12} M_{23} &= 2m - p, \\ \Sigma M_{11} M_{23} &= 2m - p^2, & \Sigma M_{11} M_{12} &= -m, & \Sigma M_{12} M_{34} &= -6m + 2p + p^2. \end{aligned}\right\} \tag{24}$$

On substituting (23) and (24) in (22) and writing S_4 and S_2 in terms of Fisher's k statistics

$$(N-1) k_2 = S_2, \quad (N-1)^{(3)} k_4 = N(N-1) S_4 - 3(N-1) S_2^2,$$

we have

$$\begin{aligned} V_P(W) &= E_P(S_0^2)/S_2^2 - \{E_P(W)\}^2 \\ &= \frac{2p(N-p-1)}{(N+1)(N-1)^2} + \frac{k_4/k_2^2}{(N-1)^2} \left\{m - \frac{p^2}{N} - \frac{2p(N-p-1)}{N(N+1)}\right\}. \end{aligned}$$

We now put $C_y = k_4/k_2^2$ and for reasons which will be apparent later we write

$$C_X = \frac{N(N-1)(N+1)}{p(N-p-1)(N-3)} \left\{m - \frac{p^2}{N} - \frac{2p(N-p-1)}{N(N+1)}\right\} \tag{25}$$

so that we have finally

$$V_P(W) = V_N(W) \left\{1 + \frac{(N-3) C_y C_X}{2N(N-1)}\right\}, \tag{26}$$

where

$$V_N(W) = \frac{2p(N-p-1)}{(N+1)(N-1)^2}$$

is the variance of W on the usual normal theory.

5. C_X AS A MEASURE OF 'NON-NORMALITY' OF THE x'S

We now show that the function C_X of the x's is analogous with the function C_y of the y's and is a measure of 'non-normality' of the x's. We first notice that M is invariant under any non-singular transformation $\mathbf{W} = \mathbf{XT}$. For

$$\mathbf{M} = \mathbf{X}(\mathbf{X}'\mathbf{X})^{-1}\mathbf{X}' = \mathbf{X}\mathbf{T}\mathbf{T}^{-1}(\mathbf{X}'\mathbf{X})^{-1}\mathbf{T}'^{-1}\mathbf{T}'\mathbf{X}' = \mathbf{W}(\mathbf{W}'\mathbf{W})^{-1}\mathbf{W}'.$$

If we now regard $m = \sum_{u=1}^{N} M_{uu}$ as a function of $m\{\mathbf{X}\}$ of the elements x_{iu} of $\mathbf{X}$, we see that its value is unchanged when every element x_{iu} is replaced by the corresponding w_{iu}.

Now let us choose $\mathbf{T}$ so that $\mathbf{W}'\mathbf{W}$ is diagonal with the ith element equal to $\sum_{u=1}^{N} w_{iu}^2$. Then

$$m = \sum_{u=1}^{N} M_{uu}^2 = \sum_{u=1}^{N} \left\{ \sum_{i=1}^{p} \left(w_{iu}^2 \Big/ \sum_{u=1}^{N} w_{iu}^2 \right) \right\}^2,$$

that is

$$m = \sum_{i=1}^{p} \Sigma_4^i/(\Sigma_2^i)^2 + \sum_{i=1}^{p} \sum_{j \neq i=1}^{p} \Sigma_{22}^{ij}/(\Sigma_2^i \Sigma_2^j) \tag{27}$$

where

$$\Sigma_2^i = \sum_{u=1}^{N} w_{iu}^2, \quad \Sigma_4^i = \sum_{u=1}^{N} w_{iu}^4, \quad \Sigma_{22}^{ij} = \sum_{u=1}^{N} w_{iu}^2 w_{ju}^2.$$

Now, because of the invariant property of m, equation (27) is still true if the corresponding power sums for the x's (which we write as $S_2^i, S_4^i, S_{22}^{ij}$) replace the power sums $\Sigma_2^i, \Sigma_4^i, \Sigma_{22}^{ij}$ for the w's.

Defining k statistics for the x's in the usual way by

$$(N-1)\,k_2^i = S_2^i, \quad (N-1)^{(3)}\,k_4^i = N(N-1)\,S_4^i - 3(N-1)\,(S_2^i)^2,$$
$$(N-1)^{(3)}\,k_{22}^{ij} = N(N+1)\,S_{22}^{ij} - (N-1)\,S_2^i S_2^j$$

we find after a little reduction that

$$m = \frac{(N-2)(N-3)}{(N+1)N(N-1)} \left[\sum_{i=1}^{p} \left\{ \frac{k_4^i}{(k_2^i)^2} \right\} + \sum_{i=1}^{p} \sum_{j \neq i=1}^{p} \left\{ \frac{k_{22}^{ij}}{k_2^i k_2^j} \right\} \right] + \frac{(N-1)\,p(p+2)}{N(N+1)}. \tag{28}$$

Substituting this expression in (25) we have finally

$$C_X = \frac{N-2}{p(N-p-1)} \left\{ \sum_{j=1}^{p} \frac{k_4^i}{(k_2^i)^2} + \sum_{i=1}^{p} \sum_{j \neq i=1}^{p} \frac{k_{22}^{ij}}{k_2^i k_2^j} \right\}. \tag{29}$$

We see that just as $C_y = k_4/k_2^2$ is a measure of non-normality (specifically a measure of kurtosis) for the y's so C_X can be regarded as its multi-variable analogue for the x's. In particular:

(*a*) If $p = 1$ then C_X is the same function of the elements of the single vector x_1 as C_y is of the elements of y.

(*b*) The expected value of C_X is zero for samples from a normal population. This is so because the ratio of k statistics which C_X contains are homogeneous functions of degree zero in the x's. All such functions of normal variates are distributed independently of scaling statistics k_2^i $(i = 1, 2, \ldots, p)$ and consequently the expected values of these ratios are the ratios of the expectations. But for a normal distribution,

$$E(k_4^i) = E\{k_{22}^{ij}\} = 0 \quad (i \neq j = 1, 2, \ldots, p).$$

(*c*) C_X is invariant under linear transformations of the x's. Any sensible measure of multi-variate non-normality would clearly need to possess this property.

Upper and lower bounds for C_X

We first show that
$$\frac{p^2}{N} \leqslant m \leqslant \frac{N-1}{N} p.$$

Because of the invariant property of m we may suppose, without loss of generality, that $\mathbf{X}'\mathbf{X} = \mathbf{I}$. Then
$$\sum_{u=1}^{N} M_{uu} = \sum_{u=1}^{N} \sum_{i=1}^{p} x_{iu}^2 = p.$$

But
$$\sum_{u=1}^{N} M_{uu}^2 - \left\{\sum_{u=1}^{N} M_{uu}\right\}^2 \Big/ N \geqslant 0.$$

Hence
$$m = \sum_{u=1}^{N} M_{uu}^2 \geqslant \frac{p^2}{N}.$$

Now suppose $N-p$ further columns are added to $\mathbf{X}$, the first of which is $N^{-\frac{1}{2}}\mathbf{1}$, to form an orthogonal matrix $\mathbf{H}$. Then since the sums of squares of the elements of each row of $\mathbf{H}$ is unity
$$M_{uu} = \sum_{i=1}^{p} x_{iu}^2 \leqslant 1 - 1/N,$$

whence
$$M_{uu}^2 \leqslant M_{uu}(1-1/N).$$

But
$$M_{uu} = p,$$

whence
$$m = \Sigma M_{uu}^2 \leqslant \{(N-1)/N\} \Sigma M_{uu} = p(N-1)/N.$$

Substituting these bounds in (25) we have finally
$$-2 \leqslant \frac{N-3}{N-1} C_X \leqslant N-1. \tag{30}$$

We shall show later that the lower bound is actually obtainable. The upper bound is approached but cannot be attained in finite samples as is clear from the manner of its derivation.

6. Approximate distribution of R

Case 1. Permutation distribution of R

Following Pitman (1937) and Welch (1937) we now approximate the permutation distribution of W with a beta distribution with degrees of freedom adjusted so as to have the correct mean and variance. If $\nu_1 = p$ and $\nu_2 = N-p-1$ are the degrees of freedom of the beta distribution appropriate on normal theory, then it is readily shown that the approximating distribution has degrees of freedom $\delta\nu_1$ and $\delta\nu_2$ where
$$\delta_1^{-1} = 1 + \frac{(N+1)\alpha_1}{N-1-2\alpha_1} \quad \text{with} \quad \alpha_1 = \frac{N-3}{2N(N-1)} C_X C_y. \tag{31}$$

Equivalently the permutation distribution of R is approximated by an F distribution with degrees of freedom $\nu_1 = \delta_1 p$ and $\nu_2 = \delta_1(N-p-1)$. In those cases where C_X and C_y are not simultaneously close to upper bounds we have to order N^{-1}
$$\delta_1^{-1} = 1 + C_X C_y / 2N. \tag{32}$$

Case 2. *Distribution of R under general non-normality of y.* **X** *fixed*

If we now take expectations of the permutation moments over all samples, then C_y in $V(W)$ is replaced by $E(C_y)$ which we denote by Γ_y. Again using the beta-distribution approximation we have that if $p(\boldsymbol{\epsilon})$ is any symmetric function of the elements of $\boldsymbol{\epsilon}$ and in particular if the ϵ's may be regarded as independent random drawings from any probability distribution whatever, R is distributed approximately as F with $\nu_1 = \delta_2 p$ and $\nu_2 = \delta_2(N-p-1)$ degrees of freedom, where

$$\delta_1^{-1} = 1+\frac{(N+1)\alpha_2}{N-1-2\alpha_2} \quad \text{and} \quad \alpha_2 = \frac{N-3}{2N(N-1)} C_X \Gamma_y \tag{33}$$

or, to order N^{-1},

$$\delta_2^{-1} = 1+C_X\Gamma_y/2N. \tag{34}$$

Case 3. *Distribution of R under general non-normality of y and x*

Finally, we may suppose that the regression variables themselves are random variables distributed independently of the y's in some p-variate distribution and that it is the deviations of these variates from their sample means that is recorded in the matrix X.

Taking expectations of the moments over all realizations of X, C_X in (31) must be replaced by $E\{C_X\}$ which we denote by Γ_X.

Once again using the beta approximation, R will be approximately distributed as F with $\nu_1 = \delta_3 p$ and $\nu_2 = \delta_3(N-p-1)$ degrees of freedom, where

$$\delta_3^{-1} = 1+\frac{(N+1)\alpha_3}{N-1-2\alpha_3} \quad \text{with} \quad \alpha_3 = \frac{N-3}{2N(N-1)} \Gamma_X \Gamma_y, \tag{35}$$

or, to order N^{-1},

$$\delta_3^{-1} = 1+\Gamma_X\Gamma_y/2N. \tag{36}$$

7. Accuracy of the approximation

It should be noticed that the above approximations do not depend for their accuracy on the fitting of an *arbitrary* distribution to the first two moments of W. The distribution of W is known to be given *exactly* by the approximation when in case 2 the observations y are normally distributed and when in case 3 *either* the observations y *or* the regression variables x, or both, are normally distributed. We might expect to be able to represent moderate departures from normality by suitable changes in the mean and variance of a system of curves which are of the right *basic* shape. Evidences that such a hope is justified are:

(*a*) For case 1, Pitman (1937) has shown for analysis of variance tests that except in very unlikely samples the third and fourth moments of W agree fairly closely with those of the approximating beta distribution. We show later that the analysis of variance test for equal groups corresponds to the general regression test when C_X attains its *lower* bound.

(*b*) For case 1, Box & Andersen showed by means of sampling experiments that the permutation distribution appropriate to the comparison of two independent variances was well represented by the approximation for distributions as non-normal as the rectangular and the double exponential. As we shall see, the test they considered can be very nearly reproduced in the general regression framework when C_X approaches its *upper* bound.

(*c*) For case 2, Box & Andersen showed the close agreement between the results of Gayen (1950) and results obtained by this approximation in analysis of variance tests. This confirms in particular the appropriateness of the general regression approximation at the lower bound of C_X.

(*d*) For case 2, Box (1953), using an argument not employing permutation theory, obtained an F approximation to the distribution of R for the comparison of two independent sample variances. In this the degrees of freedom were modified by functions which to order N^{-1} are identical with those given by the approximating F distribution derived via permutation theory. This supports the essential validity of the approximation as C_X approaches its *upper* bound.

8. Special case of the general one-way classification

We can readily check our formulae against those of Welch for the one-way classification analysis of variance with not necessarily equal groups. In the general regression model we arrange that the uth element x_{iu} of the ith x vector is $1-n_i/N$ when y_u falls in the ith group and $-n_i/N$ otherwise. As is well known the general regression test then reduces to the usual 'one way classification' analysis of variance test for the comparison of $p+1$ means. Then

$$S_0 = \mathbf{z}'\mathbf{Mz} = \mathbf{y}'\mathbf{My} = \sum_{i=1}^{p+1} n_i \bar{y}_i^2 - N\bar{y}^2,$$

where the overall sample mean is $\bar{y}$ and the sample mean for n_i observations in the ith group is $\bar{y}_i$.

Now if y_u is any one of the n_i observations in the ith group then the corresponding diagonal element M_{uu} of M is $n_i^{-1}-N^{-1}$. Hence

$$m = \sum_{u=1}^{N} M_{uu} = \sum_{i=1}^{p+1} n_i(n_i^{-1}-N^{-1})^2$$

$$= \sum_{i=1}^{p+1} n_i^{-1} - \frac{2p+1}{N}, \tag{37}$$

$$\frac{N-3}{N-1}C_X = 2\left[\frac{N(N+1)}{2p(N-p-1)}\left\{\sum_{i=1}^{p+1} n_i^{-1} - \frac{(p+1)^2}{N}\right\} - 1\right]. \tag{38}$$

Substituting this expression in (21) gives the value obtained by Welch for $V_P(W)$ and the corresponding F approximation given by Box & Andersen.

8·1. *Equal groups*

If the number of observations is the same in every group so that

$$n_i = n = N/(p+1) \quad (i = 1, 2, \ldots, N),$$

we have using equation (31)

$$m = \frac{p^2}{N} \quad \text{and} \quad \frac{N-3}{N-1}C_X = -2. \tag{39}$$

The lower bound of the inequality (30) is thus *actually attained* for the equal groups analysis of variance test.

Substituting the result in (31) we obtain correctly

$$\delta^{-1} = 1 - \frac{1}{N}\left\{\frac{N+1}{N-1+2C_y/N}\right\} C_y$$

or, to order N^{-1},

$$\delta^{-1} = 1 - C_y/N. \tag{40}$$

The corresponding result with $\Gamma_y = E(C_y)$ replacing C_y provides the appropriate correction factor for case 2 exemplifying the approximate effect of parent non-normality.

8·2. *Very unequal groups*

It is not possible to reproduce *exactly* the test for the comparison of two variances from the present general regression set-up. As we have already mentioned in § 3·2 we can, however, reproduce such a test with a slightly modified model in which the overall mean β_0 is supposed known.

The permutation approximation given by Box & Andersen for the set-up of § 3·2 with $n_1 + n_2 = N-1$ observations is

$$\delta^{-1} = 1 + \frac{1}{2}\left\{\frac{N+1}{N-2-C'_y}\right\} C'_y$$

or, to order N^0,

$$\delta^{-1} = 1 + \tfrac{1}{2}C'_y, \tag{41}$$

where C'_y is defined in equation (12). The corresponding result which gives the effect of non-normality on this comparison of variances test is obtained as before by replacing C'_y with $\Gamma'_y = E(C'_y)$. It is also possible by a different modification in which both the group means are eliminated to reproduce the usual test to compare two variances when their group means are estimated from the samples. With the present unmodified set-up we come closest to reproducing the test for the comparison of two variances by selecting the ith vector to have an element $x_{iu} = (N-1)/N$ when $u = i$ and $x_{iu} = -1/N$ otherwise. This then corresponds to the analysis of variance test discussed above with one observation in each of p groups and the remaining $N-p$ in the remaining groups. After a little manipulation we obtain for this case

$$R = \frac{\left\{\sum_{u=1}^{p} (y_u - \bar{y}')^2 + \frac{pN}{N-p}(\bar{y}' - \bar{y})^2\right\} \Big/ p}{\sum_{s=p+1}^{N} (y_s - \bar{y}'')^2/(N-p-1)}, \tag{42}$$

where $\bar{y}'$ is now used for the sample mean of the first p observations and y'' for the sample mean of the last $N-p$ observations.

The usual test criteria to compare the variances of two samples of p and $N-p$ observations would be

$$R' = \frac{\left\{\sum_{u=1}^{p} (y_u - \bar{y}')^2\right\} \Big/ (p-1)}{\sum_{s=p+1}^{N} (y_s - \bar{y}'')^2/(N-p-1)}, \tag{43}$$

which differs from R above only in that the latter contains a single extra comparison which contrasts the mean $\bar{y}'$ of the first p observations with the overall mean $\bar{y}$.

For this extreme case of very unequal groups, which nearly reproduces the comparison of variances test, we have

$$m = \frac{p}{N}\left\{N-1-\frac{(N-p-1)}{N-p}\right\} \quad \text{and} \quad \frac{N-3}{N-1}C_X = N-1-\left(\frac{N+1}{N-p}\right). \tag{44}$$

Provided that the ratio of p to N is small, C_X will approach its upper bound of $N-1$ quite closely.

If we write $\dot{N}$ for $N-1-(N+1)/(N-p)$, the modifying factor appropriate to this case is

$$\delta^{-1} = 1 + \frac{1}{2}\frac{N-1}{N}\left\{\frac{\dot{N}}{N-1-\dot{N}C_y/N}\right\} C_y, \tag{45}$$

which is very similar to the corresponding expression in equation (41) and once more, to order N^0, $\delta^{-1} = 1+\frac{1}{2}C'_y$.

8·3. *Group sizes which approximately nullify the effect of non-normality*

We have seen that particular choices of group sizes can be made so that C_X approaches a lower value of -2 and an upper value of N, giving rise to proportionate corrections. Provided the number of observations is not very small, the slight corrective increase in the degrees of freedom at the lower extreme is usually of little concern, but the considerable corrective decrease at the upper extreme is much more serious. Consideration of equations (31), (33) and (35) shows that in general if in cases 1 and 2 we choose X so that $C_X = 0$, or, in case 3, sample from a population in which $\Gamma_X = 0$, the corrective factor supplied by our approximations are all zero irrespective of the y's or their distribution. One way of exploiting this fact in experimental design theory has been noted by Box (1952).

Returning, now, to the one-way classification analysis of variance we see from equation (38) that, if we choose the group sizes n_i so that

$$N\sum_{i=1}^{p+1} n_i^{-1} - (p+1)^2 = \frac{2p(N-p-1)}{(N+1)}, \tag{46}$$

then $C_X = 0$ and the correction term is zero. Rather surprisingly therefore the effect of non-normality as measured by our approximation is *not* smallest for equal group sizes.

As an example, suppose there are just two groups of size $n_1 = rN$ and $n_2 = (1-r)N$. Substituting in (46) we obtain

$$r = \frac{1}{2}\left\{1 \pm \sqrt{\frac{N-2}{3N}}\right\} \tag{47}$$

for the optimum ratio of subgroup sizes. If $N = 12$, for example, then approximately the optimal group sizes are 9 and 3. For large N the optimal sizes are approximately in the ratio 4:1.

At first sight the idea that unequal group sizes could produce *less* sensitivity to non-normality seemed sufficiently surprising as to be almost unacceptable. In fact, as we show in the next section, it is not difficult to explain it. The result itself may be confirmed independently by study of the results of Gayen (1950). This author obtained the exact distribution of R for the one-way classification analysis of variance test for a parent population expressed by an Edgeworth series. His method of derivation is, of course, quite different from that used here. His corrective factor for kurtosis is proportional to a quantity which he denotes by

$$(\nu_{22}) = \frac{2\nu_1\nu_2 - (k^2-k'^2)(\nu_1+\nu_2+2)}{8(\nu_1+\nu_2+1)(\nu_1+\nu_2+2)}.$$

In our notation $\nu_1 = p$, $\nu_2 = N-p-1$, $k = p+1$, $k' = (p+1)^2 - N\sum_{i=1}^{p+1} n_i^{-1}$. Making the necessary substitution we see that when (46) is satisfied Gayen's quantity (ν_{22}) is zero, providing verification of our result.

We do not of course suggest that one would deliberately seek unequal sample sizes to lessen the effect of non-normality in a test to compare means. The reduction in precision with which comparison among the means could be made and the increase in sensitivity to variance inequalities which would result would certainly not be worth the small increase in robustness to non-normality.

9. ROBUSTNESS DETERMINED BY 'NORMALITY' IN THE x'S

We shall conduct the following discussion in terms of the situation (cases 2 and 3) where the error vector $\boldsymbol{\epsilon}$ is drawn from any symmetric distribution. We have seen that our modifying factor involves a measure of non-normality in the y's multiplying an analogous measure of 'non-normality' in the x's. That such factors would be involved symmetrically is to be expected from geometric considerations. The criterion R is a function of the angle between the observation vector and the plane of the x vectors and, as was first shown by Fisher, will follow its normal theory distribution if the y's *or* the x's *or* both are normally distributed. The multiplicative characteristic shows how non-normality in the y's is magnified or diminished depending on whether the x's are 'normal-looking' or not.

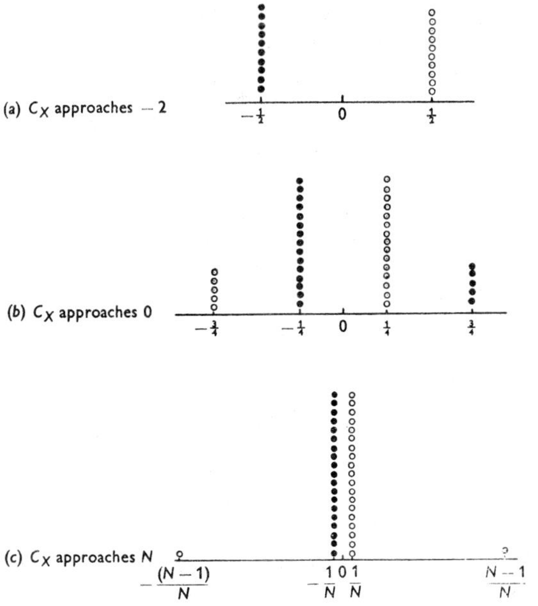

Fig. 1. Some distributions of the elements of x for $N = 20$.

The effects can be understood intuitively by means of particular examples. Consider the analysis of variance test for two equal groups. In this case the single vector x_1 has for its elements $\frac{1}{2}N$ values equal to $-\frac{1}{2}$ and $\frac{1}{2}N$ values equal to $+\frac{1}{2}$. The distribution of individual x's is like that shown in Fig. 1(a) (for $N = 20$) and represents the most 'platykurtic' distribution possible, for which the value C_X tends to its lower limit of -2. The set-up involving very unequal groups, discussed in § 8·2, comes closest to representing the comparison of two independent variances. The x vectors then each have one value equal to $(N-1)/N$ and the remaining $N-1$ values equal to $-1/N$. The distribution for $N = 20$ is that of the full circles shown in Fig. 1(c). This distribution has the same measure of kurtosis as when its mirror image, shown by open circles, is added. This represents the most leptokurtic distribution possible and the value of C_X tends to its upper limit, N. The full circles in Fig. 1 (*b*) show a distribution in which $\frac{1}{4}$ of the observations are set a distance $\frac{3}{4}$ from the origin and the remaining $\frac{3}{4}$ of the observations are set a distance $-\frac{1}{4}$. This distribution is close to that expected to minimize the effect of non-normality in the y's. Again the mirror image distribution of open circles is added. We see that in this example the distribution of the x's is doing its best to approximate a normal curve which accounts for the resulting insensitivity to non-normality in the y's.

10. Conclusion

Our results may be summarized in the simple statement that sensitivity to non-normality in the y's is determined by the extent of the 'non-normality of the x's'. The small effect in one direction experienced with the equal groups analysis of variance test and the much larger effect in the opposite direction found in the test for the comparison of independent variances provide extremes of sensitivity within which the sensitivity of the general test will be found. In the analysis of data arising from experimental designs such as factorials, a small and usually unimportant degree of sensitivity characteristic of the equal groups analysis of variance may be expected. With data in which the x's themselves are drawn from near normal distributions an even smaller degree of sensitivity is to be expected. Tests which employ x vectors in which one or two elements are very different in magnitude from the remainder may be expected to show much greater sensitivity to non-normality in the y's. In addition to the tests for comparing variances, certain tests concerned with outliers and with missing observations will show this greater sensitivity. One would expect that the usual normal theory *multi-variate* overall regression tests will have analogous corrective factors.

References

Box, G. E. P. (1952). Multifactor designs of the first order. *Biometrika*, **39**, 49–57.

Box, G. E. P. (1953). Non-normality and tests on variances. *Biometrika*, **40**, 318–35.

Box, G. E. P. & Andersen, S. (1955). Permutation theory in derivation of robust criteria and the study of departures from assumption. *J. R. Statist. Soc.* **17**, 1–26.

David, F. N. & Johnson, N. L. (1951*a*). A method of investigating the effect of non-normality and heterogeneity of variance on tests of the general linear hypothesis. *Ann. Math. Statist.* **22**, 382–92.

David, F. N. & Johnson, N. L. (1951*b*). The effect of non-normality on the power function of the *F*-test in the analysis of variance. *Biometrika*, **38**, 43–57.

David, F. N. & Kendall, M. G. (1949). Tables of symmetric functions—Part 1. *Biometrika*, **36**, 431–49.

Fisher, R. A. (1947). *The Design of Experiments*, 4th ed. Edinburgh and London: Oliver and Boyd.

Gayen, A. K. (1950). The distribution of the variance ratio in random samples of any size drawn from non-normal universes. *Biometrika*, **37**, 236–55.

Geary, R. C. (1947). Testing for normality. *Biometrika*, **34**, 209–42.

Pearson, E. S. (1931). Analysis of variance in cases of non-normal variation. *Biometrika*, **23**, 114–33.

Pitman, E. J. G. (1937). Analysis of variance test for samples from any population. *Biometrika*, **29**, 322–35.

Scheffé, H. (1953). A method for judging all contrasts in the analysis of variance. *Biometrika*, **40**, 87–104.

Welch, B. L. (1937). On the *z*-test in randomised blocks and Latin squares. *Biometrika*, **29**, 21–52.

1.6
A Further Look at Robustness Via Bayes's Theorem*

G. E. P. BOX and G. C. TIAO†
University of Wisconsin

1. Introduction and summary

In recent years, under the leadership of Savage (1954, 1959, 1960), there has been a great revival of interest in subjective probability and in the interpretation of data via Bayes's theorem. Many statisticians now feel that this provides the most satisfactory basis for a theory of statistical inference. In particular, such an approach seems necessary if one is to give explicit cognisance to the uncertainty in the assumptions which are built into many statistical procedures. Classical statistical arguments lead us to treat such assumptions as if they were in some way axiomatic and yet consideration will show that, in fact, they are conjectures which in practice may be expected to be more or less true. The mathematical expression of 'more or less true' seems to require the explicit injection of subjective probability distributions. For instance, in many problems the particular physical set-up is such that the errors involved might behave like a linear aggregate of component errors and, consequently, a central limit effect would operate. In fact, of course, the central limit theorem does not tell us that a linear aggregate of a *finite* number of component errors would be exactly normal. We are, however, entitled to expect in this physical situation that the distribution we are dealing with will be a member of a 'distribution of distributions' in which the normal curve occupies the central place. In this situation we can express our true state of mind by the use of a prior distribution of some parameter or parameters, measuring the non-normality of the parent distribution.

If we *assume* normality, we can proceed with an 'objective' classical analysis. But by making this normality assumption, however, we act in fact as if the distribution of our non-normality parameter were a delta function. As seems to be inevitably the case in other problems as well as this one, therefore, our 'objectivity' is gained by pretending to knowledge we do not have and in so doing we even ignore what the sample has to tell us about the matter in question.

On classical theory, once having assumed the form of the parent distribution, we can derive a criterion which is appropriate on this assumption. For example, on the assumption of normality, for the comparison of two means we would derive the t-statistic. It is then customary to justify the use of such a normal theory criterion in the practical circumstance in which normality cannot be guaranteed by arguing that the distribution of the criterion is but little affected by non-normality of the parent distribution—that is, it is robust under non-normality. However, this argument ignores the fact that if the parent distribution really differed from the normal, the appropriate criterion would no longer be the normal-theory statistic. It is easy to produce examples in which the distribution of the normal-theory criterion is little affected if the parent is assumed to be some distribution other than the normal; and yet, the inference to be drawn when a criterion *appropriate* to this other distribution is employed is markedly different.

* This research was supported by the United States Navy through the Office of Naval Research.
† Dr Tiao was supported by a grant from the Ford Foundation.

The Collected Works of George E. P. Box, 1984, Wadsworth, Inc., Belmont, CA 94002.
Originally published in *Biometrika*, vol. 49, parts 3 and 4 (1962), pp. 419–432.

In this paper the analysis of Darwin's paired data on the heights of self- and cross-fertilized plants quoted by Fisher in *The Design of Experiments* (1935) is reconsidered. In our development the parent distribution is not assumed to be normal, but only a member of a class of symmetric distributions which include the normal, and whose kurtosis is measured by a parameter β. In this example, the physical nature of the experimental environment is certainly such that a central limit effect would be expected. That this expectation justifies us only in supposing that the error distributions will *approach* the normal is specifically recognised in our formulation by giving a subjective prior probability distribution to β centred at its normal value. The sharpness of this subjective prior distribution can be varied so that we can represent a range of situations in which a greater and greater degree of central limit effect is injected. Finally, when the prior distribution becomes a delta function, we produce the usual formulation in which an exact assumption of normality is made. At the other extreme, we can produce the situation in which all the information about normality or the lack of it is essentially being generated from the sample itself. The extent to which the usual normal theory t-test could be approximately justified over this wide range of circumstances is illustrated and discussed.

It is believed that the injection into the model of subjective prior probability distributions to represent tentatively held 'assumptions' has general application. Extension of these ideas to other statistical procedures is being carried out.

2. Various approaches to the analysis of Darwin's data

Darwin's experiments were conducted in pairs so that on the normal assumption and on classical theory one may interpret these data by using the paired t-test. Fig. 1 shows the observed differences by circles marked along the horizontal axis. The solid curve in the same figure is an appropriately scaled t-distribution centred about the sample mean $\bar{y}$ with scale factor $s/\sqrt{n}$ where the quantity $s^2 = \Sigma(y_i - \bar{y})^2/(n-1)$ is the usual sample estimate of the variance of the differences. In what follows, for definiteness, we shall call this distribution the *reference distribution* for the population mean θ. This reference distribution may be variously interpreted. It is the fiducial distribution of θ. It can also be regarded as showing a complete set of confidence intervals for θ for all values of the 'confidence coefficient'. Finally, if we make suitable assumptions discussed later, concerning the prior distributions of θ and σ, it is the *posterior* distribution of θ. If we are interested in a significance test appropriate to the hypothesis that $\theta = 0$ against the alternate $\theta > 0$, then the associated significance level for the present example is 2·485 %.

Now suppose that instead of assuming normality for the parent distribution, we assumed it to be uniform over some unknown range $\theta - \sigma$ to $\theta + \sigma$, where here and in what follows, σ is used as a general scale parameter and does not necessarily refer to the standard deviation. This assumption would, of course, be quite ridiculous in the present example. First, we know that the many contributing errors arising from genetic differences, soil differences, and so forth, will produce a strong central limit effect so that we may expect with good reason that the heights themselves and, even more, their differences will be closely normally distributed. Second, the evidence from the sample itself does not support the uniform assumption. However, to illustrate our point, let us make the assumption of a uniform instead of a normal parent. One thing we might then consider is the effect of this extreme degree of non-normality on the distribution of the t-statistic. This can be approximately calculated using, for example, the work of Gayen (1950) or of Box & Anderson (1955).

Following these latter authors, it is readily shown that the null distribution of t^2 is approximated by an F-distribution with δ and $\delta(n-1)$ degrees of freedom where

$$\delta = 1+E(b-3)/n$$

and $$E(b-3) \sim \gamma_2' - n^{-1}(2\gamma_4' - 3\gamma_2'^2 + 11\gamma_2') + n^{-2}(3\gamma_6' - 16\gamma_4'\gamma_2' + 15\gamma_2'^3 + 38\gamma_4' + 86\gamma_2'),$$

where $$\gamma_{r-2}' = \kappa_r(y)/\{\kappa_2(y)\}^{\frac{1}{2}r}$$

are the standardized cumulants of the parent distribution of differences. In our present example, δ is found to be 0·913. Thus, t^2 is approximately distributed as F with 0·913 and 12·78 degrees of freedom. In particular, the significance level associated with the hypothesis that $\theta = 0$ against the alternative $\theta > 0$ is now 2·388 % as compared with the previous value of 2·485 %. The test of the hypothesis that the true difference is zero using the t-criterion is thus very little affected by this major departure from normality. Similarly, confidence intervals based on the t-statistic would be very little affected by this departure. The robustness of the t-statistic in this example was, in fact, demonstrated by Fisher who derived the exact randomization distribution in this case and showed that the null probability agreed very closely with that obtained from the t-criterion.

However, if we really knew that the parent distribution was uniform, we would not consider the t-criterion at all. We would be led instead to consider the function

$$W = |m-\theta|/h,$$

where $$m = \tfrac{1}{2}(y_L + y_S), \quad h = \tfrac{1}{2}(y_L - y_S)$$

and y_L and y_S are respectively the largest and the smallest of the observations—jointly sufficient statistics for θ and σ on the uniform assumption. On this same assumption (Neyman & Pearson, 1928; Carlton, 1946), the variate $(n-1)W$ is distributed as F with 2 and $2(n-1)$ degrees of freedom.

Just as the solid curve in Fig. 1 exemplifies the inferential situation with the normal assumption, so the broken curve in the same figure correspondingly exemplifies this situation with the uniform assumption. As before, we can interpret the distribution represented by the broken curve either as the fiducial distribution of θ or as defining a complete set of confidence intervals for θ or, finally (if we adopt identical assumptions about the prior distributions of θ and σ as those needed before) as the posterior distribution of θ. We notice that this reference distribution is markedly different from that we obtained from the normal assumption, especially with regard to its location. In particular, the significance level associated with the hypothesis that $\theta = 0$ against the alternative $\theta > 0$ is not 2·485 %, but 23·215 %. Thus, whichever form of derivation we favour, we see that the conclusions which we would draw if we assumed a uniform parent distribution are very different from those which would be appropriate if we assumed a normal parent distribution, even though the *t-criterion* itself is very little affected by this large departure from normality. The principal reason for this large difference is that in one case the reference distribution is centred at the sample mean (20·9) and in the other case it is centred at the sample mid-point (4·0). For this particular sample, the mean and the mid-point differ considerably, mainly because of two rather large negative differences.

As we have explained, we are not seriously suggesting that the uniform distribution is a reasonable choice for the parent. We wish only to emphasize that uncertainty in our knowledge of the parent distribution transmits itself rather forcefully into an uncertainty about the conclusion we can draw concerning θ, and the difficulty which this presents in our interpretation of the data is not avoided by our knowledge of the robustness of the t-criterion. It seems to us that this difficulty can only be resolved by explicitly including the knowledge that we have about the parent distribution into our formulation. This knowledge is of two kinds, that coming from the sample itself and that coming from knowledge of the physical set-up appropriate to this problem. We shall see that they are both taken into account in an appropriate Bayesian formulation.

At this point, it may be instructive to remind ourselves of the Bayesian justification of the t-distribution such as has been essentially given by Jeffreys (1957, 1961).

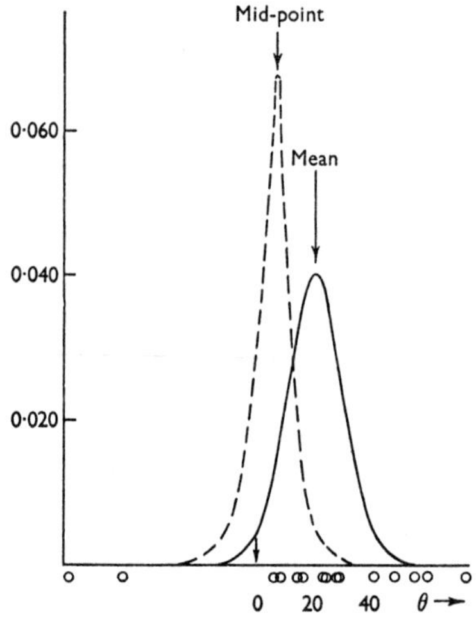

Fig. 1. In this figure, the solid curve represents the reference distribution for θ when the parent is assumed normal. The broken curve represents the reference distribution for θ when the parent is assumed rectangular.

3. Derivation of the t-test via Bayes's theorem on the principle of precise measurement

The Bayesian argument requires that we have some prior distributions for θ and σ. We assume, as it seems reasonable, that the local prior distribution of these location and scale parameters are independent. So far as the location parameter is concerned, the situation met in actual circumstances of experimentation would often permit us to assume that the prior distribution of θ was locally uniform, using what Savage (1959, 1960) calls the principle of precise measurement. This principle says, in effect, that we do not need to know exactly what the prior distribution of θ is if we can say only that in the region in which the likelihood is appreciable it does not change very much, and at no other point is it of sufficiently great magnitude as to become appreciable when multiplied by the likelihood. This principle would be applicable in situations where the likelihood dominates and is inapplicable in situations where the prior probability density dominates. What makes this principle of particular importance is that most actual experiments will be conducted only when it is expected that the likelihood will exert a much stronger influence in the final result than the prior distribution. Otherwise, there is little point in doing the experiment. For instance, suppose that the value of the gravitational constant in suitable units had been estimated as $32{\cdot}2 \pm 0{\cdot}1$ then there would be little justification for making further measurements with a method whose accuracy was, say, $\pm 0{\cdot}2$, but considerable justification for conducting further experiments using a method whose accuracy was $\pm 0{\cdot}02$.

The argument that if θ is taken as locally uniform, then $\log \theta$, $1/\theta$, etc., will not be, loses its force if we remember that unless the range of values of θ over which the likelihood is appreciable is large compared with the average magnitude of θ over the same range, then such transformations will make little practical difference *in the range considered.* In the example considered above, for instance, if the prior distribution of θ were assumed uniform from, say, $\theta = 32{\cdot}0$ to $\theta = 32{\cdot}4$, then, to a close approximation, the prior distributions of, for example, $\log \theta$ and $1/\theta$ would be uniform over corresponding ranges.

We can also demonstrate this lack of sensitivity to prior assumption when we consider the scale parameter. Suppose we merely assume that either σ or its logarithm or some power of σ is locally uniform. We have then

$$p(\theta) \propto k, \quad p(\sigma) \propto \begin{cases} \sigma^{q-1} & \text{if} \quad \sigma^q \text{ assumed uniform,} \\ \sigma^{-1} & \text{if} \quad \log \sigma \text{ assumed uniform,} \end{cases} \tag{1}$$

whence denoting $l(\theta, \sigma|\mathbf{y})$ for the likelihood function given the sample $\mathbf{y}$

$$p(\theta, \sigma|\mathbf{y}) = kl(\theta, \sigma|\mathbf{y}) . p(\theta) . p(\sigma), \tag{2}$$

where
$$k^{-1} = \iint_R l(\theta, \sigma|\mathbf{y}) . p(\theta) . p(\sigma)\, d\theta\, d\sigma.$$

On the normal assumption, then

$$p(\theta, \sigma|\mathbf{y}) = p(\theta|\sigma, \bar{y}) . p(\sigma|s), \tag{3}$$

where
$$p(\theta|\sigma, \bar{y}) = \{n/(2\pi\sigma^2)\}^{\frac{1}{2}} \exp\{-(\tfrac{1}{2}n/\sigma^2)(\bar{y}-\theta)^2\},$$

$$p(\sigma|s) = 2\{\Gamma[\tfrac{1}{2}(\nu - q)]\}^{-1} (\tfrac{1}{2}\nu s^2)^{\frac{1}{2}(\nu-q)} \sigma^{q-(\nu+1)} \exp\{-\tfrac{1}{2}\nu s^2/\sigma^2\} \quad (\nu = n-1, \text{ and } q < \nu).$$

On integrating out σ we obtain

$$p\left(\frac{\theta - \bar{y}}{s/\sqrt{n}}\bigg|\mathbf{y}\right) = p[t_{\nu-q}], \tag{4}$$

where $p[t_{\nu-q}]$ is the t-distribution with $\nu - q$ degrees of freedom. The only effect of changing the power q of σ supposed uniform, is to change the number of degrees of freedom in the final posterior t-distribution. In particular, by assuming $\log \sigma$ to be locally uniform, we obtain a posterior distribution of θ as a t-distribution with the traditional $\nu = n - 1$ degrees of freedom. If we suppose σ to be locally uniformly distributed, we will have a t-distribution with $n-2$ degrees of freedom, and if we suppose σ^2 to be locally uniform, a t-distribution with $n-3$ degrees of freedom.

If we take $\log \sigma$ as the function of σ to be regarded as locally uniform we are consistent in the sense that $\log \sigma$ is a location parameter for $\log |y - \theta|$, just as θ is a location parameter for y. We do not regard this argument as conclusive, but it is comforting to notice that from a moderate sized sample such as that from Darwin's data, rather drastic changes in the nature of the prior distribution of σ do not greatly affect the final conclusion and in what follows we make the assumption of uniform distribution for $\log \sigma$.

For our later purposes, it is perhaps worth while to consider this well-known result geometrically. The joint posterior distribution of θ and σ can be regarded as being built up from a series of normal distributions each centred at $\bar{y}$, each of which represents the conditional distribution $p(\theta|\sigma,\bar{y})$ of θ for some given σ, multiplied by $p(\sigma|s)$, the marginal posterior distribution of σ which has the form of an inverted gamma function. If we knew the value of σ, then the posterior distribution of θ would be normal about $\bar{y}$ with this known value of $\sigma = \sigma_0$. When σ is unknown, then we must average all these normal distributions, using for weights the ordinates of the marginal posterior distribution of σ. In so doing, we obtain the t-distribution. There is, of course, nothing new in the above; we recall it here only to introduce the more general argument which follows.

4. A WIDER CHOICE OF THE PARENT DISTRIBUTION

If, in the analysis of Darwin's data, we suppose that the parent distributions of self- and cross-fertilized plants are of the same form, then the distribution of the differences would certainly be symmetric. Let us, therefore, assume that our parent distribution is a member of a class of symmetric distributions which includes, in particular, the normal, together with other distributions on the one hand more leptokurtic, and on the other hand more platykurtic than the normal. A convenient choice is the class of power distributions employed in other contexts, for example, by Diananda (1949), Box (1953) and Turner (1960), where

$$p(y|\theta,\sigma,\beta) = \omega \exp\left\{-\frac{1}{2}\left|\frac{y-\theta}{\sigma}\right|^{2/(1+\beta)}\right\} \tag{5}$$

$$\omega^{-1} = \Gamma[1+\tfrac{1}{2}(1+\beta)]\, 2^{[1+\frac{1}{2}(1+\beta)]}\sigma$$

$$(-\infty < y < \infty,\ 0 < \sigma < \infty,\ -\infty < \theta < \infty,\ -1 < \beta < 1).$$

In particular, we see that when $\beta = 0$, we have the normal distribution; when β is 1, we have the double exponential; and when β tends to -1, our distribution tends to the uniform distribution.

5. DERIVATION OF THE POSTERIOR DISTRIBUTION OF θ FOR A SPECIFIC SYMMETRIC PARENT

We now derive the posterior distribution of θ, supposing that the parent distribution to be a member of the above class of distributions in which β is assumed to have a fixed value β_0. In so doing, we shall adopt the same assumptions *a priori* as are necessary to derive the traditional t-distribution when β is assumed to be zero.

We have

$$l(\theta,\sigma|\mathbf{y},\beta_0) = [\Gamma\{1+\tfrac{1}{2}(1+\beta_0)\}\, 2^{\{1+\frac{1}{2}(1+\beta_0)\}}\sigma]^{-n} \exp\left\{-\tfrac{1}{2}\sum_i\left|\frac{y_i-\theta}{\sigma}\right|^{2/(1+\beta_0)}\right\}, \quad p(\theta) \propto k',\ p(\sigma) \propto \sigma^{-1} \tag{6}$$

so that

$$p(\theta,\sigma|\mathbf{y},\beta_0) = k\sigma^{-(n+1)} \exp\left\{-\tfrac{1}{2}\sum_i\left|\frac{y_i-\theta}{\sigma}\right|^{2/(1+\beta_0)}\right\},^{*} \tag{7}$$

where

$$k^{-1} = \iint_R \sigma^{-(n+1)} \exp\left\{-\tfrac{1}{2}\sum_i\left|\frac{y_i-\theta}{\sigma}\right|^{2/(1+\beta_0)}\right\} d\theta\, d\sigma.$$

By integrating out σ as before, we finally obtain for the posterior distribution of θ for any fixed $\beta = \beta_0$ in the permissible range the remarkably simple expression

$$p(\theta|\mathbf{y},\beta_0) = k[M(\theta)]^{-\frac{1}{2}[n(\beta_0+1)]}, \tag{8}$$

where

$$M(\theta) = [\sum_i |y_i-\theta|^{2/(1+\beta_0)}]$$

* It is understood here that at least two of the observations are not equal.

and $M(\theta)/n$ is the absolute moment of order $2/(1+\beta_0)$ of the observations about θ. The integral

$$k^{-1} = \int_{-\infty}^{\infty} [M(\theta)]^{-\frac{1}{2}n(1+\beta_0)}\, d\theta$$

is merely a normalizing factor which ensures that the total area under the distribution is unity. This integral cannot usually be expressed as a simple function; it can, of course,

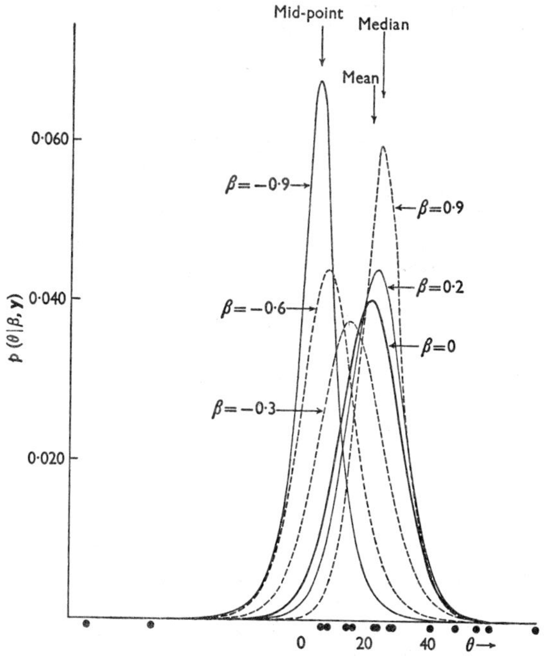

Fig. 2. Posterior distributions of θ for various choices of β.

always be computed numerically, and with the availabilty of electronic computers, this presents no particular difficulty.*

Using equation (8), posterior distributions computed from Darwin's data for various values of β_0 are shown in Fig. 2.

* It is interesting to notice here that when $\beta \leqslant 0$ and the quantity $\frac{1}{2}n(\beta+1)$ is large, we can approximate the posterior distribution by

$$p(\theta|\mathbf{y},\beta) \sim k\exp\{-\tfrac{1}{2}[\tfrac{1}{2}\{n(1+\beta)\}]h''(\theta_0)(\theta-\theta_0)^2\},$$

where

$$k = \{\tfrac{1}{2}[n(1+\beta)]h''(\theta_0)/2\pi\}^{\frac{1}{2}}, \quad h(\theta) = \log[M(\theta)]$$

and θ_0 is the value of θ at which $h(\theta)$ attains its minimum. This is a special case of a powerful method, known as 'Saddle Point Approximation', developed by Jeffreys & Jeffreys (1956) and Daniels (1954). In our case, it is equivalent to the normal approximation of the distribution $p(\theta|\mathbf{y},\beta)$ around the maximum likelihood estimate of θ. This device has been used by Lindley (1961), in a similar treatment of this problem. He discusses the effect on the posterior distribution of θ when β has a normal distribution about 0 with small variance by the steepest descent method.

6. PROPERTIES OF THE POSTERIOR DISTRIBUTION OF θ FOR FIXED $\beta = \beta_0$

Since $p(\theta|\mathbf{y},\beta_0)$ is a monotonic function of $M(\theta)$, we find (see Appendix) the following:

(1) $p(\theta|\mathbf{y},\beta_0)$ is continuous, differentiable and unimodal, although not necessarily symmetric, the mode being attained in the interval $[y_S, y_L]$.

(2) When $\beta_0 = 0$, $M(\theta) = \Sigma(y_i-\theta)^2 = (n-1)s^2+n(\bar{y}-\theta)^2$ and making necessary substitutions we obtain for the posterior distribution of θ

$$p\left(\frac{\theta-\bar{y}}{s/\sqrt{n}}\middle|\mathbf{y},\beta_0\right) = p(t_{n-1})$$

as before.

(3) When β approaches -1, $\lim_{\beta_0\to-1} [M(\theta)]^{\frac{1}{2}(\beta_0+1)} = (h+|m-\theta|)$ and making the necessary substitutions,

$$\lim_{\beta_0\to-1} p(\theta|\mathbf{y},\beta_0) = k[h+|m-\theta|]^{-n}, \tag{9}$$

where

$$k^{-1} = \int_{-\infty}^{\infty} (h+|m-\theta|)^{-n}\,d\theta$$

so that

$$\lim_{\beta_0\to-1} p\left(\frac{|\theta-m|}{h/(n-1)}\middle|\mathbf{y},\beta_0\right) = p[F_{2,2(n-1)}].$$

This is then the reference distribution shown by the broken curve in Fig. 1 but now derived as a posterior distribution.

Thus, we see that, when the parent is normal ($\beta_0 = 0$), our expression (8) yields the t-distribution as expected, and when the parent approaches the uniform ($\beta_0 \to -1$), again as expected, our expression (8) gives the double F-distribution with 2 and $2(n-1)$ degrees of freedom. In each of these cases, the posterior distribution can be expressed in terms of simple functions of the observations which provide then, of course, minimal sufficient statistics for θ and σ.

(4) When β_0 approaches 1, the distribution is not expressible in simple functions of the observations, but in the limit the mode of the posterior distribution is the median of the observations if the latter is uniquely defined; and, if not, it is some unique value between the values of the middle two observations.

(5) In certain other cases, it is possible to express the posterior distribution of θ in terms of a fixed number of functions of the observations. For instance, when

$$\beta = (1-q)/q \quad (q = 1,2,3,\ldots)$$

we have

$$p(\theta,\sigma|\mathbf{y},\beta_0) \propto \sigma^{-(n+1)}\exp\left\{-\tfrac{1}{2}\sigma^{-2q}\sum_{r=0}^{2q}(-1)^r\binom{2q}{r}\theta^r S_{2q-r}\right\} \tag{10}$$

and

$$p(\theta|\mathbf{y},\beta_0) \propto \left[\sum_{r=0}^{2q}(-1)^r\binom{2q}{r}\theta^r S_{2q-r}\right]^{-n/2q}, \tag{11}$$

where

$$S_r = \sum_i y_i^r$$

and it is readily seen that the set of $2q$ functions, $S_1, S_2, \ldots, S_{2q}$ of the observations are jointly sufficient for θ and σ.

In general, however, the posterior distribution cannot be expressed in terms of a few simple functions of the observations, the minimal sufficient statistics are the observations themselves. If we wish to think in terms of sufficiency and information as defined by Fisher, our posterior distribution always, of course, employs a complete set of sufficient statistics, and, consequently, no matter what is the value of β, no information is lost. The posterior distribution of θ for given β always has as its modal value the maximum likelihood estimate of θ, but it should be noticed that we are not concerned with the distribution of this maximum likelihood estimate; rather, we are considering the distribution of θ given each one of the observations.

From the family of distributions for various values of β_0 as shown in Fig. 2, we see that very different inferences will be drawn concerning θ, depending upon which value of β_0 is assumed. The chief reason for this wide discrepancy is the fact that in Darwin's data, the centre of the posterior distribution changes markedly as β is changed. In particular, for this sample, the median, mean and the mid-point are respectively 24·0, 20·9, 4·0, and these are the modes of the posterior distributions for the double exponential, normal and uniform parent, respectively.

7. Posterior distribution of θ and β when β is regarded as a variable parameter

Because of the wide differences which occur in the posterior distribution of θ depending on which parent distribution (that is, which value of β_0) we employ, in practice it might be thought there would be considerable uncertainty as to the nature of the valid inference that could be drawn from this data. We now show in this section that this is not the case, when we use appropriate evidence concerning the value of β. We have, in fact, two sources of information about the value of β, one from the data itself and the other from our knowledge *a priori* that a central limit effect would operate in the circumstances of the experiment. Both types of evidence can be injected into our analysis by allowing β itself to be a variable parameter associated with a prior distribution.

We can represent the central limit tendency of the errors by choosing a prior distribution for β which has a maximum value at $\beta = 0$, and which extends from -1 to $+1$. A convenient distribution for this purpose is the beta distribution having mean zero and extending from -1 to $+1$ and, consequently, possessing only one adjustable parameter which we call a. We assume that

$$p(\beta) = w(1-\beta^2)^{a-1} \quad (-1 < \beta < 1),$$

where

$$w = \Gamma(2a)\,[\Gamma(a)]^{-2}\,2^{-(2a-1)} \quad (a \geqslant 1). \tag{12}$$

When $a = 1$, this distribution is uniform. With $a > 1$, it is a symmetric distribution having its mode at the normal theory value $\beta = 0$. If we wished to represent a situation in which some value other than $\beta = 0$ occupied the central position, then this could be done in a similar way by using a beta function having two adjustable parameters.

After eliminating the scale parameter σ, we now obtain for the joint posterior distribution of θ and β

$$\begin{aligned} p(\theta,\beta|\mathbf{y}) &= k_1(1-\beta^2)^{a-1}\,\Gamma[1+\tfrac{1}{2}n(1+\beta)]\,[\Gamma\{1+\tfrac{1}{2}(1+\beta)\}]^{-n}\,[M(\theta)]^{-\frac{1}{2}n(1+\beta)} \\ &= k_2 . f(\theta,\beta|\mathbf{y})\,p(\beta), \end{aligned} \tag{13}$$

where $p(\beta)$ is given by equation (12) and k_1 and k_2 are the appropriate normalizing constants. The function $f(\theta,\beta|\mathbf{y})$ can be written

$$f(\theta,\beta|\mathbf{y}) = p(\theta|\beta,\mathbf{y})\,\phi(\beta), \tag{14}$$

where

$$\phi(\beta) = k_3\,\Gamma[1+\tfrac{1}{2}n(1+\beta)]\,[\Gamma\{1+\tfrac{1}{2}(1+\beta)\}]^{-n}\int_{-\infty}^{\infty}[M(\theta)]^{-\frac{1}{2}n(1+\beta)}\,d\theta \tag{15}$$

and $p(\theta|\beta, \mathbf{y})$ is given by equation (8). The conditional distributions $p(\theta|\beta, \mathbf{y})$ are the t-like distributions which we have already plotted in Fig. 2 and which represent the posterior distributions of θ for different specific choices of β. The function $\phi(\beta)$ which is sketched in Fig. 3 (i) can be regarded as representing *information coming from the sample concerning* β. We see that the function $f(\theta, \beta|\mathbf{y})$ is in fact built up of these t-like distributions suitably weighted with the weight function $\phi(\beta)$.

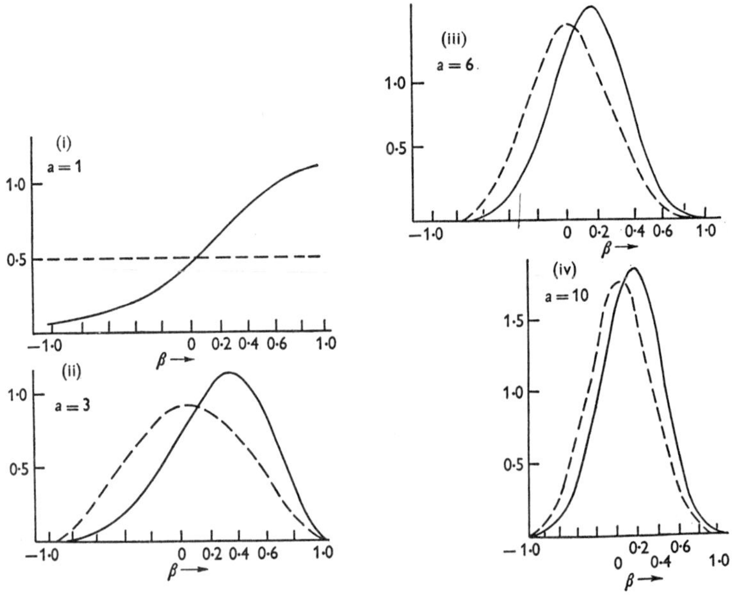

Fig. 3 (i)–(iv). Prior and posterior distributions of β for various choices of a. In each figure the broken curve represents the prior distribution $p(\beta)$ and the solid curve the posterior distribution $p(\beta|\mathbf{y})$. When $a = 1$, $p(\beta|\mathbf{y}) = \phi(\beta)$.

We can interpret $f(\theta, \beta|\mathbf{y})$ as the joint posterior distribution of θ and β for which the prior distribution of β is uniform. In this case, $\phi(\beta)$ is then the posterior distribution of β. Of course, it would usually be quite unrealistic to suppose that the distribution of β were uniform *a priori*. Since

$$p(\theta, \beta|\mathbf{y}) = kp(\theta|\beta, \mathbf{y})\,\phi(\beta)\,p(\beta)$$

we see that the joint posterior distribution $p(\theta, \beta|\mathbf{y})$ will in general be obtained by weighting $p(\theta|\beta, \mathbf{y})$ not with $\phi(\beta)$ but with the function $\phi(\beta).p(\beta)$. The parameter a in $p(\beta)$ (equation (12)) can be adjusted to allow for any desired strength of central limit effect. The case $a = 1$ giving a uniform distribution for $p(\beta)$ corresponds to 'no central limit effect'. When a tends to infinity, $p(\beta)$ becomes a delta function and represents an overwhelmingly strong central effect. This corresponds to the assumption of exact normality for the parent distribution.

8. Posterior distribution of θ

From the joint posterior distribution of θ and β

$$p(\theta, \beta|\mathbf{y}) = p(\theta|\beta, \mathbf{y})\,\phi(\beta)\,p(\beta)$$

the posterior distribution of θ is obtained by integrating out β yielding

$$p(\theta|\mathbf{y}) = \int_{-1}^{1} p(\theta, \beta|\mathbf{y})\,d\beta$$

$$= \int_{-1}^{1} p(\theta|\beta, \mathbf{y})\,\phi(\beta)\,p(\beta)\,d\beta. \tag{16}$$

In obtaining this integral, we are averaging the t-like distributions $p(\theta|\beta, \mathbf{y})$ with a weight function $\phi(\beta).p(\beta)$ which is in fact $p(\beta|\mathbf{y})$, the posterior distribution of β. The value of this weight function is seen to depend partly upon information from the sample through $\phi(\beta)$ and partly from prior information characterized by $p(\beta)$. The way in which this weight function $\phi(\beta).p(\beta)$ changes as the assumed central limit effect is increased is shown in Fig. 3 (i)–(iv). In these diagrams, the broken curve is, in each case, the prior distribution $p(\beta)$. When $a = 1$ (Fig. 3 (i)), $p(\beta)$ is uniform and $p(\beta|\mathbf{y})$ equals $\phi(\beta)$. This represents the situation where the information concerning β is essentially coming from the sample itself. The value of the parameter a is 3 in Fig. 3 (ii), 6 in Fig. 3 (iii) and 10 in Fig. 3 (iv). These three diagrams show how increasing certainty of a central limit effect tends to over-ride the information from the sample. Finally, when a tends to infinity, both $p(\beta)$ and $p(\beta|\mathbf{y})$ would approach a delta function at $\beta = 0$.

The integration

$$p(\theta|\mathbf{y}) = \int_{-1}^{1} p(\theta|\beta, \mathbf{y})\,\phi(\beta)\,p(\beta)\,d\beta$$

has been actually carried out for each of these weight functions and the results are shown in Fig. 4 together with the t-distribution which would be appropriate for the case $a \to \infty$ corresponding to an assumption of exact normality. In the diagram, the case $a = 10$ is not shown since this curve is almost indistinguishable from the t-distribution. This is interesting because, as will be seen from Fig. 3 (iv), the central limit effect implied by $a = 10$ is not an overwhelmingly strong one. For instance, it certainly leaves as acceptable *a priori* the possibility that $\beta = +0{\cdot}3$ or $\beta = -0{\cdot}3$. If we call the normal distribution a 'second power' distribution, this is equivalent to supposing *a priori* that 'third power' distributions and '$1\frac{1}{2}$ power' distributions are possible.

Fig. 4 then represents the final inference we could draw for θ depending on how strong a central limit effect would be appropriate in the physical situation. In view of the very large differences exhibited by the t-like distributions in Fig. 2, it seems remarkable how alike these distributions are. In particular, it will be seen that the tail areas which have been traditionally regarded as the most important part of the distribution are very little affected even with no 'central limit effect'. The main reason for this is that those widely discrepant t-like distributions generated by parents which approach the uniform are almost ruled out by information coming from the sample itself (see Fig. 3 (i)).

We may remark here that the precise form of the posterior distribution of θ would of course depend to some extent upon the way we parametrize the constant measuring normality. The measure β which makes the double exponential and the rectangular distributions equally discrepant from the normal seems not unreasonable. However, we might have used for example the familiar kurtosis measure $\lambda_4 = \kappa_4/\kappa_2^2$ for the class of distributions we have considered. It is easily shown, in fact, that

$$\lambda_4 = \frac{\Gamma[\frac{5}{2}(1+\beta)]\,\Gamma[\frac{1}{2}(1+\beta)]}{\Gamma[\frac{3}{2}(1+\beta)]^2} - 3.$$

On this scale, the double exponential distribution would appear as 3 and the rectangular distribution as $-1{\cdot}2$. It seems that whether β, λ_4 or any other reasonable measure of non-normality is adopted, the overall conclusions are very similar.

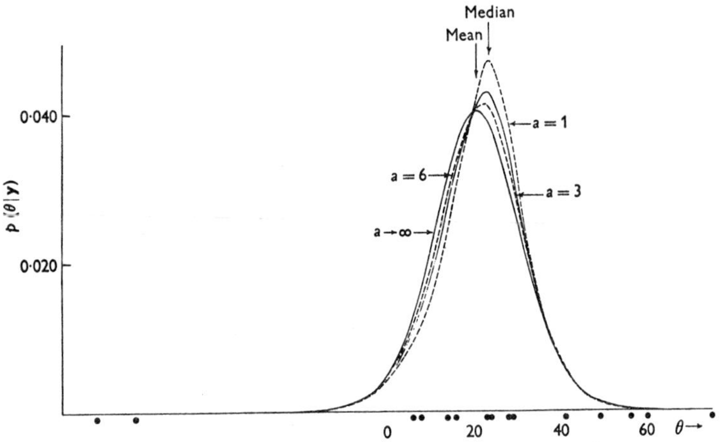

Fig. 4. Posterior distributions of θ for various choices of a.

9. Information concerning the nature of the parent distribution coming from the sample

In the past the normality or otherwise of a sample has usually been decided either by employing certain preliminary tests, for example, the χ^2 goodness-of-fit test, and the Kolmogoroff–Smirnoff test (1941) or by calculating statistics of skewness or kurtosis such as $\lambda_3 = \kappa_3/(\kappa_2)^{\frac{3}{2}}$, $\lambda_4 = \kappa_4/\kappa_2^2$. In the present instance, since we are dealing with differences in heights, it is reasonable to assume the distribution is symmetric. It would then seem that the calculation of $\phi(\beta)$, i.e. $p(\beta|\mathbf{y})$ for uniform $p(\beta)$ as shown in Fig. 3 (i) would provide a much more satisfactory way of summarizing what the data has to tell us concerning the nature of the parent distribution from which the sample is drawn. It will be noted that in our approach, we have done more than merely 'test' the assumption of normality and then, in the absence of 'a significant' result, assume it. The information concerning β coming from the sample is included in the formulation itself and as we have seen in the case of Darwin's data it plays an important role in virtually eliminating the influence of unlikely parent distributions.

APPENDIX

In § 6 we have asserted certain properties of the posterior distribution $p(\theta|\mathbf{y},\beta_0)$ in the permissible range of β. That they are so follows essentially from work on the median of a set of numbers by Jackson (1921) some 40 years ago. For the class of parent distributions given by equation (5), the maximum likelihood estimate of θ for fixed β_0, which is the mode of $p(\theta|\mathbf{y},\beta_0)$, was considered for certain specific choices of β_0 by Turner (1960), who seems to have been unaware of the much more general result obtained by Jackson. In our notation, consider the function:

$$M(\theta) = \sum_{i=1}^{n} |y_i-\theta|^{2/(1+\beta)} \quad (-\infty < \theta < \infty, \ -1 < \beta < 1).$$

For convenience, let us denote $q = 2/(1+\beta)$ so that

$$M(\theta) = \sum_{i=1}^{n} |y_i-\theta|^{q} \quad (q > 1).$$

(1) We first show that

(*a*) $M(\theta)$ is continuous and has continuous first derivative, and

(*b*) $M(\theta)$ has a unique minimum which is attained in $[y_S, y_L]$.

To see (*a*), consider

$$g_i(\theta) = |\theta-y_i|^q \quad (i = 1, 2, \ldots, n).$$

Clearly $g_i(\theta)$ is continuous everywhere. Now,

for

$$\theta < y_i,$$

$$g_i'(\theta) = -q(y_i-\theta)^{q-1},$$

for

$$\theta > y_i,$$

$$g_i'(\theta) = q(\theta-y_i)^{q-1}$$

and as θ approaches y_i from both directions,

$$\lim_{\theta\uparrow y_i} g_i'(\theta) = \lim_{\theta\downarrow y_i} g_i'(\theta) = 0$$

which implies that $g_i'(y_i) = 0$. Since $q-1 > 0$, $g_i'(\theta)$ exists and is continuous everywhere. Our assertion (*a*) is proved since $M(\theta)$ is the sum of all $g_i(\theta)$. Let us now consider $M'(\theta)$. We see that when

$$\theta < y_S,$$

$$M'(\theta) = -q \sum_{i=1}^{n} (y_i-\theta)^{q-1} < 0$$

and when

$$\theta > y_L,$$

$$M'(\theta) = q \sum_{i=1}^{n} (\theta-y_i)^{q-1} > 0.$$

Thus, by a property of a continuous function, there exists at least one θ_0, $y_S \leqslant \theta_0 \leqslant y_L$, such that

$$M'(\theta_0) = 0.$$

Further, since $M'(\theta)$ is a monotonically increasing function of θ, we conclude that $M'(\theta)$ can vanish once and only once and that the extreme value of $M(\theta)$ must be a minimum. This completes the proof of assertion (*b*).

(2) It has been shown by Jackson that when q approaches 1, in the limit the value of θ which minimizes $M(\theta)$ is the median of the y_i's, if the latter is uniquely defined; and, if not, is some unique value between the middle two of the y_i's.

(3) We now show that, when q is arbitrarily large

$$\lim_{q\to\infty} [M(\theta)]^{1/q} = (h+|m-\theta|),$$

where

$$m = \tfrac{1}{2}(y_L+y_S), \quad h = \tfrac{1}{2}(y_L-y_S).$$

Proof. Consider a finite sequence of monotone increasing positive numbers $\{a_n\}$ and a number S such that

$$S = \left\{\sum_{i=1}^{n} a_i^q\right\}^{1/q}.$$

We can write

$$S = a_n\left\{\sum_{i=1}^{n} \left(\frac{a_i}{a_n}\right)^q\right\}^{1/q},$$

where

$$a_i/a_n \leqslant 1 \quad \text{for all } i.$$

Hence
$$\left(\frac{S}{a_n}\right) = \left\{\sum_{i=1}^{n}\left(\frac{a_i}{a_n}\right)^q\right\}^{1/q},$$
so that
$$\log\left(\frac{S}{a_n}\right) = \frac{1}{q}\log\left\{\sum_{i=1}^{n}\left(\frac{a_i}{a_n}\right)^q\right\}.$$
When $q \to \infty$,
$$\lim_{q\to\infty}\log\left\{\sum_{i=1}^{n}\left(\frac{a_i}{a_n}\right)^q\right\} = \log r, \quad \text{where} \quad 1 \leqslant r \leqslant n.$$
But this implies that
$$\lim_{q\to\infty}\log\left(\frac{S}{a_n}\right) = 0,$$
whence
$$\lim_{q\to\infty} S = a_n.$$
Thus, for any given value of θ, when q is arbitrarily large,
$$\begin{aligned}\lim_{q\to\infty}[M(\theta)]^{1/q} &= \sup|y_i-\theta| \\ &= \sup[|\theta-y_S|, |\theta-y_L|] \\ &= \begin{Bmatrix} h+(m-\theta) \\ h \\ h+(\theta-m)\end{Bmatrix} \quad \text{for} \quad \begin{cases}\theta < m, \\ \theta = m, \\ \theta > m.\end{cases}\end{aligned}$$
Hence,
$$\lim_{q\to\infty}[M(\theta)]^{1/q} = (h+|m-\theta|)$$
and the assertion is proved.

References

Box, G. E. P. (1953). A note on regions for test of kurtosis. *Biometrika*, **40**, 465–8.

Box, G. E. P. & Anderson, S. L. (1955). Permutation theory in the derivation of robust criteria and the study of departures from assumptions. *J. R. Statist. Soc.* B, **17**, 1–34.

Carlton, G. A. (1946). Estimating the parameters of a rectangular distribution. *Ann. Math. Statist.* **17**, 355–8.

Daniels, H. E. (1954). Saddlepoint approximations in statistics. *Ann. Math. Statist.* **25**, 631–50.

Diananda, P. H. (1949). Note on some properties of maximum likelihood estimates. *Proc. Camb. Phil. Soc.* **45**, 536–44.

Fisher, R. A. (1935). *The Design of Experiments.* Edinburgh: Oliver and Boyd.

Gayen, A. K. (1950). The distribution of the variance ratio in random samples of any size drawn from a non-normal universe. *Biometrika*, **37**, 236–55.

Jackson, D. (1921). Note on the median of a set of numbers. *Bull. Amer. Math. Soc.* **27**, 160–4.

Jeffreys, H. (1957). *Scientific Inference* (2nd edition). Cambridge University Press.

Jeffreys, H. (1961). *Theory of Probability* (3rd edition). Oxford: Clarendon Press.

Jeffreys, H. & Jeffreys, B. S. (1956). *Methods of Mathematical Physics* (3rd edition). Cambridge University Press.

Kolmogoroff, A. (1941). Confidence limits for an unknown distribution function. *Ann. Math. Statist.* **12**, 461–3.

Lindley, D. V. (1961). The robustness of interval estimates. *Bull. Int. Statist. Inst.* **38**, 209–20.

Neyman, J. & Pearson, E. S. (1928). On the use and interpretation of certain test criteria for purposes of statistical inference, Part I. *Biometrika*, **20** A, 175–240.

Savage, L. J. (1954). *The Foundation of Statistics.* New York: John Wiley and Sons.

Savage, L. J. (1959). Subjective probability and statistical practice. *Technical Note* 59-1161, Air Force Office of Scientific Research.

Savage, L. J. (1960). The foundation of statistics reconsidered. *Proceedings of the 4th Berkeley Symposium on Mathematical Statistics and Probability*, **1**, 575–86. Berkeley: University of California Press.

Turner, M. C. (1960). On heuristic estimation methods. *Biometrics*, **16**, 299–301.

1.7

A Note on Criterion Robustness and Inference Robustness*

G. E. P. BOX and GEORGE C. TIAO
University of Wisconsin

Introduction

Since many statistical methods make specific assumptions concerning the nature of the parent distribution, it has long been a concern of statisticians to determine how far conclusions might be affected if these assumptions were false. In particular, a considerable literature exists on the effect of non-normality on analysis of variance tests to compare means and on the test to compare the variances of independent samples—see for example Geary (1936), Gayen (1950) and Box & Andersen (1955). In the majority of this work, the behaviour under non-normality of the probability distribution of a specific criterion is considered. These investigations we refer to as investigations of *criterion* robustness to non-normality.

Recently, the authors have employed Bayes's theorem to reconsider the classical problem of comparing two means (1962) and of comparing two variances (1964). In these investigations, it is being supposed that observations are drawn from a class of parent distributions which are characterized by a location parameter θ, a scale parameter σ and a third parameter β. This parameter β can be thought of as measuring the degree of non-normality of the parent. For a specific value of β, the inference to be drawn about the parameter of interest (such as difference between the location parameters or ratio of scale parameters) is determined by the relevant posterior distribution for that fixed value of β. By studying the change in this conditional distribution as β is changed, we measure the sensitivity of the inference to non-normality. The degree of this sensitivity can be said to determine the robustness of the Bayesian inference in any given example.

We may thus distinguish between two types of sensitivity to non-normality: *criterion* robustness and *inference* robustness. While most previous work has concerned the former kind of robustness, the latter is perhaps more important. The distinction can be understood without reference to Bayes's theorem and although the Bayesian formulation is most appealing to the present authors, we shall in this paper from now on discuss and illustrate this distinction principally within the Neyman–Pearson formulation. In their approach, a criterion with optimal properties is selected, and the criterion robustness refers to the changes in its probability distribution when the parent distribution deviates from the form assumed. We could also consider the robustness of *inferences* to be drawn within the Neyman–Pearson formulation. These would concern changes, for example, in the significance level when appropriate changes were made in the nature of the criterion to correspond with the changes in the parent distribution. In the Bayesian formulation, this 'change in criterion' is automatically taken account of. Equivalent sampling results are not usually obtainable except for special parent distributions for which the relevant parameters happen to have sufficient statistics. However, in the very exceptional circumstances in which sufficient

* This research was supported by the Wisconsin Alumni Foundation.

The Collected Works of George E. P. Box, 1984, Wadsworth, Inc., Belmont, CA 94002.
Originally published in *Biometrika,* vol. 51, parts 1 and 2 (1964), pp. 169–173.

statistics are available over the whole spectrum of non-normal parents, results equivalent to those obtained from Bayes's formula may be derived from classical sampling theory. We shall illustrate our point, therefore, by considering a special example of this kind.

The example we select is that which concerns the comparison of two variances when the means are known. In this case, a simple family of non-normal distributions can be found for which sufficient statistics exist for the variance over the whole range of parent distributions. We discuss our results using data quoted in a book edited by Davies (1949, pp. 67–8). Our investigation illustrates the relationship between the two kinds of robustness.

The analyst example

The data consists of 20 independent observations made by an analyst A_1 and 13 made by an analyst A_2 in their assay of carbon in a mixed powder (see Box & Tiao, 1964). The object of the experiment was to compare the spread of results obtained by the analysts. This was done originally by assuming that the parent distributions were normal and referring the sample variance ratio to the usual F table. In fact, the means were unknown, but for the purpose of this demonstration we shall assume them known and equal to the sample means. The normal theory significance level, i.e. the probability of exceeding the observed variance ratio by chance, is then 7·6 %.

Table 1 shows the significance level calculated for this example under a number of different circumstances. The details of how these calculations were made are given later.

Table 1. *Changes in significance level induced by departure of the parental β from the β_0 defining the criterion*

β_0 \ β	−0·6	−0·4	−0·2	0·0	0·2	0·4	0·6
−0·6	**0·060**	0·110	—	—	—	—	—
−0·4	·035	**·065**	0·090	0·140	—	—	—
−0·2	·030	·048	**·070**	·098	0·115	0·170	0·220
0·0	·028	·045	·060	**·076**	·095	·120	·145
0·2	·025	·040	·050	·060	**·080**	·100	·120
0·4	·024	·035	·048	·060	·080	**·086**	·095
0·6	·024	·040	·048	·050	·065	·075	**·092**

In calculating the table, we assume the parent distributions to be members of the following class of power distributions employed in our previous papers,

$$p(y|\theta,\sigma,\beta) = k\exp\left\{-\frac{1}{2}\left|\frac{y-\theta}{\sigma}\right|^{2/(1+\beta)}\right\} \quad (-\infty<y<\infty) \tag{1}$$

with

$$k = [\Gamma\{1+\tfrac{1}{2}(1+\beta)\}\,2^{\{1+\frac{1}{2}(1+\beta)\}}\sigma]^{-1}, \quad (-\infty<\theta<\infty, \quad 0<\sigma<\infty, \quad -1<\beta<1).$$

When $\beta = 0$, we have the usual normal distribution; when $\beta = 1$, we have the double exponential distribution; and as β tends to the limit -1 the distribution tends to the rectangular. For intermediate values of β varying from -1 to $+1$, we obtain intermediate symmetrical distributions, platykurtic for $\beta < 0$ and leptokurtic for $\beta > 0$.

We suppose that the parents have common value of β and that θ_1 and θ_2, the means of both populations, are known. Then, for any fixed value of β, say $\beta = \beta_0$, the uniformly most powerful similar test for testing the hypothesis H_0: $\sigma_1^2/\sigma_2^2 = 1$ against the alternative H_1: $\sigma_1^2/\sigma_2^2 > 1$ is provided by the ratio

$$F(\beta_0) = \frac{\sum^{n_1} |y_{1i} - \theta_1|^{2/(1+\beta_0)}/n_1}{\sum^{n_2} |y_{2i} - \theta_2|^{2/(1+\beta_0)}/n_2} \tag{2}$$

in which the numerator and the denominator are sufficient statistics for the scale parameters σ_1 and σ_2, respectively. In particular, when $\beta_0 = 0$, this is the usual F criterion. Referring to the row in the table for $\beta_0 = 0$, the probabilities listed are the appropriate significance levels for the analysts example for various values of β in the parent distributions. For instance, 7·6 % is the normal theory significance level and the value 9·5 % immediately to the right of this indicates the change in probability that would occur for a somewhat more leptokurtic parent distribution with *the same normal theory criterion*. Thus, this row represents the sensitivity of this criterion to this particular type of changes in the parent distributions. Similarly, if we take the values corresponding to the next row for $\beta_0 = 0{\cdot}2$, we have the corresponding probabilities for the criterion

$$\frac{n_2}{n_1}\{\sum^{n_1} |y_{1i} - \theta_1|^{2/1\cdot 2} / \sum^{n_2} |y_{2i} - \theta_2|^{2/1\cdot 2}\}$$

which would provide a uniformly most powerful similar test for the parents in which $\beta = 0{\cdot}2$. In addition, the change from 7·6 to 8 % in the diagonal gives a measure of inference robustness. That is, specifically of how much the significance level changes when the parent distributions and *the appropriate criterion* are changed correspondingly.

Thus, in Table 1, inference robustness is measured by the way the diagonal elements change their values, while criterion robustness is measured by the changes as we read horizontally across the table. For this particular example, it is noticeable that the changes which occur along the diagonal are considerably less than those occurring horizontally. That is, the appropriate inference about the ratio of variances is less affected by changes in β than the probabilities associated with a particular criterion.

Too much should not be read into a single example, but it is noticeable that whereas the probability of the error of the first kind for the normal theory criterion is changed by a factor of 5, from 2·8 to 14·5 % in changing from a rather platykurtic distribution in which $\beta = -0{\cdot}6$ to a somewhat leptokurtic distribution in which $\beta = 0{\cdot}6$, it is changed only by a factor of 1·5 from 6·0 to 9·2 % when appropriate modification is made in the criterion. It is certainly true for this set of data that we would not be led far astray by 'assuming' normality so far as inference robustness is concerned. We may here remark that, as demonstrated in our earlier work (1964), the diagonal elements are precisely the *a posteriori* probabilities that the variance ratio σ_2^2/σ_1^2 exceeds unity for the corresponding values of β_0 in the parents, when $\log \sigma_1$ and $\log \sigma_2$ are taken to be locally uniformly distributed *a priori*.

Derivation of the entries in Table 1

It was shown in our work (1964) that for any specific value of β, say $\beta = \beta_0$, the uniformly most powerful similar criterion $F(\beta_0)$ in equation (2) follows an F distribution with $n_1(1+\beta_0)$ and $n_2(1+\beta_0)$ degrees of freedom when the hypothesis H_0: $\sigma_1^2/\sigma_2^2 = 1$ is true. Equivalently, the statistic

$$W(\beta_0) = \frac{\sum^{n_1} |y_{1i} - \theta_1|^{2/(1+\beta_0)}}{\sum^{n_1} |y_{1i} - \theta_1|^{2/(1+\beta_0)} + \sum^{n_2} |y_{2i} - \theta_2|^{2/(1+\beta_0)}} \tag{3}$$

has a beta distribution with parameters $\frac{1}{2}n_1(1+\beta_0)$ and $\frac{1}{2}n_2(1+\beta_0)$. Using this result, the exact probabilities in the diagonal of Table 1, namely, $\Pr\{F_{n_1(1+\beta_0),\, n_2(1+\beta_0)} > F(\beta_0)\}$, can be readily calculated. For the off-diagonal elements, we must find the distribution of the $F(\beta_0)$ criterion when the parent β takes some value other than β_0. This can be approximated using permutation theory.

From the population distribution given in (1), it is readily shown that the variate

$$X = \left|\frac{y-\theta}{\sigma}\right|^{2/(1+\beta_0)}$$

has as its rth moment (about the origin)

$$\mu_r = \frac{\Gamma\left[(1+\beta)\left(\frac{1}{2}+\frac{r}{1+\beta_0}\right)\right]}{\Gamma\{\frac{1}{2}(1+\beta)\}}\, 2^{r(1+\beta)/(1+\beta_0)}. \tag{4}$$

On the hypothesis H_0: $\sigma_1^2/\sigma_2^2 = 1$, the permutation moments of $W(\beta_0)$ can be written

$$W(\beta_0) = \frac{\sum_{i=1}^{n_1} X_{1i}}{\sum_{i=1}^{N} X_i}, \quad (N = n_1+n_2).$$

Following the derivation in Box & Andersen (1955), we have

$$\underset{p}{E}[W(\beta_0)] = \frac{n_1}{N}, \quad \underset{p}{V}[W(\beta_0)] = \frac{2n_1 n_2}{N^2(N+2)}\left\{1+\frac{1}{2}\frac{N}{N-1}(b_2-3)\right\}, \tag{5}$$

where
$$b_2 = (N+2)\frac{\sum^N X_i^2}{(\sum^N X_i)^2}.$$

Expanding the denominator of b_2 around the mean of X and taking expectation, we find that, to order N^{-2}

$$E(b_2-3) = \frac{\mu_2}{\mu_1^2}-3+N^{-1}\left[\frac{\mu_2}{\mu_1^2}-\frac{2\mu_3}{\mu_1^3}+\frac{3\mu_2^2}{\mu_1^4}\right]$$
$$-N^{-2}\left[\frac{\mu_2}{\mu_1^2}-\frac{2\mu_3}{\mu_1^3}-\frac{3}{\mu_1^4}(\mu_4-\mu_2^2)+\frac{16\mu_3\mu_2}{\mu_1^5}-\frac{15\mu_2^3}{\mu_1^6}\right], \tag{6}$$

where μ_r are given in (4). Thus, for a specific value of β, the statistic $W(\beta_0)$ is approximately distributed as a beta variable with parameters $[\frac{1}{2}n_1(1+\beta_0)\,\delta, \frac{1}{2}n_2(1+\beta_0)\,\delta]$, where

$$\delta = [1+\tfrac{1}{2}E(b_2-3)]^{-1}$$

represents the modification due to departure of β from β_0. Equivalently, $F(\beta_0)$ is approximately distributed as an F variable with $n_1(1+\beta_0)\,\delta$ and $n_2(1+\beta_0)\,\delta$ degrees of freedom. The above serves as a method from which the off-diagonal probabilities in Table 1 can be approximately determined.

REFERENCES

BOX, G. E. P. & ANDERSEN, S. L. (1955). Permutation theory in the derivation of robust criteria and the study of departures from assumptions. *J. R. Statist. Soc.* B, **17**, 1–34.

BOX, G. E. P. & TIAO, G. C. (1962). A further look at robustness via Bayes's theorem. *Biometrika*, **49**, 419–32.

BOX, G. E. P. & TIAO, G. C. (1964). A Bayesian approach to the importance of assumptions applied to the comparison of variances. *Biometrika*, **51**, 153–67.

DAVIES, O. L. *et al.* (1949). *Statistical Methods in Research and Production—with Special Reference to the Chemical Industry*. London: Oliver and Boyd.

GAYEN, A. K. (1950). The distribution of the variance ratio in random sample of any size drawn from non-normal universe. *Biometrika*, **37**, 236–55.

GEARY, R. C. (1936). The distribution of 'Student's' ratio for non-normal samples. *J. R. Statist. Soc.*, Suppl., **3**, 178–84.

1.8
The Experimental Study of Physical Mechanisms

G. E. P. BOX and WILLIAM G. HUNTER*
University of Wisconsin

This paper is concerned with the dual problem of generating and analyzing data in experimental investigations in which the goal is to develop a suitable mechanistic model. The problem is first distinguished from that of response surface methodology. With regard to the analysis of data, topics that are discussed include the behavior of estimated constants with an inadequate model, a diagnostic technique for model-building, and the importance of visual scrutiny of data. With regard to the generation of data, the concept of placing a model in jeopardy is discussed. Designs for model discrimination and for parameter estimation are considered.

1. Introduction

1.1 *The General Problem*

The search for underlying physical mechanisms constitutes a major portion of effort in many scientific and engineering fields. To engineers, for example, basic mechanism studies are of interest principally because a deeper understanding makes it possible to cope with engineering design problems in a more successful manner than would be possible if the mechanism were entirely unknown.

Suppose that an experimenter is interested in studying a particular system for which there exists a mathematical model $\eta = f(\boldsymbol{\theta}, \boldsymbol{\xi})$, that is perhaps non-linear in the parameters $\boldsymbol{\theta}$, which relates a measureable response η to the controllable variables $\boldsymbol{\xi}$. In a typical chemical engineering situation, for example, the response might be the rate of a chemical reaction, the variables might be partial pressures of the reactants and products, and the parameters might be adsorption equilibrium constants and an overall reaction rate constant. In practice, the quantities $\boldsymbol{\theta}$ are often referred to as constants rather than parameters.

The objective of the experimenter may be (1) to obtain an estimate of the response η over some particular region of interest in the space of the variables or (2) to determine the underlying physical mechanism of the phenomenon under investigation. Mathematically, we could say, for problem (2) the object is to discover the nature of the function $\eta = f(\boldsymbol{\theta}, \boldsymbol{\xi})$. In practical situations it is unlikely that we can ever know this completely. We shall say, however, that we have an adequate theoretical model when we have derived from a con-

* This research was supported by the National Science Foundation under Contract number GP 2755. Reproduction in whole or in part is permitted for any purpose of the United States Government.

The Collected Works of George E. P. Box, 1984, Wadsworth, Inc., Belmont, CA 94002.
Originally published in *Technometrics,* vol. 7, no. 1 (1965), pp. 23–42.

sideration of the mechanism a function which closely predicts the results of actual experiments.

In problem (1), which has come to be called the response surface problem, it is useful but not essential to employ such a theoretical model [7, 9, 11, 19]. In many circumstances even though no theoretical model is available, perfectly good empirical approximations can be obtained by fitting a polynomial or some other flexible graduating function over the region of interest [6, 12, 13, 14, 31, 34, 35, 36, 38]. Empirical models are, however, of limited value when the aim is to develop a suitable mechanistic theory. This paper is not concerned directly with problem (1) but rather with problem (2). Most statistical discussions begin by assuming that a model is known even though in practice, the model is usually unknown and the main problem is to build such a suitable model. The science of model-building has been a field neglected by most statistical authors. A notable exception, however, is the pioneering work of Cox on tests of separate families of hypotheses [21, 22].

1.2 *Cooperation between Experimenter and Statistician*

In model-building it is appropriate to speak of "tentatively entertaining" a given model rather than "assuming" it. This usage describes the attitude of the experimenter who suspects that a particular model may provide an adequate description of a system but bears in mind that the model may have to be modified or abandoned.

Data themselves cannot produce information; they can only produce information in the light of a particular model. But the model itself is not a fixed thing. Experimental results often show peculiarities which although meaningless to the statistician, might suggest to the experimenter a phenomenon other than that originally anticipated, that is, a change of model. To interpret the clues, therefore, one needs the cooperation of the expert in the subject matter field with the statistician. In some cases this is achieved by having the experimenter trained to act as his own statistician. A relevant example of the interplay between statistical analysis on the one hand and chemical engineering knowledge on the other is to be found in reference [28] where an appropriate model for the catalytic oxidation of methane was developed.

Here, then we are considering the problem of improving hypotheses iteratively, special emphasis being given to catalyzing the links between determining that a given model is inadequate, finding out *how* it is inadequate, and hence constructing a more appropriate model. The traditional statistical procedure of testing hypotheses has a more limited goal.

1.3 *The Iterative Nature of Experimentation*

The process of experimental investigation is an iterative one. In practice there may be backtracking, false starts, etc., but an underlying pattern can be seen. This involves the steps of conjecture (C), design (D), experiment (E), and analysis (A) continually repeated. As a result of some intuition, hypothesis, guess or *conjecture* the experimenter decides on a particular kind of plan or *design* of one or more experiments. These runs having been performed constitute the *experiment* the data from which is mulled over from many points of view. This is the *analysis* which frequently leads to a modification of ideas. Then a new and perhaps somewhat more realistic conjecture is made which begins the iterative cycle over again. The adaptive sequence CDEACDEA ··· may be repeated many times before an investigation is complete.

Experimentation is thus essentially a dynamic process. Design leads to analysis via experiment and analysis leads to new design via conjecture. In parts 2 and 3 of this paper we will discuss design and analysis procedures developed in the context of this general framework.

2. Analysis of Data

Before discussing the generation of data we shall first consider three topics concerned with the analysis of data: (a) the behavior of estimated constants with an inadequate model, (b) a diagnostic technique for model-building, and (c) visual scrutiny of data. For a general account of statistical analysis of experimental data from mechanism studies the reader is referred to reference [9].

2.1 *Behavior of Estimated Constants with an Inadequate Model*

In the physical and biological sciences a mechanistic model is often expressed most naturally in terms of one or more time- or space-dependent differential equations. For example, we may describe some phenomenon by the time-dependent differential equation

$$\frac{d\eta}{dt} = \phi(\boldsymbol{\theta}; \eta, \boldsymbol{\xi}, t) \tag{1}$$

subject to certain boundary conditions. Upon integration this equation may yield

$$\eta = f(\boldsymbol{\theta}; \boldsymbol{\xi}, t) \tag{2}$$

In conducting experiments on such a system it is usually most convenient to set the variables $\boldsymbol{\xi}$ at some fixed set of values and make a "run" in which observations are made at m specified values of t; the values of $\boldsymbol{\xi}$ are then reset and a second run is made in which observations are taken once again at the same values of t; and so on. Let y_{ui} denote the i-th observation of the u-th run. Suppose that the method of least squares is appropriate and we can obtain a set of such estimates

$$\hat{\boldsymbol{\theta}}_u' = (\hat{\theta}_{1u}, \hat{\theta}_{2u}, \cdots, \hat{\theta}_{pu}) \tag{3}$$

for the p parameters from the m observations $y_{u1}, y_{u2}, \cdots y_{um}$ in *the u-th run.* Suppose that n runs and $N = nm$ observations are made in all.

By fitting *the complete set of data* suppose we obtain least squares estimates

$$\hat{\boldsymbol{\theta}}' = (\hat{\theta}_1, \hat{\theta}_2, \cdots, \hat{\theta}_p). \tag{4}$$

We show in Appendix A how specific inadequacies of the model may be detected by studying the dependence of the discrepancies $\hat{\boldsymbol{\theta}}_u - \hat{\boldsymbol{\theta}}$ on the levels $\boldsymbol{\zeta}_u$ of the variables.

2.2 *A Diagnostic Technique for Model-Building*

In any least squares analysis, of course, it is possible to check the lack of fit by examining the residuals and, in particular, the residual sum of squares. The object of our analysis, however, is not only to establish whether there is lack of fit but, if so, to provide some indication of its nature and so to help the experimenter decide how to modify the model.

As we have stated in the previous section, when the model is inadequate the within-run least squares estimates $\hat{\boldsymbol{\theta}}_u$ may vary as the levels of the experimental variables are changed. In order to diagnose the source of lack of fit it is important to obtain accurate information concerning which $\hat{\theta}$'s are dependent on which ξ's. To allow best estimation and separation of such effects it is useful to run a formal design such as a factorial in the ξ's.

In the simplest application of this technique, if there were no experimental error and the model were adequate, then all the estimates $\hat{\boldsymbol{\theta}}_u$ would be equal, i.e., constant from run to run. We can, therefore, proceed as follows. First, the adequacy of the model with respect to each individual run is checked by examining residuals within runs. Then a statistical analysis is applied to the estimated quantities $\hat{\boldsymbol{\theta}}_u$ to see whether there is any evidence that they are related to the level of the ξ's.

If a factorial design is employed such an analysis is particularly easy. The usual main effects and interactions are calculated treating the estimated parameters as observations. The existence of non-zero main effects and interactions would indicate via the above results the need for modification.

In other than the simplest applications, rather than no effect being expected it may be that an effect of a specific type would be expected from a particular model. For example, when an Arrhenius type model was used the logarithm of the rate constant would be expected to be linearly related to the reciprocal of the absolute temperature. Thus, a main effect would be expected in the analysis of the estimated parameter corresponding to temperature (it would, in fact, be related to the activation energy) but there would be no interaction effects with, for example, the concentrations of the reactants. This diagnostic technique for model-building has been discussed in references [15] and [28] and will here be illustrated with an example taken from the second of these references.

An Example

The catalytic oxidation of methane

$$CH_4 + 2O_2 \xrightarrow[\text{catalyst}]{} CO_2 + 2H_2O$$

was investigated. At one stage the mechanism entertained was that gaseous methane and adsorbed oxygen react on the catalyst surface to produce gaseous carbon dioxide and adsorbed water. On the assumption that oxygen is adsorbed on adjacent dual sites the following mathematical model was appropriate

$$t = \int_0^{\eta} \frac{(\theta + \theta'\eta)^2}{\xi_1(1 - \eta)(\xi_2 - 2\xi_1\eta)^{\frac{1}{2}}}\, d\eta \qquad (6)$$

where

$$\theta = \beta_0 + \beta_2\xi_2 + \beta_3\xi_3$$
$$\theta' = (\beta_3 - 2\beta_2)\xi_1 \tag{7}$$

and

$$\beta_0 = \frac{1}{\sqrt{kK_1}}$$

$$\beta_2 = \frac{1}{\sqrt{k}}$$

$$\beta_3 = \frac{K_2}{\sqrt{kK_1}}$$

with

t = (weight of catalyst)/(mass flow of gas)
ξ_1 = mole fraction of methane
ξ_2 = mole fraction of oxygen
ξ_3 = mole fraction of carbon dioxide
η = fractional conversion of methane
k = overall surface reaction rate constant
K_1 = adsorption equilibrium constant for oxygen
K_2 = adsorption equilibrium constant for carbon dioxide

It will be noted that the variable ξ_4, the mole fraction of water vapor, does not appear in model 6 at all, although this could be an important factor if certain plausible rival mechanisms were appropriate. This model, then, implies in particular that the parameter θ, although linearly related to the variables ξ_2 and ξ_3, should be unrelated to the variables ξ_1 and ξ_4. A 2^{4-1} fractional factorial design run in the variables ξ_1, ξ_2, ξ_3, and ξ_4 allowed us to discover experimentally with which of the variables the parameter θ was related. For each of the 8 sets of factor combinations of the design, observations of η were made at three distinct level of t.

It was necessary to realign the apparatus when the variable t was changed as well as when the variables ξ_1, ξ_2, ξ_3, and ξ_4 were changed so that for this investigation the 24 distinct experiments could without inconvenience be made completely randomly in time order. From these 24 experiments, eight η vs. t curves could be fitted from which eight individual estimates $\hat{\theta}_u$ could be readily obtained. Furthermore, a reasonably reliable estimate of experimental error could be obtained from the deviations from the fitted curves.

It is now a simple matter to fit

$$\tilde{\theta} = \hat{\beta}_0 + \hat{\beta}_2\xi_2 + \hat{\beta}_3\xi_3$$

which if the model were adequate, would completely account for the variation in $\hat{\theta}$. In this expression $\hat{\beta}_0$ would simply be the average of the $\hat{\theta}$'s while $\hat{\beta}_2$ and $\hat{\beta}_3$ would be one half of the "main effects" associated with ξ_2 and ξ_3. Our model implies that only these "effects" ought to be found. In particular there should be no "effects" associated with ξ_1 or ξ_4 and no interactions of any kind. In fact, the discrepancies $\hat{\theta} - \tilde{\theta}$ shown in Table 1 indicate when experimental error is taken into account that this is not the case and hence that the model is inadequate.

It will be noticed that the signs of $\hat{\theta} - \tilde{\theta}$ are perfectly correlated with the signs of variable ξ_4 , water vapor. This particular model, then, does not take the concentration of water vapor properly into account. Since it had at the beginning of this stage been supposed that the reaction yields non-adsorbed water vapor, a modified model which considers adsorbed water vapor was a logical next step. After further analysis it was shown that a model of this kind was indeed adequate.

TABLE 1

Intermediate Analysis of Data from Study on the Catalytic Oxidation of Methane.

Run	Design				Discrepancies
	ξ_1	ξ_2	ξ_3	ξ_4	$(\hat{\theta}_u - \tilde{\theta}_u) \times 10^4$
1	−1	−1	−1	+1	+3.63
2	+1	−1	−1	−1	−2.88
3	−1	+1	−1	−1	−6.68
4	+1	+1	−1	+1	+5.28
5	−1	−1	+1	−1	−5.13
6	+1	−1	+1	+1	+4.07
7	−1	+1	+1	+1	+6.27
8	+1	+1	+1	−1	−4.56

2.3 *Visual Scrutiny of Data*

The value of graphs and diagrams cannot be overemphasized. It has been our experience that research workers do not exploit visual devices as much as they should. We summarize here some of the more important kinds of graphs and diagrams that have been found helpful.

Residuals

The residuals are defined as the discrepancies

$$y_i - \hat{y}_i \qquad (j = 1, 2, \cdots, N)$$

where

y_i = observed value of response
$\hat{y}_i$ = fitted value of response.

If the model is adequate the residuals will be manifestations of completely random variation apart from restrictions imposed by the analysis. An example of this kind of restriction is that the sum of the residuals from an arithmetic average always equals zero.

All the information relating to the possible inadequacy of a tentatively entertained model is contained in the residuals. Plots of residuals, therefore, can reveal particular aspects of the model that should be improved. Consequently, as a matter of course, residuals should always be plotted in any way that might shed light on pertinent questions [2, 9]. To list some examples, the residuals might be plotted against (1) the time order in which the experiments were performed, (2) the level of each of the variables, and (3) the predicted value of the response. (Two plots illustrating case (3) are presented in Anscombe [2]). The question being asked in each case might be (1) Is there a time trend in the results? (2) Does the model fail to take any of the variables properly into account? Is there evidence of a quadratic trend? (3) Does the variability increase as the magnitude of the predicted value of the response increases? Is there a quadratic dependence between the residuals and the predicted values of the response suggesting lack of fit which can be cured by transforming y? [19A]

Dot diagrams or (if enough observations are available) histograms can be constructed to check on the distribution of the residuals to see if it has any unusual characteristics. If there is one residual that is much larger in magnitude than the others, for example, it will deserve further attention. There may have been a mistake in carrying out that particular experiment (e.g., the wrong reagent was used, the variables were set at grossly incorrect levels, or the measuring instrument was out of calibration) or there may have been a copying or an arithmetic error. On the basis of a careful check for blunders of this kind the response for that run may be accepted as is, corrected, or rejected. The literature on outliers is relevant here. (See [1], [23], [26], [27], [31], [32], [40], and the references listed therein).

Perhaps no blunder has occurred but rather the apparently anomalous observation can be explained in some other way, e.g., by a mechanism other than the one currently being considered. The value of a single observation in this way can sometimes be greater than all the other observations in model-building situations because it leads the experimenter to new and valuable conjectures.

Contour Diagrams

Plotting contour diagrams can also be helpful in summarizing the main features of a model. In such diagrams lines (or surfaces) of constant response levels are plotted as functions of the experimental variables. Contour diagrams are often useful when more than one response is of importance. In such cases the individual diagrams can be overlaid. Examples of plots showing more than one response are contained in references [34], [35], [36], and [38].

Likelihood Function

Diagrams can also be extremely helpful in gaining a better understanding of the estimation situation. Point estimates of parameters are unsatisfactory without some indication of their precision. One effective way of displaying this information is to plot the likelihood function, as was advocated by Fisher and later reemphasized by Barnard, Jenkins and Winsten [3] and Box [9].

If the errors are Normally and independently distributed with constant variance σ^2 the log likelihood $l(\boldsymbol{\theta})$ is simply a linear function of the sum of squares $S(\boldsymbol{\theta})$. That is, with m observations,

$$l(\boldsymbol{\theta}) = -\frac{m}{2} \ln 2\pi\sigma^2 - \frac{1}{2\sigma^2} S(\boldsymbol{\theta})$$

where

$$S(\boldsymbol{\theta}) = \sum_{i=1}^{m} \{y_i - f(\boldsymbol{\theta}, \xi_i)\}^2$$

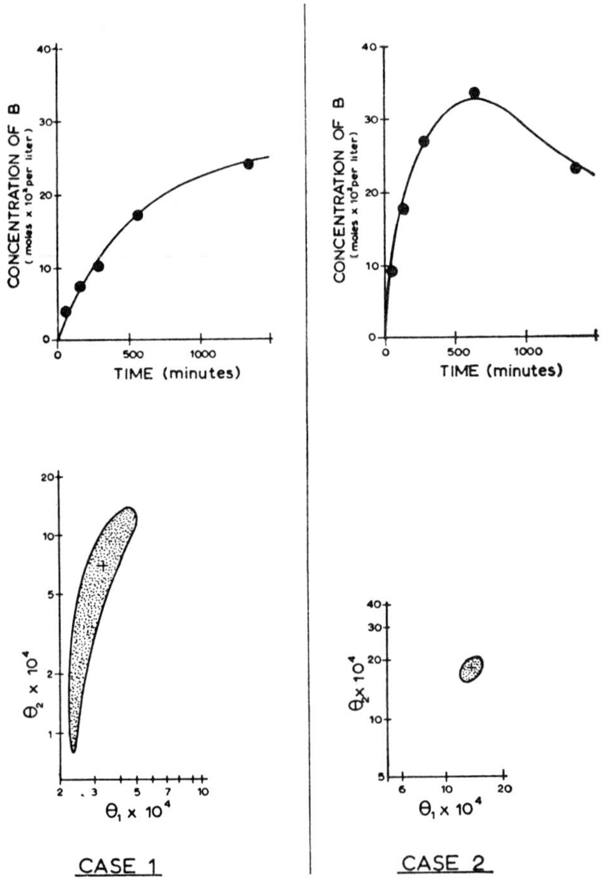

FIGURE 1

Typical Fitted Curves and Corresponding Approximate 95% Confidence Regions for Two Different Situations

Therefore, if there are only two or three parameters a contour diagram of the likelihood (or, equivalently, the sum of squares) function can be readily produced.

For example, suppose that chemical reaction of the type $A \rightarrow B \rightarrow C$ were being studied and the concentration of $[B]$ were measured at five different times t. The models being entertained might be of the form

$$\eta = \frac{[A]_0\theta_1}{\theta_1 - \theta_2}(e^{-\theta_2 t} - e^{-\theta_1 t})$$

where

η = concentration of B at time t
$[A]_0$ = initial concentration of A

and θ_1 and θ_2 are first order rate constants to be estimated. Figure 1 shows typical runs that might have been made with two different reactions of this kind. (These are, in fact, runs 1 and 13 in reference [15].)

The contours shown are those for which the sum of squares $\sum_{i=1}^{5}(y_{ui} - \eta_{ui})^2$, considered as a function of θ_1 and θ_2, is equal to the constant

$$S_{\min} + s^2 p F_{0.05}(2,48)$$

where

S_{min} = minimum value of the sum of squares $S(\boldsymbol{\theta})$
s^2 = an estimate of the within run variance associated, in this example, with 48 degrees of freedom
p = 2, the number of parameters estimated
$F_{0.05}(2,48)$ = the usual 5% value of the F criterion with 2 and 48 degrees of freedom.

If the model were linear the regions so defined would be 95% confidence regions, or, under suitable assumptions, 95% Bayesian regions, (see Appendix B).

These contours agree with common sense. They clearly show, for example, that if data are taken only over the early part of such a curve, relatively imprecise and highly correlated estimates are obtained, whereas if the data "cover" the major features of the curve, markedly better estimates can be obtained. Many choices of the parameters which can match the beginning part of the curve are eliminated when more of the curve is covered by experimental runs. This is a simple example which clearly illustrates the importance of good design in obtaining reliable estimates of parameters.

This graphical approach can be extended to three or four parameters by exhibiting a "grid" of two-dimensional (θ_1, θ_2) plots for various combinations of θ_3 and θ_4. An instrument such as the "Calcomp" plotter can produce such plots automatically from digital computer output and such plotting should be part of the normal stock in trade of the practicing statistician. The technique immediately shows up peculiarities of an estimation situation (for example: non-stationary maxima on a boundary of the permissible parameter space and excessive non-linearity evidenced by non-elliptic contours). When each of several responses contribute information about a particular set of parameters overlaid plots of this kind are particularly valuable [19B].

3. Design of Experiments

Of the two, design and analysis, the former is undoubtedly of greater importance. The damage of poor design is irreparable; no matter how ingenious the analysis, little information can be salvaged from poorly planned data. On the other hand, if the design is sound, then even quick and dirty methods of analysis can yield a great deal of information. For further discussion of this point see [8].

3.1 *Placing a Model in Jeopardy*

In many instances the experimental design is such that even when serious inadequacies exist they are undetectable. Yates [42] was referring to this fact when he stated that "nothing is easier than to 'prove' that a hypothesis is true by testing it by an experiment which is sufficiently inaccurate". It should be understood that the difficulty here is not merely associated with possible inaccuracy in the observations. Even with fairly accurate observations experiments can be of such a type that they are insensitive to departures from the model. Although they do not disprove a particular model they do not disprove a vast number of other models either.

For example, suppose a model represents the concentration η of a chemical product as a function of time t and a number of parameters $\theta' = (\theta_1, \theta_2, \cdots \theta_p)$. Suppose that $\eta = 0$ at time $t = 0$. If observations are made only for small values of t over which period the curve starting from the origin is roughly a straight line, then clearly it could represent the beginning of almost any kind of reaction. Undoubtedly, values of the θ's can be found which will fit the data with this particular type of model but this would be true for many other models as well. Specifically, the model being fitted could be wrong and yet this fact could go unrecognized because of the nature of the experimental design. In such a case we shall say that the experimental design has been such as to never *place the model in jeopardy*. The common practice in kinetic studies of analysing only initial rates of chemical reaction is particularly vulnerable to this criticism.

3.2 *Designs for Model Discrimination*

Suppose that an experimenter is studying a chemical reaction in which a reactant A is used to make a product B. Simultaneously, however, an undesired by-product C is formed. Suppose further that the experimenter knows that the reaction is one of the following:

Mechanism 1: $A \rightarrow B \rightarrow C$
Mechanism 2: $A \rightarrow B \rightleftarrows C$

Typical curves showing the concentration of the product B as a function of time are shown in Figure 2. If one of these mechanisms is actually correct and all experiments consist of measurements of the concentration of B at times less than 10 minutes, then obviously it will be virtually impossible to distinguish between Mechanism 1 and 2. Observations are needed at times of the order of 100 minutes.

Where to place experiments to discriminate most efficiently between two models will depend in general on the values of the estimated parameters. A sequential scheme, therefore, suggests itself in which at each stage the two models are refitted and a decision is made as to what experiment should be run next. The feasibilitv of such an iterative procedure is presently under investigation.[28A]

3.3 *Designs for Parameter Estimation*

If the form of the theoretical model is known, the problem which confronts the experimenter is to evaluate the physical parameters (e.g., rate constants in chemical kinetics examples). In this section, we consider the problem of the generation of data in this situation.

If experiments are not carefully planned the experimental points may be so situated in the space of the variables that the estimates which can be obtained for the parameters $\boldsymbol{\theta}$ are not only imprecise but also highly correlated. Once the data are collected a statistical analysis, no matter how elaborate, can do nothing to remedy this unfortunate situation. However, by selecting a suitable experimental design in advance these shortcomings can often be overcome. Although the problem of designing experiments in non-linear situations has received comparatively little attention, some possible approaches have been suggested. See, for example, references [17], [20], [25], [29], [30], [39] and [41]. Blakemore and Hoerl [5] made a rather strong appeal for further work in this important area.

Box and Lucas [17], whose approach has since been applied, for example, to chemical problems by Behnken [4], proceeded by attempting to choose a design D in such a manner that the volume of the approximate confidence region for $\boldsymbol{\theta}$ is minimized, or, equivalently, under suitable assumptions, trying to choose D to minimize the volume in the parameter space which contains a given percentage of the posterior distribution. If the experimental errors are approximately Normally distributed and the response relationship is approximately linear in the vicinity of the least squares estimates $\hat{\boldsymbol{\theta}}$, then the volume of this region is proportional to the square root of the reciprocal of the determinant $\Delta = X'X$ where $X = \{x_{ru}\}$ and

$$x_{ru} = \left\{\frac{\partial f(\boldsymbol{\theta}, \boldsymbol{\xi}_u)}{\partial \theta_r}\right\}_{\boldsymbol{\theta}=\hat{\boldsymbol{\theta}}}$$

Unfortunately, since we do not know the values of $\hat{\boldsymbol{\theta}}$ in advance, we do not know the derivatives x_{ru} on which the design is to be based. In most cases, however, some knowledge of the size of the parameters will be available and it was suggested [17] that preliminary guesses $\boldsymbol{\theta}^0$ should be made, and that the derivatives should be determined at these values $\boldsymbol{\theta}^0$ instead of $\hat{\boldsymbol{\theta}}$. The resulting determinant $\Delta^0 = X^{0\prime}X^0$ is an explicit function of the settings of the experimental variables $\boldsymbol{\xi}$. It is therefore possible to find (perhaps analytically but, in any event, numerically) those values for $\boldsymbol{\xi}$ which maximize the determinant Δ^0.

At first sight it may seem strange that in order to use this scheme one must initially have estimates of the parameters since, after all, it is the purpose of the experiment to obtain such estimates. Actually, however, this is merely an example of the fact that any experimental design uses the experimenter's beliefs about the situation being studied. It is thus efficient depending on whether the experimenter turns out to be nearly right. In general, the more one knows initially the better he can design experiments. As has been pointed out previously [9], if nothing is known about the experimental situation then strictly speaking no experiment can be planned.

Since the best design depends upon the parameters that are to be estimated,

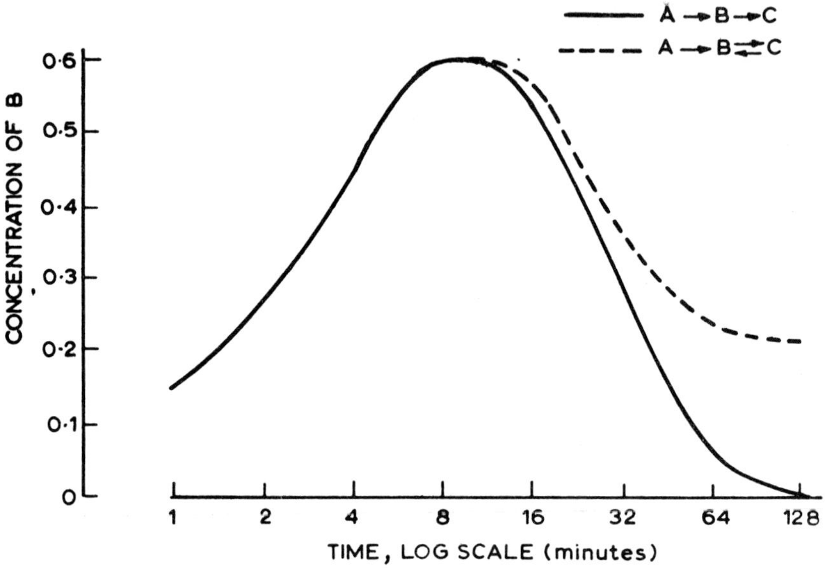

FIGURE 2

Two Curves Representing the Concentration of B as a Function of Reaction Time for the Two Mechanisms $A \rightarrow B \rightarrow C$ and $A \rightarrow B \rightleftarrows C$

it was suggested [16] that a sequential plan be employed in which at each stage the parameters be reestimated. With these current estimates a calculation would be made to determine where the next experiment should be run. The feasibility of this method was demonstrated with an example based on the catalytic dehydration of n-hexyl alcohol [33], the model being

$$\eta = \frac{\theta_3\theta_1\xi_1}{1 + \theta_1\xi_1 + \theta_2\xi_2}$$

where

η = true rate of reaction
θ_1 = adsorption equilibrium constant for alcohol
θ_2 = adsorption equilibrium constant for olefin
θ_3 = effective reaction rate constant
ξ_1 = partial pressure of alcohol
ξ_2 = partial pressure of olefin

Data were constructed on the assumption that the true parameter values were

$$\theta_1 = 2.9 \qquad \theta_2 = 12.2 \qquad \theta_3 = 0.69$$

and that

$$y = \eta + \epsilon$$

where

y = observed rate of reaction
η = true rate of reaction
ϵ = error

The error ϵ was taken to be Normally and independently distributed with zero mean and standard deviation $\sigma = 0.01$. It was further supposed that neither partial pressure could be set greater than three atmospheres so that the experimental region was that shown in Figure 3.

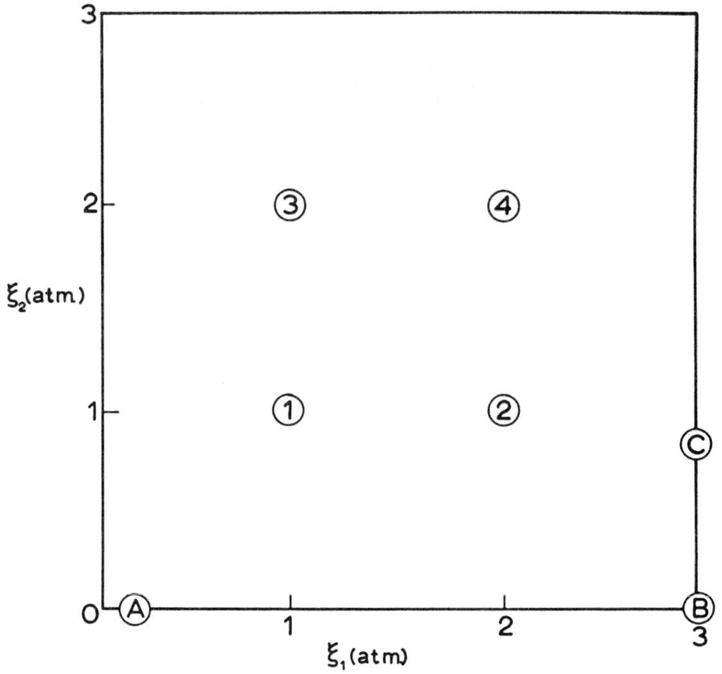

FIGURE 3

Location of Experimental Points in the Factor Space for a Sequentially Designed Set of Runs for the Model $\eta = \theta_3\theta_1\xi_1/(1 + \theta_1\xi_1 + \theta_2\xi_2)$

An initial set of four experiments was "run" according to a standard 2^2 factorial pattern. These runs are indicated by the circled numbers 1, 2, 3 and 4 in the figure. At each successive stage thereafter, the experimenter supplied a digital computer with (a) the model, (b) the data, and (c) the current least squares estimates of the parameters of the model; and the computer in turn produced (a) the new least squares estimates, (b) the value of the determinant criterion as a function of various choices of contemplated settings of the variables, and (c) the experimental conditions for the next experiment which maximized the determinantal criterion. The next nine runs fell near the three areas denoted by circled letters A, B, and C in the following order:

ABABACBAC

The procedure can be modified by considering, for example, special information about the dispersion matrix.

4. Conclusion

This paper has been concerned with some statistical aspects of the dual problem of generating and analyzing data in experimental investigations in which the goal is to develop or build a suitable mechanistic model. Such investigations consist logically of two steps: (i) establishing the form of an adequate model and then (ii) estimating precisely the values of its parameters. Fisher, incidentally, referred to these as the problems of (i) specification and (ii) estimation. For the first step one of the methods mentioned for placing the model in jeopardy might be used, and for the second step one of the suggestions discussed in Section 3.3 might be used. The aim of the first type of design would be, by straining a tentative model, to iterate towards a satisfactory formulation, and the aim of the second type would be to obtain improved estimates of the parameters once a suitable form of the function is established.

There may be better methods, however, for dealing with this two-part problem. One might suppose that a class of sequential designs could exist which would permit an attack on both problems from the outset of the investigation and would facilitate the transition from step (i) to (ii).

5. Appendix A
Behavior of Estimated Constants with an Inadequate Model

The general setup is that given in Section 2.1. Recall that the estimates of the parameters from the observations $y_{u1}, y_{u2}, \cdots, y_{um}$ in *the u-th run* are denoted by $\hat{\boldsymbol{\theta}}_u$ and the estimates obtained from fitting *the complete set of data* are denoted by $\hat{\boldsymbol{\theta}}$. Suppose that

$$r_{ui} = y_{ui} - f(\boldsymbol{\theta}^0; \boldsymbol{\xi}_u, t_i)\ {}_{i=1,2,\cdots,m}^{u=1,2,\cdots,n} \tag{A-1}$$

where $\boldsymbol{\theta}^0$ is some convenient point in the parameter space in the neighborhood of the point representing estimates and that

$$x_{ui}^{(i)} = \left[\frac{\partial f(\boldsymbol{\theta}; \boldsymbol{\xi}_u, t_i)}{\partial \theta_j}\right]_{\boldsymbol{\theta}=\boldsymbol{\theta}^0} \tag{A-2}$$

Then if linear approximation suffices over the relevant region our model may be written

$$E(r_{ui}) = \sum_{j=1}^{p} (\theta_j - \theta_j^0) x_{ui}^{(j)}. \tag{A-3}$$

All the information on the adequacy of the model is contained in the N residuals whose sum of square is

$$S_R = \sum_{u=1}^{n} \sum_{i=1}^{m} \left\{ r_{ui} - \sum_{j=1}^{p} (\hat{\theta}_j - \theta_j^0) x_{ui}^{(j)} \right\}^2 \tag{A-4}$$

which is associated with $N-p$ degrees of freedom.

Decomposition of the Residuals

Now

$$S_R = \sum_{u=1}^{n}\sum_{i=1}^{m}\left\{\left[r_{ui} - \sum_{j=1}^{p}(\hat{\theta}_{ju} - \theta_j^0)x_{ui}^{(j)}\right] - \left[\sum_{l=1}^{p}(\hat{\theta}_l - \hat{\theta}_{lu})x_{ui}^{(l)}\right]\right\}^2 \quad \text{(A-5)}$$

$$= \sum_{u=1}^{n}\sum_{i=1}^{m}\left\{r_{ui} - \sum_{j=1}^{p}(\hat{\theta}_{ju} - \theta_j^0)x_{ui}^{(j)}\right\}^2 + \sum_{u=1}^{n}\sum_{j=1}^{p}\sum_{l=1}^{p}(\hat{\theta}_j - \hat{\theta}_{ju})(\hat{\theta}_l - \hat{\theta}_{lu})c_{ujl} \quad \text{(A-6)}$$

where

$$c_{ujl} = \sum_{i=1}^{m} x_{ui}^{(j)}x_{ui}^{(l)}$$

since

$$\sum_{u=1}^{n}\sum_{l=1}^{p}(\hat{\theta}_l - \hat{\theta}_{lu})\sum_{i=1}^{m}\left\{r_{ui} - \sum_{j=1}^{p}(\hat{\theta}_{ju} - \theta_j^0)x_{ui}^{(j)}\right\}x_{ui}^{(l)} = 0 \quad \text{(A-7)}$$

The residual sum of squares may thus be decomposed into two parts S_W and S_B which add to give S_R. The within-run residual sum of squares S_W is

$$S_W = \sum_{u=1}^{n}\left\{\sum_{i=1}^{m} r_{ui} - \sum_{j=1}^{p}(\hat{\theta}_{ju} - \theta_j^0)x_{ui}^{(j)}\right\}^2 \quad \text{(A-8)}$$

This, in turn, can be split into n separate sums of squares associated with the n separate runs. The u-th such sum of squares is

$$S_{W_u} = \sum_{i=1}^{m}\left\{r_{ui} - \sum_{j=1}^{p}(\hat{\theta}_{ju} - \theta_j^0)^2x_{ui}^{(j)}\right\}^2 \quad \text{(A-9)}$$

Lack of fit in the u-th run will be indicated by an inflation of this sum of squares. The between run sum of squares is

$$S_B = \sum_{u=1}^{n}\sum_{j=1}^{p}\sum_{l=1}^{p}(\hat{\theta}_{ju} - \hat{\theta}_j)(\hat{\theta}_{lu} - \hat{\theta}_l)c_{ujl} \quad \text{(A-10)}$$

We notice that this second quadratic form focuses attention on the changes in the estimates of the θ's which occur from run to run, that is to say as the levels of the variables are changed.

Model Adequate

Let the $m \times p$ matrix of the derivatives $x_{ui}^{(j)}$ in the u-th run be denoted by X_u and the corresponding $m \times 1$ vector of the elements $r_{u1}, r_{u2}, \cdots, r_{um}$ by r_u. Then

$$(\hat{\boldsymbol{\theta}}_u - \boldsymbol{\theta}^0) = (X_u'X_u)^{-1}X_u'r_u \quad \text{(A-11)}$$

Now the expected values of these estimates are

$$E(\hat{\boldsymbol{\theta}}_u - \boldsymbol{\theta}^0) = (X_u'X_u)^{-1}X_u'E(r_u) \quad \text{(A-12)}$$

and if the model is adequate

$$E(r_u) = X_u(\boldsymbol{\theta} - \boldsymbol{\theta}^0) \quad \text{(A-13)}$$

whence

$$E(\hat{\boldsymbol{\theta}}_u - \boldsymbol{\theta}^0) = \boldsymbol{\theta} - \boldsymbol{\theta}^0. \tag{A-14}$$

Thus on our assumptions the estimates $\hat{\boldsymbol{\theta}}_u$ are unbiassed estimates of $\boldsymbol{\theta}$ and in particular they do not depend on the levels of the variables $\boldsymbol{\xi}$.

Model Inadequate

Now suppose that the true form of the function is not $\eta = f(\boldsymbol{\theta}; \boldsymbol{\xi}, t)$ but

$$\eta = \zeta(\boldsymbol{\psi}; \boldsymbol{\xi}, t) \tag{A-15}$$

Then

$$E(\hat{\boldsymbol{\theta}}_u - \boldsymbol{\theta}^0) = (X_u'X_u)^{-1}X_u'\Delta_u \qquad u = 1, 2, \cdots, n \tag{A-16}$$

where the i-th element of the $m \times 1$ vector Δ_u is

$$\zeta(\boldsymbol{\psi}; \boldsymbol{\xi}_u, t_i) - f(\boldsymbol{\theta}^0; \boldsymbol{\xi}_u, t_i) = \Delta(\boldsymbol{\psi}, \boldsymbol{\theta}^0; \boldsymbol{\xi}_u, t_i) \tag{A-17}$$

We note that now the elements in X_u and in Δ_u on the right hand side of Equation A-16 all depend on $\boldsymbol{\xi}$.

Specifically, suppose we choose a particular central point $\boldsymbol{\xi}^0$ in the space of the variables about which the levels $\boldsymbol{\xi}_u$ vary in successive experiments. Then expanding about this value we can write

$$X_u = \left[\frac{\partial f(\boldsymbol{\theta}; \boldsymbol{\xi}_u, t_i)}{\partial \theta_j}\right]_{\boldsymbol{\theta}=\boldsymbol{\theta}^0} \doteq A + \sum_{h=1}^{k} (\xi_{hu} - \xi_h^0)B_h \qquad \begin{aligned} i &= 1, 2, \cdots, m \\ j &= 1, 2, \cdots, p \\ u &= 1, 2, \cdots, n \end{aligned} \tag{A-18}$$

where

$$A = \left[\frac{\partial f(\boldsymbol{\theta}; \boldsymbol{\xi}, t_i)}{\partial \theta_j}\right]_{\substack{\boldsymbol{\theta}=\boldsymbol{\theta}^0 \\ \boldsymbol{\xi}=\boldsymbol{\xi}^0}} \tag{A-19}$$

$$B_h = \left[\frac{\partial^2 f(\boldsymbol{\theta}; \boldsymbol{\xi}, t_i)}{\partial \theta_j \, \partial \xi_h}\right]_{\substack{\boldsymbol{\theta}=\boldsymbol{\theta}^0 \\ \boldsymbol{\xi}=\boldsymbol{\xi}^0}} \tag{A-20}$$

Also

$$\Delta_u \doteq C + \sum_{h=1}^{k} (\xi_{hu} - \xi_h^0)D_h \tag{A-21}$$

where

$$C = \Delta(\boldsymbol{\psi}, \boldsymbol{\theta}^0; \boldsymbol{\xi}^0, t_i) \tag{A-22}$$

$$D_h = \left[\frac{\partial\{\Delta(\boldsymbol{\psi}, \boldsymbol{\theta}^0; \boldsymbol{\xi}^0, t_i)\}}{\partial \xi_h}\right]_{\boldsymbol{\xi}=\boldsymbol{\xi}^0} \tag{A-23}$$

Hence Equation A-16 can be written as

$$E(\hat{\boldsymbol{\theta}}_u - \boldsymbol{\theta}^0) \doteq \left[\left(A + \sum_{h=1}^{k} (\xi_{hu} - \xi_h^0)B_h\right)'\left(A + \sum_{h=1}^{k} (\xi_{hu} - \xi_h^0)B_h\right)\right]^{-1}$$

$$\cdot\left(A + \sum_{h=1}^{k} (\xi_{hu} - \xi_h^0)B_h\right)'\left(C + \sum_{h=1}^{k} (\xi_{hu} - \xi_h^0)D_h\right) = F(\boldsymbol{\xi}_u - \boldsymbol{\xi}^0) \tag{A-24}$$

Thus we find that when the model is inadequate $E(\hat{\boldsymbol{\theta}}_u - \boldsymbol{\theta}^0)$ will be a function of the levels of $\boldsymbol{\xi}$.

6. Appendix B
Summarizing the Information in the Likelihood Function

It is the likelihood *function* itself (or on the Normal assumption the sum of squares function $S(\boldsymbol{\theta})$) which, when the model is adequate, embodies all the information about the parameter supplied by the data. If the model were *linear* in the parameters then $S(\boldsymbol{\theta})$ would be quadratic in $\boldsymbol{\theta}$ and so could be described entirely in terms of its first and second derivatives with respect to $\boldsymbol{\theta}$. Equivalently, it could be described by the maximising values $\hat{\boldsymbol{\theta}}$ and the second derivatives and if the parameter space were *unrestricted* $\hat{\boldsymbol{\theta}}$ could be located by equating first derivatives to zero. In model analysis, however, the models dealt with are usually *non-linear* in the parameters and the parameter space is often *restricted* (for example certain parameters are necessarily positive). Special care is therefore needed in analysing $S(\boldsymbol{\theta})$ and in particular graphical analysis can be very informative as we have already indicated. If, however, we are satisfied that $\hat{\boldsymbol{\theta}}$ is not close to a restricting boundary and that the model can be approximated over the ranges of interest by a function linear in $\boldsymbol{\theta}$ (so that $S(\boldsymbol{\theta})$ is approximately locally quadratic over these same ranges) then with N observations and p parameters, we may employ as a measure of precision of the estimates the region R defined by

$$S(\boldsymbol{\theta}) \leq S(\hat{\boldsymbol{\theta}})\left\{1 + \frac{p}{N-p} F_\alpha(p, N-p)\right\} \tag{B-1}$$

where $F_\alpha(p, N - p)$ indicates the α significance level of the F distribution with p and $N - p$ degrees of freedom. This region enclosed by the contour on which $S = S(\boldsymbol{\theta})$ provides an approximate $100(1-\alpha)\%$ confidence region. From the Bayesian point of view, R would be a region within which $100(1-\alpha)\%$ of the posterior distribution lay, if the prior distribution of $\boldsymbol{\theta}$ could be assumed locally uniform [18, 37].

More generally suppose there exist *non-linear* transformations $\varphi_i(\boldsymbol{\theta})$, $i = 1, 2, \cdots, p$, [3A, 27A] of the p parameters $\boldsymbol{\theta}$ in terms of which an adequate representation can be obtained over the region R by the linear expansion

$$\eta_u = f(\boldsymbol{\theta}, \boldsymbol{\xi}) = f_1(\boldsymbol{\phi}, \boldsymbol{\xi}) \doteq f_1(\hat{\boldsymbol{\phi}}, \boldsymbol{\xi}) + \sum_{i=1}^{p} (\phi_i - \hat{\phi}_i)\left[\frac{\partial f_1(\boldsymbol{\phi}, \boldsymbol{\xi})}{\partial \phi_i}\right]_{\boldsymbol{\phi}=\hat{\boldsymbol{\phi}}} \tag{B-2}$$

In terms of these transformed parameters $\boldsymbol{\phi}$ the region R could alternatively be defined by

$$\frac{1}{2}\sum_{i=1}^{p}\sum_{j=1}^{p} S^{ij}(\phi_i - \hat{\phi}_i)(\phi_j - \hat{\phi}_j) \leq \frac{S(\hat{\boldsymbol{\phi}})p}{N-p} F_\alpha(p, N-p) \tag{B-3}$$

where

$$\frac{1}{2\sigma^2} S^{ij} = \frac{1}{2\sigma^2}\frac{\partial^2 S(\boldsymbol{\phi})}{\partial \phi_i\, \partial \phi_j}\bigg|_{\boldsymbol{\phi}=\hat{\boldsymbol{\phi}}} = -\frac{\partial^2 l(\boldsymbol{\phi})}{\partial \phi_i\, \partial \phi_j}\bigg|_{\boldsymbol{\phi}=\hat{\boldsymbol{\phi}}} \tag{B-4}$$

In the $\boldsymbol{\phi}$ space the region R is thus ellipsoidal.

If the condition B–2 is satisfied almost all the important information is contained in the estimates $\hat{\boldsymbol{\phi}}$(or, equivalently, in the corresponding values $\hat{\boldsymbol{\theta}}$) and in the derivatives S^{ij} which as noted in Equation B–4 are proportional to the second derivatives of the log likelihood. These second derivatives, in fact, supply estimates of the variance-covariance matrix of $\hat{\boldsymbol{\phi}}$

$$E(\hat{\phi}_i - \phi_i)(\hat{\phi}_j - \phi_j)' = 2[S^{ij}]^{-1}\sigma^2 \tag{B-5}$$

With condition B–2 satisfied the result applies to small samples but it corresponds to the well-known asymptotic result concerning the variance-covariance matrix and the second derivatives of the log likelihood function. According to this result the asymptotic variance-covariance matrix for $\hat{\boldsymbol{\theta}}$ is given by

$$E(\hat{\theta}_i - \theta_i)(\hat{\theta}_j - \theta_j)' = -\frac{1}{2}\left[\frac{\partial^2 l(\boldsymbol{\theta})}{\partial\theta_i\,\partial\theta_j}\right]^{-1}_{\boldsymbol{\theta}=\hat{\boldsymbol{\theta}}} \tag{B-6}$$

or equivalently with $\hat{\boldsymbol{\phi}}$ by

$$E(\hat{\phi}_i - \phi_i)(\hat{\phi}_j - \phi_j) = -\left[\frac{\partial^2 l(\boldsymbol{\phi})}{\partial\phi_i\,\partial\phi_j}\right]^{-1}_{\boldsymbol{\phi}=\hat{\boldsymbol{\phi}}} = 2[S^{ij}]^{-1}\sigma^2 \tag{B-7}$$

The logic of these results can be seen if we notice that as the number N increases the region over which the condition B–2 needs to apply becomes smaller and smaller, and the non-linear transformations $\phi_i(\boldsymbol{\theta})$, $i = 1, 2, \cdots, p$, between $\boldsymbol{\theta}$ and $\boldsymbol{\phi}$ will be closely represented by linear transformations which would lead from Equation B–5 to B–6 directly.

To summarize the situation with regard to typifying the likelihood function, in any particular problem one of three circumstances can occur.

(i) It happens quite frequently that an adequate linear approximation of the form of Equation B–2 is available in the original parameterization $\boldsymbol{\theta}$ in which case the above results are approximately justified in terms of the θ's themselves. Virtually all the information concerning the estimation situation is then contained in $\hat{\boldsymbol{\theta}}$ and the derivatives

$$\left.\frac{\partial^2 l(\boldsymbol{\theta})}{\partial\phi_i\,\partial\phi_j}\right|_{\boldsymbol{\theta}=\hat{\boldsymbol{\theta}}}$$

(ii) In some cases it may be that the condition B–2 can be satisfied after suitable transformations $\boldsymbol{\phi}(\boldsymbol{\theta})$ have been made on the parameters $\boldsymbol{\theta}$; the summary statistics would then be $\hat{\boldsymbol{\phi}}$ and

$$\left.\frac{\partial^2 l(\boldsymbol{\phi})}{\partial\phi_i\,\partial\phi_j}\right|_{\boldsymbol{\phi}=\hat{\boldsymbol{\phi}}}$$

(iii) In still other cases no transformation of the parameters $\boldsymbol{\theta}$ can satisfy condition B–2 to a sufficient degree of accuracy. It would then be necessary to consider higher order derivatives of the likelihood function to completely exemplify the situation. When condition B–2 cannot be satisfied we say that intrinsic non-linearity exists.

REFERENCES

1. ANSCOMBE, F. J., 1960. Rejection of outliers, *Technometrics, 2*, 2, 123–147.
2. ANSCOMBE, F. J., 1961. Examination of residuals, in the *Proceedings of the Fourth Berkeley Symposium on Mathematical Statistics and Probability,* Edited by Jerzy Neyman, Vol. I, 19 and 24.
3. BARNARD, G. A., JENKINS, G. M., and WINSTEN, C. B., 1962. Likelihood inference and time series, *J. R. Stat. Soc.* (A), *125*, 321–372.
3A. BEALE, E. M. L., 1960. Confidence regions in non-linear estimation, *J. R. Stat. Soc.* (*B*), 22, 41–76.
4. BEHNKEN, D. W., 1964. Estimation of copolymer reactivity ratios: an example of non-linear estimation, *Journal of Polymer Science*, Part A, *2*, 645–668.
5. BLAKEMORE, J. W., and HOERL, A. E., 1963. Fitting non-linear reaction rate equations to data, *Chem. Engr. Prog. Symp., Series*, Vol. 59, 42, 14–27.
6. BOX, G. E. P., 1954. The exploration and exploitation of response surfaces: some general considerations and examples, *Biometrics, 10*, 16–60.
7. BOX, G. E. P., 1958. Use of statistical methods in the elucidation of basic mechanisms, *Bull. Inst. Intern. De Statistique, 36*, 215–225.
8. BOX, G. E. P., 1959. Discussion on papers by Satterthwaite and Budne on random balance, *Technometrics, 1*, 174–180.
9. BOX, G. E. P., 1960. Fitting empirical data, *Annals of the New York Academy of Sciences, 86*, 792–816.
10. BOX, G. E. P., 1961. The effects of errors in the factor levels and experimental design, *Bull. Intern. Stat. Inst., 38*, 339–355. Also, reprinted in 1963 in *Technometrics, 5*, 247–262.
11. BOX, G. E. P., and COUTIE, G. A., 1956. Application of digital computers in the exploration of functional relationships, *Proc. I.E.E., 103*, Part B suppl. No. 1, 100–107.
12. BOX, G. E. P., and DRAPER, N. R., 1959. A basis for the selection of a response surface design, *J. Am. Stat. Assoc., 54*, 622–654.
13. BOX, G. E. P., and DRAPER, N. R., 1963. The choice of a second order rotatable design, *Biometrika, 50*, 335–352.
14. BOX, G. E. P., and HUNTER, J. S., 1957. Multifactor experimental designs for exploring response surfaces, *Ann. Math. Stat., 28*, 195–241.
15. BOX, G. E. P., and HUNTER, W. G., 1962. A useful method for model-building, *Technometrics, 4*, 301–318.
16. BOX, G. E. P., and HUNTER, W. G., 1963. Sequential design of experiments in non-linear situations, Technical Report No. 21, Department of Statistics, University of Wisconsin.
17. BOX, G. E. P., and LUCAS, H. L., 1959. Design of experiments in non-linear situations, *Biometrika, 46*, 77–90.
18. BOX, G. E. P., and TIAO, G. T., 1962. A further look at robustness via Bayes' Theorem *Biometrika, 49*, 419–432.
19. BOX, G. E. P., and YOULE, P. V., 1955. The exploration and exploitation of response surfaces: an example of the link between the fitted surface and the basic mechanism of the system, *Biometrics, 11*, 287–323.
19A. BOX, G. E. P., and COX, D. R., 1964. An analysis of transformations, *J. R. Stat. Soc.* (*B*), 26, 211–252.
19B. BOX, G. E. P., and DRAPER, N. R., 1965. The Bayesian estimation of common parameters from several responses, *Biometrika* (to appear).
20. CHERNOFF, H., 1953. Locally optimal designs for estimating parameters, *Ann. Math. Stat., 24*, 586–602.
21. COX, D. R., 1961. Tests of separate families of hypotheses, *Proc. Fourth Berkeley Symp., 1*, 105–123.
22. COX, D. R., 1962. Further results on tests of separate families of hypotheses, *J. R. Stat. Soc.* (*B*), *24*, 406–424.
23. DANIEL, CUTHBERT, 1960. Locating outliers in factorial experiments, *Technometrics, 2*, 2, 149–156.
24. DAVIES, O. L. (Ed.), 1956. *Design and Analysis of Industrial Experiments*, Hafner, New York, 495–578.

25. Elfving, G., 1952. Optimum allocation in linear regression theory, *Ann. Math. Stat.*, *23*, 255–262.
26. Ferguson, T. S., 1960. Discussion of the papers of Messrs. Anscombe and Daniel, *Technometrics*, *2*, 2, 159–160.
27. Gumbel, E. J., 1960. Discussion of the Papers of Messrs. Anscombe and Daniel, *Technometrics*, *2*, 2, 165–166.
27A. Guttman, I., and Meeter, D. A., 1964. Use of transformations on parameters in nonlinear theory II. Technical report No. 38, Dept. of Statistics, University of Wisconsin.
28. Hunter, W. G., and Mezaki, R., 1964. A model-building technique for chemical engineering kinetics, *A.I.Ch.E. Journal 10*, 3, 315–322.
28A. Hunter, W. G., and Reiner, A. M., 1964. Designs for discriminating between rival models, Technical report No. 32, Dept. of Statistics, University of Wisconsin.
29. Kiefer, J. C., 1959. Optimum experimental designs, *J.R. Stat. Soc. (B)*, *21*, 272–319.
30. Kiefer, J. C., 1961. Optimal experimental designs V, with applications to systematic and rotatable designs, *Proc. Fourth Berkeley Symp.*, *1*, 381–405.
31. Kruskal, W. H., 1960. Some remarks on wild observations, *Technometrics*, *2*, 1, 1–3.
32. Kruskal, W. H., 1960. Discussion of the papers of Messrs. Anscombe and Daniel, *Technometrics*, *2*, 2, 157–158.
33. Laible, J. R., 1959. Ph.D. Thesis, University of Wisconsin, Madison, Wisconsin.
34. Lind, E. E., Goldin, J., Hickman, J. B., 1960. Fitting yield and cost surfaces, *CEP*, *56*, 11, 66.
35. Norton, C. J., Moss, T. E., 1964. Oxidation dealkylation of alkylation of alkylaromatic hydrocarbons, *I & E C Process Design and Development*, *3*, 1, 31–32.
36. Remmers, E. G., Dunn, C. G., 1961. Process improvement of a fermentation product, *I & E C*, *53*, 9, 745.
37. Savage, L. J. and others, 1962. *The Foundations of Statistical Inference*, Wiley, 20–25.
38. Smith, H., Rose, A., 1963. Subjective responses in process investigation, *I & E C*, *55*, 7, 26–27.
39. Stone, M., 1959. Application of a measure of information to the design and comparison of regression experiments, *Ann. Math. Stat.*, *30*, 55–70.
40. Tukey, J. W., 1960. Discussion of the Papers of Messrs. Anscombe and Daniels, *Technometrics*, *2*, 2, 160–165.
41. Wald, A., 1943. On the efficient design of statistical investigations, *Ann. Math. Stat.*, *14*, 134–140.
42. Yates, F., 1959. Discussion on Dr. Kiefer's paper, *J. R. Stat. Soc. (B)*, *21*, 309.

1.9

Use and Abuse of Regression†

G. E. P. BOX
University of Wisconsin

Let us first restate the usual assumptions and conclusions for linear least squares. Gauss showed that if we have n observations $y_1, y_2, \cdots, y_n$ and *if* an appropriate model for the uth observation is

$$y_u = \beta_0 + \beta_1 x_{1u} + \beta_2 x_{2u} + \cdots + \beta_k x_{ku} + \epsilon_u \tag{1}$$

where the β's are unknown parameters, the x's known constants, and the ϵ's random variables uncorrelated and having the same variance and zero expectation, then estimates $b_0, b_1, \cdots, b_k$ of the β's obtained by minimizing $\sum (y - \hat{y})^2$ with $\hat{y} = b_0 x_0 + b_1 x_1 + b_2 x_2 + \cdots + b_k x_k$ are unbiassed and have smallest variance among all linear unbiassed estimates.

The method of least squares is used in the analysis of data from planned experiments and also in the analysis of data from unplanned happenings. The word "regression" is most often used to describe analysis of unplanned data. It is the tacit assumption that the requirements for the validity of least squares analysis are satisfied for unplanned data that produces a great deal of trouble. Whether the data are planned or unplanned the quantity ϵ, which is usually quickly dismissed as a random variable having the very specific properties mentioned above, really describes the effect of a large number of "latent" variables $x_{k+1}, x_{k+2}, \cdots, x_m$ which we know nothing about. If we suppose that it is enough to consider the linear effects of these latent variables (which would often be realistic for small variations in $x_{k+1}, \cdots, x_m$) we should have

$$\epsilon = \beta_{k+1} x_{k+1} + \beta_{k+2} x_{k+2} + \cdots + \beta_m x_m \tag{2}$$

Thus in matrix notation we can write for the column of n observations $\mathbf{y}$

$$\mathbf{y} = \mathbf{X}_1 \boldsymbol{\beta}_1 + \mathbf{X}_2 \boldsymbol{\beta}_2 \tag{3}$$

where $\mathbf{X}_1$ has for elements the n values of the k regression variables and $\mathbf{X}_2$ has for elements the n unknown values of the $m - k$ latent* variables. The situation is illustrated in Figure 1 in which the variables $x_{k+1} \cdots, x_m$ are "hidden behind the wall." In practice various kinds of linkages would occur between the variables indicated by lines. These linkages might indicate causative relations; for instance, an increase in temperature might necessarily produce an increase

Received March 1966.

† A talk prepared for the tenth conference on the Design of Experiments in Army Research, Development and Testing. Washington, D. C. November 1964.

* More dramatically described as "lurking" variables.

The Collected Works of George E. P. Box, 1984, Wadsworth, Inc., Belmont, CA 94002.
Originally published in *Technometrics,* vol. 8, no. 4 (1966), pp. 625–629.

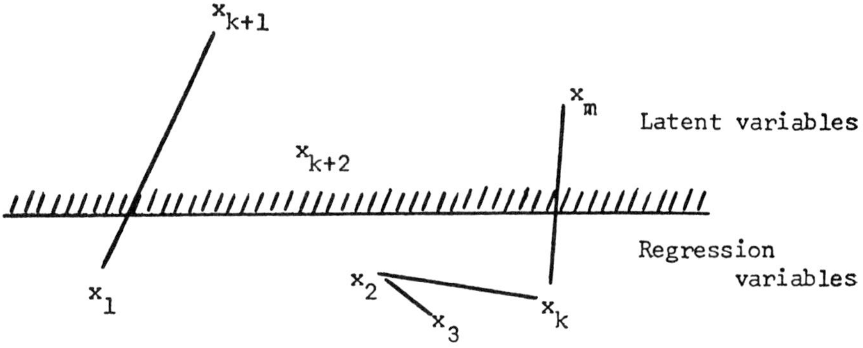

FIGURE 1 Latent variables and regression variables.

in pressure; or merely relationships due to correlation. Thus, an operator in charge of a process might *as a standard operating procedure* always reduce the flow of one of the reactants if a certain temperature was observed to be high.

We must now ask the question, "What do we wish to do with the fitted regression equation?" We might

(i) desire to predict y in the future from passive observation of $x_1 \cdots x_k$. We assume that the causal and correlative system which operated during the data taking has not been interferred with and also operates during the period when predictions are being made.

(ii) to discover how deliberate *changes* in $x_1 \cdots x_k$ will effect y with the intention of actually *modifying* the system to get a better value for y.

The position is quite different depending upon whether prediction from passive observation or improvement from active interference is in mind. This is made clear by the following example.

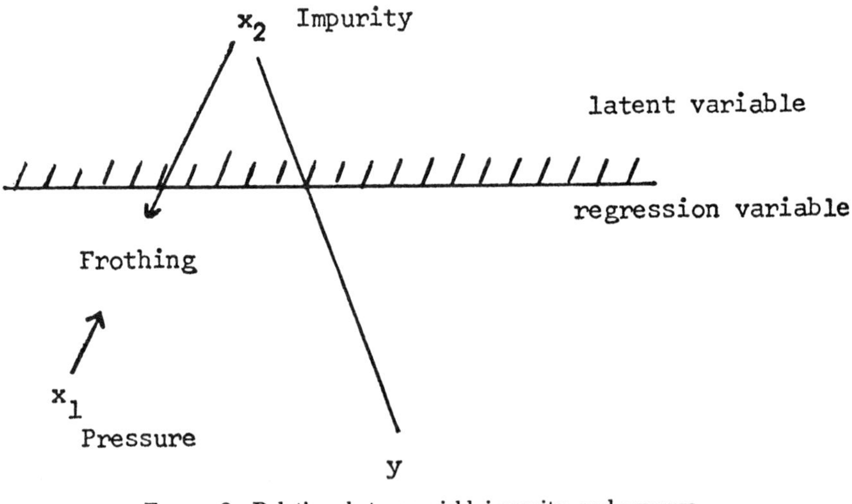

FIGURE 2 Relations between yield, impurity, and pressure.

Suppose that in a chemical process it has been found that undesirable frothing can be reduced by increasing pressure. The standard operating procedure is, therefore, to increase pressure whenever frothing appears. Suppose that the frothing in fact occurs because of an unsuspected impurity x_2 (which is, of course, not measured because it is unknown). Suppose finally that a high value of impurity x_2 not only produces frothing but also lowers yield but that yield is unaffected directly by a change in pressure.

If (with now x and y representing deviations from respective averages) a "regression of yield on pressure" $\hat{y} = b_1x_1$ is fitted by the usual least squares procedure we may well find a highly significant coefficient b_1.

This well known phenomenon of "nonsense" correlation exhibited in this example is worth studying further. Suppose there is a relationship $y = \beta_1x_1 + \beta_2x_2$ connecting y *exactly* with the two variables x_1 and x_2 (with, in the present instance, $\beta_1 = 0$). Now, of course, the actual levels of x_2 are unknown but suppose that $\hat{x}_2 = ax_1$, is the formal regression of x_2 on x_1 which would be obtained if values of x_2 *were* available. Then it is readily shown that

$$b_1 = \beta_1 + a\beta_2 \tag{4}$$

In this expression β_1 is zero and we appear to obtain a real effect only because of the influence of the bias term $(a\beta_2)$. On the other hand using (4) we see that our fitted equation $\hat{y} = b_1x_1$ which ignored x_2 can be written $\hat{y} = \beta_1x_1 + \beta_2ax_1$ or as

$$\hat{y} = \beta_1x_1 + \beta_2\hat{x}_2 \tag{5}$$

This equation which replaces x_2 by $\hat{x}_2$ is the best estimate of y we can expect to get from observing x_1 only. Provided the system *continues to be run in the same fashion as when the data were recorded* we can use pressure to indicate the level of y. Of course, if we had measured x_2 a more accurate (indeed in the present instance an exact) value of y would be deducible, but, lacking knowledge of the importance of x_2, we might nevertheless appropriately use the simple regression equation $\hat{y} = b_1x_1$.

On the other hand the value of b_1 will be utterly misleading if interpreted as the effect on the variable y of a unit *change* in x_1. If we hope to increase yield by reducing pressure we will be disappointed.

A similar argument applies for any number of variables. The true model is

$$\mathbf{y} = \mathbf{X}_1\boldsymbol{\beta}_1 + \mathbf{X}_2\boldsymbol{\beta}_2 \tag{6}$$

By including only the variables X_1 in the regression equation our prediction equation for y becomes

$$\hat{\mathbf{y}} = \mathbf{X}_1\mathbf{b}_1 = \mathbf{X}_1(\mathbf{X}_1'\mathbf{X}_1)^{-1}\mathbf{X}_1'\mathbf{y} \tag{7}$$

$$= \mathbf{X}_1(\mathbf{X}_1'\mathbf{X}_1)^{-1}\mathbf{X}_1'(\mathbf{X}_1\boldsymbol{\beta}_1 + \mathbf{X}_2\boldsymbol{\beta}_2) \tag{8}$$

i.e.

$$\hat{\mathbf{y}} = \mathbf{X}_1\boldsymbol{\beta}_1 + \hat{\mathbf{X}}_2\boldsymbol{\beta}_2 \tag{9}$$

where $\hat{\mathbf{X}}_2 = \mathbf{X}_1\mathbf{A}$ and $\mathbf{A} = (\mathbf{X}_1'\mathbf{X}_1)^{-1}\mathbf{X}_1'\mathbf{X}_2$ is the $k + 1 \times m - k$ matrix of regression coefficients of the latent variables on the regression variables.

Again we see that so far as the passive prediction of $\hat{\mathbf{y}}$ is concerned our simple regression onto the known variables $\mathbf{X}_1$ in effect replaces the unknown $\mathbf{X}_2$ by $\hat{\mathbf{X}}_2$.

On the other hand the regression coefficients $\mathbf{b}_1 = \boldsymbol{\beta}_1 + \mathbf{A}\boldsymbol{\beta}_2$ represent combinations of effects due to regression variables and latent variables and as before it is impossible to draw any valid conclusions as to how interference with the levels of the regression variables will affect the system.

In a designed experiment, we are in quite a different case. It was, of course, to overcome such difficulties as those described above that Fisher introduced the idea of designed experiments and in particular of randomization. When the levels of the regression variables are chosen in some deliberately random manner it is impossible for the levels of a regression variable to be affected by the level of a latent variable. The only cause of the particular values which the regression variables have within the design framework is the throw of an unbiassed die or other random process. Fisher makes it possible to analyze the data *as if* Gaussian assumptions were true by making $\mathbf{X}_1$ a random variable. The regression variables can, of course, still affect the latent variables and these may in turn effect y. Provided, however, we apply our results to the same system for which we obtained our data this will cause no problem. It will be genuinely true that apart from experimental error *manipulation* of regression variables will produce the predicted change in y even though it does it via some latent variable.

The basic difficulty mentioned above is by no means the only one that faces us in the analysis of unplanned data. In the operation of an industrial process past experience often shows that certain variables are of major importance. In order to control fluctuations in the process, therefore, care is taken to hold precisely these variables very close to fixed values. As the "statistical significance" of any variable is greatly affected by the range it covers there is a strong probability, therefore, that the most important variables will be dubbed "not significant" by a standard regression analysis. A further difficulty is that with unplanned data regression variables will frequently be highly correlated only because of operating policy. The operator is told to reduce x_2 whenever x_1 becomes high. With such data even if difficulties from latent variables could be ignored it may be almost impossible to discover whether changes in y are associated with x_1, with x_2, or with both. In designed experiments, of course, one normally arranges that x_1 and x_2 are uncorrelated by using an orthogonal design.

In summary the regression analysis of unplanned data is a technique which must be used with great care. However,

(i) It may provide a useful prediction of y in a fixed system being passively observed even when latent variables of some importance exist. For this application computer programs which progressively add or drop variables make some sense.

(ii) It is one of a number of tools sometimes useful in indicating variables which ought to be included in some latter planned experiment (in which randomization will, of course, be included as an integral part of the design). It ought never to be used to decide which variable should be excluded from further investigation for reasons which are obvious from the above.

To find out what happens to a system when you interfere with it you have to interfere with it (not just passively observe it).

REFERENCE

1. FISHER, R. A., 1937. *Design of Experiments*, published by Oliver and Boyd.

1.10
Discrimination among Mechanistic Models

G. E. P. BOX and W. J. HILL*
University of Wisconsin

This paper is concerned with research, the object of which is to discover the *mechanism* for a particular phenomenon leading to a specific mathematical model. Such investigations are distinguished from those in which the object is merely to estimate the output y of a process over a range of values of the input $\xi_1, \xi_2, \cdots, \xi_k$. Frequently, a number of possible mechanisms are suggested from theoretical considerations leading to a number of different mathematical models. To discriminate among these a sequential procedure is developed in which calculations made after each experiment determine the most discriminatory process conditions for use in the next experiment. The method is illustrated with examples.

1. Using Mathematical Models to Describe Physical Phenomena

The experimenter is often concerned with the study of systems in which there is a dependent variable y related to independent variables $\boldsymbol{\xi} = (\xi_1, \xi_2, \cdots, \xi_k)$. For example, he might be studying the yield y of a chemical reaction under various conditions of temperature ξ_1, pressure ξ_2, and catalyst concentration ξ_3. Any such phenomenon is theoretically capable of representation by a mathematical equation

$$E(y) = f(\boldsymbol{\theta}, \boldsymbol{\xi}), \tag{1.1}$$

which takes account of the mechanism of the process. In equation (1.1) $\boldsymbol{\theta} = (\theta_1, \theta_2, \cdots, \theta_p)'$ is the vector of p parameters which are usually physical constants of the system under study. We refer to this as the *true* mathematical model. In practice such a model accounting for the phenomenon under study in terms of its exact mechanism is seldom obtainable. Many examples occur, however, where the main characteristics of a mechanism can be elucidated and a close mechanistically based mathematical representation is possible.

It should be noted that the objective which we are presently considering is that of finding out "how a system works." The reason for this may be no more than scientific curiosity. However, if we know how the system works and can describe it by a mathematical model, then we can use this knowledge for practical aims such as predicting the behavior of the process under various experimental conditions and, in particular, in finding optimum operating conditions. This last fact leads to some confusion because if all we need to do is either to estimate the behavior of the process under various experimental conditions or to find optimum operating conditions, we do not necessarily need a mechanistic model.

Received: July, 1965; revised July, 1966.
* Presently with National Aniline, Buffalo, New York.

The Collected Works of George E. P. Box, 1984, Wadsworth, Inc., Belmont, CA 94002.
Originally published in *Technometrics,* vol. 9, no. 1 (1967), pp. 57–71.

In some circumstances, an attempt to discover the mechanism merely to develop an operable system would be needlessly time consuming. For example, it may be known that the observed strength y of an extruded plastic sheet depends on the rate of extrusion ξ_1 and the rate of cooling immediately after extrusion ξ_2. Nevertheless, a study to discover the nature of the true mechanism relating y to ξ_1 and ξ_2 might be extremely difficult and laborious. It would certainly not be needed if all that was required was to determine how the process should be adjusted to give a plastic sheet of greater strength. This more limited objective could probably be obtained from a purely empirical approach, for example, response surface methods (see [1], [2], [3], [4], [5], [6], [7]). However, when there is a good possibility of obtaining a mechanistically based model with a reasonable experimental effort, this approach and not the purely empirical one ought to be followed. This is so even when the objective is only the practical one of estimating the response y over a particular region of the experimental conditions.

The approach via a mechanistic model is particularly useful in the development of new processes. The chances of meaningful extrapolation are very much greater with such a model. Even here, however, because mechanisms are never perfectly known, we cannot hope for much more than having the model indicate a region which has not yet been investigated but is worthy of experimental study. The chance that such predictions will be borne out may be reasonably good with a mechanistic model. Although empirical models such as interpolation polynomials may be perfectly adequate to represent what is happening in the immediate region of the experiments, they will provide little basis for extrapolation.

2. Identifying the Correct Model

The process of building a mechanistic model can be thought of as involving three stages:

(i) identification
(ii) fitting
(iii) diagnostically checking the adequacy of fit.

The statistical literature has tended to concentrate on stages (ii) and (iii). Less attention has been given to (i). To illustrate what identification involves suppose we consider the catalytic reaction

$$2A \rightleftarrows B + C.$$

If we assume the dual site mechanism outlined in [10]

$$\begin{array}{lll} \text{(a)} & A + s \rightleftarrows As & \text{adsorption,} \\ \text{(b)} & 2As \rightleftarrows Bs + Cs & \text{surface reactions,} \\ \text{(c)} & Bs \rightleftarrows B + s & \text{desorption,} \\ \text{(d)} & Cs \rightleftarrows C + s & \text{desorption,} \end{array} \qquad (2.1)$$

where s is an active catalytic site, we see that there are four different possible rate controlling steps. From each of these possible rate controlling steps can be derived a reaction rate model. These candidate models are

(a) $$r = \frac{k_1\Phi\left(\xi_A - \sqrt{\frac{\xi_B\xi_C}{K}}\right)}{\left(1 + \sqrt{\frac{\theta_A^2}{K}}\,\xi_A\xi_C + \theta_B\xi_B + \theta_C\xi_C\right)},$$

(b) $$r = \frac{\gamma\theta_A^2\left(\xi_A^2 - \frac{\xi_B\xi_C}{K}\right)}{(1 + \theta_A\xi_A + \theta_B\xi_B + \theta_C\xi_C)^2},\qquad (2.2)$$

(c) $$r = \frac{k_2\theta_B K\Phi\left(\frac{\xi_A^2}{\xi_C} - \frac{\xi_B}{K}\right)}{\left(1 + \theta_A\xi_A + \theta_B K\frac{\xi_A^2}{\xi_C} + \theta_C\xi_C\right)},$$

(d) $$r = \frac{k_3\theta_C K\Phi\left(\frac{\xi_A^2}{\xi_B} - \frac{\xi_C}{K}\right)}{\left(1 + \theta_A\xi_A + \theta_B\xi_B + \theta_C K\frac{\xi_A^2}{\xi_B}\right)},$$

where r is the rate of reaction, $\boldsymbol{\theta}$, Φ, k_1, k_2, k_3, K, and γ are constants of the system, and ξ are partial pressures. If we took into account other mechanisms besides (2.1), we could postulate many more candidates which could conceivably describe the system.

It might be expected that all one needs to distinguish among rival models is to run a number of careful experiments at various experimental conditions. It might then be hoped that one candidate model would fit the data very well whereas the others would show a marked lack of fit. Unfortunately, it is easy to collect data that are well fitted by a large number of different models. Different research groups commonly claim widely varying mechanisms for the same chemical system. Each can produce data which prove that their mechanism is the right one. One such reaction is the Water Gas Shift reaction [9]

$$CO + H_2O \underset{\text{catalyst}}{\longrightarrow} CO_2 + H_2,$$

where there are at least ten or more different candidates.

The above situation can occur because the model that one group favors is never really "put in jeopardy" by the experiments they conduct. As an illustrative example of this consider the two candidate mechanisms for a chemical reaction involving components A, B, and C

(a) $A \to B \to C,$

(b) $A \to B \rightleftarrows C.$

In Figure 1 the concentration of B is plotted against time ξ for both reactions. Obviously, in order to discriminate between these two mechanisms it is not enough to observe B only over small values of ξ where the curves can almost be identical in shape. We need to observe B at higher values of ξ.

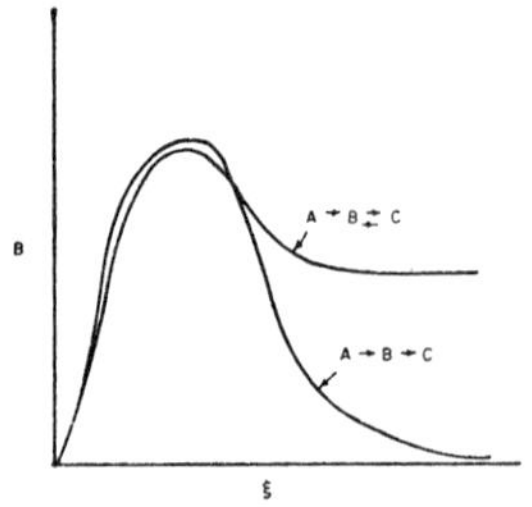

Figure 1—Two Rival Mechanisms for a Chemical Reaction Involving Components A, B, and C.

3. Sequential Designs for Discrimination

In the problem considered here, it is supposed that an experimenter is investigating a physical system in which the underlying theory suggests at the outset that any one of $m(> 1)$ models might describe the system. It is also supposed that after $n - 1$ preliminary experimental runs, the results are not conclusive as to which model is best. In order to discriminate among these m rival models the experimenter may wish to perform further experimental runs. If experiments are conducted in sequence, the experimenter can consider each result before he runs the next experiment. In particular, computations prior to each new experiment can indicate where the next experiment ought to be conducted to provide maximum discrimination among m rival models. A sequential approach for $m = 2$ was that of Hunter and Reiner [8]. Here experiments were performed at that set of conditions which essentially gave the greatest estimated difference in response. This method, however, did not take into account the magnitude of the error of the estimated difference.

In considering the general problem ($m \geq 2$), we first decide upon some relevant discrimination criterion and then choose experimental conditions which maximize this measure at each stage. In the measure of discrimination one must not only take account of the difference in response given by the models to be discriminated among but also take into consideration the variance of the estimated response. Thus, as shown in Figure 2, if the two lines represent the estimated relationship y on ξ and the crossed areas indicate the limits of error (say 95% confidence region), then it will by no means be true that the best point to discriminate between the two models will be that where there is maximum divergence between the two lines. This is so, since clearly it is the divergence relative to the limits of error to be considered rather than divergence itself at any chosen point.

4. A Discrimination Criterion

In studying the amount of information supplied by a communication system, Shannon [12] used the concept of entropy where entropy is defined as

$$S = -\sum_{i=1}^{m} \Pi_i \ln \Pi_i , \tag{4.1}$$

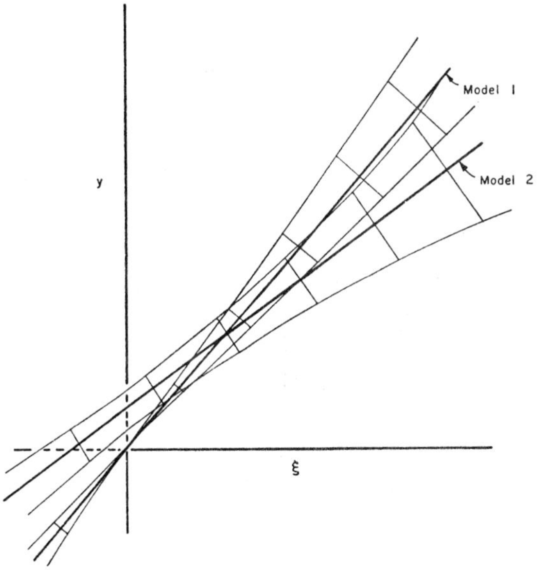

FIGURE 2—Two Rival Prediction Equations and Their Corresponding Error Regions.

where Π_i is the probability associated with a symbol i. The least possible information corresponds to maximum entropy, that is, when $\Pi_1 = \Pi_2 = \cdots = \Pi_m = 1/m$. An illustration of this situation for $m = 4$ is shown in Figure 3(a). It is more desirable to have a situation where $\Pi_j \gg \Pi_i$ where $i \neq j$. Here, the entropy is small and the amount of information is large. See, for example, Figure 3(b). In a communication system, the first situation (Figure 3(a)) usually represents the input state and the second situation ($\Pi_j \gg \Pi_i$) usually represents the output state. Therefore, in order to obtain the maximum information from the system it is desirable to have a maximum change in entropy between input and output. This occurs, of course, when for any given input the output is such that $\Pi_j = 1$ and $\Pi_i = 0$ for $i \neq j$. (In a communication system, this means that the probability of the symbol j being correctly received is unity.) This concept is now applied to the discrimination among m models where it is desired to go from a noninformative situation like that in Figure 3(a) to a more informative situation like that in Figure 3(b).

At input (after $n - 1$ observations) the prior probabilities associated with the m models are $\Pi_{1n-1}, \Pi_{2n-1}, \cdots, \Pi_{mn-1}$. The posterior probabilities found by taking the n-th observation y_n are $\Pi_{1n}, \Pi_{2n}, \cdots, \Pi_{mn}$. The *expected* change in entropy between input and output is expressed as

$$R = \text{entropy at input—expected entropy at output}$$

$$= -\sum_{i=1}^{m} \Pi_{in-1} \ln \Pi_{in-1} - (-1) \int \left(\sum_{i=1}^{m} \Pi_{in} \ln \Pi_{in}\right) q(y_n)\, dy_n\,, \qquad (4.2)$$

where

$$q(y_n) = \sum_{i=1}^{m} \Pi_{in-1} p_i\,, \qquad (4.3)$$

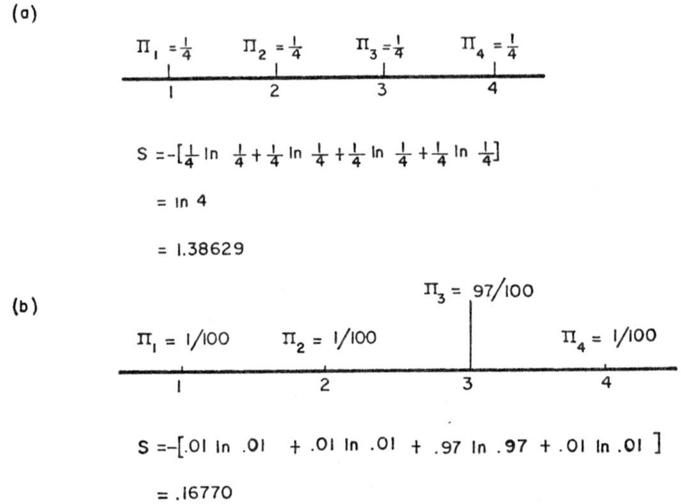

FIGURE 3—Illustration of Entropy for (a) the Least Informative Case, and (b) a More Informative Case.

where p_i is the probability density function of the n-th observation under model i. Since the posterior probability associated with the i-th model is

$$\Pi_{in} = \frac{\Pi_{in-1} p_i}{q(y_n)} , \tag{4.4}$$

substitution of (4.4) into (4.2) gives

$$R = \sum_{i=1}^{m} \Pi_{in-1} \int p_i \ln \frac{p_i}{q(y_n)} \, dy_n . \tag{4.5}$$

By substituting into (4.5) the following inequality (Corollary 3.1 and Example 3.2 of Kullback [11])

$$\sum_{j=1}^{m} \Pi_{jn-1} p_i \ln \frac{p_i}{p_j} \geq p_i \ln \frac{p_i}{q(y_n)} , \tag{4.6}$$

we have

$$R \leq D, \tag{4.7}$$

where

$$D = \sum_{i=1}^{m} \sum_{j=i+1}^{m} \Pi_{in-1} \Pi_{jn-1} \left(\int p_i \ln \frac{p_i}{p_j} \, dy_n + \int p_j \ln \frac{p_j}{p_i} \, dy_n \right) . \tag{4.8}$$

The expression in the parentheses of equation (4.8) is Kullback's total measure of information (or divergence) when discriminating between two hypotheses H_i and H_j. The function D represents the maximum change in entropy expected from y_n and is a measure of discrimination among m rival hypotheses or models.

In order to use D in a discrimination criterion, the following distribution theory is developed for the n-th observation. Under model i ($i = 1, 2, \cdots, m$) it is assumed that the observation y_n is normally distributed with expected value $\eta = E(y_n)$ and known variance σ^2. The probability density function of y^u

given η and σ is

$$p_i(y_n \mid \eta, \sigma) = \frac{1}{\sqrt{2\pi}\,\sigma} \exp\left\{-\frac{1}{2\sigma^2}(y_n - \eta)^2\right\}. \tag{4.9}$$

If η is linear in the parameters $\boldsymbol{\theta}_i$ $(i = 1, 2, \cdots, m)$ or can be approximately expressed as a linear function of $\boldsymbol{\theta}_i$ in the region of the parameter estimates $\tilde{\boldsymbol{\theta}}_i$, then from a Bayesian development η is normally distributed about $\tilde{y}_n^{(i)}$ with standard deviation σ_i. The predicted value of y_n under model i using the first $n - 1$ observations is $\tilde{y}_n^{(i)}$ and the variance of $\tilde{y}_n^{(i)}$ is σ_i^2. (See Appendix for this linear theory development and for the calculation of σ_i^2.) The probability density function of η given σ is then written as

$$p_i(\eta \mid \sigma) = \frac{1}{\sqrt{2\pi}\,\sigma_i} \exp\left\{-\frac{1}{2\sigma_i^2}(\eta - \tilde{y}_n^{(i)})^2\right\}. \tag{4.10}$$

Therefore, the probability density function of y_n under model i given σ and the first $n - 1$ observations is

$$\begin{aligned} p_i &= p_i(y_n \mid \sigma) \\ &= \int p_i(y_n \mid \eta, \sigma) p_i(\eta \mid \sigma)\, d\eta. \end{aligned} \tag{4.11}$$

Substituting equations (4.9) and (4.10) into (4.11) and integrating, we have

$$p_i = \frac{1}{\sqrt{2\pi(\sigma^2 + \sigma_i^2)}} \exp\left\{-\frac{1}{2(\sigma^2 + \sigma_i^2)}(y_n - \tilde{y}_n^{(i)})^2\right\}. \tag{4.12}$$

When the results of (4.12) are substituted in the discrimination function D of equation (4.8) we have

$$\begin{aligned} D = \frac{1}{2} \sum_{i=1}^{m} \sum_{j=i+1}^{m} \Pi_{i\,n-1}\Pi_{j\,n-1} \Bigg\{ &\frac{(\sigma_i^2 - \sigma_j^2)^2}{(\sigma^2 + \sigma_i^2)(\sigma^2 + \sigma_j^2)} \\ &+ (\tilde{y}_n^{(i)} - \tilde{y}_n^{(j)})^2 \left(\frac{1}{\sigma^2 + \sigma_i^2} + \frac{1}{\sigma^2 + \sigma_j^2}\right)\Bigg\}. \end{aligned} \tag{4.13}$$

To select the n-th experiment that attains maximum expected discrimination among the m rival models, one chooses those operating conditions that maximize D. After y_n is observed the current standing of each model can be checked by calculating its corresponding posterior probability from equation (4.4). The procedure is repeated until the posterior probabilities indicate that one model is clearly superior to the others.

5. Examples

It is usually found that it is only possible to conduct experiments in any given situation over a particular region of $\boldsymbol{\xi}$. This region will be called the operability region O. Thus, for example, with a particular apparatus it might be impossible to conduct experiments outside temperature and pressure ranges. It will be noted that the operability region O is usually very extensive and it would normally be quite impossible to conduct experiments adequately covering all of O. Cases may occur where during the course of an investigation we may be led to redesign our apparatus to extend the operability region in a particular direction of interest.

The following three examples illustrate the application of the discrimination criterion. Although polynomial models are used in the first two examples, this is not meant to suggest that this discrimination criterion is proposed for the empirical model situation. These examples are essentially intended to illustrate the characteristics of the criterion. More in keeping with the purpose of this study, that is, the elucidation of the underlying mechanism of a system, the third example includes mechanistically based models that might be considered as rivals when an experimenter is studing the reaction of the form $A \rightarrow B$.

Example 5.1

As a simple illustration of the discrimination criterion, consider the case of the $m = 2$ models

$$\begin{aligned} &\text{Model 1.} \quad \underset{1}{E(y)} = \theta_{11}\xi, \\ &\text{Model 2.} \quad \underset{2}{E(y)} = \theta_{21} + \theta_{22}\xi, \end{aligned} \tag{5.1}$$

where the variance is assumed known. See Figure 2 for a graphical representation of the prediction equations for these models and their corresponding error regions. Model 1 implies that the expected value of the response is proportional to ξ and hence is represented by a straight line passing through the origin. Model 2 implies that the expected value of the response follows a straight line not necessarily passing through the origin. For this particular example it is easy to derive theoretically the course which the discrimination procedure will follow. That is, when $-\infty < \xi < \infty$, D is maximized if the first discriminatory observation is taken at

$$\xi_n = -\sum_{u=1}^{n-1} \xi_u \, . \tag{5.2}$$

It follows from equation (5.2) that all further maximum discrimination points (taken one at a time) occur at $\xi = 0$. Therefore, as one might expect, in order to discriminate between the two first order models (5.1), future settings of ξ should be chosen at the origin after an initial discrimination point has been chosen according to expression (5.2).

Example 5.2

To illustrate the case $m > 2$, consider the discrimination among the $m = 4$ polynomial models

$$\begin{aligned} &\text{Model 1.} \quad \underset{1}{E(y)} = \theta_{11}\xi, \\ &\text{Model 2.} \quad \underset{2}{E(y)} = \theta_{21} + \theta_{22}\xi, \\ &\text{Model 3.} \quad \underset{3}{E(y)} = \theta_{31} + \theta_{32}\xi + \theta_{33}\xi^2, \\ &\text{Model 4.} \quad \underset{4}{E(y)} = \theta_{41}\xi + \theta_{42}\xi^2, \end{aligned} \tag{5.3}$$

representing respectively a straight line passing through the origin, a straight line not necessarily passing through the origin, a quadratic not constrained to pass through the origin, and a quadratic passing through the origin. The standard deviation $\sigma = 1$ is assumed known. Data are generated from model 3 where $\theta_{31} = \theta_{32} = \theta_{33} = 1$. (In an actual experimental situation, of course, the true model and the corresponding parameter values would not be known.) The operability region for ξ is $0 \leq \xi \leq 4$ and preliminary data are generated at $\xi = 0, 1, 2, 3$, and 4. When the prior probabilities are taken to be $\Pi_{10} = \Pi_{20} = \Pi_{30} = \Pi_{40} = .25$, then the posterior probabilities after five data points are $\Pi_{15} = .0024$, $\Pi_{25} = .0058$, $\Pi_{35} = .6583$, and $\Pi_{45} = .3335$. Models 3 and 4 appear to be the important rivals from these initial data. For $n = 6$, the grid ($\xi = 0(.2)4$) is searched to find where D is maximized. Maximum discrimination occurs at $\xi = 0$ and when an observation is generated at this point the posterior probabilities are $\Pi_{16} = .0009$, $\Pi_{26} = .0012$, $\Pi_{36} = .8777$, and $\Pi_{46} = .1202$. The procedure is repeated for $n = 7$, 8, and 9. The discrimination results are summarized in Figure 4. Because the important rivals (models 3 and 4) differ only by a constant, this may partially explain why maximum discrimination

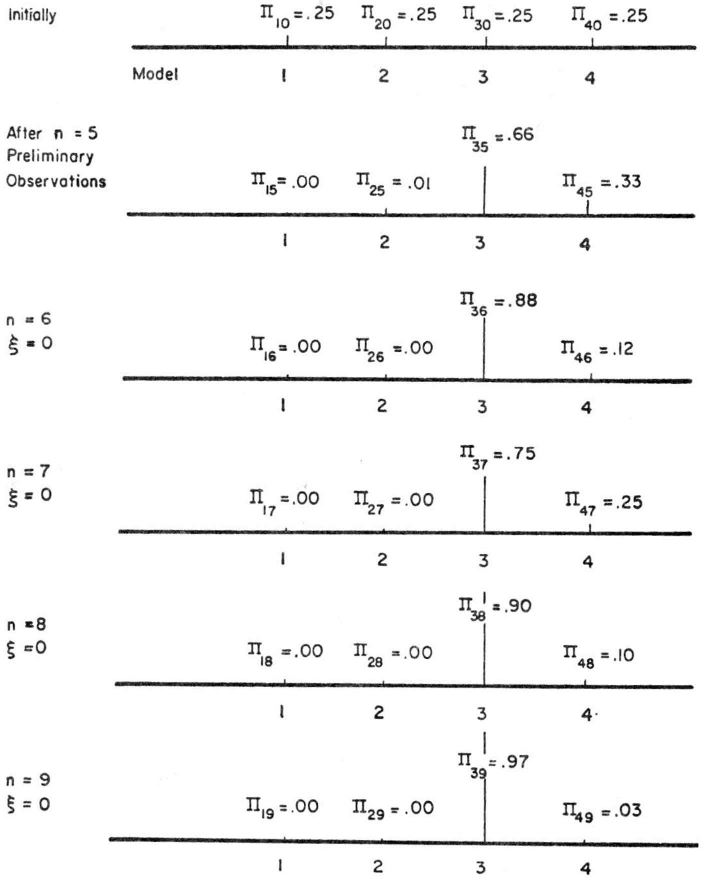

FIGURE 4—Posterior Probabilities for Four Polynomial Models (Model 3 True).

occurred at $\xi = 0$ for each run. Since after nine observations the posterior probabilities indicate that model 3 is clearly superior to the other three models, we discriminate in favor of model 3 (true model).

This example is one in which certain models are specific cases of others. For instance, by setting $\theta_{21} = 0$ in model 2, $\theta_{31} = \theta_{33} = 0$ in model 3, and $\theta_{42} = 0$ in model 4, one obtains model 1 in each case. In these circumstances it might be thought that if model 1 were the true model and since it follows that models 2, 3, and 4 would also be correct, it might be impossible to discriminate among them. In fact, the criterion does discriminate effectively in this case, essentially because the average variances of the estimates of the response, obtained from

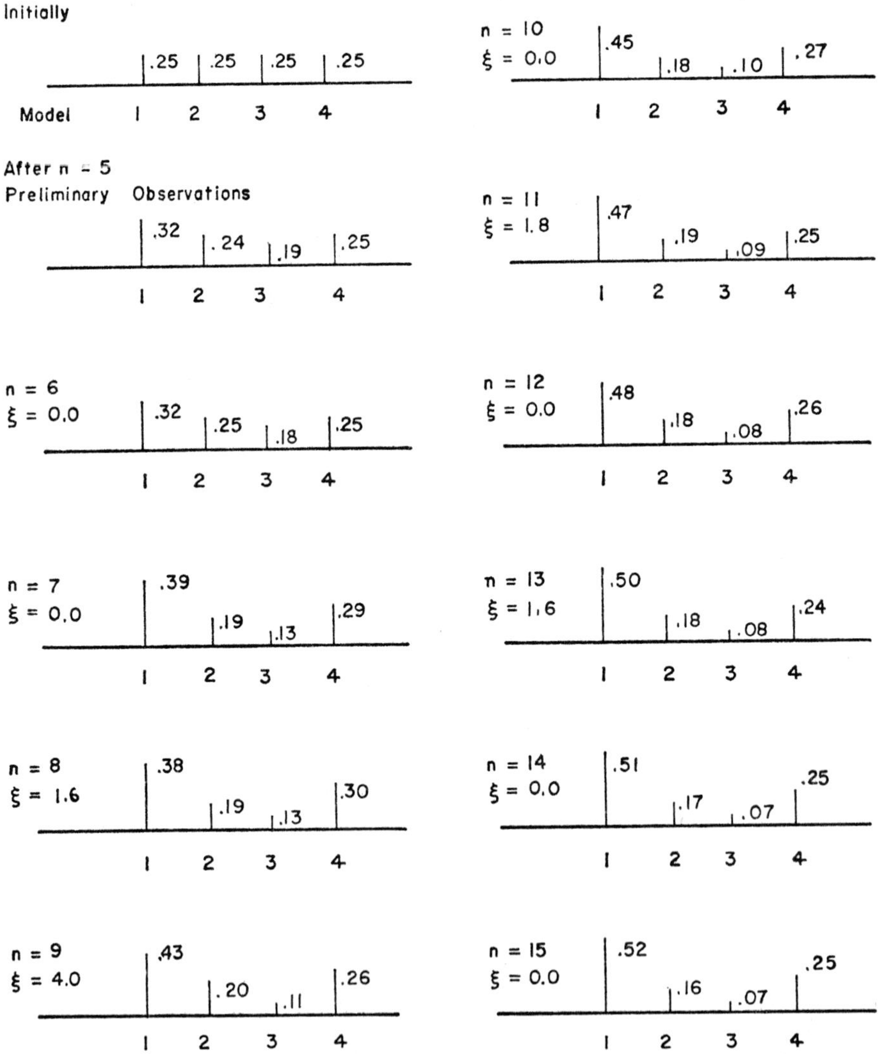

FIGURE 5—Posterior Probabilities for Four Polynomial Models (Model 1 True).

models with fewer parameters are smaller and thus the posterior probabilities become larger.

This reducible model situation is illustrated when model 1 is taken to be true with $\theta_{11} = 4$. The posterior probabilities are shown in Figure 5 at each stage up to $n = 15$ observations. Although the convergence is slow, as would be expected, and although more observations are needed to complete the analysis, there is a marked preference for the simplest model, model 1, even though all the models fit the data. Notice that model 3 which has the most parameters has the lowest probability associated with it after 15 observations.

Example 5.3

Suppose an experimenter is studying a chemical reaction of the type $A \rightarrow B$. Depending on whether the reaction is of first, second, third, or fourth order, the following $m = 4$ rival models can be considered

$$
\begin{aligned}
&\text{Model 1.} \quad \underset{1}{E(y)} = \exp\left(-\theta_{11}\xi_1 \exp\left(-\theta_{12}/\xi_2\right)\right), \\
&\text{Model 2.} \quad \underset{2}{E(y)} = 1/\left(1 + \theta_{21}\xi_1 \exp\left(-\theta_{22}/\xi_2\right)\right), \\
&\text{Model 3.} \quad \underset{3}{E(y)} = 1/\left(1 + 2\theta_{31}\xi_1 \exp\left(-\theta_{32}/\xi_2\right)\right)^{\frac{1}{2}}, \\
&\text{Model 4.} \quad \underset{4}{E(y)} = 1/\left(1 + 3\theta_{41}\xi_1 \exp\left(-\theta_{42}/\xi_2\right)\right)^{\frac{1}{3}},
\end{aligned}
\tag{5.4}
$$

where y is the concentration of A, ξ_1 is the reaction time, ξ_2 is the temperature, and θ_{i1} and θ_{i2} are the parameters of model i ($i = 1, 2, 3, 4$). In a constructed study, model 2 is chosen as the true model with $\theta_{21} = 400$ and $\theta_{22} = 5000$. The standard deviation $\sigma = .05$ is assumed known. The operability region in the (ξ_1, ξ_2) space is $0 < \xi_1 \leq 150$ and $450 \leq \xi_2 \leq 600$. It is desired to select design points in units of 25 for both ξ_1 and ξ_2. A preliminary 2^2-factorial design is chosen at levels $\xi_1 = 25$ and 125, and $\xi_2 = 475$ and 575. Data are generated from model 2 at these points and the initial priors are chosen to be $\Pi_{10} = \Pi_{20} = \Pi_{30} = \Pi_{40} = .25$. After these initial four runs, model 3 is slightly favored over

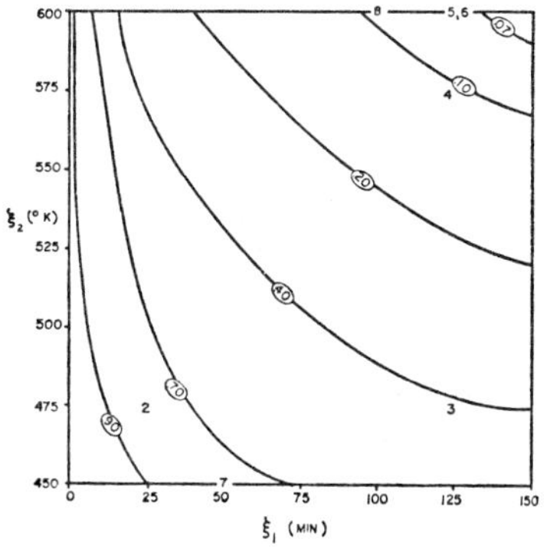

FIGURE 6—Contours of Unreacted A: 1–4 Initial Design Points; 5–8 Sequentially Chosen Points Using D.

model 2 (true model), whereas models 1 and 4 appear to have little importance. This is indicated by the posteriors $\Pi_{14} = .0069$, $\Pi_{24} = .4290$, $\Pi_{34} = .5008$, and $\Pi_{44} = .0633$ which are shown in Table 1. For $n = 5$ the discrimination function D is maximized on the grid ($\xi_1 = 25\ (25)\ 150$, $\xi_2 = 450\ (25)\ 600$) when $(\xi_1, \xi_2) = (125, 600)$. When an observation is generated at these conditions and when the posterior probabilities are calculated, model 2 is slightly favored. This is seen in Table 1 for $n = 5$. If the procedure is repeated until $n = 8$, model 2 is chosen as the best model on the basis of the posterior probabilities. The discrimination results are shown in Figure 6 and Table 1.

TABLE 1

Results when Discriminating Among $m = 4$ Rate Models where $\Pi_{10} = \Pi_{20} = \Pi_{30} = \Pi_{40} = .25$ (Model 2 True).

n	ξ_1	ξ_2	y	Π_1	Π_2	Π_3	Π_4
1	25	575	0.3961				
2	25	475	0.7232				
3	125	475	0.4215				
4	125	575	0.1297	.0069	.4290	.5008	.0633
5	125	600	0.0984	.0019	.5602	.4291	.0088
6	125	600	0.0556	.0018	.8639	.1339	.0004
7	50	450	0.7969	.0021	.9736	.0243	.0000
8	100	600	0.0325	.0032	.9956	.0012	.0000

Suppose that due to either wild observations or poor prior knowledge a wrong model is initially favored. Then, because of the weighting in D due to the presence of the prior probabilities, is there a possibility that the procedure may never recover from this incorrect weighting and continue to favor this wrong model? Consideration was given to the above question when this example was done again using an initial prior probability for model 3, seven times as great as those for models 1, 2, and 4. In reality, of course, model 2 was the true model and model 3 was the most important rival. Even with this handicap the procedure still rapidly discriminated and unmistakably revealed the true model after five additional observations. The results are shown in Table 2.

TABLE 2

Results when Discriminating Among $m = 4$ Rate Models where $\Pi_{10} = \Pi_{20} = \Pi_{40} = .1$ and $\Pi_{30} = .7$ (Model 2 True).

n	ξ_1	ξ_2	y	Π_1	Π_2	Π_3	Π_4
1	25	575	0.3961				
2	25	475	0.7232				
3	125	475	0.4215				
4	125	575	0.1297	.0018	.1071	.8752	.0159
5	125	600	0.0984	.0006	.1567	.8402	.0025
6	125	600	0.0556	.0011	.4790	.5197	.0002
7	50	450	0.7969	.0020	.8496	.1484	.0000
8	100	600	0.0600	.0035	.9884	.0081	.0000
9	75	600	0.2140	.0002	.9900	.0098	.0000

6. Conclusions

The availability of fast electronic computers ensures that calculations of the type envisaged here can be made in a reasonably short space of time. It is therefore possible to consider an important new development in the scientific method. The complexity of models which result from other than rather simple mechanisms may make it almost impossible for the experimenter to judge what constitutes a worthwhile experiment at a given stage of an investigation. Placing the computer in the iterative circuit shown in Figure 7 enormously

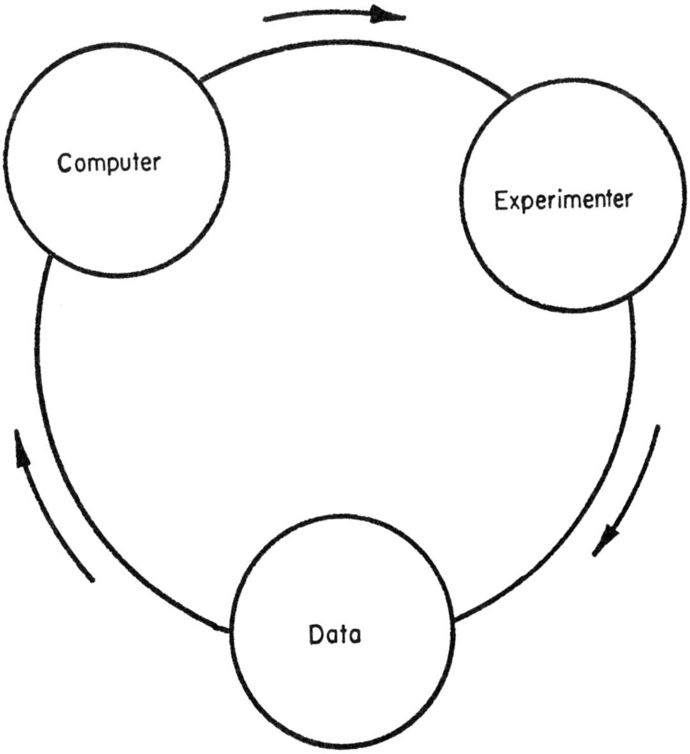

Figure 7—Iterative Experimental Procedure.

helps the experimenter to increase the discriminatory power of his experiments. The use of techniques like the above will increase the efficiency of experimentation and should reduce the frequency with which contradictory conclusions are reached in any given situation.

7. Appendix

The response function for model i ($i = 1, 2, \cdots, m$) is of the form

$$\underset{i}{E}(y) = f_i(\boldsymbol{\theta}_i, \boldsymbol{\xi}), \tag{7.1}$$

where $f_i(\boldsymbol{\theta}_i, \boldsymbol{\xi})$ depends upon p parameters $\boldsymbol{\theta}_i = (\theta_{i1}, \theta_{i2}, \cdots, \theta_{ip})'$ and k

independent variables $\boldsymbol{\xi} = (\xi_1, \xi_2, \cdots, \xi_k)$. The number of parameters p need not be the same for all models. It is supposed that for a region in the $\boldsymbol{\theta}_i$ space sufficiently close to the parameter estimates $\tilde{\boldsymbol{\theta}}_i$, the response function can be approximately expressed in the linear form

$$E_i(y_u) = f_i(\hat{\boldsymbol{\theta}}_i, \boldsymbol{\xi}_u) + \sum_{l=1}^{p} (\theta_{il} - \tilde{\theta}_{il})x_{lu}^{(i)} \qquad u = 1, 2, \cdots, n-1, \tag{7.2}$$

where $x_{lu}^{(i)} = [\partial f_i(\boldsymbol{\theta}_i, \boldsymbol{\xi}_u)/\partial\theta_{il}]_{\boldsymbol{\theta}_i = \hat{\boldsymbol{\theta}}_i}$ and where $\boldsymbol{\xi}_u = (\xi_{1u}, \xi_{2u}, \cdots, \xi_{ku})$ are the operating conditions for the u-th run.

After $n-1$ independent observations $\mathbf{y} = (y_1, y_2, \cdots, y_{n-1})'$, the likelihood function for parameters $\boldsymbol{\theta}_i$ is

$$L(\boldsymbol{\theta}_i \mid \sigma, \mathbf{y}) = \frac{1}{(\sqrt{2\pi}\,\sigma)^{n-1}} \cdot \exp\left\{-\frac{1}{2\sigma^2}(\mathbf{T}_i - \mathbf{X}_i(\boldsymbol{\theta}_i - \tilde{\boldsymbol{\theta}}_i))'(\mathbf{T}_i - \mathbf{X}_i(\boldsymbol{\theta}_i - \tilde{\boldsymbol{\theta}}_i))\right\}, \tag{7.3}$$

where

$$\mathbf{T}_i = \begin{bmatrix} y_1 - f_i(\tilde{\boldsymbol{\theta}}_i, \boldsymbol{\xi}_1) \\ y_2 - f_i(\tilde{\boldsymbol{\theta}}_i, \boldsymbol{\xi}_2) \\ \vdots \\ y_{n-1} - f_i(\tilde{\boldsymbol{\theta}}_i, \boldsymbol{\xi}_{n-1}) \end{bmatrix} \tag{7.4}$$

and

$$\mathbf{X}_i = \begin{bmatrix} x_{11}^{(i)} & x_{21}^{(i)} & \cdots & x_{p1}^{(i)} \\ x_{12}^{(i)} & x_{22}^{(i)} & \cdots & x_{p2}^{(i)} \\ \vdots & & & \\ x_{1n-1}^{(i)} & x_{2n-1}^{(i)} & \cdots & x_{pn-1}^{(i)} \end{bmatrix}. \tag{7.5}$$

Since the likelihood function (7.3) is maximized if and only if

$$\mathbf{T}_i'\mathbf{X}_i = \mathbf{0} \tag{7.6}$$

then

$$L(\boldsymbol{\theta}_i \mid \sigma, \mathbf{y}) = \frac{1}{(\sqrt{2\pi}\,\sigma)^{n-1}} \exp\left\{-\frac{1}{2\sigma^2}(\mathbf{T}_i'\mathbf{T}_i + (\boldsymbol{\theta}_i - \tilde{\boldsymbol{\theta}}_i)'\mathbf{X}_i'\mathbf{X}_i(\boldsymbol{\theta}_i - \tilde{\boldsymbol{\theta}}_i)\right\}. \tag{7.7}$$

If a locally uniform prior distribution is assumed for $\boldsymbol{\theta}_i$ then from Bayes theorem the posterior probability density function of $\boldsymbol{\theta}_i$ after $n-1$ observations is

$$p(\boldsymbol{\theta}_i \mid \sigma, \mathbf{y}) = \frac{L(\boldsymbol{\theta}_i \mid \sigma, \mathbf{y})}{\int L(\boldsymbol{\theta}_i \mid \sigma, \mathbf{y})\, d\boldsymbol{\theta}_i}. \tag{7.8}$$

Upon simplification (7.8) reduces to

$$p(\boldsymbol{\theta}_i \mid \sigma, \mathbf{y}) = \frac{|\mathbf{X}_i'\mathbf{X}_i|^{\frac{1}{2}}}{(\sqrt{2\pi}\,\sigma)^{p}} \exp\left\{-\frac{1}{2\sigma^2}(\boldsymbol{\theta}_i - \tilde{\boldsymbol{\theta}}_i)'\mathbf{X}_i'\mathbf{X}_i(\boldsymbol{\theta}_i - \tilde{\boldsymbol{\theta}}_i)\right\}. \tag{7.9}$$

Therefore, $(\boldsymbol{\theta}_i - \tilde{\boldsymbol{\theta}}_i)$ is normally distributed about zero with variance-covariance matrix $(\mathbf{X}_i'\mathbf{X}_i)^{-1}\sigma^2$.

Using the estimates of $\boldsymbol{\theta}_i$ based on the first $n - 1$ observations, we have the following linearized expression for the expected value of y_n under model i

$$\begin{aligned}\eta &= \underset{i}{E}(y_n) \\ &= f_i(\tilde{\boldsymbol{\theta}}_i, \xi_n) + \mathbf{x}_n^{(i)}(\boldsymbol{\theta}_i - \tilde{\boldsymbol{\theta}}_i),\end{aligned} \tag{7.10}$$

where $\mathbf{x}_n^{(i)} = (x_{1n}^{(i)}, x_{2n}^{(i)}, \cdots, x_{pn}^{(i)})$ is the $1 \times p$ vector of partial derivatives

$$x_{ln}^{(i)} = [\partial f_i(\boldsymbol{\theta}_i, \xi_n)/\partial \theta_{il}]_{\boldsymbol{\theta}_i = \hat{\boldsymbol{\theta}}_i}, \qquad l = 1, 2, \cdots, p. \tag{7.11}$$

Setting $\tilde{y}_n^{(i)} = f_i(\tilde{\boldsymbol{\theta}}_i, \xi_n)$, we have

$$\eta - \tilde{y}_n^{(i)} = \mathbf{x}_n^{(i)}(\boldsymbol{\theta}_i - \tilde{\boldsymbol{\theta}}_i). \tag{7.12}$$

Therefore, η is normally distributed about $\tilde{y}_n^{(i)}$ with variance

$$\sigma_i^2 = \mathbf{x}_n^{(i)}(\mathbf{X}_i'\mathbf{X}_i)^{-1}\mathbf{x}_n^{(i)\prime}\sigma^2.$$

Unless η is exactly linear in $\boldsymbol{\theta}_i$, then this distribution statement is only an approximation.

Acknowledgments

The authors wish to thank Professor Irwin Guttman for his valuable suggestions concerning the use of Kullback's information theory. The authors are also indebted to Professor D. A. S. Fraser who encouraged the authors to study the effects of poor prior knowledge on the discrimination procedure. This research was supported by the National Science Foundation under grant GP-2755.

References

1. Box, G. E. P., 1954. The exploration and exploitation of response surfaces. *Biometrics, 10,* 16–60.
2. Box, G. E. P., 1960. Some general considerations in process optimization. *J. Basic Eng., 82,* 113.
3. Box, G. E. P. and Wilson, K. B., 1951. On the experimental attainment of optimum conditions. *J. Roy. Statist. Soc., Series B, 13,* 1–45.
4. Davies, O. L., 1954. *The Design and Analysis of Industrial Experiments.* Hafner, New York, Chapter 11, 495–578.
5. Hunter, J. S., 1958. Determination of optimum conditions by experimental methods, Part II-1. *Indus. Qual. Control, 15,* 16–24.
6. Hunter, J. S., 1959. Determination of optimum conditions by experimental methods, Part II-2. *Indus. Qual. Control, 15,* 7–15.
7. Hunter, J. S., 1959. Determination of optimum conditions by experimental methods, Part II-3. *Indus. Qual. Control, 15,* 6–14.
8. Hunter, W. G. and Reiner, A. M., 1965. Designs for discriminating between two rival models. *Technometrics, 7,* 307–23.
9. Jenkins, G. M., Kisiel, A. J., Price, R. J., and Rippin, D. W., 1963. Preliminary Report, Optimization Project, Imperial College.
10. Kittrell, J. R., 1964. Preliminary Report, Department of Chemical Eng., University of Wisconsin, Madison, Wisconsin.
11. Kullback, S., 1959. *Information Theory and Statistics.* John Wiley and Sons, Inc., New York.
12. Shannon, C. E., 1948. A mathematical theory of communication. *Bell System Tech. Journal, 27,* 379–423 and 623–56.

1.11
A Bayesian Approach to Some Outlier Problems

G. E. P. BOX and G. C. TIAO
University of Wisconsin

SUMMARY

The problem of outlying observations is considered from a Bayesian viewpoint. We suppose that each of the observations in an experiment may come from either a 'good' run or a 'bad' run. By specifying the models corresponding to good and bad runs and the prior probabilities of which runs being bad, we then employ standard Bayesian inference procedures to derive the appropriate analysis. In particular, we consider the linear model and assume that a good observation is normally distributed about its mean with variance σ^2, and a bad one is normal with the same mean but a larger variance $k^2\sigma^2$. An example is given.

1. INTRODUCTION

How to bring order and yet to permit diversity is a problem for all men and not least for the statistician. Most statistical procedures are arrived under the assumption that *each one* of a given set of observations is generated by a specific stochastic model containing a modest number of adjustable parameters. That this assumption does not correspond with what the statistician really believes is shown in particular by the fact that he makes provision for the rejection of 'outliers'; an outlier being an observation which is suspected of being partially or wholly irrelevant because it is *not* generated by the stochastic model assumed (Anscombe, 1960; Daniel, 1960; Gebhardt, 1964; Grubbs, 1950; Thompson, 1935; etc.).

When no knowledge is available as to which particular observation or observations is irrelevant, an approach more descriptive of the statistician's state of mind might suppose that there was a small prior probability α that any given observation was *not* generated by the central stochastic model as well as a complementary prior probability $(1-\alpha)$ that it *was* so generated. The main object of this paper is to explore the consequences of such a supposition for the normal theory linear model.

Usually the mathematical derivation of a statistical procedure goes forward on the basis that the assumptions are certain. In practice, however, the statistician's feelings about the model are of hope rather than certainty, and having derived his procedures on fixed assumptions, he, very sensibly, examines these assumptions by analysis of residuals and other means. It is of interest to see what analysis he might be led to if his doubts were built into the original model. Doubts about the linear normal theory model, for example, might concern additivity, homogeneity of variance, independence of observations and normality. It was a desire to discover the result of building in doubts of these kinds using Bayes's theorem which was the motivation for the studies by Box & Tiao (1962, 1964, 1965), Zellner & Tiao (1964) and Tiao & Tan (1966).

Now in practice hedgings must be paid for with additional parameters in the model, but if our treatment of the data is to be worthy of the name of analysis, our final model should be as simple as possible. Additional parameters if they are to be introduced at all must therefore be added artfully and grudgingly, in a word, parsimoniously. One example of such

The Collected Works of George E. P. Box, 1984, Wadsworth, Inc., Belmont, CA 94002.
Originally published in *Biometrika,* vol. 55, no. 1 (1968), pp. 119–129.

parsimony is the improvement in additivity, homogeneity of variance and normality which sometimes results by the introduction of a single transformation parameter (see, e.g. Box & Cox, 1964).

A different type of parsimonious extension of the normal theory model which might be considered along or in combination with transformation arises from the concept of outliers. When a series of runs is made on some experimental apparatus the data analyst may feel that while most of these runs will be 'good' runs, a few may be 'bad' runs; a 'bad' run being one which was consciously or unconsciously not made in the manner intended. Different situations can occur. At one end of the scale it may be known precisely which runs were bad runs and we may know the exact experimental conditions at which they were made. In such a case exact allowance may be made for these discrepancies and the only effect of the bad runs will be some loss in efficiency resulting from departure from the desired form of design. At the other end of the scale there may be no specific information to distinguish the goodness of one run from that of another, but there is a general possibility that one or more of the runs may be bad. It is the latter case that we consider here.

In § 2, we present a general formulation of the problem. We suppose that for any given run there exist two alternative possible models: a standard model appropriate when an observation comes from a good run and an alternative model appropriate when an observation comes from a bad run. Once the prior probability of which runs being bad is specified, we can employ standard procedures of inference to derive the appropriate analysis.

We consider in § 3 the case where each observation has the common probability α of being generated from a bad run. We suppose that a good observation is normally distributed about its mean with fixed variance σ^2. A bad observation is normally distributed about the same mean but with a larger variance $k^2\sigma^2$. This type of model has been considered by Tukey (1960) in a somewhat different context. For purpose of illustration, we give a numerical example in the case of the estimation of a single mean.

2. A general formulation of the problem

Consider the familiar linear model

$$\mathbf{y} = X\boldsymbol{\theta} + \boldsymbol{\epsilon}, \tag{2·1}$$

where $\mathbf{y}$ is an $n \times 1$ vector of observations, X an $n \times p$ matrix of fixed elements with rank p, $\boldsymbol{\theta}$ a $p \times 1$ vector of regression coefficients and $\boldsymbol{\epsilon}$ an $n \times 1$ vector of random errors or disturbances. We suppose that each of the errors could independently come from either one of two distributions, a standard distribution $f(\epsilon \mid \boldsymbol{\xi}_1)$ and an alternative distribution $g(\epsilon \mid \boldsymbol{\xi}_2)$. The regression coefficients $\boldsymbol{\theta}$ are the parameters of principal interest and $(\boldsymbol{\xi}_1, \boldsymbol{\xi}_2)$ are nuisance parameters.

Let $a_{(r)}$ be the event that a particular set of r of the n ϵ's are from $g(\epsilon \mid \boldsymbol{\xi}_2)$ and the remaining $s = n - r$ from $f(\epsilon \mid \boldsymbol{\xi}_1)$. Corresponding to $a_{(r)}$, we partition the vector $\boldsymbol{\epsilon}$ into $\boldsymbol{\epsilon}_{(r)}$ and $\boldsymbol{\epsilon}_{(s)}$, the vector $\mathbf{y}$ into $\mathbf{y}_{(r)}$ and $\mathbf{y}_{(s)}$, and the matrix X into $X_{(r)}$ and $X_{(s)}$. Given the observations $\mathbf{y}$, the likelihood of $(\boldsymbol{\theta}, \boldsymbol{\xi}_1, \boldsymbol{\xi}_2, a_{(r)})$ is

$$l(\boldsymbol{\theta}, \boldsymbol{\xi}_1, \boldsymbol{\xi}_2, a_{(r)}) \propto \dot{f}(\boldsymbol{\epsilon}_{(s)} \mid \boldsymbol{\xi}_1)\, \dot{g}(\boldsymbol{\epsilon}_{(r)} \mid \boldsymbol{\xi}_2) \propto \dot{f}(\mathbf{y}_{(s)} - X_{(s)}\boldsymbol{\theta} \mid \boldsymbol{\xi}_1)\, \dot{g}(\mathbf{y}_{(r)} - X_{(r)}\boldsymbol{\theta} \mid \boldsymbol{\xi}_2), \tag{2·2}$$

where $\dot{f}$ is the product of the density functions of the elements of $\boldsymbol{\epsilon}_{(s)}$ and $\dot{g}$ that of $\boldsymbol{\epsilon}_{(r)}$. Thus, the entire likelihood function consists of 2^n expressions of the type given in (2·2) corresponding to the 2^n possible combinations of the ϵ's.

Suppose we let $p^{(r)}$ be the prior probability of the event $a_{(r)}$, so that

$$p^{(r)} \geqslant 0 \quad \text{and} \quad \sum_{(r)} p^{(r)} = 1. \tag{2·3}$$

Further, let $p(\boldsymbol{\theta}, \boldsymbol{\xi}_1, \boldsymbol{\xi}_2)$ be the prior distribution of the parameters. Then, given the observations $\mathbf{y}$, the posterior distribution of $(\boldsymbol{\theta}, \boldsymbol{\xi}_1, \boldsymbol{\xi}_2, a_{(r)})$ is

$$p(\boldsymbol{\theta}, \boldsymbol{\xi}_1, \boldsymbol{\xi}_2, a_{(r)} \mid \mathbf{y}) = \frac{p^{(r)} p(\boldsymbol{\theta}, \boldsymbol{\xi}_1, \boldsymbol{\xi}_2) \dot{f}(\mathbf{y}_{(s)} - X_{(s)}\boldsymbol{\theta} \mid \boldsymbol{\xi}_1) \dot{g}(\mathbf{y}_{(r)} - X_{(r)}\boldsymbol{\theta} \mid \boldsymbol{\xi}_2)}{\sum_{(r)} p^{(r)} h(\mathbf{y}_{(r)} \sim g;\, \mathbf{y}_{(s)} \sim f)}. \tag{2·4}$$

The expressions $\mathbf{y}_{(r)} \sim g$ and $\mathbf{y}_{(s)} \sim f$ are used to indicate respectively the event that the elements of $\boldsymbol{\epsilon}_{(r)}$ are drawn from the distribution $g(\epsilon|\boldsymbol{\xi}_2)$ and the event that the elements of $\boldsymbol{\epsilon}_{(s)}$ are drawn from the distribution $f(\epsilon|\boldsymbol{\xi}_1)$, so that

$$h(\mathbf{y}_{(r)} \sim g;\, \mathbf{y}_{(s)} \sim f) = \int_R p(\boldsymbol{\theta}, \boldsymbol{\xi}_1, \boldsymbol{\xi}_2) \dot{f}(\mathbf{y}_{(s)} - X_{(s)}\boldsymbol{\theta} \mid \boldsymbol{\xi}_1) \dot{g}(\mathbf{y}_{(r)} - X_{(r)}\boldsymbol{\theta} \mid \boldsymbol{\xi}_2)\, d\boldsymbol{\xi}_1\, d\boldsymbol{\xi}_2\, d\boldsymbol{\theta} \tag{2·5}$$

is the marginal, or unconditional, distribution of $\mathbf{y}$ under the assumption that the elements of $\boldsymbol{\epsilon}_{(r)}$ are from $g(\epsilon \mid \boldsymbol{\xi}_2)$ and the elements of $\boldsymbol{\epsilon}_{(s)}$ are from $f(\epsilon \mid \boldsymbol{\xi}_1)$. After integrating out $(\boldsymbol{\xi}_1, \boldsymbol{\xi}_2)$ from (2·4), the distribution of $(\boldsymbol{\theta}, a_{(r)})$ can be written

$$p(\boldsymbol{\theta}, a_{(r)} \mid \mathbf{y}) = p(a_{(r)} \mid \mathbf{y})\, p(\boldsymbol{\theta} \mid a_{(r)}, \mathbf{y}), \tag{2·6}$$

where

$$p(a_{(r)} \mid \mathbf{y}) = \frac{p^{(r)} h(\mathbf{y}_{(r)} \sim g;\, \mathbf{y}_{(s)} \sim f)}{\sum_{(r)} p^{(r)} h(\mathbf{y}_{(r)} \sim g;\, \mathbf{y}_{(s)} \sim f)} \tag{2·7}$$

is the marginal posterior distribution of $a_{(r)}$, and

$$p(\boldsymbol{\theta} \mid a_{(r)}, \mathbf{y}) = \frac{\int_{R_1} p(\boldsymbol{\theta}, \boldsymbol{\xi}_1, \boldsymbol{\xi}_2) \dot{f}(\mathbf{y}_{(s)} - X_{(s)}\boldsymbol{\theta} \mid \boldsymbol{\xi}_1) \dot{g}(\mathbf{y}_{(r)} - X_{(r)}\boldsymbol{\theta} \mid \boldsymbol{\xi}_2)\, d\boldsymbol{\xi}_1\, d\boldsymbol{\xi}_2}{h(\mathbf{y}_{(r)} \sim g;\, \mathbf{y}_{(s)} \sim f)} \tag{2·8}$$

is the conditional posterior distribution of $\boldsymbol{\theta}$ given that a particular r combination of the ϵ's are from $g(\epsilon \mid \boldsymbol{\xi}_2)$ and the rest from $f(\epsilon \mid \boldsymbol{\xi}_1)$. Thus, the marginal posterior distribution of $\boldsymbol{\theta}$ is a weighted average of 2^n such conditional distributions

$$p(\boldsymbol{\theta} \mid \mathbf{y}) = \sum_{(r)} p(a_{(r)} \mid \mathbf{y})\, p(\boldsymbol{\theta} \mid a_{(r)}, \mathbf{y}) \tag{2·9}$$

with the weights given by $p(a_{(r)} \mid \mathbf{y})$.

Now, the quantity $h(\mathbf{y}_{(r)} \sim g;\, \mathbf{y}_{(s)} \sim f)$ can be written as the product

$$h(\mathbf{y}_{(r)} \sim g;\, \mathbf{y}_{(s)} \sim f) = h(\mathbf{y}_{(s)} \sim f)\, h(\mathbf{y}_{(r)} \sim g \mid \mathbf{y}_{(s)} \sim f), \tag{2·10}$$

where

$$h(\mathbf{y}_{(s)} \sim f) = \int_{R_2} h(\mathbf{y}_{(r)} \sim g;\, \mathbf{y}_{(s)} \sim f)\, d\mathbf{y}_{(r)} \tag{2·11}$$

is the marginal distribution of $\mathbf{y}_{(s)}$ and $h(\mathbf{y}_{(r)} \sim g \mid \mathbf{y}_{(s)} \sim f)$ the 'predictive' distribution of $\mathbf{y}_{(r)}$ given $\mathbf{y}_{(s)}$, under the assumption that $\boldsymbol{\epsilon}_{(r)}$ are from $g(\epsilon \mid \boldsymbol{\xi}_2)$ and $\boldsymbol{\epsilon}_{(s)}$ from $f(\epsilon \mid \boldsymbol{\xi}_1)$. From (2·5) and because of the independence assumption of the ϵ's, we can write

$$h(\mathbf{y}_{(s)} \sim f) = \int_R p(\boldsymbol{\theta}, \boldsymbol{\xi}_1, \boldsymbol{\xi}_2) \dot{f}(\mathbf{y}_{(s)} - X_{(s)}\boldsymbol{\theta} \mid \boldsymbol{\xi}_1)\, d\boldsymbol{\xi}_1\, d\boldsymbol{\xi}_2\, d\boldsymbol{\theta}. \tag{2·12}$$

Note that as might be expected, this expression holds irrespective of whether $\boldsymbol{\epsilon}_{(r)}$ are assumed to be from $g(\epsilon \mid \boldsymbol{\xi}_2)$ or $f(\epsilon \mid \boldsymbol{\xi}_1)$. Making use of (2·12), we can express the weights $p(a_{(r)} \mid \mathbf{y})$ alternatively as

$$\begin{aligned} p(a_{(r)} \mid \mathbf{y}) &= C\frac{p^{(r)} h(\mathbf{y}_{(r)} \sim g;\, \mathbf{y}_{(s)} \sim f)}{p^{(0)} h(\mathbf{y}_{(r)} \sim f;\, \mathbf{y}_{(s)} \sim f)} = C\frac{p^{(r)} h(\mathbf{y}_{(s)} \sim f)\, h(\mathbf{y}_{(r)} \sim g \mid \mathbf{y}_{(s)} \sim f)}{p^{(0)} h(\mathbf{y}_{(s)} \sim f)\, h(\mathbf{y}_{(r)} \sim f \mid \mathbf{y}_{(s)} \sim f)} \\ &= C\frac{p^{(r)} h(\mathbf{y}_{(r)} \sim g \mid \mathbf{y}_{(s)} \sim f)}{p^{(0)} h(\mathbf{y}_{(r)} \sim f \mid \mathbf{y}_{(s)} \sim f)}, \end{aligned} \tag{2·13}$$

where
$$C = \frac{p^{(0)}h(\mathbf{y}_{(r)} \sim g;\, \mathbf{y}_{(s)} \sim f)}{\sum_{(r)} p^{(r)}h(\mathbf{y}_{(r)} \sim g;\, \mathbf{y}_{(s)} \sim f)}$$

is the appropriate normalizing constant. That is,

$$\begin{aligned} p(a_{(r)} \mid \mathbf{y}) &= C \times \text{prior probability ratio} \times \text{ratio of 'predictive' densities} \\ &= C \times \text{prior probability ratio} \times \text{'likelihood' ratio} \\ &= C \times \text{posterior probability ratio.} \end{aligned}$$

In other words, $p(a_{(r)} \mid \mathbf{y})$ is proportional to the ratio of the posterior probabilities of two hypotheses (i) $\boldsymbol{\epsilon}_{(r)}$ from $g(\epsilon \mid \boldsymbol{\xi}_2)$ and (ii) $\boldsymbol{\epsilon}_{(r)}$ from $f(\epsilon \mid \boldsymbol{\xi}_1)$, given that the remaining ϵ's are from $f(\epsilon \mid \boldsymbol{\xi}_1)$, a result which is intuitively appealing.

We now turn to illustrate the above approach for a specific choice of $p^{(r)}$, $f(\epsilon \mid \boldsymbol{\xi}_1)$ and $g(\epsilon \mid \boldsymbol{\xi}_2)$.

3. A special linear model

We suppose that the random error ϵ associated with each observation could have been drawn from two sources, a *central* model $N(0, \sigma^2)$ and an alternative model $N(0, k^2\sigma^2)$, with probabilities $(1-\alpha)$ and α respectively. We proceed with the analyses by further supposing that k is fixed, and the θ's and $\log\sigma$ are locally independent and uniform *a priori*. In the context of our general framework, we have

$$f(\epsilon \mid \boldsymbol{\xi}_1) = \frac{1}{\sigma\sqrt{(2\pi)}}\exp\left(-\frac{\epsilon^2}{2\sigma^2}\right), \quad g(\epsilon \mid \boldsymbol{\xi}_2) = \frac{1}{k\sigma\sqrt{(2\pi)}}\exp\left(-\frac{\epsilon^2}{2k^2\sigma^2}\right), \tag{3·1}$$

$$p^{(r)} = \alpha^r(1-\alpha)^{n-r} \tag{3·2}$$

and
$$p(\boldsymbol{\xi}_1, \boldsymbol{\xi}_2, \boldsymbol{\theta}) = p(\boldsymbol{\theta}, \sigma) \propto \frac{1}{\sigma}. \tag{3·3}$$

As mentioned earlier, we can express the posterior distribution of $\boldsymbol{\theta}$ as a weighted average of 2^n distributions

$$p(\boldsymbol{\theta} \mid \mathbf{y}) = \sum_{(r)} w_{(r)} p_{(r)}(\boldsymbol{\theta} \mid \mathbf{y}), \tag{3·4}$$

where $w_{(r)}$ corresponds to $p(a_{(r)} \mid \mathbf{y})$ and $p_{(r)}(\boldsymbol{\theta} \mid \mathbf{y})$ to $p(\boldsymbol{\theta} \mid a_{(r)}, \mathbf{y})$ in the general framework. We now sketch the derivation of the distributions and the weights.

The distribution $p_{(r)}(\boldsymbol{\theta} \mid \mathbf{y})$. Consider a particular combination (r). Under the assumptions made in (3·1)–(3·3), the corresponding posterior distribution of $\boldsymbol{\theta}$ is, from (2·8),

$$p(\boldsymbol{\theta} \mid a_{(r)}, \mathbf{y}) = p_{(r)}(\boldsymbol{\theta} \mid \mathbf{y}) \propto \int_0^\infty \sigma^{-(n+1)} \exp\left\{-\frac{1}{2\sigma^2} S_{(r)}(\boldsymbol{\theta})\right\} d\sigma, \tag{3·5}$$

where $S_{(r)}(\boldsymbol{\theta}) = (\mathbf{y}_{(n-r)} - X_{(n-r)}\boldsymbol{\theta})'(\mathbf{y}_{(n-r)} - X_{(n-r)}\boldsymbol{\theta}) + (1/k^2)(\mathbf{y}_{(r)} - X_{(r)}\boldsymbol{\theta})'(\mathbf{y}_{(r)} - X_{(r)}\boldsymbol{\theta})$.

Integrating out σ, we obtain

$$p_{(r)}(\boldsymbol{\theta} \mid \mathbf{y}) \propto \{S_{(r)}(\boldsymbol{\theta})\}^{-\frac{1}{2}n}. \tag{3·6}$$

Now it is well known that we can write

$$S_{(r)}(\boldsymbol{\theta}) = S_{(r)}(\hat{\boldsymbol{\theta}}_{(r)}) + (\boldsymbol{\theta} - \hat{\boldsymbol{\theta}}_{(r)})'\left(X'_{(n-r)}X_{(n-r)} + \frac{1}{k^2}X'_{(r)}X_{(r)}\right)(\boldsymbol{\theta} - \hat{\boldsymbol{\theta}}_{(r)}), \tag{3·7}$$

where
$$\hat{\boldsymbol{\theta}}_{(r)} = \left(X'_{(n-r)}X_{(n-r)} + \frac{1}{k^2}X'_{(r)}X_{(r)}\right)^{-1}\left(X'_{(n-r)}\mathbf{y}_{(n-r)} + \frac{1}{k^2}X'_{(r)}\mathbf{y}_{(r)}\right).$$

In particular, when $(r) = (0)$, i.e. all the errors are generated from the central model $N(0,\sigma^2)$, we have

$$\left.\begin{aligned}\hat{\boldsymbol\theta}_{(0)} = \hat{\boldsymbol\theta} = (X'X)^{-1}X'\mathbf{y}, \quad S_{(0)}(\hat{\boldsymbol\theta}_{(0)}) = \nu s^2,\\ \nu = n-p \quad\text{and}\quad s^2 = (1/\nu)(\mathbf{y}-X\hat{\boldsymbol\theta})'(\mathbf{y}-X\hat{\boldsymbol\theta}).\end{aligned}\right\} \tag{3·8}$$

It is informative to express the quantities $\hat{\boldsymbol\theta}_{(r)}$ and $S_{(r)}(\hat{\boldsymbol\theta}_{(r)})$ in terms of the more familiar least squares estimates $\hat{\boldsymbol\theta}$, sample variance s^2 and residuals $\mathbf{y}_{(r)} - X_{(r)}\hat{\boldsymbol\theta}$. Making use of the identity

$$\left(X'_{(n-r)}X_{(n-r)}+\frac{1}{k^2}X'_{(r)}X_{(r)}\right)^{-1} = (X'X-\phi X'_{(r)}X_{(r)})^{-1}$$
$$= (X'X)^{-1}+\phi(X'X)^{-1}X'_{(r)}\{I-\phi X_{(r)}(X'X)^{-1}X'_{(r)}\}^{-1}X_{(r)}(X'X)^{-1}, \quad \phi = 1-k^{-2}, \tag{3·9}$$

and after a little algebraic reduction, we obtain

$$\left.\begin{aligned}\hat{\boldsymbol\theta}_{(r)} &= \hat{\boldsymbol\theta} - \phi(X'X)^{-1}X'_{(r)}\{I-\phi X_{(r)}(X'X)^{-1}X'_{(r)}\}^{-1}(\mathbf{y}_{(r)}-X_{(r)}\hat{\boldsymbol\theta}),\\ \nu s^2_{(r)} &= \nu s^2 - \phi(\mathbf{y}_{(r)}-X_{(r)}\hat{\boldsymbol\theta})'\{I-\phi X_{(r)}(X'X)^{-1}X'_{(r)}\}^{-1}(\mathbf{y}_{(r)}-X_{(r)}\hat{\boldsymbol\theta}),\end{aligned}\right\} \tag{3·10}$$

where
$$s^2_{(r)} = \frac{1}{\nu}S_{(r)}(\hat{\boldsymbol\theta}_{(r)}).$$

Using (3·7)–(3·10) and upon normalizing, the posterior distribution in (3·6) is

$$p_{(r)}(\boldsymbol\theta\mid\mathbf{y}) = \frac{\Gamma(\tfrac12 n)\,|X'X-\phi X'_{(r)}X_{(r)}|^{\frac12}}{\Gamma(\tfrac12\nu)\,(\pi\nu s^2_{(r)})^{\frac12 p}} \times\left\{1+\frac{(\boldsymbol\theta-\hat{\boldsymbol\theta}_{(r)})'(X'X-\phi X'_{(r)}X_{(r)})(\boldsymbol\theta-\hat{\boldsymbol\theta}_{(r)})}{\nu s^2_{(r)}}\right\}^{-\frac12 n}, \tag{3·11}$$

which is a p-dimensional multivariate t-distribution with means $\hat{\boldsymbol\theta}_{(r)}$, dispersion matrix $s^2_{(r)}(X'X-\phi X'_{(r)}X_{(r)})^{-1}$, and $\nu = n-p$ degrees of freedom. In particular, for $(r) = (0)$, i.e. all the ϵ's are from $N(0,\sigma^2)$, the distribution in (3·11) reduces to the ordinary p-dimensional t-distribution with ν degrees of freedom centred at the least squares estimates $\hat{\boldsymbol\theta}$ and having dispersion matrix $s^2(X'X)^{-1}$.

The weights $w_{(r)}$. For the model considered in this section, the predictive density $h(\mathbf{y}_{(r)}\sim g\mid\mathbf{y}_{(s)}\sim f)$ in (2·13) is

$$h(\mathbf{y}_{(r)}\sim g\mid\mathbf{y}_{(n-r)}\sim f) = \int_R p(\boldsymbol\theta,\sigma)\dot f(\mathbf{y}_{(n-r)}-X_{(n-r)}\boldsymbol\theta\mid\sigma^2)\,\dot g(\mathbf{y}_{(r)}-X_{(r)}\boldsymbol\theta\mid k^2\sigma^2)\,d\boldsymbol\theta\,d\sigma \Bigg/\int_R p(\boldsymbol\theta,\sigma)\dot f(\mathbf{y}_{(n-r)}-X_{(n-r)}\boldsymbol\theta\mid\sigma^2)\,d\boldsymbol\theta\,d\sigma. \tag{3·12}$$

Making use of (3·7), it is easy to see that

$$h(\mathbf{y}_{(r)}\sim g\mid\mathbf{y}_{(n-r)}\sim f) = C_1k^{-r}|X'X-\phi X'_{(r)}X_{(r)}|^{-\frac12}(\nu s^2_{(r)})^{-\frac12(n-p)}, \tag{3·13}$$

where C_1 is some positive constant independent of $(k, \mathbf{y}_{(r)}, X_{(r)})$. In the special case when $X_{(n-r)}$ is of rank p, we have that

$$h(\mathbf{y}_{(r)}\sim g\mid\mathbf{y}_{(n-r)}\sim f) = \frac{\Gamma(\tfrac12 r)\,|A_{(r)}|^{\frac12}}{\Gamma\{\tfrac12(n-p-r)\}\{\pi(n-p-r)\,\bar s^2_{(n-r)}\}^{\frac12 r}} \times\left\{1+\frac{(\mathbf{y}_{(r)}-X_{(r)}\bar{\boldsymbol\theta}_{n-r)})'A_{(r)}(\mathbf{y}_{(r)}-X_{(r)}\bar{\boldsymbol\theta}_{(n-r)})}{(n-p-r)\,\bar s^2_{(n-r)}}\right\}^{-\frac12(n-p)} \tag{3·14}$$

with
$$\bar{\boldsymbol\theta}'_{(n-r)} = (X'_{(n-r)}X_{(n-r)})^{-1}X'_{(n-r)}\mathbf{y}_{(n-r)},$$

$$\bar s^2_{(n-r)} = \frac{1}{(n-r-p)}(\mathbf{y}_{(n-r)}-X_{(n-r)}\bar{\boldsymbol\theta}_{(n-r)})'(\mathbf{y}_{(n-r)}-X_{(n-r)}\bar{\boldsymbol\theta}_{(n-r)})$$

and
$$A_{(r)} = \frac{1}{k^2}\left\{I+\frac{1}{k^2}X_{(r)}(X'_{(n-r)}X_{(n-r)})^{-1}X'_{(r)}\right\}^{-1},$$

that is, an r-dimensional multivariate t-distribution with means $X_{(r)}\bar{\boldsymbol{\theta}}_{(n-r)}$, dispersion matrix $\bar{s}^2_{(n-r)}A^{-1}_{(r)}$ and $n-p-r$ degrees of freedom (Raiffa & Schlaifer, 1961). When the rank of $X_{(n-r)}$ is less than p, the predictive density in (3·13) is a singular distribution.

The predictive density $h(\mathbf{y}_{(r)} \sim f \mid \mathbf{y}_{(s)} \sim f)$ in (2·13) for the present model is obtained simply by setting $k=1$ in (3·13). It follows the weights $w_{(r)}$ are

$$w_{(r)} = C\left(\frac{\alpha}{1-\alpha}\right)^r k^{-r} \frac{|X'X|^{\frac{1}{2}}}{|X'X-\phi X'_{(r)}X_{(r)}|^{\frac{1}{2}}}\left\{\frac{s^2_{(r)}}{s^2}\right\}^{-\frac{1}{2}\nu}, \tag{3·15}$$

where C is the constant making the weights sum to unity.

Marginal distribution of θ_l. Using the property of the multivariate t-distribution, we obtain from (3·4), (3·11) and (3·15) the marginal distribution of the lth element of $\boldsymbol{\theta}$, $l=1,\ldots,p$, as

$$p(\theta_l \mid \mathbf{y}) = \sum_{(r)} w_{(r)} p_{(r)}(\theta_l \mid \mathbf{y}), \tag{3·16}$$

where
$$p_{(r)}(\theta_l \mid \mathbf{y}) = s^{-1}_{(r)}(v^{ll}_{(r)})^{-\frac{1}{2}}\, p\left\{t_{n-p} = \frac{\theta_l - \hat{\theta}_{l(r)}}{s_{(r)}(v^{ll}_{(r)})^{\frac{1}{2}}}\right\},$$

$\hat{\theta}_{l(r)}$ is the lth element of $\hat{\boldsymbol{\theta}}_{(r)}$ and $v^{ll}_{(r)}$ the lth diagonal element of the matrix $(X'X-\phi X'_{(r)}X_{(r)})^{-1}$. It is easy to see that the mean and standard deviation of this distribution are, respectively,

$$E(\theta_l \mid \mathbf{y}) = \bar{\theta}_l = \sum_{(r)} w_{(r)}\hat{\theta}_{l(r)}, \tag{3·17}$$

$$\text{S.D.}\,(\theta_l \mid \mathbf{y}) = \{\text{var}\,(\theta_l \mid \mathbf{y})\}^{\frac{1}{2}} \tag{3·18}$$

with
$$\text{var}\,(\theta_l \mid \mathbf{y}) = \sum_{(r)} w_{(r)}\left\{\left(\frac{n-p}{n-p-2}\right)s^2_{(z)}\, v^{ll}_{(r)} + (\hat{\theta}_{l(r)}-\bar{\theta}_l)^2\right\}.$$

Estimation of a single mean θ. A special case of the linear model (2·1) which is of particular importance is that of a single mean θ. In this case $p=1$, $\boldsymbol{\theta}=\theta$, and X is a $n\times 1$ vector of unities. It is straightforward to verify that

$$\left.\begin{aligned} \hat{\theta}_{(r)} &= \bar{y} - \frac{r\phi}{n-r\phi}(\bar{y}_{(r)}-\bar{y}),\\ \nu s^2_{(r)} &= \nu s^2 - \phi\left\{(\mathbf{y}_{(r)}-\mathbf{1}_r\bar{y})'(\mathbf{y}_{(r)}-\mathbf{1}_r\bar{y}) + \frac{\phi r^2}{n-\phi r}(\bar{y}_{(r)}-\bar{y})^2\right\},\\ X'X-\phi X'_{(r)}X_{(r)} &= n-\phi r, \quad \phi = 1-k^{-2}, \end{aligned}\right\} \tag{3·19}$$

where $\bar{y}$ is the sample average, s^2 the sample variance, $\mathbf{1}_r$ an $r\times 1$ vector of unities and $\bar{y}_{(r)}$ the average of the elements of $\mathbf{y}_{(r)}$. The posterior distribution of θ is thus

$$\left.\begin{aligned} p(\theta \mid \mathbf{y}) &= \sum_{(r)} w_{(r)}\frac{(n-\phi r)^{\frac{1}{2}}}{s_{(r)}}\, p\left\{t_{n-1} = \frac{\theta-\hat{\theta}_{(r)}}{s_{(r)}/\sqrt{(n-\phi r)}}\right\},\\ w_{(r)} &= C\left(\frac{\alpha}{1-\alpha}\right)^r k^{-r}\left(\frac{n}{n-\phi r}\right)^{\frac{1}{2}}\left\{\frac{s^2_{(r)}}{s^2}\right\}^{-\frac{1}{2}(n-1)}, \end{aligned}\right\} \tag{3·20}$$

i.e. a weighted average of 2^n scaled t-distributions with $n-1$ degrees of freedom.

Mixture of distributions. In this section, we have supposed that the error ϵ associated with each observation in the linear model (2·1) could have been from either $N(0,\sigma^2)$ or $N(0,k^2\sigma^2)$ with probabilities $(1-\alpha)$ and α respectively. Through the general results in § 2 we obtained the posterior distribution of $\boldsymbol{\theta}$ as a weighted average of 2^n multivariate t-distributions corresponding to the 2^n possible configurations of the errors. Alternatively, we can write the distribution of ϵ as

$$p(\epsilon\mid\sigma) = (1-\alpha)\frac{1}{\sigma\sqrt{(2\pi)}}\exp\left(-\frac{\epsilon^2}{2\sigma^2}\right)+\alpha\frac{1}{k\sigma\sqrt{(2\pi)}}\exp\left(-\frac{\epsilon^2}{2k^2\sigma^2}\right), \tag{3·21}$$

which is a mixture of two normal distributions containing three parameters (α, k, σ). The likelihood function of $(\boldsymbol{\theta}, \sigma, k, \alpha)$ is then

$$l(\boldsymbol{\theta},\sigma,k,\alpha\mid\mathbf{y})\propto\sigma^{-n}\prod_{i=1}^{n}\left[(1-\alpha)\exp\left\{-\frac{1}{2\sigma^2}(y_i-\mathbf{x}_i'\boldsymbol{\theta})^2\right\},\right.$$
$$\left.+\alpha k^{-1}\exp\left\{-\frac{1}{2k^2\sigma^2}(y_i-\mathbf{x}_i'\boldsymbol{\theta})^2\right\}\right] \tag{3·22}$$

where $\mathbf{x}_i'$ is the ith row of X. Adopting the prior distribution for $(\boldsymbol{\theta}, \sigma)$ in (3·3), then for *fixed* (k, α) the posterior distribution of $\boldsymbol{\theta}$ is

$$p(\boldsymbol{\theta}\mid k,\alpha,\mathbf{y}) = \gamma\int_0^\infty \sigma^{-1}l(\boldsymbol{\theta},\sigma,k,\alpha\mid\mathbf{y})\,d\sigma, \tag{3·23}$$

where
$$\gamma^{-1} = \int_R \sigma^{-1}l(\boldsymbol{\theta},\sigma,k,\alpha\mid\mathbf{y})\,d\boldsymbol{\theta}\,d\sigma.$$

This distribution is mathematically equivalent to the distribution in (3·4). In fact, if we expand the integrand in (3·22) and perform the integration term by term, we obtain exactly the expression (3·4).

Some simplification of the distribution of $\boldsymbol{\theta}$. For mathematical convenience, we have adopted the notation (r) to denote the combination that a particular set of r of the n error ϵ's are from the model $N(0, k^2\sigma^2)$ and the remaining ones are from the model $N(0, \sigma^2)$. In practical application of the posterior distribution of $\boldsymbol{\theta}$ in (3·4) and its associated expressions, it is informative to arrange the summation over the 2^n possibilities such that

$$p(\boldsymbol{\theta}\mid\mathbf{y}) = w_0p_0(\boldsymbol{\theta}\mid\mathbf{y})+\sum_{i=1}^{n}w_ip_i(\boldsymbol{\theta}\mid\mathbf{y})+\sum_{i<j}^{n}w_{ij}p_{ij}(\boldsymbol{\theta}\mid\mathbf{y})+\sum_{i<j<t}^{n}w_{ijt}p_{ijt}(\boldsymbol{\theta}\mid\mathbf{y})+\dots \tag{3·24}$$

The distribution $p_0(\boldsymbol{\theta}\mid\mathbf{y})$ in the first term on the right of (3·24) would be appropriate if all the observations were good ones, i.e. generated from the central model. The distributions $p_i(\boldsymbol{\theta}\mid\mathbf{y})$ in the next summation correspond to allowing the possibility of each observation in turn being a bad value, i.e. from the alternative model. Those in the next summation allow for the possibility of two bad values and so on. Similarly, the weight w_0 is the posterior probability that no observation is bad. Correspondingly, the weight w_i is the posterior probability that only the ith observation is bad, w_{ij} the posterior probability that both the ith and the jth are bad and so on.

Since the distribution of $\boldsymbol{\theta}$ is a weighted average of 2^n posterior distributions, exact valuation would be very burdensome except when n is small. In practical application of the procedure for moderate values of n (say $n \leqslant 20$), it would usually be sufficient to include the leading term, terms in the first summation and at most those in the second summation in (3·24). Unless the experimental set-up is extremely unreliable, one or at most two bad runs in a total of not more than twenty runs would perhaps be the worst that would usually happen. This can be injected into our formulation by giving a heavy *a priori* discounting of the possibility of a run being bad, namely assigning a small value for α. For instance, with $n = 7$ and $\alpha = 0{\cdot}05$, it would be sufficient to consider the possibility of no bad run and that of a single bad one, since the chance *a priori* that there is at most one bad run is 95·5%. Unless the data are extreme, the weights $w_{ij}, w_{ijl}, \ldots$ will be negligible so that the associated distributions can be ignored.

We can, alternatively, interpret the distribution in (3·24) in the following manner. Suppose we entertain no bad run to begin with, then we would base our inference on the distribution in the leading term alone. Suppose we entertain the possibility of only one bad run, we would then consider both the leading term and terms in the first summation, and so on. In each instance the weights can be renormalized to make them sum to unity and the posterior distribution of $\boldsymbol{\theta}$ is always a weighted average of the possibilities considered.

4. Re-examination of Darwin's data

To illustrate the above procedure, we consider Darwin's data concerning fifteen differences of the heights of cross- and self-fertilized plants quoted by Fisher (1960, p. 37). This set of data, given in Table 1 in the order of their numerical values, was employed by the authors (1962) to study from a Bayesian viewpoint the effect of non-Normality on inferences about a location parameter. A peculiar aspect of the data is that the two observations (−67, −48) look rather discrepant compared with the remaining ones. While in our previous study this led to broadening the model to include the possibility of leptokurtic and platykurtic parent distributions, we could alternatively analyse the data in the light of the present formulation by supposing that the central model is $N(\theta, \sigma^2)$ but with a small probability α that each of the observations might have come from a normal distribution with a much larger variance $N(\theta, k^2\sigma^2)$.

Table 1. *Darwin's data*

y_1	−67	y_6	16	y_{11}	41
y_2	−48	y_7	23	y_{12}	49
y_3	6	y_8	24	y_{13}	56
y_4	8	y_9	28	y_{14}	60
y_5	14	y_{10}	29	y_{15}	75

For illustration, we have computed the posterior distribution of θ with $\alpha = 0{\cdot}05, k = 5$ and the result is given by the solid curve A in Fig. 1. In this calculation, we used the expansion (3·24) and included the leading term, and terms in the first and second summation. That terms in the second summation (corresponding to the possibility of two bad runs) should be included is not only supported by the data, but also by the choice $\alpha = 0{\cdot}05$. With $\alpha = 0{\cdot}05$ and $n = 15$, the *a priori* probability of having two bad experiments is 0·1347 which is appreciable. In general, the larger the values of α and n, the more terms in (3·24) should be included by consideration of prior probability alone. As we have explained the posterior distribution of θ will be a weighted sum of posterior t densities each of which would be appropriate if certain specific observations were bad. The five largest components of this distribution which together contribute 87·9% of the weight are as follows:

Weight (%)	46·2	19·0	17·5	3·6	1·6
Bad observations	None	y_1, y_2	y_1 only	y_2 only	y_{15} only

The broken curve B in Fig. 1 represents the appropriate distribution of θ if all observations were known to be good. It is a t-distribution with 14 degrees of freedom centred at

$\bar{y} = 20{\cdot}933$ with scale factor $s/\sqrt{n} = 9{\cdot}75$. The dotted curve C in the same figure is the posterior distribution corresponding to the situation if observations y_1 and y_2 were *known* to be bad.

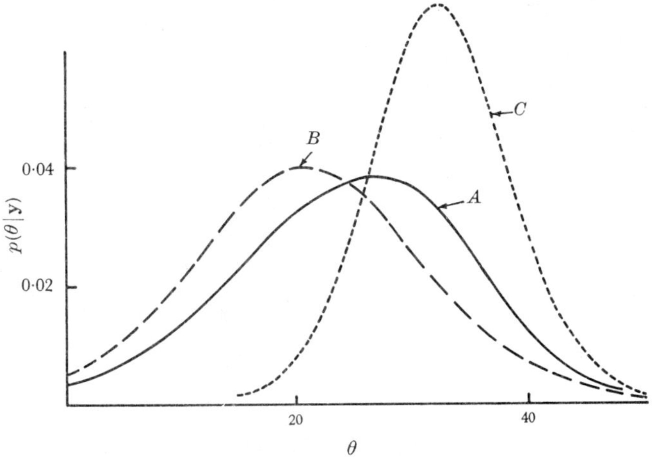

Fig. 1. Posterior distributions of θ.

A: $p(\theta \mid \mathbf{y})$, assuming possibility of outliers.

B: $(9{\cdot}75)^{-1} p\left(t_{14} = \dfrac{\theta - 20{\cdot}933}{9{\cdot}75}\right)$, assuming no outliers.

C: $(5{\cdot}83)^{-1} p\left(t_{14} = \dfrac{\theta - 32{\cdot}45}{5{\cdot}83}\right)$, assuming y_1 and y_2 are outliers.

Sensitivity of the results to changes in (α, k). In the above example, we calculated the posterior distributions of θ by setting $k = 5$ and $\alpha = 0{\cdot}05$. While the choice of values for α and k depends mainly upon the experimenter's experience and knowledge of the physical setup, it is, nevertheless, of interest to investigate how critical inferences about θ are affected by changes in (α, k). In Table 2 we show how the posterior means and standard deviations change for various choices of α and k in the ranges $0{\cdot}01 \leqslant \alpha \leqslant 0{\cdot}10$, $3 \leqslant k \leqslant 10$. The posterior means and variances are calculated from

$$\left.\begin{aligned} E(\theta \mid \mathbf{y}, k, \alpha) &= \bar{\theta} = w_0 \bar{y} + \sum_{i=1}^{n} w_i \bar{y}_i + \sum_{i<j} w_{ij} \bar{y}_{ij} + \ldots \\ \text{S.D.}\ (\theta \mid \mathbf{y}, k, \alpha) &= \{\operatorname{var}(\theta \mid \mathbf{y}, k, \alpha)\}^{\frac{1}{2}} \end{aligned}\right\} \qquad (4{\cdot}1)$$

with
$$\operatorname{var}(\theta \mid \mathbf{y}, k, \alpha) = w_0 \left\{\left(\frac{n-1}{n-3}\right)\left(\frac{s^2}{n}\right) + (\bar{y} - \bar{\theta})^2\right\} + \sum_{i=1}^{n} w_i \left\{\left(\frac{n-1}{n-3}\right)\left(\frac{s_i^2}{n-\phi}\right) + (\bar{y}_i - \bar{\theta})^2\right\}$$
$$+ \sum_{i<j} w_{ij} \left\{\left(\frac{n-1}{n-3}\right)\left(\frac{s_{ij}^2}{n-2\phi}\right) + (\bar{y}_{ij} - \bar{\theta})^2\right\} + \ldots,$$

where from (3·19) and (3·23) $\bar{y}_i$ is the sample mean given the ith observation being bad, s_i^2 the corresponding sample variance and so on. In calculating the entries in Table 2 only the leading term, and terms in the first and the second summations were included.

The posterior mean and standard deviation are very insensitive to changes in k in the range considered. This suggests that for this example at least, precise determination of the

Table 2. *Posterior mean and standard deviation of θ for various (α, k): Darwin's data*

(a) $E(\theta \mid \mathbf{y}, \alpha, k)$

$\alpha \backslash k$	3	4	5	6	7	8	9	10
0·01	21·48	21·53	21·50	21·45	21·40	21·35	21·31	21·27
·02	22·09	22·25	22·21	22·11	22·00	21·89	21·79	21·71
·03	22·70	22·99	22·97	22·84	22·67	22·50	22·35	22·22
·04	23·26	23·69	23·71	23·56	23·36	23·14	22·94	22·76
·05	23·78	24·34	24·41	24·26	24·03	23·78	23·54	23·32
·06	24·23	24·91	25·03	24·90	24·66	24·39	24·13	23·87
·07	24·63	25·42	25·59	25·47	25·24	24·96	24·68	24·41
·08	25·98	25·86	26·08	25·99	25·77	25·49	25·21	24·92
·09	25·29	26·25	26·51	26·45	26·25	25·98	25·69	25·40
·10	25·56	26·58	26·89	26·86	26·67	26·42	26·13	25·85

$E(\theta \mid \mathbf{y}, 0, k) = E(\theta \mid \mathbf{y}, \alpha, 1) = E(\theta \mid \mathbf{y}, \alpha, \infty) = 20{\cdot}93$

(b) S.D. $(\theta \mid \mathbf{y}, \alpha, k)$

$\alpha \backslash k$	3	4	5	6	7	8	9	10
0·01	10·53	10·54	10·55	10·55	10·55	10·55	10·54	10·54
·02	10·53	10·56	10·58	10·59	10·59	10·58	10·58	10·58
·03	10·51	10·55	10·59	10·60	10·61	10·61	10·61	10·61
·04	10·47	10·51	10·55	10·59	10·61	10·62	10·63	10·63
·05	10·41	10·44	10·49	10·54	10·57	10·60	10·62	10·63
·06	10·35	10·35	10·40	10·45	10·51	10·55	10·59	10·61
·07	10·29	10·25	10·29	10·35	10·42	10·48	10·53	10·57
·08	10·21	10·15	10·17	10·24	10·32	10·39	10·45	10·51
·09	10·14	10·04	10·05	10·12	10·20	10·29	10·36	10·43
·10	10·08	9·94	9·93	9·99	10·08	10·17	10·26	10·34

S.D. $(\theta \mid \mathbf{y}, 0, k)$ = S.D. $(\theta \mid \mathbf{y}, \alpha, 1)$ = S.D. $(\theta \mid \mathbf{y}, \alpha, \infty) = 10{\cdot}53$

value of k is not necessary. It does not mean, of course, that we can take k as large, or as close to unity as we please. In fact, both as $k \to 1$ and $k \to \infty$, the posterior distribution of θ approaches the t-distribution

$$\left(\frac{s}{\sqrt{n}}\right)^{-1} p\left(t_{n-1} = \frac{\theta - \bar{y}}{s/\sqrt{n}}\right)$$

appropriate to the situation that all observations were from $N(\theta, \sigma^2)$, and little insight will be gained from the present framework of analysis. In the ranges considered, the effect of the choice of α on the mean and standard deviation is somewhat more pronounced. However, as per a change in α of say, 0·01, the change in the posterior mean is still only a small fraction of the average standard deviation. In fact, Table 2 shows that the ratio

$$|E(\theta \mid \mathbf{y}, \alpha, k) - E(\theta \mid \mathbf{y}, \alpha + 0{\cdot}01, k)| / \tfrac{1}{2}\{\text{S.D.}\,(\theta \mid \mathbf{y}, \alpha, k) + \text{S.D.}\,(\theta \mid \mathbf{y}, \alpha + 0{\cdot}01, k)\}$$

is at most about 10 % and is considerably smaller than 10 % for $\alpha \geqslant 0{\cdot}03$. This suggests that, for this example at least, uncertainty about the value of α in the range $0{\cdot}05 \pm 0{\cdot}02$ would not have any appreciable effect on the inferences about θ.

It might be argued that in the above analysis our choice of the range $0{\cdot}01 \leqslant \alpha \leqslant 0{\cdot}1$ is rather arbitrary and larger values of α should be considered. However, we feel that, in practical applications of the procedure, α would usually be small and $\alpha = 0{\cdot}1$ is already too extreme a value for realistic consideration. Some insight into the reasonable range for α can be obtained by considering a 'typical' experiment involving, say, twenty runs. A fairly optimistic data analyst might perhaps expect some discrepant observations 50% of the time. A rather pessimistic analyst might expect discrepant ones 75% of the time. Using the Poisson approximation, these possibilities would correspond to values of α equal to 0·035 and 0·07 respectively. On the other hand, $\alpha = 0{\cdot}1$ implies that at least one of the runs will be bad 86% of the time, a rather unacceptable situation which is probably less often met.

It will be noticed that in our formulation we have included k and α as parameters but we have not thought it appropriate to give the parameters prior distributions and to obtain a marginal distribution for θ by integrations. Rather in our example we carry out a 'sensitivity analysis' to show the extent to which our inferences about θ are sensitive to the choice of these values. The place and importance of this type of sensitivity analysis or analysis of 'robustness' within the Bayesian framework have been discussed in some detail elsewhere (Box & Tiao, 1964).

The fact is that we probably would never know α and k very exactly. We can be certain, however, that α is not equal to zero, the value implied by an analysis which ignores the possibility of outliers. Further, in the above example, if we were to give some prior distribution for (α, k) the mass of which were concentrated roughly uniformly in the region $0{\cdot}01 \leqslant \alpha \leqslant 0{\cdot}10$, $3 \leqslant k \leqslant 10$, it seems clear from Table 2 that the resulting posterior distribution of θ would be little different from the one shown by curve A in Fig. 1.

This research was supported in part by the National Science Foundation.

References

Anscombe, F. J. (1960). Rejection of outliers. *Technometrics* **2**, 123–46.

Box, G. E. P. & Cox, D. R. (1964). An analysis of transformations. *J.R. Statist. Soc.* B **26**, 211–43.

Box, G. E. P. & Tiao, G. C. (1962). A further look at robustness via Bayes' theorem. *Biometrika* **49**, 419–32.

Box, G. E. P. & Tiao, G. C. (1964). A Bayesian approach to the importance of assumptions applied to the comparison of variances. *Biometrika* **51**, 153–67.

Box, G. E. P. & Tiao, G. C. (1965). A change in level of a non-stationary time series. *Biometrika* **52**, 181–92.

Daniel, C. (1960). Locating outliers in factorial experiments. *Technometrics* **2**, 149–56.

Fisher, R. A. (1960). *The Design of Experiments* (7th edition). Edinburgh: Oliver and Boyd.

Gebhardt, F. (1964). On the risk of some strategies for outlying observations. *Ann. Math. Statist.* **35**, 1524–36.

Grubbs, F. E. (1950). Sample criteria for testing outlying observations. *Ann. Math. Statist.* **21**, 27–58.

Raiffa, H. & Schlaifer, R. (1961). *Applied Statistical Decision Theory*. Harvard Business School.

Thompson, W. R. (1935). On a criterion for the rejection of observations and the distribution of the ratio of deviation to sample standard deviation. *Ann. Math. Statist.* **6**, 214–9.

Tiao, G. C. & Tan, W. Y. (1966). Bayesian analysis of random-effect models in the analysis of variance. II. Effect of autocorrelated errors. *Biometrika* **53**, 477–95.

Tukey, J. W. (1960). A survey of sampling from contaminated distributions. *Contributions to Probability and Statistics: Volume Dedicated to Harold Hotelling*. Stanford University Press.

Zellner, A. & Tiao, G. C. (1964). Bayesian analysis of the regression model with autocorrelated errors. *J. Am. Statist. Ass.* **59**, 763–78.

[*Received December* 1966. *Revised March* 1967]

1.12
Science and Statistics

G. E. P. BOX*

Aspects of scientific method are discussed: In particular, its representation as a motivated iteration in which, in succession, practice confronts theory, and theory, practice. Rapid progress requires sufficient flexibility to profit from such confrontations, and the ability to devise parsimonious but effective models, to worry selectively about model inadequacies and to employ mathematics skillfully but appropriately. The development of statistical methods at Rothamsted Experimental Station by Sir Ronald Fisher is used to illustrate these themes.

1. INTRODUCTION

In 1952, when presenting R.A. Fisher for the Honorary degree of Doctor of Science at the University of Chicago, W. Allen Wallis described him in these words.

> He has made contributions to many areas of science; among them are agronomy, anthropology, astronomy, bacteriology, botany, economics, forestry, meteorology, psychology, public health, and—above all—genetics, in which he is recognized as one of the leaders. Out of this varied scientific research and his skill in mathematics, he has evolved systematic principles for the interpretation of empirical data; and he has founded a science of experimental design. On the foundations he has laid down, there has been erected a structure of statistical techniques that are used whenever men attempt to learn about nature from experiment and observation.

Fisher was introduced by the title which he himself would have chosen—not as a statistician but as a scientist, and this was certainly just, since more than half of his published papers were on subjects other than statistics and mathematics. My theme then will be first to show the part that his being a good scientist played in his astonishing ingenuity, originality, inventiveness, and productivity as a statistician, and second to consider what message that has for us now.

2. ASPECTS OF SCIENTIFIC METHOD

A heritage of thought about the process of scientific learning comes to us from such classical writers as Aristotle, Galen, Grossteste, William of Occam, and Bacon who have emphasized aspects of good science and have warned of pitfalls.

2.1 Iteration Between Theory and Practice

One important idea is that science is a means whereby learning is achieved, not by mere theoretical speculation on the one hand, nor by the undirected accumulation of practical facts on the other, but rather by a motivated *iteration* between theory and practice such as is illustrated in Figure A(1).

A. The Advancement of Learning

A(1) An Iteration Between Theory and Practice

A(2) A Feedback Loop

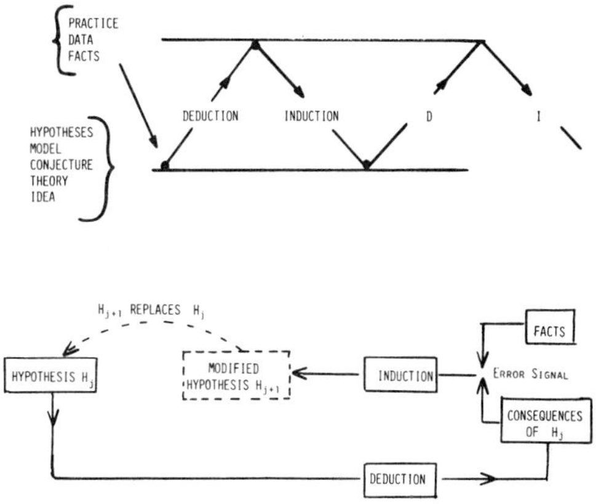

Matters of fact can lead to a tentative theory. Deductions from this tentative theory may be found to be discrepant with certain known or specially acquired facts. These discrepancies can then induce a modified, or in some cases a different, theory. Deductions made from the modified theory now may or may not be in conflict with fact, and so on. In reality this main iteration is accompanied by many simultaneous subiterations (see, e.g., [1, 2]).

2.2 Flexibility

On this view efficient scientific iteration evidently requires unhampered feedback. The iterative scheme is shown as a *feedback* loop in Figure A(2). In any feedback loop it is, of course, the *error* signal—for example, the discrepancy between what tentative theory suggests *should* be so and what practice says *is* so—that can produce learning. The good scientist must have the flexibility and courage to seek out, recognize, and exploit such errors—especially his own. In particular, using Bacon's

* George E.P. Box is R.A. Fisher Professor of Statistics, University of Wisconsin, Madison, WI 53706. Research was supported by the United States Army under Grant DAHC04-76-G-0010. This is the written version of the R.A. Fisher Memorial Lecture presented at the joint statistical meetings of the American Statistical Association and Biometric Society given at St. Louis in 1974. The author gratefully acknowledges the assistance of his wife Joan who generously shared the results of her research on her father's life and made available the manuscript of her biography of Fisher.

The Collected Works of George E. P. Box, 1984, Wadsworth, Inc., Belmont, CA 94002.
Originally published in *J. Amer. Stat. Assoc.*, vol. 71, no. 356 (1976), pp. 791-799.

analogy, he must not be like Pygmalion and fall in love with his model.

2.3 Parsimony

Since all models are wrong the scientist cannot obtain a "correct" one by excessive elaboration. On the contrary following William of Occam he should seek an economical description of natural phenomena. Just as the ability to devise simple but evocative models is the signature of the great scientist so overelaboration and overparameterization is often the mark of mediocrity.

2.4 Worrying Selectively

Since all models are wrong the scientist must be alert to what is importantly wrong. It is inappropriate to be concerned about mice when there are tigers abroad.

2.5 Role of Mathematics in Science

Pure mathematics is concerned with propositions like "given that A is true, does B *necessarily* follow?" Since the statement is a conditional one, it has nothing whatsoever to do with the truth of A nor of the consequences B in relation to real life. The pure mathematician, acting in that capacity, need not, and perhaps should not, have any contact with practical matters at all.

In applying mathematics to subjects such as physics or statistics we make tentative assumptions about the real world which we know are false but which we believe may be useful nonetheless. The physicist knows that particles have mass and yet certain results, approximating what really happens, may be derived from the assumption that they do not. Equally, the statistician knows, for example, that in nature there never was a normal distribution, there never was a straight line, yet with normal and linear assumptions, known to be false, he can often derive results which match, to a useful approximation, those found in the real world.

It follows that, although rigorous derivation of logical consequences is of great importance to statistics, such derivations are necessarily encapsulated in the knowledge that premise, and hence consequence, do not describe natural truth. It follows that we cannot know that any statistical technique we develop is useful unless we use it. Major advances in science and in the science of statistics in particular, usually occur, therefore, as the result of the theory-practice iteration.

The researcher hoping to break new ground in the theory of experimental design should involve himself in the design of actual experiments. The investigator who hopes to revolutionize decision theory should observe and take part in the making of important decisions. An appropriately chosen environment can suggest to such an investigator new theories or models worthy to be entertained. Mathematics artfully employed[1] can then enable him to derive the logical consequences of his tentative hypotheses and his strategically selected environment will allow him to compare these consequences with practical reality. In this way he can begin an iteration that can eventually achieve his goal. An alternative is to redefine such words as *experimental design* and *decision* so that mathematical solutions which do not necessarily have any relevance to reality may be declared optimal.

[1] The researcher's purely mathematical ingenuity is likely to be exercised more, not less, by the fact of his dealing with genuine problems.

3. FISHER—A SCIENTIST

With these ideas in mind let us see how Fisher qualifies as a scientist, using for illustration some of the events occurring during his stay at Rothamsted Experimental Station.

3.1 Rothamsted

In 1919, Fisher had rejected the security and prestige of working under Karl Pearson in the most distinguished statistical laboratory in Britain and at that time certainly in the world. Instead, he took up a temporary job as the sole statistician is a small agricultural research station in the country. He was then already 29 years old and he later said that he was aware that he had failed at both the jobs (teacher and actuary) that he had so far attempted.

Sir John Russell, then Director of Rothamsted, later recalled [17, p. 326]

> . . . when I first saw him in 1919 he was out of a job. Before deciding anything I wrote to his tutor at Caius college . . . about his mathematical ability. The answer was that he could have been a first class mathematician had he "stuck to the ropes" but he would not. That looked like the type of man we wanted. . . . I had only £200 and suggested he should stay as long as he thought that should suffice. . . . He reported to me weekly at tea at my house. . . . It took me a very short time to realize that he was more than a man of great ability, he was in fact a genius.

At the end of a year, Fisher, who had a wife and child, had used up twice the £200, but by that time he had been given a permanent post.

3.2 Weighing the Baby

For the theory-practice iteration to work, the scientist must be, as it were, mentally ambidextrous; fascinated equally on the one hand by possible meanings, theories, and tentative models to be induced from data and the practical reality of the real world, and on the other with the factual implications deducible from tentative theories, models and hypotheses.

Fisher had great interest in practical matters. For example, he begins the real business of his book *Statistical Methods for Research Workers* in Chapter 2, by discussing different ways of plotting data. His first example is introduced as follows [12, p. 25]: "Figure 1 represents the growth of a baby weighed to the nearest ounce at equal intervals from birth." He does not say that this is any particular baby. Recently I was fortunate to see the Fisher family records in which in Fisher's own hand are recorded the weight from birth of every one of his nine children, weighed by himself, with the results carefully graphed. Comparison shows that the child is his second son, Harry Leonard, who was born in 1923 shortly before the first edition of the book was written. The next leg of the scientific iteration is hinted at as he goes on to discuss how best to plot the data so as to make "a rough examination of the agreement of observation with any (proposed) law of increase."

3.3 Find the Lady

The extraordinary extent to which Fisher's actual every day experience was grist to the mill of his inductive mind is further illustrated in the famous opening lines of Chapter II of Fisher's book *The Design of Experiments* [11, p. 11]: "A lady declares that by tasting a cup of tea . . . she can discriminate whether the milk or the tea infusion was first added to the cup. We will consider the problem of designing an experiment by means of which this assertion can be tested." Fisher proceeds to use this example to explain and illustrate the basic principles of good statistical design.

There was, of course, a real lady. This incident happened many years before the book was written and just after Fisher came to Rothamsted. The lady was Dr. Muriel Bristol, the algologist, and she had declined the cup of tea that Fisher had offered her because he had added the tea first. Fisher declared it made no difference. To which she replied "Of course it did." Her future husband, William Roach, who was close at hand said "Let's test her," they did, and according to him she made nearly every choice correctly. In this she behaved similarly to the lady in the book who got one wrong.

3.4 From Soil Bacteria to Nonlinear Design

The tea urn was a great catalyst to iteration. There, each afternoon, Fisher conversed with members of the scientific staff and with visitors and became involved in their problems, often with dramatic consequences. One scientist who came to Rothamsted about the same time as Fisher and became his intimate friend was the bacteriologist, Gerard Thornton. It was he who first interested Fisher in improving the time consuming dilution methods for making bacterial counts. This resulted in Fisher's pioneering work on nonlinear design in 1922 mentioned by Cochran [4].

3.5 From Cotton to Extreme Values

One of the early visitors to Rothamsted was L.H.C. Tippett from the Cotton Research Institute. A matter of great practical concern to him was the strength of cotton yarn. Since the breaking strength of a piece of cotton is the strength of the weakest link, he was faced with what we should now call the extreme value problem. Tippett had first studied with Karl Pearson and had earlier approximated the distribution using the method of moments. In cooperation with Fisher the problem was tackled rather differently. The authors note [14, p. 180] that, "the limiting distribution must be such that the extreme member of a sample of n from such a distribution has itself a similar distribution." This simple but remarkable insight leads to a functional equation which yields as its solution the basic limiting forms. From these forms almost all subsequent work on the subject springs. The theory has applications in such different fields as the design of dams and the reliability of components. Like so many of Fisher's brain children this is now regarded as a distinct field of study.

I will use for further illustration work that Fisher did at Rothamsted between 1919 and 1927 which began with regression analysis and ended with a complete and elegant theory of experimental design which is still the basis for most statistically planned experiments. This work was published in a series of papers having the general title "Studies in Crop Variation" and numbered I, II, III,[2] IV, and VI [7, 13, 8, 5, 6].

3.6 From Dung to Orthogonal Polynomials and Residual Analysis

By 1919 13 plots on Broadbalk wheat fields had received thirteen different manurial treatments uniformly for 67 years. In "Studies in Crop Variation I" [7], Fisher begins by presenting a workmanlike discussion, which lasts for twelve pages, of the responses to the thirteen different manures revealed by his analysis of the Broadbalk data. In particular, he concludes that there is really nothing like plain dung. It gives a high yield with no significant diminution of its effect over the years. He then quite suddenly shifts from manure to mathematics revealing where his analysis has come from. In the next few pages he introduces orthogonal polynomials, presents formulas for their calculation from equispaced data, obtains the distributional properties of the coefficients, and shows how their significance may be judged. Without calling it that he presents the appropriate analysis of variance which he has used in fitting fifth degree polynomials to the annual yields. Most interesting of all, he discusses the properties of the residuals $y - \hat{y}$ from a fitted polynomial of any degree r allowing us to see him in the guise of what some people now call a data analyst.

Data analysis, a subiteration in the process of investigation, is illustrated here.

TENTATIVE MODEL	INFERENCE → / ← CRITICISM	TENTATIVE ANALYSIS

In the inferential stage, the analyst acts as a sponsor for the model. Conditional on the assumption of its truth he selects the best statistical procedures for analysis of the data. Having completed the analysis, however, he must switch his role from sponsor to critic.[3] Conditional now on the contrary assumption that the model may be

[2] This paper [8] was presented to the Royal Society without the general title but was mysteriously labelled III and had clearly been originally intended for this series.

[3] The apt christening of statistical criticism is due to Cuthbert Daniel.

seriously faulty in one or more suspected or unsuspected ways he applies appropriate diagnostic checks, involving various kinds of residual analysis.

In order to conduct his analysis of the residuals from the fitted polynomials, Fisher obtained

i. the average value of $V(y - \hat{y})$ as $(1 - (r + 1)/n)\sigma^2$,
ii. the individual variances of the residuals $y_j - \hat{y}_j$ for the 67 observations,
iii. the identity

$$\sigma^2 = V(y_j) = V(y_j - \hat{y}_j) + V(\hat{y}_j) \ , \quad j = 1, 2, \ldots, n \ ,$$

iv. an approximate formula for the autocorrelations of residuals from a fitted polynomial of any degree.

The average value of $V(\hat{y})$ from (i) and (iii) is $\sigma^2(r + 1)/n$. Thus, Fisher says if we want to have a small variance for $\hat{y}$ we should keep r small—a demonstration of the value of parsimony, helping to justify his use of polynomials of only fifth degree. Fisher plots the variances $V(y_j - \hat{y}_j)$ for the individual residuals against their time order j. Because of relation (iii) the graph looked at upside down is also a plot of $V(\hat{y}_j)$. Using this he notes the deceptive reduction of $V(y_j - \hat{y}_j)$ at the extremities of the scale and the corresponding increase of $V(\hat{y}_j)$ and says [7, p. 123] "it is a weakness of the polynomial form that the extreme terms should be so much affected."

Finally, mentioning that overfitting and underfitting are both to be avoided he uses the matching of theoretical and empirical *autocorrelations* of residuals to check when a polynomial of sufficiently high degree has been fitted. In particular, he compares theoretical and observed autocorrelations of residuals from polynomials of degree zero and five to show the inadequacy of the former and the satisfactory fit of the latter. This application of serial correlation of residuals to the awkward problem of deciding at what point adequacy of fit has been achieved has great freshness and interest 55 years later.

3.7 Weeds and the Education Acts

Fisher was perplexed by the shapes of his fitted yield graphs. These showed a pattern of significant slow changes *common* to all the 13 Broadbalk plots. In particular, there was a common tendency for low yields roughly in the period 1870–1880. This common pattern was not due to weather; a similar analysis he conducted for successive yields of experimental wheat at Woburn, wheat averages for the whole of Hertfordshire, and for barley and grass from experimental plots at Rothamsted, failed to show it. He speculates [7, p. 129], "Of all the organic factors which influence the yield of wheat it is probable that weeds alone change sufficiently slowly to explain the changes at Broadbalk."

He goes on to describe, as only a dedicated gardener could, all the various weeds that were found there. He notes that old records show that, in 1853, 211 man-days and 714 boy-days were spent in weeding the field. In particular, the boys probably held in check by hand weeding the slender foxtail grass *Alopecuris agrestis*. But he says [7, p. 131] "it may be remembered that the Education Acts of 1876 and 1880 made attendance at school compulsory." We are left to speculate whether the low wheat yields occurred after that time because the hands of the little boys who pulled the foxtail grass were now covered with ink and not with earth.

3.8 From Rainfall and Wheat Yield to Distributed Lags

In 1924, in the third paper of the series [8], he used the Broadbalk data to demonstrate the influence of rainfall on the wheat yield. At the beginning of the paper he seemed to fear that he might be expected to account for the effects not only of rainfall but also for such other variables as maximum and minimum temperature, dew point, and hours of bright sunshine. But he points out that allowances for the effect of each of these on the final harvested yield would need to be included at least for each month separately. And he says if so many regressors are included a very high proportion of the total variation can seem to be accounted for by chance alone. In case some dissident reader might doubt it, he thereupon outlines the derivation of the distribution of the multiple correlation coefficient in one paragraph flat using n-dimensional geometry and on the next page produces a short table of tail areas for R. He then goes on to discuss the misleading effects of selection in what would now be called step-wise regression.

Fisher's data were as follows:

i. for each of the 13 Broadbalk plots he had the harvested wheat yields for each of 60 years,[4]
ii. for each of these 60 years he had daily rainfall records and for convenience he aggregated these for each year into 61 six-day periods ($6 \times 61 = 366$) beginning immediately after the harvest.

In a remarkable demonstration of parsimonious modeling he first suggests that the yield of wheat in the jth year, w_j say, might be represented by

$$w_j = c + \sum_t^{61} a_t r_{jt} \ , \quad j = 1, 2, \ldots, 60 \ . \qquad (3.1)$$

In this model the coefficient a_t provides the average effect on eventual harvested yield of one inch of rain in the tth time period. In modern parlance (3.1) might be called a "transfer function" model expressing the "memory" of the system. Economists later called it a "distributed lag" model but they seem to have been unaware of Fisher's prior work or of his ingenious way of proceeding using orthogonal polynomials.

As it stands (3.1) is highly nonparsimonious. Fisher decided, therefore, to represent the rainfall data r_{jt} by orthogonal polynomials of fifth degree. He now notes that the coefficients a_t should also follow a smooth curve which might be represented in the same way. Thus,

$$a_t = \alpha_0 T_{0t} + \alpha_1 T_{1t} + \ldots + \alpha_5 T_{5t} \ ,$$
$$r_{jt} = \rho_{0j} T_{0t} + \rho_{1j} T_{1t} + \ldots + \rho_{5j} T_{5t} \ .$$

[4] Five years 1890, 1891, 1905, 1906, and 1915 were omitted because the plots in these years had special treatment.

But if the orthogonal functions T_{it} are chosen so that $\sum_t T_{it}^2 = 1$; then, after summing, (3.1) may be written

$$w_j = c + \alpha_0\rho_{0j} + \alpha_1\rho_{1j} + \ldots + \alpha_5\rho_{5j} \; .$$

The α's which determine the lagged weights in the transfer function can thus be obtained by regressing the w_j onto the estimated ρ's.

Having carried through the necessary heavy calculations and graphed his results Fisher conducts a very extensive discussion and comparison of the polynomial distributed lag curves for the differently manured plots from which, in particular, he adduces the predominant effect of rain in reducing soil nitrates. One feels that his love of parsimony was certainly not lessened by the fact that the computations were performed by hand by himself and his assistant. Indeed, much can still be learnt from his discussion about economical processes of calculation and appropriate checks [8, p. 111–3].

3.9 From Fertilizer and Potatoes to the Analysis of Variance

About this time Fisher was getting rather tired of analyzing old records—he later described it as "raking over the muck heap." In "Studies in Crop Variation II," jointly authored with his assistant, Miss W.A. MacKenzie, and subtitled "The Manurial Response of Different Potato Varieties," [13] he tried his hand at analyzing some *experimental* data from Rothamsted. The authors remark that it would be convenient if (contrary to some expert opinion) different varieties of plants did *not* react differently to fertilizers, or as we should say now, if there were no interaction between variety and fertilizer.

An experiment had recently been run by Thomas Eden, a crop ecologist at Rothamsted, in which each of twelve varieties of potatoes were tested with six different combinations of manure. This experiment was analyzed as if it were a thrice replicated and randomized 12 × 6 factorial. (It wasn't, but we return to that later.)

From the analysis of variance which is presented, the answer to the question, "Is there significant interaction between varieties and manures?" appears to be No!

There are some remarkable things about this paper, however:

i. The analysis of variance, hinted at earlier, appears here for the first time in its completeness. It arrives quite suddenly and unannounced in the middle of the paper after the discussion of agricultural questions. It is, of course, not even mentioned in the title.

ii. After the algebraic identity between the total sum of squares and the within and between treatments sum of squares has been written down, the statement is made [13, p. 315] "If all the plots were undifferentiated, as if the numbers had been mixed up and written down in random order, the average value of each of the two parts is proportional to the number of degrees of freedom in the variation of which it is compared." Thus, at the very beginning, randomization, an important flag under which Fisher will sail, is firmly nailed to the mast.

iii. The analysis is wrong, because in fact the trial was actually run as what is now called a split plot design. Feedback in the form of the appropriate correction came quickly in the first edition of *Statistical Methods* in 1925 (see [12, p. 238]). Using part of the same data, Fisher there gives the correct analysis and points out that it is essential to use separate error variance estimates (for between and within plot comparisons) and shows that one is indeed significantly larger than the other.

iv. In this very first paper on the analysis of variance, Fisher demonstrated the flexibility of his thought by questioning the linear model (which almost everybody else has ever since accepted as representing received truth). The authors say [13, p. 316], "the above test is only given as an illustration of the method; the summation formula for combining the effects of variety and manurial treatment is evidently quite unsuitable for the purpose. No one would expect to obtain from a low yielding variety the same actual increase in yield which a high yielding variety would give ... a far more natural assumption is that the yield should be the product of two factors one depending on the variety and one on the manure." With the possibility of transformation so much a part of Fisher's everyday thought, we might expect him now to proceed along that route but in fact he derives the appropriate nonlinear analysis, devising methods which have only recently been rediscovered [18].

3.10 Mice, Tigers, and Randomization

A man in daily muddy contact with field experiments could not be expected to have much faith in any direct assumption of *independently* distributed normal errors. While the supposition of marginal normality for the errors might be regarded as innocuous, the idea that errors from adjacent plots of land could be treated as independent would be obviously absurd and dangerous. This was one important reason for Fisher's insistence (i) on the physical act of randomization as a necessary condition for the validity of any experiment and (ii) that given that randomization had been carried out inferences should be made from the appropriate randomization distribution; to which, however, standard normal theory often provided an adequate approximation.

To guarantee the exact validity of the usual null tests made with the standard linear model it is not, of course, necessary that the density function of the error vector **e** be spherically normal, it is necessary only that it be spherically symmetric,[5] i.e., the density function be of the form $f(\mathbf{e}'\mathbf{e})$. The fact that standard normal theory often provides an adequate approximation to that given by randomization theory is not because the density for randomized errors is necessarily approximated by that of independent normal deviates. It is rather because, in the appropriate vector space, the symmetry induced by randomization is approximated by spherical symmetry.

Fisher showed some irritation with later workers who saw only a rich source of purely mathematical development in his work. In particular, workers on what has come to be called "distribution-free" tests have often failed to emphasize and sometimes perhaps even to realize the limitations imposed by the necessary assumption of symmetry of the joint error distribution. The

[5] Obviously, this must be true for any criterion which is a homogeneous function of the data of degree zero.

validity of this assumption could, of course, only be guaranteed by randomization. Otherwise, the derived procedures, far from being distribution free, would be almost as restrictive as those derived on the assumption of normal independent errors. It is true that long usage has seemed to sanctify the proposition that density functions are of the form $p(\mathbf{y}) = \prod_i f(y_i)$ or at least that $p(\mathbf{y}) = S(\mathbf{y})$, where S is some symmetric function of the elements of $\mathbf{y}$. These propositions have come to be treated almost as natural laws or at least as rules of the game that no sportsman would question.[6] In fact, of course, experiments where errors cannot be expected to be independent are very common.

These points are not new but if we are to appreciate Fisher's point of view they need to be brought together and illustrated together. For this latter purpose the results of a simple sampling experiment are shown in the table. Two samples of 10 observations from identical populations of the forms indicated were taken and subjected to a t-test (t) and a Mann-Whitney test (MW). The sampling was repeated 1,000 times and the number of results significant at the 5 percent point was recorded. Ideally, this number should be 50 (that is, 5 percent of the total) but it has a standard deviation of about 7 because of sampling errors. More accurate results may be obtained by taking larger samples or by analytical procedures, however, since there is no practical difference between a significance level of say 4 percent and 6 percent, the present investigation suffices for illustration. Autocorrelation between adjacent values was introduced by generating observations from a moving average model of the form $y_t = u_t - \theta u_{t-1}$. In this model the u_t were

Frequency in 1,000 Trials of Significance at the 5 Percent Level Using the t-Test (t) *and the Mann-Whitney Test* (MW) *with No Randomization* (NR) *and Randomization* (R)

ρ_1	Test	Rectangular NR	Rectangular R	Normal NR	Normal R	Chi-square[a] NR	Chi-square[a] R
		Independent observations					
0.0	t	56	60	54	43	47	59
	MW	43	58	45	41	43	44
		Autocorrelated successive observations					
−0.4	t	5	48	3	55	1	63
	MW	5	43	1	49	2	56
+0.4	t	125	50	105	58	114	54
	MW	110	46	96	53	101	43

[a] The parent chi-square distribution has four degrees of freedom and is thus highly skewed.

independently and identically distributed about zero in the forms indicated in the table. Values of θ were chosen so that ρ_1, the first serial correlation, had values of -0.4 and $+0.4$.

[6] Except in the study of time series.

The frequencies shown under NR are those obtained for a nonrandomized test. The frequencies under R are those obtained when the observations were randomly allocated to the two groups.

As is to be expected the significance level of the t-test is affected remarkably little by the drastic changes made in the marginal parent distribution—changes for which the distribution-free test provides insurance. Unfortunately, of course, both tests are equally impaired by error dependence unless randomization is introduced when they do about equally well. The point is, of course, that it is the act of randomization that is of major importance here not the introduction of the distribution-free test function.

3.11 From Muck Raking to Group Theory

Eden's potato data served to illustrate the method of analysis of variance but Fisher appears to have had no hand in planning that experiment. The design is not randomized nor blocked and its very deficiencies call for appropriate remedies. When Fisher's friend Gosset saw the paper, he wrote to Fisher [15, Letter No. 29], "The experiment seems to me to be quite badly planned, you should give them a hand in that ...." Fisher later notes Gosset's "suggesting that I should start designing experiments" [15, summary of Letter No. 29]. This he proceeded to do. The iterative process including the design aspect is sketched in Figure B.

B. Data Analysis and Data Getting in the Process of Scientific Investigation[a]

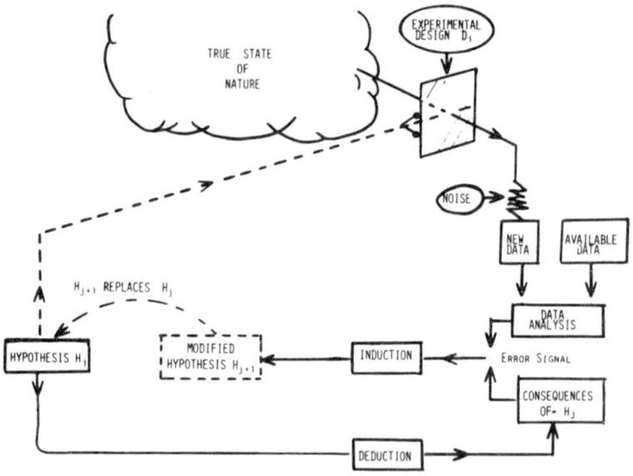

[a] The experimental design is here shown as a movable window looking onto the true state of nature. Its positioning at each stage is motivated by current beliefs, hopes, and fears.

Between 1919 and 1928 an iterative sequence occurred that went through three main stages, each leading logically to the next via interaction of theory and practice. The analysis of existing records led to the analysis of *experimental trials* which then led to the *design* of experimental trials.

There were different but interactive aspects to this development. We can see (i) sequential evolution of the new methods in response to unfolding realizations of need, (ii) the persuading of practitioners to try the new techniques, and (iii) the changing role of the statistician implied by the development.

3.12 Evolution of the New Methods

Fisher's attempts to analyze experimental data quickly led him to the essential principles of experimental design. The need for randomization to achieve validity; for replication to provide a valid estimate of error; for blocking extraneous sources of disturbance to achieve accuracy. Blocking in two directions simultaneously (by randomized Latin squares) was particularly appealing. Fisher would have been brought to see the enormous advantages of the unorthodox factorial arrangements as an economical way to assess the effects of variables in combination by, for example, his early attempts to impart meaning to the differences associated with the 13 differently manured Broadbalk plots to which fertilizers had been applied in a highly nonbalanced manner. However, while the efficiency of factorial designs could be increased by packing in more factors, larger factorial designs required bigger blocks and hence produced greater inhomogeneity in the experimental material, giving larger experimental errors. The answer which quickly followed was confounding.

3.13 Persuading Practitioners

The blessings of feedback were only available if scientists would try out his designs but, not surprisingly, Fisher at first did not have an easy job selling his revolutionary ideas at Rothamsted. Indeed, the first design run to his specification (in 1924) was not done at Rothamsted at all. It was a randomized Latin Square design run at Bagshot for the Forestry Commission who had asked for and acted on his advice. But between 1924 and 1929, as described in "Studies in Crop Variation IV and VI" [5, 6], there is a rapid development of ideas which were quickly put into practice. It is clear that Eden had become a convinced disciple during this period and it is refreshing, but alas unfamiliar, to see publication of new designs simultaneous with data obtained from their successful use. By the end of this period data were being collected from designs of great accuracy and beauty which included all of Fisher's ideas.

In spite of all this in 1926 the Director of Rothamsted, Sir John Russell, wrote a paper [16] in the *Journal of the Ministry of Agriculture* about agricultural experimentation which almost totally ignored the ideas of his protegé. However, in the next issue [9] in a paper notable for its brevity and clarity, Fisher outlined *his* philosophy on the subject, setting his boss to rights and anyone else who would listen.

3.14 A New Heritage for Statisticians

The original concept that the research station needed a statistician was revolutionary, but certainly the role initially envisaged in 1919 for the statistician was a passive and possibly even a temporary one. Russell wondered if anything more could be extracted from the existing records.

Fisher's work gradually made clear that the statistician's job did not begin when all the work was over—it began long before it was started. The statistician was not a curator of dusty relics. His responsibility to the scientific team was that of the architect with the crucial job of ensuring that the investigational structure of a brand new experiment was sound and economical. The latter role is much more fun than the former. He himself relished it and we should thank him for bequeathing it to us. It calls for abilities of a high order. It requires among other things the wit to comprehend complicated scientific problems, the patience to listen, the penetration to ask the right questions, and the wisdom to see what is, and what is not, important. Finally, it requires from the statistitian the courage to wager his reputation each time an experiment is run. For the time must come when all the data are in and conclusions must be drawn; at this stage oversights in the design, if they exist, will become embarrassingly evident.

4. PERILS OF THE OPEN LOOP

We have seen some examples of the extraordinary progress made in our science over a brief ten-year period as a result of feedback between theory and practice. Feedback requires a closed loop. By contrast, when for any reason the loop is open, progress stops. Such stagnation can occur with the (normally iterative) cycle stuck either in the practice mode or in the theory mode.

4.1 Cookbookery and Mathematistry

The maladies which result may be called *cookbookery* and *mathematistry*. The symptoms of the former are a tendency to force all problems into the molds of one or two routine techniques, insufficient thought being given to the real objectives of the investigation or to the relevance of the assumptions implied by the imposed methods. Concerning the latter, Fisher's apparently bivalent attitude towards mathematicians has often been remarked and has been the cause of perplexity and annoyance. He himself was an artist in the use of mathematics and emphasized the importance of mathematical training for statisticians—the more mathematics known the greater the potential to be a good statistician. Why then did he sometimes seem to refer so slightingly to mathematicians? The answer I think is that his real target was "mathematistry." It is to make the distinction that the word is introduced here.

Mathematistry is characterized by development of theory for theory's sake, which since it seldom touches down with practice, has a tendency to redefine the problem rather than solve it. Typically, there has once been a statistical problem with scientific relevance but this has long since been lost sight of. Fisher felt strongly about this last point, particularly when he himself had produced the originally useful idea. I have cited already the development of distribution-free tests which, he felt, misused ideas initiated in Chapter III of his book *Design of Experiments* [11, p. 48]. Another annoyance was the generalization to what he felt was absurdity of his applications of group theory and combinatorial mathematics to experimental design.

The penalty for scientific irrelevance is, of course, that the statistician's work is ignored by the scientific community. But this does not come to the notice of a statistician who has no contact with that community. It is sometimes alleged that there is no actual harm in mathematistry. A group of people can be kept quite happy, playing with a problem that may once have had relevance and proposing solutions never to be exposed to the dangerous test of usefulness. They enjoy reading papers to each other at meetings and they are usually quite inoffensive. But we must surely regret that valuable talents are wasted at a period in history when they could be put to good use.

Furthermore, there is unhappy evidence that mathematistry is not harmless. In such areas as sociology, psychology, education, and even, I sadly say, engineering, investigators who are not themselves statisticians sometimes take mathematistry seriously. Overawed by what they do not understand, they mistakenly distrust their own common sense and adopt inappropriate procedures devised by mathematicians with no scientific experience.

An even more serious consequence of mathematistry concerns the training of statisticians. We have recently been passing through a period where nothing very much was expected of the statistician. A great deal of research money was available and one had the curious situation where the highest objective of the teacher of statistics was to produce a student who would be another teacher of statistics. It was thus possible for successive generations of teachers to be produced with no practical knowledge of the subject whatever. Although statistics departments in universities are now commonplace there continues to be a severe shortage of statisticians competent to deal with real problems. But such are needed.

4.2 Meeting the Challenge

As long ago as 1950, Fisher, delivering the Eddington Memorial Lecture at Cambridge, said [10, p. 22]

> For the future, so far as we can see it, it appears to be unquestionable that the activity of the human race will provide the major factor in the environment of almost every evolving organism. Whether they act consciously or unconsciously human initiative and human choice have become the major channels of creative activity on this planet. Inadequately prepared we unquestionably are for the new responsibilities, which with the rapid extension of human control over the productive resources of the world have been, as it were, suddenly thrust upon us.

One by one, the various crises which the world faces become more obvious and the need for hard facts on which to take sensible action becomes inescapable. The demand for competent statisticians who can tease out the facts by analyzing data, planning investigations, and developing the necessary new theory and techniques will, therefore, continue to increase.

4.3 Training of Statisticians

Competent statisticians will be front line troops in our war for survival—but how do we get them? I think there is now a wide readiness to agree that what we want are neither mere theorem provers nor mere users of a cookbook. A proper balance of theory and practice is needed and, most important, statisticians must learn how to be good scientists; a talent which has to be acquired by experience and example. To quote Fisher once more, in 1952, in a letter concerning a proposed Statistics Center to be set up in Scotland he said: "I have no hesitation in advising that such a centre as you have under discussion should plan to integrate teaching closely with project work in which practical experience can be gained by those who are capable of learning from it; in contradistinction to the ruinous process of segregating the keener minds into a completely sterile atmosphere" [3]. It is encouraging that at more and more statistical centers such advice is now being taken seriously.

5. CONCLUSION

We may ask of Fisher

Was he an applied statistician?
Was he a mathematical statistician?
Was he a data analyst?
Was he a designer of investigations?

It is surely because he was all of these that he was much more than the sum of the parts. He provides an example we can seek to follow.

[*Received May 1976.*]

REFERENCES

[1] Box, G.E.P. and Tiao, G.C., *Bayesian Inference in Statistical Analysis*, Reading, Mass.: Addison-Wesley Publishing Co., 1973.

[2] ——— and Youle, P.V., "The Exploration and Exploitation of Response Surfaces: An Example of the Link Between the Fitted Surface and the Basic Mechanism of the System," *Biometrics*, 11, No. 3 (1955), 287–323.

[3] Box, Joan Fischer, *Fisher, The Life of a Scientist*, New York: John Wiley & Sons, Inc. In press.

[4] Cochran, W.G., "Experiments for Nonlinear Functions," *Journal of the American Statistical Association*, 68, No. 344 (1973), 771–81.

[5] Eden, T. and Fisher, R.A., "Studies in Crop Variation IV. The Experimental Determination of the Value of Top Dressings with Cereals," *Journal of Agricultural Science*, 17 (1927), 548–62.

[6] ——— and Fisher, R.A., "Studies in Crop Variation VI. Experiments on the Response of the Potato to Potash and Nitrogen," *Journal of Agricultural Science*, 19 (1929), 201–13.

[7] Fisher, R.A., "Studies in Crop Variation I. An Examination of the Yield of Dressed Grain from Broadbalk," *Journal of Agricultural Science*, 11 (1921), 107–35.

[8] ———, "Studies in Crop Variation III. The Influence of Rainfall on the Yield of Wheat at Rothamsted," *Philosophical Transactions of the Royal Society of London*, B, No. 213 (1924), 89–142.

[9] ———, "The Arrangement of Field Experiments," *Journal of the Ministry of Agriculture*, 33 (1926), 503–13.

[10] ———, "Creative Aspects of Natural Law," The Eddington Memorial Lecture, Cambridge, Eng.: Cambridge University Press, 1950.

[11] ———, *The Design of Experiments*, (8th Ed.), Edinburgh: Oliver & Boyd, Ltd., 1966.

[12] ———, *Statistical Methods for Research Workers*, (14th Ed.), Edinburgh: Oliver & Boyd, Ltd., 1970.

[13] ——— and MacKenzie, W.A., "Studies in Crop Variation II. The Manurial Response of Different Potato Varieties," *Journal of Agricultural Science*, 13 (1923), 311–320.

[14] ——— and Tippett, L.H.C., "Limiting Forms of the Frequency Distribution of the Largest and Smallest Member of a Sample," *Proceedings of the Cambridge Philosophical Society*, 24 (1928), 180–90.

[15] Gosset, W.S., *Letters from W.S. Gosset to R.A. Fisher, 1915–1936*, with summaries by R.A. Fisher and a foreword by L. McMullen, (2nd Ed.), Privately circulated, 1970.

[16] Russell, E. John, "Field Experiments: How They Are Made and What They Are," *Journal of the Ministry of Agriculture*, 32 (1926), 989–1001.

[17] ———, *A History of Agricultural Research in Great Britain*, London: Allyn and Unwin, Ltd., 1966.

[18] Wold, H., "Nonlinear Estimation by Iterative Least Squares Procedures," in F.N. David, ed., *Research Papers in Statistics, Festschrift for J. Neyman*, New York: John Wiley & Sons, Inc., 1966, 411–44.

1.13
Some Problems of Statistics and Everyday Life

G. E. P. BOX*

When Fred Leone, our Executive Director, was explaining to me what my presidential duties were, he told me that one of the "perks" associated with this job is that I get to give the annual address to the Association, and I have a captive audience for as long as they are prepared to sit there. Fred said to me, "George, don't give them anything too technical because this is a light occasion and there will be a lot of people that the statisticians have dragged along—husbands, wives, friends—who have had about all the statistics they can take."

Well, imagine my disappointment. I had prepared a 200-page draft of my talk. It was called "The Present Status of the One-Armed Secretary Problem: A Decision-Theoretic Approach," and it made free use of σ-fields, Hilbert spaces, and all kinds of squiggly letters with dots on. This I reluctantly set aside. (I don't think any of you would have understood it anyway.) I have had to look for an alternative. I toyed for some time with the title, "Whither Statistics?," subtitled "Perhaps We Shouldn't Start from Here," but in the end abandoned that too. Eventually it struck me that many of the issues that we face as members of the American Statistical Association are really not very different from those we face as ordinary human beings. This is what my talk is about.

THE BEST IS OFTEN NOT VERY GOOD

Some of us have had a preoccupation with optimal or best procedures. But the best, of course, is not necessarily very good. For instance, to bring in the aspect of everyday life, if ever I *had* to decide between cutting my throat with a razor blade or with a rusty nail, I suppose I would choose the razor blade. But, although not strictly relevant to the problem as posed, one question that might cross my mind would be, "Have I considered all my options?"

A principle that is being given more attention these days is that of "robustification." Here one doesn't attempt to guarantee that things will be optimal over some tractable, but perhaps very narrow, set of circumstances. Instead one tries to ensure that they will be fairly good over a wide range of possibilities *likely to happen in practice*. Look at the human hand, for example. I doubt if there is any single thing that it does that could not be done better by some special instrument, but it is very good at doing a very large number of things that come up in facing the world as it actually is.

Another way to say this is that there is really nothing wrong with optimization per se, but that we ought to try to optimize over *that distribution of circumstances which the world really presents to us*. The mistake is choosing the best over too narrow a set of alternatives, suboptimization. It is sometimes argued that by doing simplified exercises, we can at least obtain useful pointers. However, I feel that such pointers are very likely to indicate the *wrong* direction, as might be true in the case of the razor blade and the rusty nail.

WHAT IS THE REAL WORLD LIKE?

The difficulty in taking the wider robustification approach is that we cannot expect to get good results unless we are really prepared to engage in the hazardous undertaking of finding out *what the world is really like*. It requires us, as statisticians, to have some knowledge of reality.

I believe we do have members, we may have ASA fellows, possibly we have even had ASA Presidents who really do not care what the world is really like. Some years ago a friend of mine told me about his daughter who was then at Oxford University. She was a very bright girl, but she got interested in politics (it was in the 1960s); she got behind in her studies, and the time of graduation was approaching. You may know that in the English system, there are many different grades of bachelor's degree. The young lady started to worry: was she going to get a "pass" degree (which is almost like the University spitting at you), or was it to be a third class, lower second class, upper second class, or a first class honours degree? She decided to ask her tutor about it. Finding him buried somewhere in the dust of one of the Oxford colleges, she eventually got around to asking him the delicate question, "Would it matter in the outside world if I didn't get a very good degree?" He looked very startled and said, "*Outside world*? What do *I* know about the outside world?"

When the statistician looks at the outside world, he cannot, for example, rely on finding errors that are independently and identically distributed in approximately normal distributions. In particular, most economic and business data are collected serially and can be expected, therefore, to be heavily serially dependent. So is much

* George E.P. Box is R.A. Fisher Professor of Statistics, University of Wisconsin, Madison, WI 53706. This article is the text of the Presidential Address delivered at the 138th Annual Meeting of the American Statistical Association, August 15, 1978, in San Diego.

The Collected Works of George E. P. Box, 1984, Wadsworth, Inc., Belmont, CA 94002.
Originally published in *J. Amer. Stat. Assoc.*, vol. 74, no. 365 (1979), pp. 1–4.

of the data collected from the automatic instruments which are becoming so common in laboratories these days. Analysis of such data, using procedures such as standard regression analysis which assume independence, can lead to gross error. Furthermore, the possibility of contamination of the error distribution by outliers is always present and has recently received much attention. More generally, real data sets, especially if they are long, usually show inhomogeneity in the mean, the variance, or both, and it is not always possible to randomize.

To find out what the world is really like, we must spend more time looking for ourselves at real sets of data. For example, David Cox says that deviations from normality often occur in the direction of light-tailed distributions as well as heavy-tailed ones. Let us discover if he's right. If he is, it could seriously affect some proposed robust methods.

In order to better confront the realities of the outside world, some have urged us to abandon classical methods of estimation, such as employ likelihood and Bayes' Theorem, and resort to a *new* empiricism for each problem that arises, and for each author that writes about it. I think that notion is wrong-headed. The imperfection, of course, lies not with the estimation method, but with the model that we put into it. For example, it is true that the sample average, which is the maximum likelihood estimate of the mean on standard normal assumptions, could be a very poor estimate if we believed that the data were generated by a *contaminated* normal distribution. Nevertheless, if our model took account, not of what we did *not* believe, but rather of what we *did* believe, we could obtain excellent estimates by standard methods. The great advantage of the model-based over the ad hoc approach, it seems to me, is that at any given time we know what we are doing.

Models, of course, are never true, but fortunately it is only necessary that they be useful. For this it is usually needful only that they not be grossly wrong. I think rather simple modifications of our present models will prove adequate to take account of most realities of the outside world. The difficulties of computation which would have been a barrier in the past need not deter us now.

CHOOSING THE "BEST" DOESN'T MAKE SENSE WITH OPTIONS THAT ARE NOT ALTERNATIVES

Another difficulty with optima is a tendency to want to choose the best *one* of a set of items that are not really alternatives. The relevant question then is not, "Which is best?," but "Do these different entities have a role, and if so, what is it?" For example, it turns out that it is much better to have two sexes than one—and this not merely for hedonistic reasons. Again, if one mentions Bayesian analysis and sampling theory analysis in the same breath, one not only hears the question, "Which is best?," but also the question, "Which is right?," and religious passions are quickly aroused. Yet, to my mind Bayes theory and sampling theory are not alternatives at all.

It is widely recognized that the advancement of learning does not proceed by conjecture alone, nor by observation alone, but by an iteration involving both. Certainly, scientific investigation proceeds by such iteration. Examination of empirical data inspires a tentative explanation which, when further exposed to reality, may lead to its modification. This modified explanation is again put in jeopardy by further exposure to reality, and so on, in a continued alternation between induction and deduction.

I am continually surprised that statisticians, even good ones, still seem to ignore this iterative aspect of investigation and talk as if the movement from an initial (perhaps ill-posed) question, to design, to data collection, to analysis of the data, to "the answer" were a one-shot affair. The wise investigator expends his effort not in one grand design (necessarily conceived at a time when he knows least about unfolding reality), but in a series of smaller designs, analyzing, modifying, and getting new ideas as he goes. This iterative aspect of research has a profound influence on almost everything the investigator and the statistician do, and it has been the source of much misunderstanding. Just as the rules that govern mathematical iteration are very different from those that govern solutions in closed form, so the rules that ought to apply to the statistics of most real scientific investigations are different, broader, and vaguer than those that might apply to a single decision or to a single test of hypothesis.

Now, since scientific advance, to which all statisticians must accommodate, takes place by the alternation of *two* different kinds of reasoning, we would expect also that *two* different kinds of inferential process would be required to put it into effect.

The first, used in estimating parameters from data *conditional* on the truth of some tentative model, is appropriately called *Estimation*. The second, used in checking whether, in the light of the data, *any* model of the kind proposed is plausible, has been aptly named by Cuthbert Daniel *Criticism*.

While estimation should, I believe, employ Bayes' Theorem, or (for the fainthearted) likelihood, criticism needs a different approach. In practice, it is often best done in a rather informal way by examination of residuals or other suitable functions of the data. However, when it is done formally, using tests of goodness of fit, it must, I think, employ sampling theory for its justification.

Bayes and likelihood inferences are necessarily conditional and, therefore, should not be used alone for the same reason that the statement, "If the moon was made of green cheese, it would be a great place for mice," should not tempt a mouse to hang around Cape Kennedy.

INAPPROPRIATE DIVISION OF AN ENTITY

While we can make a mistake by looking for *one* answer when we should be looking for two or more, we can make another mistake by *dividing* an entity inappropriately. You will recall the story of Solomon, who determined the true mother of a child of disputed parentage by offering to cut it in two. One slicing of our subject which I think can be harmful is that into Applied Statistics and Theoretical Statistics.

I hear people saying things like, "Of course I'm a theoretical statistician myself, but I agree there should be some applied statisticians and there should even be applied statistics departments; in fact, some of my *best friends* are applied statisticians." Now, in my opinion, that isn't any good, because, if you imagine the theoretical statisticians distributed about a point on the right of a scale and the applied statisticians distributed about a point on the left, you will end up with a bimodal distribution with low density in the center. Now the people most needed are, in my opinion, those in the middle, and perhaps that's why they seem to be in such short supply. If, alternatively, we aimed at a central target, then we might achieve a single unimodal distribution. This would still, of course, allow diversity. We would have some highly theoretical people in one tail and some highly applied people in the other. But the majority, while having proper theoretical training, might also possess ability and experience in applying what they knew to the solution of scientific problems.

TRAINING OF STATISTICIANS

This suggests the question of how statisticians should be trained. It's fairly easy to see how we should *not* train them. I will make an analogy with swimming.

Swimming could be taught by lecturing the student swimmers in the classroom three times a week on the various kinds of strokes and the principles of buoyancy and so forth. Some might believe that on completing such a course of study, the graduates would all eagerly run down to the pool, jump in, and swim at once. But I think it's much more likely that they would want to stay in the classroom to teach a fresh lot of students all that they had learned.

Let me mention another distinction which is now needed, and which threatens to become an unnecessary and harmful slicing. Statistical practitioners have known for a long time that, prior to using the methods that most textbooks emphasize, there is a very important and largely neglected[1] phase of activity which Fisher called specification and which has also been called model identification. This involves informal techniques of analysis of data, many of them graphical, aimed at looking at the data in a preliminary and exploratory way in order to help understand what questions should be asked and what tentative models might be entertained. Until recent years, however, this process was regarded by the majority as not entirely respectable. Like the black art, it was widely felt that it should be conducted, if at all, only behind closed doors.

It was a stroke of genius to realize that to render "a deed without a name" respectable, you should name it (or perhaps I should say rename it), and we are all grateful for the name "Data Analysis." This important part of our subject can now be studied without apology or shame, and courses on it are taught and may be attended by consenting adults. The elevation of Data Analysis to its proper place as a subject meriting serious study makes me as happy as I would be if some neglected but important activity of the carpenter, such as the use of the saw or the chisel, had at last received proper recognition and study. But my enthusiasm for the naming of Data Analysis does not extend to the renaming of Statisticians as "Data Analysts," any more than I should be happy to hear a carpenter described as a sawyer or a chiseler. Indeed, I am as appalled by the appearance of Data Analysts *as entities* as I would be at contemplating one half of the baby over which Solomon adjudicated, and for the same reason. There can be no feedback between the parts of a once-living thing cut in two.

Please can Data Analysts get themselves together again and become whole Statisticians before it is too late? Before they, their employers, and their clients forget the other equally important parts of the job statisticians should be doing, such as designing investigations and building models? By invention of the concept of Experimental Design, Fisher promoted the statistician from a curator of dusty relics to a valued member of a scientific team, responsible for planning and taking part in the conduct of an investigation. Let us not allow him to be relegated to his previous passive and inferior role by an injudicious choice of a name. "Our Data Analyst" is too close for my liking to "Our Tame Statistician," a poor thing if that is all he is.

THE AMERICAN STATISTICAL ASSOCIATION

Finally, I want to talk a bit about *our* Association because it is ours and it can be as good or as bad as we make it.

During my time as president I have received a number of letters, all of them interesting and some of them critical. Some members feel that we should be doing things we are not doing, some feel that we are doing things we should not be doing, some feel that the articles in our journals are not on the subjects they would like, or are not written with sufficient clarity. The suggestions made in such letters are, of course, given careful consideration not only by your president but by your Board and its committees. If you have ideas on these or other subjects, however, and want to see something more done about them, I do urge you, if you have not done so already, to volunteer for active duty. The Association is always seeking new faces and new ideas for its committees. And there are other things you can do too.

Suppose, for example, as a statistical practitioner, you feel that the journal *Technometrics* is not adequately

[1] An early exception was the second chapter of *Statistical Methods for Research Workers* (Fisher 1925), first published in 1925, in which Fisher discussed the use of preliminary graphical techniques.

fulfilling that part of its mandate that says it will publish

> papers illustrating the application of known statistical method to new or novel environments, expository or tutorial papers on particular statistical methods, and papers dealing with the philosophy and problems of applying statistical methods to research, development, design and performance.

I have it on good authority that editors have two problems in carrying out such a mandate: in the first place, articles of this kind are extremely hard to come by, and in the second place, referees tend to reject such articles for the wrong reasons (perhaps because they have not read the mandate). I urge you, therefore, to consider one or both of the following courses:

1. If you have suitable material, please write it up and send it in.
2. Please volunteer to act as a referee.

The editors are in desperate need of good referees for articles of this and every other sort. They need people who will go carefully through a paper, say encouraging words about good things, suggest how ideas could be clarified and how imperfections can be put right and who, when necessary, will firmly reject unsuitable manuscripts.

In closing, I want to say how much I have enjoyed being your president, especially because of the kindness, consideration, and help I have received from the members, board, and officers. In particular, I wish to thank Fred Leone, Ed Bisgyer, Jean Smith, and the rest of the Washington staff. The Association is indeed fortunate to be served by such accomplished and dedicated people.

REFERENCE

Fisher, R.A. (1925), *Statistical Methods for Research Workers*, Edinburgh: Oliver & Boyd.

1.14

Bayesian Analysis of Some Outlier Problems in Time Series

BOVAS ABRAHAM
University of Waterloo, Ontario

G. E. P. BOX
University of Wisconsin

SUMMARY

Two models, the aberrant innovation model and the aberrant observation model, are considered to characterize outliers in time series. The approach adopted here allows for a small probability α that any given observation is 'bad' and in this set-up the inference about the parameters of an autoregressive model is considered.

Some key words: Aberrant innovation; Autoregression; Posterior distribution.

1. INTRODUCTION

Since time series often contain discrepant observations, it is appropriate to employ models which reflect this fact. In a different context, Dixon (1958), Tukey (1960) and Box & Tiao (1968) suggested set-ups in which a small probability α exists that any observation is bad. We apply this idea to time series models in two different ways.

Consider a familiar time series model $\pi(B)y_t = a_t$, where $\pi(B) = 1-\pi_1 B-\pi_2 B^2-\ldots$, B is a backward shift operator such that $By_t = y_{t-1}$, and $\{a_t, t = \ldots, 1, 2, \ldots\}$ is a sequence of independent identically distributed normal random variables with mean zero and variance σ^2. The function $\pi(B)$ is often expressed as a ratio $\Phi(B)/\Theta(B)$ of finite autoregressive and moving average polynomial operators.

If it is supposed that any given innovation has a small probability α of being aberrant then we can write

$$\pi(B)y_t = \delta x_t + a_t, \tag{1·1}$$

where $x_t = 1$ if there is an aberrant innovation at t and $x_t = 0$ otherwise. This will be referred to as the aberrant innovation model.

Alternatively, it might be that the aberration affects the observation itself rather than the innovation. In that case, we may write

$$\pi(B)z_t = a_t, \tag{1·2}$$

where $z_t = y_t + \delta x_t$, $\pi(B)$ and a_t are as defined before, and $x_t = 1$ if the tth observation is aberrant and $x_t = 0$ otherwise. We refer to this as the aberrant observation model.

In §§ 2, 3 and 4 of this paper we discuss how to make inferences about the parameters of an autoregressive model with the possibility of aberrant innovations, how to assess the evidence that any innovation is an outlier, and how to estimate its deviation from expectation when the identity of the outliers may or may not be known. Section 5 treats a special case of the aberrant observation model.

The Collected Works of George E. P. Box, 1984, Wadsworth, Inc., Belmont, CA 94002. Originally published in *Biometrika,* vol. 66, no. 2 (1979), pp. 229–236.

2. ABERRANT INNOVATION MODEL AND INFERENCES ABOUT THE PARAMETERS

2·1. *Specification of model*

The pth order autoregressive aberrant innovation model may be defined as

$$y_t = V^T \Phi + \delta x_t + a_t, \tag{2·1}$$

where $V_t^T = (y_{t-1}, \ldots, y_{t-p})$, $\Phi^T = (\phi_1, \ldots, \phi_p)$, and x_t and a_t are as defined in (1·1). The parameters Φ, δ and σ are unknown. We also define $X^T = (x_1, \ldots, x_n)$ as a vector of r unities and $(n-r)$ zeros, where r and hence X are unknown. Assuming X to be known, Fox (1972) considered a likelihood ratio criterion for this set-up of the autoregressive model. The observations y_t are assumed to be deviations from the mean and in the case when the mean is unknown we can include that also in the expression for the likelihood and often to a sufficient approximation we can take y_t as the deviations from the sample mean.

2·2. *Posterior distribution of Φ given Y and X*

The posterior density of Φ given $Y^T = (y_1, \ldots, y_n)$ and X is

$$P(\Phi \mid Y, X) \propto \int P(\Phi, \delta, \sigma \mid X) P(Y \mid \Phi, \delta, \sigma, X)\, d\delta\, d\sigma,$$

where $P(\Phi, \delta, \sigma \mid X)$ is the prior density of $(\Phi, \delta, \sigma \mid X)$ and $P(Y \mid \Phi, \delta, \sigma, X)$ is the joint density of Y given $(\Phi, \delta, \sigma, X)$. This joint density may be written as

$$P(Y \mid \Phi, \delta, \sigma, X) = P(V_{p+1} \mid \Phi, \delta, \sigma, X) \prod_{t=p+1}^{n} P(y_t \mid V_t, \Phi, \delta, \sigma, X).$$

Also we can write

$$P(y_t \mid V_t, \Phi, \delta, \sigma, X) = P(y_t \mid V_t, \Phi, \delta, \sigma, x_t = 0) \exp\left[-\frac{1}{2\sigma^2}\{\delta^2 x_t^2 - 2(y_t - V_t^T \Phi) x_t \delta\}\right].$$

Suppose now that $x_1 = \ldots = x_p = 0$, i.e. that the first p observations are not outliers. Then $(V_{p+1} \mid \Phi, \delta, \sigma, X)$ is multivariate normal with mean zero and covariance matrix $\sigma^2 M_p^{-1}$, say. Hence we obtain

$$P(Y \mid \Phi, \delta, \sigma, X) = P(Y \mid \Phi, \delta, \sigma, X = \bar{0}) \exp\left[-\frac{1}{2\sigma^2}\{r\delta^2 - 2(Y - V^T \Phi)^T X \delta\}\right], \tag{2·2}$$

where $r = \Sigma_t x_t^2$, $\bar{0} = (0, \ldots, 0)^T$, and $V = (0, \ldots, 0, V_{p+1}, \ldots, V_n)^T$. Note here that since $x_1 = x_2 = \ldots = x_p = 0$ the first p rows of V may be taken as zero without loss of generality. If we follow Box & Jenkins (1976, p. 274)

$$P(Y \mid \Phi, \delta, \sigma, X = \bar{0}) = (2\pi\sigma^2)^{-\frac{1}{2}n} |M_p|^{\frac{1}{2}} \exp\left\{-\frac{1}{2\sigma^2} S(\Phi)\right\}, \tag{2·3}$$

where

$$S(\Phi) = Y^T Y - 2d^T \Phi + \Phi^T D \Phi, \quad d^T = (d_{01}, \ldots, d_{0p}), \quad D = ((d_{ij})),$$

a $p \times p$ symmetric matrix and

$$d_{ij} = y_{i+1} y_{j+1} + y_{i+2} y_{j+2} + \ldots + y_{n-j} y_{n-i} \quad (i, j = 0, 1, \ldots, p).$$

Then substituting (2·3) in (2·2) and simplifying, when $r \neq 0$ we obtain

$$P(Y \mid \Phi, \delta, \sigma, X) = (2\pi\sigma^2)^{-\frac{1}{2}n} |M_p|^{\frac{1}{2}}$$

$$\times \exp\left[-\frac{1}{2\sigma^2}\{(\Phi - \Phi^*)^T B (\Phi - \Phi^*) + r(\delta - \delta^*)^2 + Y^T(I - r^{-1} X X^T) Y - \Phi^{*T} B \Phi^*\}\right], \tag{2·4}$$

where

$$B = D - r^{-1} V^{\mathrm{T}} X X^{\mathrm{T}} V, \quad \Phi^* = B^{-1}(d - r^{-1} V^{\mathrm{T}} X X^{\mathrm{T}} Y), \quad \delta^* = r^{-1}(Y - V\Phi)^{\mathrm{T}} X.$$

We now assume that δ, σ and Φ are *a priori* independent of each other and of X, and that δ is locally uniform. Also, if we do not have any prior information about σ and Φ and since information about σ would supply no information about Φ it might be sensible (Box & Tiao, 1973, pp. 41–3) to employ a prior distribution of the form

$$P(\Phi, \sigma) \propto |I(\Phi)|^{\frac{1}{2}} \sigma^{-1},$$

where $I(\Phi)$ is the information matrix of Φ. It is shown by Box & Jenkins (1976, pp. 280–1) that for an autoregressive model of order p, $I(\Phi) \propto M_p^{-1}$, so that

$$P(\Phi, \delta, \sigma | X) = P(\Phi, \sigma) \propto |M_p|^{-\frac{1}{2}} \sigma^{-1}. \tag{2·5}$$

Now multiplying (2·4) by (2·5) and integrating over δ and σ, we get

$$P(\Phi | Y, X) \propto \left\{1 + \frac{(\Phi - \Phi^*)^{\mathrm{T}} B (\Phi - \Phi^*)}{\nu S^2}\right\}^{-\frac{1}{2}(\nu + p)}, \tag{2·6}$$

where $\nu = n - p - 1$, $\nu S^2 = Y^{\mathrm{T}}(I - r^{-1} X X^{\mathrm{T}}) Y - \Phi^{*\mathrm{T}} B \Phi^*$ which does not involve Φ. Similarly it follows for $r = 0$ that

$$P(\Phi | Y, X) \propto \left\{1 + \frac{(\Phi - \hat{\Phi})^{\mathrm{T}} D (\Phi - \hat{\Phi})}{\nu_0 S_0^2}\right\}^{-\frac{1}{2}(\nu_0 + p)},$$

where $\hat{\Phi} = D^{-1} d$, $\nu_0 = n - p$ and $\nu_0 S_0^2 = Y^{\mathrm{T}} Y - d^{\mathrm{T}} D^{-1} d$. These results should be considered approximate in as much as the priors on δ and σ are improper.

Hence it follows that if it were given that a specific set of r observations was spurious the approximate posterior distribution of Φ would be a p-dimensional multivariate t distribution with mean vector Φ^*, dispersion matrix $S^2 B^{-1}$ and degrees of freedom $\nu = n - p - 1$. Note that if the Φ's are subject to stationarity conditions the above distribution is appropriate provided that the probability outside the region defined by the stationarity conditions is negligible.

2·3. *Posterior distribution of Φ given Y for unknown X*

Suppose now that there is a small prior probability α that any element of X is unity and a complementary probability that it is 0. If u_r denotes an $n \times 1$ vector with a specific set of r unities and $(n - r)$ zeros then the prior probability is

$$\mathrm{pr}(X = u_r | \alpha) = \alpha^r (1 - \alpha)^{n-r} \quad (r = 0, 1, \ldots, n). \tag{2·7}$$

Hereafter α is assumed known and conditioning on α will be implicit. Also to simplify notation we write X for $X = u_r$.

Given the observations Y the posterior distribution of $(\Phi, \delta, \sigma, X)$ is

$$P(\Phi, \delta, \sigma, X | Y) = \frac{P(X) P(\Phi, \delta, \sigma | X) P(Y | \Phi, \delta, \sigma, X)}{\Sigma_r P(X) P(Y | X)}, \tag{2·8}$$

where $P(\Phi, \delta, \sigma | X)$ is the prior distribution given in (2·5), $P(Y | \Phi, \delta, \sigma, X)$ is as expressed in (2·4) and

$$P(Y | X) = \int P(\Phi, \delta, \sigma | X) P(Y | \Phi, \delta, \sigma, X) \, d\delta \, d\Phi \, d\sigma. \tag{2·9}$$

The summation in (2·8) takes into account all 2^n possible values of X. Integrating (2·8) over δ and σ, we get $P(\Phi, X \mid Y) = P(X \mid Y) P(\Phi \mid Y, X)$, where $P(\Phi \mid Y, X)$ is the posterior distribution in (2·6) and

$$P(X \mid Y) = P(X) P(Y \mid X) / \Sigma_r P(X) P(Y \mid X)$$

which may be rewritten as

$$P(X \mid Y) = C\{P(X)/P(X = \bar{0})\}\{P(Y \mid X)/P(Y \mid X = \bar{0})\},$$

where

$$C = P(X = \bar{0}) P(Y \mid X = \bar{0}) / \Sigma_r P(X) P(Y \mid X).$$

Now from (2·3), (2·4), (2·5), (2·9) and with some algebraic manipulation it can be shown that

$$\frac{P(Y \mid X)}{P(Y \mid X = \bar{0})} = \beta(\tfrac{1}{2}, \tfrac{1}{2}\nu)\{|D| \nu_0 S_0^2/(r|B|)\}^{\frac{1}{2}}\{\nu_0 S_0^2/(\nu S^2)\}^{\frac{1}{2}\nu},$$

where $\beta(a, b)$ is the beta function. Now summing $P(\Phi, X \mid Y)$ over all possible values of X, we get

$$P(\Phi \mid Y) = \Sigma_r w_{(r)} P(\Phi \mid Y, X), \tag{2·10}$$

where

$$w_{(r)} = P(X \mid Y) = C\{\alpha/(1-\alpha)\}^r \{P(Y \mid X)/P(Y \mid X = \bar{0})\}.$$

Note that (2·10) is a weighted average of multivariate t distributions and for computational purposes it can be written as

$$P(\Phi \mid Y) = w_0 P(\Phi \mid Y, X = \bar{0}) + \sum_{i=p+1}^{n} w_i P_i(\Phi \mid Y, X = u_1)$$

$$+ \sum_{i>j=p+1}^{n} w_{ij} P_{ij}(\Phi \mid Y, X = u_2) + \ldots, \tag{2·11}$$

where the subscripts $0, i, ij$ refer to the possibilities respectively that no innovations, the ith innovation, the ith and jth innovations are outliers. The exact evaluation of (2·11) would be computationally difficult. However, in practice α will usually be small and it will often be possible without undue approximation to ignore the possibility of more than two and in some cases one outlier.

2·4. *Autoregressive model of order* 1

Consider the model $y_t = \phi y_{t-1} + \delta x_t + a_t$, where x_t and a_t are as defined before. Suppose that $\bar{y}_q$ and $\bar{y}_{q-1}$ respectively, denote the average of the suspected observations and that of the observations just previous to the suspected ones. Then

$$D = \sum_{t=3}^{n} y_{t-1}^2, \quad d = \sum_{t=2}^{n} y_{t-1} y_t, \quad B = D - r\bar{y}_{q-1}^2, \quad \phi^* = B^{-1}(d - r\bar{y}_q \bar{y}_{q-1}), \quad \hat{\phi} = D^{-1} d,$$

$$\nu S^2 = \sum_{t=1}^{n} y_t^2 - r\bar{y}_q^2 - \phi^{*2} B, \quad \nu_0 S_0^2 = \sum_{t=1}^{n} y_t^2 - \hat{\phi}^2 D.$$

When $r = 1$, the qth observation being suspected, say, $\phi^* = (d - y_q y_{q-1})/(D - y_{q-1}^2)$, which is very like the regression estimate of ϕ except that relevant quantities allowing for the effect of the outlier are subtracted from the numerator and denominator.

It can be seen that $P(\phi \mid Y)$ is a weighted average of scaled t distributions with mean ϕ^*, scaling factor $SB^{-\frac{1}{2}}$ and degrees of freedom $n-2$. This distribution gives us all the information about ϕ. In particular, the posterior mean and variance are given by

$$\bar{\phi}_m = E(\phi \mid Y) = \Sigma_r w_{(r)} \phi^*,$$

$$\operatorname{var}(\phi \mid Y) = w_0 \left\{ \frac{n-1}{n-3} \frac{S^2}{D} + (\hat{\phi} - \bar{\phi}_m)^2 \right\} + \sum_{r \neq 0} w_{(r)} \left\{ \frac{n-2}{n-4} \frac{S^2}{B} + (\phi^* - \bar{\phi}_m)^2 \right\}. \tag{2·12}$$

3. Example

The data given in Fig. 1(a) consist of 70 observations generated from the model $y_t = \frac{1}{2} y_{t-1} + 5x_t + a_t$, where $x_{50} = 1$, $x_t = 0$ $(t \neq 50)$, and $\{a_t\}$ is a 'white noise' sequence with $\sigma^2 = 1$. We analyse these data supposing that the magnitude and identity of the outlier are unknown. If we assume that there is a prior probability that it is from $N(\delta, 1)$, the posterior distribution of Φ given Y is computed for $\alpha = 0{\cdot}0001$, 0·001, 0·01, 0·03, and 0·05, and one of them is shown in Fig. 1(b). In these computations the leading term and the next two summations in the expansion (2·11) were used. The weights after these are very small and hence the associated distributions are ignored.

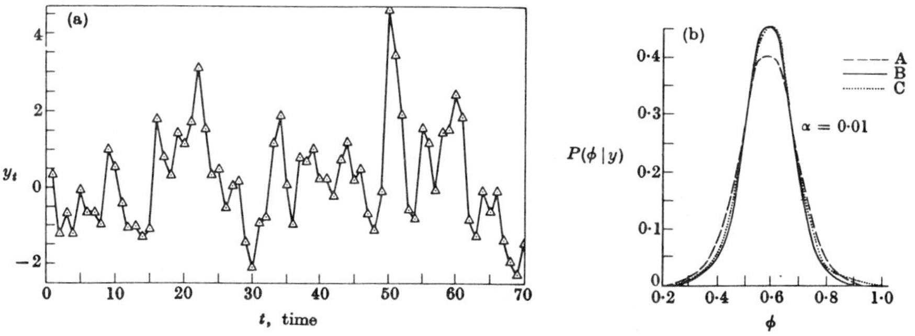

Fig. 1. (a) Data generated from a model $y_t = \frac{1}{2} y_{t-1} + a_t$ with a discrepancy ($\delta = 5$) introduced at $t = 50$. (b) Posterior distribution of ϕ given Y: A, assuming no outliers, $t(0{\cdot}582, 0{\cdot}099^2, 69)$; B, assuming that y_{50} is an outlier $t(0{\cdot}584, 0{\cdot}086^2, 68)$; C, assuming the possibility of outliers, $\alpha = 0{\cdot}01$.

The posterior probabilities w_t that the tth observation is an outlier and the others are not can be computed. For $\alpha = 0{\cdot}01$, $w_{50} = 0{\cdot}8922$, $w_{16} = 0{\cdot}0004$, $w_j = 0{\cdot}0002$ $(j = 22, 55, 62)$ and $w_j \simeq 0{\cdot}0001$ $(j \neq 16, 22, 50, 55, 62)$. This shows that these probabilities lie very close to zero except for the discrepant 50th observation which stands out dramatically. In fact, w_{50} is more than 2200 times the next biggest one and about 9000 times the smallest one. Very similar results are obtained also for other values of α. These posterior probabilities can be plotted as illustrated by Abraham & Box (1978). The possibility for two outliers can also be graphically represented by a two way diagram as has been demonstrated by Abraham & Box (1978). For this example, as might be expected, these posterior probabilities are negligible and distributed uniformly over the space except for the pairs which include the observation y_{50}.

Figure 1(b) shows respectively posterior distributions for ϕ, A assuming no outliers, B assuming that t_{50} is an outlier, and C assuming the possibility of outliers with $\alpha = 0{\cdot}01$.

The situation is very insensitive to the choice of α. In particular, values of α of 0·0001, 0·001, 0·01, 0·03 and 0·05 give very similar graphs.

Table 1 shows the mean and variance of the posterior distribution of ϕ for various values of α. For comparison we show the corresponding mean and variance when it is assumed that

Table 1. *Posterior mean and variance of* Φ *given* Y

α	$E(\phi \mid Y)$	var $(\phi \mid Y)$
0·0	0·5820	0·0101
0·0001	0·5841	0·0076
0·001	0·5845	0·0077
0·01	0·5855	0·0078
0·03	0·5873	0·0079
0·05	0·5887	0·0089

there are no outliers. Again it is found that the conclusions are not sensitive to moderate changes in α. However, as might be expected there is a dramatic difference between the assumption of no possibility of outliers ($\alpha = 0$) and the assumption of some such possibility, even a very remote one ($\alpha = 0{\cdot}0001$).

4. Inference about δ

4·1. *Posterior distribution of* δ

Using the results of § 2, we obtain

$$P(\Phi, \delta, \sigma \mid Y, X) \propto \sigma^{-(n+1)} \exp\left\{-\frac{1}{2\sigma^2}(\Phi - \bar{\Phi})^{\mathrm{T}} D(\Phi - \bar{\Phi}) + C_X(\delta - \hat{\delta})^2 + \nu S_d^2\right\}, \tag{4·1}$$

where

$$\bar{\Phi} = D^{-1}(d - V^{\mathrm{T}} X\delta), \quad C_X = r - X^{\mathrm{T}} V D^{-1} V^{\mathrm{T}} X,$$

$$\hat{\delta} = C_X^{-1}(Y^{\mathrm{T}} X - d^{\mathrm{T}} D^{-1} V X), \quad \nu S_d^2 = Y^{\mathrm{T}} Y - d' D^{-1} d - C_X \hat{\delta}^2.$$

Integrating (4·1) over Φ and σ, we obtain

$$P(\delta \mid Y, X) \propto \left\{1 + \frac{C_X(\delta - \hat{\delta})^2}{\nu S_d^2}\right\}^{-\frac{1}{2}(\nu+1)},$$

where $\nu = n - p - 1$. Hence $(\delta \mid Y, X)$ is approximately a scaled t distribution with mean $\hat{\delta}$, scaling factor $S_d/\sqrt{C_X}$ and degrees of freedom $\nu = n - p - 1$. Therefore, proceeding as in § 2·2, we get

$$P(\delta \mid Y) = \Sigma_r w_{(r)} P(\delta \mid Y, X),$$

where $w_{(r)}$ is the weight function shown in (2·10).

4·2. *Special case*

Consider again the model given in § 2·3. In this case

$$\hat{\delta} = r(\bar{y}_q - \hat{\phi}\bar{y}_{q-1})/C_X, \quad \nu S_d^2 = \nu_0 S_0^2 - C_X \hat{\delta}^2,$$

where $\hat{\phi}$, $\bar{y}_q$ and $\bar{y}_{q-1}$ are all given in § 2·3, and

$$C_X = r\left(1 - \bar{y}_{q-1}^2 \Big/ \sum_{t=3}^{n} y_{t-1}^2\right).$$

When $r = 1$, $\hat{\delta} = (y_q - \hat{\phi} y_{q-1})/C_X$. Notice that $\hat{\phi} y_{q-1}$ serves as an estimate of y_q and hence $y_q - \hat{\phi} y_{q-1}$ is a measure of the discrepancy.

4·3. *Example*

With the data of § 3·1, the posterior means and variances of δ given Y are shown in Table 2. Changes in α over the range considered seem once more to have little effect on inferences about δ.

Table 2. *Posterior means and variances of δ given Y*

α	$E(\delta \mid Y)$	var $(\delta \mid Y)$
0·001	4·56	1·02
0·01	4·50	0·97
0·03	4·31	0·90
0·05	4·11	0·86

5. Autoregressive aberrant observation model

We now consider the aberrant observation model described in (1·2). In the autoregressive case the model becomes $\Phi(B) z_t = a_t$, where $z_t = y_t + \delta x_t$, and a_t and x_t are all as defined before in (1·2).

To find the posterior distribution of δ given $Y^T = (y_1, \ldots, y_n)$ and $X^T = (x_1, \ldots, x_n)$, we proceed as follows. The joint probability distribution of $Z^T = (z_1, \ldots, z_n)$ can be given exactly as in (2·3) with Z replacing Y. Then, the joint probability distribution of Y can be obtained by making a transformation from Z to Y and this transformation has unit Jacobian. Hence we have

$$P(Y \mid \Phi, \delta, \sigma) = P(Y \mid \Phi, \delta, \sigma, X = \bar{0}) \exp\left\{-\frac{1}{2\sigma^2}(\delta - \hat{\delta})^2 C_X(\Phi) - \hat{\delta}^2 C_X(\Phi)\right\}, \quad (5\cdot1)$$

where

$$\hat{\delta} = (Y^T X - Y^T U\Phi - X^T V\Phi + \Phi^T V^T U\Phi)/C_X(\Phi), \quad C_X(\Phi) = r - 2X^T U\Phi + \Phi^T U^T U\Phi,$$

$$U = (\bar{0}, \ldots, \bar{0}, U_{p+1}, \ldots, U_n)^T, \quad U_t^T = (x_{t-1}, \ldots, x_{t-p}),$$

and V is as defined in (2·2). Now the posterior distribution $P(\Phi, \delta, \sigma \mid Y, X)$ can be obtained by multiplying the prior distribution given in (2·5) with (5·1). Integrating this posterior distribution over σ, we get the joint posterior of Φ and δ to be

$$P(\Phi, \delta \mid Y, X) \propto \{(\delta - \hat{\delta})^2 C_X(\Phi) + S(\Phi) - \hat{\delta}^2 C_\Phi(X)\}^{-\frac{1}{2}n}$$

and inference about (Φ, δ) can be made through this distribution.

It seems difficult, at present, to obtain the marginals of Φ or δ analytically and one has to resort to numerical methods. However, conditional posterior distributions of δ or Φ can be obtained analytically. In particular, it can be shown that

$$P(\delta \mid \Phi, Y, X) \propto \left\{1 + \frac{C_X(\Phi)(\delta - \hat{\delta})^2}{S(\Phi) - \hat{\delta}^2 C_X(\Phi)}\right\}^{-\frac{1}{2}n}, \quad (5\cdot2)$$

and this is a scaled t distribution.

Now if we proceed as in § 2·2 it is possible to get an expression for the posterior distribution of δ given Φ and Y, for unknown X. However, we feel that instead of repeating that procedure it is interesting to consider some simple cases of (5·2) for illustration.

Consider an autoregressive model of order one with $r = 1$ and suppose that the qth observation is suspected as an outlier. Then $\hat{\delta} = y_q - \hat{y}_q$, where $\hat{y}_q = \phi(1+\phi^2)^{-1}(y_{q-1} + y_{q+1})$ is an estimate of y_q giving equal weights to y_{q-1} and y_{q+1}. In the case of an autoregressive model of order p with $r = 1$, $\delta = y_q - \hat{y}_q$, where

$$\hat{y}_q = \sum_{j=1}^{p} \psi_j (y_{q-j} + y_{q+j}), \quad \psi_j = \left(\phi_j - \sum_{i=1}^{p-j} \phi_i \phi_{i+j}\right) \Big/ \left(1 + \sum_{i=1}^{p} \phi_i^2\right) \quad (j = 1, \ldots, p).$$

Here the estimate of y_q is a weighted sum of y_{q-j} and y_{q+j} $(j = 1, \ldots, p)$ with equal weights being given to observations which are equidistant from y_q. It can be shown, in both the cases considered, that $\hat{y}_q$ is the predictive mean of y_q given Φ and the other observations and hence δ is a measure of the discrepancy.

6. CONCLUDING REMARKS

The model used in this paper implies that the effect of what has happened to the innovation at a given time stays in the system. This model assumes that the aberrant observations all belong to the same population. Obviously, there are physical situations where this is likely to be true but in general it will not be. However, the model discussed gives us access to the most interesting practical case which is that where there is a single aberrant observation.

We thank a referee for helpful comments. The first author was partially supported by a grant from the National Research Council of Canada and the second author was supported by a U.S. Army Grant.

REFERENCES

ABRAHAM, B. & BOX, G. E. P. (1978). Linear models and spurious observations. *Appl. Statist.* **27**, 131–8.

BOX, G. E. P. & JENKINS, G. M. (1976). *Time Series Analysis Forecasting and Control*, revised edition. San Francisco: Holden Day.

BOX, G. E. P. & TIAO, G. C. (1968). A Bayesian approach to some outlier problems. *Biometrika* **55**, 119–29.

BOX, G. E. P. & TIAO, G. C. (1973). *Bayesian Inference in Statistical Analysis*. Reading, Mass: Addison-Wesley.

DIXON, W. J. (1958). Processing data for outliers. *Biometrics* **9**, 74–89.

FOX, A. J. (1972). Outliers in time series. *J. R. Statist. Soc.* B **34**, 350–63.

TUKEY, J. W. (1960). A survey of sampling from contaminated distributions. In *Contributions to Probability and Statistics*, Essays in Honor of Harold Hotelling, Eds I. Olkin, S. G. Churye and others, pp. 448–85. Stanford University Press.

[*Received April* 1978. *Revised November* 1978]

1.15

Sampling and Bayes' Inference in Scientific Modelling and Robustness

G. E. P. BOX
University of Wisconsin

[Read before the ROYAL STATISTICAL SOCIETY at a meeting organized by the South Wales Local Group on Thursday, May 15th, 1980, the President SIR CLAUS MOSER in the Chair]

SUMMARY

Scientific learning is an iterative process employing Criticism and Estimation. Correspondingly the formulated model factors into two complementary parts—a predictive part allowing model criticism, and a Bayes posterior part allowing estimation. Implications for significance tests, the theory of precise measurement and for ridge estimates are considered. Predictive checking functions for transformation, serial correlation, bad values, and their relation with Bayesian options are considered. Robustness is seen from a Bayesian viewpoint and examples are given. For the bad value problem a comparison with M estimators is made.

Keywords: ITERATIVE LEARNING; MODEL BUILDING; INFERENCE; BAYES THEOREM; SAMPLING THEORY; PREDICTIVE DISTRIBUTION; DIAGNOSTIC CHECK; TRANSFORMATIONS; SERIAL CORRELATION; BAD VALUES; OUTLIERS; ROBUST ESTIMATION

0. INTRODUCTION

No statistical model can safely be assumed adequate. Perspicacious criticism employing diagnostic checks must therefore be applied. But while such checks are always necessary, they may not be sufficient, because some discrepancies may on the one hand be potentially disastrous and on the other be not easily detectable. In addition therefore it is often pertinent to make the developing model robust against contingencies to which it is currently judged sensitive.

The object of this paper is to review the complementary roles in the model building process of the predictive distribution and of the posterior distribution; the former in producing diagnostic checks of parametric as well as residual features of the model, the latter in providing a general basis for robust estimation.

1. SCIENTIFIC LEARNING AND STATISTICAL INFERENCE

Much of statistics is concerned with extending knowledge by building empirico-mechanistic models that involve probability. A theory about such scientific model building ought to explain what good statisticians and scientists actually do. It seems that scientific knowledge advances by a practice–theory iteration. Known facts (data) suggest a tentative theory or model, implicit or explicit, which in turn suggests a particular examination and analysis of data and/or the need to acquire further data; analysis may then suggest a modified model that may require further practical illumination and so on. I shall suppose that data are acquired from a designed experiment, but the same argument would apply if data acquisition was from a sample survey or even from a visit to the library. New knowledge thus evolves by an interplay between *dual* processes of induction and deduction in which the model is not fixed but is continually developing. The statistician's role is to assist this evolution (see, for example, Box and Youle, 1955; Box, 1976). In doing so he employs two inferential devices: *Criticism*† and *Estimation.*

Suppose that at some stage i of an investigation, model M_i is being entertained.

Criticism can induce model modification. It involves a confrontation of M_i with available data $\mathbf{y}$ (old as well as newly acquired), and asks whether M_i is consonant with $\mathbf{y}$ and, if not, how

† The apt naming of inferential *criticism* is due to Cuthbert Daniel, see also Popper (1959).

The Collected Works of George E. P. Box, 1984, Wadsworth, Inc., Belmont, CA 94002.
Originally published in *J. Roy. Stat. Soc.,* Series A, vol. 143, part 4 (1980), pp. 383–430.

not. It employs a process of diagnostic checking (see, for example, Box and Jenkins, 1970), which is often done informally using plots of various kinds of residual quantities, or more formally, with tests of goodness of fit or "tentative overfitting" procedures. When a modification to M_{i+1} has been made, this new model, in addition to confronting the same data, will in some cases be checked against new data generated by a design D_{j+1}. This new design will be chosen to explore those shadowy regions whose illumination is judged currently to be important in view of the nature of the modified model and the needs of independent verification.

Estimation. When the iteration leads to a model worthy to be entertained it may be used to estimate parameters conditional on its truth. In practice such estimation is used not only at the termination of the model building sequence but at many stages throughout it. This is because, to conduct criticism of a model, it is often necessary to estimate provisionally parameters at intermediate stages.

In any such enterprise many subjective choices are made, conscious or unconscious, good or bad. They determine for instance which plots, displays and checks of data and residuals are looked at; and what treatments and variables are included at which levels, over what experimental region, in which transformation, in what design, to illuminate which models. The wisdom of these choices over successive stages of development is the major determinant of how fast the iteration will converge or of whether it converges at all, and distinguishes good scientists and statisticians from bad. It is in this context that theories of inference need be considered. While it is comforting to remember that a good scientific iteration is likely to share the property of a good numerical iteration—that mistakes often are self-correcting, this also implies that the investigator must worry particularly about mistakes which are likely not to be self-correcting.

1.1. *Rival Theories of Inference*

The distinction between model criticism and parameter estimation has not always been made and proponents both of sampling inference and Bayesian inference have long sought for a single comprehensive theory.

I believe that, subject to some overlap discussed later, sampling theory is needed for exploration and ultimate *criticism* of an entertained model in the light of current data, while Bayes' theory is needed for *estimation* of parameters conditional on the adequacy of the entertained model. On this view (see also Box and Tiao, 1973) both processes would have essential roles in the continuing scientific iteration just as the two sexes are required for human reproduction. Attempts to choose between two entities which were not alternative but complementary could certainly be expected to lead to contention, paradox and confusion of the kind we have been experiencing. The view that more than one mode of statistical reasoning can be useful is not, of course, new and was advanced (however with a different emphasis and conclusions) by R. A. Fisher. See also in particular Dempster (1971).

1.2. *The Need for Prior Distributions*

In the past, the need for probabilities expressing prior belief has often been thought of, not as a necessity for all scientific inference, but rather as a feature peculiar to Bayesian inference. This seems to come from the curious idea that an outright assumption does not count as a prior belief. The interconnection between model assumptions and prior distributions becomes clear when it is remembered that every model can be imagined as embedded in a more complex one. For example, an outright assumption of normality can be modelled by a suitable parametric family of distributions indexed by a parameter β, which has a sharp prior at the normal value. I believe that it is impossible logically to distinguish between model assumptions and the prior distribution of the parameters. The model *is* the prior in the wide sense that it is a probability statement of all the assumptions currently to be tentatively entertained *a priori*. On this view, traditional sampling theory was of course not free from assumptions of prior knowledge. Instead it was as if only two states of mind had been allowed—complete certainty or complete uncertainty.

One illustration of how implied prior knowledge which is *implausibly imprecise* can lead to trouble in sampling theory is the famous discovery by Stein (1956) of the inadmissibility of the multivariate sample mean. Consider for example the usual one-way analysis of variance set-up. The prior assumption which justifies the shrinkage estimator (see, for example, Box and Tiao, 1968a; Lindley and Smith, 1972) that the group means μ_j are random samples from some normal super-population having unknown mean and variance might, in appropriate circumstances, be eminently reasonable. It is easy, however, to miss the lesson which is to be learned from such examples. Notice that there are many circumstances in which this "Model II" assumption would not be sensible either. For example, if the μ's were daily batch yields from some production process, it might be much more reasonable to postulate *a priori* that they followed some time series model such as a stationary autoregressive process. The estimators (Tiao and Ali, 1971) then derived from Bayesian means are not Stein's shrinkage estimators, but alternative estimators allowing incorporation of relevant sample information about the *autocorrelation* of the batch means. Thus while for this example, except as a numerical approximation, we ought not to use the sample means as estimates, we ought not to use Stein's shrinkage estimates either. There seems no logical way to avoid trouble except by the explicit prior statement of the model we wish to entertain.

1.3. *Two Complementary Factors from Bayes' Formula*

If the prior probability distribution of parameters is accepted as essential, then a complete statement of the entertained model at any stage of an investigation is provided by the joint density for potential data $\mathbf{y}$ and parameters $\boldsymbol{\theta}$ calculated from

$$p(\mathbf{y},\boldsymbol{\theta}\mid A) = p(\mathbf{y}\mid\boldsymbol{\theta},A)\,p(\boldsymbol{\theta}\mid A). \tag{1.1}$$

In these expressions A is understood to indicate conditionality on all or some of the assumptions in the model specification. This model (1.1) means to me that current belief about the outcome of contemplated data acquisition would be calibrated with adequate approximation by a *physical simulation* involving random sampling from the distributions $p(\mathbf{y}\mid\boldsymbol{\theta},A)$ and $p(\boldsymbol{\theta}\mid A)$.

The model can also be factored as

$$p(\mathbf{y},\boldsymbol{\theta}\mid A) = p(\boldsymbol{\theta}\mid\mathbf{y},A)\,p(\mathbf{y}\mid A). \tag{1.2}$$

In particular the second factor on the right, which can be computed before any data become available,

$$p(\mathbf{y}\mid A) = \int p(\mathbf{y}\mid\boldsymbol{\theta},A)\,p(\boldsymbol{\theta}\mid A)\,d\boldsymbol{\theta} \tag{1.3}$$

is the *predictive* distribution. It is the distribution of the totality of all possible samples $\mathbf{y}$ that could occur if the assumptions were true.

When an actual data vector $\mathbf{y}_d$ becomes available

$$p(\mathbf{y}_d,\boldsymbol{\theta}\mid A) = p(\boldsymbol{\theta}\mid\mathbf{y}_d,A)\,p(\mathbf{y}_d\mid A) \tag{1.4}$$

and the first factor on the right is Bayes' posterior distribution of $\boldsymbol{\theta}$ given $\mathbf{y}_d$

$$p(\boldsymbol{\theta}\mid\mathbf{y}_d,A) \propto p(\mathbf{y}_d\mid\boldsymbol{\theta},A)\,p(\boldsymbol{\theta}\mid A). \tag{1.5}$$

But of equal importance is the second factor

$$p(\mathbf{y}_d\mid A) = \int p(\mathbf{y}_d\mid\boldsymbol{\theta},A)\,p(\boldsymbol{\theta}\mid A)\,d\boldsymbol{\theta}, \tag{1.6}$$

the predictive density associated with the particular data $\mathbf{y}_d$ actually obtained. Fig. 1 illustrates for a single parameter θ and a sample $\mathbf{y}_d$ of $n = 2$ observations.

If the model is to be believed, the posterior distribution $p(\boldsymbol{\theta}\mid\mathbf{y}_d,A)$ allows all relevant estimation inferences to be made about $\boldsymbol{\theta}$. However, if $\mathbf{y}_d$ were such as would be very unlikely to

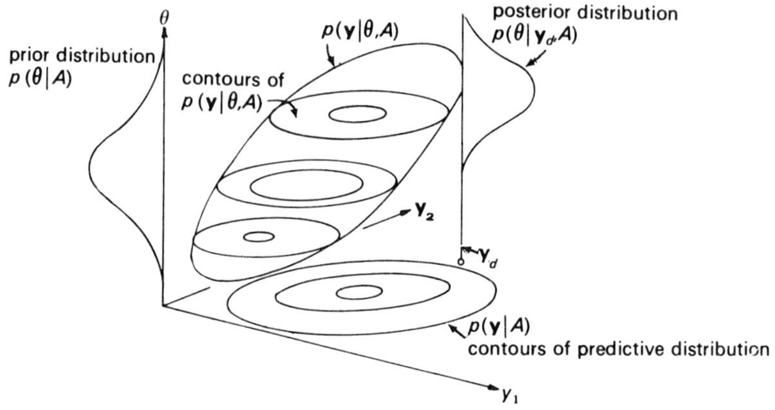

FIG. 1. A representation of the prior distribution, the posterior distribution and the predictive distribution for a single parameter θ and sample $\mathbf{y}_d$ of two observations.

be generated by the model, this could not be shown by any abnormality in this factor, but *could* be assessed by reference of the density $p(\mathbf{y}_d \mid A)$ to the predictive reference distribution $p(\mathbf{y} \mid A)$ or of the density $p\{g_i(\mathbf{y}_d) \mid A\}$ of some relevant checking function $g_i(\mathbf{y}_d)$ to its predictive distribution. The importance of the predictive distribution and the possibility of using it in some way as a model checking device has been discussed by a number of authors. See in particular Roberts (1965), Guttman (1967), Geisser (1971, 1975), Dempster (1971, 1975), Geisser and Eddy (1979) and Kadane *et al.* (1979). Also measures of surprise other than that discussed here have been proposed, for example by Good (1956).

2. ESTIMATION OF THE MEAN OF A NORMAL DISTRIBUTION

As an example consider a sample of n observations drawn randomly from a normal distribution with unknown mean θ and known variance σ^2 with uncertainty about the mean expressed by supposing that, *a priori*, θ is distributed normally about θ_0 with known variance σ_θ^2. Then *conditional* on the adequacy of the model, θ is estimated by combining data and prior information in the normal posterior distribution

$$p(\theta \mid \mathbf{y}, A) \propto (I_{\bar{y}} + I_\theta)^{\frac{1}{2}} \exp\{-\tfrac{1}{2}(I_{\bar{y}} + I_\theta)(\theta - \bar{\theta})^2\}, \tag{2.1}$$

where $I_{\bar{y}} = n\sigma^{-2}$, $I_\theta = \sigma_\theta^{-2}$ and $\bar{\theta} = w\bar{y} + (1-w)\theta_0$ is an appropriately weighted average of $\bar{y}$ and θ_0, with $w = I_{\bar{y}}/(I_{\bar{y}} + I_\theta)$ the proportion of information coming from the data.

The predictive distribution allowing criticism of the model by contrasting data and prior information is

$$p(\mathbf{y} \mid A) \propto \sigma^{-(n-1)}\left(\frac{\sigma^2}{n} + \sigma_\theta^2\right)^{-\frac{1}{2}} \exp\left\{-\frac{1}{2}\left[\frac{(n-1)s^2}{\sigma^2} + \frac{(\bar{y}-\theta_0)^2}{n^{-1}\sigma^2 + \sigma_\theta^2}\right]\right\}. \tag{2.2}$$

An overall predictive check is supplied by calculating

$$\alpha = \Pr\{p(\mathbf{y} \mid A) < p(\mathbf{y}_d \mid A)\} = \Pr\{\chi_n^2 > g(\mathbf{y}_d)\}, \tag{2.3}$$

where

$$g(\mathbf{y}_d) = \frac{(\bar{y}_d - \theta_0)^2}{n^{-1}\sigma^2 + \sigma_\theta^2} + \frac{(n-1)s_d^2}{\sigma^2}. \tag{2.4}$$

As an example suppose the sample consists of $n = 4$ analytical tests of yield $\mathbf{y}_d' = (77, 74, 75, 78)$ performed on a single batch from an industrial process for which it is believed that the testing variance $\sigma^2 = 1$, the process mean $\theta_0 = 70$ and the batch to batch variance is $\sigma_\theta^2 = 2$. We wish to estimate the mean θ for *this particular batch*.

In this example $\bar{y}_d = 76$, $\theta_0 = 70$, $s_d^2 = 3{\cdot}33$, $I_{\bar{y}} = 4$, $I_\theta = 0{\cdot}5$, $w = 0{\cdot}89$; so that, given the appropriateness of the model previously discussed, θ is estimated by the normal distribution $N(\bar{\theta}, \bar{\sigma}^2)$ with $\bar{\theta} = (0{\cdot}89 \times 76) + (0{\cdot}11 \times 70) = 75{\cdot}3$, $\bar{\sigma}^2 = (4 + 0{\cdot}5)^{-1} = 0{\cdot}22$.

However, from the predictive check

$$g(\mathbf{y}_d) = \frac{(76-70)^2}{2{\cdot}25} + \frac{3 \times 3{\cdot}3}{1} = 26 \tag{2.5}$$

and

$$\alpha = \Pr\{\chi_4^2 > 26\} < 0{\cdot}001. \tag{2.6}$$

Thus for this example the model, and hence the estimate of θ supplied by the posterior distribution $N(75{\cdot}3, 0{\cdot}22)$, is discredited by the predictive check.

Notice the following: (a) While the posterior distribution *combines* information from data and prior in a manner which is entirely appropriate if the model is to be believed, the predictive distribution *contrasts* these two sources of information and checks their compatibility.

(b) The predictive check formalizes questions that any competent statistician would raise having been presented with the supposed form of the model and the data. The components of $g(\mathbf{y}_d)$, $\{(n-1)s_d^2\}/\sigma^2$ and $(\bar{y}_d - \theta_0)^2/\{n^{-1}\sigma^2 + \sigma_\theta^2\}$ are the standard checking functions for contrasting an estimate of variance with a prior value and contrasting two estimates of the same mean.

(c) In making this predictive check it was not necessary to be specific about an alternative model. This issue is of some importance for it seems a matter of ordinary human experience that an appreciation that a situation is unusual does not necessarily depend on the immediate availability of an alternative.

(d) Whereas in estimating θ assuming the model to be true the posterior distribution makes use only of the single data vector $\mathbf{y}_d$ that has actually occurred, by contrast, an assessment of whether the sample $\mathbf{y}_d$ is likely to have occurred at all is necessarily achieved by relating $\mathbf{y}_d$ to a relevant reference set of *all* data vectors $\mathbf{y}$ which could have occurred with the model true.

Inspection of the global function $g(\mathbf{y}_d)$ alone would rarely ensure adequate checking of the model. In this example, for instance, it would be natural to consider the individual contributions from $\bar{y}_d$ and s_d^2 not only so that they could be separately considered, but also because unusually small values of $(n-1)s_d^2/\sigma^2$ as well as unusually large ones could point to model inadequacy. Also if n were larger, we might wish to consider other functions $g_i(\mathbf{y}_d)$ of the data such as moment coefficients and serial correlation coefficients which could reveal model inadequacies believed important in the current experimental situation. This could be done by referring $p\{g_i(\mathbf{y}_d) \mid A\}$ to the predictive distribution $p\{g_i(\mathbf{y}) \mid A\}$ derived by appropriate integration of $p(\mathbf{y} \mid A)$. Associated with these more specific checks are (possibly vague) model alternatives, the logical consequences of which are discussed in Section 4.6.

In practice, criticism of the model is often conducted by visual inspection of residual displays and other more sophisticated plots. But such a process, although it is informal, seems to me to fall within the logical framework described above. The plots are designed to make manifest certain "features" in the data that would rarely be extreme, if the model were true. If such a feature can be described by a function $g_i(\mathbf{y}_d)$, its unusualness, if formalized, would be measured appropriately by reference to $p\{g_i(\mathbf{y}) \mid A\}$.

For the above example obvious functions for checking individual features of the model are $\bar{y}$, s^2 and suitably chosen functions of standardized residuals $\mathbf{r} = (r_1, \ldots, r_n)'$ with $r_i = (y_i - \bar{y})/s$, $i = 1, \ldots, n$. These would usually include the individual residuals themselves plus other functions which, depending on the context, might include checks for needed transformation, heteroscedasticity, serial correlation, "bad values", skewness and kurtosis. See, for example, Anscombe (1961), Anscombe and Tukey (1963), Andrews (1971a, b).

The standardized residuals can be expressed more conveniently in terms of $n-2$ independently distributed functions obtained by making an orthogonal transformation from $\mathbf{y}$ to $\mathbf{Y} = (Y_1, Y_2, \ldots, Y_n)'$ with $Y_n = \sqrt{(n)}\,\bar{y}$ and then transforming to $\bar{y}$, s^2 and $\mathbf{u}$ where $\mathbf{u}$ is a vector of $n-2$ residual quantities $\mathbf{u} = (u_1, u_2, \ldots, u_{n-2})'$ such that

$$u_j = Y_{j+1} \Big/ \left\{ \sum_{i=1}^{j} Y_i^2 \Big/ j \right\}^{\frac{1}{2}}. \tag{2.7}$$

The Jacobian of the transformation from $\mathbf{y}$ to $\bar{y}$, s^2, $\mathbf{u}$ is proportional to

$$(s^2)^{\frac{1}{2}(n-1)-1} \prod_{j=1}^{n-2} \{1 + u_j^2/j\}^{-\frac{1}{2}(j+1)}.$$

After transformation, the predictive distribution contains n elements distributed independently

$$p(\bar{y}, s^2, \mathbf{u} \mid A) = p(\bar{y} \mid A)\, p(s^2 \mid A)\, p(\mathbf{u} \mid A), \tag{2.8}$$

where

$$p(\bar{y} \mid A) \propto (\sigma_\theta^2 + \sigma^2/n)^{-\frac{1}{2}} \exp\{-\tfrac{1}{2}(\bar{y} - \theta_0)^2/(\sigma_\theta^2 + \sigma^2/n)\}, \tag{2.9}$$

$$p(s^2 \mid A) \propto (\sigma^2)^{-\frac{1}{2}(n-1)} \{s^2\}^{\frac{1}{2}(n-1)-1} \exp\{-\tfrac{1}{2}(n-1)s^2/\sigma^2\}, \tag{2.10}$$

$$p(\mathbf{u} \mid A) \propto \prod_{j=1}^{n-2} \left\{1 + \frac{u_j^2}{j}\right\}^{-\frac{1}{2}(j+1)} \tag{2.11}$$

The unusualness of $g_1 = \bar{y}$, $g_2 = s^2$ and of any residual functions of interest $g_3, g_4, \ldots, g_q$ can then be assessed by computing

$$\Pr\{p(g_j \mid A) < p(g_{jd} \mid A)\}, \quad j = 1, 2, \ldots, q. \tag{2.12}$$

which for unimodal distributions will be tail area probabilities. For this example these would be obtained by referring

(i) $(\bar{y}_d - \theta_0)/(\sigma_\theta^2 + \sigma^2/n)^{\frac{1}{2}}$ to the Normal table;

(ii) $(n-1)s_d^2/\sigma^2$ to the χ^2 table;

(iii) $g_{3d}, \ldots, g_{qd}$ to reference distributions obtained by appropriate integration of $p(\mathbf{u} \mid A)$.

These probabilities are of course affected by transformation. Thus the answer will be a little different depending for example on whether we ask a question about s or about s^2. I do not find this particularly disturbing. Slightly different questions can be expected to have slightly different answers. We now illustrate some implications.

2.1. *Significance Tests*

Suppose σ_θ^2 is assumed small compared with σ^2/n, so that w, the relative amount of information supplied by the data, is close to zero. Then, *if this model can be relied upon*, the posterior distribution will be essentially the same as the prior, sharply centered at θ_0. A practical context is one where the statistician is told that the process mean is known to be θ_0 and the batch to batch variance σ_θ^2 is negligible compared with testing variance σ^2. If he believed this model, then any data $\mathbf{y}$ could do very little to change his belief that $\theta \doteq \theta_0$. However, *it could deny the relevance of this model.* In particular $g_1(\mathbf{y}_d)$ now involves essentially the reference of $(\bar{y}_d - \theta_0)/(\sigma/\sqrt{n})$ to normal tables; the failure of this check means that the model is discredited and therefore the Bayes calculation that leads to a sharp posterior distribution at θ_0 may not logically be undertaken.

The above most satisfactorily explains to me the rationale of a significance test.

(a) The tentative model (null hypothesis) implies that θ is close to θ_0.

(b) A check on the compatibility of this model and the data, so far as the mean is concerned, is provided by reference of $(\bar{y}_d - \theta_0)/(\sigma/\sqrt{n})$ to the Normal Table.

(c) If the tail area probability is not small we do not question the model. The *application of Bayes' theorem* then produces a posterior distribution which is sharply centred at θ_0. We have "no reason to question the null hypothesis".

(d) If the tail area probability is small we conclude that the model which postulated that $\theta \doteq \theta_0$ is discredited by the data, i.e., the "null hypothesis is discredited".

(e) Notice too that although the failure of this check would most immediately proscribe the use of Bayes' theorem, the failure of other checks (and of that based on s^2 in particular) would also suggest the need for model modification before proceeding further.

A difficulty that this removes for me is that, as usually formulated, significance tests had seemed to provide *no basis for belief.* On this formulation, however, the significance test provides a means of discrediting a model which *if* accepted would inevitably imply acceptance of the belief that θ lay close to θ_0. It is admitted that this formulation does not cover all possible circumstances in which significance tests have been used (see in particular Cox, 1977), but it is arguable that other applications are best dealt with in other ways.

2.2. *Precise Measurement and Improper Priors*

Suppose now that σ_θ^2 is assumed large compared with σ^2/n, so that $1-w$, which measures the proportion of the information about θ coming from the prior, is close to zero. Then σ_θ^2 dominates the denominator in the predictive checking function $(\sigma_\theta^2+\sigma^2/n)^{-\frac{1}{2}}(\bar{y}-\theta_0)$ implying that the model would not be called into question by sets of data having widely different sample averages. This is the situation where we can invoke what L. J. Savage called the "theory of precise measurement" to justify the very useful numerical approximation of the posterior distribution by $N(\bar{y}, \sigma^2/n)$. Now since the predictive distribution for $\bar{y}$ does not exist at the limit $1-w=0$ when this limiting posterior distribution is obtained, it might be argued that, when precise measurement theory is appropriate, we have a license to apply Bayes' theorem without any restraining checks on the model. Obviously, however, in any imaginable experimental situation there *would* be values of $\bar{y}$ which would rightly be regarded as implausible given the investigator's current beliefs. Thus what is really being verified is that a non-informative prior must, to make practical sense, always be proper, even though the appropriate posterior distribution can, in suitable circumstances, be *numerically approximated* by substituting an improper prior.

3. The Normal Linear Model

Suppose

$$\mathbf{y} \sim N(\mathbf{1}\mu+\mathbf{X\theta}, \mathbf{I}_n\sigma^2) \tag{3.1}$$

with $\mathbf{1}$ a vector of unities and $\mathbf{X}$ of full rank k such that $\mathbf{X}'\mathbf{1}=\mathbf{0}$ and suppose that prior densities are locally approximated by

$$\mu \sim N(\mu_0, c^{-1}\sigma^2),\ \mathbf{\theta} \sim N(\mathbf{\theta}_0, \mathbf{\Gamma}^{-1}\sigma^2),\ \{\sigma^2/\nu_0 s_0^2\} \sim \chi^{-2}(\nu_0) \tag{3.2}$$

with μ and $\mathbf{\theta}$ independent but conditional on σ^2, and $\chi^{-2}(\nu_0)$ the inverted χ^2 distribution.

Given a sample $\mathbf{y}_d$, special interest attaches to $\mathbf{\theta}$ and $\mathbf{\sigma}^2$ which given the assumptions are estimated by $p(\mathbf{\theta}, \sigma^2 \mid \mathbf{y}_d, A)$ with marginal distributions

$$p(\mathbf{\theta} \mid \mathbf{y}_d, A) \propto \left\{1+\frac{(\mathbf{\theta}-\bar{\mathbf{\theta}}_d)'(\mathbf{X}'\mathbf{X}+\mathbf{\Gamma})(\mathbf{\theta}-\bar{\mathbf{\theta}}_d)}{(n+\nu_0)\hat{\sigma}_d^2}\right\}^{-\frac{1}{2}(n+\nu_0+k)} \tag{3.3}$$

$$p(\sigma^2 \mid \mathbf{y}_d, A) \propto \sigma^{-(n+\nu_0+2)}\exp\{-\tfrac{1}{2}(n+\nu_0)\hat{\sigma}_d^2/\sigma^2\} \tag{3.4}$$

with

$$\bar{\mathbf{\theta}}_d = (\mathbf{X}'\mathbf{X}+\mathbf{\Gamma})^{-1}(\mathbf{X}'\mathbf{X}\hat{\mathbf{\theta}}_d+\mathbf{\Gamma\theta}_0),\quad \hat{\mathbf{\theta}}_d=(\mathbf{X}'\mathbf{X})^{-1}\mathbf{X}'\mathbf{y}_d,\quad \nu=n-k-1,$$

$$(n+\nu_0)\hat{\sigma}_d^2 = \nu s_d^2+\nu_0 s_0^2+(\hat{\mathbf{\theta}}_d-\mathbf{\theta}_0)'\{(\mathbf{X}'\mathbf{X})^{-1}+\mathbf{\Gamma}^{-1}\}^{-1}(\hat{\mathbf{\theta}}_d-\mathbf{\theta}_0)+(n^{-1}+c^{-1})^{-1}(\bar{y}-\mu_0)^2.$$

Now let s_p^2 be the pooled estimate

$$(\nu+\nu_0)^{-1}(\nu s^2+\nu_0 s_0^2). \tag{3.5}$$

Then the predictive distributions for $(\hat{\mathbf{\theta}}-\mathbf{\theta}_0)/s_p$, s^2 and the $\nu-1$ elements of the residual vector $\mathbf{u}$, defined in an analogous manner to that previously employed in (2.7), are independent and are given by

$$p\{(\hat{\boldsymbol\theta}-\boldsymbol\theta_0)/s_p \mid A\} \propto \left\{1+\frac{(\hat{\boldsymbol\theta}-\boldsymbol\theta_0)'\{(\mathbf{X}'\mathbf{X})^{-1}+\boldsymbol\Gamma^{-1}\}^{-1}(\hat{\boldsymbol\theta}-\boldsymbol\theta_0)}{(\nu+\nu_0)\,s_p^2}\right\}^{-\frac{1}{2}(n+\nu_0-1)}, \tag{3.6}$$

$$p(s^2/s_0^2 \mid A) \propto F^{\frac{1}{2}\nu-1}\left\{1+\frac{\nu}{\nu_0}F\right\}^{-\frac{1}{2}(\nu+\nu_0)}, \quad F = s^2/s_0^2, \tag{3.7}$$

$$p(\mathbf{u} \mid A) \propto \prod_{j=1}^{\nu-1} \{1+(u_j^2/j)\}^{-\frac{1}{2}(j+1)}. \tag{3.8}$$

The predictive check derived from (3.6)

$$\Pr\{p((\hat{\boldsymbol\theta}-\boldsymbol\theta_0)/s_p \mid A) < p((\hat{\boldsymbol\theta}_d-\boldsymbol\theta_0)/s_{pd} \mid A)\}$$
$$= \Pr\left\{F_{k,\nu+\nu_0} > \frac{(\hat{\boldsymbol\theta}_d-\boldsymbol\theta_0)'\{(\mathbf{X}'\mathbf{X})^{-1}+\boldsymbol\Gamma^{-1}\}^{-1}(\hat{\boldsymbol\theta}_d-\boldsymbol\theta_0)}{k s_{pd}^2}\right\} \tag{3.9}$$

is the standard analysis of variance check for compatibility of two estimates $\hat{\boldsymbol\theta}_d$ and $\boldsymbol\theta_0$. It was earlier proposed as a check for compatibility of prior and sample information by Theil (1963). The predictive check derived from (3.7) $\Pr\{p(s^2 \mid A) < p(s_d^2 \mid A)\}$ yields the F test having ν and ν_0 degrees of freedom appropriate to check whether the two estimates s_d^2 and s_0^2 are compatible. Residual checks derived from (3.8) are obtainable as before.

3.1. *Ridge Estimates*

Now suppose the **X** matrix to be in correlation form and assume $\boldsymbol\theta_0 = \mathbf{0}, \boldsymbol\Gamma = \mathbf{I}_k\gamma_0, \nu_0 \to 0$ so that $s_p^2 \to s^2$. Then the estimates $\bar{\boldsymbol\theta}_d$ are the ridge estimators of Hoerl and Kennard (1970) which, given the assumptions, appropriately combine information from the prior with information from the data. The predictive check (3.9) now yields

$$\alpha = \Pr\left\{F_{k,\nu} > \frac{\hat{\boldsymbol\theta}_d'\{(\mathbf{X}'\mathbf{X})^{-1}+\mathbf{I}\gamma_0^{-1}\}^{-1}\hat{\boldsymbol\theta}_d}{k s_d^2}\right\} \tag{3.10}$$

allowing any choice of γ_0 to be criticized.

For example, in their original analysis of the data of Gorman and Toman (1966), Hoerl and Kennard (1970) chose a value $\gamma_0 = 0{\cdot}25$. However, substitution of this value in (3.10) yields $\alpha = \Pr\{F_{10,25} > 3{\cdot}59\} < 0{\cdot}01$ which discredits this choice. More recently it has been pointed out (Lindley and Smith, 1972; Hoerl, Kennard and Baldwin, 1975) that given the model, γ can be estimated from the data. If we do this, much smaller values of γ are obtained which of course are not in conflict with the wider model. The two kinds of analysis further illustrate the overlap between predictive checking and Bayesian estimation later discussed in Section 4.6.

The Bayes approach to ridge estimators has the characteristic advantage that the somewhat arbitrary prior assumptions, which have to be made even for compatible values of γ, are uncovered for criticism (see also Draper and Van Nostrand, 1977). If $\gamma_0^{-1} \to 0$, (3.10) yields the standard ANOVA significance test which has a detailed interpretation parallel to that set out in Section 2.1.

4. Diagnostic Checks

It is useful to distinguish two kinds of checks which may be called respectively Overall or Multidirectional checks and Specific or Unidirectional checks. An example of the first would be a general inspection of residuals and the second a Durbin–Watson test for first-order serial correlation. This distinction is made, for example, by Box and Jenkins (1970) in their discussion of the general philosophy of diagnostic checking. Concerning these two kinds of checks these authors say "... although [overall checks] can point out unsuspected peculiarities ... [they] may not be particularly sensitive. Tests for specific departures ... are more sensitive, but may fail to warn of trouble other than that specifically anticipated." The two alternatives ought properly to be regarded as extremes on some scale of dependence of checking procedures on specific alternatives. For example, consider the fitting of a parametric time series model. While residuals themselves should always be inspected there are a number of way-stations between this overall but insensitive check and the device called "overfitting" in which a model is tentatively elaborated in a *specific* direction. Thus inspection of, and application of overall tests to, the autocorrelation function and the periodogram of the residuals while still non-specific is less general than the first device and much less specific than the second.

The model checking problem is comparable to that faced by a nation which fears aerial attack that might come from any direction but with certain rather wide zones more likely than others and certain specific directions believed especially likely. How should limited radio detection devices, which are less sensitive the less they are focused, be deployed? The best solution obviously involves some combination of wide and more specific searches, and theoretically could be achieved knowing prior probabilities and expected losses. Correspondingly, the competent statistician must, in a variety of contexts, be able to make intelligent guesses not only of what discrepancies are particularly likely, but which are potentially disastrous, and to allocate his effort accordingly. In practice this is done informally and is part of what an adequate training in statistics achieves.

4.1. *Checking Parametric Features of the Model*

In the examples considered above where sufficient statistics were available parameter preferences evidenced by proper priors were directly challenged, leading without a direct statement of alternatives to appropriate checks. When a specific set of assumptions A_1 alternative to A_0 are in mind then an appropriate checking function might also be obtained from the predictive ratio

$$p(\mathbf{y}_d \mid A_1)/p(\mathbf{y}_d \mid A_0). \tag{4.1}$$

We shall not explore this possibility further here, except to note that this ratio is a component in the direct assessment of Bayesian odds to which we refer briefly in Section 4.6.

4.2. *Checking Residual Features of the Model*

Residual checking functions are sometimes chosen on an *ad hoc* basis and sometimes using specific models. I think the best course is again to employ an iteration—this time between theory and intuition. An empirical procedure that works well invites the question: What kind of model would be needed for its justification? Such a model can then be considered for use in a wider context. For instance, exponential smoothing and the "three term" controller were both empirically developed techniques found to be practically effective. ARIMA time series models are generalizations of the stochastic processes that could justify these methods (Box and Jenkins, 1970). In a similar way the practical usefulness of such things as the *jackknife* and *cross-validation* implies the existence of corresponding models which are worthy of further analysis.

The distinction between parametric features of the model and residual features is of course arbitrary and a matter of convenience. In practice the needs of parsimony urge us to settle for reasonably simple models and to consider possible deviations from them. Consider now therefore an interesting but by no means unique method for obtaining an appropriate function of the data for informal or formal checking for a particular kind of deviation from a current model parametrized by a discrepancy parameter β.

Suppose the predictive distribution conditional on some specific choice of β is $p(\mathbf{y} \mid \beta)$. Then a scaleless function of the data alone, appropriate to measure discrepancies from the value β_0 taken in the current model, is provided by Fisher's score function

$$g_\beta(\mathbf{y}) = \left. \frac{\partial \ln p(\mathbf{y} \mid \beta)}{d\beta} \right|_{\beta=\beta_0}. \tag{4.2}$$

We illustrate by considering some possible discrepancies from the standard normal linear model. First consider the model when there is no discrepancy so that $\beta = \beta_0$, and using the structure of (3.1) write

$$\mathbf{\Theta}' = (\mu \vdots \mathbf{\theta}'), \quad \mathscr{X} = (\mathbf{1} \vdots \mathbf{X}). \tag{4.3}$$

For simplicity we here suppose that the distributions of $\mathbf{\Theta}$ and $\ln \sigma$ are locally flat *a priori* so that $p(\mathbf{\Theta}, \sigma) \doteq \text{const}\, \sigma^{-1}$. Then $p(\mathbf{y} \mid \beta_0)$ is locally approximated by the singular distribution

$$p(\mathbf{y} \mid \beta_0) \doteq \text{const}\, S^{-\nu}, \tag{4.4}$$

where $S^2 = \Sigma_{i=1}^n (y_i - \hat{y}_i)^2 = \mathbf{y}'\mathbf{R}\mathbf{y}$ and $\mathbf{R} = \mathbf{I} - \mathbf{M}$ with $\mathbf{M} = \mathscr{X}(\mathscr{X}'\mathscr{X})^{-1}\mathscr{X}'$. If we transform to $\hat{\mathbf{\Theta}}$, S and $\mathbf{u}$ then the standardized residuals $\mathbf{u}$ which are functions of $\nu - 1$ angles are distributed as in (3.8) and,

$$p(\hat{\mathbf{\Theta}}, S, \mathbf{u} \mid \beta_0) \doteq \text{const}\, S^{-1}\, p(\mathbf{u} \mid \beta_0). \tag{4.5}$$

To see the reasonableness of this set-up notice that by invocation of the linear model the investigator in effect predicts that the sample point $\mathbf{y}$ will lie somewhere close to a hyperplane $\mathscr{h}_{\mathscr{X}}$ spanned by the columns of $\mathscr{X}$. The formulation above interprets "somewhere close to" as follows. Consider a future sample $\mathbf{y}$ in relation to $(\hat{\mathbf{\Theta}}, S)$ where $\hat{\mathbf{\Theta}}$ are the $k+1$ coordinates of the projection $\dot{\mathbf{y}}$ of $\mathbf{y}$ on $\mathscr{h}_{\mathscr{X}}$, and S is the perpendicular distance of $\mathbf{y}$ from $\mathscr{h}_{\mathscr{X}}$. Equation (4.5) says that locally any value of $\hat{\mathbf{\Theta}}$ is equally acceptable but that the density for the distance S will fall off inversely with S.

To obtain $g_\beta(\mathbf{y})$ we need to determine how $p(\mathbf{y} \mid \beta)$ depends on the discrepancy parameter β in the neighbourhood of $\beta = \beta_0$.

4.3. *A Check for Needed Power Transformation*

Especially when $y_{\max}/y_{\min}$ is large some transformation of the data, for example $y^{(\lambda)} = (y^\lambda - 1)/\lambda$, might permit closer representation. Following the approximate argument of Box and Cox (1964), with λ the discrepancy parameter and $\dot{y}$ the geometric mean of the y's, and for λ close to 1,

$$p(\mathbf{y} \mid \lambda) \mathrel{\dot{\propto}} \dot{y}^{\nu(\lambda-1)} Q_\lambda^{-\frac{1}{2}\nu}, \tag{4.6}$$

where the omitted constant does not depend on $\mathbf{y}$ or on λ and where $Q_\lambda = \mathbf{y}^{(\lambda)'}\mathbf{R}\mathbf{y}^{(\lambda)}$,

$$g_\lambda(\mathbf{y}) = \left.\frac{\partial \ln p(\mathbf{y} \mid \lambda)}{\partial \lambda}\right|_{\lambda=1} \doteq \mathbf{z}'\mathbf{R}\mathbf{y}/s^2 = s^{-1} \sum_{i=1}^{n} z_i r_i \tag{4.7}$$

where $z_i = y_i\{1 - \ln(y_i/\dot{y})\}$, $s^2 = \mathbf{y}'\mathbf{R}\mathbf{y}/\nu$ and $r_i = (y_i - \hat{y}_i)/s$. The predictive check may thus be performed by regressing the residuals $y - \hat{y}$ on the residuals $z - \hat{z}$ of the constructed variable $z = y\{1 - \ln(y/\dot{y})\}$, which accords with a proposal of Atkinson (1973). The check can be made informally by plotting one set of residuals versus the other. More formally the distribution of $g_\lambda(\mathbf{y})$ is not precisely known although an approximate level can be obtained by computer simulation.

Relation to other proposed checks

Related checks proposed by Tukey (1949) and by Andrews (1971a) correlate the original residuals with those from the constructed variables $(\hat{y} - \bar{y})^2$ and $\hat{y} \ln \hat{y}$ respectively. Both possess the advantage of having exactly known sampling distributions.

For illustration we consider:

(a) the biological data of Box and Cox (1964), for which they recommend a reciprocal transformation;

(b) the trapping data of Snedecor and Cochran (1967), for which they recommend a log transformation.

Figures 2(a) and (b) show plots of residuals $y-\hat{y}$ against residuals from $y\{1-\ln(y/\dot{y})\}$ and $-(\hat{y}-\bar{y})^2$. The correlation coefficient for the latter transforms directly to give Tukey's one degree of freedom for non-additivity. Plots for the constructed variable $\hat{y}\ln\hat{y}$ are not shown since they are essentially identical to the Tukey plot. The relationship between these various procedures can be seen by noting that $z = y\{1-\ln(y/\dot{y})\}$ may be closely approximated by

$$z \doteq \dot{y} - B(y-\dot{y})^2. \tag{4.8}$$

Thus after writing

$$\frac{y_i-\dot{y}}{s} = \frac{y_i-\hat{y}_i}{s} + \frac{\hat{y}_i-\bar{y}}{s} + \frac{\bar{y}-\dot{y}}{s} = r_i + Y_i + d, \tag{4.9}$$

$$g_\lambda(\mathbf{y}) \mathrel{\dot{\propto}} -\Sigma(r_i+Y_i+d)^2\, r_i = -\{\Sigma r_i^3 + 2\,\Sigma r_i^2\, Y_i + \Sigma r_i\, Y_i^2 + 2\,\Sigma r_i^2\, d\} \tag{4.10}$$

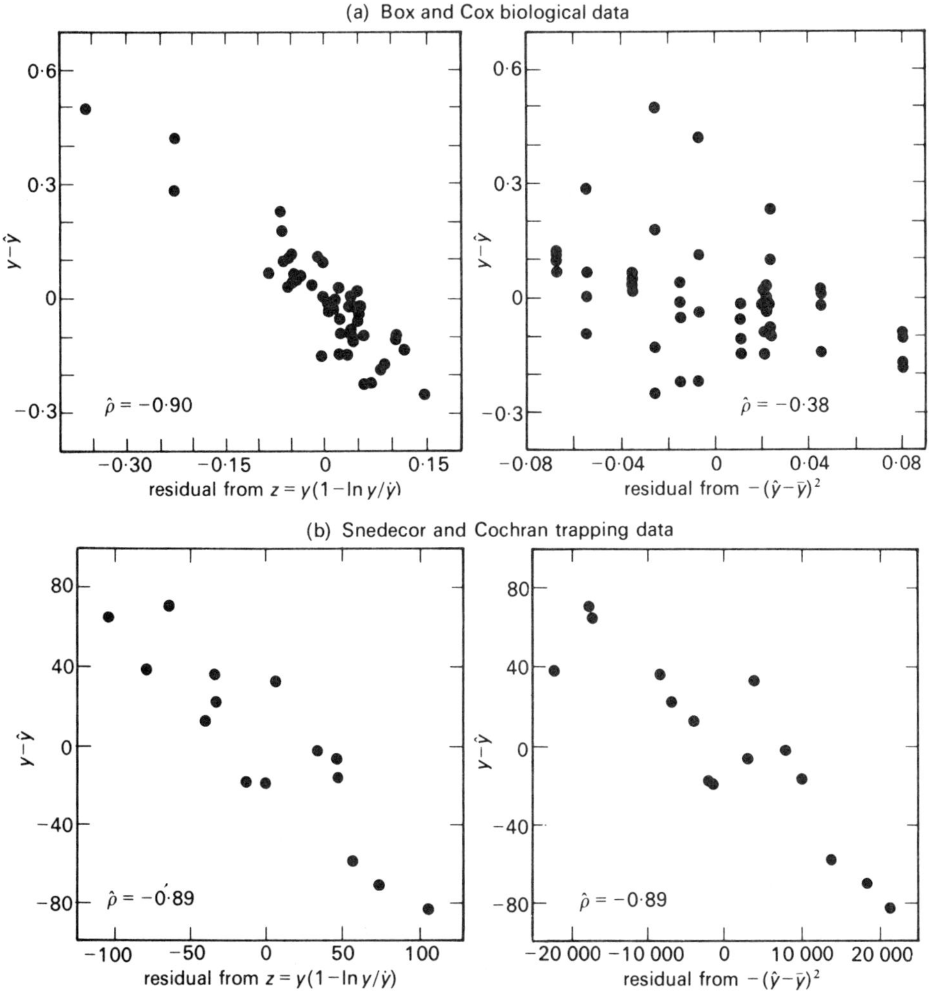

FIG. 2. Plots of residuals $y-\hat{y}$ against residuals from two constructed variables $z = y(1-\ln y/\dot{y})$ and $-(\hat{y}-\bar{y})^2$, with correlation coefficient $\hat{\rho}$ to indicate strength of association.

and

$$g_\lambda(\mathbf{y}) \,\dot{\propto}\, -(T_{30}+2T_{21}+T_{12}+2vd), \tag{4.11}$$

where the T_{ij} are checking functions proposed by Anscombe (1961) and Anscombe and Tukey (1963). See also Box and Cox (1964). In particular T_{12} is the component associated with Tukey's one degree of freedom for non-additivity. The approximation shows how $g_\lambda(\mathbf{y})$ jointly employs skewness (T_{30}), dependence of variance on level (T_{21}), as well as transformable non-additivity (T_{12}) to indicate the need for transformation.

The Box and Cox data were generated by a 3×4 factorial with four-fold replication supplying a good deal of information about the variance as a function of location. It is not surprising therefore (see also Atkinson, 1973) that for this example $g_\lambda(\mathbf{y})$ is considerably more sensitive than T_{12} (or almost equivalently, than Andrew's criterion) as a measure of the need for transformation. By contrast the Snedecor and Cochran data are from an unreplicated 3×5 arrangement where most of the information comes from T_{12} measuring non-additivity.

4.4. *A check for Serial Correlation*

For data known or suspected to have been taken in a specific serial order in time or space, a model that permitted the errors to follow a first-order autoregressive process with parameter $|\phi|<1$ might provide an improved approximation. The dispersion matrix for the n-dimensional vector of errors $\mathbf{e}$ would then be $\mathbf{W}_\phi^{-1}\sigma^2$ where $\mathbf{W}_\phi$ is a symmetric continuant with principal diagonal $\{1, 1+\phi^2, 1+\phi^2, \ldots, 1+\phi^2, 1\}$ and with all the elements of super- and sub-diagonals equal to $-\phi$. Thus in particular $\mathbf{W}_0 = \mathbf{I}$. Then

$$p(\mathbf{y}\,|\,\phi) \,\dot{\propto}\, (1-\phi^2)^{\frac{1}{2}}\, Q_\phi^{-\frac{1}{2}v}, \tag{4.12}$$

where $Q_\phi = \mathbf{y}'\mathbf{R}_\phi\mathbf{y}$ and $\mathbf{R}_\phi = \mathbf{W}_\phi - \mathbf{M}_\phi$ with $\mathbf{M}_\phi = \mathbf{W}_\phi\mathscr{X}(\mathscr{X}'\mathbf{W}_\phi\mathscr{X})^{-1}\mathscr{X}'\mathbf{W}_\phi$. Then with $\partial\mathbf{W}_\phi/\partial\phi\,|_{\phi=0} = -\mathbf{C}$ where $\mathbf{C}$ is $n\times n$ with unities in super- and sub-diagonals and zeros elsewhere, after some algebraic manipulation,

$$g_\phi(\mathbf{y}) = \left.\frac{\partial \ln p(\mathbf{y}\,|\,\phi)}{\partial\phi}\right|_{\phi=0} \doteq \tfrac{1}{2}(\mathbf{y}'\mathbf{RCRy})/s^2, \tag{4.13}$$

where $R = R_0$. Thus

$$g_\phi(\mathbf{y}) \doteq \sum_{i=1}^{n-1} r_i r_{i+1} \tag{4.14}$$

which is a multiple of the sample first lag autocorrelation of the residuals from the fitted model. This points to the sensitive graphical diagnostic procedure of plotting residuals r_{i+1} against r_i and yields the standard checking function of Durbin and Watson (1950).

4.5. *A Check for Bad Values*

Competent investigators have over the centuries treated data as possibly containing atypical values, see for example Stigler (1973). This imples that they would not really have believed standard textbook models of the kind $y_i = f(\boldsymbol{\theta}, \mathbf{x}_i)+e_i$ $(i = 1, 2, \ldots, n)$ which state that the same structure is appropriate for *every one* of a sample of n observations.

When it is unknown which observations are dubious a more credible "contaminated" model proposed by Jeffreys (1932), Dixon (1953) and by Tukey (1960) supposes that there is a probability α that any given observation is "bad" (cannot be represented by the ideal model). Given α, let $p(\mathbf{y}\,|\,\alpha)$ be the predictive distribution and let $p(b\,|\,\alpha)$ denote the probability of getting b bad values, then (Box and Tiao, 1968b; Bailey and Box, 1980a)

$$p(\mathbf{y}\,|\,\alpha) = \sum_{b=0}^{n}\binom{n}{b}\alpha^b(1-\alpha)^{n-b}\,p(\mathbf{y}\,|\,b) \tag{4.15}$$

and

$$g_\alpha(\mathbf{y}) = \left.\frac{\partial \ln p(\mathbf{y} \mid \alpha)}{\partial \alpha}\right|_{\alpha=0} = n\left\{\frac{p(\mathbf{y} \mid b = 1)}{p(\mathbf{y} \mid b = 0)} - 1\right\}. \tag{4.16}$$

Now let z_i indicate that the ith observation is bad, then

$$p(\mathbf{y} \mid b = 1) = n^{-1} \sum_{i=1}^{n} p(\mathbf{y} \mid z_i) \tag{4.17}$$

so

$$g_\alpha(\mathbf{y}) = \left(\sum_{i=1}^{n} p(\mathbf{y} \mid z_i) \Big/ p(\mathbf{y} \mid b = 0)\right) - n.$$

Depending on experimental circumstances, there are a variety of ways in which bad values might be modelled. In particular, contamination could come from increased error variance, unknown bias, and mistaken sign. The last possibility was suggested by Barnard (1978) to account for two suspiciously large outliers in Darwin's data on cross- and self-fertilized plants, quoted by Fisher (1935).

For illustration consider the first possibility. With one bad value, suppose the error covariance matrix $W_i^{-1}\sigma^2$ has all elements equal to σ^2, except for the ith element which is equal to $\kappa^2\sigma^2 (\kappa > 1)$. Then

$$g_\alpha(\mathbf{y}) = \kappa^{-1}\left(\frac{n}{n-q}\right)^{\frac{1}{2}} \sum_{i=1}^{n} \left(\frac{s_i^2}{s^2}\right)^{-\frac{1}{2}\nu} - n, \tag{4.18}$$

where

$$\nu s^2 = \mathbf{y}'\mathbf{R}\mathbf{y}, \quad \nu s_i^2 = \mathbf{y}'\mathbf{R}_i\mathbf{y} = \mathbf{y}'\{\mathbf{W}_i - \mathbf{W}_i\mathscr{X}(\mathscr{X}'\mathbf{W}_i\mathscr{X})^{-1}\mathscr{X}'\mathbf{W}_i\}\,\mathbf{y}, \tag{4.19}$$

where $\mathbf{W}_i = \mathbf{I} - q\mathbf{G}_i$, $q = 1 - \kappa^{-2}$ and $\mathbf{G}_i$ is an $n \times n$ matrix with a single unity for the ith diagonal element and all other elements zero. Now

$$\nu s_i^2 = \nu s^2 - \frac{q}{1 - qv_i}(y_i - \hat{y}_i)^2,$$

where $v_i = \operatorname{var}(\hat{y}_i)/\sigma^2 = x_i'(\mathscr{X}'\mathscr{X})^{-1}x_i$ and $y_i - \hat{y}_i$ is the ith residual from the ideal model fit, $\mathbf{y} - \hat{\mathbf{y}} = \mathbf{R}\mathbf{y}$.

Thus finally $g_\alpha(\mathbf{y}) = \kappa^{-1}\{n/(n-q)\}^{\frac{1}{2}} D - n$ where

$$D = \Sigma\left\{1 - \frac{q}{\nu(1 - qv_i)} r_i^2\right\}^{-\frac{1}{2}\nu} \tag{4.20}$$

This is the simplest form for computation. The nature of this checking function D can however be more clearly seen by writing it in terms of the weighted residuals $\tilde{r}_i = (y_i - \tilde{y}_i)/s_i$ where $\mathbf{y} - \tilde{\mathbf{y}} = \mathbf{R}_i\mathbf{y}$. Thus

$$D = \Sigma\left\{1 + \frac{q(1 - qv_i)}{\nu}\tilde{r}_i^2\right\}^{\frac{1}{2}\nu}.$$

Thus D is proportional to the sum of the reciprocals of the n residual t ordinates obtained by downweighting (omitting as $q \to 1$) each observation in turn and recomputing the fitted value $\tilde{y}_i$ and the standard deviation s_i.

The situation of most interest is when κ is large (say $\kappa \geqslant 5$). Then q approaches unity and the check may be carried out by calculating

$$D = \Sigma\left(1 - \frac{r_i^2}{\nu(1 - v_i)}\right)^{-\frac{1}{2}\nu} = \Sigma\left\{1 + \frac{(1 - v_i)}{\nu}\tilde{r}_i^2\right\}^{\frac{1}{2}\nu}. \tag{4.21}$$

Equation (4.16) brings out a feature of the checking function (4.2) which can be a disadvantage. Differentiation at $\alpha = 0$ on the boundary of the parameter space ensures that only the possibility of one bad value is taken account of. Thus as is clear from (4.21), D in its present form would not be expected to be sensitive to the occurrence of two or more bad values. Thus with $\kappa = 5$, we obtain the value $D = 59{\cdot}05$ for Darwin's data. A Monte Carlo study with 5000 samples of 15 observations shows that this value would be exceeded by chance about 14 per cent of the time, which hints only mildly at inadequacy in the standard model, confirming D's insensitivity for this example.

4.6. *Bayesian Options for Specific Alternatives*

When concrete alternatives are in mind, Bayesian options are available. In particular, the predictive ratio $p(\mathbf{y} \mid A_1)/p(\mathbf{y} \mid A_0)$, mentioned earlier, is a component in the posterior odds ratio which, with suitable priors, might be used to assess directly the relative evidence for one model versus another. Also $g_\beta(\mathbf{y})$ of (4.2) has a Bayesian interpretation for, if corresponding to some discrepancy parameter β, the prior distribution $p(\beta)$ was locally flat then the posterior distribution $p(\beta \mid \mathbf{y}_d)$ would be proportional to the predictive density $p(\mathbf{y}_d \mid \beta)$ regarded as a function of β. Furthermore, if that posterior distribution was approximated by a normal distribution, then

$$g_\beta(\mathbf{y}) \doteq \{-(\beta_0 - \hat{\beta})/\sigma_\beta^2\} \tag{4.22}$$

and a second differentiation would produce a standardized variate.

The relation shows how any specific predictive check $g_\beta(\mathbf{y})$ is linked to a posterior distribution. In particular, considering the illustrative examples of the last section, the marginal posterior distribution for λ was given by Box and Cox (1964), for ϕ by Zellner and Tiao (1964), for the ridge regression parameters by Lindley and Smith (1972) and that for α may be obtained using the results of Box and Tiao (1968b) and Bailey and Box (1980a).

Before leaving the topic of diagnostic checks two final points need to be made:

(i) The above discussion illustrates the "overlap" previously mentioned when specific alternatives are in mind. It does not, however, establish the omnipotence of purely Bayesian inference. However far the process of model elaboration is taken by Bayesian methods the final model involving say the mth set of assumptions A_m can still be factored

$$p(\mathbf{y}, \boldsymbol{\theta} \mid A_m) = p(\boldsymbol{\theta} \mid \mathbf{y}, A_m)\, p(\mathbf{y} \mid A_m) \tag{4.23}$$

thus there always remains an unexplored n-dimensional predictive distribution $p(\mathbf{y} \mid A_m)$ in relation to which a small relative value for $p(\mathbf{y}_d \mid A_m)$ could, on a sampling theory argument, discredit the assumptions on which the Bayes analysis was conditional. The same is true of the more plausible of two models chosen using a posterior odds ratio.

(ii) In addition to possible discrepancies to which we have been alerted by experience, other features may appear pointing to inadequacies of a kind not previously suspected. This possibility has sometimes proved perplexing, for while on the one hand the truly unexpected could point the way to precious new knowledge, on the other, associated probabilities would be indeterminate because of the uncountable character of other features that might also have been regarded as surprising. I think the calculation which ignores this difficulty of indeterminate selection is still worth making, for at least it helps to correct a misjudgement of something that appears unusual but really is not. For example, Feller (1968) shows that for a random group of 30 people, the probability that at least two have coincident birthdays is over 70 per cent; this tells us we need look no further for an explanation when we are surprised to find two such people at a party. While the proposed policy will lead to the too frequent pursuit of non-existent assignable causes, the iterative process will quickly terminate this chase.

5. Robust Estimation

Efficient model building requires both *diagnostic checking* and *model robustification*, where by robustification I mean judicious and grudging elaboration of the model to ensure against particular hazards (see also Box, 1979). Robustification becomes necessary when it is known that likely, but not easily detectable, model discrepancies can yield badly misleading analyses. It is well known, for example, that least squares analysis can be dramatically affected by moderate serial correlation of errors.

Recently the serious consequences of bad values on standard least squares analysis has been especially emphasized and numerous authors have proposed methods which rely on abandonment or modification of classical estimation methods. In discussing the rationale for this approach Huber (1977) says "The traditional approach to theoretical statistics was and is to optimize an idealized model and then to rely on a continuity principal: what is optimal at the model should be optimal nearby. Unfortunately, this reliance on continuity is unfounded: the classical optimal procedures tend to be discontinuous in the statistically meaningful topologies."

He then quotes a motivating example given by Tukey (1960), who pointed out for example that if a normal distribution were very mildly contaminated with another which is centrally located but of larger variance, then the sample standard deviation could be a very poor estimate of scale. Tukey's contribution was remarkable because it had previously gone unnoticed that the assumption that the same structure must apply to *every* observation y_i $(i = 1, 2, ..., n)$ *with absolute certainty* $(1 - \alpha = 1)$, not only was unrealistic (since no responsible investigator would make the claim that inadvertent bad values were impossible), but also could have serious consequences. While Huber goes on to say that typical "good data" samples in the physical sciences appear to be well modelled by this contaminated normal model, he does not develop methods based on this more realistic set up. This is presumably because his objection would apply equally to the new as well as to the old model.

I do not agree that the example would support a thesis of the need to abandon model-based procedures. A model that omits the parameter α is, of course, the same as one that includes it but sets its value exactly to zero. A value of $\alpha = 0$, which allows no possibility whatever for bad values, and a value of $\alpha = 0{\cdot}001$ are, I think, *not close* in any statistically meaningful topology. Although 0·001 may look close to zero, an odds ratio of $0{\cdot}999/0{\cdot}001 = 999$ for a "good" to a "bad" value is obviously very different from one of infinity. Such differences in probability distinguish, for example, a lifeless world in which no evolution could possibly occur from the one we live in.

The proper conclusion to draw from Tukey's example is, I think, that for many practical situations in which occasional bad values are to be expected the standard linear model provides an inadequate approximation that is potentially misleading and therefore the model should be appropriately changed to approximate what is believed rather than what is not. The situation is logically the same for a model that implicitly insists there can be no serial correlation, when data have in fact been collected serially, or that no transformation of y could be needed when $y_{\max}/y_{\min}$ is large. As in the classical Stein problem if we know something *a priori* it may be disastrous to omit it. On this view for robust estimation of the parameters of interest we should modify the model which is at fault, rather than the method of estimation which is not.

5.1. *Bayesian Robust Estimation*

As was argued for example by Box and Tiao (1964), all relevant aspects of the problem are brought out in an appropriate Bayes analysis. Supposing that $\boldsymbol{\theta}$ has the same physical interpretation for all β then estimation of $\boldsymbol{\theta}$ which is robust relative to the discrepancy parameter β is supplied by the posterior distribution

$$p(\boldsymbol{\theta} \mid \mathbf{y}) = \int p(\boldsymbol{\theta} \mid \beta, \mathbf{y})\, p_u(\beta \mid \mathbf{y})\, p(\beta)\, d\beta. \tag{5.1}$$

This expression contains three key elements that repay individual study:

(a) the sensitivity of inferences about $\boldsymbol{\theta}$ to changes in β is reflected by $p(\boldsymbol{\theta} \mid \beta, \mathbf{y})$ considered as a function of β;

(b) the information about β coming from the data themselves is reflected in the pseudo-likelihood

$$p_u(\beta \mid \mathbf{y}) = p(\beta \mid \mathbf{y})/p(\beta) \propto p(\mathbf{y} \mid \beta); \tag{5.2}$$

(c) the probability of occurrence of different values of β in the real world is represented by $p(\beta)$ which can be chosen to approximate what is believed or feared.

This route was used to explore deviations from the standard normal model for a particular class of heavy-tailed distributions by Box and Tiao (1962, 1964); for the contaminated model of Section 4.5 by Jeffreys (1932) and Box and Tiao (1968b); for a serial correlation model by Zellner and Tiao (1964); for a transformation problem by Box and Cox (1964). Notice that using this approach the parameters $\boldsymbol{\theta}$ of interest are completely estimated in the sense that their distribution rather than merely a point estimate is available. Also the various elements of $p(\boldsymbol{\theta} \mid \mathbf{y})$ which can be studied individually can provide a deep understanding of each robustness problem. A particularly informative display shows contours of the joint distribution $p_u(\theta, \beta \mid \mathbf{y})$ for some parameter θ of interest and the discrepancy parameter β together with the marginal distribution $p_u(\beta \mid \mathbf{y})$. When a less prodigal display is necessary the mean and standard deviation of $p(\theta \mid \beta, \mathbf{y})$ may be shown with $p_u(\beta \mid \mathbf{y})$. For illustration we consider some serial data analysed by Coen, Gomme and Kendall (1969). They regressed quarterly values of the Financial Times Share Index y on detrended lagged values of UK car production X_1, and of the Financial Times Commodity Index X_2 using a model† which could be written (Box and Newbold, 1971) as

$$y_t = \beta_0 + \beta_1 t + \theta_1 X_{1,t-6} + \theta_2 X_{2,t-7} + e_t \quad \text{with } e_t = \phi e_{t-1} + a_t \tag{5.3}$$

with a_t white noise, and ϕ constrained to be equal to zero. Fig. 3 illustrates an analysis made by Pallesen (1977) in which ϕ is unconstrained. It shows the joint posterior distribution for θ_1 and ϕ and the marginal distribution for ϕ assuming locally flat priors for $\boldsymbol{\theta}$, $\ln \sigma$ and ϕ. Although for this example serial correlation could have been easily detected by diagnostic checks, notice the enormous shift (about five standard deviations) of the conditional distribution $p(\theta_1 \mid \phi, \mathbf{y})$ which occurs as ϕ changes from zero to more plausible values. This illustrates the point that smaller serial correlation, of a magnitude difficult to detect with diagnostic checks, could disastrously invalidate estimates of $\boldsymbol{\theta}$.

A second example discussed more fully in Bailey and Box (1980b) further illustrates this approach for the "bad value" problem using the contaminated normal model of Section 4.5. The data were used originally by Box and Behnken (1960) to illustrate the analysis of a balanced incomplete four factor three-level design with $n = 27$ observations arranged in three blocks of nine. A residual plot suggests the possibility of two bad values (y_{10} and y_{13}). However, the small number of residual degrees of freedom and the nature of this particular design would induce large correlations yielding potentially misleading residual patterns.

Table 1 gives Bayesian means and standard deviations for coefficients in the fitted model

$$y = \beta_0 + \sum_{i=1}^{4} \beta_i x_i + \sum_{i=1}^{4} \beta_{ii} x_i^2 + \sum_{i=1}^{4} \sum_{j>1}^{4} \beta_{ij} x_i x_j + e. \tag{5.4}$$

In this analysis κ was set equal to 5 and the values of α varied over the range 0 to 0·091. It has been shown by Chen and Box (1979) that for $\kappa \geqslant 5$ the posterior distribution is mainly a function of $\varepsilon = \alpha/(1-\alpha)\kappa$ so the results are also labelled in terms of this dominant discrepancy parameter ε. It will be noticed:

(a) The large change in each estimated effect and standard deviation occurs when no possibility whatever of bad values ($\varepsilon = 0$) is replaced by a small possibility ($\varepsilon = 0{\cdot}001$). For good

† For the present purpose we retain the model structure of Coen, Gomme and Kendall. However, its relevance seems dubious, for example, a multivariate time series analysis by Tiao and Box (1980) for these three series shows the stock prices $\mathbf{y}$ acting as a weak *leading* indicator for the commodity index X_2.

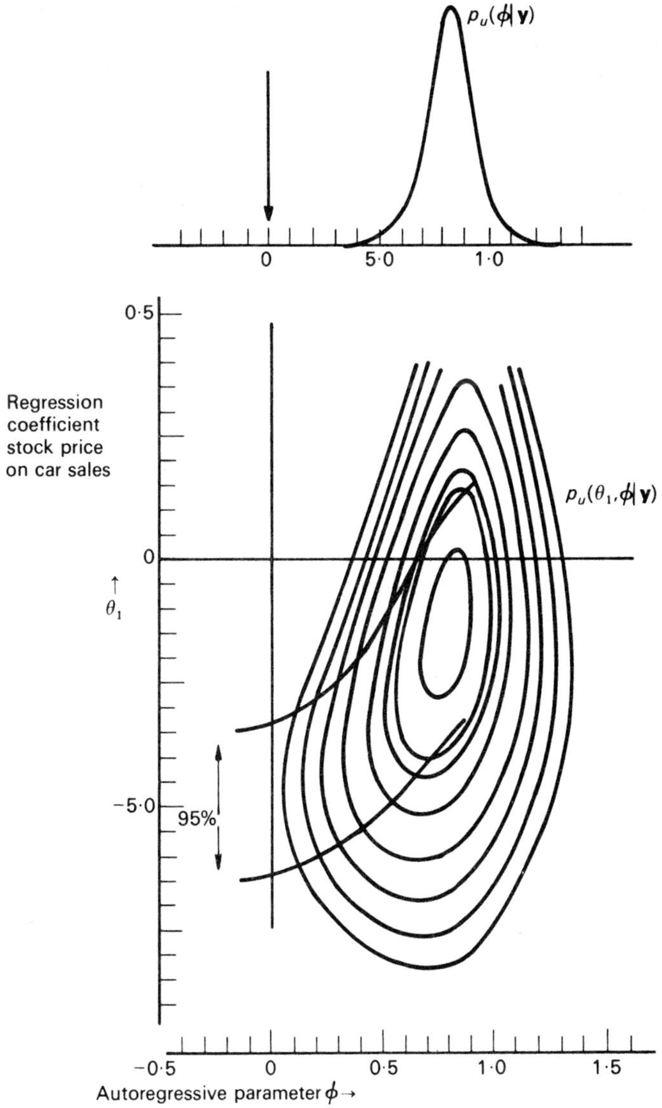

FIG. 3. Joint posterior distribution of θ_1 and ϕ and marginal posterior distribution of ϕ. Note shift in approximate 95 per cent interval as ϕ is changed.

data the typical behaviour of a table of this kind is that only very minor changes in mean and standard deviation occur as ε is changed over the plausible range.

(b) For all the estimates except β_{14} the standard deviations of effects are about halved. Thus for these effects the use of the more appropriate model is equivalent to a four-fold increase in the size/sensitivity of the experiment. This may be compared for example with a parallel analysis by Box and Cox of their biological data where a three-fold increase in sensitivity resulted from the use of an appropriate transformation.

(c) The analysis can be further illuminated by considering other available quantities. In particular a plot of the probability that the ith value is bad, given that one value is bad (see, for example, Abraham and Box, 1978), results in a plot with 94 per cent of the probability associated with the tenth observation and the remainder spread among the remaining 26 observations. It is likely therefore that y_{10} alone is a bad value. It is a deficiency of the design being used here that least squares estimates of interactions employ only four of the 27 observations and so lack robustness to bad observations (see, for example, Box and Draper, 1975). In particular $\hat{\beta}_{14} = 0{\cdot}25\,(y_{10} - y_{11} - y_{12} + y_{13})$ so that the Bayesian down-weighting of y_{10} accounts for the large change in this estimate and the *increase* in the standard deviation.

(d) We saw in the case of ridge regression how failure to take account of observational information could lead to an unrealistic choice of the discrepancy parameter γ. To complete the

TABLE 1

Bayesian means and (standard deviations) for polynomial coefficients using the contaminated model of Section 4.5 with $\kappa = 5$ $(\varepsilon = \alpha/(1-\alpha)\kappa)$

ε	0	0·001	0·005	0·010	0·015	0·020
α	0	0·005	0·024	0·048	0·070	0·091
β_0	90·60	90·60	90·60	90·60	90·60	90·60
	(0·94)	(0·45)	(0·41)	(0·41)	(0·41)	(0·41)
β_1	1·93	2·46	2·49	2·49	2·49	2·49
	(0·47)	(0·28)	(0·23)	(0·22)	(0·22)	(0·22)
β_2	−1·96	−1·96	−1·96	−1·96	−1·96	−1·96
	(0·47)	(0·22)	(0·20)	(0·20)	(0·20)	(0·20)
β_3	1·13	1·13	1·13	1·13	1·13	1·13
	(0·47)	(0·22)	(0·20)	(0·20)	(0·20)	(0·20)
β_4	−3·68	−3·15	−3·12	−3·12	−3·12	−3·12
	(0·47)	(0·28)	(0·23)	(0·22)	(0·22)	(0·22)
β_{11}	−1·42	−1·88	−1·90	−1·90	−1·89	−1·89
	(0·70)	(0·44)	(0·41)	(0·41)	(0·42)	(0·42)
β_{22}	−4·33	−4·10	−4·09	−4·09	−4·09	−4·09
	(0·70)	(0·36)	(0·34)	(0·34)	(0·34)	(0·34)
β_{33}	−2·24	−2·01	−2·00	−2·00	−2·00	−2·00
	(0·70)	(0·38)	(0·34)	(0·34)	(0·34)	(0·34)
β_{44}	−2·58	−3·05	−3·06	−3·06	−3·06	−3·05
	(0·70)	(0·44)	(0·41)	(0·41)	(0·42)	(0·42)
β_{12}	−1·67	−1·67	−1·67	−1·67	−1·67	−1·67
	(0·81)	(0·39)	(0·35)	(0·35)	(0·34)	(0·34)
β_{13}	−3·83	−3·82	−3·82	−3·82	−3·82	−3·82
	(0·81)	(0·39)	(0·35)	(0·35)	(0·34)	(0·34)
→ β_{14}	0·95	−0·45	−0·51	−0·50	−0·49	−0·48
	(0·81)	(0·95)	(0·92)	(0·93)	(0·95)	(0·95)
β_{23}	−1 67	−1·67	−1·67	−1·67	−1·67	−1·67
	(0·81)	(0·39)	(0·35)	(0·35)	(0·35)	(0·35)
β_{24}	−2·62	−2·62	−2·62	−2·62	−2·62	−2·62
	(0·81)	(0·39)	(0·35)	(0·35)	(0·35)	(0·35)
β_{34}	−4·25	−4·25	−4·25	−4·25	−4·25	−4·25
	(0·81)	(0·39)	(0·35)	(0·34)	(0·34)	(0·34)

picture, therefore, a plot of the marginal distribution of the discrepancy parameter ε should be made in conjunction with Table 1 (compare also with the serial correlation example in Fig. 3). For these data the distribution $p_u(\varepsilon \mid \mathbf{y})$ has its mode close to $\varepsilon = 0{\cdot}010$.

The Bayes approach to robust estimation has the advantage of generality; furthermore it clearly reveals at any given stage, on precisely what assumptions the analysis is conditional. With the increased speed of computers and availability of visual display equipment a general Bayesian computer program, that can analyse any model we wish to entertain, seems a much more attractive prospect than the fresh devising of semi *ad hoc* procedures for each new possibility.

Some parallels in the two approaches are briefly considered below for the "bad value" problem.

5.2. *Robust Estimation for the "bad value" Problem*

For the "bad value" problem a wide variety of semi-empirical estimators have been proposed. Among these are the M, L and R, and various kinds of adaptive estimators. In turn among the M estimators a number of different "ψ" functions have been suggested leading to different ways of downweighting extreme observations.

Now consider the model of Section 4.5 for the simple location structure $E(y_i) = \mu$. Then (see, for example, Box and Tiao, 1968b) the Bayesian mean may be written

$$\hat{\mu} = \sum_{b=0}^{n} p(b \mid \mathbf{y}, \alpha)\bar{y}^{(b)}, \tag{5.5}$$

where $p(b \mid \mathbf{y}, \alpha)$ is the posterior probability that there are b bad values and $\bar{y}^{(b)}$ is the corresponding conditional posterior mean. Consider in particular $\bar{y}^{(1)} = \Sigma\, w_i\, y_i$. Then Chen and Box (1979) show that for $\kappa \geqslant 5$

$$w_i \doteq (n-1)^{-1}(1 - D_i/D), \tag{5.6}$$

$$D_i = \left\{1 - \frac{nr_i^2}{(n-1)^2}\right\}^{-\frac{1}{2}(n-1)} = \{1 + n^{-1}\tilde{r}_i^2\}^{\frac{1}{2}(n-1)}, \tag{5.7}$$

where r_i and $\tilde{r}_i$ are unweighted and weighted residuals defined in Section 4.5. Fig. 4(a) and (b) show plots of $w = w_1$ against r_1 and $\tilde{r}_1$ for three random normal samples of ten observations from a normal distribution when a multiple $0, 1, 2, \ldots,$ of σ is added to the first observation in each sample. Empirical approximations for these weighting curves are provided by the functions

$$w = 0{\cdot}1 \exp\{-|0{\cdot}49\, r_1|^7\} \quad \text{and} \quad w = 0{\cdot}1 \exp\{-|0{\cdot}3\,\tilde{r}_1|^{3{\cdot}5}\}.$$

Also shown in Fig. 4(b) for comparison is Tukey's biweight function $w = 0{\cdot}1\{1-(\tilde{r}/c)^2\}^2$ for $c = 5{\cdot}3$ (chosen to roughly match the curve). Although the Bayesian weights are sample dependent they remain remarkably stable as is indicated (a) by the smooth manner in which the remaining weight is evenly spread throughout the non-discrepant observations; (b) by the closeness with which points from different samples follow the same curve.

The estimate $\hat{\mu}$ is sample adaptive in another more striking way however. For illustration consider the case where the $p(b \mid \mathbf{y}, \alpha)$ are negligible for $b \geqslant 2$. Then writing $p = p(1 \mid \mathbf{y}, \alpha)$ (5.5) becomes

$$\hat{\mu} = (1-p)\,\bar{y} + p\bar{y}^{(1)} \tag{5.8}$$

and the Bayesian mean is an interpolation between $\bar{y}$ and the "robustified" $\bar{y}^{(1)}$. In this expression the value of p is determined by the posterior odds ratio for one *vs* no bad values

$$p/(1-p) \doteq \varepsilon\{n/(n-1)\}\, D, \quad \varepsilon = \alpha/(1-\alpha)\,\kappa \tag{5.9}$$

and D is the checking function encountered earlier.

Sample adaptivity is evidenced as follows. For a sample with no outliers $\bar{y}$ and $\bar{y}^{(1)}$ are not very different so that $\hat{\mu}$ is close to $\bar{y}$. But in the presence of an outlier of larger and larger size two

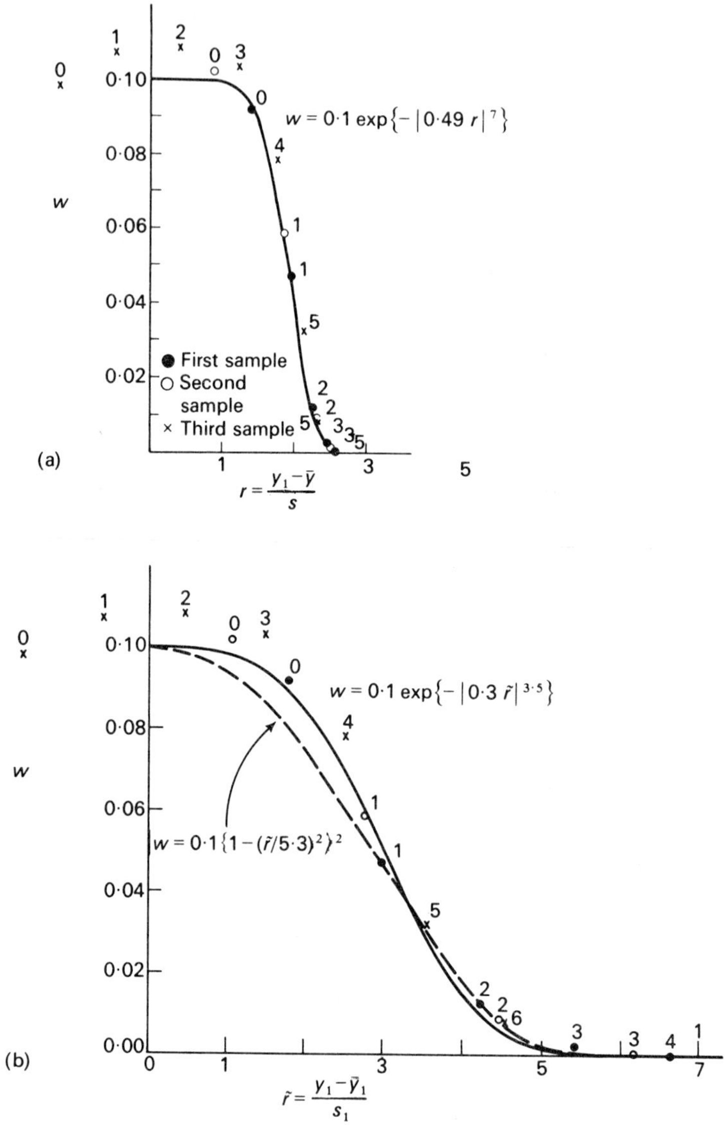

FIG. 4. Weight w applied to y_1 for three samples from a Normal Distribution. Numbers 0, 1, 2, ... indicate that $0, \sigma, 2\sigma, \ldots$ has been added to y_1.

things happen: the outlier is downweighted in $\bar{y}^{(1)}$ which becomes more and more different from $\bar{y}$ and also p becomes larger placing more and more emphasis on $\bar{y}^{(1)}$.

The purpose of this discussion is to show that sensible solutions which appropriately downweight suspected bad values may be obtained directly from an appropriate model. From the viewpoint of the traditional M estimator, the weight function (W say) for $\hat{\mu}$ itself is an interpolation $W = (1-p)n^{-1} + pw$ between $1/n$ and w. Thus W will descend to the value $(1-p)/n$ for large $\tilde{r}$. For a sample containing a large outlier, $1-p$ will be negligible and W will approach w plotted in Fig. 4(b) and will descend like Tukey's biweight. However for a more normal-looking sample W will flatten out to some moderate non-zero value and will more closely resemble the weighting originally proposed by Huber.

In choosing robust estimators there is room for empiricism but I think that some of its inspiration should be applied to the choice, study, and consequences of appropriate parsimonious models. The structure of the resulting Bayesian analysis should in each case be carefully analysed, for the great strength of such a model-based approach is that the exact consequences of whatever goes into the model must come out. These consequences will either agree with "common sense" or they will not. If they do not then we know either that what went in was inappropriate in a way we had failed to forsee, or else, as happens quite frequently, that our common sense was too shortsighted. In either case we learn something.

ACKNOWLEDGEMENTS

This research was sponsored by the United States Army under Contract No. DAAG29–75–C–0024. My thanks are due to Steve Bailey, Gina Chen, Conrad Fung, Kevin Little and Ruey-Shiong Tsay for their help in preparing this manuscript, and to Tom Leonard, a referee, and many other friends for helpful suggestions and comments.

REFERENCES

ABRAHAM, B. and BOX, G. E. P. (1978). Linear models and spurious observations. *Appl. Statist.*, **27**, 131–138.

ANDREWS, D. F. (1971a). A note on the selection of data transformations. *Biometrika*, **58**, 249–254.

—— (1971b). Significance tests based on residuals. *Biometrika*, **58**, 139–148.

ANSCOMBE, F. J. (1961). Examination of residuals. In *Proc. 4th Berkeley Symp. Math. Statist. Prob.*, **1**, 1–36. Berkeley and Los Angeles: University of California Press.

ANSCOMBE, F. J. and TUKEY, J. W. (1963). The examination and analysis of residuals. *Technometrics*, **5**, 141–160.

ATKINSON, A. C. (1973). Testing transformations to normality. *J. R. Statist. Soc.* B, **35**, 473–479.

BAILEY, S. P. and BOX, G. E. P. (1980a). Modeling the nature and frequency of outliers. Technical Report 2085 Math. Research Center, University of Wisconsin at Madison.

—— (1980b). The duality of diagnostic checking and robustification in model building: Some considerations and examples. Technical Report 2086, Math. Research Center, University of Wisconsin at Madison.

BARNARD, G. (1978). Personal communication.

BOX, G. E. P. (1976). Science and statistics. *J. Amer. Statist. Ass.*, **71**, 791–799.

—— (1979). Robustness in the strategy of scientific model building. In *Robustness in Statistics*, pp. 201–236. New York: Academic Press.

BOX, G. E. P. and BEHNKEN, D. W. (1960). Some new three-level designs for the study of quantitative variables. *Technometrics*, **2**, 455–475.

BOX, G. E. P. and COX, D. R. (1964). An analysis of transformations. *J. R. Statist. Soc.* B, **26**, 211–243.

BOX, G. E. P. and DRAPER, N. R. (1975). Robust designs. *Biometrika*, **62**, 347–352.

BOX, G. E. P. and JENKINS, G. M. (1970). *Time Series Analysis: Forecasting and Control.* San Francisco: Holden-Day.

BOX, G. E. P. and NEWBOLD, P. (1971). Some comments on a paper by Coen, Gomme and Kendall. *J. R. Statist. Soc.* A, **134**, 229–240.

BOX, G. E. P. and TIAO, G. C. (1962). A further look at robustness via Bayes' theorem. *Biometrika*, **49**, 419–432.

—— (1964). A Bayesian approach to the importance of assumptions applied to the comparison of variances. *Biometrika*, **51**, 153–167.

—— (1968a). Bayesian analysis of means for the random effect model. *J. Amer. Statist. Ass.*, **63**, 174–181.

—— (1968b). A Bayesian approach to some outlier problems. *Biometrika*, **55**, 119–129.

—— (1973). *Bayesian Inference in Statistical Analysis.* Reading, Mass.: Addison-Wesley.

BOX, G. E. P. and YOULE, P. V. (1955). The exploration and exploitation of response surfaces: an example of the link between the fitted surface and the basic mechanism of the system. *Biometrics*, **11**, 287–323.

CHEN, G. G. and BOX, G. E. P. (1979). Further study of robustification via a Bayesian approach. Technical Report 1998, Math. Research Center, University of Wisconsin at Madison.

COEN, P. J., GOMME, E. E. and KENDALL, M. G. (1969). Lagged relationships in economic forecasting. *J. R. Statist. Soc.* A, **132**, 133–152.

COX, D. R. (1977). The role of significance tests. *Scand. J. Statist.*, **4**, 49–70.

DEMPSTER, A. P. (1971). Model searching and estimation in the logic of inference (with Discussion). In *Foundations of Statistical Inference*, pp. 56–81. Toronto: Holt, Rinehart and Winston.

—— (1975). A subjectivist look at robustness. *I.S.I. Bulletin*, **46**, 349–374.

DIXON, W. J. (1953). Processing data for outliers. *Biometrics*, **9**, 74–89.

DRAPER, N. R. and VAN NOSTRAND, R. C. (1977). Ridge regression: is it worthwhile? Technical Report 501, Department of Statistics, University of Wisconsin at Madison.

DURBIN, J. and WATSON, G. S. (1950). Testing for serial correlation in least square regression I. *Biometrika*, **37**, 409–428.
FELLER, W. (1968). *An Introduction to Probability Theory and its Applications*, Vol. 1. New York: Wiley.
FISHER, R. A. (1935). *The Design of Experiments*. Edinburgh: Oliver and Boyd.
GEISSER, S. (1971). The inferential use of predictive distributions (with Discussion). In *Foundations of Statistical Inference*, pp. 458–469. Toronto: Holt, Rinehart and Winston.
—— (1975). The predictive sample reuse method with applications. *J. Amer. Statist. Ass.*, **70**, 320–328.
GEISSER, S. and EDDY, W. F. (1979). A predictive approach to model selection. *J. Amer. Statist. Ass.*, **74**, 153–160.
GOOD, I. J. (1956). The surprise index for the multivariate normal distribution. *Ann. Math. Statist.*, **27**, 1130–1135.
GORMAN, J. W. and TOMAN, R. J. (1966). Selection of variables for fitting equations to data. *Technometrics*, **8**, 27–51.
GUTTMAN, I. (1967). The use of the concept of a future observation in goodness-of-fit problems. *J. R. Statist. Soc.* B, **29**, 83–100.
HOERL, A. E. and KENNARD, R. W. (1970). Ridge regression : applications to non-orthogonal problems. *Technometrics*, **12**, 69–82.
HOERL, A. E., KENNARD, R. W. and BALDWIN, K. F. (1975). Ridge regression : some simulations. *Comm. in Statist.*, **4**, 105–124.
HUBER, P. J. (1977). Robust statistical procedures. Society for Industrial and Applied Mathematics, **27**, Philadelphia.
JEFFREYS, H. (1932). An alternative to the rejection of observations. *Proc. Roy. Soc.* A, **CXXXVII**, 78–87.
KADANE, J. B., DICKEY, J. M., WINKLER, R. L., SMITH, W. S. and PETERS, S. C. (1979). Interactive elicitation of opinion for a normal linear model. Technical Report 150, Department of Statistics, Carnegie Mellon University.
LINDLEY, D. V. and SMITH, A. F. M. (1972). Bayes' estimates for the linear model (with Discussion). *J. R. Statist. Soc.* B, **34**, 1–41.
PALLESEN, L. C. (1977). Studies in the analysis of serially dependent data. Ph.D. thesis, Department of Statistics, University of Wisconsin at Madison.
POPPER, K. R. (1959). *The Logic of Scientific Discovery*. New York: Harper and Row.
ROBERTS, H. V. (1965). Probabilistic prediction. *J. Amer. Statist. Ass.*, **60**, 50–62.
SNEDECOR, G. W. and COCHRAN, W. G. (1967). *Statistical Methods*. Ames : Iowa State University Press.
STEIN, C. (1956). Inadmissibility of the usual estimator for the mean of a multivariate normal distribution. *Proceedings of the Third Berkeley Symposium*, **1**, 197–206. Berkeley and Los Angeles: University of California Press.
STIGLER, S. M. (1973). Simon Newcomb, Percy Daniel and the history of robust estimation 1885–1920. *J. Amer. Statist. Ass.*, **68**, 872–879.
THEIL, H. (1963). On the use of incomplete prior information in regression analysis. *J. Amer. Statist. Ass.*, **58**, 401–414.
TIAO, G. C. and ALI, M. M. (1971). Analysis of correlated random effects: linear model with two random components, *Biometrika*, **58**, 37–51.
TIAO, G. C. and BOX, G. E. P. (1980). An introduction to applied multiple time series analysis. Technical Report 582, Department of Statistics, University of Wisconsin at Madison.
TUKEY, J. W. (1949). One degree of freedom for non-additivity, *Biometrics*, **5**, 232–242.
——(1960). A survey of sampling from contaminated distributions. In *Contributions to Probability and Statistics: Essays in Honor of Harold Hotelling*, pp. 448–485. Stanford, Stanford University Press.
ZELLNER, A. and TIAO, G. C. (1964). Bayesian analysis of the regression model with autocorrelated errors. *J. Amer. Statist. Ass.*, **59**, 763–778.

DISCUSSION OF PROFESSOR BOX'S PAPER

Professor G. A. BARNARD (University of Waterloo): I very much welcome this important paper as a further indication that views on the foundations of statistical inference are converging and the worrying prospect that seemed to be opening itself up four or five years ago of hopelessly irreconcilable attitudes being adopted amongst experts in the field can now be thought less probable.

Although I shall follow tradition and emphasize my difference from the author, I must first of all say how strongly I agree with what he has said concerning "robust" estimation. The conditional approach which he adopts—and which follows naturally from a Bayesian approach—is surely the only sound one. We ought not to look for robust *procedures*, but rather for robust *samples*. For a robust sample varying assumptions about the shape of the distribution will have little effect on the inference to be drawn. For a non-robust sample these assumptions have an effect; and in that case the *existence of this effect must form part of the inference*. We must tell our clients that unless they can find out something concerning the shape of the distribution, the message of their sample is ambiguous. Alternatively, by taking further observations, they may convert what was a non-robust sample into a robust sample and so avoid the necessity for investigations about the population shape. Insofar as the Bayesian approach bases itself essentially on using the likelihood function, of course, we would expect a likelihood approach and a Bayesian approach to agree in this respect. It has sometimes been said that the trouble with likelihood is that we need to know the form of the distribution with more precision than is commonly available. Such a view overlooks the fact that if we have doubts about the form of the distribution we can always make a range of suppositions concerning this, and derive a corresponding range of likelihood functions. It will very often be the case that the range of likelihood functions so obtained will be quite narrow, showing that our sample is a robust one and our ignorance of the precise shape of the distribution will do no harm. In the contrary case, where the shape of the distribution does matter it is actually misleading to fail to point this out. It will be no comfort to our current client if we should say that cases such as his rarely occur—he is concerned with what has happened in his particular case, and if the inference drawn proves to be wholly incorrect, he will be entitled to say that he ought to have been warned.

When the conditions for the simple-minded application of the method of maximum likelihood apply—that is when the log likelihood function is nearly parabolic—the sensitivity or otherwise of a particular sample to changes in distribution form can be indicated by the change in the position of the maximum likelihood estimate, primarily, and secondarily by changes in the second derivative of the score function, the information. If the maximum likelihood estimate does not change very much then for "point estimate" purposes our sample is robust, and very often a modest allowance for variations in the information will give the protection that is needed.

Now to return to the main topic. Professor Box had adopted a position very close to that of Emile Borel. Borel's view of the nature of probability was very close to that of de Finetti—the strongest advocate of the personalistic view—but Borel was criticized, I think wrongly, by de Finetti for adopting "Cournot's principle" according to which we behave as if events of small probability do not occur. Another version of Cournot's principle is the so-called empirical law of large numbers, according to which probabilities must agree with long run frequencies.

The fact is that if an individual went through life continually being amazed to find that the events for which his subjective probability had been extremely small regularly turned out to be those that occurred, the individual in question ought to ask himself whether his modes of assessment of probabilities were reliable. The fallacy in the personalistic argument appears to me to arise from the fact that a personalistic view of probability makes no allowance for the limitations of human imagination. In Box's sequence of model criticism, revised model and so on, the pure personalistic view has to treat the propositions "model M_i is true" and "model M_i is false" as propositions on the same footing. Yet this is obviously absurd. If we take as given that model M_i is true we have a very clear picture of the possible events to be observed and the parameter values to be associated with them. But to say simply that model M_i is false give us absolutely no idea of what sort of model might be true—we might or might not be required to introduce further parameters, we might or might not think of a great multiplicity of shapes of distribution that might arise and so on. Observations of incompatible events could well lead us to feel that M_i could not be true, and yet the intellectual effort involved in constructing a model that we think would apply to the situation in hand could form as large a part of the whole intellectual exercise as any other. To take one example, Michelson and Morley gained well-deserved fame for demonstrating the falsity of the simple minded notion of an ether. But their credit is small compared with that of Einstein whose tremendous imagination was needed to set up an alternative model which accounts for their data. To put the point formally, the pure theory of personal probability treats H and not-H as propositions of the same kind. But the problem of calculating the probability of E given H may be quite trivial, whereas the problem of calculating the probability of E given not-E could be quite insoluble.

It is perhaps worthwhile to indicate how the de Finetti theory could be modified to take account of the points that Box is making. To simplify a little, we can regard the model M_i as being our (tentative) view of what constitutes the totality of possibilities before we perform the experiment. If we then observe a result y for which the probability density on the predictive distribution is low, we need to reconsider M_i and ask whether it really does encompass all the possibilities. When y presents a feature which, in the words of Borel., is "en quelque sort remarquable" the fact that it is remarkable points us in a certain direction to try to use our imagination to conjure up possibilities which previously we had not recognized. M_{i+1} will then embody these possibilities and the iteration process begins again. Thus as I see it a strict adherent to de Finetti's views need only admit the obvious fact that we can never think of everything to accept the force of Box's argument.

The modified personalistic view of probability put forward by Box and Borel is, I think, inadequate for statisticians basically because statisticians work for clients. They are therefore not concerned with personal probabilities but with what might be called "agreed probabilities". And whereas personal probabilities may be said always to exist in principle, *agreed* probabilities need not exist. We may have a parameter involved in the statistical problem for which an agreed prior probability distribution does not exist and in such a case I think it will often be wise for a statistician to regard the parameter as simply unknown, capable of taking on a number of values, and for the statistician to see his job as pointing to the probabilistic consequences, in the light of the data, of assuming that the parameter takes various values in its range. This is *not* equivalent to a prior concentrated on a single value. This is my major disagreement with Professor Box.

As an extension of the general theory of likelihood, I have in recent years been developing a theory of what I call "pivotal inference" concerning which I hope to present a paper to the Society shortly. In this theory we start from quantities, called pivotals, which do have agreed probability distributions. Such pivotals will usually be functions both of observations and of parameters. But some of them may also be, or be transformed into, quantities which are functions of the observations only—these then become ancillaries, upon which we should condition when we have the observations—or they may be functions of the parameters only, in which case they represent prior distributions for the functions of the parameters involved. By applying the standard rules of probability, concerned with marginalization and with conditioning in the light of the values of the observations when these become known, we can infer distributions for functions of the basic pivotals which will sometimes give posterior distributions, if the necessary information for a full Bayesian inference is available, but more often will enable us only to make statements to the effect that an assumption that a parameter has a value in some specified range will entail that an event has occurred whose probability is small. We shall be disinclined to swallow such improbabilities.

The pivotal approach does not require us to take a general position on the question, whether or not unknown parameters are required to be endowed with probability distributions. In each specific case, we can exercise judgement as to whether to ascribe such distributions to some or all of the parameters and we can explore the final effects—often small—of adding or removing such assumptions. In any case the basic inferential procedure is always the same—to condition on known, or approximately known quantities, whose distribution is known, to arrive, if we can, at invertible pivotals for the quantities of interest, which enable us to say that accepting the notion that a parameter value lies in a certain range entails accepting that an event of a specifically low probability has occurred. This is all we can derive from a Bayesian approach, unless we have—as is very rarely the case—a well-specified loss function which we aim to minimize.

The neo-Bayesian movement has purged statistical inference of a great many stupidities which arise from neglect of proper conditioning. I hope this paper will come to represent a major step towards a situation where such absurdities no longer plague us, and we are much closer to general agreement on foundations than we have been in the recent past.

It gives me great pleasure to propose the vote of thanks to Professor Box.

Professor A. P. DAWID (The City University): This Society has traditionally recognised the importance of both *Estimation* and *Criticism* as principal features of a vote of thanks. It is my very agreeable task to bring these twin criteria to bear on tonight's paper.

First, then, to register my esteem. Professor Box has given us a compelling account of his search for the guiding principles of scientific learning, and illuminated it with practical examples which are both interesting and important. We are all the richer for having him share his experience and insights with us. Whilst, as he admits, his views are not new, his paper is a valuable and forceful reminder of an important lesson: that there are at least two distinct fundamental functions of statistical reasoning, Criticism and Estimation, and "never the twain shall meet". We should constantly bear the distinction in mind when we construct, select, or teach students about statistical techniques. A hypothesis test, for example, can be used for model testing (Criticism) or model simplification (Estimation). The purpose, interpretation and relevance are quite different for the two different cases.

Whether or not one agrees with Professor Box that Bayesian conditioning is the way to go about Estimation, one must agree that something else is needed for Criticism, where no fully specified alternative model is given. Professor Box makes a valuable contribution in proposing the predictive distribution as the basis of Criticism, but the question of *how* to use it is not clarified, and we are left with familiar *ad hoc* devices such as tail-area tests and (in a framework that makes some concession to Estimation) score statistics. This is not to belittle the usefulness of *ad hoc* solutions in the absence of underlying principles, merely to point to a gaping hole in all our current theoretical formulations.

At a very general level, Box's dualistic view of statistical reasoning is crudely analogous to Thomas Kuhn's account of the progress of scientific theories. In periods of "normal science", a particular paradigm, or model, is taken for granted, and scientists work at refining it and apply it ("Estimation"). But the predictions of the current paradigm are always open to confrontation with the real world ("Criticism"), and if and when discrepancies become unacceptable a "scientific revolution" topples the old paradigm and puts a new one in its place. However, the old paradigm can be discredited even when no workable alternative is in sight.

Kuhn's ideas have made an enormous impact on the philosophy of Science, and it is valuable to be reminded that similar reasoning can, and should, be applied to statistical investigations, however mundane.

Let me now switch to Criticism mode and consider some details of tonight's paper. Box distinguishes between checks on parametric and residual features of the model. I believe that the former, which is basically a check on the prior distribution, as in (3.9), will be the most useful practical contribution of this paper, but it may not be much appreciated by non-Bayesians who do not accept, with Box, that a model cannot exist in isolation from a prior distribution. Moreover, I do not find the attempted Bayesian justification of significance tests (Section 2.1) any advance on the non-Bayesian interpretation. As for residual checks, these have long been available in a non-Bayesian setting. Indeed, with the exception of Section 4.3 (to which I shall return), Box's analysis has merely reproduced classical tests. These examples all involve models of the form

$$p(\hat{\mathbf{\Theta}}, S, \mathbf{u} \mid \mathbf{\Theta}, \sigma, \beta) = S^{-2} \,.\, p((\hat{\mathbf{\Theta}} - \mathbf{\Theta})/\sigma, S/\sigma \mid \mathbf{u}; \beta) \,.\, p(\mathbf{u} \mid \beta).$$

Box takes his prior for $(\mathbf{\Theta}, \sigma)$ given β to be "locally uniform":

$$p(\mathbf{\Theta}, \sigma \mid \beta) \doteqdot c(\beta) \,.\, \sigma^{-1}.$$

(We must of course allow the general level, determined by $c(\beta)$, to depend on β). Integrating out $(\mathbf{\Theta}, \sigma)$ gives

$$p(\hat{\mathbf{\Theta}}, S, \mathbf{u} \mid \beta) \doteqdot S^{-1} \,.\, c(\beta)\, p(\mathbf{u} \mid \beta),$$

and so the score-statistic $g_\beta(\mathbf{y})$, from (4.2), is just $\{c'(\beta_0)/c(\beta_0)\} + h_\beta(\mathbf{u})$, a simple transform of $h_\beta(\mathbf{u})$, the score statistic derived from the purely classical approach of considering the standardized residuals only, and having a known null distribution. A generalization of this remark holds for arbitrary group-invariant models where the underlying pivot has a distribution governed by the discrepancy parameter β.

At least the above theory leads us to hope that the Box approach to checking residual features, when applied (as it should be) with a genuinely informative prior distribution, may lead to a test statistic approximating $h_\beta(\mathbf{u})$, with null distribution not over-dependent on the prior. The situation is not so clear for Section 4.3, however. If we again take a "locally uniform" prior of the form $p(\mathbf{\Theta}, \sigma \mid \lambda) \doteqdot c(\lambda)\, \sigma^{-1}$, we find

$$p(\mathbf{y} \mid \lambda) \,\dot{\propto}\, c(\lambda)\, \dot{y}^{n(\lambda-1)} Q_\lambda^{-\frac{1}{2}\nu}.$$

(The slightly different expression (4.6) results from an interesting attempt by Box and Cox to specify $c(\lambda)$ reasonably; for present purposes, this can be avoided.) Then we find

$$g_\lambda(\mathbf{y}) \doteqdot \{c'(1)/c(1)\} + (k+1) \log \dot{y} + s^{-1} \textstyle\sum r_i z_i.$$

The constant term is irrelevant to the test statistic, leaving the extra term $(k+1) \log \dot{y}$ as a correction to (4.7).

Even if we stick to (4.7), what are we to do with it? Box suggests an informal graphical analysis, but we have to know what features of the diagram to look for. It seems implicit that a substantial sample correlation between $z - \hat{z}$ and $y - \hat{y}$ is to be regarded as evidence against $\lambda = 1$. In fact, (4.7) is $\nu \times$ the slope of the regression of $z - \hat{z}$ on $y - \hat{y}$, and is not a function of correlation. And is zero the appropriate "null value" for (4.7)? What departures are "significant"? These problems do not disappear on "actual reference of $p(g_\lambda(\mathbf{y}_d))$ to its sampling distribution", since the null sampling distribution depends on the unknown $(\mathbf{\Theta}, \sigma)$. We could marginalize it with respect to the prior distribution (for $(\mathbf{\Theta}, \sigma)$ given $\lambda = 1$), but the answer would be critically dependent on the (proper, locally uniform) prior chosen, and cannot be approximated by using an improper prior. Nevertheless, this marginalization would be the right course for the Bayesian. The moral is that the need for transformation can only be assessed in the light of prior knowledge.

Box's representation (4.11) is in error, a factor $(s/\dot{y})$ being omitted ($B = 1/\dot{y}$ in (4.8)). For the simplest case $k = 0$, we get $g_\lambda(\mathbf{y}) \doteqdot -\frac{1}{2}(s/\bar{y})\, T_{30}$, and the variance of this is approximately proportional to σ^2/μ^2, making assessment of significance impossible without prior information about (μ, σ). If we adjust by omitting the $-\frac{1}{2}(s/\bar{y})$, we get T_{30}, a function of the normalized residuals with a known null distribution; but after all these *ad-hockeries*, have we gained anything from a pseudo-Bayesian approach?

Finally, let me remove my critic's hat and welcome Box's whole-hearted Bayesian account of robust estimation (Section 5). This captures all the common-sense features one would want: for example, there is no point in guarding against model departures of a size which the data themselves suggest is implausible. Even non-Bayesians could learn useful lessons from this analysis.

Altogether I regard tonight's paper as of the greatest importance, and second the vote of thanks to its author, George Box, most warmly.

The vote of thanks was passed by acclamation.

Dr A. O'HAGAN (University of Warwick): Like all Professor Box's work, this paper is suffused with common-sense and insight. I applaud all that he has done in identifying and illuminating very clearly a difficult problem. I have two complaints, one small and one I think rather larger. Both concern his curious belief that Bayesian methods are appropriate to the selection of sensible parameter values but not to the selection of models. The obvious Bayesian solution to the latter is to compute the posterior probability of model M_i via

$$P(M_i \mid y_d) = \frac{p(y_d \mid M_i)\, P(M_i)}{\sum_j p(y_d \mid M_j)\, P(M_j)} \qquad (*)$$

Of course this is conditional on one of the stated M_j being the true model, but for the moment let us suppose that this is so. Professor Box argues that if $p(y_d \mid M_i)$ is sufficiently small, that is if the model M_i completely fails to fit the data y_d, then M_i is discredited. But the appearance of the prior probabilities $P(M_j)$ in (*) shows that these are also relevant. We cannot say that $P(M_i \mid y_d)$ is small unless we can find a model M_j which fits the data well *and* is credible *a priori*. Otherwise the denominator in (*) will also be small and $P(M_i \mid y_d)$ may even be near to one. Professor Box on his first page tells us that "A theory about scientific model building ought to explain what good statisticians and scientists actually do". Practising statisticians when criticizing a model look at things like residual plots and sample autocorrelations, which Professor Box has shown us carry implicit ideas of alternative models. Every way the statistician chooses to look at the data corresponds to a feeling, which perhaps never becomes conscious and explicit, for a specific kind of alternative. Furthermore, his implied alternative is *a priori credible*—if he couldn't conceive of autocorrelation then he wouldn't bother to look for it.

When good statisticians *compare* models they should and effectively do employ (*). The new idea in Professor Box's approach is that we can criticize a model without having any alternatives in mind. This is very tempting because of course we can never think of more than a few of the countless possible models. Certainly if $p(y_d \mid M_i)$ is small it sounds a warning to me and I start to look for sensible alternatives, but I do not feel that I can gauge that smallness or reject M_i until I've found a better alternative. Professor Box judges the size of $p(y_d \mid M_i)$ by means of a "tail-area", but this is wrong in this context for the same reasons that it is wrong in other stages of the analysis—the role played by $p(y_d \mid M_i)$ in (*) is that of the likelihood. Studying $p(y \mid M)$ as y varies is useless because we are interested in it only as M varies. Of course, in other contexts sampling-theory methods based on tail-areas have often been found to approximate closely to Bayesian procedures. But tail-areas are also known to be unreliable indicators, and I feel sure that uncritical use of them to compare models will lead to familiar pitfalls. Just as in parameter estimation, there is no substitute for a proper Bayesian analysis. We must consider all the alternatives which occur to us, and if our original model M_i is not bettered then we cannot in any sense reject it yet, however "small" $p(y_d \mid M_i)$ may appear to be. Nor should we forget that if we are satisfied with our current model we still cannot rule out the possibility of something better being proposed.

In conclusion, I am sure that Professor Box is a far better practical statistician than I, and if he tells me that he can persuade a thing for taking stones out of horses' hooves to turn screws I believe him. *I* will stick to my screwdriver.

Dr A. C. ATKINSON (Department of Mathematics, Imperial College, London): I can still recall the intellectual excitement of first reading Chapter 11 of "Big Davies" (Davies, 1956). More than any other, this was the experience which inspired me to become a statistician. It is a pleasure to be able to thank Professor Box in person for the initial and continuing stimulus of his books and papers.

My comments on tonight's paper concern the diagnostic checking of regression models. If outlying values of the carriers x are suspected, residual plots should be augmented by functions which respond to the influence of the individual observations. Fig. D1 shows a half normal plot, for the Box and Behnken data, of a modification of a statistic due to Cook (1977) which I call T_i. For this designed experiment the plot is similar to a residual plot, with identity for a D optimum design. There is clearly something strange about one of the observations.

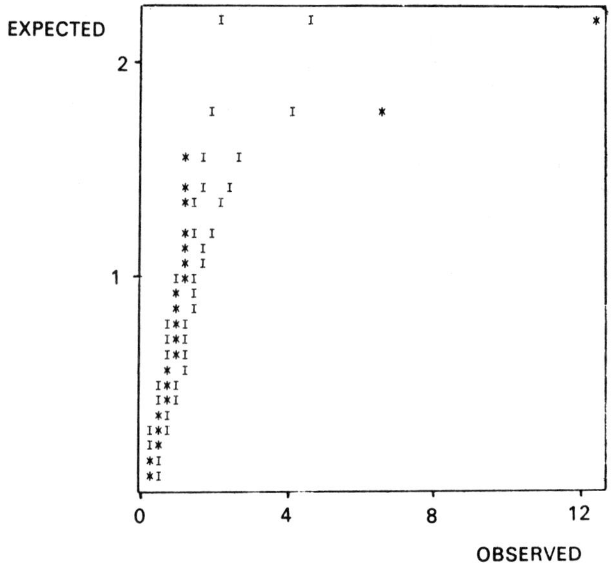

FIG. D1. Half normal plot of modified Cook statistic T_i for 27 observations from Box and Behnken. * Observed. I, Envelope from 19 simulations.

In several standard examples (Atkinson, 1980, 1981) transformation of the data provides an alternative to rejection of outliers. In this case the asymptotically standard normal score statistic for transformations, T_D, has the value $-0{\cdot}360$, so no transformation is needed. Incidentally, I do not understand the remark after (4.7) about the distribution of this statistic. Perhaps Professor Box was thinking of the apparently anomalous results of Schlesselman (1973) which are due to a programming error (Fuchs, 1979). Other transformations of the Box and Behnken data, such as considering $(100-y)^\lambda$ and replacement of

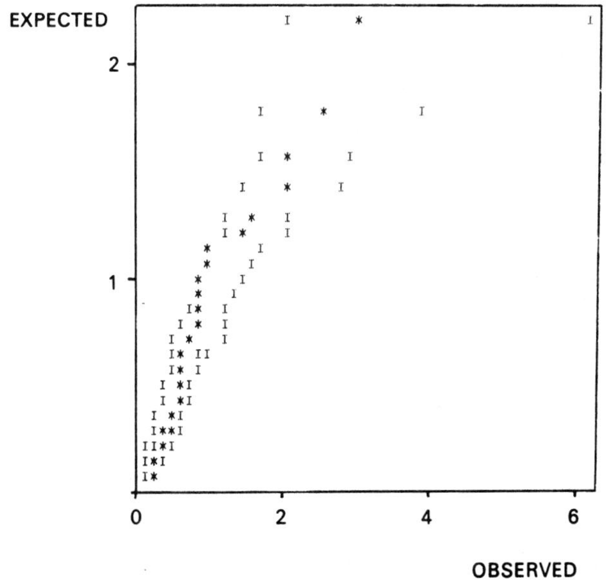

FIG. D2. Plot of T_i with y_{10} deleted. * Observed. I, Envelope from 19 simulations.

$y_{10} = 89{\cdot}4$ by 84·9, also fail to remove the outlier. If the observation is rejected the plot of T_i for the remaining 26 observations, Fig. D2, shows no further features of interest.

Table D1 gives the least squares estimates of the coefficients and the associated standard errors for the full second-order model fitted to the 26 observations. If only two factor interactions are considered, with the exception of β_{14}, the estimated coefficients agree to 3 significant figures with those in the last column of Table 1. Complete agreement can be achieved by interchanging the two rows for β_{14} in Table 1, which also brings the standard errors more nearly into line. I wonder whether the observed behaviour for β_{14} may not be due to a transcription error for at least some of the columns of these two rows.

In conclusion I would like to echo Professor Box's remarks about the importance of computer graphics in model checking. It is ironical that at a time when we are technically able to produce interesting plots with great ease, economic pressures are such as to discourage the publication of figures and graphs.

TABLE D1

Parameter estimates and associated standard deviations for Box and Behnken's data with y_{10} deleted

Parameter	*Least squares estimate*	*Standard deviation*
β_1	2·568	0·124
β_2	−1·958	0·111
β_3	1·133	0·111
β_4	−3·041	0·124
β_{11}	−2·051	0·175
β_{22}	−4·012	0·168
β_{33}	−1·924	0·168
β_{44}	−3·241	0·175
β_{12}	−1·675	0·192
β_{13}	−3·825	0·192
β_{14}	−0·953	0·252
β_{23}	−1·675	0·192
β_{24}	−2·625	0·192
β_{34}	−4·250	0·192

Professor D. R. Cox (Department of Mathematics, Imperial College, London): I admire the paper for its combination of important general discussion with interesting examples.

Equation (2.12) derives a tail area by integration over all sample points with a density equal to or smaller than that of the observed point. While, as stated just below the equation, relatively minor changes such as from s to s^2 (or $s^{2/3}$ or $\log s$) will make relatively minor changes to the answer, in fact adjusting the relation between two tails, it is not clear why they should make any difference at all: or to put the point differently, how do we decide which is appropriate in a given instance? There is, of course, also the theoretical possibility that radically non-linear transformations, or the use of quantities with very spiky distributions, would lead to strange regions of integration. Would it not be better to define the test quantity so as to order the sample points in order of increasing discrepancy in some respect and to integrate over large values of the test quantity? I appreciate that this involves an implied qualitative specification of alternatives which it might be desirable to avoid, but is some such specification really avoidable?

The section on robust estimation is very appealing. The idea of modelling suspected complications and exploring the consequences by general theory is excellent for broad guidance, but if undue complication is to be avoided, this idea would presumably have to be severely restrained in applications. We all recall what Lord Kelvin said.

Professor M. STONE (University College London): The emphasis, in this authoritative and readable paper, on the concept of compatibility between data and their marginal distribution is very welcome. I hope that Professor Box's message will be studied by those who have up to now ignored the value of the compatibility concept in constructive criticism of a Bayesian model and elucidation of any paradoxical aspects it may have when improper priors are used.

In Section 2.2, Professor Box rightly advises us that impropriety of the marginal data distribution does not dispense with the need for criticism but he does not give much guidance to the potential critic. One analytical method available to deal with this troublesome corner is roughly as follows. Find, if you can, a sequence of proper priors such that, for every $\varepsilon > 0$, the (proper) marginal probability that the (proper) posterior differs by "more than ε" from the improper one of the model tends to unity down the sequence. You can then say that the improper posteriors are justifiable in marginal probability but you cannot conclude that the improper posterior for fixed data (i.e. what you wish to use) is also justified. Whether or not it is acceptable may be determined by (a) the way in which the posterior is used (b) whether the (fixed) data are in some relevant way asymptotically incompatible with the proper marginals. In some cases, the test of asymptotic incompatibility is dramatically simple: any fixed data are unacceptable!

In a paper so rich in suggestions, I should not be surprised that the author has touched on another problem that has worried me for some time. That is the question of the possible link between cross-validatory procedures and the Bayesian models to whose output they bear a striking resemblance. The similarity is most striking for Bayesian models that incorporate flexible priors of the multistage variety and that are therefore rendered data-adaptive. By construction, cross-validatory procedures are data-adaptive within the ambit prescribed for choice. However, I have been unable to obtain any technical insight into problems of real significance. It is tempting to try to relate the cross-validatory weights in Modelmix to posterior probabilities of the component "models". A simple example with discontinuous weights, (1, 0) or (0, 1), is that of estimating the true mean μ from a random sample of size n when it is known that μ is either μ_1 or μ_2 with $\mu_1 < \mu_2$. If the cross-validatory prescription is to say $\hat{\mu} = \mu_1$ if $\bar{y} < \alpha$ and $\hat{\mu} = \mu_2$ if $\bar{y} \geqslant \alpha$ and if the "loss function" is quadratic between $\hat{\mu}$ and an observation, a cross-validatory choice is to take $\alpha^{\dagger} = \infty$ or $-\infty$ according to whether $\bar{y} < \frac{1}{2}(\mu_1 + \mu_2)$ or $\bar{y} \geqslant \frac{1}{2}(\mu_1 + \mu_2)$, that is $\hat{\mu} = \mu_1$ if and only if $\bar{y} < \frac{1}{2}(\mu_1 + \mu_2)$. A model for which this would be the posterior modal estimate of μ would be that of normality and a prior with $P(\mu = \mu_1 \mid \sigma^2) \equiv P(\mu = \mu_2 \mid \sigma^2)$. Is the normality implied by the use of $\bar{y}$ in the prescription and are the equiprior probabilities mildly expressive of maximal data-adaptivity? In some sense, I believe they are and that it ought to be possible to develop better illustrations of the relationship

Professor A. F. M. SMITH (University of Nottingham): Professor Box is a wise, practical statistician and tonight's distillation of his wisdom merits careful consideration by everyone genuinely concerned with statistical methodology. In particular, it should be required reading for all those statistical Yahoos who stridently proclaim the Bayesian approach to be misguided or irrelevant without bothering to study in detail what it has to offer. Perhaps there *are* ad hoc "non-Bayesian" analogues of everything Box mentions (and no doubt many of these *ad hockeries* "came first"), but Box's *unified perspective* is surely more satisfying and suggestive as a framework for overall understanding and further advance—even if, as various other, basically sympathetic, discussants have indicated, there are still many issues to be clarified *within* the framework.

So far as the details of the paper are concerned, I shall confine myself to a brief comment on the Bayesian approach to Robust Estimation (Section 5). The author's discussion concentrates on robustness against bad values, where his approach is shown to have close links with non-Bayesian procedures. In fact, an even more direct association with ideas like "M-estimates" and "influence functions" can be established using the fact that if $y = \theta + \varepsilon$, $p_\varepsilon(\cdot)$ arbitrary, with $\theta \sim N(m, c)$, then, using Masreliez (1975),

$$E(\theta \mid y) = m + c\left[-\frac{\partial}{\partial y}\log p(y)\right]$$

$$\approx m + c\left[-\frac{\partial}{\partial y}\log p_\varepsilon(y - m)\right],$$

where $p(y) = \int p_\varepsilon(y - \theta)\, p(\theta)\, d\theta$. Using the approximate form, we see that the way in which the "innovation" is used to update the prior mean depends on the choice of $p_\varepsilon(\cdot)$ through the score (or "influence") function. Interesting families of error distributions—such as the t-family, exponential power family, normal centre/Laplace tails family—can be investigated for use as (model-based) robust estimation procedures. The approach can be extended to provide robust Bayesian sequential learning within the Kalman filter framework, with both location and covariance structure unknown. An account of these ideas is being written as a Ph.D. thesis by Michael West at Nottingham University. Finally, it is worth remarking that these and other robustifying and checking devices tend to lead to some tricky *numerical* problems if integration over several (perhaps highly correlated) parameters is required in order to isolate a marginal feature. An account of some recent progress in this area will be given in a paper currently being written by myself and John Naylor, based on material which will form part of a Ph.D. thesis at Nottingham.

Professor JOSE M. BERNARDO (Department of Biostatistics, University of Valencia, Spain): Although I certainly agree with Professor Box in the cyclical nature of scientific experimentation through model criticism and parameter estimation, I do not understand his claim that "sampling theory is needed for ultimate criticism of a model". Indeed, when for this purpose he computes (2.3), that is $\alpha = \Pr\{p(\mathbf{y}\,|\,A) < p(\mathbf{y}_d\,|\,A)\}$, he is making use of a *predictive* distribution, which is not defined unless one has a prior. The (interesting) mathematical accident by which the use of a reference (non-informative) prior often leads to the same numerical manipulations as classical tests should not obscure the fact that one is arguing within a Bayesian framework. Incidentally, reference should be made to the use of the *surprise index* (2.3) by Barnard (1967, p. 28) and by Aitchison and Dunsmore (1975, p. 224).

At a more concrete level, I find unsatisfactory the use of Fisher's score function (4.2) to measure discrepancies from β_0 from a current model $p(\mathbf{y}\,|\,\beta)$. Indeed, as its Bayesian interpretation (4.22) openly shows, (i) $g_\beta(\mathbf{y})$ is *not* invariant under scale changes in β, as one might wish, and (ii) it assumes implicitly a uniform prior for β, that would only be appropriate if β were a location parameter. An alternative definition could be

$$g'_\beta(\mathbf{y}) = \frac{E(\beta\,|\,\mathbf{y}) - \beta_0}{D(\beta\,|\,\mathbf{y})}.$$

Here, $E(\beta\,|\,\mathbf{y})$ and $D(\beta\,|\,\mathbf{y})$ are the mean and standard deviation of the *reference posterior* distribution (Bernardo, 1979) $\pi(\beta\,|\,\mathbf{y})$ defined as $\pi(\beta\,|\,\mathbf{y}) \propto p(\mathbf{y}\,|\,\beta)\,\pi(\beta)$ where $\pi(\beta)$ is a reference (non-informative) prior. If β is a continuous one-dimensional variable and $p(\mathbf{y}\,|\,\beta)$ is well behaved, then the appropriate choice for $\pi(\beta)$ is Jeffreys' prior

$$\pi(\beta) \propto \left\{ -\int p(\mathbf{y}\,|\,\beta) \frac{\partial^2}{\partial \beta^2} \log p(\mathbf{y}\,|\,\beta)\, d\mathbf{y} \right\}^{\frac{1}{2}}.$$

If only distance from β to β_0 matters, I would probably use $\{g'_\beta(y)\}^2$ rather than $g'_\beta(y)$ as a measure of discrepancy.

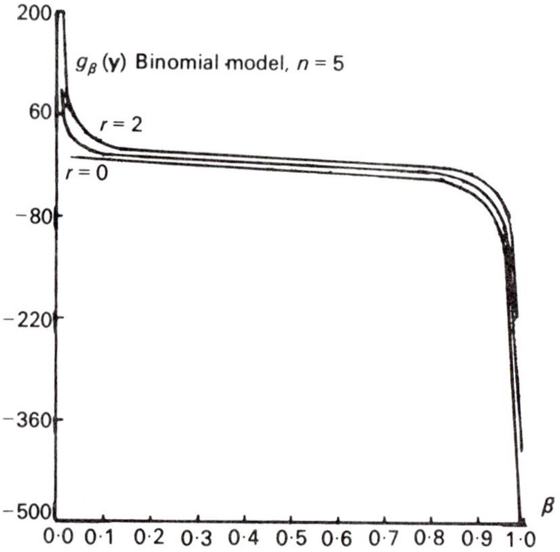

Fig. D3. Box's discrepancy.

In Figs D3 and D4 the behaviour of Box's $g_\beta(\mathbf{y})$ and the proposed measure $\{g'_\beta(\mathbf{y})\}^2$ is shown for the simple situation in which the value β_0 as the proportion of elements in a population which possess a given feature, is to be tested using a binomial model $p(\mathbf{y}\,|\,\beta) = p(r\,|\,\beta, n) \propto \beta^r(1-\beta)^{n-r}$. Here, one has

$$g_\beta(\mathbf{y}) = \frac{r}{\beta} - \frac{n-r}{1-\beta}$$

$$\pi(\beta) \propto \beta^{-\frac{1}{2}}(1-\beta)^{-\frac{1}{2}}, \quad \pi(\beta\,|\,r) \propto \beta^{r-\frac{1}{2}}(1-\beta)^{n-r-\frac{1}{2}},$$

$$E(\beta\,|\,r) = (r+1/2)/(n+1), \quad D^2(\beta\,|\,r) = \{(r+1/2)(n-r+1/2)\}/\{(n+1)^2(n+2)\},$$

$$\{g'_\beta(\mathbf{y})\}^2 = \frac{(n+1)(n+2)}{(r+1/2)(n-r+1/2)}\left(\frac{r+1/2}{n+1}-\beta\right)^2.$$

I believe most people would prefer the behaviour of $g'_\beta(\mathbf{y})$ to that of $g_\beta(\mathbf{y})$.

An even more attractive possibility which does **not** invoke approximate normality would be to test the data against an assumed *distribution* $p_0(\beta)$ (maybe centred in β_0). A good measure of discrepancy would then be the (positive, invariant) directed divergence

$$\int \pi(\beta\,|\,\mathbf{y}) \log \frac{\pi(\beta\,|\,\mathbf{y})}{p_0(\beta)}\, d\beta$$

where $\pi(\beta\,|\,\mathbf{y})$ is the reference posterior mentioned before.

I would like to finish by congratulating Professor Box by this very interesting, thought-provoking paper.

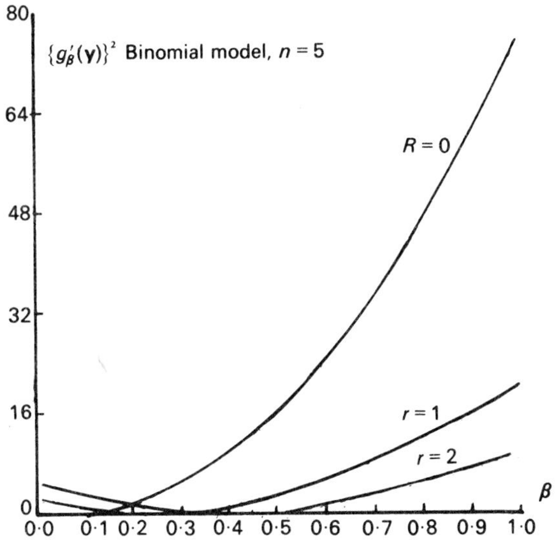

Fig. D4. Alternative discrepancy.

Professor Bovas Abraham (University of Waterloo, Canada): I wish to congratulate Professor Box for this excellent paper in which he discusses how Bayesian and Sampling Theory methods can complement each other in the scientific modelling process. It seems implied in this paper as well as in many others in the literature that the Bayesians do not perform model checking. I submit that any practically minded, and realistic Bayesian would perform diagnostic checks (residual plots, etc.) before using the appropriate posterior distributions obtained. After all, these distributions are derived under certain model assumptions. It is, therefore, comforting to see that this paper, using the concept of predictive distributions and checks, has formally justified the use of significance tests for model checking.

I agree fully with the author in his preference to parametric models over the *ad hoc* procedures suggested in the literature for "robustification". The parametric analysis has the flexibility of presenting the analyst with various alternatives. For instance, a parametric model containing a contamination parameter gives the analyst the opportunity to examine the analysis corresponding to various degrees (including zero) of contamination. Very often there is a dramatic difference in the analyses between models with extremely small and zero amounts of contamination. Of course this difference will not be revealed in an analysis of a model which excludes this possibility. This latter model is the same as the one which contains this parameter but with the value zero. It should be noted that this type of model including the contamination parameter also lends itself for further predictive checking.

I am sure that this paper will contribute significantly towards the progress of statistical science.

Professor PETER J. BICKEL (University of California): I very much agree with the formulation in this paper of the process of scientific inference and even with some of the critique of traditional sampling theory—the lack of recognition that prior assumptions are prior assumptions and specifying the class of distributions which the data are presumed to come from (the parametric model) may be a giant step compared to the further specification of prior distributions on the parameters.

Where I differ from Professor Box is in his next step of requiring that prior beliefs on the parameters are made precise in terms of prior distributions and that subsequent inference be expressed in terms of corrections to such prior beliefs. It seems to me there is middle ground between what he calls implausibly imprecise prior knowledge i.e., Haar priors or global minimax criteria and proper Bayesian priors and posterior inference. These types of compromises have been explored by Hodges and Lehmann (1952) and Efron and Morris (1971) among others. I am also exploring such compromises.

I find his use of tests based on the predictive distribution very stimulating though perhaps a little inconsistent with what I think of as a purist Bayesian point of view which should try to find a prior on the space of all possible models. While the predictive distribution approach is very useful from a Bayesian point of view it is not clear to me what advantages it possesses over the classical tests of model adequacy. It can fail even as they do in detecting departures which can lead to overconservatism in inferences about parameters of interest.

Suppose, for example, in the model of Section 2 that the assumptions are off in two respects.

1. The observations are not quite normal say they are better approximated by a long-tailed symmetric density f.

2. The true variance σ_0^2 (say) is less than σ^2, the assumed variance.

Then the significance test based on (2.6) leads to large (non-significant) values of α. To see this note

$$(n-1)\frac{s_d^2}{\sigma^2} \simeq \sum_{i=1}^{n}\left\{\frac{y_i-\theta}{\sigma}\right\}^2$$
$$= n\frac{\sigma_0^2}{\sigma^2}+O_p(\sqrt{n}).$$

Since $\chi_{n-1}^2 = (n-1)+O_p(\sqrt{n})$, $\sigma_0^2 < \sigma^2$ implies that α will tend to be larger than $\frac{1}{2}$ for n large.

On the other hand, the true posterior distribution of θ will be approximately normal $(\hat{\theta}, I(f)/n)$ where $\hat{\theta}$ is the MLE for f and $I(f)$ is the Fisher information.

A posterior probability such as

$$P\left[\bar{Y}+\frac{\sigma}{\sqrt{n}} \leqslant \theta \leqslant \bar{Y}+\frac{\sigma}{\sqrt{n}} \,\middle|\, Y_1, \ldots, Y_n\right]$$

will tend to a limit law, that of,

$$V = \Phi((A+\sigma)\sqrt{(I(f))}) - \Phi((A-\sigma)\sqrt{(I(f))}),$$

where $A \sim$ normal $(0, \sigma_0^2 - I^{-1}(f))$ (in this case).

Clearly inference is still conservative on the average (Box's robustness of validity more or less)

$$E(U) = 2\Phi(\sigma/\sigma_0)-1$$

but intervals centred on a robust estimate with length proportional to its estimated standard deviation could be much shorter and also accurate.

I realize that the procedure discussed in 5.2 in which the prior takes other models into account should do better (though I am a bit worried about masking of bad values by one another). However, the point I am making is that these methods have the same shortcomings as purely frequentist methods do. In fact they can do worse. In the example I just discussed a standard goodness of fit test of normality with variance σ^2 would have (for very large sample sizes!) detected the lack of fit of the model. The approach in Section 5.2 is a help with the "bad value—long tails" problem but so is Huber's and I guess I do not see why one "fix" is preferable to the other.

Professor BRUNO DE FINETTI (University of Rome, Mathematical Institute): I have always felt myself to be on the whole in agreement with the point of view of G. E. P. Box, and this paper seems to confirm strongly such a feeling.

Its illustration of "Bayesianism"—interpreted not simply as a systematic (and maybe sometimes careless) application of a formula—should make clear that what is essential is an exploratory way of thinking and of weighing the evidence. This involves considering when to use or not to use theorems and formulae rather than ideas expressible in words or simply at the level of instinctive propensities.

The Bayesian way of thinking must be followed everywhere, but this happens also under the guidance of unconscious mental processes (often better than conscious ones). Mathematical machinery may be a proper tool in some cases (those, for example, with a high degree of complexity, but which permit exploration), but not in others: in conditions of lack of knowledge, of haste, or of pain, its use might be disastrous.

Professor A. P. DEMPSTER (Harvard University): Professor Box displays an important trail of evolution over the 25 years since R. A. Fisher's *Statistical Methods and Scientific Inference.* The major change of principle is that scattered varieties of likelihood and fiducial methods are subordinated to Bayesian inference as the central principle of good estimation. The major practical change is a greatly heightened concern for potentially damaging effects of model-dependence, while holding firm on the necessity of models. I am in basic agreement with Box.

In the next 25 years we need much more emphasis on what a range of models tells us about the real world as partially reflected in a data set, and less on parameter estimation. The computing revolution will gradually but substantially expand the tool kit of models for which *a posteriori* analysis is available. I am pessimistic about the prospects for robustification, since an extended range of available models may show that nonrobustness is too often built into the limitiations of data and design. Checking sensitivity of results to additional parameters is often inadequate, for there may be sensitivity to simple models rather differently parametrized. For example, the extended discussion of robustness of location estimates for symmetric populations has distracted attention from the less tractable but more important task of estimating the population mean from asymmetric long-tailed populations. It hardly behoves us to be "grudging" in the pursuit of such problems.

If statisticians are to address real questions they will need formal structures to assess and incorporate sources of knowledge outside the current data set, whether from related data sets or rational arguments based on tentatively accepted science. I look to extended development of such formal structures over the next quarter century.

In summary, I applaud Professor Box's report of progress, but am impatient for more revolutionary changes ahead.

Mr R. GATHERCOLE (University College London): In general, the use of predictive checks is to be commended. Certainly, any statistical analysis should contain a review of the prior assumptions, with the aim of better understanding of the process in hand. It is useful to see the use of the Predictive Distribution in this paper, and I would like to add some remarks, together with some further ideas.

With reference to the example of Section 2, the batch mean, the preoccupation with the batch mean implies that the decision-maker has a quadratic loss function. Thus, the actual loss incurred in using the prior mean is of magnitude 36, and the expected loss is of magnitude 2·25. This as a ratio appears in the first half of the test statistic $g(y_d)$. With the same air of vagueness that surrounds the test statistic, we can use the discrepancy in the two losses to show the weakness of the prior as a predictive model. Now consider the case where the data are a random permutation of the sequence 70, 70, 70, 74. The posterior distribution is $N(70{\cdot}8, 0{\cdot}22)$, which is (informally) quite reasonable in its proximity to the prior. The test $g(y_d)$ yields a value of 12·44. Considering that $\Pr(X^2(4)\,11{\cdot}07) = 0{\cdot}05$, what inference do we draw now about the prior?

For those of us who want to test our assumptions against completely specified alternatives it may be of interest to consider using a loss function of the sort;

$$L(d,\theta) = 1, \quad \text{if } |d-\theta| \leqslant b,$$
$$= 0, \quad \text{otherwise.}$$

This is a step loss function, gauge b, and it is only a convenience that the upper bound of the loss is one. This is not dissimilar to establishing a confidence region. In particular, a "$1-z$" confidence region may be interpreted as a decision rule with constant risk, z. This loss function has the bonus that it may be used as an indicator of how well a model performs. If we have two competing prior formulations, we may want to choose a value of the gauge that will maximize the difference in their respective expected losses. As an illustration, take two models which are univariate normal, with respective variances, c_1 and c_2. Under the stated criterion, the optimal choice of gauge is

$$b^* = [\{c_1 c_2 \,.\, \ln(c_1/c_2)\}/(c_1-c_2)]^{\frac{1}{2}}.$$

If we have n models with ranked variances $c_1, \ldots, c_n$, then the same formula applies, with c_n substituted for c_2.

Although this may be of little worth in "one off" cases, such as the batch mean, in sequential sampling, there will be established a string of losses (binary) for each model, which can be used in conjunction with the familiar, conventional odds ratio etc. to make an informed decision about the appropriateness of a model. No such analogy exists with validation under unbounded loss.

Professor SEYMOUR GEISSER (University of Minnesota): Professor Box considers a variety of model checking functions on the joint predictive (marginal) distribution of the entire data set. One can carry this further and calculate

$$p(\mathbf{y} \mid A) = p(\mathbf{y}_2 \mid \mathbf{y}_1, A)\, p(\mathbf{y}_1 \mid A), \tag{1}$$

where $\mathbf{y} = (\mathbf{y}_1, \mathbf{y}_2)$ for subsets $\mathbf{y}_1$ and $\mathbf{y}_2$. Suppose it is supected that the discrediting of the model is due to some outliers or "bad" observations say, the set $\mathbf{y}_2$. One can check first whether the set $\mathbf{y}_1$ adheres to the model in Boxist fashion and assuming it does, then conditional on $\mathbf{y}_1$, we can now test whether $\mathbf{y}_2$ would have been an appropriate "prediction" from the model after observing $\mathbf{y}_1$, without necessarily a discernible alternative in mind. For example, if there is one observation at issue, e.g. $\mathbf{y}_2 = y_j$ and the model is as given in Section 2, then Y_j conditional on $\mathbf{Y}_{(j)} = \mathbf{y}_{(j)}$ which represents the set $\mathbf{y}$ with y_j deleted, is normally distributed with

$$\mathrm{E}(Y_j \mid \mathbf{y}_{(j)}) = \frac{(n-1)\,\sigma_\theta^2\, \bar{y}_{(j)} + \sigma^2\, \theta_0}{(n-1)\,\sigma_\theta^2 + \sigma^2} = \mu_j, \tag{2}$$

$$\mathrm{var}(Y_j \mid \mathbf{y}_{(j)}) = \frac{\sigma^2(\sigma^2 + n\sigma_\theta^2)}{\sigma^2 + (n-1)\,\sigma_\theta^2} = \sigma_j^2, \tag{3}$$

where $\bar{y}_{(j)}$ is the mean with y_j deleted. Note that in this distribution, as σ_θ^2 grows, μ_j tends to $\bar{y}_{(j)}$ and σ_j^2 to $\sigma^2(1+1/(n-1))$. So analysis based on this distribution may still be useful in the so-called "precise measurement" situation.

Now suppose the four observations were given as

$$y_1 = 71, \quad y_2 = 68, \quad y_3 = 69, \quad y_4 = 76 \quad \text{with} \quad \sigma^2 = 1, \quad \theta_0 = 70, \quad \sigma_\theta^2 = 1.$$

The Boxist computation

$$\bar{y}_4 = 71, \quad 3s_4^2 = 38, \quad g(\mathbf{y}) = \frac{(71-70)^2}{1{\cdot}25} + 38 = 38{\cdot}8, \tag{4}$$

presumably discredits the model. Notice that all of the so-called discreditation accrues to the second part of (4)—the component with three degrees of freedom. On the first glance one might be tempted to patch up the model by increasing σ^2. But an increase in σ^2 while reducing the second component also reduces the first component to such a degree as to make it unbelievably small unless one displaced θ_0. One could, in many instances, vary θ_0, σ_θ^2 and σ^2 simultaneously so that a compatible model resulted. However, the danger of deluding oneself by such Procrustean activities is obvious. In looking at the data however, one is immediately struck by y_4 which appears to be somewhat discrepant from the other three observations.

Checking to see if the first three observations discredit the model, we obtain

$$g(y_1, y_2, y_3) = \frac{(69{\cdot}33 - 70)^2}{1{\cdot}33} + 2{\cdot}43 = 2{\cdot}8,$$

which is certainly in line with a χ_3^2 random variable. Now notice that Y_4 conditional on y_1, y_2, y_3 has

$$E(Y_4 \mid y_1, y_2, y_3) = 69{\cdot}5, \quad V(Y_4 \mid y_1, y_2, y_3) = 1{\cdot}25,$$

so that the observation $y_4 = 76$ is more than 5 standard deviations from the centre of its conditional predictive distribution. So perhaps the model may not be wrong but y_4 may be aberrant. One then might look for reasons why y_4 appears aberrant if one had confidence in the original model. This type of analysis could be of some value in the "precise measurement" situation where the model was assumed true but one or more observations were made under suspicious circumstances and then aberrancy needed checking. Since Box is not averse to probabilities affected by transformations, he could also entertain a diagnostic

(suppressing A)

$$d_j = \frac{p(\mathbf{y})}{p(\mathbf{y}_{(j)})} = p(y_j \mid \mathbf{y}_{(j)}), \tag{5}$$

to hunt for an observation which may be aberrant from the assumed model. In the "precise measurement" case, neither numerator or denominator need exist whilst the ratio often does.

Another diagnostic put forth by Johnson and Geisser (1979, 1980) compares the predictive distribution $p(z \mid \mathbf{y})$ of a future observation z based on $\mathbf{y}$ with the predictive distribution $p(z \mid \mathbf{y}_{(j)})$ based on $\mathbf{y}_{(j)}$. One natural measure for how these distributions differ is the Kullback–Leibler–Good–Turing measure of divergence:

$$I(p, p_j) = \int p(z \mid \mathbf{y}) \log \frac{p(z \mid \mathbf{y})}{p(z \mid \mathbf{y}_{(j)})} dz, \quad j = 1, \ldots, n. \tag{6}$$

A single "bad" observation should yield an $I(p, p_j)$ which stands out from the rest. The first diagnostic d_j stresses how poorly will the "bad" observation be predicted given the remaining data; the second $I(p, p_j)$ can be considered an overall measure of the influence of y_j for predicting future observations from the data. It reflects how the predictive distributions differ with and without the "bad" observation—which is usually the more critical issue in regard to the ultimate use of the data. Sometimes even if $I(p, p_j)$ is an order of magnitude above the others it still may be so small that the discrepancy in the probability that a future observation lies in a particular region of interest with and without the offending observation is negligible enough so that its inclusion is of little or no consequence. How it affects what you are going to conclude, infer or decide is the final arbiter of the observation's influence.

Box notes that for the Bayesian, the "bad" value problem is in a sense manageable by modelling. For one who is predictively oriented the "bad" value problem depends on what the predictive intent is. For example, bad values may come about because of circumstances that may be considered *sui generis* in a particular experiment so that no modelling apparatus is appropriate and the intent is to make a prediction that does not permit the recurrence of such circumstances. Such observations then are to be excised from influencing the predictive distribution of a future observation. Contaminated models are of a different nature and two views are possible towards them. For example, if you assume the sampling in the experiment was from

$$f(y \mid \alpha, \beta, \theta) = (1-\alpha) f_1(y \mid \beta) + \alpha f_2(y \mid \theta),$$

and there is no reason to suppose that your future observations will not be from this model, then you calculate

$$P(y_{n+1} \mid y_1, \ldots, y_n) \propto \int \prod_{i=1}^{n+1} f(y_i \mid \alpha, \beta, \theta)\, g(\alpha, \beta, \theta)\, d\alpha\, d\beta\, d\theta,$$

where $g(\alpha, \beta, \theta)$ is the prior density, and that is an end to the matter.

On the other hand, as seems to be implicit in many situations if it is your intention to predict a value from $f_1(y \mid \beta)$ then the appropriate density is

$$P_1(y_{n+1} \mid y_1, \ldots, y_n) = \int f_1(y_{n+1} \mid \beta)\, p(\beta \mid y_1, \ldots, y_n)\, d\beta$$

where $p(\beta \mid y_1, \ldots, y_n)$ is the posterior marginal density of β. As other models arise they can be dealt with as long as the predictive intent is clear.

With regard to robustness, I believe that Table 1 is incomplete without a demonstration of the effect α or ε has on the predictive distribution of future observables at appropriate values of the independent variables.

Formula (4.1) stressed by Geisser (1969, 1971) for comparing alternatives was modified by Geisser and Eddy (1979) to

$$\prod_{j=1}^{n} p(y_j \mid \mathbf{y}_{(j)}, A_1) \Big/ \prod_{j=1}^{n} p(\mathbf{y}_j \mid y_{(j)}, A_0)$$

for "precise measurement" cases and is also viewed as a sample reuse procedure.

Professor Box has allowed us a glimpse at the contents of his artful and eclectic box of techniques and though purists of various persuasions may perceive a similarity to the legendary Box, those less apprehensive will enthusiastically welcome further revelation.

Professor V. P. GODAMBE and Mr P. FERREIRA (University of Waterloo, Canada): We are thankful to Professor Box for explaining in so many details how the predictive distribution could be used for *model criticism.* His "model" includes a *prior distribution* (Section 1.2) and "criticism" includes *model modification* (Section 1). Thus criticism must also include estimation of the prior distribution, partly or fully. This points to the following problematic area concerning Bayesian logic and practice.

With a notation slightly extended from the author's, let a model $M \equiv (X, \Omega, \mathscr{P}, \xi)$ where $X = \{x\}$ is the sample space, $\Omega = \{\theta\}$ the parameter space, $\mathscr{P} = \{p_\theta : \theta \in \Omega\}$ the class of distributions and ξ is a (prior) distribution on Ω. Assuming all distributions here to be discrete we can write the predictive distribution $p_\xi(x) = \Sigma_\Omega p_\theta(x)\xi(\theta) \ldots (1)$. If $x = x_0$ is the (present) data, one may estimate ξ by maximizing $p_\xi(x_0)$ in (1) for the variations of ξ. Similarly we can estimate θ from $p_\theta(x_0)$. If $\hat{\xi}$ and $\hat{\theta}$ $(\hat{\theta} \in \Omega)$ are the corresponding estimates we have

$$[p_{\hat{\xi}}(x_0) \geqslant p_\xi(x_0), \forall \xi \text{ and } p_{\hat{\theta}}(x_0) \geqslant p_\theta(x_0), \theta \in \Omega] \Rightarrow [\hat{\xi}(\hat{\theta}) = 1].$$

That is, the maximum likelihood estimate $\hat{\xi}$ of the prior distribution ξ will have all its mass concentrated at the maximum likelihood estimate $\hat{\theta}$ of θ. It then follows that $p_{\hat{\xi}}(\theta = \hat{\theta} \mid x_0) = 1$. Thus one may as well use just the maximum likelihood estimate of θ ignoring the Bayesian prior distribution and the methodology completely.

The above extreme example illustrates the general problem. There is nothing in Bayesian logic which can tell in any given situation how to distinguish the "prior or past knowledge" on the one hand and the "present data" on the other (Godambe, 1974, 1980). Yet such a distinction surely underlies all the conventional Bayesian applications; the past knowledge is used (informally) to construct a prior distribution and the present data to compute (formally) the posterior distribution. Worse still, it is not unusual to see a model constructed after (what is judged rightly or wrongly to be) the present data, are at hand. The data suggest some interesting investigation leading to a construction of an appropriate model.

Another related question concerns the *Bayesian options* (Section 4.6), for choosing between the alternative models. What should one do if the significance testing based on the predictive distribution (Section 2.1) rejects model M_1 but not M_2 while the posterior odds ratio based on some assumed (Bayesian) prior probabilities for the models M_1 and M_2 prefers M_1 to M_2? Such a situation can certainly arise in practice. Then one could not avoid the question, which comes first, the result of significance testing or the assumed prior probabilities?

Professor Box's suggestion for the use of the predictive distribution is unquestionably very persuasive. But its satisfactory implementation, obviously depends on how one resolves the problematic situation mentioned in the above paragraphs.

Professor PETER J. HUBER (Harvard University): The first and main part of Professor Box's paper is a masterful exposition of the learning process in applied statistics, explaining how knowledge is extended by a spiralling move through modelling, design, data acquisition, data analysis, inference, then again model modification, with various minor loops, and so on. It is certainly the best such article I have ever seen.

I cannot extend this praise without reservations to the last section of the paper, on robustness from a Bayesian viewpoint. I have often wondered why Bayesian robustness did not develop as vigorously as its non-Bayesian counterpart; after all, the term "robust" was coined some 27 years ago by Professor Box, and the early childhood of robustness had quite some Bayesian flavour. For me (and also for Frank Hampel, see Hampel (1973), p. 95) some open robustness problems seem to be ideally fit for a Bayesian approach; I can only surmise they were let lie fallow because some of the goals of robustness may clash with some of the Bayesian dogmas—if the latter are interpreted narrowly. The present paper seems to offer some interesting clues on these "theological" difficulties.

The first issue is rather deeply seated in the foundations and concerns a curious misunderstanding about the "frequentist" probability interpretation. Professor Box alludes to it in Section 1.2, but I think he misses the real point. It is not that the frequentist is unaware of the subjective nature of his probability interpretation, he is only more restrictive than the Bayesian and admits a subjective interpretation only for probabilities which are sufficiently close to zero or one. For intermediate probabilities, where he does not have a direct interpretation, he takes recourse to the weak law of large numbers. This is tersely and lucidly enunciated in Kolmogorov's basic "Ergebnisheft" (1933) and the passage is worth quoting in full (from the English translation (1956), p. 4):

"Under certain conditions, which we shall not discuss here, we may assume that to an event A which may or may not occur under conditions C, is assigned a real number $P(A)$ which has the following characteristics:

(a) One can be practically certain that if the complex of conditions C is repeated a large number of times, n, then if m be the number of occurrences of event A, the ratio m/n will differ very slightly from $P(A)$.

(b) If $P(A)$ is very small, one can be practically certain that when conditions C are realized only once, the event A would not occur at all."

Thus, in a certain sense, the frequentists constitute a most austere fraction of Bayesians!

As a consequence, modelling by frequentist dogma is restricted to choosing either a single point, or more generally, a point set in a space of probability distributions, in which the unknown true element lies with practical certainty.

At the bottom level, the Bayesian has more freedom: he can choose both a pointset and a prior distribution supported by it. But by Bayesian dogma, he must formalize his ignorance in terms of a prior distribution. This can create problems on the next higher level with robustness, since it is well-nigh impossible to condense some beliefs ("I expect up to about 5 per cent gross errors, which could be anywhere") into a single prior distribution, without making very arbitrary choices; these choices might matter more than what intuition tells us.

Just as in topology, it is not possible to specify smallness in absolute terms. The widespread use of 5 per cent levels in statistical testing shows that for many purposes a probability 0·05 is already close enough to zero for an investigator to accept something as a practical certainty (at least provisionally, as a working hypothesis). Sometimes, 10^{-9} is not small enough. But from the point of view of interpretation, it is absolutely essential that sufficiently small probability values are considered to be topologically close to zero. The statements made by Box in the first part of Section 5 (on $\alpha = 0{\cdot}001$) in essence negate Kolmogorov's interpretation of probability. If we take them literally, they would even knock off the props underneath the practical translation of probabilities into actions (does the probability of a fatal accident have to be exactly zero for you to risk a plane trip?).

The second issue is related and is specific to the foundations of robustness. It takes off from the very opening line of the paper: "No statistical model can safely be assumed adequate." Among other things, this dictum (with which I wholeheartedly agree) implies that the infinite regress of model improvement has to be cut short somewhere (by a judicious application of Occam's razor), and that one has to rely on robustness to take care of the remaining inadequacies.

Here Bayesian robustness fails by too strictly adhering to the dogma that uncertainty has to be formalized by a prior distribution, pulled from one's mind by introspection. This is curious, because otherwise Bayesians are much less strict and often choose priors out of mathematical convenience (e.g. the so-called conjugate priors).

Typically, Bayesians try to achieve robustness with a last model, which contains some contamination or tail-length parameters, together with some prior on these parameters. Let us call this the "super model". The super model is somewhat arbitrary, because typically neither the data nor the prior knowledge offer adequate information at this late modelling stage. Estimation then proceeds in the pious but unwarranted hope that the super model provides robustness. If the statistician has not tired yet, and still is keeping up the good work, he should note a possible lack of robustness in the next criticism stage and he will only waste a few iterations of the learning process.

Such an approach may have been the best available until about 10–12 years ago, but in between the demands and potentialities of robustness have progressed a stage further. Essentially, by now the Bayesian approach should be concerned not with the *ad hoc* construction of super models, but with deriving reliable guide-lines on how to choose the super model (within the inherent arbitrariness) so as to guarantee robustness, and how to do so in a best possible fashion. The paper by Rubin (1977) is a first small step toward such a goal.

All this has to do with modelling within a last speck of probability, just before both the frequentist and the Bayesian would switch to practical certainty, and I believe that in this region some ideas that ordinarily are anathema to Bayesians (like minimax strategies) may be much preferable to picking arbitrary distribution out of one's zone of indifference. There may lie the cause of my disagreement with Professor Box: while the $\alpha = 0{\cdot}001$ in the beginning of Section 5 may be too large to be lumped with zero, it is too small for a reliable, data- or intuition-based modelling. If we cannot change the model to approximate what is believed, because our belief is too fuzzy to stand a 1000-fold magnification, we better change the model into the worst case compatible with our belief (namely, the belief that there might be a fraction of up to about α unspecified bad values). It is curious that the Bayesian version of this (i.e. formalizing the belief about α in terms of a non-degenerate prior) to my knowledge has never been seriously explored.

Professor W. G. HUNTER (University of Wisconsin): I am reminded of something said some time ago by J. Williard Gibbs: "One of the principal objects of theoretical research in any department of knowledge is to find the point of view from which the subject appears in its greatest simplicity." I believe that Professor Box has taken an important step in that direction. He has found a vantage point from which it seems sensible essentially to divide statistical real estate between those who use Bayesian methods and those who use sampling theory. (It will be interesting to see to what extent this proposal actually leads to a reduction in territorial disputes that have enlivened the statistical landscape over the years.) Note that adjoining lands are occupied by experimenters and other research workers. I would like to comment on the value of randomization as perceived by these three different groups.

Leaving aside refinements, the gist of the story is as follows. Suppose the experimenter gives the Bayesian the data necessary for estimating the parameter $\boldsymbol{\theta}$ in a cause-and-effect model A that purportedly connects $\mathbf{x}$ to $\mathbf{y}$. Suppose, as is usually the case, that the variance σ^2 is unknown. The Bayesian's job is to produce the posterior distribution $p(\boldsymbol{\theta} \mid \mathbf{x}, \mathbf{y}, A)$ for the experimenter, and the residuals $\mathbf{y}-\hat{\mathbf{y}}$ for the critic. The critic is provided, by the experimenter, with additional information about how the data $(\mathbf{x}, \mathbf{y})$ were collected, including measurements on variables not contained in model A. The critic's job is to report to the experimenter whether any defects can be detected in A. The Bayesian uses (1.5), the critic (2.3).

Given $\mathbf{x}$, $\mathbf{y}$, and A, the added knowledge about whether randomization was used would not influence any of the Bayesian's calculations. To the Bayesian, therefore, randomization has no value. By contrast, randomization stands between the critic and unemployment and so, to the critic, it has high value. The critic's job simply cannot be done if the data have not arisen from a randomized experiment. The critic must consider whether the residual vector $\mathbf{y}-\hat{\mathbf{y}}$ resembles random error, in particular, whether it can reasonably be regarded as an approximation to a random sample from the error distribution $p(\boldsymbol{\varepsilon})$. Unlike the Bayesian, the critic does not adhere to the likelihood principle. In calculating a significance level, which is the critic's stock-in-trade, a reference distribution is required. Consequently, the critic must consider information other than $\mathbf{x}$ and $\mathbf{y}$, basically for the same reason that Samuel Johnson said that "among the works of Nature no man can properly call a river deep, or a mountain high, without the knowledge of many mountains, many rivers".

The experimenter wants information about $\boldsymbol{\theta}$ and A, and the statistical task is finished when the Bayesian supplies the posterior distribution for $\boldsymbol{\theta}$ and the critic reports that no defects can be found in A. But the Bayesian and the critic are not infallible; even if they make no mistakes in calculation and conclude that A does not appear to be inadequate, A in fact may be grossly inadequate for future use because of the presence of an undetected lurking variable x_0. Note, in this eventuality, that randomization helps the experimenter in two ways.

(1) It tends to produce an orthogonal "design" in the sense of making the lurking variable vector perpendicular to the space defined by the vectors of the x's in the model. Suppose measurements on x_0 are available but they have thus far been ignored in the analysis of the data. If they are later brought forward, then randomization having been used will improve the experimenter's chances of detecting x_0's importance. In practice, if randomization is not used, the vector $\mathbf{x}_0$ will often tend to be close to the space defined by the vectors of the x's in the model (because of internal feedback or regulatory mechanisms or other linkages within the system being studied); accordingly, to discover the effect of x_0 will be extremely difficult, if not impossible.

(2) Even if the existence of x_0 remains undetected and it thereby biases the estimate of $\boldsymbol{\theta}$, proper randomization will tend to reduce the amount of this bias. The price paid will be to increase the variance of the posterior distribution of $\boldsymbol{\theta}$. This is a desirable trade-off. Note that the Bayesian and the critic are blind to advantages (1) and (2).

Thus, randomization should be used by the experimenter, even though the Bayesian cannot see any sense in it. Although the critic knows it is a good idea, the critic does not realize how good it is. Rather than use either Bayesian (B) or sampling theory (ST) methods, a statistician should use both. A statistician will be even more effective by learning at least some elements of the subject matter field from which the data arise, thereby becoming somewhat of an experimenter (E), too. This combination is clearly best.

Mr P. H. JACKSON (University College of Wales, Aberystwyth): I wonder whether Professor Box could be persuaded to substitute the expression "the model is called into question" for "the model is discredited" throughout Section 2 of the paper?

Practising statisticians of any school of inference will react to surprising data by asking "Have I overlooked something?", and the formal checks for surprising features proposed will be valuable in provoking this question. Usually the next questions to ask oneself are "Have I misunderstood the data?" and "Do the data contain gross recording errors?", followed by "Am I using an inappropriate model?" In considering the last question the Bayesian will recall that his theory requires him to assign prior probabilities to *all* states of nature (models in the present application), whereas his human finiteness has led him to assign zero probability to models which should really have received at least epsilon. He will therefore enquire whether there are models for which the likelihood ratio calculated from the data would have increased even an epsilon's worth of prior probability to a posterior probability greater than that for the model employed. This is especially likely to be the case when, as in many examples given in the paper, parsimony rather than genuine prior belief was the reason for selecting the model.

The objection to saying that "the model is discredited" is that it implies a decision, not a reconsideration. It will be widely interpreted to mean that the model, regarded as a hypothesis, is rejected; Professor Box himself encourages this interpretation in 2.1(d). Suppose that in my first season as captain of a cricket team I lose the toss at all twelve matches of the season. An event has occurred for which almost any relevant tail-area calculation, frequentist or Bayesian, will give a probability less than 0·001. But what is discredited? The toss as a fair way to start a match? My suitability as captain? Certainly there are models which make the data less surprising: telekinesis, "bad vibes", or the ever-available divine intervention. Believers in any of these theories might consider it wise to relieve me of the captaincy. Most of us, however, having satisfied ourselves that there was nothing fishy about the way the tosses were conducted, would still conclude that I had simply had a run of bad luck; that is, we would assign such a small probability to these alternative models *a priori* that even data as extreme as this would not give them a large probability *a posteriori*.

Professor JOSEPH B. KADANE (Carnegie–Mellon University): I am in sympathy with much of what Professor Box has proposed in this paper, and yet there are parts I cannot accept. His claim for the need for both model estimation and model criticism is well taken. His discussion of examples is illuminating, and his remarks on robustness in the Bayesian context are important and insightful.

My difficulty with this paper lies in his proposal to resurrect significance testing, now to be done with respect to the predictive distribution, as a method of model criticism. My difficulty does not have to do with the use of predictive distribution itself. After all, an equivalent Bayesian analysis can be performed without mentioning parameters at all, using Bayes' Theorem to update the predictive distribution,

$$p(y_{n+1}, y_{n+2}, \ldots, | y_1, y_2, y_n) = \frac{p(y_1, y_2, \ldots, y_n, y_{n+1}, \ldots)}{P(y_1, y_2, \ldots, y_n)}.$$

Rather, my difficulty comes in interpreting his equation (2.3). How shall we choose an appropriate level α, below which we decide that the model is discredited? Apparently, from equation (2.6), 0·001 is too small. Why is that? What coherent theory can justify the use of such a critical value α_0, without reference to the size of the likely discrepancies, their impact on the conclusions drawn from the analysis, etc., in other words, without a full decision-theoretic treatment?

A mathematical way of putting the same question is to point out that α, as computed in (2.3) is not invariant to changes in the underlying measure μ with respect to which the density p is a Radon–Nikodym derivative. Thus if $\mu(x)$ is taken to be Lebesgue measure if $x<0$ and k times Lebesgue measure if $x \geqslant 0$, a different interval $p(y|A)<p(y_d|A)$ results for each k, and hence a different α. Which is to be used? Generalizations of this device can lead to α's arbitrarily close to 0 and 1, by changing μ. Does Professor Box wish to argue that only densities with respect to Lebesgue measure are legitimate? By contrast, the predictive ratio (4.1) is invariant to such changes, which suggests to me that it is on a more solid footing.

Dr RON KENETT (University of Wisconsin): A testimony to the illuminating nature of this work is that while reading it I kept asking myself why it is that this natural compromise between pure Bayesian inference and pure Sampling inference was not already widely recognized.

My comments are on the general nature of model building and specifically on paragraph 1.2 describing the need for prior distributions. I tend to agree with the author that for most problems attacked by statistical means "there seems no logical way to avoid trouble except by the explicit prior statement of the model we wish to entertain" but this might sometimes be impossible to achieve.

The alternative I have in mind stands half-way between Tukey's exploratory data analysis and Box's proposals. In other words, it seems to me that there are situations when no prior elicitation (even of a model) is possible but still there is some relevant prior information that can be used. If the model is the prior in the wide sense we might have a very "diffuse" prior. One might look at an analysis in such circumstances as a structural exploratory data analysis.

To illustrate my point let me describe an analysis in which I was involved (Karlin, Kenett and Bonne'-Tamir, 1979). Professor Bonne'-Tamir of the Human Genetics Department at Tel-Aviv University collected frequencies of various biochemical genetic traits in various Jewish populations living now in Israel and sharing a common origin such as Jews from Iraq, Poland, etc.... An investigative look at these data for similarities and differences between populations and between and within standard demographic–anthropological classifications of these populations can provide clues relevant to the study of genetic diseases and set a basic framework for successful genetic counselling. We had available information on the history of these populations, where they lived, cultural exchanges, migrations, admixture and more. A reasonable analysis of such data should incorporate this information but 2000 years of an eventful history cannot be summarized in a meaningful model. This situation does not give an initial stage for criticism to start iterating on.

My point therefore is that in some situations one will have to use *ad hoc* techniques incorporating features appropriate to the data, such as measures of particular meaning in genetics, as was done in the analysis mentioned above, and making use of prior information, such as historical knowledge, without putting it in an analytical model.

Professor TOM LEONARD (University of Wisconsin): I would like to add my congratulations to Professor Box for this highly creative landmark paper which pioneers the unification of the Bayesian and frequentist elements of parametric statistical methodology. It is by now fairly widely accepted that Bayes is very good when the statistician conditions upon the truth of his assumed sampling model; for example the dualities between Bayes and admissible procedures lead us to some of the best properties available under a frequentist philosophy. However, practical statistics is primarily concerned with aspects of modelling; whilst standard Bayesian approaches, based upon finite mixtures of specified models, are helpful, they do not seem to provide the final answer (e.g. the statistician needs to specify accurately several priors corresponding to several sampling distributions). Two alternative choices involve (a) proceeding pragmatically in the manner recommended by Professor Box, or (b) referring to a non-parametric procedure.

Some Bayesian non-parametric procedures (e.g. Ferguson, 1973; Leonard, 1978) parallel Professor Box's Bayes/non-Bayes compromise; they fit in very naturally with his important philosophy of iterative model-building. A hypothesized model is introduced as a prior estimate; the posterior estimate then indicates both local and overall differences from the hypothesized model and also suggests how the hypothesized model could be revised. One of the further prior parameters measures the degree of belief in the hypothesized model and parallel's Professor Box's significance level; it may itself be estimated from the data by either an empirical or a hierarchical Bayesian procedure.

A beneficial conclusion to be drawn by intermingling the ideas in (a) and (b) is that it might be a bit overambitious for a statistician to try to check out his model against a finite data set unless either (i) he also refers to the scientific background of the data (e.g. by interacting with a client), or (ii) he makes some particular assumptions (e.g. independence and homogeneity of errors) about the true model (or equivalently about available alternative models). In the context of non-parametrics suppose that the true density of the observation vector **y** is given by

$$f(\mathbf{y}) = \frac{\exp\{g_0(\mathbf{y})+\varepsilon(\mathbf{y})\}}{\int \exp\{g_0(\mathbf{u})+\varepsilon(\mathbf{u})\}\, d\mathbf{u}}, \tag{1}$$

where g_0 is the logistic transform of the hypothesized density

$$f_0(\mathbf{y}) = \frac{\exp\{g_0(\mathbf{y})\}}{\int \exp\{g_0(\mathbf{u})\}\, du}, \tag{2}$$

and the multi-dimensional function ε measures the departure of f from f_0.

The expression in (1) also provides the likelihood functional of ε and therefore concisely summarizes all information contained in the data about departures of f from f_0. It seems (e.g. by trying to maximize (1) with respect to ε) that this information can only be sensibly utilized on its own by making some specific assumptions about ε which effectively reduce its dimensionality. We could, for example, assume ε to take the form

$$\varepsilon(\mathbf{y}) = \sum_{i=1}^{n} \eta(y_i),$$

and then estimate the one-dimensional function η, e.g. by maximum likelihood or a Bayesian smoothing procedure or an approximation based upon a polynomial or a linear combination of basic splines. This involves an assumption analogous to independence of error terms of the true model. Without a simplifying assumption like this, on the true or alternative models, it seems that we would need to bring in information external to the likelihood functional of ε in order to reach a viable conclusion. This parallels Professor Box's choice of diagnostic statistics, which may be based upon background considerations; also his choice of significance level may be made pragmatically.

In summary, either a Bayes/non-Bayes compromise or a suitably chosen non-parametric procedure is useful in modelling situations. George Box's brilliant approach will prove historically to be a splendid addition to this area.

Professor D. V. Lindley (Somerset): Professor Box would have us abandon the likelihood principle when it is a question of testing the adequacy of fit of a statistical model. Even outside the Bayesian paradigm, the arguments in support of the principle by Birnbaum (1962) and Basu (1975) are most convincing and it would be interesting to know why they have been implicitly rejected tonight. *Any* test of a model surely requires some consideration of alternatives as the following example illustrates. A statistician judges a sequence of 12 0's and 1's to be Bernoulli (the model). On observing the sequence he sees 010101010101 and a plausible alternative immediately suggests itself. Suppose, however, he had observed 111010100010, then the alternative that trials of prime (composite) order always give a 1(0) is not seriously entertained because it has low probability. Yet both these sequences have the same probability on the model.

A curious feature of abandoning the likelihood principle is the need to appeal to aspects of the data previously thought to be irrelevant, namely the data values that were not obtained originally but might have been. For example, (2.3) implicitly assumes the sample size was fixed. Curiously, were the alternatives themselves to be parameterized, this information would again not be necessary.

The bulk of the paper is not, however concerned with such issues, but with the task of developing workable tests for the adequacy of a model and, master craftsman that he is, Box succeeds admirably. It is interesting to notice that the valuable procedures derived from the $g_\beta(y)$ criterion scarcely depend on the viewpoint adopted. Only casually, as at the end of Section 4.5 when a Monte Carlo study is mentioned, does the sampling attitude surface. And even there, under rather special assumptions, the procedure has a Bayesian interpretation given by (4.22). (Incidentally, in that argument, would it be reasonable to suppose $p(\beta)$ flat since the use of β_0 rather suggests high probability for that value?) The sampling approach is typically quite satisfactory in suggesting a criterion to consider—for, after all, even it admits that the only solutions worth considering are Bayesian solutions—where it fails is in saying what to do with the criterion, or with the plots. Tail-areas are preferred to probability ratios and it is there that the sampling argument is dangerous. Let us use the valuable results derived in tonight's paper but let us judge them by coherent standards.

Dr Robert B. Miller (University of Wisconsin): Our purpose as scientists is to draw conclusions about the world from data. Scientists engage in modelling in order to guide the data analysis and data collection processes, which in turn inform the process of drawing conclusions. Thus a model must be viewed as an expression of a scientist's thinking at a particular moment and not as an objective entity. The fact that a model is widely accepted among scientists makes it neither objective nor true, only widely accepted. The only objective entity is data, and even data are corrupted by measurement error, selection bias, processing error, etc.

Professor Box rightly reminds us that the predictive distribution provides a formal mechanism for checking conclusions against objective evidence. I have reservations about equating this mechanism with sampling theory. Whereas contemplation of hypothetical, identical repetitions of an experiment is very useful in model conception, it does not enter so naturally into model validation. At least I cannot think of a natural sampling theory way to deal with structural shifts in either a model or its parameters or, at a more elementary level, with even the validation of a stationary time series model.

I believe a very important principle is embodied in Professor Box's definition of robustification as "judicious and grudging elaboration of the model to ensure against particular hazards". The principle is that robustness is more in the scientist's frame of mind than in any particular model. While statisticians will inevitably speak of robust models and robust procedures, they will, consciously or unconsciously, be speaking about the one who guides the application of these models and procedures toward conclusions about the world.

Perhaps robustness should be defined as a scientists' ability to ferret simple, lasting structures from data that are subject to myriad sources of variation. In keeping with this point of view it is well to remember that a surprisingly large variety of data patterns is consistent with the assumption of a fixed (and relatively simple) model structure if the parameters are allowed to fluctuate randomly over groups or over time or both. If we make a parameter process a part of our model, then Bayesian analysis becomes hierarchical. Some will say heretical, but I believe this point of view is the best hope for robust data analysis in such volatile fields as business, economics, and environmental modelling. Significantly, random coefficient models already have wide currency in these fields.

Finally, I wish to thank Professor Box for his very useful contribution to both the philosophical and the practical sides of statistics.

Dr D. J. SPIEGELHALTER (University of Nottingham): Professor Box has suggested the use of Fisher's score function to measure discrepancies from a specific model assumption. One application is within the context of testing for the shape of a univariate distribution, when, for example, normality may be embedded in the exponential power family. For known location and assuming $p(\sigma) \propto \sigma^{-1}$, Fisher's score has a simple form, and for unknown location the statistic may be closely approximated by $\Sigma\, z_i^2 \ln z_i^2$, where $z_i = (x_i - \bar{x})/s$.

Similarly, the exponential shape may be embedded in the gamma or Weibull family and simple "locally", in a specific sense, most powerful invariant tests obtained. The sampling characteristics of these tests are currently under investigation.

Another Bayes/sampling theory combination involves using the posterior probability of a specific model, embedded in a finite family of alternatives, as a test statistic. This has been shown to be successful as a small sample test for normality (Spiegelhalter, 1977, 1980).

Professor S. M. STIGLER (University of Chicago): George Box has made a bold, and I think largely successful, attempt to spell out the compatible and complementary roles Bayesian inference and significance tests may have in scientific investigation. There is one point that I think is implicit in his *tour de force* that, I feel, deserves greater emphasis. He notes the importance to model criticism of the predictive distribution,

$$p(\mathbf{y} \mid A) = \int p(\mathbf{y} \mid \boldsymbol{\theta}, A)\, p(\boldsymbol{\theta} \mid A)\, d\boldsymbol{\theta}.$$

But the predictive distribution is, of course, not unique. Let $\mathbf{z} = \psi(\mathbf{y})$. Then the predictive distribution of $\mathbf{z}$ may be very different from that of $\mathbf{y}$, even though if ψ is $1-1$, the data given by $\mathbf{z}$ are equivalent to those given by $\mathbf{y}$. Different choices of ψ will render the significance test sensitive to different departures from the model. In the example that begins Section 2, we have a likelihood $p(\mathbf{y} \mid \boldsymbol{\theta})$ that is a function of a sufficient statistic $T(\mathbf{y}) = (\bar{y}, s^2)$; $p(\mathbf{y} \mid \theta) = h(\mathbf{y}) \cdot g(T(\mathbf{y}) \mid \boldsymbol{\theta})$, where here $h(\mathbf{y}) \equiv 1$. This yields (2.2), a predictive distribution that is also a function of $T(\mathbf{y})$, and thus a significance test that is only sensitive to model departures that perturb the distribution of $T(\mathbf{y})$. Many other formulations are possible, such as the one Professor Box cleverly exploits later in Section 2: in effect, he works in (2.7)–(2.12) with the predictive distribution of $\mathbf{z} = \psi(\mathbf{y}) = (T(\mathbf{y}), \mathbf{u})$. He is then led to a different set of significance tests, remarking that while the test (and its outcome) is affected by transformation, this is not particularly disturbing. I expect many of his readers will be disturbed, but I am not. I think he is quite correct in expecting different answers to different questions. Still, I think he leaves us with a dilemma, namely what question should we ask?

If the predictive distribution and the resulting significance test are affected in important ways by transformation (and they are), we need guidance in the choice of transformation. Professor Box has shared his experience and considerable intuition in providing some of the needed guidance, but I feel he should give more emphasis to a formal consideration of alternative hypotheses. The tests discussed are introduced without specific (only vague) alternatives in mind, but they are in fact likelihood ratio tests for specific families of alternative hypotheses (as can be seen from De Groot, 1973, for example). I would suggest that we could borrow a clue from Professor Box's last section, where he increases our understanding of Tukey's Biweight by relating it to a Bayesian procedure, and gain some of the needed guidance in choice of transformation by studying the classes of alternatives for which the suggested tests are likelihood ratio tests. Whether or not such an approach will be both feasible and enlightening remains to be seen, but I do think further guidance in choice of transformation (or, equivalently, choice of significance test) is needed. Otherwise, we risk asking either the wrong questions, or too many questions. In many important situations involving social data, non-stationary aspects of the underlying processes will prohibit any practical appeal to the iteration that Professor Box correctly notes (at the end of Section 4) will "quickly terminate" the chase in much industrial or laboratory experimentation.

Two centuries have passed since the first statistician, Laplace, embraced both Bayesian inference and significance tests. It is a pleasure to learn at last that this was not an adulterous relationship, but one that can be defended on principle.

Professor A. ZELLNER (University of Chicago): This stimulating paper prompts the following observations. First, I believe that a scientific model is better represented by $p(\mathbf{y}, \boldsymbol{\theta} \mid A) = p(\mathbf{y} \mid \boldsymbol{\theta}, A)\, p(\boldsymbol{\theta} \mid A)$ than by $p(\mathbf{y} \mid \boldsymbol{\theta}, A)$, the likelihood function since the form and content of $p(\boldsymbol{\theta} \mid A)$, the prior distribution are significant parts of any scientific model. Second, sampling theory considerations appear relevant before we observe the data in establishing sampling properties of Bayesian estimation and significance testing procedures, for example admissibility and average risk properties of Bayesian estimators and operating characteristics of Bayesian significance testing procedures based on posterior odds ratios; on this latter topic see, for example, Dyer (1973), Jeffreys (1967, p. 396) and Zidek (1969). After data are observed, posterior distributions and odds ratios provide a basis for inference about parameter values and hypotheses, sharp or non-sharp. Significance testing based on posterior odds ratios, mentioned briefly by Box in Sections 4.1 and 4.6 is a basic Bayesian procedure for model criticism when specific alternative hypotheses are available, as they usually are. Jeffreys (1967, Ch. V and Ch. VI) provides posterior odds ratios for many problems including hypotheses about a normal mean and standard deviation that are relevant for the normal mean problem that Box analysed in Section 2. Third, for any given sample of data, there will be some function of the observations, including the whole set that will be improbable or unusual given the model, a circumstance that led Jeffreys to write (1967, p. 385), "If mere improbability of the observations, given the hypothesis, was the criterion, any hypothesis whatever would be rejected." Then, without an alternative hypothesis, we are left with no model at all. Also, his critique of the rationale for the use of tail areas or P-values that Box employs warrants attention. Last, the basic idea of significance testing or model criticism is of great importance as Box emphasizes and thus it is fortunate that Bayesian posterior odds ratios are available for this important aspect of inference and that they generally have very good sampling properties.

The AUTHOR replied later, in writing, as follows.

I need hardly say how happy I am at the reception afforded my paper which I was particularly anxious to present here because of the unique vitality of this Society and its well known willingness to entertain and criticize ideas.

To clear up some misunderstandings and to set my reply in context, let me first make clear what I regard as the proper role of a statistican. This is not as the analyst of a single set of data, nor even as the designer and analyser of a single experiment, but rather as a colleague working with an investigator throughout the whole course of iterative deductive–inductive investigation. As a general rule he should, I think, not settle for less. In some examples the statistician is a member of a research team. In others the statistician and the investigator are the same person but it is still of value to separate his dual functions. Also I have tended to set the scene in the physical sciences where designed experiments are possible. I would however argue that the scientific process is the same for, say, an investigation in economics or sociology where the investigator is led along a path, unpredictable *a priori*, but leading to (a) the study of a number of different sets of already existing data and/or (b) the devising of appropriate surveys.

The objective taking precedence over all others is that the scientific iteration converge as surely and as quickly as possible. In this endeavour the statistician has an alternating role as sponsor and critic of the evolving model. Deduction, based on the temporary pretense that the current model is true, is attractive because it involves the statistician in "exact" estimation calculations which he alone controls. By contrast induction resting on the idea that the current model may not be true is messy and the statistician is much less in control. His role is now to present analyses in such a form, both numerical and graphical, as will accurately portray the current situation to the investigator's mind, and appropriately stimulate his colleague's imagination, leading to the next step. Although this inductive jump is the only creative part of the cycle and hence is scientifically the most important, the statistician's role in it may appear inexact and indirect.

If he finds these facts not to his liking, or if his training has left him unfamiliar with them, the statistician can construct an imaginary world consisting of only the clean deductive half of the scientific process. This has undoubted advantages. A model dubbed true remains so, all alternative models are known *a priori*, the likelihood principle and the principle of coherence reign supreme. It is possible to devise rigid "optimal" rules with known operating characteristics which aspire to elevate the statistician from a mere subjective artist to an objective automaton. But there are disadvantages. Deduction alone is sterile—by cutting the iterative process in two you kill it. What is left can have little to do with the never-ending process of model evolution which is the essence of Science.

My object then is to suggest a theory which can fully explain both the inductive and deductive statistical aspects of investigation. I argue that for this we need look *no further* than the factorization of the model $p(\mathbf{y}, \boldsymbol{\theta} \mid A_i)$, expressing all aspects of currently held belief at some stage i, into its Bayesian and predictive parts. In particular the predictive distribution $p(\mathbf{y} \mid A_i)$, since it is the distribution of all possible samples generated by the current model, provides, free from nuisance parameters $\boldsymbol{\theta}$, the appropriate reference set for $\mathbf{y}_d$ or for any diagnostic checking function $g(\mathbf{y}_d)$.

Some contributors have found confusing my speaking of a "sampling theory argument". I mean by this, reference of a function of the data to an appropriate reference distribution implied by the model as in a significance test. The problem of what *is* the appropriate reference set is completely resolved in the present context: it is the set defined by the predictive distribution. Notice that on this basis a "sampling theory argument" is no less so because a prior distribution is involved in the generation of the (predictive) reference set.

Although I do not think that Professor Barnard and I are far apart in our basic philosophy, I fear that his final remarks about subjectivity and relations with clients (or as I would say investigators) may be misunderstood. Surely what he says applies only to perhaps the last 5 per cent of the experimental effort when it is to be demonstrated that the final destination reached is where it is claimed to be. The other 95 per cent—the wandering journey that has finally led to that destination—involves, as I have said, many heroic subjective choices (what variables? what levels? which scales? etc., etc.) at every stage. So far as this major effort is concerned then, since we must swallow the subjective camel, why strain at the subjective gnat?

Professor Dawid complains that much of my analysis in Section 4 has reproduced classical tests, and asks what has been gained. The answer is that my proposals come from a theory having general application, in which the Predictive Eve is no longer separated from the Bayesian Adam. While highly specialized structure is not a requirement, when it exists it is appropriately used, and then reproduces sensible classical tests. So far as it goes I suspect my readers will find this more reassuring than if the contrary had been the case. For ordinary well-behaved predictive distributions zero will be the appropriate value for the score functions of Section 4 since this will imply that β_0 maximizes the predictive density.† In particular examination of the plots in Box and Cox (1964) and elsewhere verifies that, for all examples I have seen, this is sensible for $g_\lambda(\mathbf{y})$ of (4.7). The phrase "reference of $p\{g_\lambda(\mathbf{y}_d)\}$ to its sampling distribution" does not appear in the paper presented to this Society. It appeared in an earlier draft which Professor Dawid saw but was changed precisely because the earlier version did not make it sufficiently clear that the sampling involved was from the predictive reference set. It is quite evident in Sections 1, 2 and 3 that a formal check is made by referring the predictive densities $p(\mathbf{y}_d)$ and $p\{g_\beta(\mathbf{y}_d)\}$ to their predictive distributions, where necessary by computer simulation. These distributions cannot of course contain nuisance parameters.

Concerning (4.8), trials with actual samples showed that although local quadratic approximation of z was good, the approximation $B = \dot{y}^{-1}$ was exceptionally poor. It was not used therefore. Of course B is data dependent but the only purpose of the analysis leading to (4.11) is to show that (making only the assumption that z is locally quadratic in y), $g_\lambda(\mathbf{y})$ is a function not only of Tukey's statistic T_{12} but of T_{21} and T_{30} as well; as common sense says it ought to be. This analysis also sets Andrews' criterion in its proper context.

Suppose, following Professors Bernardo and Lindley my proposals are applied to the binomial, and for illustration, using Bernardo's notation, assume a beta function prior

$$p(\beta \mid A) = \beta^{m\beta_0 - 1}(1-\beta_0)^{m(1-\beta_0)-1} / B\{m\beta_0, m(1-\beta_0)\}.$$

† Which, with some advantages, replaces likelihood.

Then the predictive distribution is

$$p(r \mid A) = \binom{n}{r} B\{m\beta_0 + r, m(1-\beta_0)+n-r\} \Big/ B\{m\beta_0, m(1-\beta_0)\}.$$

Referral of $p(r_d \mid A)$ to this distribution can, without specification of alternatives, discredit the model and subsequent Bayesian analysis. In particular, as $m \to \infty$, $p(\beta \mid A)$ becomes concentrated at β_0 and the predictive check consists of the standard binomial significance test in which r_d is referred to

$$p(r \mid A) = \binom{n}{r} \beta_0^r (1-\beta_0)^{n-r}.$$

This is despite Lindley's remark that any test of a model requires alternatives. However, as I say, although such an analysis can discredit the model, it cannot adequately check it. In particular suspected deviations, such as that implied by Lindley's example, from binomial sampling would require checking functions which, I grant, would imply specific alternatives. In particular a suitable function of the data $g_\delta(r)$ with δ measuring some discrepancy from binomial sampling could be obtained using the ideas in Section 4 of the paper and would then be referred to its predictive distribution.

The analysis via $p(r \mid A)$ discussed above, and not that fathered on me by Professor Bernardo, follows the suggestions made in this paper. If, however, we do look at

$$g_\beta(y) = n\left(\frac{r}{n} - \beta_0\right) \Big/ \beta_0(1-\beta_0)$$

this simply confirms the point (in this case trivial) that the appropriate function of the data to consider is r itself. Reference to the predictive distribution brings us once again to the previous result. Concerning Professor Bernardo's Figs D3 and D4, I can only say that outside the famous classic by Darrell Huff (1954) it is unusual to see comparisons made between unstandardised unsquared quantities and standardised squared ones.

Returning to Professor Lindley's remarks, there is in my mind no question of abandoning the likelihood principle so far as estimation is concerned. When we are asking what can be said of a variable $\boldsymbol{\theta}$ in relation to a single vector $\mathbf{y}_d$, I do not find it surprising that the sampling rule should be irrelevant. But if we aim to judge whether samples resembling in some relevant respect the one we have observed are or are not rare, it seems to me essential to know (as part of the model) the rule by which the samples were generated. In most cases this information is available. But if it is not, then I believe no absolute check on the model is possible. In particular as Professor Hunter points out this allows an appropriate role for randomization.

Concerning tail areas and probability ratios, my prescription is that low predictive density, not location in tail areas *per se*, casts doubt on the model. It is not difficult to produce examples where the predictive distribution of a sensible checking function could have two humps, with extreme *and intermediate* values appropriately suspect. I acknowledge that I prefer to sit on a different horn of this dilemma than the one favoured by Professor Cox. My difficulty with probability ratios (and likelihood ratios and predictive ratios) is of course that while I grant that it may be useful to know that it is a million times more probable that the first man I meet when I walk down the street will be called John Smith rather than Jeremiah Hezekiah Bramblebottom, this in itself tells me little about the chance of meeting a man called John Smith. Invent a few more names and one sees the difficulty with the formula which Dr O'Hagan displays and with whose uses and difficulties I am not unfamiliar (see, for example, Box and Hill, 1967). I do find it astonishing, however, that he regards as trivial the assumption that all possible models $M_1, M_2, \ldots, M_k$ are known *a priori*. If he lacks personal experience of scientific investigation, he would need only to read any moderately accurate account of one such (e.g. Watson's *The Double Helix*, 1968) to know that models *evolve*. Dr O'Hagan is welcome to his screwdriver; I am saying that it is an unsuitable instrument for driving in a nail. Once more let me say that the difficulty with any attempt to use the Bayesian half of the model alone, is that it is eternally conditional. We can move the conditionality around but we cannot lose it. The buck stops when we cease to ignore the other half of the model $p(\mathbf{y} \mid A)$.

Dr Atkinson's generous attribution to me of some part in his becoming a statistician is very flattering. In response to his further remarks, Steve Bailey and I have also found that on suitable transformation apparent outliers can vanish away (Bailey and Box, 1980b), but, as he says, this does not seem to be true for the Box and Behnken data. Dr Atkinson has doubts about the accuracy of Table 1; these calculations have now been carefully rechecked and apart from very minor discrepancies they seem not to be in error but remarkably sensible as evidenced by Table D2.

TABLE D2

	β_1	β_2	β_3	β_4	β_{11}
No omissions	1·93 (0·42)	−1·96 (0·42)	1·13 (0·42)	−3·68 (0·42)	−1·42 (0·63)
$\varepsilon = 0{\cdot}001$	2·46 (0·28)	−1·96 (0·22)	1·13 (0·22)	−3·15 (0·28)	−1·88 (0·44)
$\varepsilon = 0{\cdot}020$	2·49 (0·22)	−1·96 (0·20)	1·13 (0·20)	−3·12 (0·22)	−1·89 (0·42)
y_{10} deleted	2·57 (0·12)	−1·96 (0·11)	1·13 (0·11)	−3·04 (0·12)	−2·05 (0·18)
	β_{22}	β_{33}	β_{44}	β_{12}	β_{13}
No omissions	−4·33 (0·63)	−2·24 (0·63)	−2·58 (0·63)	−1·67 (0·73)	−3·82 (0·73)
$\varepsilon = 0{\cdot}001$	−4·10 (0·36)	−2·01 (0·38)	−3·05 (0·44)	−1·67 (0·39)	−3·82 (0·39)
$\varepsilon = 0{\cdot}020$	−4·09 (0·34)	−2·00 (0·34)	−3·05 (0·42)	−1·67 (0·34)	−3·82 (0·34)
y_{10} deleted	−4·01 (0·17)	−1·92 (0·17)	−3·21 (0·18)	−1·68 (0·19)	−3·82 (0·19)
	β_{14}	β_{23}	β_{24}	β_{34}	
No omissions	0·95 (0·73)	−1·68 (0·73)	−2·62 (0·73)	−4·25 (0·73)	
$\varepsilon = 0{\cdot}001$	−0·45 (0·95)	−1·67 (0·39)	−2·62 (0·39)	−4·25 (0·39)	
$\varepsilon = 0{\cdot}020$	−0·48 (0·95)	−1·67 (0·35)	−2·62 (0·35)	−4·25 (0·34)	
y_{10} deleted	−0·95 (0·25)	−1·68 (0·19)	−2·62 (0·19)	−4·25 (0·19)	

The four rows of the table show (i) least squares estimates with no omissions; the Bailey and Box analysis for (ii) $\varepsilon = 0{\cdot}001$; (iii) $\varepsilon = 0{\cdot}020$ taken from Table 1 and (iv) the least squares estimates with y_{10} omitted. Comparison of the estimates shows that over the very wide range $\varepsilon = 0{\cdot}001$ to $\varepsilon = 0{\cdot}020$ ($\alpha = 0{\cdot}005$ to $\alpha = 0{\cdot}091$) the Bayesian means and standard deviations do not change very much. Also that when the four estimates differ appreciably the Bayesian means occupy a position between, but sharply different from estimates which assume for certain that there are no bad values, and from those which assume for certain that y_{10} is bad. The Bayes analysis thus produces a stable compromise which correctly acknowledges that since we do not know for sure whether y_{10} is bad or good, we should neither accept nor reject it but only downweight it. In fact it is a little more subtle than this because of course it allows appropriately for other possible bad values. In this case this mostly means that y_{13} also is downweighted slightly. The large standard deviation associated with Bayesian means for β_{14} occurs because two of the four observations essentially involved (y_{10} and y_{13}) are the major suspects.

If I understand Dr Gathercole correctly, he believes that the results from the analysis in Section 2 would be anomalous if the data had been 70, 70, 70, 74. I do not see why. In that case

$$g(\mathbf{y}_d) = \frac{(\bar{y}_d - \theta_0)^2}{n^{-1}\sigma^2 + \sigma_\theta^2} + \frac{(n-1)\,s_d^2}{\sigma^2} = 0{\cdot}44 + 12{\cdot}00$$

correctly indicating that the supposition that the test results can be treated as independently and identically distributed random variables with $\sigma^2 = 1$ is called into question, but that the other structure is not. Incidentally, responding to Mr Jackson, I agree that to speak of a model being "called into question" rather than of its being "discredited" would have reduced the chance of my intention being misunderstood. We certainly need (as Fisher clearly intended by his use of the term "discredit") to distance ourselves as far as possible from the terminology "*reject* the model/hypothesis".

Since I am not proposing to estimate the prior distribution in the manner proposed by Professor Godambe, I do not see the relevance of the first part of his contribution to this discussion. Concerning the second part, while, as I have said, I do not "reject" models on the basis he suggests and am suspicious of probability ratios, I acknowledge that it is possible to construct conundrums of the kind he mentions. But when he talks of how these should be "resolved" I assume he does not mean as some kind of mathematical puzzle but in the context of a real scientific investigation. In that case if at first there seemed to be inconsistencies in the quantities I had calculated or the plots I had made, I would try to see why, by reviewing the assumptions behind and the meanings of my various quantities. Then in cooperation with the investigator and taking into account many other vexing indeterminacies which were perhaps more relevant, I would help him to decide what to do next.

My response to Professor Kadane about choice of "significance levels" follows similar lines. In (2.6) I did not mean that 0·001 was to be taken as a critical value but only that the approximate probability should be recorded and in practice would be considered in relation to (among other things) "the size of likely discrepancies, their impact on the conclusions . . ." which he mentions. I believe however that in practice this has to be done informally and not by formal appeal to decision theory. I believe that the question: how small must α be to be small? unrealistically selects one aspect of necessary scientific subjectivity for criticism in the mistaken belief that scientific research can be made wholly objective. It seems to me that the significance test idea is natural and indeed a necessary part of the conduct and management of everyday life, and I find it hard to understand the horror with which it is sometimes greeted nowadays. The process of modification of belief occurs in two stages: (a) the recognition that the data do not fit with the presently entertained model of the world, (b) the later consideration of what are alternative models that might better explain the data. For example, suppose I have an office that looks onto, say, Oxford Street in London, normally thronged with people. One day I look out of the window at 11 o'clock in the morning and notice that there are only two people in the whole street. My initial reaction surely is that on the null (*status quo*) model this is an unusual event possibly worthy of further investigation. Alternative models that might explain the phenomenon come later. These might posit that the street has been blocked off for a ceremonial occasion, that there is a bomb scare, or that it is a Sunday, etc. But notice that the basis of the initial reaction, which requires no alternatives, is surely that I have (or could have) looked out of the window on many previous occasions and rarely have I (would I have) seen as few as two people in the street. The motivation is economy of effort and is employed by all of us hundreds of times in our daily lives—when the null model is plausible I will not worry, but when data make it implausible perhaps I should be concerned. Quality control charts and the principle of "management by exception" also employ this concept, thus ensuring that we are not often distracted except when we should be.

Professor Huber asks why Bayesian robustness has not developed as rapidly as its non-Bayesian counterpart. I think that this is a temporary situation arising from inferior publicity, lack of computer programs and (if we allow "develop" to mean proliferation of theory rather than use) to the fact that non-Bayesian robustness lends itself to mathematistry. Indeed, the problem of arbitrary choice which he lays at the Bayesian door would, I should have thought, have been a greater embarrassment to non-Bayesians. Even for the single problem of robustification against systematic heavy-tailed distributions they have already produced a bewildering plethora of solutions and the end seems nowhere in sight.

I find Professor Huber's objections to my statements in the first part of Section 5 incomprehensible, and my head is unbowed even after the invocation of Kolmogorov and the weak law of large numbers. In relation to the small probability of mishap on a plane trip "the practical translation of probabilities into actions" takes the form before each flight of my fastening my seat belt and attending to the cabin demonstration of the safety equipment, not of my refusing to fly. In the same way, the knowledge that there is a small finite probability that a sample could contain one or more bad values results in my using a robust procedure, not in my refusal to analyse the data.

The idea of using a prior distribution for α is discussed in Bailey and Box (1980a); see also (d) after (5.4) in the present paper. The results are not very different from those presented here. I believe Professor Huber's fear about the possibility of my suggestion's wasting iterations is groundless. My experience has been that to produce a satisfactory model it is not necessary that it be exactly right (no model/procedure is perfect) but rather that it not be grossly wrong in the context in which it is to be used. Examples of models that can be grossly wrong in specific contexts are the standard linear model when data are collected serially (because the model totally discounts the possibility of serial correlation), the same model in the common situation where bad values can occur (because it says they cannot happen), the normal model for the comparison of variances (because it uses, directly or indirectly, the fact that for exact normality $\mu_4 = 3\mu_2^2$). Simple and obvious repairs for these and similar cases can, I believe, be extremely effective. We do not need to throw away our traditional approaches to estimation and statistical models in an orgy of *ad hockery*.

Concerning Dr Kenett's interesting observation, I think it necessary to recognise that the scientific iteration is not necessarily completed at one location (in a wide sense it is never completed). The results from his publication based on vague models may well make possible a course of investigation, perhaps by others, resulting in models that are much more precise.

I thank the remaining contributors for their many kind remarks. Where we differ I find I have already stated my side of the case as well as I can. It therefore only remains for me to thank the Society for tentatively entertaining me and my paper in such a generous spirit.

References in the Discussion

Aitchison, J. and Dunsmore, I. R. (1965). *Statistical Prediction Analysis.* Cambridge University Press.

Atkinson, A. C. (1980). Examples showing the use of two graphical displays for the detection of influential and outlying observations in regression. In *COMPSTAT 1980* (M. M. Baritt and W. Wishart, eds). Vienna: Physica Verlag.

—— (1981). Two graphical displays for outlying and influential observations in regression. *Biometrika*, **68**, in press.

BARNARD, G. A. (1967). The use of the likelihood function in statistical practice. *Proc. 5th Berkeley Symp. Math. Statist. Prob.*, **1**, pp. 27–40.

BASU, D. (1975). Statistical information and likelihood. *Sankhyā* A, **37**, 1–71.

BERNARDO, J. M. (1979). Reference posterior distributions for Bayesian inference (with Discussion). *J. R. Statist. Soc.* B, **41**, 113–147.

BIRNBAUM, A. (1962). On the foundations of statistical inference. *J. Amer. Statist. Ass.*, **57**, 269–306.

BOX, G. E. P. and HILL, W. J. (1967). Discrimination among mechanistic models. *Technometrics*, **9**, 57–71.

COOK, R. D. (1977). Detection of influential observations in linear regression. *Technometrics*, **19**, 15–18.

DAVIES, O. L. (1956). *The Design and Analysis of Industrial Experiments.* London: Oliver and Boyd.

DE GROOT, M. H. (1973). Doing what comes naturally: interpreting a tail area as a posterior probability or as a likelihood ratio. *J. Amer. Statist. Ass.*, **68**, 966–969.

DYER, A. R. (1973). Discrimination procedures for separate families of hypotheses. *J. Amer. Statist. Ass.*, **68**, 970–974.

FERGUSON, T. S. (1973). A Bayesian analysis of some non-parametric problems. *Ann. Statist.*, **1**, 209–230.

FUCHS, C. (1979). Comments on a criterion of transformation proposed by Schlesselman. *J. Amer. Statist. Ass.*, **74**, 238–239.

GEISSER, S. (1969). Invited discussion on "The Bayesian outlook and its application" by J. Cornfield. *Biometrics*, **4**, 643–647.

GODAMBE, V. P. (1974). Review of de Finetti's *Probability Induction and Statistics. J. Amer. Statist. Ass.*, **69**, 578–580.

HAMPEL, F. R. (1973). Robust estimation: a condensed partial survey. *Z. Wahrscheinlichkeitstheorie und Verw. Gebiete*, **27**, 87–104.

HUFF, D. (1954). *How to Lie with Statistics.* New York: Atheneum.

JEFFREYS, H. (1967). *Theory of Probability*, 3rd revised edition. London: Oxford University Press.

JOHNSON, W. and GEISSER, S. (1979). Assessing the predictive influence of observations, University of Minnesota Technical Report No. 355 (to be published in the C. R. Rao Birthday volume).

—— (1980). A predictive view of the detection and characterization of influential observations in regression analysis, University of Minnesota Technical Report No. 365.

KARLIN, S., KENETT, R. S. and BONNE'-TAMIR, B. (1979). Analysis of biochemical genetic data on Jewish populations II. Results and interpretations of heterogeneity indices and distance measures with respect to standards. *Amer. J. Hum. Genet.*, **31**, 341–365.

KOLMOGOROV, A. N. (1956). *Foundations of the Theory of Probability.* New York: Chelsea Publications Co.

LEONARD, T. (1978). Density estimation, stochastic processes and prior information (with Discussion). *J. R. Statist. Soc.* B, **40**, 113–146.

MASRELIEZ, C. J. (1975). Approximate non-Gaussian filtering with linear state and observation relations. *IEEE Trans. Autom. Control*, **AC-20**, 107–110.

RUBIN, H. (1977). Robust Bayesian estimation. In *Statistical Decision Theory and Related Topics, II* (S. S. Gupta and D. S. Moore, eds). New York: Academic Press.

SCHLESSELMAN, J. J. (1973). Data transformation in two way analysis of variance. *J. Amer. Statist. Ass.*, **68**, 369–378.

SPIEGELHALTER, D. J. (1977). A test for normality against symmetric alternatives. *Biometrika*, **64**, 415–418.

—— (1980). An omnibus test for normality for small samples. *Biometrika*, **67**, 493–496.

WATSON, J. D. (1968). *The Double Helix.* New York: Atheneum.

ZIDEK, J. V. (1969). A representation of Bayes invariant procedures in terms of Haar measure. *Ann. Inst. Statist. Maths*, **21**, 291–308.

1.16
Linear Models and Spurious Observations

BOVAS ABRAHAM
University of Waterloo, Ontario

G. E. P. BOX
University of Wisconsin

[Received February 1977. Revised November 1977]

SUMMARY

A Bayesian approach is adopted here to make inferences about the parameters of a linear model in the possible presence of one or more spurious observations. The method proposed is illustrated by analysing a classical set of data.

Keywords: SPURIOUS OBSERVATIONS; LINEAR MODELS; POSTERIOR DISTRIBUTIONS; *t*-DISTRIBUTIONS

1. INTRODUCTION

IN this paper we consider (i) how to make inferences about the parameters of a linear model in the possible presence of one or more spurious observations (ii) how to assess the evidence that any particular observation is spurious and to estimate its deviation from expectation.

2. INFERENCES ABOUT THE PARAMETERS OF A LINEAR MODEL

Consider the following linear model in which some of the observations are not generated in the manner intended:

$$\mathbf{y} = X\boldsymbol{\beta} + \delta\mathbf{Z} + \boldsymbol{\varepsilon}, \tag{2.1}$$

where $\mathbf{y}$ is an $n \times 1$ vector of observations, X an $n \times p$ nonstochastic matrix with rank p, $\boldsymbol{\beta}$ a $p \times 1$ vector of unknown parameters and $\boldsymbol{\varepsilon}$ an $n \times 1$ vector of independent identically distributed random variables with approximate $N(0, \sigma^2)$ (normal distribution with mean zero and variance σ^2) distributions. $\mathbf{Z}' = (Z_1, Z_2, \ldots, Z_n)$ is a vector of r (unknown in general) unities and $(n-r)$ zeros such that

$$Z_i = \begin{cases} 1 & \text{if the } i\text{th observation is spurious,} \\ 0 & \text{otherwise.} \end{cases}$$

In Section 2.1 we regard $\mathbf{Z}$ as known, i.e. the identities of the suspected observations are known and the more interesting case of unknown $\mathbf{Z}$ will be considered in Section 2.2.

2.1. *Posterior Distribution of* $\boldsymbol{\beta}$ *given* $\mathbf{y}$ *and* $\mathbf{Z}$

$$P(\boldsymbol{\beta} \mid \mathbf{y}, \mathbf{Z}) = \int P(\boldsymbol{\beta}, \delta, \sigma \mid \mathbf{Z}) P(\mathbf{y} \mid \boldsymbol{\beta}, \delta, \sigma, \mathbf{Z})\, d\delta\, d\sigma, \tag{2.2}$$

where $P(\boldsymbol{\beta}, \delta, \sigma \mid \mathbf{Z})$ is the prior of $(\boldsymbol{\beta}, \delta, \sigma \mid \mathbf{Z})$ and $P(\mathbf{y} \mid \boldsymbol{\beta}, \delta, \sigma, \mathbf{Z})$ is the joint distribution of $\mathbf{y}$ given $(\boldsymbol{\beta}, \delta, \sigma, \mathbf{Z})$. This joint density may be written as

$$P(\mathbf{y} \mid \boldsymbol{\beta}, \delta, \sigma, \mathbf{Z}) = (2\pi\sigma^2)^{-\frac{1}{2}n} \exp [(2\sigma^2)^{-1}\{(\mathbf{y} - X\boldsymbol{\beta})'(\mathbf{y} - X\boldsymbol{\beta}) + r\delta^2 - 2(\mathbf{y} - X\boldsymbol{\beta})'\mathbf{Z}\delta\}], \tag{2.3}$$

where $r = \mathbf{Z}'\mathbf{Z}$. Now assuming that $r \neq 0, n$ and that $X'VX$ is nonsingular (2.3) may be rearranged (see Appendix 1) as

$$P(\mathbf{y} \mid \boldsymbol{\beta}, \delta, \sigma, \mathbf{Z}) = (2\pi\sigma^2)^{-\frac{1}{2}n} \exp [-(2\sigma^2)^{-1}\{(\boldsymbol{\beta} - \boldsymbol{\beta}^*)'(X'VX)(\boldsymbol{\beta} - \boldsymbol{\beta}^*) + r(\delta - \delta_\beta)^2 + \nu S_r^2\}], \tag{2.4}$$

The Collected Works of George E. P. Box, 1984, Wadsworth, Inc., Belmont, CA 94002.
Originally published in *Applied Statistics,* vol. 27, no. 2 (1978), pp. 131–138.

where $V = I - r^{-1}\mathbf{Z}\mathbf{Z}'$, $\delta_\beta = r^{-1}(\mathbf{y} - X\boldsymbol{\beta})'\mathbf{Z}$, $\boldsymbol{\beta}^* = (X'VX)^{-1}X'V\mathbf{y}$ and

$$\nu S_r^2 = \mathbf{y}'V(I - X(X'VX)^{-1}X')V\mathbf{y}.$$

We now assume that $\boldsymbol{\beta}$, δ, $\log\sigma$ are locally independent and uniform *a priori*, i.e.

$$P(\boldsymbol{\beta}, \delta, \sigma \mid \mathbf{Z}) \propto \sigma^{-1}. \tag{2.5}$$

Multiplying (2.5) with (2.6) and integrating over δ and σ we get

$$P(\boldsymbol{\beta} \mid \mathbf{y}, \mathbf{Z}) \propto \left(\frac{1 + (\boldsymbol{\beta} - \boldsymbol{\beta}^*)'(X'VX)(\boldsymbol{\beta} - \boldsymbol{\beta}^*)}{\nu S_r^2}\right)^{-\frac{1}{2}(\nu+p)}, \tag{2.6}$$

where $\nu = n - p - 1$. Hence it follows that if a specific set of r observations were given to be spurious the posterior distribution of $\boldsymbol{\beta}$ would be a p-dimensional t-distribution with mean $\boldsymbol{\beta}^*$, dispersion matrix $S_r^2(X'VX)^{-1}$ and degrees of freedom (d.f.) $\nu = n - p - 1$. In fact $\boldsymbol{\beta}^*$ is a weighted least squares estimator of $\boldsymbol{\beta}$ with the weighting matrix V depending on the identity of the spurious observations. When $r = 0$, $P(\boldsymbol{\beta} \mid \mathbf{y}, \mathbf{Z})$ is p-dimensional t with mean $\hat{\boldsymbol{\beta}} = (X'X)^{-1}X'\mathbf{y}$, dispersion matrix $S^2(X'X)^{-1}$ and d.f. $\nu_0 = n - p$ where

$$\nu_0 S^2 = \mathbf{y}'\{I - X(X'X)^{-1}X'\}\mathbf{y}.$$

2.2. *Posterior Distribution of* $\boldsymbol{\beta}$ *given* $\mathbf{y}$ *(*$\mathbf{Z}$ *unknown)*

Suppose now that there is a small prior probability α (known) that any element of $\mathbf{Z}$ is unity and a complementary probability that it is zero. If $\mathbf{U}_r$ denotes an $n \times 1$ vector with a specific set of r unities and $(n - r)$ zeros we have, *a priori*,

$$P(\mathbf{Z} = \mathbf{U}_r \mid \alpha) = \alpha^r(1 - \alpha)^{n-r}, \quad r = 0, 1, 2, \ldots, n. \tag{2.7}$$

Hereafter, to simplify notation, we will write $\mathbf{Z}$ for $\mathbf{Z} = \mathbf{U}_r$ and conditioning on α will be avoided.

Given $\mathbf{y}$, the posterior distribution of $(\boldsymbol{\beta}, \delta, \sigma, \mathbf{Z})$ is given by

$$P(\boldsymbol{\beta}, \delta, \sigma, \mathbf{Z} \mid \mathbf{y}) = \frac{P(\mathbf{Z})P(\boldsymbol{\beta}, \delta, \sigma \mid \mathbf{Z})P(\mathbf{y} \mid \boldsymbol{\beta}, \delta, \sigma, \mathbf{Z})}{\Sigma P(\mathbf{Z})P(\mathbf{y} \mid \mathbf{Z})}, \tag{2.8}$$

where $P(\mathbf{Z})$ is the prior distribution given in (2.7), $P(\mathbf{y} \mid \boldsymbol{\beta}, \delta, \sigma, \mathbf{Z})$ given in (2.4) and

$$P(\mathbf{y} \mid \mathbf{Z}) = \int P(\boldsymbol{\beta}, \delta, \sigma \mid \mathbf{Z}) P(\mathbf{y} \mid \boldsymbol{\beta}, \delta, \sigma, \mathbf{Z})\, d\delta\, d\boldsymbol{\beta}\, d\sigma. \tag{2.9}$$

The summation in (2.8) takes into account all the 2^n possible values of $\mathbf{Z}$. Integrating (2.8) over δ and σ we get

$$P(\boldsymbol{\beta}, \mathbf{Z} \mid \mathbf{y}) = P(\mathbf{Z} \mid \mathbf{y}) P(\boldsymbol{\beta} \mid \mathbf{y}, \mathbf{Z}), \tag{2.10}$$

where $P(\boldsymbol{\beta} \mid \mathbf{y}, \mathbf{Z})$ is the posterior distribution given in (2.6) and

$$P(\mathbf{Z} \mid \mathbf{y}) = P(\mathbf{Z})P(\mathbf{y} \mid \mathbf{Z}) / \Sigma P(\mathbf{Z})P(\mathbf{y} \mid \mathbf{Z}), \tag{2.11}$$

which may be rewritten as

$$P(\mathbf{Z} \mid \mathbf{y}) = C\{P(\mathbf{Z})/P(\mathbf{Z} = \mathbf{0})\}\{P(\mathbf{y} \mid \mathbf{Z})/P(\mathbf{y} \mid \mathbf{Z} = \mathbf{0})\}, \tag{2.12}$$

where

$$C = P(\mathbf{Z}=\mathbf{0})P(\mathbf{y}|\mathbf{Z}=\mathbf{0})/\sum P(\mathbf{Z})P(\mathbf{y}|\mathbf{Z}).$$

Now summing (2.10) w.r.t. r we obtain

$$P(\boldsymbol{\beta}|\mathbf{y}) = \sum P(\mathbf{Z}|\mathbf{y})P(\boldsymbol{\beta}|\mathbf{y},\mathbf{Z}), \tag{2.13}$$

where $P(\boldsymbol{\beta}|\mathbf{y},\mathbf{Z})$ is given in (2.6) and $P(\mathbf{Z}|\mathbf{y})$ is in (2.12) which simplifies, using (2.9) and Appendix 2 (see Appendix 3 for some additional simplifications), to

$$P(\mathbf{Z}|\mathbf{y}) = C\left(\frac{\alpha}{1-\alpha}\right)^r \left(\frac{r|X'VX|}{\nu_0 S^2|X'X|}\right)^{-\frac{1}{2}} \left(\frac{\nu S_r^2}{\nu_0 S^2}\right)^{-\frac{1}{2}\nu} \beta(\tfrac{1}{2},\tfrac{1}{2}\nu), \tag{2.14}$$

where C is the normalizing constant which makes $\sum P(\mathbf{Z}|\mathbf{y}) = 1$. Hence the posterior distribution of $\boldsymbol{\beta}|\mathbf{y}$ is a weighted average of multivariate t-distributions, with the weights being equal to the posterior probability of a particular set of observations being spurious. For computational purposes (2.13) may be written as

$$P(\boldsymbol{\beta}|\mathbf{y}) = w_0 P(\boldsymbol{\beta}|\mathbf{y},\mathbf{Z}=\mathbf{0}) + \sum_{i=1}^{n} w_i P_i(\boldsymbol{\beta}|\mathbf{y},\mathbf{Z}=\mathbf{U}_1) + \sum_{\substack{i=1\\i<j}}^{n}\sum_{j=1}^{n} w_{ij} P_{ij}(\boldsymbol{\beta}|\mathbf{y},\mathbf{Z}=\mathbf{U}_2) + \ldots, \tag{2.15}$$

where the subscripts 0, i, ij refer to the possibilities, respectively, that no observations, the ith observation, and the ith and jth observations, are spurious and the w's are weights $P(\mathbf{Z}|\mathbf{y})$ corresponding to the different possibilities. The exact evaluation of (2.15) would be computationally difficult. However, in practice α will be small and for moderate n it will often be possible without undue approximation to ignore the possibility of more than three (or even two) spurious observations.

2.3. *Inference about the Mean of a Normal Distribution*

In this special case $X' = \mathbf{1}_n' = (1,1,\ldots,1)$ and $\beta^* = \bar{y}(g)$, where $g = n-i_1-i_2-\ldots-i_r$ implying that $\bar{y}(g)$ is the average after deleting the spurious observations $y_{i_1}, y_{i_2}, \ldots, y_{i_r}$. Also $\nu_0 S^2 = \sum(y_i-\bar{y})^2$, $\nu S_r^2 = \sum y_i^2 - (n-r)\bar{y}^2(g) - r\bar{y}^2(b)$ where $\bar{y}(b)$ is the average of the spurious observations, and $X'VX = n-r \neq 0$ when $r \neq n$. $P(\beta|\mathbf{y})$ is a weighted average of scaled t-distributions and the posterior mean and variance of $\beta|\mathbf{y}$ are given by

$$\left.\begin{aligned} E(\beta|\mathbf{y}) &= w_0\bar{y} + \sum_{i=1}^{n} w_i \bar{y}(n-i) + \sum_{i<j}\sum w_{ij}\bar{y}(n-i-j) + \ldots = \phi \quad \text{(say)},\\ V(\beta|\mathbf{y}) &= w_0\left(\frac{n-1}{n-3}\frac{S^2}{n} + (\bar{y}-\phi)^2\right) + \sum_i w_i\left(\frac{n-2}{n-4}\frac{S_1^2}{n-1} + \{\bar{y}(n-i)-\phi\}^2\right) + \ldots. \end{aligned}\right\} \tag{2.16}$$

Also $\text{S.D.}(\beta|\mathbf{y}) = \sqrt{\{V(\beta|\mathbf{y})\}}$.

3. Example: Darwin's Data

The data given in ascending order in Table 1 (Fig. 1(b)) consist of the differences in heights of 15 self-fertilized and cross-fertilized plants grown in the same plot (see Box and Tiao, 1973).

Table 1

−67,	−48,	6,	8,	14,	16,	23,	24,
28,	29,	41,	49,	56,	60,	75	

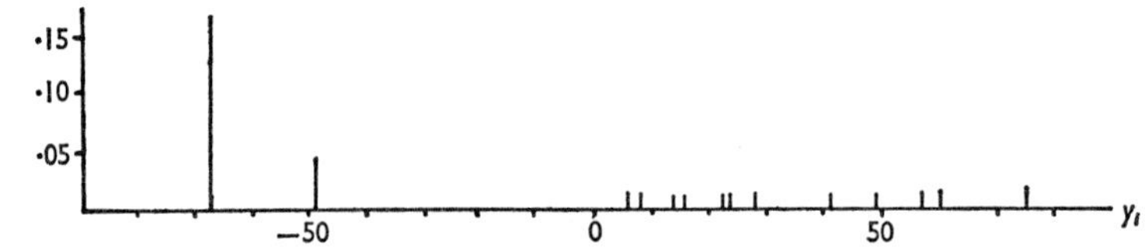

FIG. 1(a). The weights w_i plotted corresponding to each observation for α = ·05; w_i = posterior probability that the ith observation is spurious and the others are not.

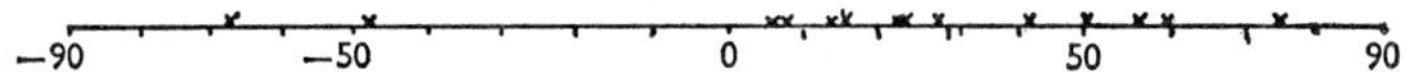

FIG. 1(b). Darwin's data.

The first two observations are rather discrepant. Assuming that the central model is $N(\beta, \sigma^2)$ but with a small probability α that any observation could have been generated by an alternative model $N(\beta+\delta, \sigma^2)$, the posterior distributions of β given **y** were computed for α = ·001, ·01 (·01), ·1, ·15 and these are shown in Fig. 2. In these computations the leading term and the next three summations in (2.15) are used. The weights after these are very small and hence the associated distributions are ignored.

Fig. 2 shows respectively posterior distributions for β, assuming no spurious observations (A), assuming that y_1 and y_2 are spurious (B) and assuming the possibility of spurious

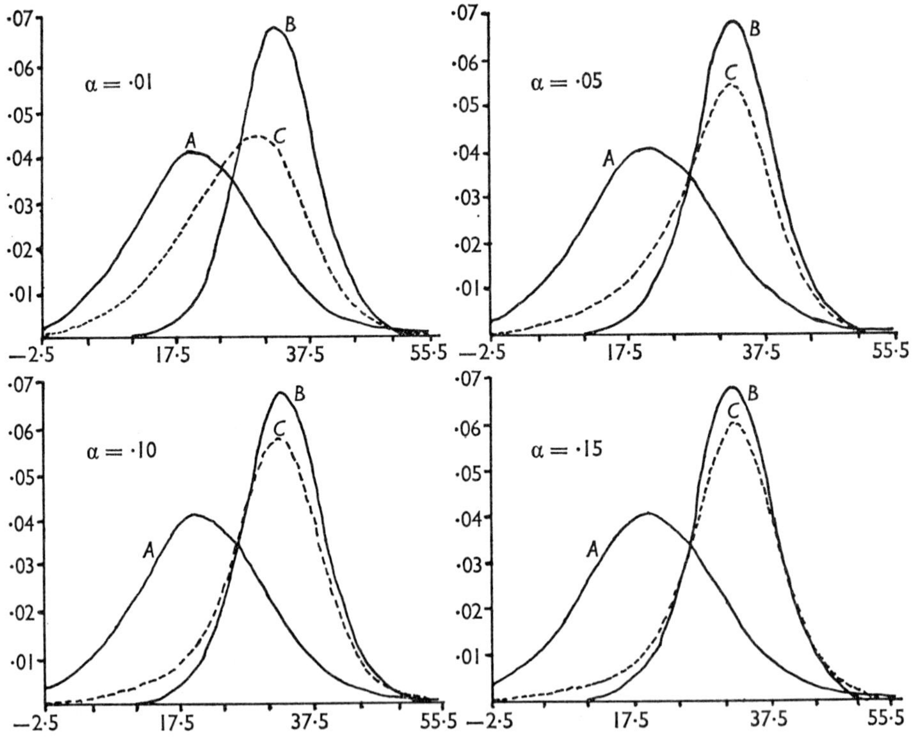

FIG. 2. Posterior distribution of β. A: Assuming no spurious observations, t (20·93, 9·75^2, 14); B: Assuming y_1 and y_2 to be spurious, t (32·45, 5·78^2, 13); C: Assuming the possibility of spurious observations.

observations (C). It is seen that curves B and C are closely similar for $\alpha = \cdot 05$, $\cdot 10$, $\cdot 15$ and they are considerably different from curve A.

Fig. 1(a) is the plot of the posterior probability that the ith observation is spurious given that the others are not, for $\alpha = \cdot 05$. Similar graphs were obtained for other values of α also. Table 2 gives the posterior probabilities that the ith and jth observations are spurious given that the others are not, for $\alpha = \cdot 05$. These can be graphically represented by a two-way diagram on which circles are drawn with areas proportional to the posterior probabilities as has been demonstrated in Abraham and Box (1975). However, here (in the lower triangle of Table 2) some symbols are employed to distinguish between weights of different magnitudes. The remarkable point is that these weights are very close to zero except the one corresponding to the discrepant pair (1, 2) which is more than 300 times the next biggest one and more than 3,000 times the smallest one.

Table 3 shows the mean and standard deviation of β given $\mathbf{y}$ for various values of α. For comparison we also include the corresponding quantities when it is assumed that there are no spurious ones. We find that the conclusions are not sensitive to moderate changes in α. However, as might be expected, there is a dramatic difference between the assumption of no possibility of spurious observations ($\alpha = 0$) and the assumption of some such possibility even a very remote one ($\alpha = \cdot 001$).

4. Inference about δ

The model we have been entertaining so far is a mixture of two normals where only one of them is regarded as the right one. This model has applications in the case where "bad" observations are suspected and one is interested to find out how much these observations deviate from the mean of the "good" observations.

4.1. *Posterior Distribution of* δ *given* $\mathbf{y}$

Proceeding as in Section 2, using (2.3), (2.5), Appendix 1 and assuming that $r \neq 0, n$ and that $\mathbf{Z}'M\mathbf{Z} \neq 0$

$$P(\boldsymbol{\beta}, \delta, \sigma \mid \mathbf{y}, \mathbf{Z}) = (2\pi)^{-\frac{1}{2}n}\,\sigma^{-(n+1)} \exp[-(2\sigma^2)^{-1}\{(\boldsymbol{\beta}-\boldsymbol{\beta}_\delta)'X'X(\boldsymbol{\beta}-\boldsymbol{\beta}_\delta) + (\mathbf{Z}'M\mathbf{Z})(\delta-\delta^*)^2 + \nu S_d^2\}], \tag{4.1}$$

where $\boldsymbol{\beta}_\delta = (X'X)^{-1}(\mathbf{y}-\mathbf{Z}\delta)$, $\delta^* = (\mathbf{Z}'M\mathbf{Z})^{-1}\mathbf{Z}'M\mathbf{y}$, $M = I - X(X'X)^{-1}X'$ and

$$\nu S_d^2 = \mathbf{y}'M\{I - \mathbf{Z}(\mathbf{Z}'M\mathbf{Z})^{-1}\mathbf{Z}'\}M\mathbf{y}.$$

Now integrating with respect to $\boldsymbol{\beta}$ and σ we get

$$P(\delta \mid \mathbf{y}, \mathbf{Z}) \propto \left(1 + \frac{(\mathbf{Z}'M\mathbf{Z})(\delta-\delta^*)^2}{\nu S_d^2}\right)^{-\frac{1}{2}(\nu+1)}, \tag{4.2}$$

which is a scaled t-distribution. Following the procedure in Section 2 one can obtain $p(\delta \mid \mathbf{y})$ and this will be a weighted average of scaled t-distributions.

4.2. *Special Case*

Again we examine the case considered in Section 2.3, i.e. when $X = \mathbf{1}_n$. In this case $\mathbf{Z}'M\mathbf{Z} = r(n-r)/n \neq 0$ for $r \neq 0, n$,

$$\delta^* = \left(\frac{n}{n-r}\right)\{\bar{y}(b) - \bar{y}\}, \quad \nu S_d^2 = \nu_0 S^2 - \frac{nr}{n-r}\{\bar{y}(b)-\bar{y}\}^2. \tag{4.3}$$

Notice that $\bar{y}(b)$ is the average of the spurious observations, $\bar{y}(b) - \bar{y}$ is a measure of the discrepancy and, as might be expected, $\delta^* = E(\delta \mid \mathbf{y}, \mathbf{Z})$ is proportional to this discrepancy. Also it is very interesting to observe that the posterior mean $E(\delta \mid \mathbf{y})$, which can be computed using an equation very similar to (2.16), is a weighted average of these δ^*'s.

TABLE 2

Posterior probability that the ith *and* jth *observations are spurious* (α = ·05)

	1	2	3	4	5	6	7	8	9	10	11	12	13	14	15
1		·6455	·0021	·0019	·0014	·0013	·0010	·0009	·0008	·0008	·0005	·0004	·0003	·0003	·0003
2	■		·0009	·0008	·0007	·0006	·0005	·0005	·0004	·0004	·0003	·0003	·0003	·0002	·0003
3	+	×		·0002	·0002	·0002	·0002	·0002	·0002	·0002	·0002	·0002	·0002	·0002	·0003
4	+	×	×		·0002	·0002	·0002	·0002	·0002	·0002	·0002	·0002	·0002	·0002	·0003
5	+	×	×	×		·0002	·0002	·0002	·0002	·0002	·0002	·0002	·0002	·0002	·0003
6	+	×	×	×	×		·0002	·0002	·0002	·0002	·0002	·0002	·0002	·0002	·0003
7	+	×	×	×	×	×		·0002	·0002	·0002	·0002	·0002	·0003	·0003	·0004
8	×	×	×	×	×	×	×		·0002	·0002	·0002	·0002	·0003	·0003	·0004
9	×	×	×	×	×	×	×	×		·0002	·0002	·0003	·0003	·0003	·0004
10	×	×	×	×	×	×	×	×	×		·0002	·0003	·0003	·0003	·0004
11	×	×	×	×	×	×	×	×	×	×		·0003	·0004	·0004	·0006
12	×	×	×	×	×	×	×	×	×	×	×		·0004	·0005	·0008
13	×	×	×	×	×	×	×	×	×	×	×	×		·0006	·0011
14	×	×	×	×	×	×	×	×	×	×	×	×	×		·0013
15	×	×	×	×	×	×	×	×	×	×	×	×	+	+	

× : ·0002–·0009; + : ·0010–·0021; ■ ⩾ ·6455.

TABLE 3

$E(\beta\,|\,\mathbf{y})$ *and* S.D.$(\beta\,|\,\mathbf{y})$

α	0	·001	·01	·02	·03	·04	·05	·06	·07	·08	·09	·10	·15
$E(\beta\,|\,\mathbf{y})$	20·93	24·50	27·05	28·36	29·16	29·70	30·08	30·37	30·59	30·77	30·91	31·03	31·33
S.D.$(\beta\,|\,\mathbf{y})$	10·53	10·40	10·10	9·76	9·47	9·24	9·06	8·82	8·81	8·72	8·64	8·59	8·47

5. Concluding Remarks

For making inferences about $\boldsymbol{\beta}$ and δ an alternate approach is to consider the joint distribution of $\boldsymbol{\xi} = \begin{pmatrix}\boldsymbol{\beta}\\ \delta\end{pmatrix}$. Following the procedure in Section 2 it is not difficult to show that $P(\boldsymbol{\xi}\mid\mathbf{y})$ is a weighted average of $(p+1)$-dimensional t-distributions. Hence the marginal of $\boldsymbol{\beta}$ will be a weighted average of p-dimensional t-distributions and that of δ will be a weighted average of one-dimensional t-distributions.

The model used in this paper assumes that the spurious observations all belong to the same population. Obviously, there are physical situations where this is likely to be true but in general it may not be. The more general case which is not discussed here would be to consider spurious observations as coming from different populations. In any case, the model discussed here gives us access to the most interesting practical case which is that where there is one spurious observation (or even two belonging to the same population).

Tables and plots of the kind produced here could be made a routine part of the regression and analysis of variance programmes and would be of considerable assistance to the data analyst in deciding how to proceed.

Acknowledgement

We would like to thank the referees for their helpful comments. This project was supported in part by a Grant A3079 from the National Research Council of Canada.

References

Abraham, B. and Box, G. E. P. (1975). Outliers in linear models. Technical Report No. 437, Department of Statistics, University of Wisconsin at Madison.

Anscombe, F. J. (1960). Rejection of outliers. *Technometrics*, **2**, 123–146.

Box, G. E. P. and Tiao, G. C. (1968). A Bayesian approach to some outlier problems. *Biometrika*, **55**, 119–129.

—— (1973). *Bayesian Inference in Statistical Analysis*. Reading, Mass.: Addison-Wesley.

Guttman, I. (1973). Care and handling of univariate or multivariate outliers in detecting spuriocity—a Bayesian approach. *Technometrics*, **15**, 723–738.

Jeffreys, H. (1932). An alternative to the rejection of observations. *Proc. Roy. Soc.* A, **137**, 78–87.

Newcomb, S. (1886). A generalized theory of the combination of observations so as to obtain the best result. *Amer. J. Math.*, **8**, 343–366.

Pearson, E. S. and Chandrasekar, C. (1936). The efficiency of statistical tools and a criterion for the rejection of outlying observations, *Biometrika*, **28**, 308–320.

Pierce, B. (1852). Criterion for the rejection of doubtful observations. *Astron. J.*, No. 45, **2**, No. 21, 161–163.

Appendix

1. (i) $Q = (\mathbf{y}-X\boldsymbol{\beta})'(\mathbf{y}-X\boldsymbol{\beta})+r\delta^2-2(\mathbf{y}-X\boldsymbol{\beta})'\mathbf{Z}\delta$: Assuming $r\neq 0, n$, this expression can be rewritten as

$$(\mathbf{y}-X\boldsymbol{\beta})'(I-r^{-1}\mathbf{ZZ}')(\mathbf{y}-X\boldsymbol{\beta})+r(\delta-\delta_\beta)^2$$

where $\delta_\beta = r^{-1}(\mathbf{y}-X\boldsymbol{\beta})'\mathbf{Z}$. Now writing $I-r^{-1}\mathbf{ZZ}'$ as V, rearranging the terms and assuming that $X'VX$ is nonsingular we get

$$(\boldsymbol{\beta}-\boldsymbol{\beta}^*)'(X'VX)(\boldsymbol{\beta}-\boldsymbol{\beta}^*)+r(\delta-\delta_\beta)^2+y'Vy-\mathbf{y}'VX(X'VX)^{-1}X'V\mathbf{y},$$

where $\boldsymbol{\beta}^* = (X'VX)^{-1}X'V\mathbf{y}$. Defining $\nu S_r^2 = y'V\{I-X(X'VX)^{-1}X'\}V\mathbf{y}$, the above expression can be written as

$$Q = (\boldsymbol{\beta}-\boldsymbol{\beta}^*)'X'VX(\boldsymbol{\beta}-\boldsymbol{\beta}^*)+r(\delta-\delta_\beta)^2+\nu S_r^2.$$

(ii) Consider again the first expression for Q. This may be rewritten as

$$(\boldsymbol{\beta}-\boldsymbol{\beta}_\delta)'X'X(\boldsymbol{\beta}-\boldsymbol{\beta}_\delta)-(\mathbf{y}-\mathbf{Z}\delta)'X(X'X)^{-1}X'(\mathbf{y}-\mathbf{Z}\delta)+\mathbf{y}'\mathbf{y}+r\delta^2-2\mathbf{y}'\mathbf{Z}\delta,$$

where $\boldsymbol{\beta}_\delta = (X'X)^{-1}X'(\mathbf{y}-\mathbf{Z}\delta)$. Denoting $I-X(X'X)^{-1}X'$ by M this can be written as

$$(\boldsymbol{\beta}-\boldsymbol{\beta}_\delta)'X'X(\boldsymbol{\beta}-\boldsymbol{\beta}_\delta)+\mathbf{y}'M\mathbf{y}+\delta^2\mathbf{Z}'M\mathbf{Z}-2\delta\mathbf{y}'M\mathbf{Z}.$$

Now assuming that $\mathbf{Z}'M\mathbf{Z}\neq 0$, and completing the squares for δ, we get

$$Q=(\boldsymbol{\beta}-\boldsymbol{\beta}_\delta)'X'X(\boldsymbol{\beta}-\boldsymbol{\beta}_\delta)+(\mathbf{Z}'M\mathbf{Z})(\delta-\delta^*)^2+\nu S_d^2,$$

where $\delta^*=(\mathbf{Z}'M\mathbf{Z})^{-1}\mathbf{Z}'M\mathbf{y}$ and $\nu S_d^2 = y'M\{I-\mathbf{Z}(\mathbf{Z}'M\mathbf{Z})^{-1}\mathbf{Z}'\}M\mathbf{y}$.

2. (i) $p(\mathbf{y}|\mathbf{Z})=\int(2\pi\sigma^2)^{-\frac{1}{2}(n+1)}\exp\{-(2\sigma^2)^{-1}Q\}\,d\delta\,d\boldsymbol{\beta}\,d\sigma$ (see also Appendix 1)

$$\propto r^{-\frac{1}{2}}|X'VX|^{-\frac{1}{2}}\int_0^\infty \sigma^{-(n-p)}\exp\{-\nu S_r^2/2\sigma^2\}\,d\sigma$$

$$\propto r^{-\frac{1}{2}}|X'VX|^{-\frac{1}{2}}(\nu S_r^2)^{-\frac{1}{2}\nu},$$

where $\nu = n-p-1$.

(ii) $p(\mathbf{y}|\mathbf{Z}=\mathbf{0})=\int(2\pi\sigma^2)^{-\frac{1}{2}(n+1)}\exp[(2\sigma^2)^{-1}\{(\boldsymbol{\beta}-\hat{\boldsymbol{\beta}})'X'X(\boldsymbol{\beta}-\hat{\boldsymbol{\beta}})+\nu_0 S^2\}]\,d\boldsymbol{\beta}\,d\sigma$,

where $\hat{\boldsymbol{\beta}}=(X'X)^{-1}X'\mathbf{y}$, $\nu_0 S^2=(\mathbf{y}-X\hat{\boldsymbol{\beta}})'(\mathbf{y}-X\hat{\boldsymbol{\beta}})$ and $\nu_0=n-p$. Hence we get

$$p(\mathbf{y}|\mathbf{Z}=0)\propto|X'X|^{-\frac{1}{2}}[\nu_0 S^2]^{-\frac{1}{2}\nu_0}.$$

3. Using some matrix algebra it can be shown that

(i) $|X'VX|/|X'X| = \mathbf{Z}'M\mathbf{Z}/r$.

(ii) $\nu S_r^2 = \nu S_d^2$.

2
Experimental Design and Response Surface Methodology

Contents

2.9 "Some New Three Level Designs for the Study of Quantitative Variables" (with D. W. Behnken). *Technometrics,* vol. 2, no. 4 (1960), pp. 455–475 517

2.10 "The 2^{k-p} Fractional Factorial Designs, Part 1" (with J. S. Hunter). *Technometrics,* vol. 3, no. 3 (1961), pp. 311–351 539

2.11 "The 2^{k-p} Fractional Factorial Designs, Part II" (with J. S. Hunter). *Technometrics,* vol. 3, no. 4 (1961), pp. 449–458 581

2.12 "The Choice of a Second Order Rotatable Design" (with N. R. Draper). *Biometrika,* vol. 50, parts 3 and 4 (1963), pp. 335–352 591

2.13 "Sequential Design of Experiments for Nonlinear Models" (with W. G. Hunter). *Proceedings of IBM Scientific Computing Symposium on Statistics* (1963), pp. 113–137 609

2.14 "Some Aspects of Randomization" (with I. Guttman). *J. Roy. Stat. Soc.,* Series B, vol. 28, no. 3 (1966), pp. 543–558 635

2.0
Introduction

BARRY H. MARGOLIN
National Institute of Environmental Health Sciences

It is generally acknowledged that R. A. Fisher created the field of statistically designed experiments more than half a century ago. Presumably, Fisher was motivated by the need he perceived for greater efficiency and broader applicability of results obtained in the type of bio-agricultural research then being conducted at Rothamsted Agricultural Station. It was George E. P. Box, however, starting in the early 1950s, who provided the necessary intellectual catalyst that allowed Fisher's creation to fully attain its current position of importance in chemical, physical, and engineering research. The fourteen papers reprinted here, with their many and varied interconnections, exhibit the breadth and depth of Box's contribution to the design of experiments.

At the times that Fisher and Box made their initial fundamental contributions to the design of experiments, each was employed in an interdisciplinary experimental research setting. Fisher's colleagues at Rothamsted were agronomists, biologists, and geneticists, while Box collaborated at Imperial Chemical Industries, Ltd. (ICI), with chemists and chemical engineers. Experimentation at ICI was quite distinctive from that at Rothamsted in three key aspects. At ICI, compared to Rothamsted, (1) the relative magnitude of random observational errors was considerably smaller, (2) the time involved in producing a single observation was considerably shorter, and (3) the interest in quantitative factors such as concentration of a catalyst or temperature at which a chemical reaction is conducted, as opposed to qualitative factors such as varieties of wheat, was considerably greater.

It was Box's creative genius that perceived clearly the manner in which to exploit these characteristics of ICI experimentation. At that time, factorial and fractional factorial designs were the classical approach to statistically designed experiments, and departures from them were viewed with distaste (see Tocher's discussion in 2.1). Box, however, was keenly aware of the inefficiencies and limitations of factorial-based experiments for the research conducted at ICI. The first paper published by Box on the design of experiments (2.1), coauthored in 1951 with chemist K. B. Wilson, is one of those rare, truly pioneering papers that completely shatters the existing paradigm of how to solve a particular problem. The paper presents "results of a study extending over the past few years by a chemist and a statistician" regarding strategies for the determination of optimum conditions in a chemical process.

The design strategy that is formulated by Box and Wilson to attain an optimum condition is brilliant in both its logic and its simplicity. Rather than exploring the entire continuous experimental region in one fell swoop, one explores a sequence of subregions. Two distinct phases of such a study are discernible. First, in each subregion a classical two-level fractional factorial design is employed to obtain reasonable estimates of the first order partial derivatives of the underlying response surface. The method of steepest ascent then guides the experimenter to the next promising subregion. This first-phase process is iterative and is terminated when a region of near-stationarity is reached. At this point, a new phase is begun, one necessitating radically new designs for the successful culmination of the research effort.

It is in the exploration of the nearly stationary region that Box most clearly exhibits his superb geometrical talents and insights. Whereas algebraic group theory served Finney well in his development of fractional factorial designs, Box's approach to the design and analysis of response surface experiments involves contour plots, reductions to canonical form, and composite designs (i.e., designs consisting of two-level fractional factorials augmented with axial points to permit estimation of quadratic effects). Numerous ideas developed in later papers by Box and coauthors are evident in this paper with Wilson, the most notable being emphasis on the important role of bias and its tradeoff with variance in the selection of a response surface design, and discussion of methods for constructing 2^{k-p} fractional factorial designs.

The creation of composite designs and other "multi-factor" designs, Box's imprecise term for designs freed from the standard factorial framework with its few factor levels, opened an entirely new and important research direction for design creators. In 2.2 Box exhibited the full beauty and flexibility of multi-factor designs when he proved that the minimum variance property of an optimal design is invariant to rotation, and so a given optimal design can be rotated to reduce bias or to eliminate systematic effects deriving from blocking or time trends. This flexibility is well illustrated by the design proposal of Box and Hay (2.3) for the removal of systematic effects when one is sequentially comparing two response curves and a time trend is suspected or known to exist. In their example, which involved a bioassay to determine the relative potency of a test compound to a standard, the experimental object was known to exhibit fatigue with repeated testing. The ease with which this time trend is eliminated should be compared to difficulties with this elimination when the design structure is factorial (see Cox, 1958, and Daniel and Wilcoxon, 1966). The infinite flexibility of rotational randomization discussed in 2.2 later served, in part, to motivate Box and Guttman (2.14) to examine how to achieve the desired effects of Fisherian randomization when the assumption of a finite randomization set is relaxed.

In the years immediately following the Box-Wilson publication (2.1), response surface methodology for the attainment of optimal conditions was applied to a variety of investigations, many outside the chemical industry. One of the more intriguing studies involved factors that affect the motility and viability of bull sperm. In 1954, Box (2.4) published a highly lucid exposition of the main ideas that were distilled from these various investigations, reiterating earlier themes of first order designs and steepest ascent followed by composite designs and exploration of near-stationary regions, and summarizing the gains in understanding achieved by the reduction of fitted surfaces to their canonical forms. Box stressed graphical illustrations as aids to both the reader of the paper and the decision makers who need to understand the estimated response surface obtained in any particular investigation as well as the resulting conclusions. Particularly impressive is the photograph of a hand-constructed three-dimensional model illustrating one such surface. Anyone reading this paper will want to take note of two secondary themes: (1) the roles of the statistician and the experimenter (pp. 26–27), and (2) the risk of one-factor-at-a-time experimentation grounding itself on a ridge surface and seriously missing the channel leading to the true optimum.

Although the early proposals for response surface designs apparently served reasonably well, Box returned to the construction of these designs in 1957 and with J. S. Hunter (2.5) produced a most comprehensive treatment of the topic. Box and Hunter list five desirable properties for a response surface design, a list that would inflate with time and experience to fifteen by the time of the publication of the paper "Robust Designs" with N. R. Draper in 1975. A most important contribution of the paper (2.5) is the introduction of the variance function of a design, which is proportional to the variance of the estimated response at a specified point x. This permits one to construct variance contours for alternative designs and, because the orientation of the response surface is initially unknown, leads naturally to consideration of rotatable designs, that is, designs whose variance contours are spherical. The class of rotatable designs includes the orthogonal first order designs commonly used in early stages of a response surface investigation. When second order designs are considered, however, this class offers substantial improvement over the standard factorial and fractional factorial designs, whose orthogonality is dependent upon a particular orientation of the response surface. A major portion of 2.5 is devoted to the construction of designs that are rotatable or admit orthogonal blocking. More than twenty-five years later, this paper is still a tour de force on the design of multi-factor experiments.

A further method to construct second order rotatable designs is presented in 2.8. Box and Behnken, using results for the moments of means of samples from a k-variate finite population of n elements, demonstrate that such designs can always be derived from first order simplex designs by using a simplex-sum method.

By virtue of working in areas of research grounded in scientific theory, Box knew the importance of models that are nonlinear in their parameters. In 2.6, he and Lucas adopted as their design criterion the minimization of the determinant of the crossproduct matrix of first partial derivatives with respect to the parameters of the nonlinear response function. As in 2.1, the assumptions are that experimentation is sequential and that design variables are continuous. Despite the lack of a general analytical solution, Box and Lucas, via a set of practical examples and a sharing of their geometrical insights, point the way to design solutions for low dimensional problems and prepare the reader for grappling with higher dimensional problems in a more intelligent fashion. The value of preliminary parameter estimates $\underset{\sim}{\theta}^*$ is stressed, and a hint of things to come is evident in the suggestion of using as a measure of the reliability of $\underset{\sim}{\theta}^*$ "some rough estimate of the prior probability distribution."

This shyness about matters Bayesian evaporated completely by the time Box, with W. G. Hunter as coauthor (2.13), resumed the struggle with designs for nonlinear response functions. That Bayesian methods are applicable to experimental design in a practical way may seem surprising to some readers, for most of what is considered to be the core of design of experiments evolved independently of formal Bayesian reasoning. The situation is quite different for nonlinear problems, where preliminary parameter estimates are needed. As Box and Hunter (2.13) comment:

> At first sight, [this] may seem strange that . . . one must initially have estimates of the parameters; after all, it is the purpose of the experiment to obtain estimates. Actually, however, this paradox is merely an example of the fact that any experimental design uses the experimenter's beliefs about the situation being studied. . . . In general, the more one initially knows, the better he can design experiments.

In short, experimental design with nonlinear models will always remain in part an art, dependent upon the skill and knowledge of the experimenter. What better framework for Bayesian methods could possibly exist! Nevertheless, the non-Bayesian reader can rest easy. Box is first and foremost a pragmatist; the design criterion that he and Hunter arrive at via their new formulation is the same determinantal criterion adopted earlier by Box and Lucas (2.6), although newly interpreted as maximizing the posterior density for the most probable value.

From the first work with Wilson (2.1) on designs for the study of response surfaces, it was clear that Box, far more than any other statistician, appreciated the importance for design of bias resulting from a systematic lack of fit of a postulated model. It was not until the publication of his two papers with Draper (2.7, 2.12) on the selection of response surface designs, however, that Box formally incorporated bias into a design criterion. In this pair of papers, the authors make three points relevant to the response surface problem. First, if attention is focused solely on variance or sampling error, this may readily lead to a design that is at odds with both common sense and an experimenter's intuition. Typically, this might involve experimenting with as large a design as possible within the operating region *O,* and thereby employing a design that is poorly prepared to obtain evidence for model inadequacy within the region of interest *R,* frequently a subset of *O.* Second, integrated mean square error, an attractive criterion for design selection, can be represented as the sum of two components of error, each averaged over *R,* the first reflecting integrated squared bias (B) and the second measuring the integrated variance function (V).

The third point is the most crucial for their subsequent development: In the cases they considered, designs obtained by minimizing the bias term B alone are very nearly the same as those obtained by minimizing the integrated mean square error. Thus, it might often be desirable to ignore sampling variability rather than bias in selecting a design. For the polynomial response function, this can be achieved by setting the appropriate design moments equal to the corresponding moments of a uniform distribution over R. In the first paper (2.7), Box and Draper focus mainly on situations where the fitted or graduating polynomial function is linear and the true functional relationship is quadratic. Later they returned to the problem, and in the second paper (2.12) they provided the necessary extensions to the more difficult problem where the graduating function is quadratic and the true relationship is cubic. In this second paper, Box and Draper also introduced as a generalization the possibility of a weighting function that could emphasize or de-emphasize specific subregions of the operating region.

Although Box's major insight was to free himself and other researchers from the narrow limitations of factorial designs, he did not reject factorials. On the contrary, his proposals employ factorial and fractional factorial designs in crucial ways, such as in the use of 2^{k-p} fractions in the first phase of a response surface investigation. It is then no surprise at all to find that in a pair of papers with J. S. Hunter (2.10, 2.11) Box produced what is still the most lucid exposition on the structure, construction, and blocking of regular fractions of 2^k factorials. This includes discussions of alias patterns, fold-over construction and combinations of fractions, and the introduction of the important concept of resolution. Students of design theory would be hard put to find a better entree to the study of regular 2^{k-p} fractional factorial designs.

Nor does Box restrict himself within the factorials solely to the 2^k class. In 2.9, he and Behnken discovered a means to employ incomplete block designs and two-level factorials in the construction of a class of rotatable or near-rotatable incomplete three-level factorial designs suitable for use in response surface investigations. These designs frequently admit orthogonal blocking and, in marked contrast to regular 3^{k-p} fractional factorial designs, are parsimonious in their use of observations to estimate the parameters in a second degree polynomial graduating function.

A careful reading of these reprinted works of George E. P. Box on experimental design and response surface methodology makes it clear that such a thorough exploration of this topic might have kept a very fine research statistician occupied for a quarter of a century. What is impressive is that for Box, as any reader of this work can readily see, this pursuit represented merely one of five major areas of his statistical research over a thirty-year period. It is difficult to describe just how important Box's introduction of methods for the exploration of response surfaces was. Perhaps it is best conveyed by a quotation borrowed from T. S. Eliot:

We shall not cease from exploration
And the end of all our exploring
Will be to arrive where we started
And know the place for the first time.

T. S. Eliot, *Four Quartets*

References Other Than Those of George E. P. Box

Cox, D. R., 1958. *Planning of Experiments,* Chapter 14. New York: J. Wiley and Sons.

Daniel, C., and Wilcoxon, F. 1966. Factorial 2^{p-q} plans robust against linear and quadratic trends. *Technometrics* 8:259–278.

2.1
On the Experimental Attainment of Optimum Conditions

G. E. P. BOX and K. B. WILSON
Imperial Chemical Industries Ltd.

[Read before the RESEARCH SECTION OF THE ROYAL STATISTICAL SOCIETY, November 29th, 1950, Professor M. G. KENDALL in the Chair]

1. INTRODUCTION

THE work described is the result of a study extending over the past few years by a chemist and a statistician. Development has come about mainly in answer to problems of determining optimum conditions in chemical investigations, but we believe that the methods will be of value in other fields where experimentation is sequential and the error fairly small.

1.1 *Statement of the Problem*

The problem of experimental attainment of optimum conditions has been stated and discussed by Hotelling (1941) and by Friedman and Savage (1947). We define the problem as follows:

A *response* η is supposed dependent on the levels of k quantitative *factors* or *variables* $x_1, \ldots x_2, \ldots x_k$ capable of exact measurement and control. Thus for the uth combination of factor levels ($u = 1, \ldots N$)

$$\eta_u = \varphi(x_{1u}, \ldots x_{ku})$$

The Collected Works of George E. P. Box, 1984, Wadsworth, Inc., Belmont, CA 94002.
Originally published in *J. Roy. Stat. Soc.,* Series B, vol. XIII, no. 1 (1951), pp. 1–45.

Owing to unavoidable uncontrolled factors, the observed response y_u varies in repeated observations, having mean η_u and variance σ^2.

In the whole k dimensional factor space, there is a region R, bounded by practical limitation to change in the factors, which we call the *experimental region.* The problem is to find, in the smallest number of experiments, the point $(x_1^0, \ldots x_t^0, \ldots x_k^0)$ within R at which η is a maximum or a minimum. In our field, yield, purity or cost of product are the responses which have to be maximized (or minimized in the case of cost). The factors affecting these responses are variables such as temperature, pressure, time of reaction, proportions of the reactants.

2. Outline of the Methods Adopted

2.1 *Experimental Strategy*

The following circumstances influence the strategy of the experimenter:

(i) The magnitude of experimental error,

(ii) The complexity of the response surface,

(iii) Whether or not experiments may be conducted sequentially so that each set may be designed using the knowledge gained from the previous sets.

A sure way of finding optimum conditions would be to explore the whole experimental region. In practice this would have to be done by carrying out experiments on a grid of points extending through R, and would in principle be possible whatever were the circumstances i, ii, iii, above. Now for a response surface of given complexity there must exist a grid of minimum density to allow adequate approximation, but it is easy to show that the number of points on such a grid would usually be far too large, in the investigations with which we have been concerned. On the other hand, in these investigations the experimental error is usually rather small and the experiments are conducted sequentially. A shorter method is therefore required which will exploit these advantages.

Since the experimental error is small, small changes can be determined accurately, and the experimenter may explore adequately a small *sub-region* of the whole region R with only a few experiments. Also since the experiments are sequential the possibility arises of using the results obtained in one sub-region to move to a second in which the response is higher. By successive application of such a procedure, a maximum or at least a near-stationary point of high response should be reached. By restricting the region in which experiments are conducted, we may find only a local maximum and miss a higher ultimate maximum; however, where fuller exploration is impracticable this possibility must be accepted, and we have not found the implied risk troublesome.

2.2 *Sequential Procedures for Attaining Higher Responses*

A rule is required, therefore, which will lead from one sub-region to another where the response is improved. The most obvious procedure of this sort is the "one factor at a time" method. It has in one form or another been used for a long time, and is formalized by Friedman and Savage. A more efficient procedure is developed below.

2.2.1 *Maximum Gain Methods and Steepest Ascent*

We shall assume that within the region considered the derivatives of the response function are continuous. It is desired to proceed from a point O in the k-dimensional space to a point P distant r from O, at which the gain in response is a maximum. For convenience make O the origin so that the response there is $\varphi(O)$ and at P is $\varphi(P) = \varphi(x_1 \ldots x_k)$. Since OP is to be equal to r,

$$r^2 = \Sigma x_t^2 \qquad (1)$$

We require $\varphi(P) - \varphi(O)$ to be a maximum subject to condition (1). Using Lagrange's method of undetermined multipliers we construct the function:

$$\psi = \varphi(P) - \varphi(O) - \tfrac{1}{2}\mu \, \Sigma x_t^2 \qquad (2)$$

The required maximum is where $\partial\psi/\partial x_t$ are all zero, that is at the point where the k equations

$$\mu x_t = \varphi_t(P) \qquad t = 1, 2, \ldots k \qquad (3)$$

are satisfied. (The notation $\varphi_t(P)$ denotes that the function is differentiated with respect to x_t and the value at P inserted.) From (1) and (3)

$$\mu = \pm \{\Sigma [\varphi_t(P)]^2\}^{\frac{1}{2}}/r \quad . \quad . \quad . \quad . \quad . \quad . \quad . \quad (4)$$

The equations show that for P to be the point distance r from O at which the gain is a maximum, the co-ordinates at P must be proportional to the first order derivatives at P (assumed not all zero). This is equivalent to saying that the point of maximum gain will be one of the points at which a hypersphere radius r and centre O touches a contour surface. Now derivatives at P are in general unknown. Assuming however that φ can be represented about the origin by its Taylor's series in which terms of degree greater than d are ignored, the derivatives at P may be expressed in terms of those at O by the equation

$$\varphi_t(P) = [D_t \sum_{s=0}^{d-1} \{(D_1 x_1 + D_2 x_2 + \ldots + D_k x_k)^s/s!\}]\varphi(O), \quad . \quad . \quad (5)$$

where D_t denotes differentiation with respect to x_t and the expression within the square brackets is expanded and allowed to operate on φ, the values of the derivatives at O being inserted in the equation. Formulae which specify the co-ordinates of the point P of maximum gain distance r from O in terms of the derivatives at O may now be obtained by substituting (5) in (3) and choosing μ so that the solutions of (3) give a maximum. In particular if in the region considered second and higher degree terms may be ignored we have k equations,

$$\mu x_t = \varphi_t(O) \qquad t = 1, 2, \ldots k \quad . \quad . \quad . \quad . \quad . \quad (6)$$

These are the equations of "steepest ascent" or greatest slope from O. By varying the factors in proportion to their first order partial derivatives at O, we move in a direction at right angles to contour planes assumed to be locally parallel and equidistant. If, therefore, the first order derivatives at O can be determined, we may use them to move to a better response at P, provided that the distance r is not taken so large as to seriously strain the linear approximation. Fig. 1 illustrates for two factors x_1 and x_2 the "one factor at a time" approach via the path $OPQR$ and the steepest ascent approach along the path $O'P'$. The curved lines are response contours.

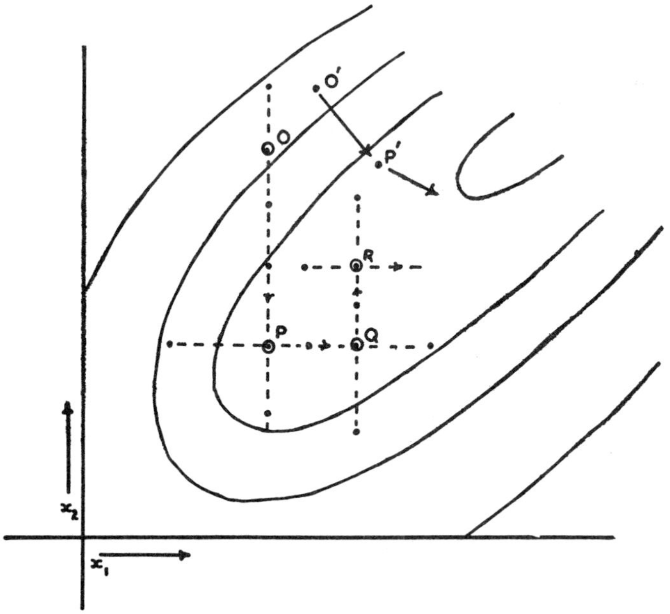

FIG. 1.—The "one factor at a time" path – – – and the steepest ascent path —.

If the experimental error were larger and experimentation could therefore be conducted more economically by covering a wider subregion with rather more experimental points, the second degree approximation could be used. The locus of P would then be obtained by solving the k linear equations,

$$\mu x_t = \varphi_t + x_1\varphi_{1t} + x_2\varphi_{2t} + \ldots + x_k\varphi_{kt} \quad . \quad . \quad . \quad . \quad (7)$$

for a suitable set of values of μ, the values of the derivatives being taken at the point O.

2.2.2 *Relative Scales of Measurement of the Factors*

The formulae of the last section are clearly not scale-invariant, and it is only after the relative scales of measurement have been agreed that "distance" in the factor space can have any real meaning. The general experimental implications can easily be seen. In practice an experimental design is carried out around O, and the choice of factor levels there tacitly fixes the relative scale units. For example, an experimenter might propose a temperature change of say 5° C., but a change in concentration of a catalyst of as much as 100 per cent. to match it. The direction of steepest ascent calculated would then be at right angles to contour surfaces drawn on *this scale*. Experience is necessary, and clearly the choice can affect the amount of subsequent work. It is shown in §(4) how a grossly wrong choice can be rectified.

2.2.3 *Application of Steepest Ascent Formula*

If the derivatives could be determined without experimental error it would be possible by successive application of the steepest ascent formula (6) to move to points of higher and higher response till eventually a stationary point was reached. Because of experimental error, the more successful we were in reducing the derivatives, the more difficult would it become in the next set of experiments to determine them with sufficient accuracy to proceed further. In practice, therefore, the steepest ascent technique is employed to move from a point remote from a stationary value at which the surface slopes are large, to a point closer to it at which they are small compared with the experimental error.

2.3 *Exploration of the Surface in a Near-Stationary Region*

When the experimenter is in a near-stationary region (either as the result of successive application of the steepest ascent procedure, or because he is working to improve a process which has already received much previous attention), he will wish to conduct more detailed experiments in the limited region in order to determine the local nature of the surface.

In particular for example he will desire to know whether it probably contains a true maximum (in which case he will wish to estimate its position), a minimax or col (in which case he will wish to know how to "climb out of it"), or a ridge (when he will wish to know its direction and slope). By determining effects of second and possibly higher order in the region by suitably designed experiments it will be possible to answer these questions.

2.3.1 *Arrangement of Sections* 3, 4 *and* 5

In what follows, we shall be concerned with: (i) the application of the steepest ascent technique to move in the space to a near-stationary point; (ii) the examination of a near-stationary region.

For (i) we shall mainly require designs for the determination of first-order effects, and for (ii) designs to determine effects of higher order. Section (3), which follows, is concerned with the derivation of these designs, (4) with the application of the first order steepest ascent process, and (5) with the examination of a near-stationary region.

3. EXPERIMENTAL DESIGNS TO DETERMINE DIFFERENTIAL COEFFICIENTS

We assume that the response η at any point $(x_1, \ldots x_t, \ldots x_k)$ in the factor space can be represented by a regression equation of the form—

$$\eta = \beta_0 + \beta_1 x_1 + \beta_2 x_2 + \ldots + \beta_{11} x_1^2 + \ldots + \beta_{12} x_1 x_2 + \ldots + \beta_{111} x_1^3 + \ldots \quad . \quad (8)$$

In general the coefficient of $x_1^\alpha x_2^\gamma \ldots x_k^\pi$ is denoted by $\beta_{1^\alpha 2^\gamma \ldots k^\pi}$, the notation implying that the subscript 1 is to be repeated α times; the subscript 2, γ times, etc.

Now suppose that *all* terms of degree d and less are included and within a given region a perfect fit is thus obtained. Then equation (8) will correspond with the Taylor expansion of the response function $\varphi(x_1, x_2, \ldots x_k)$ about the origin O of the variables and the derivatives at O will be simple multiples of the β's. For example, $\varphi_0 = \beta_0$, $\varphi_1 = \beta_1$, $\varphi_{11} = 2\beta_{11}$, $\varphi_{12} = \beta_{12}$, $\varphi_{111} = 6\beta_{111}$ and in general—

$$\varphi_{1^\alpha 2^\gamma \ldots k^\pi} = \alpha!\,\gamma!\,\ldots\,\pi!\,\beta_{1^\alpha 2^\gamma \ldots k^\pi}, \quad \ldots \quad (9)$$

where the quantity on the left denotes the value of the derivative obtained by differentiating the function α times with respect to x_1, γ times with respect to x_2, etc., at the origin O.

If responses are observed at a set of points suitably numerous and suitably placed within the region concerned, estimates b_0, b_1, b_2, etc., of β_0, β_1, β_2, etc., can be obtained by fitting the regression equation. Using (9) to obtain the appropriate multiplier, the b's will provide estimates f_0, f_1, f_2, etc., of the derivatives at O; $\varphi_0, \varphi_1, \varphi_2$, etc. We define an $N \times k$ matrix **D** called the *design matrix* whose elements are the levels of the k variables employed in each of the N trials. The elements $x_{1u}, x_{2u}, \ldots, x_{ku}$ of the uth row of **D** are the co-ordinates of a point in the k dimensional factor space. The N points thus defined constitute the *experimental design*. The N elements of a vector **Y** are observations of the responses obtained at these points and $\boldsymbol{\eta} = E(\mathbf{Y})$ is the corresponding vector of expected values. Suppose equation (8) contains L terms then it may be written—

$$\eta = \sum_{i=1}^{L} \beta_i X_i, \quad \ldots \quad (10)$$

where the X_i are called the independent variables. In the case we are considering they are products and powers of the co-ordinates of the experimental points. Denoting by **X** the $N \times L$ *matrix of independent variables*, showing the N values of the L functions of the co-ordinates, (10) becomes in matrix notation

$$\boldsymbol{\eta} = \mathbf{X}\boldsymbol{\beta}, \quad \ldots \quad (11)$$

where $\boldsymbol{\beta}$ is the $L \times 1$ vector of unknown constants. To fit the surface we employ the method of least squares. Assuming that the observational errors have constant variance σ^2 and are uncorrelated and also that **X** has rank L it is well known (Gauss, 1821; Markoff, 1912; Aitken, 1935; David and Neyman, 1938; Plackett, 1949) that

(i) Unbiased estimates of the elements of the vector $\boldsymbol{\beta}$, linear in the observations, with smallest variance are provided by the $L \times 1$ vector of estimates $\mathbf{B} = \mathbf{TY}$ where the *transforming matrix* **T** is $(\mathbf{X}'\mathbf{X})^{-1}\mathbf{X}'$.

(ii) The $L \times L$ matrix of variances and covariances of these estimates is $\mathbf{C}^{-1}\sigma^2$ where $\mathbf{C} = \mathbf{X}'\mathbf{X}$ is the matrix of sums of squares and products of the independent variables. (We call $\mathbf{C}^{-1}$ the *precision matrix*.)

(iii) An unbiased estimate of $(N - L)\sigma^2$ is provided by the residual sum of squares

$$(N - L)s^2 = (\mathbf{Y} - \mathbf{XB})'(\mathbf{Y} - \mathbf{XB}) = \mathbf{Y}'\mathbf{Y} - \mathbf{B}'\mathbf{CB} = \mathbf{Y}'\mathbf{Y} - \mathbf{Y}'\mathbf{XB}.$$

We note that once a design for determining derivatives up to a given order has been decided on, the matrices **T** and $\mathbf{C}^{-1}$ can be calculated and tabled once for all. The computations necessary to calculate the estimates and their standard errors from any given set of observations are then extremely simple.

3.1 *An Example*

Suppose there were two variables only and it was desired to determine the derivatives up to the second order in a neighbourhood where terms of higher order could be ignored. We should fit the equation

$$\eta = \beta_0 + \beta_1 x_1 + \beta_2 x_2 + \beta_{11} x_1^2 + \beta_{22} x_2^2 + \beta_{12} x_1 x_2 \quad \ldots \quad (12)$$

to an arrangement of six or more points in the two-dimensional space. One suitable arrangement would be a factorial design in which each of the two factors was varied at three levels. This design is rather specialized, however, and as a more general illustration we consider a design not of the factorial type.

One such arrangement consists of five points in the shape of a regular pentagon with one point in the centre. If the figure chosen had its uppermost side horizontal and the distance of any point from the centre was one unit, the co-ordinates of the point would be given by the design matrix

$$\mathbf{D} = \begin{array}{c} 1 \\ 2 \\ 3 \\ 4 \\ 5 \\ 6 \end{array}\begin{array}{c} \\ \\ \\ \end{array}\left[\begin{array}{rr} x_1 & x_2 \\ \hline 0\cdot5878 & 0\cdot8090 \\ 0\cdot9511 & 0\cdot3090 \\ 0\cdot0000 & -1\cdot0000 \\ -0\cdot9511 & -0\cdot3090 \\ -0\cdot5878 & 0\cdot8090 \\ 0\cdot0000 & 0\cdot0000 \end{array}\right] \qquad (13)$$

The matrix $\mathbf{X}$ of independent variables has six columns giving the corresponding values of x_0, x_1, x_2, x_1^2, x_2^2 and x_1x_2 at the experimental points (x_0 is the variable corresponding with β_0 and is always unity). From $\mathbf{X}$ we calculate in turn* $\mathbf{X'X} = \mathbf{C}$, $\mathbf{C}^{-1}$ and $\mathbf{T'}$; the last two matrices are given below:

$$\mathbf{C}^{-1} = (\mathbf{'XX})^{-1} = \begin{array}{c|cccccc} & b_0 & b_1 & b_2 & b_{11} & b_{22} & b_{12} \\ \hline b_0 & 1 & . & . & -1 & -1 & . \\ b_1 & . & 0\cdot4 & . & . & . & . \\ b_2 & . & . & 0\cdot4 & . & . & . \\ b_{11} & -1 & . & . & 1\cdot6 & 0\cdot8 & . \\ b_{22} & -1 & . & . & 0\cdot8 & 1\cdot6 & . \\ b_{12} & . & . & . & . & . & 1\cdot6 \end{array} \qquad (14)$$

$$\mathbf{T'} = \mathbf{X(X'X)}^{-1} = \begin{array}{c|cccccc} & b_0 & b_1 & b_2 & b_{11} & b_{22} & b_{12} \\ \hline 1 & 0 & 0\cdot2351 & 0\cdot3236 & 0\cdot0764 & 0\cdot3236 & 0\cdot7608 \\ 2 & 0 & 0\cdot3804 & -0\cdot1236 & 0\cdot5236 & -0\cdot1236 & -0\cdot4702 \\ 3 & 0 & 0\cdot0000 & -0\cdot4000 & -0\cdot2000 & 0\cdot6000 & 0\cdot0000 \\ 4 & 0 & -0\cdot3804 & -0\cdot1236 & 0\cdot5256 & -0\cdot1236 & 0\cdot4702 \\ 5 & 0 & -0\cdot2351 & 0\cdot3236 & 0\cdot0764 & 0\cdot3236 & -0\cdot7608 \\ 6 & 1 & 0\cdot0000 & 0\cdot0000 & -1\cdot0000 & -1\cdot0000 & 0\cdot0000 \end{array} \qquad (15)$$

To use the design, observations would be made at levels of the variables proportional to the elements in $\mathbf{D}$. The sums of products of the observations with the columns of the matrix $\mathbf{T'}$ supply the estimates b_0, b_1, etc., of the β's. The variances and covariances of the estimates are given by the elements of $\mathbf{C}^{-1}$ multiplied by σ^2 (assumed known).

The relations $f_0 = b_0$, $f_1 = b_1$, $f_2 = b_2$, $f_{11} = 2b_{11}$, $f_{22} = 2b_{22}$, $f_{12} = b_{12}$, would supply unbiased estimates f_0, f_1, etc., of the derivatives φ_0, φ_1, etc., at the origin of the variables, providing it could be assumed that over the region considered the second degree approximation was adequate.

3.2 *Effect of Lack of Fit of Assumed Equation. General Theory of Aliases*

In fitting an equation like (12) to a set of experimental points the estimates of the β's differ from the true values on account of: (i) experimental error in determining the responses; (ii) biases arising when it is impossible to represent the function by an equation of the type fitted.

An experimental design should therefore be judged, partly by the apparent precision with which the desired constants are estimated (as shown by the matrix $\mathbf{C}^{-1}$), and partly by the magnitude of the possible biases in the estimates.

The nature of the biases for any given design may be determined as follows. Suppose that, within a given region of the factor space, the response function may be represented exactly by an equation such as (8) involving L constants; but that the experimenter assumes an adequate

* In this particular example there are as many constants to estimate as there are observations, consequently $\mathbf{X}$ is a square matrix and $\mathbf{T} = \mathbf{X}^{-1}$ can be obtained directly.

fit to be possible using an equation involving only $M < L$ of these constants, and performs $N \geqslant M$ experiments to estimate them. Then, it is wrongly assumed that—

$$\boldsymbol{\eta} = \mathbf{X}_1\boldsymbol{\beta}_1, \qquad (16)$$

when in fact—

$$\boldsymbol{\eta} = \mathbf{X}_1\boldsymbol{\beta}_1 + \mathbf{X}_2\boldsymbol{\beta}_2, \qquad (17)$$

and $\mathbf{X}_1$ is $(N \times M)$; $\mathbf{X}_2$, $(N \times S)$; $\boldsymbol{\beta}_1$, $(M \times 1)$; $\boldsymbol{\beta}_2$, $(S \times 1)$ and $M + S = L$. The least squares estimates obtained assuming (16) is true are

$$\mathbf{B}_1 = (\mathbf{X}_1'\mathbf{X}_1)^{-1}\mathbf{X}_1'\mathbf{Y} \qquad (18)$$

and will in general be biased, since

$$E(\mathbf{B}_1) = (\mathbf{X}_1'\mathbf{X}_1)^{-1}\mathbf{X}_1'\boldsymbol{\eta} \qquad (19)$$

$$= (\mathbf{X}_1'\mathbf{X}_1)^{-1}\mathbf{X}_1'\mathbf{X}_1\boldsymbol{\beta}_1 + (\mathbf{X}_1'\mathbf{X}_1)^{-1}\mathbf{X}_1'\mathbf{X}_2\boldsymbol{\beta}_2 \qquad \text{(using (17))}$$

and consequently

$$\mathbf{B}_1 \rightarrow \boldsymbol{\beta}_1 + \mathbf{T}_1\mathbf{X}_2\boldsymbol{\beta}_2 \qquad (20)$$

The arrow notation is used to indicate that the quantity on the left is an unbiased estimate of the quantity on the right. The $M \times S$ matrix $\mathbf{T}_1\mathbf{X}_2$ will be called the *alias matrix* and be denoted by $\mathbf{A}$. The expression (20) defines M relations of the type

$$b_i \rightarrow \beta_i + \sum_{j=1}^{S} \alpha_{ij}\beta_j, \qquad (21)$$

where a_{ij} is the element of the ith row and j^{th} column of $\mathbf{A}$. Thus if S extra constants β_j are needed accurately to describe the function, these may bias the estimates of the M constants which the experimenter is attempting to estimate. The extent of the biases will depend on the magnitude of the elements a_{ij} of the matrix $\mathbf{A}$ which themselves depend upon the arrangement of points chosen. In particular, if a certain a_{ij} is zero then the estimate b_i will not be biased by β_j.

As an illustration the nature of the biases arising from third-order effects is found for the pentagonal design. The four third-order derivatives are proportional to β_{111}, β_{222}, β_{112}, β_{122}, which are the coefficients of x_1^3, x_2^3, $x_1^2x_2$ and $x_1x_2^2$ respectively. Therefore

$$\mathbf{X}_2 = \begin{array}{c} \\ 1\\2\\3\\4\\5\\6 \end{array} \begin{array}{cccc} x_1^3 & x_2^3 & x_1^2x_2 & x_1x_2^2 \\ \end{array}$$

$$\mathbf{X}_2 = \begin{bmatrix} 0\cdot2031 & 0\cdot5295 & 0\cdot2795 & 0\cdot3847 \\ 0\cdot8602 & -0\cdot0295 & -0\cdot2795 & 0\cdot0908 \\ 0\cdot0000 & -1\cdot0000 & 0\cdot0000 & 0\cdot0000 \\ -0\cdot8602 & -0\cdot0295 & -0\cdot2795 & -0\cdot0908 \\ -0\cdot2031 & 0\cdot5295 & 0\cdot2795 & -0\cdot3847 \\ 0\cdot0000 & 0\cdot0000 & 0\cdot0000 & 0\cdot0000 \end{bmatrix} \qquad (22)$$

(rows 1–6; columns x_1^3, x_2^3, $x_1^2x_2$, $x_1x_2^2$)

Multiplying this matrix by $\mathbf{T}$ in (15) we obtain the alias matrix

$$\mathbf{A} = \mathbf{T}\mathbf{X}_2 = \begin{array}{c} 0\\1\\2\\11\\22\\12 \end{array} \begin{bmatrix} . & . & . & . \\ 0\cdot75 & . & . & 0\cdot25 \\ . & 0\cdot75 & 0\cdot25 & . \\ . & 0\cdot25 & -0\cdot25 & . \\ . & -0\cdot25 & 0\cdot25 & . \\ -0\cdot50 & . & . & 0\cdot50 \end{bmatrix} \qquad (23)$$

(columns 111, 222, 112, 122)

whence it follows that—

$$\left.\begin{aligned} b_0 &\rightarrow \beta_0 \\ b_1 &\rightarrow \beta_1 + 0\cdot75\beta_{111} + 0\cdot25\beta_{122} \\ b_2 &\rightarrow \beta_2 + 0\cdot75\beta_{222} + 0\cdot25\beta_{112} \\ b_{11} &\rightarrow \beta_{11} + 0\cdot25\beta_{222} - 0\cdot25\beta_{112} \\ b_{22} &\rightarrow \beta_{22} - 0\cdot25\beta_{222} + 0\cdot25\beta_{112} \\ b_{12} &\rightarrow \beta_{12} - 0\cdot50\beta_{111} + 0\cdot50\beta_{122} \end{aligned}\right\} \qquad (24)$$

In general the coefficients in **A** and consequently the magnitude of the possible biases will depend upon the choice of experimental design. It can be shown, however (Appendix 1), that if more constants are needed to represent the function than there are experiments, the extra constants will either bias the estimates or else appear in the residual degrees of freedom and bias the error estimate. The only exception to this rule occurs when a design is such that one or more columns of $\mathbf{X}_2$ consists entirely of zeros. In this case the corresponding "extra" constants will not appear.

3.3 *Comparison of Designs*

The matrices $\mathbf{C}^{-1}$ and **A** depend only on the arrangement of the experimental points, and not on any particular set of observations **Y** which might be made at these points. They supply therefore an objective basis for comparing designs for precision and bias. If the number of experiments in two designs which it is desired to compare are not the same, the precision of the estimates may be brought to a common basis by considering the relative variance of the estimates *per observation.* Thus we compare $N_1\mathbf{C}_1^{-1}$ with $N_2\mathbf{C}_2^{-1}$.

A more difficult problem arises because both $\mathbf{C}^{-1}$ and **A** are dependent on scale factors. (For the exact nature of this dependence see Appendix 2.) If therefore the comparison between designs is to have any meaning they must first be brought to the same "size". The conclusions we draw will depend to some extent on how we define "size". A design is of course a k dimensional distribution of points and the measures for size adopted in this paper are the marginal second moments of these points about their means. Thus if there are k factors and N observations s_t is called the *spread* for the tth variable where

$$s_t^2 = \{\sum_{u=1}^{N} (x_{tu} - \bar{x}_t)^2\}/N,$$

and two designs are regarded as of comparable size when they are measured so that the spread for each of the factors is the same in the two designs. For designs in which each linear effect is uncorrelated with every other effect the convention amounts to choosing the units so that the linear effects are determined with equal accuracy in the designs to be compared. Other conventions could of course be used, and would possibly be more appropriate for certain types of arrangement. The equating of second moments does in the cases so far studied seem to provide reasonable results. In any case changes of scale have opposite effects on $\mathbf{C}^{-1}$ and **A**, for example choosing wider ranges for the factors in a design will result in the reduction of the variances of the estimates but an increase in the coefficients of the aliases. It is unlikely therefore that we should be seriously misled by this convention.

For example, suppose the merits of the 3^2 factorial and pentagonal design in determining the derivatives up to second order are to be compared and the levels of the factorial design are $-1, 0,$ and 1 and the levels for the pentagonal design are those shown in (13). Then for the factorial design, $s_1^2 = s_2^2 = 2/3$ and for the pentagonal, $s_1^2 = s_2^2 = 5/12$, consequently the dimensions of the pentagonal design must be increased by the factor—

$$g = s(F)/s(P) = (2/3 \times 12/5)^{\frac{1}{2}} = 1{\cdot}265 \qquad (25)$$

The appearance of the two designs scaled on this basis are shown in Fig. 2 (see p. 9)—

The relative precision of the estimates when the designs are scaled on this basis may be judged from the matrices $N^{-}\mathbf{C}^{1}$, the elements of which are proportional to the variances and covariances per observation.

FACTORIAL

$$9\mathbf{C}_F^{-1} = \begin{array}{c} \\ b_0 \\ b_1 \\ b_2 \\ b_{11} \\ b_{22} \\ b_{12} \end{array} \begin{array}{cccccc} b_0 & b_1 & b_2 & b_{11} & b_{22} & b_{12} \\ \left[\begin{array}{cccccc} 5{\cdot}0 & . & . & -3{\cdot}0 & -3{\cdot}0 & . \\ . & 1{\cdot}5 & . & . & . & . \\ . & . & 1{\cdot}5 & . & . & . \\ -3{\cdot}0 & . & . & 4{\cdot}5 & . & . \\ -3{\cdot}0 & . & . & . & 4{\cdot}5 & . \\ . & . & . & . & . & 2{\cdot}25 \end{array}\right] \end{array}$$

PENTAGONAL

$$6\mathbf{C}_P^{-1} = \begin{array}{c} \\ \end{array}\begin{bmatrix} 6\cdot0 & . & . & -3\cdot75 & -3\cdot75 & . \\ . & 1\cdot5 & . & . & . & . \\ . & . & 1\cdot5 & . & . & . \\ -3\cdot75 & . & . & 3\cdot75 & 1\cdot875 & . \\ -3\cdot75 & . & . & 1\cdot875 & 3\cdot75 & . \\ . & . & . & . & . & 3\cdot75 \end{bmatrix}. \quad (26)$$

(columns: b_0, b_1, b_2, b_{11}, b_{22}, b_{12})

The biases due to possible third order effects are found from the alias matrices:

FACTORIAL (columns: *111*, *222*, *112*, *122*; rows: *0*, *1*, *2*, *11*, *22*, *12*)

$$\mathbf{A}_F = \begin{bmatrix} . & . & . & . \\ 1\cdot0 & . & . & . \\ . & 1\cdot0 & . & . \\ . & . & . & . \\ . & . & . & . \\ . & . & . & . \end{bmatrix}$$

PENTAGONAL (columns: *111*, *222*, *112*, *122*)

$$\mathbf{A}_P = \begin{bmatrix} . & . & . & . \\ 1\cdot2 & . & . & 0\cdot4 \\ . & 1\cdot2 & 0\cdot4 & . \\ . & 0\cdot32 & -0\cdot32 & . \\ . & -0\cdot32 & 0\cdot32 & . \\ -0\cdot63 & . & . & 0\cdot63 \end{bmatrix}. \quad (27)$$

There is little to choose in the relative precision of the designs. The pentagonal design estimates the quadratic effects rather more accurately and interaction effects rather less accurately. The quadratic estimates are, however, correlated in the pentagonal design. The alias matrices show that if third order terms were not negligible the estimates would be more heavily biased in the case of the pentagonal design. This is to be expected in view of the fewer experiments used.

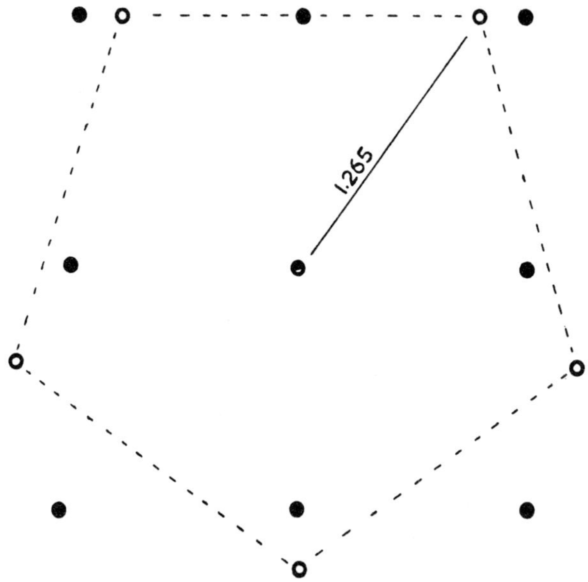

FIG. 2.—The pentagonal design and the 3^2 factorial with marginal second moments equated.

Two types of design are of particular importance, and will be referred to as designs of types A and B. Designs of type A of order d will be defined as those which give unbiased estimates of all derivatives of order 1 to d, providing the assumption is strictly true that all terms of higher degree may be ignored. In such designs N can be as small as M, the number of constants to be determined. Designs of type B of order d will be defined as those which provide unbiased estimates of all derivatives of order 1 to d, even though terms of order $d + 1$ exist. In such designs the number of experiments N must be larger than the number of constants M to be estimated, so that $N - M$ residual degrees of freedom exist in which to accommodate the aliases of order $d + 1$. Such designs need not, however, allow for separate estimation of these higher order effects, and S the number of extra constants accommodated may be considerably greater than $N - M$, the number of residual degrees of freedom.

Although φ_0, the value of the function at the point (0, 0, . . . 0) which can be regarded as the derivative of order zero, appears in the set-up, we do not need to know its value in order to apply the formula for steepest ascent or to study the nature of the response surface. From this point of view bias in its estimate will be immaterial.

3.4 *Designs for Determining First Order Differential Coefficients*

If higher order differential coefficients can be ignored, the first order differential coefficients at a point O, the origin for the variables, can be obtained by fitting the plane

$$\eta = \beta_0 + \beta_1 x_1 + \beta_2 x_2 + \ldots + \beta_k x_k \quad . \quad . \quad . \quad . \quad (28)$$

to a set of points arranged in a suitable manner about O. The estimates $b_0, b_1, \ldots b_k$ of the β's will also be estimates $f_0, f_1, \ldots f_k$ of $\varphi_0, \varphi_1, \ldots, \varphi_k$, the required derivatives at the point O. Designs suitable for this purpose can be obtained by varying each of the factors $x_1, x_2, \ldots x_k$ at only two levels. For convenience these levels are taken to be -1 and 1.

If all possible combinations of levels are included, the design will be a complete factorial design, and the 2^k points in space will be at the vertices of a k dimensional hypercube of side 2 units with its centre at the point (0, 0, . . . 0). If not all the 2^k experiments are performed, the design will be called an *incomplete factorial.* Particular classes of incomplete factorial designs are the fractional factorial designs introduced by Finney (1945) and the multifactorial designs of Plackett and Burman (1946).

First order designs of type A are provided by the multifactorial designs of Plackett and Burman. These authors supply designs for $k = 3, 7, 11, 15, \ldots, 4m - 1, \ldots 99$ factors in $N = 4, 8, 12, 16, \ldots 4m \ldots, 100$ experiments. For intermediate values of k the next higher design may be used and the appropriate number of columns omitted from the design matrix. When N is a power of two these arrangements are identical with the fractional factorial designs.

First order designs of type B can always be obtained by duplicating the appropriate Plackett and Burman design of type A and order 1, with reversed signs. This means that if a design of type A had been completed it would always be possible to make a partial check on the assumptions by modifying it in this simple way to a design of type B. Any variable held constant in the first set of experiments could be held at a different level* on duplicating the design. Thus corresponding to a type A design for $N - 1$ factors in N experiments we would obtain a type B design for N factors in $2N$ experiments. The above assertion is a special case of a more general theorem proved in Appendix (2). All these designs are orthogonal (in the sense that $\mathbf{X}'\mathbf{X} = N\mathbf{I}$). The transforming matrix $\mathbf{T}$ is therefore $N^{-1}\mathbf{X}'$, and the estimates f_0, f_1, f_2, etc., can be obtained very simply and are uncorrelated. The alias matrix is of the form $N^{-1}\mathbf{X}_1'\mathbf{X}_2$, and since for each factor the levels $+1$ and -1 occur equal numbers of times none of the first-order estimates are biased by quadratic derivatives $\varphi_{11}, \varphi_{22}$, etc. Thus, if the curvature of a surface, however great, could be expressed in terms of quadratic effects alone, the estimates of the first order effects would remain completely unaffected. On the other hand, the existence of mixed second order derivatives φ_{12}, etc., corresponding to the two-factor interactions, may bias these estimates.

* The intention here is to indicate the method for constructing the design rather than the practical advisability of introducing an extra factor in this way.

3.4.1 *Two Level Factorial and Fractional Factorial Designs*

Most of the designs of order 1 used in this work have been two level factorial and fractional factorial arrangements. It is not our intention to describe their construction and properties in detail. They have been the subject of study by Fisher (1935), Yates (1935 and 1937), Hotelling (1944), Plackett and Burman (1946), Rao (1947), Kempthorne (1947), Davies and Hay (1950), and others. We illustrate here certain features which are of particular interest in the study of the present problem. Consider for simplicity, the two-level factorial design for the variables x_1, x_2, x_3.

The design matrix **D** consists of the vertices of a 3-dimensional cube. Suppose the function

$$\eta = \beta_0 x_0 + \beta_1 x_1 + \beta_2 x_2 + \beta_3 x_3 + \beta_{12} x_1 x_2 + \beta_{13} x_1 x_3 + \beta_{23} x_2 x_3 + \beta_{123} x_1 x_2 x_3$$

is fitted to the responses found at these eight points.

The matrix of independent variables is

$$\begin{array}{cccccccc} & (1) & (2) & (3) & (5) & (6) & (7) & (4) \\ x_0 & x_1 & x_2 & x_3 & x_1x_2 & x_1x_3 & x_2x_3 & x_1x_2x_3 \end{array}$$

$$\mathbf{X}_1 = \begin{bmatrix} 1 & -1 & -1 & -1 & 1 & 1 & 1 & -1 \\ 1 & -1 & -1 & 1 & 1 & -1 & -1 & 1 \\ 1 & -1 & 1 & -1 & -1 & 1 & -1 & 1 \\ 1 & -1 & 1 & 1 & -1 & -1 & 1 & -1 \\ 1 & 1 & -1 & -1 & -1 & -1 & 1 & 1 \\ 1 & 1 & -1 & 1 & -1 & 1 & -1 & -1 \\ 1 & 1 & 1 & -1 & 1 & -1 & -1 & -1 \\ 1 & 1 & 1 & 1 & 1 & 1 & 1 & 1 \end{bmatrix} \quad . \quad . \quad . \quad (29)$$

The matrix **T** is $\frac{1}{8}\mathbf{X}'_1$. The estimate of each β is found, therefore, by forming the sum of products of the observed responses with the appropriate column of matrix $\mathbf{X}_1$ and dividing the result by 8. The estimates b_0, b_1, b_2, b_3, b_{12}, b_{13}, b_{23}, b_{123} thus obtained are the mean $\bar{y}$, and one half of the quantities usually referred to as main effects and interactions (see for example Yates, 1937).

From the approach of the previous section, the nature of possible biases in the estimates for this design may now be determined. We assume that the response function can be represented exactly by a Taylor series, including all terms* up to degree d, involving the fitting of L constants. The matrix $\mathbf{X}_1$ of independent variables then corresponds to $\mathbf{X}_1$ of equation (16), whilst $\mathbf{X}_2$ is the matrix of powers and products of degree d and less not included in $\mathbf{X}_1$. Now (Fisher, 1942; Finney, 1945) the columns of the matrix $\mathbf{X}_1$ may be regarded as the elements of a finite Abelian group. In particular they form a set closed with respect to multiplication (in the sense that if the elements of the columns are multiplied by themselves or by the elements of any other column, the new column resulting will always be identical with one or other of those already contained in $\mathbf{X}_1$). It follows that every column of the matrix $\mathbf{X}_2$ must coincide with one or other of the columns in $\mathbf{X}_1$. The alias matrix, $\mathbf{A} = \frac{1}{8}\mathbf{X}_1'\mathbf{X}_2$, is particularly simple therefore. Every column in $\mathbf{X}_2$ will have a non-zero inner product with one and only one column in $\mathbf{X}_1$, and consequently each of the extra constants will be associated with one and only one of the estimates b_0, b_1, etc. The group property arises because for any row of the matrix (29) $x_0 = 1 = x_1^2 = x_2^2 = x_3^2$; consequently, for example, $x_1^2 x_2 = x_2$, $x_1^4 x_2^3 x_3 = x_2 x_3$, etc. It follows from the nature of **A** that β_{11}, β_{22}, β_{1111}, β_{1122}, and in fact all the constants containing an even number of subscripts will appear in the expected value of b_0 and nowhere else. Similarly β_{112} will be associated with b_2, $\beta_{11112223}$ with b_{23} and so on. The defining relation for the group may be written

$$I = 11 = 22 = 33 \quad . \quad . \quad . \quad . \quad . \quad . \quad (30)$$

* The logic of including all derivatives up to a given degree d, rather than only those which correspond to interaction terms, is seen if it is supposed that experiments are being performed in the neighbourhood of a maximum. Since, for a maximum the matrix of second order derivatives φ_{st}, must be negative definite—and this in turn implies that $\varphi_{ss}\varphi_{tt} > \varphi_{st}^2$—the geometric mean of any pair of quadratic terms would be greater than the corresponding interaction term. Thus if the interaction terms could not be ignored then certainly the corresponding quadratic terms could not be negligible.

The identity I corresponds with the column x_0 and consequently with b_0. Since this also implies that $I = 1111 = 1122$, etc., the nature of the biases associated with a particular estimate may be obtained by multiplying the defining relation by the subscripts of the estimate. For example, $1 = 111 = 122 = 133 = 11111 = 11122$, etc., and consequently

$$b_1 \to \beta_1 + \beta_{111} + \beta_{122} + \beta_{133} + \beta_{11111} + \text{etc.} \qquad (31)$$

or

$$b_1 \to \varphi_1 + \tfrac{1}{6}\varphi_{111} + \tfrac{1}{2}\varphi_{122} + \tfrac{1}{2}\varphi_{133} + \tfrac{1}{120}\varphi_{11111} + \text{etc.} \qquad (32)$$

where the multipliers of the derivatives are given by (9). The expected values, including terms up to order 3, for the eight estimates from this complete factorial design are thus—

$$\left.\begin{aligned}
b_0 &\to \varphi_0 + \tfrac{1}{2}\varphi_{11} + \tfrac{1}{2}\varphi_{22} + \tfrac{1}{2}\varphi_{33}\\
b_1 &\to \varphi_1 + \tfrac{1}{6}\varphi_{111} + \tfrac{1}{2}\varphi_{122} + \tfrac{1}{2}\varphi_{133}\\
b_2 &\to \varphi_2 + \tfrac{1}{6}\varphi_{222} + \tfrac{1}{2}\varphi_{112} + \tfrac{1}{2}\varphi_{233}\\
b_3 &\to \varphi_3 + \tfrac{1}{6}\varphi_{333} + \tfrac{1}{2}\varphi_{113} + \tfrac{1}{2}\varphi_{223}\\
b_{12} &\to \varphi_{12}\\
b_{13} &\to \varphi_{13}\\
b_{23} &\to \varphi_{23}\\
b_{123} &\to \varphi_{123}
\end{aligned}\right\} \qquad (33)$$

Rao (1947) obtained fractional factorial designs with the properties he desired by associating further factors with the independent variables corresponding with the interactions in the full factorial design, a method used previously by Yates (1935). Thus to obtain the design of type B and order 1 from which unbiased estimates of the "main effects" could be derived even though two-factor interactions existed, Rao associated the factor x_4 with the elements in the column corresponding to $x_1x_2x_3$. This device is also used in the derivation of fractional factorial designs by Davies and Hay (1950), who show in addition how the association will determine the alias relationships. Thus x_4 may be put equal to $x_1x_2x_3$ or to $-x_1x_2x_3$. If the former relationship is used we have in shortened notation

$$4 = 123,$$

whence since $44 = I$ we can multiply both sides by 4 and we have

$$I = 1234.$$

This relation indicates that if the elements of the columns corresponding to the variables x_1, x_2, x_3 and x_4 are multiplied together the result will be a column of $+1$'s. Thus the particular factor combinations used correspond to that half of a full 2^4 factorial design for which the elements of the independent variable $x_1x_2x_3x_4$ are all equal to $+1$. The design under investigation is thus a one-half replicate of the full 2^4 factorial "split" along the interaction having subscript 1234. (The relationship $-4 = 123$ defines the other half.) Therefore the defining relation is

$$I = 11 = 22 = 33 = 44 = 1234 \; (= 1111 = 1122 = 111234, \text{etc.}) \qquad (34)$$

and consequently to terms of third order

$$\left.\begin{aligned}
b_0 &\to \varphi_0 + \tfrac{1}{2}\varphi_{11} + \tfrac{1}{2}\varphi_{22} + \tfrac{1}{2}\varphi_{33} + \tfrac{1}{2}\varphi_{44}\\
b_1 &\to \varphi_1 + \tfrac{1}{6}\varphi_{111} + \tfrac{1}{2}\varphi_{122} + \tfrac{1}{2}\varphi_{133} + \tfrac{1}{2}\varphi_{144} + \varphi_{234}\\
b_2 &\to \varphi_2 + \tfrac{1}{6}\varphi_{222} + \tfrac{1}{2}\varphi_{112} + \tfrac{1}{2}\varphi_{233} + \tfrac{1}{2}\varphi_{244} + \varphi_{134}\\
b_3 &\to \varphi_3 + \tfrac{1}{6}\varphi_{333} + \tfrac{1}{2}\varphi_{113} + \tfrac{1}{2}\varphi_{223} + \tfrac{1}{2}\varphi_{344} + \varphi_{124}\\
b_{12} &\to \varphi_{12} + \varphi_{34}\\
b_{13} &\to \varphi_{13} + \varphi_{24}\\
b_{23} &\to \varphi_{23} + \varphi_{14}\\
b_{123} &\to \varphi_4 + \tfrac{1}{6}\varphi_{444} + \tfrac{1}{2}\varphi_{114} + \tfrac{1}{2}\varphi_{224} + \tfrac{1}{2}\varphi_{334} + \varphi_{123}
\end{aligned}\right\} \qquad (35)$$

This is a design of type B for, if third and higher order effects were absent but second order effects were not, b_1, b_2, b_3 and b_4 would supply estimates of φ_1, φ_2, φ_3 and φ_4 unbiased by the effects of second order.

3.4.2 *Fractional Factorial Designs of Type* A

For 3 factors in 4 experiments a third factor may be associated with the independent variable x_1x_2 in the 2^2 factorial arrangement. The matrix of independent variables is thus—

$$\mathbf{X} = \begin{bmatrix} 1 & 1 & 1 & 1 \\ 1 & 1 & -1 & -1 \\ 1 & -1 & 1 & -1 \\ 1 & -1 & -1 & 1 \end{bmatrix} \quad \text{columns: } x_0,\ x_1,\ x_2,\ x_3 = x_1x_2 \qquad (36)$$

and since $I = 11 = 22 = 33 = 123$

$$\left.\begin{aligned} b_0 &\rightarrow \varphi_0 + \tfrac{1}{2}\varphi_{11} + \tfrac{1}{2}\varphi_{22} + \tfrac{1}{2}\varphi_{33} \\ b_1 &\rightarrow \varphi_1 + \varphi_{23} \\ b_2 &\rightarrow \varphi_2 + \varphi_{13} \\ b_3 &\rightarrow \varphi_3 + \varphi_{12} \end{aligned}\right\} \qquad (37)$$

For 7 factors in 8 experiments to order 2, the design is formed by associating each of the interaction columns in (29) with new factors. If we put—

$$x_4 = x_1x_2x_3,\ x_5 = x_1x_2,\ x_6 = x_1x_3,\ x_7 = x_2x_3$$

$$\left.\begin{aligned} b_0 &\rightarrow \varphi_0 + \tfrac{1}{2}\varphi_{11} + \tfrac{1}{2}\varphi_{22} + \ldots + \tfrac{1}{2}\varphi_{77} \\ b_1 &\rightarrow \varphi_1 + \varphi_{25} + \varphi_{36} + \varphi_{47} \\ b_2 &\rightarrow \varphi_2 + \varphi_{15} + \varphi_{37} + \varphi_{46} \\ b_3 &\rightarrow \varphi_3 + \varphi_{16} + \varphi_{27} + \varphi_{45} \\ b_{12} &\rightarrow \varphi_5 + \varphi_{12} + \varphi_{34} + \varphi_{67} \\ b_{13} &\rightarrow \varphi_6 + \varphi_{13} + \varphi_{24} + \varphi_{57} \\ b_{23} &\rightarrow \varphi_7 + \varphi_{14} + \varphi_{23} + \varphi_{56} \\ b_{123} &\rightarrow \varphi_4 + \varphi_{35} + \varphi_{26} + \varphi_{17} \end{aligned}\right\} \qquad (38)$$

When the factors are intermediate in number between 3, 7, etc., Rao, and Plackett and Burman, suggest that the next higher design be used, treating the remaining factors as dummies. The aliases may then be obtained from the table of aliases for the full design by omitting all terms containing dummy subscripts. Some caution is required here, since in some cases not all such designs are equally satisfactory. If four factors are to be tested in eight experiments it would usually be most desirable to omit a set of columns such as (*5*), (*6*), (*7*) in the design for seven variables. If other sets of 3 subscripts were omitted the designs would of course be of type A, but not necessarily of type B. In general the design would only be of type B if the 4 columns selected in (29) were such that, for every point on the design, the product of the co-ordinates was equal to unity, thus ensuring a sub-group relation of type $I = 1234$.

With 5 or 6 factors the arrangements are of the same type whichever subscripts are omitted. For 5 factors one of the first order effects has two second order aliases and the remaining four have only one, while the two dummy comparisons each have two second order aliases. For 6 factors each of the first order effects have two second-order aliases and the dummy comparison has three second-order aliases.

The location of the aliases is often of considerable importance both in planning the sets of experiments and in their subsequent interpretation. Thus the experimenter might feel that if appreciable interactions did occur they would be more likely, between certain pairs of factors, than between other pairs. In such instances this information could be used in arranging the experiment so that, as far as possible, these more suspect second-order effects were not associated with first order effects, but were isolated in comparisons corresponding to "dummy" factors. The possibilities can be examined by altering the identity of the dummy factors and rearranging the subscripts among themselves.

Certain limitations will be found to exist. For example in the design for determining first order effects for 5 factors in 8 experiments, it is not possible to arrange that two interactions which do not have one subscript in common should appear one in each of the two residual comparisons. From the nature of the alias relations it is clear that type *A* designs must always be used with extreme caution.

3.4.3 *Fractional Factorial Designs of Type* B

The design of type *B* for 4 factors in 8 experiments has already been derived (35). It could also have been obtained by duplicating with reversed signs the matrix of independent variables (36) for the design of type *A*, associating a fourth factor x_4 with x_0. It is easy to see that the new defining relation is obtained by reproducing terms containing an even number of elements as they stand, and multiplying the terms containing an odd number of elements by the subscripts of the newly introduced variable. In this case $I = 1234$.

If any one of the subscripts is omitted the type *B* design for 3 factors is obtained, which is of course the complete factorial. This design is one of a class which (on the assumption that effects of higher order than the second may be ignored) allows unbiased estimates to be obtained not only of effects of first order but also of mixed derivatives φ_{12}, φ_{13}, etc. To distinguish these designs they will be said to be of type B' and order 1.

The design for 8 factors in 16 experiments is obtained by duplicating with reversal of signs the matrix of independent variables (29) for the type *A* design for 7 factors in 8 experiments. The relations below, (39), may be compared with those obtained for the type *A* design. The groups of second order effects which were associated with first order effects in the original design, keep their identity, but are associated with the extra degrees of freedom arising from the duplication.

$$\left.\begin{aligned}
b_0 &\rightarrow \varphi_0 + \tfrac{1}{2}\varphi_{11} + \tfrac{1}{2}\varphi_{22} + \ldots + \tfrac{1}{2}\varphi_{88}\\
b_1 &\rightarrow \varphi_1\\
b_2 &\rightarrow \varphi_2\\
b_3 &\rightarrow \varphi_3\\
b_4 &\rightarrow \varphi_4\\
b_5 &\rightarrow \varphi_5\\
b_6 &\rightarrow \varphi_6\\
b_7 &\rightarrow \varphi_7\\
b_8 &\rightarrow \varphi_8\\
b_{12} &\rightarrow \varphi_{12} + \varphi_{34} + \varphi_{58} + \varphi_{67}\\
b_{13} &\rightarrow \varphi_{13} + \varphi_{24} + \varphi_{57} + \varphi_{68}\\
b_{14} &\rightarrow \varphi_{14} + \varphi_{23} + \varphi_{56} + \varphi_{78}\\
b_{15} &\rightarrow \varphi_{15} + \varphi_{28} + \varphi_{37} + \varphi_{46}\\
b_{16} &\rightarrow \varphi_{16} + \varphi_{27} + \varphi_{38} + \varphi_{45}\\
b_{17} &\rightarrow \varphi_{17} + \varphi_{26} + \varphi_{35} + \varphi_{48}\\
b_{18} &\rightarrow \varphi_{18} + \varphi_{25} + \varphi_{36} + \varphi_{47}
\end{aligned}\right\} \quad . \quad . \quad . \quad . \quad . \quad (39)$$

The designs for 7 and for 6 factors may be obtained as before by omitting subscripts in the design for 8 factors or by duplicating the corresponding type *A* design with reversed signs. A type *B* design for 5 factors could be obtained in a similar way, the design being a half replicate of the full 2^5 factorial typified by the sub-group relation $I = 1234$. In this case, however, a design of type B' exists, a half replicate with defining relation $I = 12345$, which would normally be used in practice. It is easily obtained by writing down the full 2^4 factorial design for four factors x_1, x_2, x_3, x_4, and adding the column $x_5 = x_1x_2x_3x_4$ obtained by multiplying these elements together. The sub-group relation $I = 12345$ together with the group relations $I = 11 = 22 = 33 = 44 = 55$ allow the aliases to be easily written down. If third and higher order effects are ignored, the 15 estimates corresponding with the 5 main effects and 10 two-factor interactions are all unbiased.

3.5 *Designs to Estimate Effects of First and Second Order*

First order designs of both types A and B were supplied by two level factorial and incomplete factorial arrangements. When it is desired to estimate effects of higher order, factorial and fractional factorial designs are less satisfactory, (1) because the number of experiments necessary is usually very much larger than the number of effects it is desired to estimate, and (2) because the relative precision with which the effects are estimated is often not that desired.

3.5.1 *Effects Estimated using Factorial Designs*

As an example, suppose that $k = 3$ variables are being considered. A table is shown below setting out the subscripts of effects up to order 6 (to save space the subscripts are not fully enumerated for $d = 4$, $d = 5$ and $d = 6$).

Subscripts for Terms of Order d *when* k = 3

$d = 0$	$d = 1$	$d = 2$	$d = 3$	$d = 4$	$d = 5$	$d = 6$
0	1	12	123	1123	11223	112233
	2	13	122	1223	11233	111111
	3	23	112	1233	12233	222222
		11	133	1122	11111	333333
		22	113	1133	22222	122222
		33	223	2233	33333	111112
			233	1111	12222	133333
			111	2222	11112	111113
			222	3333	13333	222223
			332	1222	11113	233333
				1112	23333	112222
				to 15 terms	to 21 terms	to 28 terms

Consider a factorial experiment with each factor varied at three levels. With a basic equation in which the independent variables consisted of all combinations of powers up to the second of x_1, x_2 and x_3, 27 coefficients $b_0, b_1, b_2, \ldots b_{112233}$ could be estimated from the 27 observations. Their subscripts would be those enclosed by the dotted lines above. b_1, b_2, b_3 would be proportional to what are usually called the linear effects, b_{11}, b_{22}, b_{33} to the quadratic effects, and the remaining coefficients to the interactions between these effects.

If, however, it was reasonable to approximate to the response function by a series containing *all* terms up to some specified degree d, then with $d = 2$, the ten coefficients of order 2 and less would provide unbiased estimates of the derivatives of order 2 and less, the 17 remaining coefficients being estimates of experimental error only. On the other hand, if d were higher than 2, these remaining coefficients would provide estimates of some but not all of the extra constants, and the constants not included would inevitably bias some at least of the estimates.

In general, from a complete factorial design in which each of the k factors is varied at p levels, it is possible to form estimates from the k^p observations of k^p quantities. The quantities may be chosen to be the coefficients of the k^p independent variables formed by expanding the expression

$$\prod_{t=1}^{k} (x_t^0 + x_t^1 + \ldots + x_t^{p-1}) \quad . \quad . \quad . \quad . \quad . \qquad (40)$$

The estimates will include all the b's of order $p - 1$ and less together with certain of those of order $p, p + 1, \ldots, k(p - 1)$; the latter will have subscripts repeated not more than $p - 1$ times for any single variable. On the assumption that only the $(k + p - 1)! / k!(p - 1)!$ effects of order $p - 1$ and less exist, the design supplies unbiased estimates of these effects and an estimate of experimental error based on $p^k - (k + p - 1)!/k!(p - 1)!$ degrees of freedom. If it is necessary to use a series with all terms up to order d greater than $p - 1$ it will be possible to estimate those of the extra constants in which no subscript is repeated more than $p - 1$ times; but the remaining "extra constants" will bias some of the estimates.

In practice it would often be useful to employ designs which, while supplying estimates of all effects up to order d, also estimated "sample members" of the hierarchy of higher order effects. For if such effects were not small then the experimenter would be warned of the danger of proceeding on that assumption. However, unless p and k were small the number of experiments required by the full factorial design would be unreasonably large compared with the number of constants to be estimated and the number of "extra" effects determined in the complete factorial design would often be far larger than was necessary to check the fit of the equation. These designs could not be used efficiently therefore when experimental error was small. Fractional replication, in which the comparisons measuring effects of high order are used to estimate the lower order effects of additional factors, can help to remedy this defect. Unfortunately, except when the number of levels is 2, the comparisons in the fractional factorials do not directly correspond with derivatives which it is desired to estimate and, although some saving is possible, even these designs would often require an excessive number of experiments. For example, for the 3^k designs, no useful fractional factorial exists when $k = 3$, so that it would be necessary to carry out all the 27 experiments of the complete factorial design in order to determine the 10 effects of second order or less. Similarly when $k = 4$, 81 experiments are needed to determine 15 effects. When $k = 5$ a one-third replicate may be used to determine the 21 effects of order 2 and less; but even this involves the carrying out of 81 experiments, a number which would be excessive if the experimental error were small.

3.5.2 *Relative Precision of Effects in Factorial Designs*

Consider the designs to supply estimates of effects up to the second order. These estimates may be used to indicate a path along which maximum increase might be expected; more usually, when the design has been performed in the neighbourhood of a stationary point, they may be used to estimate the position of that point by solving equations (7) with $\mu = 0$. Now the coefficients of the derivatives φ_{ts} in (7) are of the same degree, irrespective of whether $t = s$ or $t \neq s$ it would seem reasonable therefore to require that all second order effects (i.e., both the quadratic effects and interaction effects) should be determined with about the same precision. Since f_{st}, $s \neq t$, is estimated by b_{st} but f_{ss} is estimated by $2b_{ss}$, it is easily shown that 3^k factorial designs provide estimates of the quadratic derivatives having variances eight times as great as those for the interaction derivatives.

3.5.3 *Composite Designs*

The problem of finding "best" designs of types A and B when d is greater then unity is being investigated. In the meantime, designs, which did not require an excessively large number of experiments and which would estimate effects with reasonably high precision and low bias, were needed for use in current investigations. In particular, designs of type A and order 2 were required.

Factorial or fractional designs at two levels are easily found which allow the efficient estimation, without confusion, of first order derivatives and mixed derivatives not higher than the second order. These are what we have called type B' designs, and a list of such designs is given by Rao (1947). In constructing an arrangement which will estimate *all* derivatives up to the second order, we can start with a design of this sort, as a basis, and add further points which in conjunction with the type B' design make possible the estimation of the quadratic effects also. These designs have the advantage that they can be performed in stages. The first order design can first be completed. Then, if fairly large first order effects are found, while the effects of the two factor interactions, which can be estimated, are small, the experimenter may proceed to a new base by means of the first order steepest ascent formula. But if the relative magnitude of effects of first order and the two factor interactions show that it will be necessary to determine all second order derivatives, the extra points may be added to form the composite design. A design of this sort is illustrated for three variables in Fig. 3 below. The type B' arrangement forming the nucleus

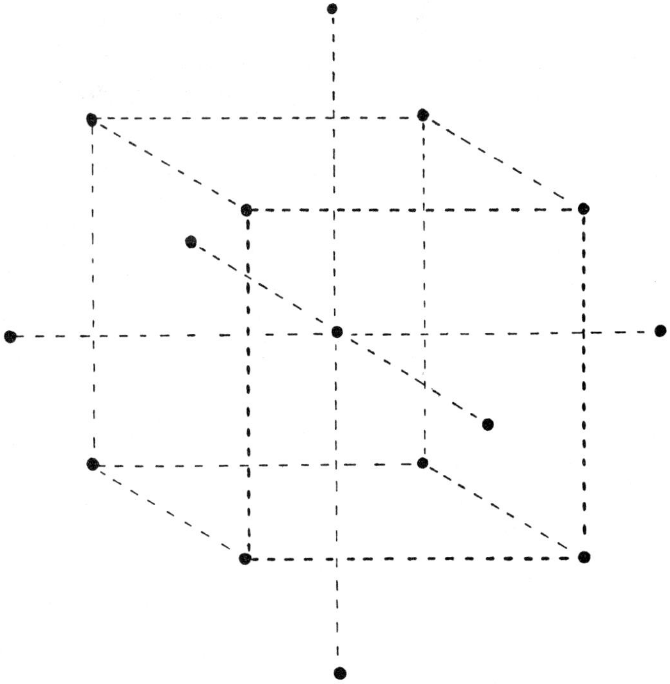

FIG. 3.—A three factor composite design.

is in this case the two level factorial design. The design matrix **D** for such an arrangement can be written

$$\mathbf{D} = \begin{array}{r} \\ \mathit{1} \\ \mathit{2} \\ \mathit{3} \\ \mathit{4} \\ \mathit{5} \\ \mathit{6} \\ \mathit{7} \\ \mathit{8} \\ \mathit{9} \\ \mathit{10} \\ \mathit{11} \\ \mathit{12} \\ \mathit{13} \\ \mathit{14} \\ \mathit{15} \end{array} \begin{array}{c} \begin{array}{ccc} x_1 & x_2 & x_3 \end{array} \\ \left[\begin{array}{ccc} 1 & 1 & 1 \\ 1 & 1 & -1 \\ 1 & -1 & 1 \\ 1 & -1 & -1 \\ -1 & 1 & 1 \\ -1 & 1 & -1 \\ -1 & -1 & 1 \\ -1 & -1 & -1 \\ \alpha & 0 & 0 \\ -\alpha & 0 & 0 \\ 0 & \alpha & 0 \\ 0 & -\alpha & 0 \\ 0 & 0 & \alpha \\ 0 & 0 & -\alpha \\ 0 & 0 & 0 \end{array} \right] \end{array} \quad . \quad . \quad . \quad . \quad . \quad . \quad (41)$$

where α is the distance of the axial points from the centre. It is shown in Appendix 4 that designs built up in this way can be very effective. In particular α may be chosen so that the design is orthogonal or, alternatively, so that the derivatives of second order are all determined with equal precision. The precision of the estimates and the extent of the possible biases show that the designs compare favourably with 3 level factorials, when the fewer experiments required is taken into account.

In examples where the first order design has indicated the probability of an imminent stationary point, in a particular direction the extra points may be added in that direction. In favour of this is the fact that the distance between the centroid of the design and the stationary point will probably be reduced and the fit of the second degree approximating equation improved. A number of interesting possibilities arise, one of which is demonstrated in the example in section 5.

4. Application of Steepest Ascent Procedure to Move to a Near-stationary Region

We have seen (Section 2) that if the experimental error is small compared with the expected gain, the experimenter can determine the differential coefficients about a base point O, confining his experiments to a small sub-region of the factor space. In such cases it is often possible, at least initially, to ignore all derivatives of order higher than the first. This situation is met quite frequently in our work, particularly in the investigation of new chemical routes for manufacture, the experiments being conducted in the first place on the laboratory scale. It is then possible to proceed by successive application of the steepest ascent formula (6).

This usually results in rapid progress, shown by the increase in the average response in successive sets of experiments and a falling off in the magnitude of the first order derivatives. A point is inevitably reached, however, at which first order effects are so reduced that they can no longer be regarded as large compared with effects of higher order, and no further progress is possible. At this stage the experimenter will usually have been brought to a near-stationary region which, without employing an unreasonably large number of experiments, can be further investigated by the methods described in the next section.

4.1 *Procedure when Some Factors Produce only Small Effects*

When a second group of observations is to be carried out the experimenter will be at liberty to alter the units for the variables if he thinks this desirable. In particular, if certain factors have shown only small effects, this might be because the units for the factors were disproportionately small. If this were so, it would result in very slow convergence for these factors, and a possibility that the small effects would be indistinguishable from experimental error.

This difficulty may be overcome as follows: If the effect corresponding to a particular factor is small compared with the other effects, then it may be that—

(i) the average level chosen for the factor is near a conditional maximum;
(ii) the unit adopted for the factor is disproportionately small;
(iii) the system is independent of the factor level.

To guard against the danger which would arise when (ii) is true, the average level for the factor should be changed away from the calculated path, and a larger unit used in the next set of experiments. Then if (iii) is true the factor will again be found to be without effect, while if (i) or (ii) is true a real effect should be found. In what follows, the change in level of a factor corresponding with a change from 0 to 1 in the design space will be called the *unit*, the level associated with the value zero in the design space will be called the *base level.*

Thus if the experimenter decided to vary the temperature of reaction at the levels 60 and 65° C., associating the lower level with -1 and the upper level with $+1$ in a two-level factorial design, the *unit* for temperature would be fixed at the value $+2{\cdot}5$° C. The *base level* used would be $62{\cdot}5$° C.

4.2 *An Example*

To illustrate the procedure the following example is given. The investigation was on a laboratory scale and concerned a chemical reaction which was of the type $A + B + C \rightarrow D +$

other products. The reaction took place in the presence of a solvent E; the object of the experiment was to maximize the yield of the product D for a given amount of A, the most expensive of the three starting materials. The amount of A was kept constant throughout. Conditions were known which gave about 45 per cent. of the theoretical yield (i.e., the yield which would be obtained if all of A were changed to D), but for a reaction of this type at least 75 per cent. was thought to be possible. The experimental error of a single preparation was expected to be rather less than 1 per cent. Five factors were varied: the amount of $E(x_1)$, the proportion of C to A (x_2), the concentration of C* (x_3), the time of reaction (x_4) and the proportion of B to A (x_5). After some deliberation and consultation between the chemist and the statistician, the following levels for x_1, x_2, x_3, x_4 and x_5 were fixed for the first group of experiments. The base levels are close to those which were known to give about 45 per cent. of the theoretical yield.

	-1	$+1$	
x_1 Amount of solvent E	200	250	c.c.
x_2 Proportion of C to A	4·0	4·5	mol./mol.
x_3 Concentration of C	90	93	%
x_4 Time of reaction	1	2	hours
x_5 Proportion of B to A	3·0	3·5	mol./mol.

In view of the large gain possible (and consequently the probable remoteness of the starting conditions from the optimum levels), it was thought that first order effects would be dominant and that higher order effects could be ignored, at least in the first stages of the investigation. After some consideration of the theoretical background of the reaction, the chemist decided that the effects most likely to produce complexity would be x_1 and x_3. Provision was made, in the design, for the isolation of any interaction between these variables; no extra factor was associated with x_1x_3. The design chosen and the yields obtained, when the experiments were performed in random order, are set out below:

$$\begin{array}{c} \\ \\ 1\\2\\3\\4\\5\\6\\7\\8 \end{array}\quad \mathbf{D}_1 = \begin{array}{ccccc} x_1 & x_2 & x_3 & x_4 & x_5 \\ & & & (x_1x_2x_3) & (-x_2x_3) \end{array} \begin{bmatrix} -1 & -1 & -1 & -1 & -1 \\ -1 & -1 & 1 & 1 & 1 \\ -1 & 1 & -1 & 1 & 1 \\ -1 & 1 & 1 & -1 & -1 \\ 1 & -1 & -1 & 1 & -1 \\ 1 & -1 & 1 & -1 & 1 \\ 1 & 1 & -1 & -1 & 1 \\ 1 & 1 & 1 & 1 & -1 \end{bmatrix} \qquad \mathbf{Y}_1 = \begin{bmatrix} 34{\cdot}4 \\ 51{\cdot}6 \\ 31{\cdot}2 \\ 45{\cdot}1 \\ 54{\cdot}1 \\ 62{\cdot}4 \\ 50{\cdot}2 \\ 58{\cdot}6 \end{bmatrix} \quad . \quad . \quad (42)$$

The columns of the design correspond with those of (29), as follows:

Variable	x_1	x_2	x_3	x_4	$-x_5$
Column in (29)	1	2	3	4	7

The second order aliases are obtained from (38) by dropping the subscripts 5 and 6 and writing 5 for 7. Since $-x_5$ is associated with the variable 7, all derivatives bearing the subscript 5 an odd number of times must be written with reversed signs. Alternatively the arrangement of aliases could be obtained directly from the relations $4 = 123$, $5 = -23$. We find—

$$\begin{array}{llcr} b_0 & \rightarrow + \varphi_0 (+ \tfrac{1}{2}\varphi_{11}{}^* + \tfrac{1}{2}\varphi_{22} + \tfrac{1}{2}\varphi_{33}{}^* + \tfrac{1}{2}\varphi_{44} + \varphi_{55}) & = & 48{\cdot}5 \pm 0{\cdot}4 \\ b_1 & \rightarrow + \varphi_1 (- \varphi_{45}) & = & 7{\cdot}9 \pm 0{\cdot}4 \\ b_2 & \rightarrow + \varphi_2 (- \varphi_{35}) & = & -2{\cdot}2 \pm 0{\cdot}4 \\ b_3 & \rightarrow + \varphi_3 (- \varphi_{25}) & = & 6{\cdot}0 \pm 0{\cdot}4 \\ b_{123} & \rightarrow + \varphi_4 (- \varphi_{15}) & = & 0{\cdot}4 \pm 0{\cdot}4 \\ b_{23} & \rightarrow - \varphi_5 (+ \varphi_{14} + \varphi_{23}) & = & -0{\cdot}4 \pm 0{\cdot}4 \\ b_{13} & \rightarrow \quad (+ \varphi_{13}{}^* + \varphi_{23}) & = & -1{\cdot}8 \pm 0{\cdot}4 \\ b_{12} & \rightarrow \quad (+ \varphi_{12} + \varphi_{34}) & = & 0{\cdot}2 \pm 0{\cdot}4 \end{array}$$

* C was added as an aqueous solution the concentration is that of C in this solution.

The estimates are followed by their standard errors which are based on the assumption that σ is about 1 per cent. The expected values are shown with second order terms bracketed. Asterisks indicate a prior suggestion that the corresponding derivatives may not be small. The estimate due to φ_{13}, which probably corresponds to a real effect, is however, not large compared with first order effects and the steepest ascent formula (6) was applied ignoring possible effects of higher orders. The calculation of the direction in which maximum gain would be expected, if the response surface is plotted in the scale of the design is set out below.

	x_1	x_2	x_3	x_4	x_5
	c.c.	*mol.*	%	*hour*	*mol.*
(1) Base level	225	4·25	91·5	1·5	3·25
(2) Unit	25	0·25	1·5	0·5	0·25
(3) Estimated derivatives f_t (change in yield per unit)	7·9	−2·2	6·0	0·4	0·4
(4) Unit $\times f_t$	197·5	−0·55	9·0	0·2	0·1
(5) Change in level per 10 c.c. change in x_1	10	−0·028	0·456	0·011	0·005
(6) Path	x_1	x_2	x_3	x_4	x_5
	225	4·25	91·5	1·5	3·25
	235	4·22	92·0	1·5	3·25
	245	4·19	92·4	1·5	3·26
	255	4·17	92·9	1·5	3·26
	265	4·14	93·3	1·5	3·27
Experiment (9)	275	4·11	93·8	1·6	3·27
	285	4·08	94·2	1·6	3·28
Experiment (10)	295	4·06	94·7	1·6	3·28
	305	4·03	95·1	1·6	3·29

The average yield was 48·5 per cent., that of experiment 9, 80 per cent., and of 10, 79·4 per cent.

In the table, line (1) shows the base levels for the factors. Line (2) shows the unit, i.e., the change in level for the variable corresponding to the change from 0 to 1 in the design space. (For two level designs, this is half the range over which the factor is varied.) Line (3) shows the estimated derivatives calculated on the assumption that second and higher order effects are small enough to be ignored. According to equation (6) the factors must be varied in proportion to these derivatives; this is to be done *in units of the design.* Thus for each 7·9 units by which x_1 is increased, x_2 should be decreased by 2·2 units, x_3 increased by 6·0 units and so on; in terms of the original units of measurement for each 25 $\times$ 7·9 c.c. that x_1 is varied, x_2 should be changed by 0·25 $\times$ − 2·2. These changes are shown in line (4) and define the direction of steepest ascent at the base O, when the surface is scaled in the units of the design. The "path" leading from O in this direction can be set out by taking a convenient increment in one of the variables, (10 c.c. in x_1 in the present case), and calculating the proportionate changes to be made in the other variables. These are given in line (5). These quantities may then be added in succession to the average level to give the path (6). Experiments (9) and (10) were performed along this path corresponding with levels for x_1 of 275 and 295 c.c. respectively. The yields obtained were 80·0 per cent. and 79·4 per cent. Although the second value was apparently somewhat lower than the first no real significance could be attributed to this, and a second set of experiments was carried out using the conditions corresponding to experiment (10) as base point.

The procedure adopted is dictated by the results of the experiments themselves. The tentative prior assumption concerning the dominance of first order effects tended to be confirmed in this instance by the evidence of the first set of experiments. We were justified, therefore, in acting on it. If, after these experiments were completed we had found (from the smallness of the first order effects compared with those aggregates of second order effects which we were able to estimate) that the assumptions were untenable, then the method of attack would have been changed to that described in §(5), where second order effects are taken into account. It is of course theoretically possible to be misled completely, when fairly high order partial factorial designs are being used. Thus a major part of the effects attributed to first order derivatives in (43) could have been due to the second order derivatives φ_{45}, φ_{35}, etc. On the basis of the prior evidence available this would have seemed unlikely. However (and this is a final safeguard), if our assumptions were so far from the truth as to be valueless, then, when experiments were performed in the direction calculated, *the expected increase would not in fact occur*. In the above experiment, as a result of following the path of steepest ascent a large increase in yield was obtained. If this had not been so, we should have concluded that either the estimates were heavily biased by ignored effects, or the error of the observations was much greater than had at first been supposed, or both. It would then have been necessary to augment the group of observations already made. New levels would have been chosen so as to allow the unbiased estimation of first order effects (by converting the design to one of type *B*), or possibly to allow the estimation of all first and second order effects.

In the first group of experiments neither of the variables x_4 or x_5 gave appreciable effects. In the next group of experiments, therefore, average levels for both of these factors were moved away from the calculated direction of steepest ascent and larger unit lengths were used. When considering suitable step lengths for the factors x_1, x_2, x_3, in the second group of experiments, it was necessary to bear in mind that the first order effects would become smaller as a stationary point was approached. Progress by this method would become more and more difficult because—

(i) the estimates of first order effects would be relatively more in error;
(ii) the estimates of the first order effects would be relatively more biased by second order aliases.

If the units were increased to reduce the first difficulty, the second would be aggravated. In the first group of experiments there was some evidence of the existence of appreciable second order effects probably due to φ_{13}. But the estimated derivatives for x_1, x_2, x_3 were all large compared with their standard errors. Thus, if further progress was to be possible without taking effects of second order into account, the best chance of success lay in making a small reduction of step lengths for these three factors. Somewhat narrower ranges, therefore, were chosen for them in the second group of experiments, although the relation between the units was kept about the same.

Levels for the factors in the second group of experiments were:

		−1	+1	
x_1	Amount of solvent *E*	280	310	c.c.
x_2	Proportion of *C* to *A*	3·85	4·15	mol./mol.
x_3	Concentration of *C*	94	96	%
x_4	Time of reaction	2	4	hours
x_5	Proportion of *B* to *A*	3·5	5·5	mol./mol.

The design was similar to that used in the first group of experiments; but the columns omitted from the 7 variable design were (*6*) and (*7*) instead of (*6*) and (*5*).*

$$\mathbf{D}_2 = \begin{array}{c} \\ \\ 11 \\ 12 \\ 13 \\ 14 \\ 15 \\ 16 \\ 17 \\ 18 \end{array} \begin{array}{c} x_1 \\ \\ \end{array}\!\!\!\!\begin{bmatrix} \begin{array}{ccccc} x_1 & x_2 & x_3 & x_4\ (x_1x_2x_3) & x_5\ (x_1x_2) \\ -1 & -1 & -1 & -1 & 1 \\ -1 & -1 & 1 & 1 & 1 \\ -1 & 1 & -1 & 1 & -1 \\ -1 & 1 & 1 & -1 & -1 \\ 1 & -1 & -1 & 1 & -1 \\ 1 & -1 & 1 & -1 & -1 \\ 1 & 1 & -1 & -1 & 1 \\ 1 & 1 & 1 & 1 & 1 \end{array} \end{bmatrix} \quad \mathbf{Y} = \begin{bmatrix} 77\cdot1 \\ 69\cdot0 \\ 75\cdot5 \\ 72\cdot6 \\ 67\cdot9 \\ 68\cdot4 \\ 71\cdot5 \\ 65\cdot9 \end{bmatrix} . \qquad (43)$$

* It is worth while changing the arrangement of the aliases in this way to shed some light on any interactions previously associated with main effects. In the present instance it is important that b_{13} should not be used for the estimation of any first order derivative, for an appreciable effect was found associated with the comparison in the first group of experiments. The effect b_{12}, however, contains no real effect (unless of course φ_{34} and φ_{12} cancel each other out). In the second group of experiments this comparison was used for the estimation of φ_5.

Whence we obtain the estimates:

$$\begin{array}{llr}
b_0 & \rightarrow +\varphi_0(+\tfrac{1}{2}\varphi_{11}+\tfrac{1}{2}\varphi_{22}+\tfrac{1}{2}\varphi_{33}+\tfrac{1}{2}\varphi_{44}+\tfrac{1}{2}\varphi_{55}) & = 70\cdot 7 \pm 0\cdot 4 \\
b_1 & \rightarrow +\varphi_1(+\ \varphi_{25}) & = -2\cdot 8 \pm 0\cdot 4 \\
b_2 & \rightarrow +\varphi_2(+\ \varphi_{15}) & = 0\cdot 1 \pm 0\cdot 4 \\
b_3 & \rightarrow +\varphi_3(+\ \varphi_{45}) & = -2\cdot 3 \pm 0\cdot 4 \\
b_{123} & \rightarrow +\varphi_4(+\ \varphi_{35}) & = -1\cdot 7 \pm 0\cdot 4 \\
b_{12} & \rightarrow +\varphi_5(+\ \varphi_{12}+\varphi_{34}) & = -0\cdot 4 \pm 0\cdot 4 \\
b_{13} & \rightarrow \quad (+\ \varphi_{13}+\varphi_{24}) & = 0\cdot 5 \pm 0\cdot 4 \\
b_{23} & \rightarrow \quad (+\ \varphi_{23}+\varphi_{14}) & = -0\cdot 4 \pm 0\cdot 4
\end{array}$$

The average yield here is higher than in the first set, though not so high as the value 79·4 per cent. of experiment (10). One cause of this is the curvature of the surface. φ_{11}, φ_{22} etc. would be negative near a maximum; thus b_0, the mean, would be a heavily biased estimate of φ_0 the yield at the centre of the design. Another cause is the loss of yield due to using longer times. The latter is shown by the fairly large negative differential now associated with x_4. This indicates that the small effect, found for the factor before, was an underestimate because it was then near a conditional maximum. The sign of this effect indicates that the level for the factor should be changed back. In spite of the larger unit and different average level used, x_5 has still failed to show any appreciable effect. Therefore, over the levels considered (which cover most of the practical range for this factor) its value is not important. The derivatives for the first 3 effects have all changed signs; this suggests that we have moved too far from the first base, and that it will be necessary to move back.

In terms of the new units the path of steepest ascent calculated as before was:

	x_1 *c.c.*	x_2 *mol.*	x_3 *%*	x_4 *hours*	x_5 *mol.*	*Yield*
	295	4·0	95·0	3·0	4·5	
Experiment (19)	285	4·0	94·5	2·6	4·4	80·8%
	275	4·0	93·9	2·2	4·3	
Experiment (20)	265	4·0	93·4	1·8	4·2	84·0%
Experiment (21)	255	4·0	92·8	1·4	4·1	81·5%

Experiments (19), (20) and (21) performed along the path gave yields of 80·8 per cent., 84·0 per cent. and 81·5 per cent. respectively. Four further experiments in which x_1 and x_3 were varied in a 2^2 factorial arrangement, about (20) as base point all gave lower yields. The calculated differentials gave values little different from zero. Experiment (20) defined conditions therefore at which the yield was nearly stationary. Accordingly it was decided not to carry out further laboratory experiments, but to take the process to the pilot plant stage and to make further adjustments there. It was clear that little further progress was possible by this method, and that in further work higher order derivatives should be taken into account.

The values of the derivatives (brought to the same units) in the two sets of experiments are compared below:

	Average Level	*Differential Coefficient*	*Average Level*	*Differential Coefficient*
x_1	225 c.c.	3·2%/10 c.c.	295 c.c.	−1·9%/10 c.c.
x_2	4·25 mol.	−0·9%/0·1 mol.	4·0 mol.	0·1%/0·1 mol.
x_3	91·5%	4·0%/%	95·0%	−2·3%/%
x_4	1·5 hr.	0·8%/hr.	3·0/hr.	−1·7%/hr.

The table shows the close analogy between our procedure and the relaxation methods of successive approximation developed by Southwell (1940, 1946). Southwell's problem is the solution of simultaneous equations. For a wide variety of types of functions this method of successive liquidation of derivatives (or residuals) results in rapid progress to a solution, especially

(Temple, 1939; Booth, 1949) when the procedure is accelerated by the employment of steepest ascent methods.

4.2.1 *Discussion of Steepest Ascent Technique*

The technique described in this section will not, by itself, precisely locate a local maximum. It is of use in problems where the starting conditions are probably fairly remote from the optimum and the experimental error is small. Under these circumstances, a few cycles of this procedure have usually been found sufficient to allow the experimenter to move rapidly through the factor space to a near-stationary region. Such a region cannot of course contain a minimum, since from one direction at least a rising path leads towards it. It need not, however, contain a true maximum. Once the experimenter has located such a region its nature can be determined without an excessively large number of experiments.

The technique described here has been used in a number of investigations, most of them similar in nature to the example given in detail above. One investigation of particular interest was on a reaction of the type—

$$A + B + C \rightarrow D + E \text{ and other products.}$$

The original object was to maximize the yield of D and minimize that of E; if more than a small amount of E were present in the product it would not be satisfactory. In this investigation the yields for both D and E were examined; the path followed was a compromise between one of steepest ascent for D and one of steepest descent for E. For most of the variables, similar changes were indicated on either basis. From a first estimate of 64 per cent. for D and 15 per cent. for E, 92 per cent. was obtained for D after two cycles of the procedure; the product contained less than $\frac{1}{2}$ per cent. of E. At this stage it was suggested that the same reaction might be used for obtaining E and a second investigation was begun, in this case maximizing E. Three cycles of the technique led to conditions which gave a yield of about 70 per cent. for E.

5. Exploration of the Surface in a Near-Stationary Region

In this section it is assumed that the experimenter is using a sub-region of the factor space, within which first order derivatives are small. He may have arrived at this region by successive application of the steepest ascent technique, or already have found it at the commencement of his investigation. Examples of the latter kind commonly occur when the problem is further to improve some well established process; for often, large first order effects will have been eliminated in previous work. In either case, initially at any rate, only the immediate neighbourhood need be explored and this may be done without an excessively large number of experiments.

Here we shall be concerned only with approximating equations of the second degree. We assume that a suitable design has been performed, and that estimates $f_0, f_1, f_2, \ldots f_{11}, {}_{22}, \ldots f_{12}$, etc., of reasonable precision are available for second and lower orders.

5.1 *Analysis of the Fitted Equation*

In the neighbourhood of the design the response surface is given approximately by a $k + 1$ dimensional paraboloid in the variables $y, x_1, x_2 \ldots x_k$.

$$y = f_0 + f_1x_1 + f_2x_2 + \ldots + \tfrac{1}{2}f_{11}x_1^2 + \tfrac{1}{2}f_{22}x_2^2 + \ldots + f_{12}x_1x_2 + f_{13}x_1x_3 + \ldots \quad (44)$$

Associated with the surface are k dimensional contour surfaces such as

$$y_c = f_0 + f_1x_1 + f_2x_2 + \ldots + \tfrac{1}{2}f_{11}x_1^2 + \tfrac{1}{2}f_{22}x_2^2 + \ldots + f_{12}x_1x_2 + f_{13}x_2x_3 + \text{etc.} \ldots \quad (45)$$

on which the response is equal to y_c.

Differentiating (44) with respect to $x_1, x_2 \ldots x_k$ in turn, we obtain k linear simultaneous equations, such as (7) but with $\mu = 0$. On solving these we obtain $x_1^0, x_2^0, \ldots x_k^0$, the co-ordinates of S the stationary point on the fitted surface. On substituting these values in (44) we obtain the predicted response at S,

$$y^0 = f_0 + \tfrac{1}{2}(f_1x_1 + f_2x_2 + \ldots + f_kx_k). \quad (46)$$

The system of conics corresponding to the contour surfaces may take a large variety of forms; it would usually be quite impossible to appreciate the nature of the fitted surfaces by inspection of the values of the coefficients in (45). The nature of the system is, however, readily made apparent by reduction of the conic to canonical form.

This consists essentially of shifting the origin to the stationary point S, the centre of the system of curves representing the contour surfaces, and rotating the co-ordinate axes so that they correspond to the axes of these conics. Equation (45) then reduces to

$$y_c - y^0 = \lambda_1 X_1^2 + \lambda_2 X_2^2 + \ldots + \lambda_k X_k^{2}* \quad . \quad . \quad . \quad . \quad (47)$$

The expression shows the loss of response on moving from S, so that for example, if λ_1 is negative, S is a maximum for the response curve drawn along X_1. Large values of λ correspond with rapid changes in the response whilst small values indicate slow changes.

(i) If $\lambda_1, \lambda_2, \ldots \lambda_k$ were all negative, the fitted surface would have a true maximum at the stationary point, and the fitted contour surfaces would be ellipsoids.

(ii) If one or more of the λ's were positive, there would be a col or minimax with the contour surfaces elliptic hyperboloids.

(iii) If one or more of the λ's approached zero, the curves would be attenuated along their corresponding axes and the resulting surfaces would approach elliptic or hyperbolic cylinders; or the fitted response surface would possess a ridge.

Suppose that in a particular example the λ's and the directions of $X_1, X_2 \ldots X_k$ had been determined. Suppose also that on differentiating (44) a stationary point had been found in the immediate neighbourhood of the design. In case (i), above, the co-ordinates of the stationary point would provide an estimate of the position of the maximum. In case (ii) the directions corresponding to the positive λ's would indicate how the experimenter should move from the col to points of higher yield; further experiments would then be made to explore these possibilities. In case (iii) the directions of the line, plane, or space, in which the response was nearly constant could be determined. This is very important in practice; for it is thus possible to indicate alternative sets of conditions at which almost equal responses would be expected and so to find the most satisfactory compromise for auxiliary responses in the chosen process. The analysis of the surface in this way will also serve to indicate where further experiments are needed. Thus where some of the λ's are near to zero, it may be important to discover whether the true values are positive or negative. In such cases extra experiments may best be performed along the corresponding estimated axes; hence re-estimates of the constants and λ's can be made. In some cases strong divergencies between the values obtained at these extra points and the values already found may indicate the necessity for fitting an equation containing higher order terms.

Sometimes the stationary point of the fitted function will be remote from the neighbourhood of the design near which it was expected. In this remote region the fitted surface could not provide any accurate information about the true surface. The fitted contour surface in the immediate neighbourhood of the design will, however, supply valuable information. In a commonly occurring case of this kind, the near-stationary region to which the experimenter has been led is not in fact close to a stationary point, but is in the neighbourhood of a slowly rising ridge which leads to higher responses. One axis (say x_1) of the fitted conic will lie in the direction of the ridge and the corresponding λ_1 will be small. Referred to a local origin on X_1, the equations of the fitted contour surfaces are

$$y_c - y^0 = B_1 X_1 + \lambda_1 X_1^2 + \lambda_2 X_2^2 + \ldots \lambda_k X_k^2, \quad . \quad . \quad . \quad . \quad (48)$$

where y° is the response at the origin. The limiting case is when $\lambda_1 = 0$. The contour surfaces are then paraboloids with their axes along X_1 and B_1 measures the slope of the ridge. In practice

* Theoretically some of the λ's may be exactly zero, in which case no unique stationary point S will exist. With experimental data the λ's may approach zero; the behaviour of the contour surfaces, in these important limiting cases, are mentioned in the text. For a full account of the reduction of a conic to canonical form, the reader is referred to text-books such as Turnbull and Aitken (1932). An example of the reduction is given in the next section.

the experimenter would explore the axis X_1 with further experiments; if the slope was confirmed he would then follow this direction.

5.2 *The Inclusion of Additional Observations*

Thus it is frequently necessary by the use of additional points to explore further the possibilities suggested by the analysis of the second degree equation. The matrix $\mathbf{C}^{-1}$ would be known for the basic design but not for the arrangement of points including the additional observations; the calculation of a new set of constants *ab initio* would be laborious. Fortunately, however, a complete recalculation is unnecessary. Plackett (1950) shows how "corrections" can be obtained, which allow new estimates to be formed from the old with a minimum of recalculation.

In our notation Plackett's results may be stated as follows: In an experiment with a design $\mathbf{D}_1$ let the N_1 resulting observations be given by the vector $\mathbf{Y}$; let $\mathbf{X}$ be the $N_1 \times L$ matrix of rank L for the L independent variables (functions of the co-ordinates in $\mathbf{D}$). Suppose that the matrix $(\mathbf{X'X})^{-1} = \mathbf{C}^{-1}$ is known; and that the vector $\mathbf{B}$ of the estimates of the elements of $\boldsymbol{\beta}$, as well as the residual sums of squares $\mathbf{S}$ (based on $N - L$ degrees of freedom) have been calculated. Let a further N_2 observations $\mathbf{Z}$ become available as a result of experiments performed at a set of points whose co-ordinates are given by $\mathbf{D}_2$ and for which the matrix of independent variables is $\mathbf{W}$. The new matrices $\mathbf{C}_0{}^{-1}$, $\mathbf{B}'_0$ and $\mathbf{S}_0$ for the whole of the $N_1 + N_2$ points are given by—

$$\mathbf{C}_0{}^{-1} = \mathbf{C}^{-1} - \mathbf{J'GJ} \qquad (49)$$

$$\mathbf{B}_0{}' = \mathbf{B}' + \boldsymbol{\Delta}'\mathbf{GJ} \qquad (50)$$

$$\mathbf{S}_0 = \mathbf{S} + \boldsymbol{\Delta}'\mathbf{G}\boldsymbol{\Delta} \qquad (51)$$

where

$$\mathbf{J} = \mathbf{WC}^{-1}; \quad \mathbf{G} = (\mathbf{I} + \mathbf{R})^{-1}; \quad \mathbf{R} = \mathbf{WC}^{-1}\mathbf{W}'; \quad \boldsymbol{\Delta} = \mathbf{Z} - \mathbf{WB}. \qquad (52)$$

In each case the new values are obtained by adding a matrix of correction terms to the old. For formulae (50) and (51) these correction terms contain the vector $\mathbf{Z} - \mathbf{WB}$. The elements of this vector are the discrepancies between the values which the equation first fitted predicts for the responses at the additional points and the responses actually found. Each of the correction terms contains the matrix $(\mathbf{I} + \mathbf{R})^{-1}$. It is thus necessary to invert a $N_2 \times N_2$ matrix only; providing N_2, the number of additional observations, is not too great, the inverse can be found fairly readily. For systematic methods for inverting matrices the reader is referred to papers by Dwyer (1942), Hotelling (1943), and Fox *et al.* (1948).

5.3 *An Example*

This example is chosen not because the results obtained were at all spectacular; but because it provides a good illustration of the devices discussed in previous sections and, since only three variables are involved, it is possible to demonstrate the conclusions geometrically.

The investigation concerned one stage of a particular chemical process in which a reaction of type,

$$A + B \rightarrow C + \text{other products},$$

was carried out. This process was already well established on the manufacturing scale, but it was thought that slightly better conditions might possibly be found. (The experiments were carried out in the laboratory. The reaction, however, was of the type in which plant conditions were almost exactly reproducible, and it was fairly certain that any changes in the laboratory scale would be applicable in the larger scale operation.) The yield of the product C could be accurately determined and the experiments were very reproducible. They were not easy to perform, however, and were time-absorbing; each one engaged a chemist and his assistants for a number of days. It was essential therefore to reduce the number of experiments to a minimum.

The factors varied were temperature (x_1), concentration (x_2), and the molar ratio of B to A (x_3). The most economic process was not necessarily that corresponding with the highest yield; the response considered was the calculated cost per pound of final product. This was a function not only of the yield obtained but also of the experimental conditions employed. The object of the experiments was to estimate the point at which the cost would be a *minimum*.

For convenience in calculation an arbitrary quantity has been subtracted from the cost per pound of product assessed in tenths of a penny; in what follows the resulting amount y is termed "cost". There was some evidence that the standard deviation of individual observations would be about 1 unit.

The normal conditions were:

Temperature °C.	*Concentration %*	*Molar Ratio B/A*
145	20	6 .

The region of practical variation for the factors was fairly extensive. However, from the plant-handling point of view at the next stage of the process, it was essential that the concentration should not fall below 18 per cent.

When the first set of experiments was begun, it was uncertain whether the known conditions would be sufficiently far from the optimum to make necessary the preliminary application of the technique of §4 to move to the region of a stationary point. Therefore a complete 2^3 factorial was performed. This is a design of type B' which allows all the first order derivatives and mixed derivatives of second order to be estimated. The levels used were:

Temperature °C.		*Concentration %*		*Molar Ratio A/B*	
140	145	20	24	6	7
x_1		x_2		x_3	
− 1	1	− 1	1	− 1	1

The requirement that the concentration should not be less than 18 per cent. corresponds with the condition that x_2 should not be less than − 2. The co-ordinates of the experimental points and corresponding costs were (1, 1, 1) 17; (1, 1, − 1) 9; (1, − 1, 1) 12; (1, − 1, − 1) 15; (− 1, 1, 1) 24; (− 1, 1, − 1) 11; (− 1, − 1, 1) 7; (− 1, − 1, − 1); 5. The effects and their expected values, including terms up to third order, are shown below:

$$\left.\begin{array}{lll}
b_0 & \rightarrow \varphi_0 + \tfrac{1}{2}\varphi_{11} + \tfrac{1}{2}\varphi_{22} + \tfrac{1}{2}\varphi_{33} & = 12\cdot 50 \pm 0\cdot 4 \\
b_1 & \rightarrow \varphi_1 \ (+\tfrac{1}{6}\varphi_{111} + \tfrac{1}{2}\varphi_{122} + \tfrac{1}{2}\varphi_{133}) & = 0\cdot 75 \pm 0\cdot 4 \\
b_2 & \rightarrow \varphi_2 \ (+\tfrac{1}{6}\varphi_{222} + \tfrac{1}{2}\varphi_{112} + \tfrac{1}{2}\varphi_{233}) & = 2\cdot 75 \pm 0\cdot 4 \\
b_3 & \rightarrow \varphi_3 \ (+\tfrac{1}{6}\varphi_{333} + \tfrac{1}{2}\varphi_{113} + \tfrac{1}{2}\varphi_{223}) & = 2\cdot 50 \pm 0\cdot 4 \\
b_{12} & \rightarrow \varphi_{12} & = -3\cdot 00 \pm 0\cdot 4 \\
b_{13} & \rightarrow \varphi_{13} & = -1\cdot 25 \pm 0\cdot 4 \\
b_{23} & \rightarrow \varphi_{23} & = 2\cdot 75 \pm 0\cdot 4 \\
b_{123} & \rightarrow \varphi_{123} & = 0\cdot 00 \pm 0\cdot 4
\end{array}\right\} \quad . \quad . \quad (53)$$

The estimates are followed by their standard errors based on the assumption that σ the experimental error is about 1 unit. From the relative size of the interaction terms, it is clear that the steepest descent* relaxation process is unlikely to be very effective here; the design must be augmented to determine all effects of second order. This example is interesting, because owing to a misunderstanding the steepest descent path was in fact calculated and followed.

The steepest descent path will be followed when the factors are varied in proportion to the first order effects with reversed signs, i.e., − 0·75, − 2·75, − 2·50.

Two experiments were performed on this path. The levels for the factors are given below, in units of the design, together with the estimated costs:

$$\begin{array}{c} \textit{Experiment} \\ (9) \\ (10) \end{array} \quad \begin{array}{c} \quad x_1 \qquad\quad x_2 \qquad\quad x_3 \\ \begin{bmatrix} -0\cdot 54 & -2\cdot 00 & -1\cdot 82 \\ -0\cdot 81 & -3\cdot 00 & -2\cdot 73 \end{bmatrix} \end{array} \quad \begin{array}{c} y \\ \begin{bmatrix} 13 \\ 27 \end{bmatrix} \end{array} \quad . \quad . \quad . \quad (54)$$

* "Descent", since we are seeking a minimum.

As might have been expected, no reduction in cost was in fact found in these experiments. However, since the first order effects for each of the factors is positive, it seemed likely that if a minimum existed it would be towards the vertex $(-1\ -1\ -1)$. Therefore the factorial design was augmented with three further points in the manner discussed at the end of §3.53. The levels used in these experiments, together with the costs found, are shown below:

Experiment	x_1	x_2	x_3	y
(11)	−3	−1	−1	19
(12)	−1	−3	−1	12
(13)	−1	−1	−3	12

(55)

The matrix T' for the composite design is:

	b_0	b_1	b_2	b_3	b_{11}	b_{22}	b_{33}	b_{12}	b_{13}	b_{23}
1	0·2656	0·1250	0·1250	0·1250	−0·0468	−0·0468	−0·0468	0·1250	0·1250	0·1250
2	−0·0156	0·1250	0·1250	−0·1250	0·0468	0·0468	0·0468	0·1250	−0·1250	−0·1250
3	−0·0156	0·1250	−0·1250	0·1250	0·0468	0·0468	0·0468	−0·1250	0·1250	−0·1250
4	0·1406	0·1250	−0·1250	−0·1250	0·0781	−0·0468	−0·0468	−0·1250	−0·1250	0·1250
5	−0·0156	−0·1250	0·1250	0·1250	0·0468	0·0468	0·0468	−0·1250	−0·1250	0·1250
6	0·1406	−0·1250	0·1250	−0·1250	−0·0468	0·0781	−0·0468	−0·1250	0·1250	−0·1250
7	0·1406	−0·1250	−0·1250	0·1250	−0·0468	−0·0468	0·0781	0·1250	−0·1250	−0·1250
8	0·7344	−0·1250	−0·1250	−0·1250	−0·2031	−0·2031	−0·2031	0·1250	0·1250	0·1250
11	−0·1250	.	.	.	0·1250	.	.	.	.	.
12	−0·1250	.	.	.	.	0·1250	.	.	.	.
13	−0·1250	.	.	.	.	.	0·1250	.	.	.

(56)

It will be noted from the above matrix that the estimates of the linear effects and first order interactions remain the same as before. Their standard errors and aliases are also undisturbed. For b_0 and the quadratic effects the estimates obtained were:

$$\left.\begin{aligned}
b_0 &\to \varphi_0 + \tfrac{1}{2}(\varphi_{111} + \varphi_{222} + \varphi_{333}) + \tfrac{1}{2}(\varphi_{122} + \varphi_{133} + \varphi_{112} \\
&\qquad + \varphi_{233} + \varphi_{113} + \varphi_{223}) + \tfrac{9}{8}\varphi_{123} = 6\cdot750 \pm 0\cdot8 \\
b_{11} &\to \tfrac{1}{2}\varphi_{11} - \tfrac{1}{2}(\varphi_{111} + \varphi_{112} + \varphi_{113}) - \tfrac{3}{8}\varphi_{123} = 3\cdot000 \pm 0\cdot3 \\
b_{22} &\to \tfrac{1}{2}\varphi_{22} - \tfrac{1}{2}(\varphi_{222} + \varphi_{123} + \varphi_{223}) - \tfrac{3}{8}\varphi_{123} = 1\cdot625 \pm 0\cdot3 \\
b_{33} &\to \tfrac{1}{2}\varphi_{33} - \tfrac{1}{2}(\varphi_{333} + \varphi_{133} + \varphi_{233}) - \tfrac{3}{8}\varphi_{123} = 1\cdot125 \pm 0\cdot3
\end{aligned}\right\} \qquad (57)$$

On the assumption that third and higher order effects might be ignored, the estimated equation of best fit, in the neighbourhood of the design, was therefore:

$$\begin{aligned}
y = {} & 6\cdot750 + 0\cdot750x_1 + 2\cdot750x_2 + 2\cdot500x_3 + 3\cdot000x_1^2 + 1\cdot625x_2^2 + 1\cdot125x_3^2 \\
& - 3\cdot000x_1x_2 - 1\cdot250x_1x_3 + 2\cdot750x_2x_3 \qquad (58)
\end{aligned}$$

The residual sum of squares S based on one degree of freedom was zero. For the composite design the matrix $\mathbf{C}^{-1}$ is:

	0	*1*	*2*	*3*	*11*	*22*	*33*	*12*	*13*	*23*
2	0·7168	−0·0781	−0·0781	−0·0781	−0·1816	−0·1816	−0·1816	0·1094	0·1094	0·1094
1	−0·0781	0·1250	.	.	0·0468	0·0156	0·0156	.	.	.
0	−0·0781	.	0·1250	.	0·0156	0·0468	0·0156	.	.	.
3	−0·0781	.	.	0·1250	0·0156	0·0156	0·0468	.	.	.
11	−0·1816	0·0468	0·0156	0·0156	0·0762	0·0449	0·0449	−0·0468	−0·0468	−0·0156
22	−0·1816	0·0156	0·0468	0·0156	0·0449	0·0762	0·0449	−0·0468	−0·0156	−0·0468
33	−0·1816	0·0156	0·0156	0·0468	0·0449	0·0449	0·0762	−0·0156	−0·0468	−0·0468
12	0·1094	.	.	.	−0·0468	−0·0468	−0·0156	0·1250	.	.
13	0·1094	.	.	.	−0·0468	−0·0156	−0·0468	.	0·1250	.
23	0·1094	.	.	.	−0·0156	−0·0468	−0·0468	.	.	0·1250

(59)

Thus the estimates of the quadratic effects were rather strongly correlated. The point at which the two steepest ascent experiments were performed was reasonably close to the design and the information from these was now included. By substituting the independent variables corresponding to the experiments (9) and (10) in the fitted equation (58), the predicted values 12·94 and 28·51 were obtained; these may be compared with the values found, 13 and 27. The agreement is remarkably good, and we are encouraged to hope that the assumptions made are satisfactory within the region considered. With the aid of Plackett's equations, discussed in §5.2, these discrepancies, between the values predicted and those actually found, were now used to modify the estimates of the b's. A new matrix $\mathbf{C}_0^{-1}$ appropriate for the whole set of 13 points was also obtained. The matrix $\mathbf{W}$, for the two new sets of independent variables, is—

$\mathbf{W} =$

	0	*1*	*2*	*3*	*11*	*22*	*33*	*12*	*13*	*23*
(9)	1	−0·5400	−2·0000	−1·8200	0·2916	4·0000	3·3124	1·0800	0·9828	3·6400
(10)	1	−0·8100	−3·0000	−2·7300	0·6561	9·0000	7·4529	2·4300	2·2113	8·1900

(60)

The vector $\mathbf{Z}$, which gives the costs found at these points, is $\mathbf{Z} = \begin{bmatrix} 13 \\ 27 \end{bmatrix}$. Whence $\mathbf{J} = \mathbf{WC}^{-1}$ is given by

$\mathbf{J} =$

(9)	0·3000	−0·0177	−0·0843	−0·0833	−0·0695	−0·0823	−0·0951	−0·0085	0·0008	0·2170
(10)	−0·4765	0·1084	0·0054	0·0808	0·1344	0·1399	0·1068	−0·1560	−0·1349	0·3516

(61)

and $\mathbf{R} = \mathbf{WC}^{-1}\mathbf{W}' = \mathbf{JW}' =$

	(9)	*(10)*
(9)	0·7469	1·0582
(10)	1·0582	3·2742

(62)

$\mathbf{I} + \mathbf{R}$ is obtained by adding unity to each of the diagonal elements in $\mathbf{R}$; and $(\mathbf{I} + \mathbf{R})^{-1} = \mathbf{G}$, by solving the equations $(\mathbf{I} + \mathbf{R})\,\mathbf{G} = \mathbf{I}$. Thus,

$$\left\{\begin{array}{l} 1{\cdot}7469g_{11} + 1{\cdot}0582g_{12} = 1 \\ 1{\cdot}0582g_{11} + 4{\cdot}2742g_{12} = 0 \end{array}\right\} \quad \left.\begin{array}{l} 1{\cdot}7469g_{12} + 1{\cdot}0582g_{22} = 0 \\ 1{\cdot}0582g_{12} + 4{\cdot}2742g_{22} = 1 \end{array}\right\}. \qquad (63)$$

Solving for the g's we have

$$\mathbf{G} = \begin{array}{c} \\ (9) \\ (10) \end{array} \begin{array}{cc} (9) & (10) \\ \left[\begin{array}{r} 0{\cdot}6734 \\ -0{\cdot}1667 \end{array}\right. & \left.\begin{array}{r} -0{\cdot}1667 \\ 0{\cdot}2752 \end{array}\right] \end{array}. \qquad (64)$$

We now pre-multiply $\mathbf{J}$ by $\mathbf{G}$ and obtain—

$\mathbf{GJ} =$

	0	*1*	*2*	*3*	*11*	*22*	*33*	*12*	*13*	*23*
(9)	0·2815	−0·0300	−0·0727	−0·0696	−0·0692	−0·0787	−0·0819	0·0202	0·0230	0·0875
(10)	−0·1812	0·0328	0·0403	0·0361	0·0486	0·0522	0·0453	−0·0415	−0·0373	0·0606

(65)

The new values for the b's can now be calculated simply from (50).
The vector of the discrepancies is—

$$\boldsymbol{\Delta} = \mathbf{Z} - \mathbf{WB} = \begin{array}{c} (9) \\ (10) \end{array} \left[\begin{array}{r} 0{\cdot}0623 \\ -1{\cdot}5137 \end{array}\right]. \qquad (66)$$

Thus, for example, the value for b_0 is—

$$6{\cdot}7500 + \{0{\cdot}0623 \times 0{\cdot}2815\} + \{(-\,1{\cdot}5137) \times (-\,0{\cdot}1812)\} = 7{\cdot}0418$$

The calculations are carried out as a single operation on the machine. In this way, we find—

$$\begin{array}{lll} & b_0 = 7{\cdot}0418 & \\ b_1 = 0{\cdot}6985 & b_{11} = 2{\cdot}9221 & b_{12} = -\,2{\cdot}9359 \\ b_2 = 2{\cdot}6844 & b_{22} = 1{\cdot}5410 & b_{13} = -\,1{\cdot}1921 \\ b_3 = 2{\cdot}4410 & b_{33} = 1{\cdot}0510 & b_{23} = 2{\cdot}6637 \end{array}$$

The correction for $\mathbf{C}^{-1}$ may now be obtained by pre-multiplying $\mathbf{GJ}$ by $\mathbf{J}'$. Again the new matrix can be obtained from the matrices $\mathbf{C}^{-1}$, $\mathbf{GJ}$, and $\mathbf{J}$, by a series of operations on the machine. For example, the new first element in the matrix is—

$$0{\cdot}7168 - \{0{\cdot}3000 \times 0{\cdot}2815\} - \{(-\,0{\cdot}4765) \times (-\,0{\cdot}1812)\} = 0{\cdot}5460.$$

The new matrix $\mathbf{C}_0^{-1}$ obtained in this way is:

	0	*1*	*2*	*3*	*11*
0	0·5460	−0·0535	−0·0371	−0·0400	−0·1377 . . . . . .
1	−0·0535	0·1209	−0·0057	−0·0051	0·0404
2	−0·0371	−0·0057	0·1150	−0·0093	0·0051
3	−0·0400	−0·0051	−0·0093	0·1163	0·0059
11	−0·1377	0·0404	0·0051	0·0059	0·0648
22	−0·1331	0·0086	0·0352	0·0048	0·0324
33	−0·1355	0·0093	0·0044	0·0364	0·0331
12	0·0835	0·0048	0·0057	0·0050	−0·0399
13	0·0847	0·0044	0·0055	0·0049	−0·0403
23	0·1120	−0·0050	0·0016	0·0024	−0·0177 . . . . . .

22	33	12	13	23
−0·1331	−0·1355	0·0835	0·0847	0·1120
0·0086	0·0093	0·0048	0·0044	−0·0050
0·0352	0·0044	0·0057	0·0055	0·0016
0·0048	0·0064	0·0050	0·0049	0·0024
0·0324	0·0331	−0·0399	−0·0403	−0·0177
0·0624	0·0319	−0·0394	−0·0085	−0·0481
0·0318	0·0636	−0·0093	−0·0407	−0·0450
−0·0394	−0·0093	0·1187	−0·0056	0·0102
−0·0085	−0·0407	−0·0056	0·1199	0·0081
−0·0481	−0·0450	0·0102	0·0081	0·0847

(67)

The additional points have slightly reduced the variance for the estimates (cf. (67) and (59)). (The matrix (67) is required later, otherwise it might not have been worthwhile to calculate it.)

Using (51), the new value for the residual sum of squares, now based on three degrees of freedom, is—

$$S_0 = 0\cdot00 + 0\cdot66 = 0\cdot66 \qquad (68)$$

Whence, if the fit were perfect, the estimated standard deviation would be—

$$s = (0\cdot66/3)^{\frac{1}{2}} = 0\cdot47.$$

In order to study the nature of the fitted surface we must reduce it to standard form, in the manner discussed in §5.1.

The discriminating cubic is—

$$\begin{vmatrix} 2\cdot9221 - \lambda & -1\cdot4679 & -0\cdot5960 \\ -1\cdot4679 & 1\cdot5410 - \lambda & 1\cdot3318 \\ 0\cdot5960 & 1\cdot3318 & 1\cdot0510 - \lambda \end{vmatrix} = 0 \qquad (69)$$

i.e.,

$$\lambda^3 - 5\cdot5141\lambda^2 + 4\cdot9114\lambda + 0\cdot9290 = 0$$

which has the roots

$$\lambda_1 = -0\cdot1597 \qquad \lambda_2 = 1\cdot3434 \qquad \lambda_3 = 4\cdot3304 \qquad (70)$$

The equations for the co-ordinates of the centre are:

$$\begin{aligned} 5\cdot8842x_1 - 2\cdot9359x_2 - 1\cdot1921x_3 &= -0\cdot6985 \\ -2\cdot9359x_1 + 3\cdot0820x_2 + 2\cdot6637x_3 &= -2\cdot6844 \\ -1\cdot1921x_1 + 2\cdot6637x_2 + 2\cdot1020x_3 &= -2\cdot4410 \end{aligned}$$

the solutions of which give the co-ordinates of the stationary point S.

$$x_1^0 = -0\cdot3365 \qquad x_2^0 = 0\cdot2411 \qquad x_3^0 = -1\cdot6576 \qquad (71)$$

By substituating these values in the fitted equation, or using (46), the predicted value for the cost at this point is found to be $y^0 = 5\cdot2247$. The equation of the contour surfaces may therefore be written in some set of orthogonal co-ordinates as

$$y_c - 5\cdot2247 = -0\cdot1597\, X_1^2 + 1\cdot3434\, X_2^2 + 4\cdot3304\, X_3^2. \qquad (72)$$

The fitted contour surfaces are hyperboloids of one sheet; thus the sections by the planes $X_2 = 0$ and $X_3 = 0$ are hyperbolas, and that by X_1 is an ellipse. From the smallness of λ_1 compared with the other coefficients, it is clear that the surface is attenuated along the X_1 axis, i.e., that there is a ridge running in this direction.

To find the direction of the new axes we require the orthogonal matrix which transforms the old variables to the new, that is the matrix **M** for which—

$$\mathbf{X} = \mathbf{M}(x - x^0). \qquad (73)$$

If $\mathbf{M}_t$ is the vector corresponding to the tth row of **M**, it is proved in text books that—

$$\mathbf{M}_t(\tfrac{1}{2}\mathbf{F} - \mathbf{I}\lambda_t) = \mathbf{O}, \qquad (74)$$

where **F** is the matrix $\{f_{st}\}$ of second order derivatives. Substituting λ_1 in this equation we obtain the set of homogeneous equations:

$$\begin{aligned} 3\cdot0818\, M_{11} - 1\cdot4679\, M_{12} - 0\cdot5960\, M_{13} &= 0 \\ -1\cdot4679\, M_{11} + 1\cdot7007\, M_{12} + 1\cdot3318\, M_{13} &= 0 \\ -0\cdot5960\, M_{11} + 1\cdot3318\, M_{12} + 1\cdot2107\, M_{13} &= 0 \end{aligned}$$

the solutions of which are proportional to the elements of the first row of **M**. Putting $M_{11} = 1$ we have a consistent set of three equations in two unknowns. Solving any two of these we obtain the values $M_{11} = 1$, $M_{12} = 3\cdot4318$, $M_{13} = -3\cdot2813$. Since **M** is orthogonal, the sum of squares of the elements in any row or column is unity. Thus division of the elements M_{11}, etc., by the square root of their sum of squares gives the first row of **M**.

$$m_{11} = 0\cdot2061 \qquad m_{12} = 0\cdot7073 \qquad m_{13} = -0\cdot6763.$$

Substituting λ_2 in (74) we find in a similar way the elements

$$m_{21} = 0\cdot6385 \qquad m_{22} = 0\cdot4265 \qquad m_{23} = 0\cdot6406.$$

Finally substituting λ_3 we obtain

$$m_{31} = 0\cdot7416 \qquad m_{32} = -0\cdot5638 \qquad m_{33} = -0\cdot3636.$$

The co-ordinates of the centre and any point along the three principle axes are thus given by—

New Variables			*Old Variables*		
0	0	0	$-0\cdot3365$	$0\cdot2411$	$-1\cdot6756$
X_1	0	0	$-0\cdot3365 + 0\cdot2061X_1$	$0\cdot2411 + 0\cdot7073X_1$	$-1\cdot6576 - 0\cdot6763X_1$
0	X_2	0	$-0\cdot3365 + 0\cdot6385X_2$	$0\cdot2411 + 0\cdot4265X_2$	$-1\cdot6576 + 0\cdot6406X_2$
0	0	X_3	$-0\cdot3365 + 0\cdot7416X_3$	$0\cdot2411 - 0\cdot5638X_3$	$-1\cdot6576 - 0\cdot3636X_3$

. . . (75)

The appearance of the fitted surface can be appreciated from Fig. 4, which shows the arrangement of the points in space, looking down the axis X_1. The contour lines for costs of 10 and 20 are drawn as they would be seen on the plane $X_1 = 0$. Owing to the small value of λ_1, the estimated contour surfaces in the neighbourhood of the design are approximately cylinders having the elliptic cross section shown (see Fig. 4, p. 32).

From the figure it will be seen that the agreement between the contours of the fitted surface and the actual experimental points is good. In the circumstances considered this arrangement of points would be expected to supply fairly accurate estimates of λ_2 and λ_3. However, a range of hyperboloids of one sheet and ellipsoids attenuated along the X_1 axis having this elliptic section at $X_1 = 0$ would give almost as satisfactory a fit. In order to determine λ_1 more accurately, it was decided to carry out two further experiments at points where the X_1 axis intersected the two planes $x_3 = 1$, $x_3 = -3$.

The co-ordinates for these two points may be obtained from (75). In the new system they are $(-3\cdot9296, 0, 0)$ and $(1\cdot9849, 0, 0)$. The co-ordinates in the old system, and the costs found when experiments were performed at these points, are given below:

	x_1	x_2	x_3	y
(14)	$-1\cdot1465$	$-2\cdot5383$	1	6
(15)	$0\cdot0727$	$1\cdot6450$	-3	19

The first point confirms the fairly stationary nature of the cost moving up the axis X_1, but the second point suggests that the contours do in fact close in the opposite direction.

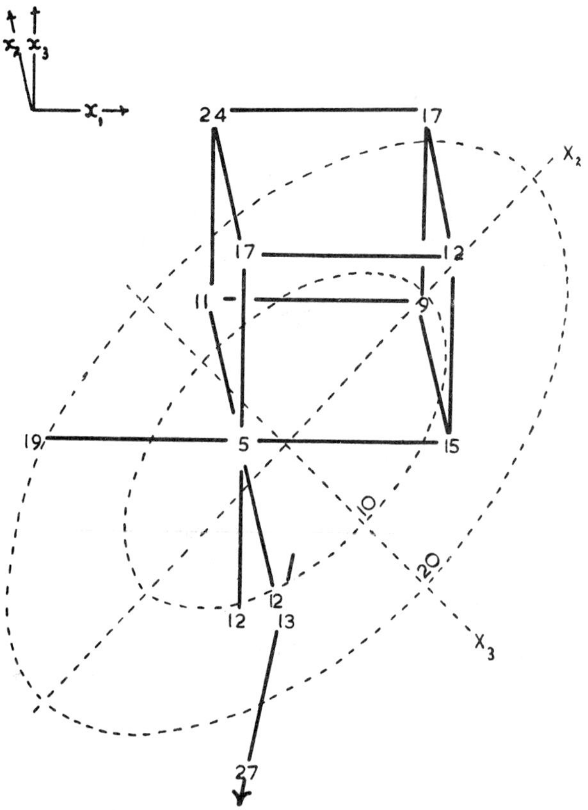

Fig. 4.—The arrangement of experiments and fitted contours "looking down".

The information supplied by these two additional points was now included, as described before. The adjusted values for the constants were:

$$b_0 = 5\cdot227$$

$$b_1 = 0\cdot874 \quad b_{11} = 3\cdot339 \quad b_{12} = -3\cdot223$$

$$b_2 = 3\cdot220 \quad b_{22} = 2\cdot239 \quad b_{13} = -1\cdot561$$

$$b_3 = 2\cdot177 \quad b_{33} = 1\cdot778 \quad b_{23} = 1\cdot7451$$

and the new residual sum of squares, based now on 5 degrees of freedom, was 19·93. Whence if the fit were perfect, the estimated standard deviation would be—

$$s = (19\cdot93/5)^{\frac{1}{2}} = 2\cdot00.$$

This is significantly greater than the value of one unit which we assumed for the standard deviation; this may mean that our preliminary value was an underestimate, or alternatively that the fit of the second degree equation was imperfect. The latter explanation is a likely one, but the departure from expectation is small enough to suggest that the second degree equation is at least a useful approximation. From the b's a new set of λ's was calculated as follows:

$$\lambda_1 = 0\cdot9461 \qquad \lambda_2 = 1\cdot4999 \qquad \lambda_3 = 4\cdot9113$$

and the new centre was—

$$x_1^0 = -0 \cdot 7601 \qquad x_2^0 = -1 \cdot 1093 \qquad x_3^0 = -0 \cdot 4014$$

at which the predicted cost was

$$y^0 = 2 \cdot 6719.$$

The new transformation was—

$$\mathbf{M} = \begin{bmatrix} 0 \cdot 3786 & 0 \cdot 7933 & -0 \cdot 4768 \\ 0 \cdot 5424 & 0 \cdot 2271 & 0 \cdot 8088 \\ 0 \cdot 7500 & -0 \cdot 5649 & -0 \cdot 3442 \end{bmatrix}. \quad . \quad . \quad . \quad (76)$$

The directions of the axes and the values of λ_2 are not greatly changed but the value for λ_1 is now positive, indicating ellipsoidal contours with the centre a true maximum. Two further experiments were performed at this centre, the costs found being 4 and 3 respectively. Four additional experiments, two performed on the newly calculated axis of X_1 and two elsewhere, confirmed the general nature of the surface; by including these the constants of the surface were not greatly changed. The final conclusion drawn was that a minimum for cost occurred at about the levels 141° C., 20 per cent. concentration and molar ratio (A/B) 6·2; here the estimated cost was about 3·2. The final estimate of the standard deviation based on the 11 residual degrees of freedom was 1·7; thus the fitted equation was probably a satisfactory approximation.

6. Summary and Discussion

The general problem is discussed of finding experimentally the levels of a number of quantitative variables at which some dependent response has a maximum value. The problem can be solved by exploring the whole experimental region; but the number of experiments necessary to do this would usually be prohibitively large. When the experimental error is small and experiments are conducted sequentially, the derivatives determined in a given sub-region may be used to locate another in which the response is greater. By repetition of this process a near-stationary region is found. This region may then be explored by determining derivatives of higher order. The derivatives are deduced by performing experiments in the sub-region and fitting an equation of suitable degree to the experimental points. Two possible sources of error arise, that due to errors of observation, and that due to bias which might occur if the response function were not capable of representation by an equation of the type assumed. The extent to which the estimates will be affected by these errors is completely determined by two matrices which depend only on the arrangement of the experimental points. Using these matrices it is possible to judge the suitability of possible designs and to compare different types of design.

When it is desired to estimate derivatives of first order only, the two-level factorial and incomplete factorial designs will provide efficient estimates. When effects of higher order are needed, multi-level factorial and incomplete factorial designs are less satisfactory and new types of design are proposed. Particular examples of these are "composite" designs which are used to determine the effects up to order 2 and which are formed by adding further points to two-level designs.

When an equation of second degree has been fitted in a particular region the nature of the local surface can be deduced approximately, by reducing the conic, of which the fitted surface is part, to canonical form. This will usually suggest points at which confirmatory experiments should be performed; the information from these may be incorporated without undue recalculation. The methods discussed below have been used in a number of investigations with considerable success; parts of two of these investigations are used as examples.

The procedure described can be regarded as an extension and rationalization of the methods of the experienced experimenter in situations where the experimental error is fairly small and sequential experimentation is possible. When confronted with problems of the sort discussed he will, as a rule, consider carefully the result of one set of experiments to plan the next, move the levels of factors in the direction in which increases are indicated, and carry out extra experiments at points where the information is inadequate. Any rigid systemization of his procedure may lose more by hindering power of manoeuvre than it gains by increased precision, and will (quite rightly in our opinion) not appeal to him. Much work remains to be done, in particular in the study of designs of second and higher orders, and in the determining of confidence regions for a predicted stationary point. Work is proceeding on both these problems, and it is hoped to discuss them in later communications. In conclusion we wish to acknowledge our indebtedness to many colleagues whose helpful criticisms and suggestions have been of the greatest value, in particular to Dr. O. L. Davies, Dr. T. S. Kenney, Dr. P. V. Youle and Dr. R. J. Benzie.

References

AITKEN, A. C. (1935), *Proc. Roy. Soc. Edin.*, **55**, 42.
—— (1937), *ibid.*, **57**, 269.
BANERJEE, K. S. (1949), *Ann. Math. Statist.*, **20**, 300.
BARNARD, G. A. (1946), *J.R. Statist. Soc. Suppl.*, **8**, 1.
BOOTH, A. D. (1949), *Quart. J. Appl. Math.*, **1**, 237.
BROWNLEE, K. A., KELLY, B. K., and LORAINE, P. K. (1948), *Biometrika*, **35**, 268.
DAVID, F. N., and NEYMAN, J. (1938), *Statist. Res. Mem.*, **2**, 105.
DAVIES, O. L., and HAY, W. (1950), *Biometrics*, **6**.
DWYER, P. S. (1942), *J. Amer. Statist. Assn.*, **37**, 441.
FINNEY, D. J. (1945), *Ann. Eugen. Lond.*, **12**, 291.
FISHER, R. A. (1935), *The Design of Experiments*. Edinburgh: Oliver and Boyd.
—— (1942), *Ann. Eugen. Lond.*, **11**, 341.
FOX, L., HUSKEY, H. O., and WILKINSON, J. H. (1948), *Quart. J. Mech. Appl. Math.*, **1**, 149.
FRIEDMAN, M., and SAVAGE, L. J. (1947), *Selected Techniques of Statistical Analysis*. New York: McGraw-Hill.
GAUSS, C. F. (1821), *Werke*, 4 Göttingen.
HOTELLING, H. (1941), *Ann. Math. Statist.*, **12**, 20.
—— (1944), *ibid.*, **15**, 297.
KEMPTHORNE, O. (1947), *Biometrika*, **34**, 255.
—— (1948), *Ann. Math. Statist.*, **19**, 238.
KENDALL, M. G. (1946), *The Advanced Theory of Statistics*, Vol. II. London: Griffin.
MARKOFF, A. A. (1912), *Wahrscheinlichkeitsrechnung*. Leipzig: Tembner.
PLACKETT, R. L., and BURMAN, J. P. (1946), *Biometrika*, **33**, 305.
PLACKETT, R. L. (1949), *Biometrika*, **36**, 458.
—— (1950), *ibid.*, **37**, 149.
RAO, C. R. (1947), *J.R. Statist. Soc., Suppl.*, **9**, 128.
SOUTHWELL, R. V. (1946), *Relaxation Methods in Theoretical Physics*. Oxford.
—— (1940), *Relaxation Methods in Engineering Science*. Oxford.
SYNGE, J. L. (1944), *Quart. J. Appl. Math.*, **1**, 237.
TEMPLE, G. (1939), *Proc. Roy. Soc.*, A, **169**, 476.
YATES, F. (1935), *J.R. Statist. Soc., Suppl.*, **2**, 181.
—— (1937), *Imp. Bur. Soil Sci. Tech. Comm.*, No. 35.
WALD, A. (1945), *Ann. Math. Statist.*, **16**, 117.

Appendix (1): *Occurrence of Bias due to Extraneous Constants*

Consider the case where N experiments are performed and N constants fitted, but S extra terms are required to obtain a perfect fit, i.e.,

$$\eta = \sum_i^N \beta_i X_i + \sum_j^S \beta_j X_j$$

and the $L = N + S$ functions X_i, X_j are linearly independent. Suppose also that the design is chosen so that $\mathbf{X}_1$ the $N \times N$ matrix of the first N independent variables is non-singular and that none of the X_j consist of zero values only. Then if the estimates b_i of the β_i are calculated, each of the extra constants β_j will occur in the expectations of one or more of the b_i whatever design is chosen.

Proof.—Suppose this is not so for a particular one of the extra constants β_j. Then from (20) the jth column of $\mathbf{T}_1\mathbf{X}_2 = X_1^{-1}X_2$ would consist entirely of zeros; this in turn implies that X_1^{-1} is singular, for it pre-multiplies a non-zero vector in $\mathbf{X}_2$, to give a zero vector. This is contrary to hypothesis hence the above assertion is true.

Appendix (2): *Generation of First Order Type B Designs from those of Type A*

Suppose we use a particular design for the estimation of the k linear effects (not necessarily an incomplete factorial or even an orthogonal design) with $N \times k$ matrix $\mathbf{D}_1$; also suppose that $\mathbf{X}$ is the corresponding $N \times (k+1)$ matrix of independent variables. Then $\mathbf{D}_2 = \begin{bmatrix} \mathbf{X} \\ -\mathbf{X} \end{bmatrix}$ will provide a design matrix for the estimation of the linear effects of $K+1$ factors and, however biased by second order effects were the estimates of first order effects using $\mathbf{D}_1$, the first order estimates using $\mathbf{D}_2$ will be completely free of bias due to this cause.

Proof.—The matrix of independent variables for the second order effects obtained by multiplying the elements of the columns of $\begin{bmatrix} \mathbf{X} \\ -\mathbf{X} \end{bmatrix}$ must be of the form $\begin{bmatrix} \mathbf{Z} \\ \mathbf{Z} \end{bmatrix}$. The alias matrix for first and second order effects is, therefore,

$$\tfrac{1}{2}(\mathbf{X}'\mathbf{X})^{-1}(\mathbf{X}'\mathbf{Z} - \mathbf{X}'\mathbf{Z}),$$

which is null.

Appendix (3): *Dependence of* $\mathbf{C}_1^{-1}$ *and* $\mathbf{A}$ *on Scale Factors*

Consider the changes produced in the precision matrix $\mathbf{C}_1^{-1}$ and the alias matrix $\mathbf{A}$ when the scales for the k factors $x_1, x_2, \ldots x_k$ are changed. Suppose the co-ordinates in the design matrix for the factor x_t are multiplied by a constant K_t ($t = 1, 2, \ldots k$). The equation (17) will become

$$\eta = \mathbf{X}_1\mathbf{K}_1\boldsymbol{\beta}_1 + \mathbf{X}_2\mathbf{K}_2\boldsymbol{\beta}_2. \qquad (1)$$

Where $\mathbf{K}_1$ $(M \times M)$ and $\mathbf{K}_2$ $(S \times S)$ are diagonal matrices which are such that if a given column in $\mathbf{X}_1$ or $\mathbf{X}_2$ corresponds with the independent variable $x_1^{\alpha}x_2^{\gamma} \ldots x_k^{\pi}$, the corresponding element in $\mathbf{K}_1$ or $\mathbf{K}_2$ is $K_1^{\alpha}K_2^{\gamma} \ldots K_k^{\pi}$. We have then instead of $\mathbf{C}_1^{-1}$ the matrix

$$\mathbf{K}_1\mathbf{C}_1^{-1}\mathbf{K}_1^{-1}, \qquad (2)$$

and instead of $\mathbf{A}$ we have

$$\mathbf{K}_1^{-1}\mathbf{A}\mathbf{K}_2. \qquad (3)$$

Thus (i) if originally the covariance* between two effects having subscripts $1^{\alpha}2^{\gamma} \ldots k^{\pi}$ and $1^{\alpha'}2^{\gamma'} \ldots k^{\pi'}$ was c, then the covariance after changing the ranges of the factors would be

$$c \times \{K_1^{-(\alpha+\alpha')}K_2^{-(\gamma+\gamma')} \ldots K_k^{-(\pi+\pi')}\} \qquad (4)$$

(ii) If originally the expected value for a particular estimate b with subscript $1^{\alpha}2^{\gamma} \ldots k^{\pi}$ contained a term $a\beta$ where β had subscript $1^{\alpha'}2^{\gamma'} \ldots k^{\pi'}$, then after changing the ranges of the factors the new multiplier corresponding with a would be—

$$a \times \{K_1^{\alpha'-\alpha}K_2^{\gamma'-\gamma} \ldots K_k^{\pi'-\pi}\}. \qquad (5)$$

For example, if the ranges were halved for all the factors then the variance of b_0 would be unchanged, the variance of the first order estimates $b_1, b_2 \ldots, b_k$ would be multiplied by the factor 4, those of second order b_{11}, b_{22}, b_{12}, etc., would be multiplied by the factor 16 and so on. For an estimate b of order d the aliases of order $d+1$ would be halved, those of order $d+2$ divided by 4 and so on.

Appendix (4): *Composite Designs of Type A and Order* 2

To the 2 level factorial or fractional factorial design of type B' observations at $2k+1$ extra points are added, one at the origin and the remaining $2k$ at a distance α units from it, in pairs, equally spread along the k axes of the design. Suppose there are F observations in the type B' design and T extra observations, and allow the possibility that the extra observations are

* In this context the term covariance is to be understood to include the variance, which is regarded as the covariance of a particular variable with itself.

replicated n times so that $T = n(2k + 1)$. We have to fit an equation including all terms up to second degree

$$\eta = \beta_0 x_0 + \beta_1 x_1 + \ldots + \beta_{11} x_1^2 + \ldots + \beta_{12} x_1 x_2 + \text{etc.} \qquad (1)$$

If $\bar{\eta}$ is subtracted from (1) we have the equivalent equation—

$$\eta = \bar{\eta} x_0 + \beta_1 (x_1 - \bar{x}_1) + \ldots + \beta_{11}(x_1^2 - \overline{x_1^2}) + \ldots + \beta_{12}(x_1 x_2 - \overline{x_1 x_2}) + \text{etc.}, \qquad (2)$$

and since the means of all independent variables except the quadratic variables are zero,

$$\beta_0 = \bar{\eta} - \beta_{11}\overline{x_1^2} - \beta_{22}\overline{x_2^2} - \text{etc.} \qquad (3)$$

The matrix **X** of independent variables corresponding to (2) may now be written down in terms of α. For all composite designs of this sort the sum of products between the columns of the matrix **X** (that is, the non-diagonal elements of **C**) are all zero except those between the columns corresponding to quadratic variables. If we denote the $k \times k$ sub-matrix corresponding to the quadratic variables by **Q**, the inverse matrix $(\mathbf{C})^{-1}$ will be of the same pattern as **C** with a matrix $(\mathbf{Q})^{-1}$ corresponding to the quadratic variables and the remaining non-diagonal elements zero. The diagonal elements not in $\mathbf{Q}^{-1}$ will be reciprocals of those in **C**.

The sub-matrix **Q** has equal diagonal elements p corresponding to the sums of squares of the quadratic variables, and equal non-diagonal elements q corresponding to their sums of products, and—

$$FT - 4nF\alpha^2 - 4n^2\alpha^4 + 2n(F + T)\alpha^4 = (F + T)p. \qquad (4)$$

$$FT - 4nF\alpha^2 - 4n^2\alpha^4 = (F + T)q. \qquad (5)$$

The inverse of a $k \times k$ matrix of this type is easily found; the diagonal elements in the inverse are all equal to

$$c = \{p + (k - 2)q\}/[(p - q)\{p + (k - 1)q\}], \qquad (6)$$

and the non-diagonal elements to

$$d = -q/[(p - q)\{p + (k - 1)q\}]. \qquad (7)$$

Orthogonal Designs

If α is chosen so that the covariances d are zero, the designs will be completely orthogonal. Now if d is zero, q is zero, whence equating (5) to zero and taking the positive root we find that for orthogonality

$$\alpha = \{\rho F/(4n^2)\}^{\frac{1}{4}} \quad \text{where} \quad \rho = \{(F + T)^{\frac{1}{2}} - F^{\frac{1}{2}}\}^2. \qquad (8)$$

Since with this value of α the design is orthogonal, it can be used very simply. Each of the estimates $\bar{y}$, b_s, b_{ss}, b_{st} is obtained by calculating the sum of products between the observations y and the appropriate independent variable in **X** and dividing the result by the sum of squares of the elements of the independent variable. The reciprocal of the latter sum of squares multiplied by σ^2 gives the variance of the estimate. The estimate of b_0 is provided by substituting the estimates $\bar{y}$, b_{11}, etc., in (3).

The spread for each of the variables is $\{F/(F + T)\}^{\frac{1}{2}}$. For the factorial and fractional factorial designs with levels $-1, 0, 1$, the spread is $(2/3)^{\frac{1}{2}}$. For comparison with the factorial design therefore the composite design must have levels $-g\alpha, -g, O, g, g\alpha$, where $g = \{4(F + T)/(9F)\}^{\frac{1}{2}}$.

Composite Designs in which the Variances of the Estimated Second Order Derivatives are All Equal

As pointed out on p. 18, for our purpose it would be desirable to employ designs in which the variances for quadratic and interaction derivatives were more nearly equal. Equality would be attained if $V(b_{ss}) = \frac{1}{4}V(b_{st})$. Now irrespective of the value of α the variance of the interaction estimate b_{st} is σ^2/F. Writing $A = \alpha^2$, substituting (4) and (5) in (6), and equating the resulting expression to $1/(4F)$ we obtain the quartic equation—

$$2n(F + n)A^4 - 4nkFA^3 + F\{nk(2k + 1) - 4(F + 3n)\}A^2 + 8(k - 1)F^2A - 2(k - 1)(2k + 1)F^2 = 0 \quad . \quad (9)$$

On substituting the appropriate values of n, F and k and solving the resulting equations we find a real positive root from which $\alpha = A^{\frac{1}{2}}$ is obtained. It is then easy to show that the variances and covariances for the effects are—

$$V(\bar{y}) = \sigma^2/(F + T), \quad V(b_s) = \sigma^2/(F + 2nA), \quad V(b_{ss}) = \sigma^2/4F, \quad V(b_{st}) = \sigma^2/F,$$

$$\text{Cov}\,(b_{ss}b_{tt}) = \sigma^2\{(1/4F) - (1/2nA^2)\}.$$

These values divided by σ^2 are the elements of the precision matrix $\mathbf{C}^{-1}$, the remaining elements all being zero. Knowing $\mathbf{C}^{-1}$ the estimates $\bar{y}$, b_s, b_{ss}, b_{st} are of course easily found, b_0 being obtained from (3) as before.

In the following table the construction of composite designs for $k = 2$, 3, 4 and 5 is illustrated and the precision of the estimates compared. The comparison is made between the factorial design at levels $-1, 0, 1$, and the composite designs at levels $-g\alpha$, $-g$, O, g, $g\alpha$, g being chosen in each case so that the spread of variables is the same for the composite design and the factorial design. The columns marked O refer to orthogonal composite designs, whilst those marked E refer to the composite designs in which the variances for second order effects are made equal.

Comparison of Two Types of Composite Design with a Three-level Factorial Design

Number of factors		$k = 2$		$k = 3$		$k = 4$		$k = 5$	
Type B' design		2^2		2^3		2^4		$\frac{1}{2}2^5(I=12345)$	
F		4		8		16		16	
n		1		1		1		1	
$T = n(2k + 1)$		5		7		9		11	
Total number of observations $F + T$		9		15		25		27	
Total number of constants estimated		6		10		15		21	
		O	E	O	E	O	E	O	E
Value of α		1·000	2·090	1·215	2·432	1·414	2·799	1·547	2·872
Efficiency* of composite designs compared with factorial. (Comparison of variances adjusted for number of observations and scale)	b_s	1·00	1·00	1·00	1·00	1·00	1·00	1·00	1·00
	b_{ss}	1·00	1·78	1·07	2·44	1·00	3·19	1·43	3·27
	b_{st}	1·00	0·22	1·00	0·31	1·00	0·40	1·00	0·41
Coefficients of correlation between quadratic effects		0·00	0·58	0·00	0·54	0·00	0·48	0·00	0·53
Scale factor (g) for composite design		1·000	0·686	0·955	0·710	0·913	0·726	0·931	0·744

* A value greater than unity denotes that the composite design gives *greater* precision.

In the row marked "Type B' design" the nature of the design is indicated. Thus $\frac{1}{2}2^5(I = 12345)$ indicates that the design used is a half replicate of the 2^5 factorial with the defining relation $I = 12345$. The two factor orthogonal composite design is identical with the 3^2 factorial, but is included for completeness. In these designs the number of experiments to be performed is not greatly in excess of the number of constants to be estimated. Thus the design for $k = 5$ estimates the 21 effects of zero, first and second order in only 27 experiments, as compared with 81 experiments required by the 1/3rd replicate of the 3^5 factorial. To obtain designs of this sort which compared favourably with 3 level factorials when k was large, n would have to be greater than 1. Thus, it would be necessary to replicate the $2k + 1$ extra points. Intuitively, this would not seem to be the most satisfactory arrangement, since, by carrying out the further experiments at new points rather than merely replicating, one might expect to be able to reduce the possible bias in the effects. This point requires further investigation.

The designs cannot, of course, be judged solely on the precision of the estimates; it is necessary to consider the possible bias which might arise if terms of higher order occurred. For $k = 2, 3, 4, 5$ the coefficients in these alias matrices are fairly small. This will be illustrated for $k = 2$, and $k = 5$. The comparisons are made as usual between designs in which the scale factors are adjusted to make the spreads equal. For $k = 2$ the orthogonal composite design is the same as the 3^2 factorial for which—

$$b_1 \to \beta_1 + \beta_{111} \qquad b_2 \to \beta_2 + \beta_{222},$$

and the remaining effects are unbiased by third order terms. For the composite design in which the second order terms have equal variance,

$$b_1 \to \beta_1 + 1\cdot56\beta_{111} + 0\cdot15\beta_{112} \qquad b_2 \to \beta_2 + 1\cdot56\beta_{222} + 0\cdot15\beta_{112}$$

and the remaining effects are unbiased by third order terms. For $k = 5$, the 3^5 factorial gives—

$$b_1 \to \beta_1 + \beta_{111},$$

and similar expressions for other first order effects. The second order effects are unbiased by third order terms.

For the orthogonal composite designs we find—

$$b_1 \to \beta_1 + 1\cdot14\beta_{111} + 0\cdot67\,(\beta_{122} + \beta_{133} + \beta_{144} + \beta_{155}),$$

and similar expressions for the other linear effects,

$$b_{12} \to \beta_{12} + 0\cdot93\beta_{345},$$

and similar expressions for other mixed second order effects, whilst b_{11}, b_{22}, etc., are unbiased by third order terms.

For the composite designs in which the second order effects have equal variance,

$$b_1 \to \beta_1 + 2\cdot59\beta_{111} + 0\cdot27\,(\beta_{122} + \beta_{133} + \beta_{144} + \beta_{155})$$
$$b_{12} \to \beta_{12} + 0\cdot74\beta_{345},$$

and b_{11}, b_{22}, etc., are unbiased by effects of third order.

In the latter type of design, when $k = 5$, the possible bias due to third order effects is becoming rather large. Both bias and correlation between quadratic effects can be reduced by replicating the $2k + 1$ "extra" points, that is by putting $n > 1$.

Discussion on Paper by Mr. Box and Dr. Wilson

Professor S. Bartlett: It is a pleasure for me to propose the vote of thanks on this paper by Mr. Box and Dr. Wilson, who are neighbours of mine at Manchester, and with whose division at I.C.I. I have had pleasant friendly contacts.

Their paper appears so comprehensive that I rather wondered whether I should find anything further to say on the problem they have raised. But there is one point on which, as far as I can determine, there is a lacuna in their discussion, and that is that the errors involved in determining the optimum conditions do not always seem to me to be adequately treated. It is now customary to assess if possible the accuracy in our final result and arrange the experiment accordingly. The problem of biases seems to me especially important here, and I may be able to indicate some of the difficulties by referring to a problem with which some of us had to deal during the war. When we were concerned with the estimation of velocity or higher order derivative of a moving projectile whose successive positions were recorded by cinefilm, radar or other means, we sometimes had to decide whether to neglect higher order derivatives in assessing one of particular order, and hence introducing an unknown bias, or to estimate the higher order derivatives also. This problem is obviously allied to the simple case of the authors' where they have only one variable, for their problem is essentially one of estimating a derivative at a point where it is zero and, even if in the war problem as well as in theirs the basic random error is relatively small, it is somewhat remarkable how large the errors in the estimation of high order derivatives can be, especially if these have to be estimated near the end of the observed range. In this case, if we are prepared to put an upper bound to their value, it may be preferable to ignore them, as I will demonstrate.

Suppose we are fitting an orthogonal polynomial series

$$y = f(x) = f_0(x) + \alpha f_1(x),$$

where $\alpha f_1(x)$ is the doubtful term. (Notice that if we are equating

$$z = \frac{\partial y}{\partial x} = f_0'(x) + \alpha f_1'(x) = z_0 + \alpha z_1, \text{ say}$$

to zero, then the error in x is connected with that in z by the relation

$$\delta z \sim \frac{\partial z}{\partial x} \delta x).$$

Here we have

$$\delta z = \delta z_0 + \alpha z_1 \qquad . \quad . \quad . \quad . \quad . \quad . \quad . \quad (1)$$

if α is neglected, and

$$\delta z = \delta z_0 + z_1 \delta\alpha \qquad . \quad . \quad . \quad . \quad . \quad . \quad . \quad (2)$$

if α is also estimated. But, taking roughly as our confidence interval $\pm 2x$ standard deviation (plus an allowance for any bias), we have a *larger* interval in the latter case if

$$| \alpha z_1 \pm 2\sigma_0 | < 2\sqrt{\sigma_0^2 + z_1^2 \sigma_a^2}, \qquad . \quad . \quad . \quad . \quad . \quad . \quad (3)$$

where $\sigma_0^2 = E(\delta z_0)^2$, $\sigma_a^2 = E(\delta\alpha)^2$; or, for α small,

$$| \alpha z_1 \sigma_0 | < z_1^2 \sigma_a^2.$$

or

$$\left| \frac{\alpha}{\sigma_0} \right| < \frac{\sigma_a^2}{\sigma_0^2} | z_1 | . \qquad . \quad . \quad . \quad . \quad . \quad . \quad . \quad (4)$$

There is a further point which emerges also in connection with the authors' discussion. They deal at some length with the useful theory of "aliases" (rather a sinister word), but they seem to assume that, if the coefficients of the derivatives they have agreed to consider are unconfounded with the coefficients of the neglected terms, then all is well; but, as I have just shown, there will be a bias in the result due to the neglected coefficients. The further error, introduced by confounding with this neglected coefficient, is necessarily of the same order as its value, but may apparently be in either direction, and therefore it is not certain that the bias will be reduced by the absence of confounding; it may be increased.

Thus let now

$$z = f_0'(x) + \alpha f_1'(x) + \beta f_2'(x)$$
$$= z_0 + \alpha z_1 + \beta z_2,$$

where α is to be estimated, but the further term is to be neglected. For our orthogonal arrangement and (for simplicity) negligible experimental error,

$$\delta z = \beta z_2. \qquad . \quad . \quad . \quad . \quad . \quad . \quad . \quad . \quad (4)$$

Suppose alternatively we alter the values of x in the design so that z_2 is no longer orthogonal to z_1, but otherwise the accuracy is similar. The estimated coefficient α now has the bias (alias)

$$\beta \sum_x z_1 z_2 / \sum_x z_1^2 = \lambda\beta$$

say, where λ depends on the design, and hence

$$\delta z = \beta(z_2 + \lambda z_1). \qquad . \quad . \quad . \quad . \quad . \quad . \quad (5)$$

Error (4) is not necessarily less than error (5), and in some situations may be greater.

There is no doubt, however, in spite of these queries, of the great value of the authors' investigation and its record in this paper, and again let me say how much pleasure I have in proposing this vote of thanks to the authors.

Mr. K. D. TOCHER: I think most of us regard the subject of design of experiments as something of a mystery. Any new approach, such as in this paper, to the problem may unravel the mystery for some of us, and is, therefore, to be welcomed.

It is more than 25 years since Fisher began his campaign for the use of factorial designs in experimentation, rather than the classical methods then in vogue. At the present time there is a considerable section of the scientific world which regards the factorial method as the classical method, and views any departure from it with almost the same distaste as their predecessors viewed a change to it.

In those early days the main difficulty was to expose the prevailing illusion that if all the factors governing some experimental set-up were varied simultaneously, then the effects of the various factors would be hopelessly intermingled. This was achieved by showing how to choose a set of comparisons between the experimental results, which could easily be identified with the separate main effects, and their interactions.

However, once the principle of allowing all factors to vary at once and dissentangling the effects from one another by analysis of the results is accepted, then it by no means follows that this disentanglement is best achieved in terms of the set of "factorial" comparisons. This, as I understand it, is the main contention of the authors.

Most other systematic methods of analysis, however, involve some assumption of the form of the response function, involving for its complete specification less parameters than observations. In the analysis of the observations into the various main effects and interactions used in the factorial method, it is still possible at the conclusion of the analysis to recover the values of the original observations. Nothing has been lost by the analysis. With other methods, however, this is not possible, and this fact is the cause of the introduction of the bias, which plays such a prominent part in the theory developed in this paper. I think it is important to notice that it is the introduction of the form of the function, and not the abandoning of the factorial contrasts, which gives rise to this bias.

The problem of designing an experiment consists of selecting an experimental arrangement that minimizes both types of error—both the bias and the random experimental error. This problem reminds one forcibly of the similar problem arising in the Neyman-Pearson theory of testing hypotheses; in both cases we have two conflicting aims. Any attempts to decrease one source of error leads to an increase in the other. It is very tempting to try to solve the design problem in the same way as the test problem. Suppose we hold the level of the error due to bias at some fixed level (thus selecting a sub-class of all possible designs), and choosing that design (from the sub-class) which minimizes the random error. Now the bias is not a single number (in contrast with the probability of error of the first or second kind) but is a set of numbers, each of which is a function of several parameters, and similarly the random error is measured, not by a single number, but by the variance matrix. Thus the method fails. This difficulty of dealing with aggregates of functions rather than single numbers is bound to arise in any experiment which is intended to give information on more than one question, for it is clear that we can adjust our arrangement to distribute our information among the different questions in many different ways.

But in the problem under review the experiment is intended to answer just one question: "Where, in the factor space, does the point of optimum response lie?" True, it may be convenient to divide the inquiry into several stages, but each stage can be considered as asking a question about the location of the optimum point. Thus, at the first stage in the inquiry, we might ask to determine the line of steepest ascent with minimum error in direction. We could fix the allowable size or cost of the experiment, and choose that arrangement which minimizes some convenient measure of the error in the estimated direction of steepest ascent. This is the sole criterion to measure the efficiency of an experiment, and consideration of relative dispersion of the experimental points in the factor space seem to me to be irrelevant.

If we advance along our estimated line of steepest descent a unit distance, the displacement of the estimated point from the true one will be given by

$$R^2 = (\boldsymbol{\delta\varphi} + \mathbf{A}\,\boldsymbol{\beta}_2)' \quad (\boldsymbol{\delta\varphi} + \mathbf{A}\,\boldsymbol{\beta}_2)/\boldsymbol{\varphi}'\boldsymbol{\varphi},$$

where the quantities $\mathbf{A}$, $\boldsymbol{\beta}_2$ are as defined in the paper, and $\boldsymbol{\delta\varphi}$ is a vector of the random errors in the estimates of φ.

We easily see that

$$E(R^2) \propto \text{trace } \mathbf{C}^{-1} + \boldsymbol{\beta}_2'\ \mathbf{A}'\ \mathbf{A}\ \boldsymbol{\beta}.$$

Now, by allowing the elements of $\boldsymbol{\beta}_2$ to increase, we can increase $E(R^2)$ beyond all limit. We shall restrain these elements by the condition $\boldsymbol{\beta}_2'\ \boldsymbol{\beta}_2 = 1$. The choice of this particular quadratic form corresponds roughly to assuming that the scales of measuring the various factors have been chosen to give equal weight to the effects of their biases.

Subject to this limitation on the elements of $\boldsymbol{\beta}_2$, we maximize the possible mean error, and use this as a measure of the efficiency of any given design. With this we can compare any two proposed experiments of equal size, but it seems unlikely that it will be possible to determine the arrangement of points that will minimize this quantity. The measure can be shown to be

$$\text{trace } \mathbf{C}^{-1} + \text{latent root of greatest modulus of } \mathbf{A}'\mathbf{A}.$$

Although it may not be so in this case, it is rather an annoying fact that most attempts to determine an optimum design produce the result that the design matrix (or as the authors call it, the matrix of independent variables) shall be orthogonal. In many cases it is combinatorially impossible to achieve this. This indicates that the mathematical tools used are hardly adequate, since if they were, only the combinatorially possible cases would have been admitted. Until improvements in mathematical techniques take place, rather crude methods of resolving the difficulty must be used. The usual method in experiments involving multiple estimates is to restrain the designs to those with a certain form of variance matrix. For example, it is not usually possible to arrange an orthogonal design for the estimation of the responses of a set of different treatments, if the experimental observations are grouped in blocks, and it is desired to eliminate entirely the effect of this grouping. Since it is the differences between the responses that are of interest it is quite reasonable, if the treatments all stand in the same relation to each other, to restrict consideration to designs which have the same accuracy for all possible comparisons. If this is done, it is found that combinatorial considerations dominate the problem, and there is no possibility of minimizing the variance of any comparison. These are of course the so-called balanced incomplete block designs.

Another possible approach to the problem of determining the optimum point in the factor space can be simply illustrated by the case of the one-factor space. Suppose we wish to determine the optimum level of some factor measured by x; we must perform some experiments at different levels of x, which we can assume for convenience are equally spaced. Without assuming a form for the response curve, it is obviously impossible to do more than choose one of the levels represented in the experiment as the best. In making the decision which is the best, comparisons must be made between the different responses. Responses corresponding to close values of x are likely to be more nearly equal than those corresponding to widely separated values of x. Consequently, the comparisons of near neighbours should be more accurate than far neighbours. If the range of x represented is sensible, the optimum is likely to lie at any of the points.

Consequently, the comparisons of every pair of immediate neighbours should have the same accuracy; the comparison of next to immediate neighbours should have a common accuracy, lower than that of immediate neighbours, and so on. We can regard the first value as the immediate neighbour of the last value to cover the possibility of a U-shaped response curve with nearly equal extremities. If we impose these restraints on the design, its variance matrix must be a circulix. If block effects have to be eliminated, this implies that all immediate neighbours must occur together in the same number of blocks, while all next-to-immediate neighbours also occur together in the same number of blocks, but in less than immediate neighbours. Once again we obtain a combinatorial problem. This approach can be extended to several factors.

Plackett's result, which is quoted in this paper, is very interesting, and of course, of great practical importance, but the derivation of it in the original paper may seem a little obscure. It can be derived immediately from a very useful but little known result concerning the inversion of certain kinds of matrices. This is $(\mathbf{I} + \mathbf{AB})^{-1} = \mathbf{I} - \mathbf{A}(\mathbf{T} + \mathbf{BA})^{-1}\mathbf{B}$.

To derive Plackett's result we have

$$\mathbf{C}_0^{-1} = (\mathbf{X}'\mathbf{X} + \mathbf{W}'\mathbf{W})^{-1} = \mathbf{C}^{-1}(\mathbf{T} + \mathbf{W}'\,\mathbf{W}\mathbf{C}^{-1})^{-1} = \mathbf{C}^{-1}\{\mathbf{I} - \mathbf{W}'(\mathbf{T} + \mathbf{W}\mathbf{C}^{-1}\,\mathbf{W}^{1})^{-1}\,\mathbf{W}\mathbf{C}^{-1}\}.$$

The matrix $\mathbf{Q}$ with $q_u = pq_{ij} = q^{i+j}$, mentioned in the paper can be immediately inverted, as it can be written as $(p - q)\mathbf{I} + q\mathbf{1}\mathbf{1}'$ where $\mathbf{1}$ is a column vector of 1's of length K. Whence we obtain

$$\mathbf{Q}^{-1} = \frac{1}{p-q}\left(\mathbf{I} + \frac{q}{p-q}\mathbf{1}\mathbf{1}'\right)^{-1} = \frac{1}{p-q}\left\{\mathbf{I} - \frac{q}{p-q}\mathbf{1}\left(1 + \frac{kq}{pq}\right)^{-1}\mathbf{1}'\right\}$$

$$= \frac{1}{p-q}\mathbf{I} - \frac{q}{(p-q)(p+k-iq)}\mathbf{1}\mathbf{1}'.$$

Some pragmatists claim that one of the prime purposes of designing experiments is to enable the calculations to be easily performed. It so happens that many of the designs with desirable properties are also easily analysed, but the excuse for using this property as the *aim* of designing will soon be swept away with the establishment of modern automatic computing machines as computing aids. These machines would be able to analyse the most complicated experiments in a fraction of an hour.

I have found this paper most stimulating, and I take great pleasure in seconding the vote of thanks to the authors.

Mr. D. R. READ: My own general reaction on reading this paper has been one of decided enjoyment, as well as interest, at seeing the various cogent points that the authors have made, and the profitable way in which they have been developed. Regarding some of the points that remain at issue, I do not pretend to be able to provide any definite answers to those that I shall raise; but I would like to present the following further discussion:

The authors have directed attention to the performance of experiments in small groups when the experimental error is small compared with the expected effects, and they have given examples where the standard error was about 1 per cent. The alternative situation, however, where the effects may be small and the error large is also of practical interest. When working on the production scale, as opposed to the laboratory scale, the factor levels must be altered cautiously in order to avoid any undue financial loss, so that the authors' suggestion of working within a small sub-region of the factor space is necessary rather than optional, and one may need to detect effects of the order of 2–3 per cent.; but in biochemical processes one may be faced with a standard error of 10–15 per cent. (It is not necessarily to be presumed, because one is on the production scale, that conditions should already be near the optimum for all factors. This may be merely a result of inadequate previous laboratory and pilot scale investigation; but it may alternatively be due to the introduction of new factors peculiar to the production scale.) A considerable amount of replication, either absolute or hidden, is required in such cases, and it will be desirable to make efficient use of this replication; but since it is not possible to explore the whole region R, the replication will have to be used instead to explore the sub-region around the base point more intensively. The question that remains to be answered is: How can one decide the optimum size of the first experiment? If this first experiment is too small, though still large enough to give at least as many points as are required to estimate the function, the estimate of the path of steepest ascent or of greatest gain will be so ill-determined as to be possibly misleading for the design of the next experiment (which experiment must also comprise many replications, and not merely one as in Expt. (9) in the authors' example in Section 4.2); and if it is too large, one is wasting effort in exploring the initial small sub-region too thoroughly. I cannot see any definite solution to this type of problem, and at present it appears to remain as a matter of considered judgment and opinion on the part of the statistician in close consultation with other persons such as the chemist or biochemist.

Reverting to the case of small experimental error, there appears also to be a matter for judgment in the authors' first example (Section 4.2). On p. 20 they give the expected path of steepest ascent from the initial base point, described between $x_1 = 225$ and $x_1 = 305$ ignoring second order effects, and they performed Expt. (9) at $x_1 = 275$. It would clearly have been undesirable to do this experiment at some low value little greater than 225, or at some very high value which might well be far the other side of the optimum. I should be interested to hear of any particular reasons that led them to take $x_1 = 275$, and I should like to know whether they have any arbitrary rule, such as that one should take the conditions corresponding (as in the present case) to 2 units increase in the variable having the greatest effect, or to an expected yield, on the basis of linear extrapolation, equal to the lowest estimate of the maximum yield that was previously thought to be possible (i.e., 75 per cent.). Perhaps the authors will confess that this is merely part of the *art* of experimentation, similar to the art in applying relaxation methods.

As regards the general utility of their method, the authors have made it clear that it may only determine a local maximum, instead of the absolute maximum, when the yield surface is not unimodal; though they say (Section 2.1) that "where fuller exploration is impracticable this possibility must be accepted, and we have not found the implied risk troublesome". From a somewhat vague sort of intuitive reasoning I am inclined to feel that the risk will in general be small, or non-existent because the surface is unimodal; but I would like to ask the authors whether they have anything more definite to say on this point, at least in connection with some particular type of system. I imagine that the surface is more likely to be unimodal in pure chemical systems than in biochemical ones; but this is only a "hunch".

Dr. HARTLEY: I should like to make only a few remarks* concerning the posing of the general problem of finding optimum conditions. The authors have clearly split their task into two stages. In their initial stage they are a good distance from the top of the mount they are trying to climb. They determine the path of steepest ascent and get, after a few steps, "within striking distance", as they call it, of the summit. This stage has a well-defined design problem: the

* The sentiment of some of these was based on the original version of the paper.

task is to find the direction of steepest ascent from the starting-point, and this is identical with that of estimating all 1st derivatives at the starting-point with equal precision. Some of the modern fractional replication layouts designed to estimate all main effects are found by the authors to be useful in this task.

In their final stage they are already within striking distance of the summit, and in unfortunate cases this summit may be a flat plateau with small curvatures so that the task of obtaining the exact *position* where the summit occurs may be awkward and exacting, and, in general, is much more difficult. Clearly all the first and second differentials must be estimated, and designs are suggested in which they can all be estimated with varying precision and degree of bias from the higher differentials which were neglected. But the question arises—what is the relative precision required for these derivatives to estimate the position of the maximum with greatest accuracy?

This appears to be a difficult problem, and the authors only give some general ideas about it.

They split their task into two parts: They first give designs from which the required derivatives can be estimated with a known precision and bias. They then state merely a *procedure* of using these derivatives to obtain the position of the maximum, without going into the question of the precision of the final answer.

In view of these difficulties one wonders whether in many practical problems it is really worth while to proceed with this final stage. When one is so close to the maximum would it not often be adequate to estimate the *amount* of the maximum and not its position, and if that amount is only a negligible gain on the amount of yield already attained, not to proceed any further.

The "*amount*" of the maximum is clearly easier to estimate than its *position*, and even if it is found that the further gains to be made are worth while, one may prefer to determine the *response surface* (near the position reached and approximately in the direction of the maximum) rather than the *exact position* of the maximum.

Mr. M. J. R. Healy: In experiments of this kind it is important to determine when to stop as well as how to go on. For this reason more consideration might be given to a preliminary rough coverage of a large region of the factor space, followed by more minute exploration of the part near the maximum. With the general knowledge of the region thereby gained, it should be possible to avoid extensive experimentation when the ultimate gain is small.

There are several directions in which this interesting investigation could be extended. In the agricultural field, with which I am chiefly concerned, experiments are usually comparative, in the sense of Mr. Anscombe's recent paper; the absolute level of response is not important, and may change from one set of readings to another. There is the added complication that the position is only partly sequential; at each stage a number of plots are used simultaneously, each plot forming one "experiment" in the authors' terminology.

With regard to the final section of the paper, it may be worth mentioning that the evaluation of latent roots and vectors can be carried out more efficaciously when the number of factors is large by the methods described by Aitken, Hotelling and others.

Dr. N. L. Johnson: I will confine my remarks to the case of initial "remoteness" from the optimum described in Section 2.2.1. The authors might have emphasized, rather more than they have, the importance of the choice of units referred to in Section 2.2.2. If the unit in which a particular factor is measured is replaced by a new unit k times as big as the first, then the new "direction of steepest ascent" will be such that the actual change in level of that factor per unit change in level of any of the other factors is increased in the ratio k^2:1. The actual relative magnitudes (but not the signs) of the required changes in level of the various factors are therefore effectively determined by the choice of units rather than by the estimated magnitudes of the partial derivatives.

While it is likely, as the authors claim, that in practice the units chosen will be reasonably appropriate, the comparative unimportance of the exact magnitudes of the partial derivatives may have certain practical implications. If it is the *signs* of these n partial derivatives that are of primary importance, then these signs (and also approximate magnitudes) could be quite well determined (assuming, as do the authors, sufficiently small experimental errors) by a set of $(n + 1)$ experiments, one at 0 and each of the other n varying one factor at a time.

The authors say that the units should be chosen so that, as far as possible, the function φ should be "symmetrical" in the n factors. Does this mean that the values of $\partial\varphi/\partial\chi_i$, should be of approximately equal magnitude? If so, it would appear that units could be adjusted after the first set of experiments. The required changes in level of the factors would then be proportional to (estimated partial derivative)$^{-1}$ in terms of the original units.

A further point of interest is the choice of actual (as opposed to relative) differences in levels of the various factors for the first set of experiments. Should these be as small as is compatible with the experimental error, or should rather larger steps be taken?

Mr. R. L. PLACKETT: This seems to me a valuable contribution to industrial experimentation. In view of the exploitation of sequential methods, and the need for a small experimental error, it must be doubtful whether applications can be made in the classical field of agricultural experiments, or in any subject where the accumulation of results takes some time. This is not to say that the problem does not exist elsewhere, but only that other approaches may then be necessary or that the optimum is already achieved.

As they wish to build up a function from its Taylor expansion, the authors have broken entirely with the usual factorial *representation* in the form of main effects and interactions. Burman and I make some reference to this subject in our section 2, but we were unable to give much heed to derivatives above the first as there were more than 20 factors in our application and the apparatus would not work with less. In connection with the example of section 3.1, I feel entitled to ask, "When is an experiment not a factorial experiment?" The design which the authors consider is one with 2 factors, one of which is varied at 5 levels and the other at 4, although only 6 of the 20 possible combinations are present. It can appropriately be described as an incomplete two-way layout, like a balanced incomplete block, but to say that it is not of the factorial type can only imply that there are no factors in it.

Another break made by the authors is that with a diagonal, or unit, covariance matrix. It has been recognized in the past that designs with a covariance matrix of this kind lead to maximum precision in the estimation of unknown parameters, and to a particularly simple analysis of variance. The range of possible designs is great, and not yet fully explored, but Box and Wilson have shown that convenient designs exist not subject to the diagonal limitation. This, of course, is an immense widening of the field of study, but results in no definite boundaries, so that for all the theorist can tell, practically any design is useable.

When two-level designs are used to estimate the coefficients in a Taylor expansion, the authors have clearly demonstrated that one can rapidly assess which differential coefficients are the aliases of those being estimated, using the same properties as form the basis of confounding in powers of two and the associated series of partial replicates. When three-level factorials are used, the assumption of main effects plus interactions has split from the Taylor expansion and although factorial designs may provide a definite first approximation to what is required, I would ask the authors whether any design matrix with columns reasonably uncorrelated would not serve their purpose as well, in view of the fact that they have abandoned the unit covariance matrix.

I should like to conclude with two remarks on the numerical analysis. "Relaxation" methods for solving linear equations have been known since the time of Gauss, and Southwell's contribution must be limited to what Aitken has called an "admixture of opportunism". It is unnecessary to invert a 2×2 matrix by solving equations (63), as all the elements of G are obtained on dividing the elements of $I + R$ by its determinant. The comments above must not obscure the fact that the authors have made a most important advance in their subject.

Mr. BOX and Dr. WILSON subsequently replied as follows:

Mr. BOX: I shall confine myself to statistical considerations and leave chemical matters to Dr. Wilson. On the question of confidence intervals for the maximum, I agree with Dr. Hartley that the most important practical problem is how far the response at the estimated maximum may fall short of that at the true maximum, rather than how far we are in error in its position. It appears, at least in chemical application, that clearly defined point maxima in many variables are rather rare, and often the most important practical problem is to determine the nature of the local ridge system. Here the local fitting of second-degree equations followed by canonical analysis usually provides a very strong lead. Where stationary ridges occur, the possibility of a range of alternative near-optimal factor combinations is indicated, and the final process would often be chosen because it brought other auxiliary responses (e.g., physical form) to more satisfactory levels rather than because it corresponded to precisely the highest value of the main response (e.g., yield).

In the estimation of derivatives, as Professor Bartlett demonstrates, the effect of bias may be large. His equation (3) allows some useful conclusions on the value of including higher order terms to be drawn, but I do not think that the ignoring of terms in α^2 which leads to (4) can be justified. If we write $t = \alpha/\sigma_a$ and $r = \sigma_a/\sigma_0$ his equation (3) tells us to ignore the doubtful term if

$$| tz_1 \pm 2/r | < 2 (1/r^2 + z_1^2)^{\frac{1}{2}}.$$

If a second degree polynomial were fitted to evenly spaced ordinates between $x = -1$ and $x = +1$, it is easy to show that $r < 2$ and $z_1 = 2x$. This leads to the conclusion that the second order term should be ignored only if $| t | < 2$, i.e., this term should be included *whatever the value of x* if it exceeds twice its standard error. Equation (4), on the other hand, leads to the erroneous and much more alarming conclusion that the term should be ignored if $|t| < |4x|$.

With regard to Professor Bartlett's second point, if one design results in a suspect higher order effect appearing as an alias of one of the estimates, whilst another gives an unbiased estimate and forces the suspect effect into the residual where its presence may be detected and the need for a more elaborate equation demonstrated, then surely the latter design is the better.

Mr. Tocher's comments are most interesting, and encouraging. I would differ with him, however, on the merits of the factorial type analysis. Nothing is lost in *any* analysis, factorial or otherwise, in which the number of linear independent estimates is equal to the number of unknowns (see for example the pentagonal design). In the problem we discuss the experimenter is interested in the responses at the experimental points only so far as they tell him what to expect at other points (that is, so far as they tell him about the surface). Surfaces generated by equations in which terms are included corresponding to all the factorial constants are often very strange creatures indeed.

I too was struck by the analogy between the design problem and the Neyman and Pearson theory. For example, it can be shown that optimum designs of type A and order 1 for N factors are provided by the $N + 1$ vertices of the corresponding regular N dimensional simplex. These designs provide uncorrelated estimates with precision independent of the orientation of the arrangement. However, as such a design is rotated so the alias relationships change, and the problem is to decide on the best direction to "point" the design to get smallest bias if the planar approximation is not justified. The answer depends on the *type of alternative* to the plane which is in mind. These considerations throw some light on the question raised by Dr. Johnson on the merits of the "one factor at a time" design. This design is not a regular simplex, and it is not very satisfactory either from the point of view of precision or bias. It has the practical disadvantage that if the response at the origin is in error, then all the effects are wrongly estimated.

On the question of what are factorial experiments, Yates (1937) defines them as "experiments which include all combinations of several different treatments or factors"; if we adopt Mr. Plackett's implied definition then all experiments with factors are factoral experiments. There is a great deal of work to be done on the question of best types of design, certainly to obtain estimates which are reasonably uncorrelated is one requirement, but the question of bias and relative accuracy of effects of different orders are equally important. These questions have been receiving attention, and it is hoped to publish something fairly soon.

Dr. Wilson: I am pleased that Mr. Read has drawn attention to the element of judgment required in the application of these techniques. We mentioned this, of course, but since little can be clearly said on the subject it has perhaps tended to be overwhelmed by those matters on which explicit statements can be made. It should be emphasized that the practical application of these methods is not automatic, that judgment is required, and that a "bad" experimenter (that is one who is not fully aware of what he does and who applies an insufficient intellect to his results) suffers here as in any other method of experiment. There is no more difficulty in the matter of judgment than is covered by this, and there are no hard and fast rules to be applied.

These remarks also cover in part Dr. Johnson's suggestion that we might have laid greater emphasis on the importance of choice of units. In fact we found two schools of thought here, the other being that the matter was obvious. This we were in fact compelled to follow for reasons of space, although we agree rather with Dr. Johnson. In practice, as we have said, this is a question of judgment, and a discussion of how it should best be exercised would lead to a discourse on what constitutes a good experimenter. Dr. Johnson's suggestion that units might be changed after a first set of experiments must be treated with reserve. Thus the experimenter chooses units in the light of his experience; for example 10° C. of temperature commonly has the same order of effect as a doubling of concentration. If it is found that in terms of such units effects are small, it would normally mean not that the units were wrong but that the effects were small *in fact*. This would imply some kind of stationary region and we have suggested suitable tests in 4.1. In addition of course it is always possible, especially in a new field of experiment, to make an unfortunate selection of units, and again it is solely a question of judgment.

Dr. Hartley and Mr. Healey both raise the question of how it should be decided to stop experimenting, and we are in complete agreement on the importance of this. Ultimately, of course, the answer depends on an estimate of return for work done. It will naturally vary with the circumstances, and may even depend on personal opinion. It would be valuable to be able to approach this problem with greater certainty. Here Dr. Hartley's suggestion that the response at the maximum be determined before its position is an interesting one.

In conclusion we wish to express our thanks to those who contributed to an illuminating and helpful discussion.

2.2
Multi-Factor Designs of First Order

G. E. P. BOX
Imperial Chemical Industries Ltd.

The problem discussed arises when it is possible to choose in advance the N combinations of levels at which a set of quantitative factors are to be held in a set of N experiments to determine the slopes of a regression surface (assumed planar). It is shown that the minimum variance property of an 'optimum' design arises from the shape of the design pattern and is independent of its orientation. This fact may be utilized as follows:

(1) When prior knowledge of the response surface exists the design may be rotated to reduce possible bias.

(2) The design may be rotated so that systematic effects, such as polynomial time trends and block effects, are eliminated without loss of efficiency.

(3) Subject to the conditions imposed by (2), the orientation of the design may be chosen at random. This has the effect of making the usual normal theory tests exact and completely independent of the distribution of the observations.

1. Suppose that the effect of k quantitative variables or factors $X_1, \ldots, X_i, \ldots, X_k$ (such as time, temperature, concentration) on some measurable response (such as yield of product) is being studied in a region of the response surface that can be represented to a sufficient degree of accuracy by a polynomial equation of degree d. We define a design of order d as an arrangement of experiments which will allow all the coefficients in this polynomial to be separately determined. In this paper it is assumed that d is equal to 1, i.e. that a planar approximation is adequate. It is also assumed that the variables can be controlled exactly at levels decided in advance and that the observed response y differs from η due to experimental error having variance σ^2:

$$E(y) = \eta, \quad E(y-\eta)^2 = \sigma^2. \tag{1}$$

We can perform $N > k$ trials; the problem is to decide which N combinations of levels to use so that the constants defining the plane are estimated with maximum accuracy. We shall not apply the usual limitation that the design is to consist of combinations of a few fixed levels of the factors, but as a means of specifying the extent of variation for a given factor X_i we define the *unit* S_i for this variable as $S_i = \left\{\sum_{u=1}^{N} \frac{(X_{iu}-\bar{X}_i)^2}{N}\right\}^{\frac{1}{2}}$, and write the design in terms of the standardized variables $x_{iu} = (X_{iu}-\bar{X}_i)/S_i$. It will be noted, therefore, that for the standardized variables

$$\sum_u^N x_{iu} = 0, \tag{2}$$

$$\sum_u^N x_{iu}^2 = N. \tag{3}$$

A discussion of the problem of scaling and comparing experimental designs will be found in a recent paper (Box & Wilson, 1951), where there is an account of the planning of experiments to attain maxima in connexion with which this investigation was undertaken. The design matrix $\mathbf{D}$ is an $N \times k$ matrix providing a programme of experiments to be performed. The k elements $x_{1u}, \ldots, x_{iu}, \ldots, x_{ku}$ of the uth row are the levels of the standardized variables to be used in the uth trial. They can also be regarded as defining the k co-ordinates of the uth *experimental point* in the k-dimensional factor space. To use the design the

The Collected Works of George E. P. Box, 1984, Wadsworth, Inc., Belmont, CA 94002.
Originally published in *Biometrika,* vol. 39 (1952), pp. 49–57.

experimenter must decide on suitable average levels $\bar{X}_1, \bar{X}_2, \ldots, \bar{X}_k$ and units $S_1, S_2, \ldots, S_k$ for the variables. The level to be used for the ith variable in the uth trial will then be $X_{iu} = \bar{X}_i + S_i x_{iu}$.

2. Suppose the true regression plane in the region considered is

$$\eta = \beta_0 + \beta_1 x_1 + \ldots + \beta_k x_k. \tag{4}$$

In an obvious matrix notation the N equations (4) at the N experimental points may be written

$$\boldsymbol{\eta} = \mathbf{X}_1 \boldsymbol{\beta}_1, \tag{5}$$

where $\mathbf{X}_1 = [\mathbf{U} \vdots \mathbf{D}]$ and each element of the column vector $\mathbf{U}$ is unity. If $\mathbf{Y}$ is a column vector of observations $y_1, \ldots, y_u, \ldots, y_N$ made at these experimental points, then providing $\mathbf{X}_1$ is of rank $k+1$ (which implies that $\mathbf{D}$ is of rank k), separate linear estimates $b_0, b_1, \ldots, b_k$ of each of the β's may be calculated. For a particular design $\mathbf{D}$, linear estimates having smallest variances are provided by the method of least squares and are given by $\mathbf{B}_1 = (\mathbf{X}_1'\mathbf{X}_1)^{-1}\mathbf{X}_1'\mathbf{Y} = \mathbf{T}_1\mathbf{Y}$, and it is well known that the matrix of variances and covariances for the estimates is $(\mathbf{X}_1'\mathbf{X}_1)^{-1}\sigma^2$. We have then to choose $\mathbf{D}$ so that the diagonal elements of $(\mathbf{X}_1'\mathbf{X}_1)^{-1}$ are minimized.

Consider the symmetrical determinant of sums of squares and products $\Delta = |c_{ij}| = |\mathbf{X}_1'\mathbf{X}_1|$, and denote by C_{hh} the cofactor of c_{hh} in Δ and by $C_{ij.hh}$ the cofactor of c_{ij} in C_{hh}. Using Cauchy's expansion we have

$$\Delta = c_{hh} C_{hh} - Q, \tag{6}$$

where Q is a quadratic form in the k variables c_{hi} $(i = 0, 1, \ldots, h-1, h+1, \ldots, k)$ and

$$Q = \sum_{ij \neq h} c_{hi} c_{hj} C_{ij.hh}. \tag{7}$$

Now since $\mathbf{X}_1$ is of rank $k+1$, Q is necessarily positive definite and Δ is positive. Also $c_{hh} = N$ (from (3)) and $V(b_h) = \sigma^2 C_{hh}/\Delta$, where $V(b_h)$ is the variance of b_h. Consequently, re-arranging (6),

$$V(b_h) = N^{-1}\sigma^2\{1 + \Delta^{-1} \text{ (positive definite quadratic form in the } c_{hi})\}.$$

Thus $V(b_h)$ is a minimum only when each of the c_{hi} (the k sums of products of the hth variable with each of the remaining variables) is zero. When this is so all the C_{hi} must be zero also, and consequently b_h is uncorrelated with each of the other estimates and has variance σ^2/N. For maximum efficiency for all the coefficients then $\{c_{ij}\} = N\mathbf{I}$ and a suitable design $\mathbf{D}$ is supplied by any k columns after the first, of a matrix $N^{\frac{1}{2}}\mathbf{O}$, where $\mathbf{O}$ is orthogonal with the elements of its first column all equal. This result was arrived at by Plackett & Burman (1946). They postulated, however, that the x's in their first-order optimum designs should take only the values $+1$ and -1. With this limitation designs existed only when N was a multiple of 4 and they obtained arrangements for $k = 3, 7, 11, \ldots, 99$ factors using $N = 4, 8, 12, \ldots, 100$ trials. Such designs may be used with qualitative or quantitative variables; in our case where the variables are essentially quantitative this restriction is not introduced and N can have any value.

3. For our problem, therefore, designs of optimum precision for up to $k = N-1$ factors in N experiments may be obtained from *any* orthogonal matrix $\mathbf{O}$ with elements in the first column all equal and $\mathbf{X}_1 = N^{\frac{1}{2}}\mathbf{O}$. We now assume $k = N-1$ and consider the geometrical implications of the above result. Since $\mathbf{D}$ is of rank $k = N-1$, the N experimental points are the vertices of an $N-1$ dimensional simplex. Write the uth row of $\mathbf{X}_1$ as $\mathbf{x}_u'$ and denote

the angle which the uth and sth experimental points make with the origin by θ_{us}, then since $X_1 = [U : D]$ the distance of each experimental point from the origin is $(N-1)^{\frac{1}{2}}$ and

$$\mathbf{x}_u'\mathbf{x}_s = 0 = 1 + (N-1)\cos\theta_{us}, \tag{8}$$

i.e.

$$\cos\theta_{us} = -(N-1)^{-1} \quad (\text{all } u \text{ and } s,\ u \neq s). \tag{9}$$

Consequently this design is formed by the vertices of the *regular* $N-1$-dimensional simplex. If two factors are tested in three trials the experimental points should be at the vertices of an equilateral triangle; for three factors tested in four trials the experimental points should form a regular tetrahedron and so on. It should be noted that no restriction is necessary on the orientation of the design. We can turn the regular figure in any direction; this will correspond simply to a different choice of the orthogonal matrix **O**, and the variance covariance matrix for the b's will remain unchanged.

As an example, Fig. 1 shows two particular orientations of the optimum design for $N = 4, k = 3$.

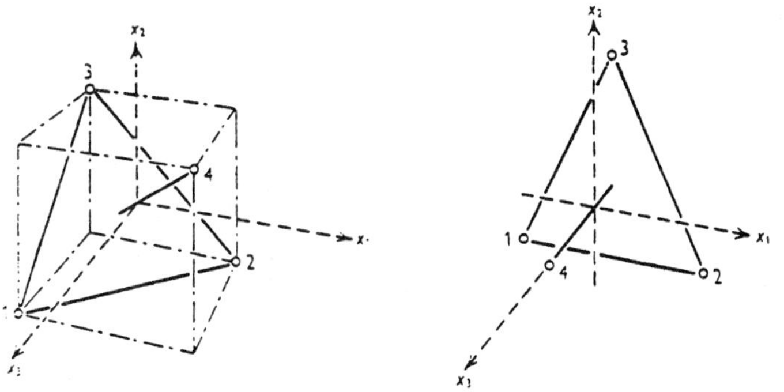

Fig. 1. Orientations of the optimum design for $N=4$, $k=3$.

The design matrices are

$$\mathbf{D}_a = \begin{matrix} (1) \\ (2) \\ (3) \\ (4) \end{matrix} \overset{\begin{matrix} x_1 & x_2 & x_3 \end{matrix}}{\begin{bmatrix} -1 & -1 & 1 \\ 1 & -1 & -1 \\ -1 & 1 & -1 \\ 1 & 1 & 1 \end{bmatrix}}, \qquad \mathbf{D}_b = \begin{matrix} (1) \\ (2) \\ (3) \\ (4) \end{matrix} \overset{\begin{matrix} x_1 & x_2 & x_3 \end{matrix}}{\underset{(\sqrt{2}) \quad (\sqrt{2}/\sqrt{3}) \quad (1/\sqrt{3})}{\begin{bmatrix} -1 & -1 & -1 \\ 1 & -1 & -1 \\ . & 2 & -1 \\ . & . & 3 \end{bmatrix}}}. \tag{10}$$

$\mathbf{D}_a$ is the familiar half-replicate of the 2^3 factorial; the other half replicate is obtained by rotation of the first and completes the cube. $\mathbf{D}_b$ is also obtained by orthogonal rotation of $\mathbf{D}_a$, so that the line joining the points (1) and (2) is parallel to the axis of x_1 and (1), (2) and (3) fall on a plane parallel to the plane of x_1 and x_2. $\mathbf{D}_b$ is seen to have elements proportional to Helmert's orthogonal matrix (for clarity the elements are given as whole numbers with the necessary multiplier shown below). This latter design has rather a curious property, for it is a 'one factor at a time design' although not of the orthodox pattern. To use it the experimenter would first perform a 'blank' experiment with all factors at the lower levels; in the second experiment the level of the first factor only would be changed; in all subsequent experiments this would then be held at the average of these two levels. In the third experiment the level of a second factor would be raised, and in all subsequent experiments this factor would be held at the average level of the three experiments. This procedure could be continued for any number of experiments and factors. The estimates would be uncorrelated, and on the convention we have adopted concerning the units for the factors, the variance of the estimates would be the same with both designs.

4. Although the variances and covariances under orthogonal rotation of the design remain constant, the magnitude and arrangement of the possible biases which might occur if the planar approximation was inadequate do not. Suppose that, contrary to assumption, to obtain a perfect fit it was necessary to include S extra terms $\mathbf{X}_2\boldsymbol{\beta}_2$, so that instead of (5) we had

$$\eta = \mathbf{X}_1\boldsymbol{\beta}_1 + \mathbf{X}_2\boldsymbol{\beta}_2, \tag{11}$$

then (Box & Wilson, 1951) $\mathbf{B}_1$ would no longer supply unbiased estimates of $\boldsymbol{\beta}_1$ but instead

$$E(\mathbf{B}_1) = \boldsymbol{\beta}_1 + \mathbf{A}\boldsymbol{\beta}_2, \tag{12}$$

where $\mathbf{A}$ is a $(k+1)\times S$ matrix of coefficients of the biases called the alias matrix and given by $\mathbf{A} = (\mathbf{X}_1'\mathbf{X}_1)^{-1}\mathbf{X}_1'\mathbf{X}_2$. With the orthogonal designs discussed here this simplifies to $\mathbf{A} = N^{-1}\mathbf{X}_1'\mathbf{X}_2$. In judging first-order designs we shall consider possible biases due to terms of second order. Now there are two varieties of second-order terms: those which are coefficients of square terms x_1^2, x_2^2, etc., sometimes called quadratic effects, and those which are the coefficients of product terms x_1x_2, x_1x_3, etc., sometimes called linear $\times$ linear interactions. In what follows it is mathematically convenient to define the effect β_{11} as the coefficient of x_1^2 whilst β_{12} is defined as the coefficient of $x_1x_2\sqrt{2}$. Equation (11) may then be written

$$\eta = \beta_0 + \beta_1x_1 + \ldots + \beta_kx_k + \beta_{11}x_1^2 + \ldots + \beta_{kk}x_k^2 + \beta_{12}(x_1x_2\sqrt{2}) + \ldots + \beta_{k-1k}(x_{k-1}x_k\sqrt{2}), \tag{13}$$

and the matrices of bias coefficients corresponding to $\mathbf{D}_a$ and $\mathbf{D}_b$ are found to be

$$\mathbf{A}_a = \begin{array}{c} \\ 0 \\ 1 \\ 2 \\ 3 \end{array}\begin{array}{c} \begin{array}{cccccc} 11 & 22 & 33 & 12 & 13 & 23 \end{array} \\ \begin{bmatrix} 1 & 1 & 1 & \cdot & \cdot & \cdot \\ \cdot & \cdot & \cdot & \cdot & \cdot & \sqrt{2} \\ \cdot & \cdot & \cdot & \cdot & \sqrt{2} & \cdot \\ \cdot & \cdot & \cdot & \sqrt{2} & \cdot & \cdot \end{bmatrix} \end{array}, \quad \mathbf{A}_b = \begin{array}{c} \\ 0 \\ 1 \\ 2 \\ 3 \end{array}\begin{array}{c} \begin{array}{cccccc} 11 & 22 & 33 & 12 & 13 & 23 \end{array} \\ \begin{bmatrix} 1 & 1 & 1 & \cdot & \cdot & \cdot \\ \cdot & \cdot & \cdot & -2 & -\sqrt{2} & \cdot \\ -\sqrt{2} & \sqrt{2} & \cdot & \cdot & \cdot & -\sqrt{2} \\ -1 & -1 & 2 & \cdot & \cdot & \cdot \end{bmatrix} \end{array}\begin{array}{c} \\ \\ (1/\sqrt{3}) \\ (1/\sqrt{3}) \\ (1/\sqrt{3}) \end{array}. \tag{14}$$

The figures in brackets are multipliers of the rows of $\mathbf{A}_b$. Using $\mathbf{D}_b$, for example, the expected value of b_3 when second-order terms were not all zero would be

$$E(b_3) = \beta_3 - (\beta_{11} + \beta_{22} - 2\beta_{33})/\sqrt{3}.$$

5. If we could, we would choose to orient the optimum design so that the bias coefficients were as small as possible. In this way both random and systematic errors might be simultaneously minimized. Consider the $(k+1)\times(k+1)$ matrix $\mathbf{AA}'$. The sums of squares of bias coefficients for the $k+1$ estimates $b_0, b_1, \ldots, b_k$ are given by the diagonal elements, and the magnitude of these would provide one indication of the efficacy of any particular orientation. We find somewhat unexpectedly, however, that for these optimum first-order designs $\mathbf{AA}'$ is invariant for any orthogonal rotation of the design. This is proved as follows.

Denote by $\mathbf{x}'_1, \ldots, \mathbf{x}'_u, \ldots, \mathbf{x}'_N$ the N rows of the matrix $\mathbf{X}_1$ and by $\mathbf{X}_3$ a matrix whose N rows are $(\mathbf{x}'_1)^{[2]}, \ldots, (\mathbf{x}'_u)^{[2]}, \ldots, (\mathbf{x}'_N)^{[2]}$ the derived power vectors of degree 2 (Aitken, 1948). Then $\mathbf{X}_3 = [\mathbf{U} \vdots \mathbf{D}\sqrt{2} \vdots \mathbf{X}_2]$ and $N^{-2}\mathbf{X}'_1\mathbf{X}_3\mathbf{X}'_3\mathbf{X}_1 = \mathbf{A}\mathbf{A}' + \mathbf{J}$, where $\mathbf{J}$ is a diagonal matrix in which the first diagonal element is 1 and each of the remaining k diagonal elements is 2. Now suppose the design is submitted to orthogonal rotation and denote the new matrices by $\dot{\mathbf{D}}, \dot{\mathbf{X}}_1, \dot{\mathbf{X}}_3$ and $\dot{\mathbf{A}}$. Then $\dot{\mathbf{D}} = \mathbf{D}\mathbf{G}$, where $\mathbf{G}$ is some $k \times k$ orthogonal matrix, and $\dot{\mathbf{X}}_1 = \mathbf{X}_1\mathbf{H}$, where $\mathbf{H}$ is a $(k+1) \times (k+1)$ orthogonal matrix consisting of $\mathbf{G}$ bordered by a first row $\mathbf{r}' = (1\,0\,0 \ldots 0)$ and a first column $\mathbf{r}$. Now $\mathbf{H}$ transforms the vector $\mathbf{x}'_u$; denote by $\mathbf{H}^{[2]}$ the matrix which correspondingly transforms the vector $(\mathbf{x}'_u)^{[2]}$.

Then
$$\dot{\mathbf{A}}\dot{\mathbf{A}}' + \mathbf{J} = N^{-2}\mathbf{H}'\mathbf{X}'_1\mathbf{X}_3\mathbf{H}^{[2]}\mathbf{H}'^{[2]}\mathbf{X}'_3\mathbf{X}_1\mathbf{H}, \tag{15}$$

and since $\mathbf{H}$ is orthogonal so is $\mathbf{H}^{[2]}$. Now the jth diagonal element of $\mathbf{X}_3\mathbf{X}'_3$ is

$$(\mathbf{x}'_j)^{[2]}(\mathbf{x}_j)^{[2]} = (\mathbf{x}'_j\mathbf{x}_j)^2 = N^2$$

and the ijth non-diagonal element is $(\mathbf{x}'_i)^{[2]}(\mathbf{x}_j)^{[2]} = (\mathbf{x}'_i\mathbf{x}_j)^2 = 0$. Consequently the right-hand side of (15) reduces to $N\mathbf{I}$. We find in consequence that $\mathbf{A}\mathbf{A}' = N\mathbf{I} - \mathbf{J}$, whatever the orientation of the design. That is, the sum of squares of the coefficient of the biases for b_0 is $N-1$, and for each of the effects $b_1 \ldots b_k$ it is $N-2$. The result is of course only true for the particular relative weighting of quadratic and interaction terms which has been adopted. This relative weighting is, however, a reasonable one, and the important conclusion emerges that if we have no prior knowledge concerning the relative importance of particular second-order terms no arrangements which are dramatically worse or better than others can be expected to arise as a result of rotation of the designs. In particular, if $k = N-1$, it is not possible to keep a selected estimate clear of bias.

6. When, on the other hand, something is known of the type of approximating second-degree equation to be expected it might be possible to reduce bias by suitable rotation of the design. Consider a particular class of designs which are such that *only* b_0 is biased by quadratic terms. For this to happen each of the $N-1$ column vectors in $\mathbf{X}_1$ corresponding to first-order effects must have zero inner product with the $N-1$ column vectors in $\mathbf{X}_2$ corresponding to quadratic effects. This can only happen if the latter have all elements equal to $+1$ which in turn implies that elements in $\mathbf{D}$ consist entirely of $+1$'s and -1's. These designs are those obtained by Plackett & Burman. Now the response contours generated by the second degree approximating equation are a set of conics. Suppose the direction of the principal axes of the system were known. Then because of the property mentioned above, if any of the designs of Plackett & Burman were rotated so that their axes were parallel to these principal axes the effects $b_1, \ldots, b_k$ would be unbiased, since in the new variables the second-degree equation contains no product terms. This may have practical application in the exploration of 'ridge' systems. Such systems occur, for example, when a line or plane of near maxima rather than a single point maximum is found. The probable existence and direction of such systems can sometimes be deduced from theoretical considerations, in which case it might be an advantage to rotate axes of the design so as to be parallel to these suspected 'ridges'.

It is worth noting (Box & Wilson, 1951) that by replication of any design with change of signs a design of 'Type B' is obtained. That is to say, one in which the first-order estimates are unbiased by terms of second order. This of course applies to all the designs discussed here.

Elimination of systematic variation

7. If the planar approximation is adequate we have seen that first-order designs of maximum efficiency are obtained by writing down any k mutually orthogonal column vectors each containing N elements and each orthogonal to U, a column vector of unit elements. The latter requirement allows for the elimination of the mean. Now it is easy to show that if any p column vectors are arbitrarily taken we can always write down further $N-p$ column vectors which are mutually orthogonal and each of which is orthogonal to each of the original p vectors. If therefore a first-order design is required for k variables which is such that, not only the mean but also a number of other systematic effects, such as time trends and block effects, are to be eliminated, this may be done by choosing the k orthogonal vectors of the design such that they are orthogonal to the mean and also to a set of $p-1 \leqslant N-k-1$ vectors representing these additional systematic effects.

A field in which there is a need for designs of this sort, and in which the extra trouble in the design and execution of the experiment is amply justified, is that of large-scale plant experiments. Here to avoid the possibility of serious loss of purity or yield, only small changes in the factor levels could usually be tolerated. This would tend to ensure the relative predominance of first-order effects, but a sensitive experiment would be required to detect such effects which would be necessarily small in magnitude. Furthermore, daily alterations to plant conditions would usually not be practicable. On some types of process if a series of trials were to be made it would be necessary to run each for a week at least. This replication would reduce random errors, but long-term trends which are of common occurrence would not be reduced. The usual way of overcoming this difficulty, of course, would be to make comparisons within blocks of weeks, using partial confounding if necessary to reduce block size. An interesting alternative procedure employs the principle discussed above.

Fig. 2 shows a set of fourteen consecutive weekly averages for a certain quality characteristic; it is obvious that a systematic trend with time typical of this particular process is occurring in the individual results. The following analysis is obtained by fitting orthogonal polynomials up to thirteenth order:

Order of term	Sum of squares	Order of term	Sum of squares
1	0·08	11	1·77
2	45·00	12	0·68
3	8·56	13	0·58
4	2·11		
5	0·80		
		Total	65·25
6	0·87		
7	0·13		
8	4·06	Terms up to fifth order	56·55
9	0·00		
10	0·61	Remainder	8·70

(A useful table giving the orthogonal polynomials for equally spaced ordinates and $N \leqslant 26$ has recently been published by De Lury (1950).)

Analysis of other sets of data from this process covering periods of the same length showed that terms up to fifth order usually accounted for a large proportion of the variation. Suppose now that it was desired to plan a set of fourteen trials on this process each lasting a week, by means of which the first-order effects of three quantitative factors were to be determined with maximum precision and the trend effect eliminated. On the assumption that the 'trend' could be represented by a polynomial of fifth degree plus *independently distributed error terms*, the levels in the 14×3 design matrix could be taken with column elements proportional to any three of the orthogonal polynomials higher than the fifth order or to any three orthogonal linear combinations of these; the remaining five degrees of freedom corresponding to high-order polynomial effects not used up would supply an internal estimate of experimental error appropriate for calculating confidence limits for the effects.

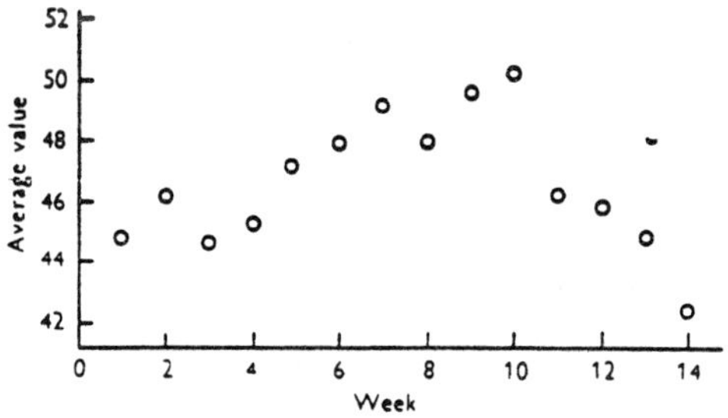

Fig. 2. Fourteen consecutive weekly averages for a quality characteristic.

The assumption that the trend would be represented by a polynomial on time plus independently distributed error terms and consequently that the residual variation would be distributed evenly through the higher order terms would often not be justifiable however. In particular, negative correlation between successive errors which might occur due to incomplete segregation of individual batches could lead to inflated higher order terms. This difficulty may be overcome by the device of randomization. In fact, we choose the directions of the three design vectors in the eight-dimensional residual subspace entirely at random subject only to the restriction that they should be mutually orthogonal. 'Angular' randomization of this sort is achieved conveniently by using a table of random normal deviates such as that of Wold (1948). Any n values of such a table provides a vector $\mathbf{z}_1$ having equal probability of lying in any direction in the n-space. A second vector $\mathbf{z}_{2.1}$ whose direction is random, subject only to the restriction that it is orthogonal to the first, may now be constructed by choosing a second set of n values $\mathbf{z}_2$ from the table and calculating

$$\mathbf{z}_{2.1} = \mathbf{z}_2 - b_{2.1}\mathbf{z}_1, \tag{15}$$

where $$b_{2.1} = \mathbf{z}_1'\mathbf{z}_2/\mathbf{z}_1'\mathbf{z}_1.$$

In a similar way a third vector $\mathbf{z}_{3.21}$ orthogonal to the first two may be constructed from

$$\mathbf{z}_{3.21} = \mathbf{z}_3 - b_{3.21}\mathbf{z}_{2.1} - b_{3.1}\mathbf{z}_1, \tag{16}$$

where $$b_{3.21} = \mathbf{z}_3'\mathbf{z}_{2.1}/\mathbf{z}_{2.1}'\mathbf{z}_{2.1} \quad \text{and} \quad b_{3.1} = \mathbf{z}_3'\mathbf{z}_1/\mathbf{z}_1'\mathbf{z}_1,$$

and so on. Having obtained $\mathbf{z}_1, \mathbf{z}_{2.1}, \mathbf{z}_{3.21}$, etc., it is convenient to standardize them by dividing the elements of each vector by the square root of the sum of squares of its elements.

These standardized vectors may be denoted by $\mathbf{Z}_1, \mathbf{Z}_{2.1}, \mathbf{Z}_{3.21}$, etc.; their elements are the direction cosines of lines which are mutually perpendicular but whose orientation in the n-space is otherwise random.

In the example quoted the three sets of eight random normal deviates taken from the table were

$$\mathbf{z}'_1 = [-0{\cdot}69 \quad 1{\cdot}40 \quad 1{\cdot}79 \quad -0{\cdot}83 \quad 0{\cdot}34 \quad 0{\cdot}19 \quad 1{\cdot}46 \quad 0{\cdot}21],$$
$$\mathbf{z}'_2 = [\ 0{\cdot}73 \quad -1{\cdot}00 \quad 0{\cdot}81 \quad 0{\cdot}01 \quad -1{\cdot}13 \quad -1{\cdot}53 \quad -0{\cdot}07 \quad -0{\cdot}22],$$
$$\mathbf{z}'_3 = [-0{\cdot}50 \quad -1{\cdot}59 \quad 0{\cdot}82 \quad 0{\cdot}71 \quad 1{\cdot}37 \quad 0{\cdot}90 \quad 0{\cdot}74 \quad -0{\cdot}83].$$

From which were obtained

$$\mathbf{Z}'_1 \ = [-0{\cdot}235 \quad 0{\cdot}476 \quad 0{\cdot}608 \quad -0{\cdot}282 \quad 0{\cdot}116 \quad 0{\cdot}065 \quad 0{\cdot}496 \quad 0{\cdot}071],$$
$$\mathbf{Z}'_{2.1} = [\ 0{\cdot}265 \quad -0{\cdot}332 \quad 0{\cdot}453 \quad -0{\cdot}046 \quad -0{\cdot}453 \quad -0{\cdot}630 \quad 0{\cdot}063 \quad -0{\cdot}080],$$
$$\mathbf{Z}'_{3.21} = [-0{\cdot}133 \quad -0{\cdot}645 \quad 0{\cdot}310 \quad 0{\cdot}270 \quad 0{\cdot}429 \quad 0{\cdot}241 \quad 0{\cdot}241 \quad -0{\cdot}313].$$

These were taken as the direction cosines of the three design vectors in an eight-dimensional subspace whose axes of reference in the fourteen-dimensional sample space were the eight vectors given by the elements of orthogonal polynomials of order 6, 7, ..., 13.

The design vectors in the original co-ordinate system were then obtained by calculating linear combinations of the orthogonal polynomials using the elements of $\mathbf{Z}_1, \mathbf{Z}_{2.1}$ and $\mathbf{Z}_{3.21}$ as coefficients. Thus the first vector is given by

$$\mathbf{x}_1 = -0{\cdot}235\boldsymbol{\xi}_6 + 0{\cdot}476\boldsymbol{\xi}_7 + \ldots + 0{\cdot}071\boldsymbol{\xi}_{13}, \tag{17}$$

where $\boldsymbol{\xi}_s$ is the vector of elements of the sth orthogonal polynomial standardized so that $\boldsymbol{\xi}'_s\boldsymbol{\xi}_s$ 14. The design* thus obtained is given below.

Observation	Factor levels		
	x_1	x_2	x_3
1	−0·2	0·5	0·3
2	0·1	−1·5	−1·1
3	1·2	1·0	1·1
4	−1·6	0·5	0·2
5	−0·4	0·5	0·0
6	0·0	−1·6	−1·5
7	1·6	−0·1	0·5
8	0·6	1·4	0·0
9	−2·0	−0·2	1·4
10	0·1	−1·4	0·2
11	−0·1	1·7	−2·3
12	1·3	−0·7	1·1
13	−0·9	−0·1	0·4
14	0·2	0·1	−0·3

To use the design the three factor levels are varied in proportion to these design elements in suitable units S_i in the manner described in § 1. Any trend occurring during the course of the experiment which can be represented by a polynomial of up to fifth degree may be

* Only a single place of decimals is given. More accurate values could of course be used if it were possible to control the levels of the factors more precisely.

eliminated without loss of efficiency. The significance of the effects may be conveniently assessed by means of the analysis of variance:

Source	Degrees of freedom
Due to fifth-order polynomial	5
x_1	1
x_2	1
x_3	1
Residual (error)	5
Total	13

Since the normal theory significance tests depend only on the angles between the vector of observations and the design vectors, angular randomization ensures that these tests are *exact* whatever the nature of the residual variation. The use of these designs is therefore justified whenever the procedure is likely to reduce the residual variation. Because of randomization no assumption need now be made that the trend can be *accurately* represented by the formal mathematical model.

8. In general, when quantitative factors are not restricted to a few fixed levels the scope and flexibility of experimental design is greatly increased. It is hoped to show in a later paper how these ideas may be applied with designs of higher order, and to discuss in more detail the interesting implications of 'angular' randomization.

The drawings were executed by Mr P. S. Ward, to whom I wish to express my best thanks.

REFERENCES

Aitken, A. C. (1948). *Determinants and Matrices*, 5th ed. Edinburgh and London: Oliver and Boyd.
Box, G. E. P. & Wilson, K. B. (1951). *J.R. Statist. Soc.*, Series B, 13, 1.
De Lury, D. B. (1950). *Values of the Integrals of the Orthogonal Polynomials up to n = 26*. University of Toronto Press.
Plackett, R. L. & Burman, J. P. (1946). *Biometrika*, 33, 305.
Wold, H. (1948). *Tracts for Computors*, no. 25. Cambridge University Press.

2.3

A Statistical Design for the Efficient Removal of Trends Occurring in a Comparative Experiment with an Application in Biological Assay

G. E. P. BOX and W. A. HAY
Imperial Chemical Industries Ltd.

SUMMARY

The method of construction of a certain class of designs for use with quantitative factors by means of which a trend occurring during a comparative experiment may be eliminated without loss of efficiency is indicated. The design and analysis is illustrated with an example from biological assay.

1. INTRODUCTION

In recent work (Box and Wilson, 1951; Box, 1952) it has been emphasised that, when dealing with quantitative factors, (i. e. factors like temperature, time, dose, which can be varied on a continuous scale) it is frequently advantageous to employ statistical designs which depart from the "factorial" principle. It has long been appreciated that it is the *orthogonal* property of designs which produces efficiency. The factorial method, in which are tested all (or a selected fraction) of possible combinations of a few levels of each of the factors, is often a convenient way of obtaining orthogonality but it is, of course, by no means the only way, nor is it, in all circumstances, the 'best' way. Designs which may be used to determine the effects of a number of factors but which are not necessarily factorial designs or parts of factorial designs may conveniently be called *multi-factor* designs.

When we are dealing with quantitative factors if we do not limit ourselves to designs involving a few equally spaced levels of the factors, a great gain in flexibility results and a widening of the possibilities for developing designs for specialised needs. Furthermore, there is no

The Collected Works of George E. P. Box, 1984, Wadsworth, Inc., Belmont, CA 94002.
Originally published in *Biometrics*, vol. 9, no. 3 (1953), pp. 304–319.

very great increase in labour in the analysis of the experiment, and arrangements involving fewer observations often result. For example, in the first of the papers referred to above "composite" multi-factor designs of second order, that is to say designs which allowed the determination of all the first and second order effects (linear effects, linear × linear interaction and quadratic effects) were developed which were at least as efficient as the orthodox three-level factorials but required many fewer experiments. In the second paper first order multi-factor designs of maximum efficiency for determination of the linear effects of any number of factors were described. These latter designs allowed simultaneous elimination of trends in the observations (e.g. time trends) without loss of efficiency, whilst a procedure called "angular randomisation" rendered subsequent statistical tests exact whatever the distribution of the observations and whether the assumed mathematical model was exactly realised or not.

Work on a related problem has been published recently by Cox (1951) whose designs retain the factorial principle, but in certain circumstances still allow the efficient elimination of time trends. The purpose of the present paper is to develop a further design of the multi-factor type which makes possible the efficient comparison of two response curves, when a time trend in the results is simultaneously occurring and to show its application in a problem of biological assay.

2. EXAMPLE

A design of this type was used in the biological assay of *d*-tubocurarine chloride. It was required to compare the estimated positions, slopes and curvatures of the dosage-response curves for a test material and a standard material whilst simultaneously eliminating, without loss of efficiency, any time trend which occurred in the course of the assay.

2.1 *Description of the Assay*

For this assay the phrenic nerve of a rat is dissected out and connected to a pointer which records on a smoked drum. This nerve which is immersed in Ringer Solution is given an electric stimulus every 12 seconds which produces a contraction. To obtain an accurate measurement the apparatus is allowed to operate under the same conditions for a period, the average recorded contraction being the value used. When a suitable dose of *d*-tubocurarine is added, a reduction of the contraction occurs. This reduction measured as a percentage is called the *percent inhibition* and is the response used to assay the drug. The apparatus is thoroughly washed out and a second dose is tested. In this way series of graded doses are given for the test and standard materials and the corresponding responses are measured. From these observations dosage response lines may be constructed from which the relative potency of the test material in terms of the standard may be calculated.

2.2 *Occurrence of Time Trend*

Unfortunately, as the test proceeds it is observed that the response (i.e. the % inhibition) becomes steadily larger due to tiring of the phrenic nerve, incomplete washing, or some other external cause not fully understood. Partial elimination of this trend could be attained, of course, by the use of a randomised block or an incomplete block design, the blocks being successive periods of time. Unless very small blocks could be employed, however, the error variance would be inflated by trends within the blocks, and an alternative method seems desirable.

2.3 *Principle of the Present Design*

In the procedure we adopt, the block size is reduced to two, that is, the experiments are performed in pairs. In the first pair a dose D_1^* of the test preparation is assayed against the same dose D_1 of the standard preparation. In the second pair of experiments the test preparation is assayed against the standard preparation at a dose D_2 and so on until k pairs of doses have been given. The order in which the preparations are tested *within* each pair is random.** We assume that the time trend which occurs during the experiment can be represented by a polynomial of degree p fitted to the block means. The problem is to choose the dose levels so that this trend can be eliminated without any loss of efficiency in the estimation of the effects. This is accomplished by so arranging the doses that the effects to be estimated are orthogonal to the estimated coefficients of the fitted polynomials. If desired, the trend fitted to the pair-means may then be used to eliminate trend effects within the pairs which might otherwise inflate the error.

For the sake of clarity we shall show first how such the design may be used in practice, leaving the method of derivation to §4.

---

*In practice of course the test preparation is suitably diluted so that equal quantities of test and standard preparations give approximately equal responses. It would, therefore, be more exact to say that the doses in each pair were proportional rather than equal.

**Here we have assumed that the order is completely random. An alternative procedure which may be adopted when k is even and which has certain advantages is to randomise subject to the condition that in $(\frac{1}{2})k$ of the pairs the test preparation is assayed first and in the remainder the standard preparation is assayed first.

2.4 *Levels of Dose Used*

In the design used for this particular assay eight pairs of doses were used as shown in Table 1. It was expected from previous experience that the response/log dose relationship would be linear for both test and standard, that the lines would be parallel, and the increase in response with time would be roughly linear and could be adequately represented by, at most, a quadratic polynomial. Provision was made in the design used to check these postulates, and less restrictive assumptions than those implied above were actually adopted. In fact effects one degree higher than those expected were allowed for, and it was arranged that:

i) Only four distinct dose levels were employed.
ii) Quadratic response/log dose lines could be fitted if necessary.
iii) A cubic polynomial could be fitted to the time trend.
iv) Differences between test and standard preparations in the dose and time trend curves could be detected. (i.e. interaction effects between samples and the dose and time effects could be estimated).
v) All the effects listed above were orthogonal one with another.

It is shown in §4 that, assuming the doses to be given at equally spaced intervals of time, one suitable arrangement is obtained if the log dose level are arranged so that the deviations from the mean log dose levels are proportional to the numbers, 0.55500, −1.50206, 1.17615, −0.22909, −0.22909, 1.17615, −1.50206, 0.55500, the doses being administered in the order indicated. This means that the actual log levels should be $X_1 = m + 0.55500\ \sigma$, $X_2 = m - 1.50206\ \sigma$, $\cdots$, $X_8 = m + 0.55500\ \sigma$ where m and σ are arbitrary constants so chosen that the design covers the required region of dose levels. It will be noted that, as mentioned above, only four distinct dose levels are used.

Although, of course, the dose of drug cannot be administered in practice to the five-decimal accuracy given above, the error in administering the doses should be no more than that involved in attempting to administer the integer levels usually required by the orthodox type of design and we shall retain the full five-decimal accuracy in our calculations to avoid introducing computational inaccuracies.

In the particular assay described the dosage region of interest lay between 200γ and 270γ. The corresponding region of log dosage is from 2.301 to 2.437. It is found by trial that by taking $m = 2.370$ and $\sigma = 0.050$ a suitable coverage is obtained and we have for the four log dose levels: 2.398, 2.295, 2.429, 2.359 corresponding to actual dose levels of 250γ, 197γ, 268γ and 228γ to the accuracy attainable. These doses of test and standard were, therefore, administered at equally spaced time intervals in the order shown in Table 1, the order within pairs of test and standard preparations at each dose level being decided by the toss of a coin.

TABLE 1. EXPERIMENTAL PLAN SHOWING ORDER IN TIME OF DOSE LEVELS AND RESPONSES OBSERVED.

Experiment No.	Dose	Test t Standard s	Response Observed (% inhibition)
1 2	250	s t	50.7 56.4
3 4	197	s t	30.9 37.5
5 6	268	t s	63.1 59.9
7 8	228	t s	52.1 48.7
9 10	228	s t	47.6 54.3
11 12	268	t s	67.8 64.3
13 14	197	t s	41.4 37.2
15 16	250	t s	63.3 57.6

3. ANALYSIS OF THE RESULTS

The constants required in the calculation are set out in Table 2 whilst the analysis of variance for the data is shown in Table 3. The total crude sum of squares based on sixteen degrees of freedom is entered in the first row of Table 3 and this is then split into two parts each based on eight degrees of freedom and accounting for the variation 'between pairs' and the variation 'within pairs' respectively. The former is the crude sum of squares calculated from the pair-sums, denoted by y_s , shown in column vi of Table 2, and the latter the crude sum of squares calculated from the pair-differences (test minus standard) denoted by y_d, shown in column vii of Table 2. The correction for the mean is next calculated in the usual way for both sums and differences. In the case of the differences the correction for the mean is in fact the sum of squares due to the difference in average response between the test samples and the standard samples and is called "samples" in Table 3. The corresponding entry in the effect column is the difference in mean response for test and standard samples. Sums of squares calculated direct from the sums or differences must, of course, be divided by two before entering in Table 3.

The two sets of seven degrees of freedom remaining after elimination of the means are now further analysed using the sets of constants given in Table 2.

In columns i), ii), and iii) of Table 2 are shown the values (as they are given by Fisher and Yates, 1942) of the orthogonal polynomials for calculating the linear, quadratic, and cubic time effects. In our notation these are denoted by t_1, t_2, and t_3. Columns iv) and v) show the orthogonal polynomials d_1 and d_2 for calculating the linear and quadratic dose effects. The method of arriving at the levels d_1 and d_2 is explained in §4. They are such that besides having zero sum of products with each other they also have zero sum of products with t_1, t_2, t_3, thus all the effects to be calculated are orthogonal and may be computed independently.

For example, by taking $\sum d_1 y_s$, the sum of products between d_1 and y_s and dividing by $\sum d_1^2 = 8$, the sum of squares for d_1, we have an estimate of the *sum* (test + standard) of the slopes of the response/log dose curves. By taking the sum of products $\sum d_1 y_d$ of d_1 with the differences y_d and dividing by $\sum d_1^2 = 8$, we have an estimate of the *differences* (test − standard) of these slopes. If, as is expected, the slopes for the test and the standard preparation are the same, then half the former quantity will provide the estimate for the common slope. On the other hand, if a discrepancy in slopes is found then the individual slopes for test and standard may be found from the estimates of the sum and differences. The corresponding sums of squares for the analysis of variance are given by half the square of the sum of products of d_1 and y_s (or y_d) divided by the sum of squares of d_1. The factor of a half appears because y_s and y_d are each calculated from two observations. In general, we may calculate all the effects and the appropriate sums of squares in this way from the two simple formulae

i) Sum (or difference) of effects $= \sum xy / \sum x^2$

ii) Sum of squares $= (\sum xy)^2 / 2\sum x^2$

where x is t_1, t_2, t_3, d_1 or d_2 and y is y_s or y_d.

TABLE 2. ORTHOGONAL POLYNOMIALS FOR TIME TREND AND DOSE, WITH SUMS, DIFFERENCES AND CORRECTED DIFFERENCES OF PAIRS OF OBSERVATIONS.

(i)	(ii)	(iii)	(iv)	(v)	(vi)	(vii)	(viii)
Time Trend			Dose		Response		
Linear t_1	Quadratic t_2	Cubic t_3	Linear d_1	Quadratic d_2	y_s Sums (Test + standard)	y_d Uncorrected Differences (Test − standard)	y_c Corrected Differences (Test − standard)
−7	7	−7	0.55500	−0.46956	107.1	5.7	4.65
−5	1	5	−1.50206	0.65424	68.4	6.6	5.55
−3	−3	7	1.17615	0.85467	123.0	3.2	4.25
−1	−5	3	−0.22909	−1.03933	100.8	3.4	4.45
1	−5	−3	−0.22909	−1.03933	101.9	6.7	5.65
3	−3	−7	1.17615	0.85467	132.1	3.5	4.55
5	1	−5	−1.50206	0.65424	78.6	4.2	5.25
7	7	7	0.55500	−0.46956	120.9	5.7	6.75
Divisor 168	168	264	8	4.91837			

TABLE 3. ANALYSIS OF VARIANCE (SUMS)

Source of Variation	Effect	Sum of Squares	D/F	Mean Squares
Grand Total		45,136.26	16	
BETWEEN PAIRS	(Effect Sums) $t + s$			
Total		45,033.70	8	
Correction for mean		43,347.24	1	
Time Linear	1.048	92.19	1	92.19***
Time Quadratic	−0.213	3.81	1	3.81
Time Cubic	−0.081	0.87	1	0.87
Dose Linear	19.917	1586.75	1	1586.75***
Dose Quadratic	−0.718	1.27	1	1.27
Residual		1.57	2	0.79
Error (1)		3.71	4	0.93

TABLE 3. ANALYSIS OF VARIANCE (DIFFERENCES)

WITHIN PAIRS (uncorrected)		(Effect Differences) $t - s$			
Total			102.56	8	
Samples		4.875	95.06	1	95.06***
Samples × Time	Linear	−0.046	0.18	1	0.18
	Quadratic	0.119	1.19	1	1.19
	Cubic	0.000	0.00	1	0.00
Samples × Dose	Linear	−0.541	1.17	1	1.17
	Quadratic	−0.622	0.95	1	0.95
Residual			4.01	2	2.01
Error (2)			4.96	4	1.24
WITHIN PAIRS (Corrected for trend)		(Effect Differences) $t - s$			
Total			104.16	8	
Samples		5.005	100.20	1	100.20***
Samples × Time	Linear	0.022	0.04	1	0.04
	Quadratic	0.100	0.85	1	0.85
	Cubic	0.046	0.28	1	0.28
Samples × Dose	Linear	−0.388	0.61	1	0.61
	Quadratic	−0.441	0.48	1	0.48
Residual			1.70	2	0.85
Error (3)			2.46	4	0.62

Proceeding in this way the sums and differences of effects due to linear, quadratic and cubic time trends, and linear and quadratic dose effects and their accompanying sums of squares are calculated and are entered in Table 3. Two residual degrees of freedom for the between pairs comparison and two residual degrees of freedom for the within pairs comparisons remain. On comparing the effect mean squares in the table with the appropriate residual mean square it is at once apparent that there is no reason to doubt the postulates, (in mind when the experiment was designed) that a quadratic time trend and linear response/log dose curves would prove adequate to represent the time and dose effects. The sum of squares due to cubic time effects and quadratic dose effects were, therefore, combined with the residuals to give error (1) and error (2) each an appropriate estimate of error for the effects in the corresponding part of the table and each having four degrees of freedom. Significance of the effects when compared with the appropriate error mean square is denoted by asterisks. (Three asterisks denote significance at the 0.1% level).

From the between pairs analysis it is seen that there is a large time trend effect almost completely accounted for by the linear component. There is also a slight suggestion of a quadratic effect but this is not sufficiently large to be definitely established.

From the within pairs analysis we see that there is a real difference in potency between the test and standard preparation, but there is no evidence of difference in slope or curvature for the time trend lines or response curves.

3.1 *Elimination of Trend Effect Within Pairs*

We have seen that a large linear time tread is occurring and that consequently some of the error within groups is due to the change in the level of response occurring in the interval between successive tests within pairs. The doses are spaced (see column (i) of Table 2) so that one unit of time elapses between the first test and the second, and the difference (second test minus first test) is therefore overestimated by an amount β, where β denotes the regression coefficient of the common linear time trend for test and standard preparations. It follows that the variance of the difference (test minus standard) is $2\sigma^2 + \beta^2$ and therefore that error (2) estimates $\sigma^2 + \frac{1}{2}\beta^2$. We may correct for the trend within-pairs therefore by adding or subtracting an amount b from the eight differences, where $b = 0.524$ is an estimate of β and is given by $\frac{1}{2} \times (1.048)$ taken from Table 3. It is subtracted from those pairs where the test preparation was assayed after the standard preparation and added in the contrary case. The "corrected" difference computed in this way to two-decimal accuracy and shown in column (viii) of Table 2 is denoted by y_c. After correction the difference (second test minus first test) is overestimated by an amount $\beta - b$. It follows that the variance of the corrected difference (test minus standard) is $2\sigma^2 + \sigma^2(b)$, where $\sigma^2(b)$ is the variance of b. Consequently, error (3) estimates $\sigma^2 + \frac{1}{2}\sigma^2(b)$. Comparing this with the previous expression we see that it will probably be advantageous to correct for the discrepancy due to trend within pairs if $b > \sigma(b)$. The analysis of the corrected differences given in Table 2, is carried in a manner exactly similar to that already described for the uncorrected differences.

Now $\sigma^2(b) = \sigma^2/(2 \sum t_1^2) = \sigma^2/336$. Hence the mean square for error (3) calculated from corrected differences estimates $(1 + 1/672)\sigma^2$. The mean square for error (3) multiplied by 672/673 is thus an unbiased estimate of σ^2. It will be seen (as would be expected if the mathematical model assumed were adequate) that estimates of σ^2 from error (1) and from error (3) are compatible and a final combined estimate s^2 based on eight degrees of freedom may be calculated from the combined sums of squares.

$$\{\text{sum of squares for error (1)}\} + \{672/673 \times \text{sum of squares for error (3)}\}$$

Dividing by eight we have the estimate of the combined error variance $s^2 = 0.70$.

3.2 *Calculation of Relative Potency*

If l is the estimated difference (in units of the design) between dose levels of test and standard preparations giving the same response, e is the estimated common slope of the log dose/response curves and a is the average corrected difference in response then

$$-l = \frac{a}{e} = \frac{5.085}{9.959} = 0.503$$

In log dose units, that is $0.503 \times 0.050 = 0.025$. Now antilog $0.025 = 1.059$ whence the estimated relative potency of test to standard preparation is 105.9%.

3.3 *Fiducial Limits for the Relative Potency*

Fiducial limits for the relative potency may now be found following Feiller (1940). We have $a = \bar{y}_c = \bar{y}_d + 2b/8$ (since the correction is subtracted in three cases and added in the remaining five) and $\bar{y}_d$ and b are distributed independently, whence

$$V(a) = V(\bar{y}_d) + \frac{1}{16} V(b) = \left(\frac{1}{4} + \frac{1}{16} \cdot \frac{1}{336}\right)\sigma^2 = \frac{1345}{5376}\sigma^2,$$

$$\text{also } V(e) = \frac{1}{16}\sigma^2.$$

Since a and e are distributed independently and (we shall assume) normally with variances $V(a)$ and $V(e)$ it follows that in repeated sampling

$$z = a + \lambda e$$

is distributed normally about zero with variance

$$V(a) + \lambda^2 V(e) = \left\{\frac{1345}{5376} + \lambda^2 \frac{1}{16}\right\}\sigma^2 = \left\{\frac{1345 + 336\lambda^2}{5376}\right\}\sigma^2$$

and that

$$t = \frac{a + \lambda e}{\left\{\frac{1345 + 336\lambda^2}{5376}\right\}^{1/2} \times s}$$

is distributed as Student's t with the same number of degrees of freedom as s (eight in this example).

Any suggested hypothetical value for λ would, therefore, be rejected at the 100 $\alpha\%$ level of significance unless

$$(a + \lambda e)^2 < \left\{\frac{1345 + 336\lambda^2}{5376}\right\} s^2 t_\alpha^2$$

where t_α is the 100 $\alpha\%$ level of significance of the t distribution. The 95% fiducial limits are provided, therefore, by those values of λ which just make the above inequality untrue i.e. which make it an equality.

Substituting the numerical values for a, e, s, and t_α and multiplying both sides of the equality by 5376 we have the quadratic equation in λ

$$531{,}895\lambda^2 - 535{,}904\lambda + 129{,}660 = 0$$

which has the solutions

$$\lambda = 0.6038 \qquad \text{and} \qquad \lambda = 0.4037$$

Multiplying these solutions by 0.050 to reduce them to log dose units and taking antilogs we have the 95% fiducial units for potency of 107.2% and 104.8%.

4. CONSTRUCTION OF DESIGNS

Suppose that we are to give k pairs of doses and that a polynomial in time (t) of p-th degree will adequately represent the time-trend and a polynomial in dose level (d) of q-th degree will adequately represent the dosage-response curve in the absence of the time-trend; then provided $p + q$ is less than k we may take as the model.

$$\eta = \beta_0' + \beta_1' t + \cdots + \beta_i' t^i + \cdots + \beta_p' t^p + \alpha_1 d + \cdots + \alpha_j d^j + \cdots + \alpha_q d^q \quad (1)$$

where η is the true response at dose d and time t.

Now instead of considering powers of t consider the orthogonal polynomials (see for example Fisher and Yates, 1942) which will be denoted by $t_1, .., t_i, .., t_p$ corresponding to $\xi_1, .., \xi_i, .., \xi_p$ in the Fisher and Yates notation. Equation (1) can then be written in the form

$$= \beta_0 + \beta_1 t_1 + \cdots + \beta_i t_i + \cdots + \beta_p t_p + \alpha_1 d + \cdots + \alpha_j d^j + \cdots + \alpha_q d^q \quad (2)$$

The orthogonal polynomials for equally spaced intervals* and $k = 8$ are given in Table 4.

TABLE 4. ORTHOGONAL POLYNOMIALS FOR $k = 8$.

t_1	t_2	t_3	t_4	t_5	t_6	t_7
−7	7	−7	7	−7	1	−1
−5	1	5	−13	23	−5	7
−3	−3	7	−3	−17	9	−21
−1	−5	3	9	−15	−5	35
1	−5	−3	9	15	−5	−35
3	−3	−7	−3	17	9	21
5	1	−5	−13	−23	−5	−7
7	7	7	7	7	1	1
$\Sigma t^2 = 168$	168	264	616	2184	264	343

The problem now is to choose the levels of the dose d so that the estimates of the α's are always uncorrelated with the estimates of the β's. This problem is solved if

$$\sum t_i d^j = 0 \qquad \text{for all values of } i \text{ and } j \quad (3)$$

where the sign $\sum$ is used to denote summation over the k values of the polynomial.

*Fisher and Yates tables give these polynomials up to fifth order only, but supply a recurrence formula from which the high polynomials are readily obtained. Alternatively an extended table giving the polynomials of all orders to $k = 26$ has recently been published by De Lury (1951). A design could, of course, be developed for values of t which are not equally spaced, but in this case, the values of the orthogonal polynomials would have to be calculated. This could be done either in the straightforward manner as described for example, by Kendall (1946) or by the Choleski method of matrix inversion recently described by Rushton (1951).

Since the polynomials up to t_p are used to represent the time trend and the α's are to be uncorrelated with the β's, d must be a linear function of the extra polynomials in time of order higher than p, namely $t_{p+1}, \cdots, t_{k-1}$.

We have, therefore,

$$d = \gamma_{p+1}t_{p+1} + \cdots + \gamma_{p+r}t_{p+r} + \cdots + \gamma_{k-1}t_{k-1} \tag{4}$$

where the constants $\gamma_{p+1}, \ldots, \gamma_{k-1}$ are chosen to satisfy the relations represented by equation (3).

4.1 *Derivation of the Design Used in the Example*

To illustrate the method the design used in the example above, will be derived. In this example experience suggested that a polynomial of second degree would adequately represent the time trend whilst the log-dose/response lines would be linear. In the design it was decided to allow for the estimation of effects of one order higher than those actually expected. Allowance was made, therefore, to fit a polynomial of up to *third* degree to represent the time trend and up to *second* degree to represent the log dose/response lines.

A preliminary examination shows that to obtain a satisfactory solution and allow a reasonable number of degrees of freedom for the estimation of error at least 8 pairs of doses would be required. We have then in the notation above,

$$p = 3, \qquad q = 2, \qquad k = 8$$

Equation (4) yields

$$d = \gamma_4 t_4 + \gamma_5 t_5 + \gamma_6 t_6 + \gamma_7 t_7 \tag{5}$$

and because of the orthogonal property of the polynomials it is true whatever the values of the γ's that

$$\sum d = \sum dt_1 = \sum dt_2 = \sum dt_3 = 0. \tag{6}$$

In addition we wish to arrange that the quadratic dose effect is also orthogonal to the time effects, so that we have to satisfy the three equations

$$\sum d^2t_1 = 0, \qquad \sum d^2t_2 = 0, \qquad \sum d^2t_3 = 0 \tag{7}$$

Now

$$\sum d^2t_1 = \sum \{(\gamma_4 t_4 + \gamma_5 t_5 + \gamma_6 t_6 + \gamma_7 t_7)^2 t_1\}$$

$$= \gamma_4^2(144) + \gamma_5^2(155) + \gamma_6^2(166) + \gamma_7^2(177) + 2\gamma_4\gamma_5(145)$$

$$+ 2\gamma_4\gamma_6(146) + \cdots \text{ etc.} \tag{8}$$

where for example (144) means $\sum t_1 t_4^2$. Similar expressions may be obtained for $\sum d^2 t_2$, and $\sum d^2 t_3$.

In practice most of the terms in these expanded expressions vanish since, for equally spaced values of t, the orthogonal polynomials of odd order are odd functions and those of even order are even functions. It follows that quantities like (157), (177), (144) involving three odd order polynomials or one odd and two even order polynomials are zero. Making use of the fact, the three equations derived from (7) are

$$\gamma_4\gamma_5(145) + \gamma_4\gamma_7(147) + \gamma_5\gamma_6(156) + \gamma_6\gamma_7(167) = 0 \qquad (9)$$

$$\gamma_4^2(244) + \gamma_5^2(255) + \gamma_6^2(266) + \gamma_7^2(277) + 2\gamma_4\gamma_6(246) + 2\gamma_5\gamma_7(257) = 0 \qquad (10)$$

$$\gamma_4\gamma_5(345) + \gamma_4\gamma_7(347) + \gamma_5\gamma_6(356) + \gamma_6\gamma_7(367) = 0 \qquad (11)$$

to which a wide variety of solutions may be found. In particular if d can be taken as a linear function of the *even* order polynomials only (i.e. if γ_5 and γ_7 can be put equal to zero) we shall obtain a design in which only four distinct levels of dosage are used instead of eight. Since it was of considerable practical advantage to reduce the number of dose levels, it was decided to adopt a solution of this kind.

Putting γ_5 and γ_7 equal to zero all the terms in equations (9) and (11) become zero and (10) becomes

$$\gamma_4^2(244) + 2\gamma_4\gamma_6(246) + \gamma_6^2(266) = 0 \qquad (12)$$

and the values of the constants are readily found from Table 4 of the orthogonal polynomials. In fact,

$$(244) = 160, \qquad (246) = 840, \qquad (266) = -672 \qquad (13)$$

whence writing $\gamma = \gamma_4/\gamma_6$ we have

$$160\gamma^2 + 1680\gamma - 672 = 0 \qquad (14)$$

$$\gamma = 0.385{,}823 \qquad \text{or} \qquad \gamma = -10.885{,}823 \qquad (15)$$

Hence, the levels for a design fulfilling all the requirements are provided either by

$$d = c(0.385{,}823 t_4 + t_6) \text{ of by } d = c'(-10.885{,}823 t_4 + t_6) \qquad (16)$$

where c and c' are scale constants at our choice. In practice it is convenient to choose the scale constant so that the 'standard deviation'

$(\sum d^2/k)^{\frac{1}{2}}$ of the levels is unity. The dose levels in these units we denote by d_1, so we have $\sum d_1^2 = k$. This choice of units enables the experimenter to assess readily the scaling-up factor which will be required so that the design covers the range of levels desired.

We require then

$$\sum d_1^2 = c^2(\gamma^2 \sum t_4^2 + \sum t_6^2) = k \tag{17}$$

whence

$$c = \left\{\frac{k}{\gamma^2 \sum t_4^2 + \sum t_6^2}\right\}^{\frac{1}{2}} \tag{18}$$

Substituting numerical values in (18) we find $c = 0.149{,}970{,}1$ $c' = 0.010{,}449{,}8$ and the required levels are

$0.057, 862t_4 + 0.149, 970t_6$			$-0.113, 755t_4 + 0.010, 450t_6$	
(1)	0.555,00		(1)	−0.785,84
(2)	−1.502,06		(2)	1.426,57
(3)	1.176,15		(3)	0.435,31
(4)	−0.229,09		(4)	−1.076,05
(5)	−0.229,09	or	(5)	−1.076,05
(6)	1.176,15		(6)	0.435,31
(7)	−1.502,06		(7)	1.426,57
(8)	0.555,00		(8)	−0.785,84

In the numerical example discussed above the first of the two alternative designs was used.

In the calculation of the quadratic dose effect it is convenient to use the orthogonal quadratic polynomial which is calculated from the formula (see for example Kendall 1946)

$$d_2 = d_1^2 - (\sum d_1^3/k)d_1 - 1 \tag{19}$$

In the present example we find,

$$d_2 = d_1^2 + 0.400{,}748d_1 - 1 \tag{20}$$

which yields the values for the quadratic polynomial given in column (v) of Table 2.

5. DISCUSSION

From the above example the general procedure will be apparent and designs may be developed for the solution of other similar problems. As might be expected if $p + q$ is nearly as large as k no solution will be possible. On the other hand, if $p + q$ is small compared with k then a great variety of solutions may be possible.

When as in the example above a linear time trend is adequate the device of angular randomisation described by Box (1952) may be used for random selection of the design. The statistical analysis will then be exact whether the mathematical model precisely fits the situation or not.

REFERENCES

Box, G. E. P. and Wilson, K. B. On the Experimental Attainment of Optimum Conditions. *J. R. Stat. Soc., B*, 13: 1, 1951.

Box, G. E. P. Multi-factor Designs of First Order. *Biometrika*, 39: 49, 1952.

Cox, D. R. Some Systematic Experimental Designs. *Biometrika*, 38: 312, 1951.

Feiller, E. C. The Biological Standardization of Insulin. *J. R. Stat. Soc. Suppl.*, 7: 1, 1940.

Fisher, R. A. and Yates, F. *Statistical Tables for use in Biological, Agricultural, and Medical Research*, 2nd edn., Edinburgh, Oliver and Boyd, 1942.

De Lury, D. B. *Values of the Integrals of the Orthogonal Polynomials up to* $n = 26$. University of Toronto Press, 1950.

Kendall, M. G. The Advanced Theory of Statistics, London. Charles Griffin and Co., Ltd., 1946.

Rushton, S. On Least Squares Fitting by Orthonormal Polynomials Using the Choleski Method, *J. R. Stat. Soc. B*, 13: 92, 1951.

2.4
The Exploration and Exploitation of Response Surfaces: Some General Considerations and Examples*

G. E. P. BOX
Imperial Chemical Industries Ltd. and University of North Carolina

Some three years ago Dr. K. B. Wilson and the author read a paper [1] before the Royal Statistical Society concerning the experimental attainment of optimum conditions. The methods there discussed grew out of experience acquired in a number of chemical investigations. Since that time many such investigations have been conducted in England along the lines we suggested, and more recently these procedures have been utilised at North Carolina, State College in experiments concerning such widely different topics as the flooding capacity of pulse columns (work performed under contract with the Atomic Energy Commission), and the motility and viability under various storage conditions of bull sperms. The object of this paper is to discuss and to illustrate with examples certain ideas which have arisen from the work and which it is believed may be of value in a wider field than that of chemical research. We shall concern ourselves with general principles. Details of the methods and calculations have been explained in [1] and more recently have been presented in a simplified form in the final chapter of a book on the design of experiments authored by a team of Imperial Chemical Industries' Chemists and Statisticians [2].

AN OUTLINE OF THE PROBLEM

1. *Response Surfaces*

We suppose that the experimenter is concerned with elucidating certain aspects of a functional relationship

$$\eta = \phi(x_1, x_2, \cdots, x_k) \tag{1}$$

connecting a *response* η such as yield, with the levels $x_1, x_2, \cdots, x_k$ of a group of k quantitative *variables* or *"factors"* like temperature, time, and pressure of reaction, concentration of reactants and speed of agitation. More generally he will be concerned not with a single

*Sponsored by the Office of Ordnance Research, United States Army under contract DA-36-034-ORD-1177.

The Collected Works of George E. P. Box, 1984, Wadsworth, Inc., Belmont, CA 94002. Originally published in *Biometrics,* vol. 10, no. 1 (1954), pp. 16–60.

response but with a number of responses which he will wish to bring to the most satisfactory levels possible. For example he will often be seeking conditions which maximize a major response η_1 such as yield (or more often cost of manufacturing unit quantity of product), while maintaining some auxiliary response η_2 such as purity, at the best level possible. To begin with we confine our discussion to a single response η_1 which for convenience we assume is the yield of product. When only one factor x (such as temperature) is studied the response function (1) could frequently be described by a graph like that in Figure 1

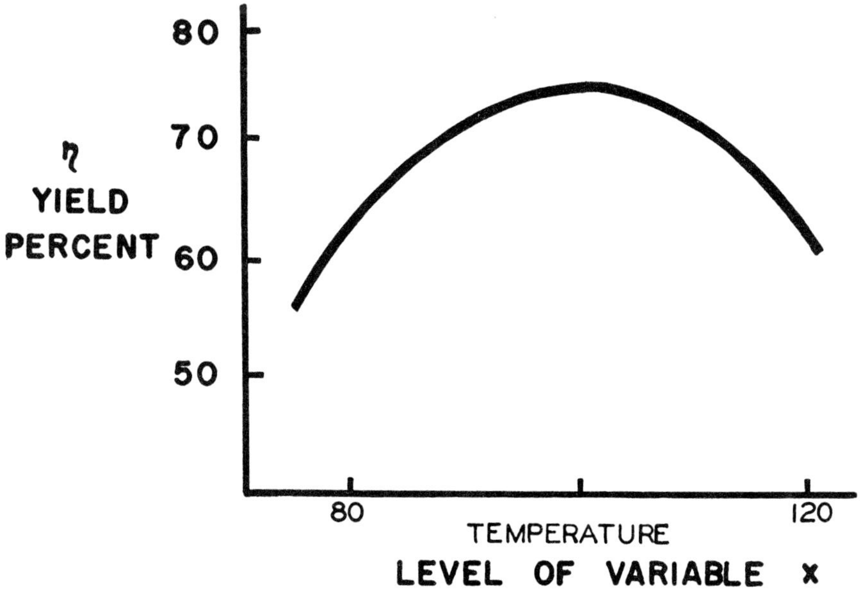

FIGURE 1. A RESPONSE CURVE FOR ONE FACTOR

With more than one factor it is tempting to generalize this by supposing that the response function could be represented by a surface like a more or less symmetrical mound, the contour representation of which would be like that shown in Figure 2.

That such a generalization is entirely inadequate in practice is found as soon as we begin to carry out experiments in which actual response surfaces can be roughly plotted. It is then clear not only that surfaces are frequently attenuated in the neighborhood of maxima as in Figure 3, but also that ridge systems like that of Figure 4 and 5 are of common occurrence.

It will be noted that any section of these figures taken parallel to either axis will yield a curve like that in Figure 1, so that these surfaces are entirely built up from such curves.

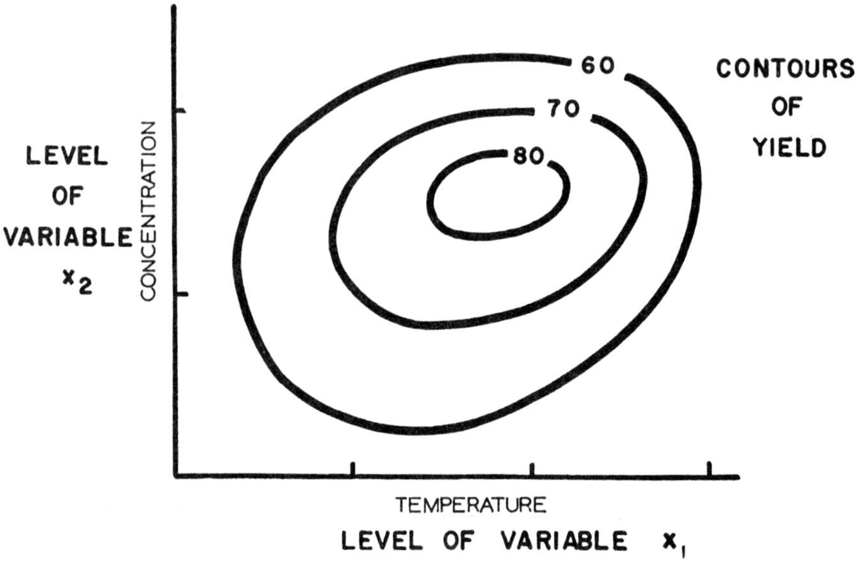

FIGURE 2. CONTOURS OF A TWO-FACTOR RESPONSE SURFACE

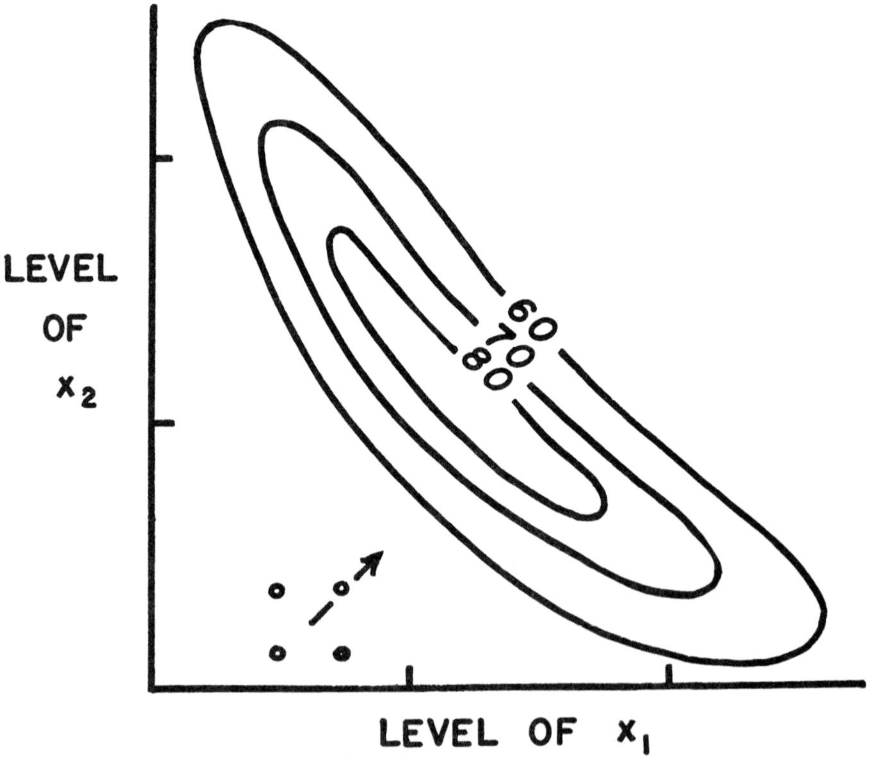

FIGURE 3
ATTENUATED MAXIMUM

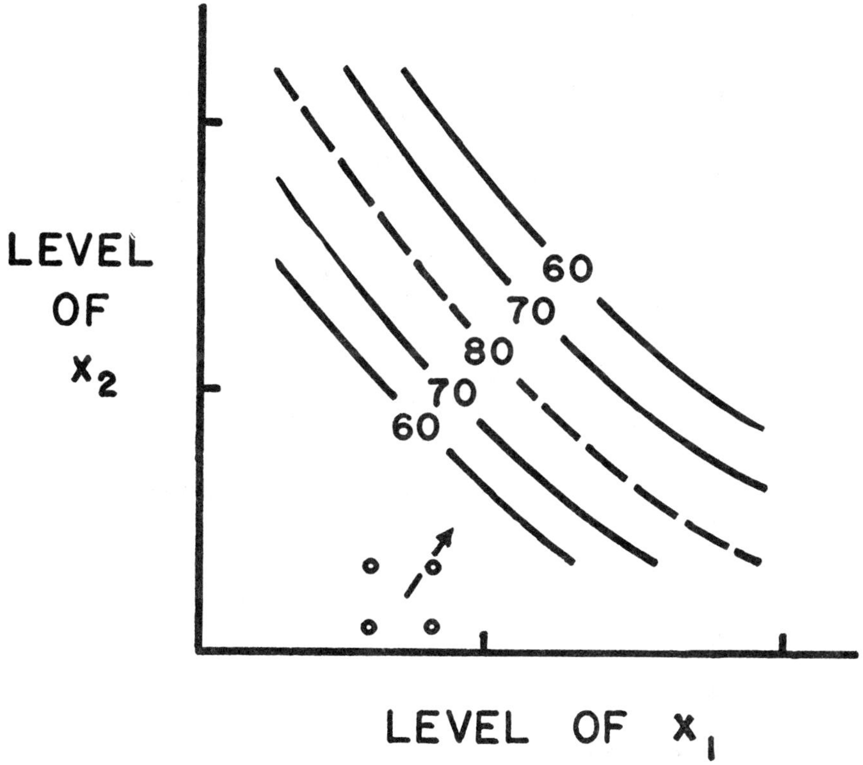

FIGURE 4
STATIONARY RIDGE

2. *Factor Dependence.*

The reason for the occurrence of such systems can be seen when it is remembered that factors like temperature, time, pressure, concentration, etc., are only regarded as "natural" variables because they happen to be quantities that can be conveniently measured separately. A more fundamental variable not directly measured, but in terms of which the behaviour of the system could be described more economically (e.g. frequency of a particular type of molecular collision), will often be a function of two or more natural variables, (e.g., temperature, concentration, pressure). For this reason many combinations of natural variables may correspond to the best level of a fundamental variable. To quote a simple example suppose that in the region of interest the response was most economically described in terms of some fundamental variable the level of which was proportional to the product $w = x_1x_2$ of two measured variables x_1 and x_2. Thus $w = x_1x_2$ and $\eta = f(w)$. Suppose that the latter relationship was that represented in Figure 6a.

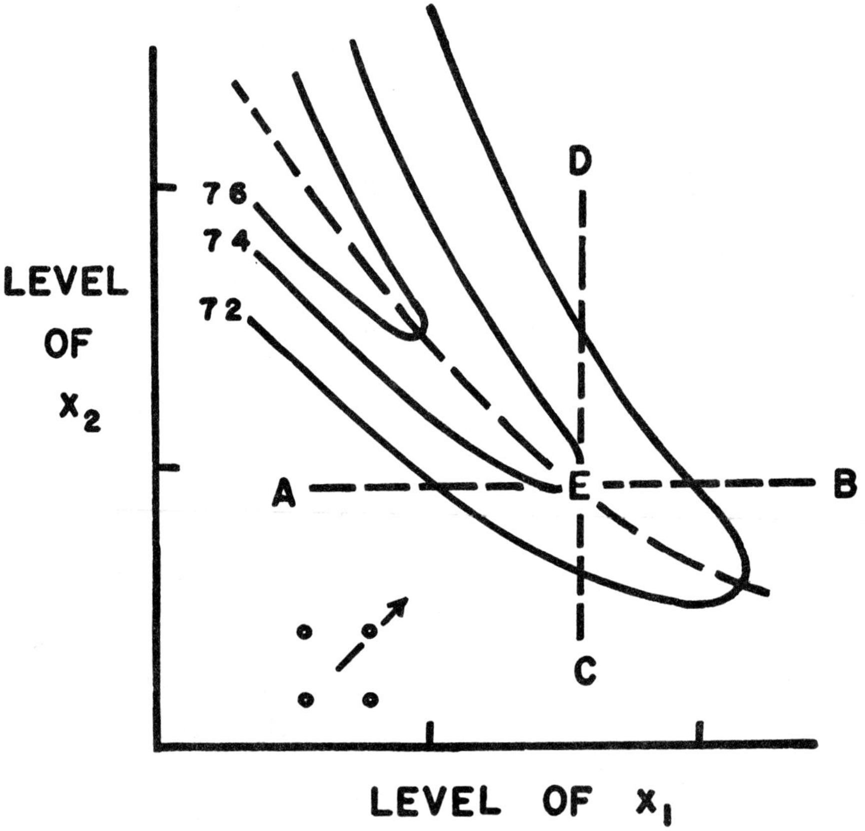

FIGURE 5
RISING RIDGE

If the experimenter did not know that the system could adequately be described in terms of the compound variable w and carried out experiments in which x_1 and x_2 were varied separately, he would be exploring a system for which the yield surface was that shown in Figure 6b, which is like that in Figure 4. Figure 6b which describes a commonly occurring practical situation was constructed by drawing contour lines through those points giving a constant product x_1x_2, the appropriate yield being read off from Figure 6b.

Instead of the function $w = x_1x_2$ we might have considered other functions $w = f(x_1, x_2)$. It will be found for example that the functions $w = a + bx_1 + cx_2$, $w = ax_1^bx_2^c$, $w = ax_1^b \exp\{-(c/x_2)\}$ all produce diagonal ridge systems running from the top left to the bottom right of the diagram like that in Figures 4 and 6b, while the functions $w = a + bx_1 - cx_2$, $w = ax_1^b/x_2^c$, $w = ax_1^b \exp\{-(cx_2)\}$ all produce ridges running in the contrary sense. These ridge systems are of course associated with

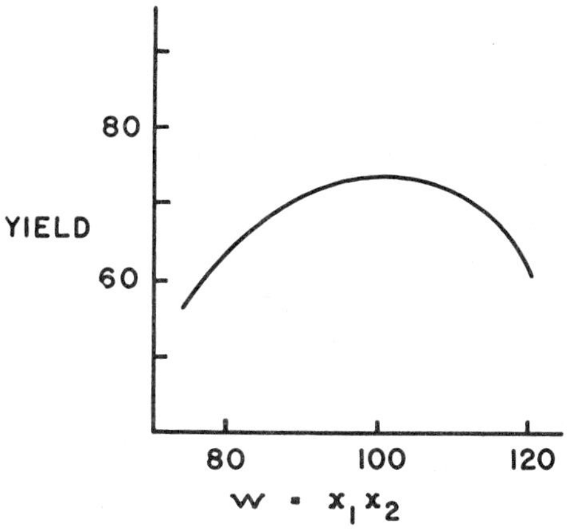

FIGURE 6a

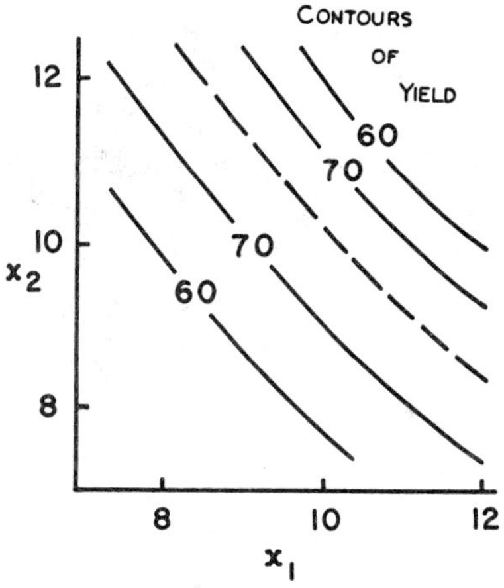

FIGURE 6b

GENERATION OF A RIDGE SYSTEM

interaction. The former type is associated with a negative two-factor interaction between x_1 and x_2 whilst the latter is associated with a positive two factor interaction.

The simplest type of ridge system is that produced by the linear relationship $w = a \pm bx_1 \pm cx_2$. The ridge system generated by such a function will have parallel straight-line contours running in a direction determined by the relative magnitudes of b and c. A section at right angles to these contours will reproduce the original graph of η on w. We note (for future use) that over limited ranges we should expect all ridge systems to be capable of approximate representation by this simple linear type. For expanding $w = f(x_1, x_2)$ in a Taylor's Series about a local origin, we have $w = a + bx_1 + cx_2 +$ (terms of second and higher order) where a is the value of the function at the origin, and b and c are the values of the partial derivatives with respect to x_1 and x_2 at the origin.

We could say in the examples quoted above that there is a 'redundancy' of one variable. The apparently two-variable system can be expressed in terms of a single fundamental 'compound' variable w. The physical and biological sciences abound with examples where, over suitable ranges of the variables, relationships exist similar to the above.

Frequently the surface which describes the system while not of the form shown in Figure 4, nevertheless contains a marked component of this type together with an additional component resulting in an elongated maximum like that shown in Figure 3, or a system like that shown in Figure 5, where the ridge is steadily rising to higher yields.

In the latter case the best practical combination may be the most extreme point on the ridge that can be attained by the experimenter.

The situations illustrated in Figure 3, 4, and 5 may be said to show *factor dependence* in the sense that the response function for one factor is not independent of the levels of the other factors. The idea of factor dependence is analogous to that of stochastic dependence and diagrams like Figure 3 call to mind the familiar contour representation of a bivariate probability surface. Again there exists the same analogy between the 'fundamental variables' we have discussed and the 'factors' in factor analysis (see for example Thurstone [3]).

These analogies are helpful but it must be emphasized that the two types of dependence are quite distinct and care should be taken to differentiate them. We are concerned here with the relationship between a response like yield and the levels of a set of variables which can be varied at our choice like temperature, time, etc. We do not need any ideas of probability to define this relationship. In the analogous stochastic situation probability takes the place of response and stochastic variables such as 'test scores' take the place of the fixed variables like temperature, time etc.

The investigation of factor dependence is of considerable practical importance.

(i) Where the surface is like that in Figure 4, not one, but a whole range of alternative optimum processes corresponding to points along the crest of the ridge will be available from which to choose. In practice some of these processes may be far less costly, or more convenient to operate, than are others. The factors in this system may be said to be *compensating* in the sense that departure from the maximum response due to change in one variable can be compensated by a suitable change in another variable. The direction of the ridge indicates how much one factor must be changed to compensate for a given change in the other.

If we imagine the contours for some auxiliary response η_2 superimposed on those for the major response η_1 we see that (provided the contours of the two systems are not exactly parallel) we could select for our optimum process that point on the crest of the ridge for the major response which maximized the auxiliary response. Thus both the major response and the auxiliary might simultaneously be brought to their best levels.

(ii) When the surface is like that in Figure 3 the direction of attenuation of the surface indicates those directions in which departures can be made from the optimum process with only small loss in response.

(iii) The detection of a rising ridge in the surface like that in Figure 5 supplies the knowledge that if the variables are changed *together* in the direction of the axis of the ridge then yield improvement is possible even though no improvement is possible by changing any *single* variable.

(iv) Of by no means least importance is the fact that discovery of factor dependence of a particular type may, in conjunction with the experimenter's theoretical knowledge lead to a better understanding of the basic mechanism of the reaction. Thus if experimentation with the variables x_1 and x_2 produced a surface like that of Figure 6b he would be led to expect that some more fundamental variable of the type $w = x_1x_2$ existed.

So far we have illustrated our discussion with examples in which there were only two variables x_1 and x_2. Such examples can be illustrated geometrically either by means of a three dimensional diagram in which two dimensions are used to accomodate the variables x_1 and x_2 and the third to accommodate the response η, or by two dimensional diagrams like Figures 2, 3, 4, and 5 in which the response η is represented by contour lines.

The situation which can occur in many variables becomes progressively more complicated and there is a distinct danger that by getting into the habit of thinking in only 2 variables we may over-simplify the problem. This danger is lessened if we become familiar with a method for representing the relationship between 3 variables x_1, x_2, x_3 and a response η by a three dimensional diagram which the contour surfaces for η are shown. Figure 7 shows such representation.

This diagram can be regarded as being built up from two dimensional contour diagrams. For example suppose that a series of two dimensional yield-contour diagrams for the variables temperature (x_1) and concentration (x_2) were made for various levels of time (x_3). If these were drawn on sheets of transparent material and the sheets were then placed one behind the other at the appropriate points on the time axis, on joining up corresponding contour lines to form contour surfaces, we would obtain a diagram like Figure 7.

In this representation in the neighborhood of a symmetrical point maximum (analogous to that in two dimensions shown in Figure 2) the contours would enclose the maximum point like the skins of an onion around its center. Insensitivity to change in conditions when variables were changed together in a certain way would correspond to attenuation of the contours in the direction of the compensatory changes. As an extreme form of this attentuation we could imagine a line maximum in the space (i.e. a line such that all sets of conditions on it gave the same high yield) surrounded by cylindrical contours of falling yield.

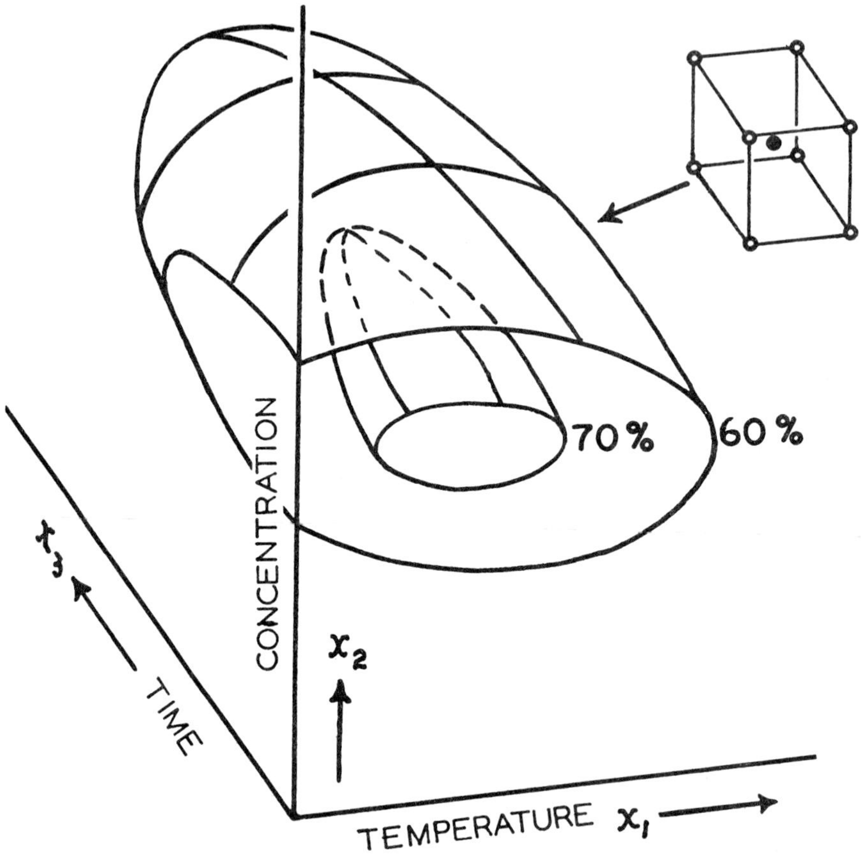

FIGURE 7. CONTOUR REPRESENTATION OF THREE-FACTOR SYSTEM

We would call this a line stationary ridge since it is analogous to the two dimensional case illustrated in Figure 4. Many other possibilities occur but we shall leave their fuller discussion till later.

It should not be thought that factor dependence is only to be expected in connection with well defined physical and chemical phenomena, nor that 'single variable' redundancy is all that need concern us.

Suppose, for example, the problem concerned the making of a cake, the object being to bring some desirable property such as 'texture' (which is our response η and which we suppose can be measured on some numerical scale) to the 'best' level. The natural variables, k in number, whose levels we denote by $x_1, x_2, \cdots, x_k$ might be numerous and involve among other things the amount of such substances as baking powder (x_1), flour (x_2) egg white (x_3) and citric acid (x_4).

Suppose it had happened that the 'texture' η depended in reality on only two fundamental variables, the *consistency* of the mix w_1 and its *acidity* w_2, and that optimum texture was attained whenever w_1 and w_2 were at their optimum levels w_{10}, w_{20}. For simplicity we will make the somewhat unreal assumption that w_1 and w_2, measured in some suitable way, were linear functions of the amounts $(x_1, x_2, \cdots, x_k)$ of the various ingredients i.e.,

$$\text{consistency} \qquad w_1 = a_1 + b_1x_1 + c_1x_2 + d_1x_3 + \cdots + p_1x_k$$

$$\text{acidity} \qquad w_2 = a_2 + b_2x_1 + c_2x_2 + d_2x_3 + \cdots + p_2x_k$$

Each of the coefficients b_1, c_1, $\cdots$, p_1 in the first equation would measure the change in consistency due to adding a unit amount of the corresponding ingredient. Thus the coefficient would be positive for solid substances like flour and negative for liquid substances like water. The optimum consistency would thus be obtained on the $k - 1$ dimensional planar surface in the k dimensional space of the factors

$$w_{10} = a_1 + b_1x_1 + c_1x_2 + d_1x_3 + \cdots + p_1x_k$$

Similarly, each of the coefficients b_2, c_2, $\cdots$, p_2 in the second equation would measure the change in acidity due to the addition of a unit of the corresponding ingredient and would be positive for acid substances and negative for alkaline substances. The optimum acidity would thus be obtained on a second $k - 1$ dimensional planar surface

$$w_{20} = a_2 + b_2x_1 + c_2x_2 + d_2x_3 + \cdots + p_2x_k$$

The intersections of the two planes would give all the levels for which both acidity and consistency were at their best levels, and hence for which optimum texture was attained. Optimum texture would thus be attained on a plane of $k - 2$ dimensions. That is to say there would be $k - 2$ directions at right angles to each other in which we could move while still maintaining optimum texture for the cake. We can express the k variable system in terms of two fundamental 'compound' variables w_1 and w_2; there is thus a redundancy of $k - 2$ variables corresponding to the $k - 2$ dimensions in which the maximum is attained. Were it not for the danger of confusion involved in using the phrase in the present context we would say that the maximum had $k - 2$ 'degrees of freedom'.

The surface of this example is a generalisation of the stationary ridge of figure 4. Rising ridges, attenuated maxima and combinations of these phenomena will all have generalisations in many dimensions. We shall discuss these generalisations later.

It is clear that the experimenter should have adequate investigational tools to explore response surfaces. For if he can appreciate at least approximately the main features of the surface with which he is dealing, this may lead not only to process improvement but also to his obtaining a more fundamental understanding of basic process mechanisms.

3. *The Roles of the Statistician and the Experimenter.*

We have been discussing from a geometrical standpoint the response function which underlies a given system and certain features of which it is desired to elucidate experimentally. We can perhaps avoid some misunderstandings by using our geometrical analogy to delineate more clearly the problems in experimentation which are statistical and those which are essentially non-statistical and in which this science cannot be of help. We shall speak in what follows of the experimenter (by which we mean the chemist, biologist, or engineer who is conducting the experiments), and the statistician as two individuals. Occasionally the experimenter will be his own statistician in which case the terms will indicate the questions which will require his statistical skill and those which require his other knowledge. The statistician's role is not, and cannot be, to design experiments in any absolute sense. The statistician's function is to advise the experimenter on the best positioning of experimental points in a space which the experimenter *must of necessity construct for him*, and construct purely on the basis of the experimenter's expert background knowledge of the subject in which he is experimenting. Were this not so there would be little point in training chemists, biologists, physicists etc., only statisticians.

The experimenter decides,

(i) *which* factors should be varied: Because the amount of effort which can be exerted on any given problem is in practice not unlimited he must often choose a few factors which he believes will be important out of a larger number which might be.

(ii) *in what way* the factors should be varied: They might be varied directly or as ratios or products of the "natural" variables as we have discussed above.

(iii) *by how much* the factors should be varied: He decides both absolutely and relatively the amounts by which the factors should be varied, and hence tacitly the units in which the surface is imagined to be drawn.

He thus decides *which* particular transform of *which* particular factor space shall be considered.

In a given problem different experimenters might well include or fail to include different variables and vary them in different ways by different amounts. The statistician may help to some extent by general words of advice on these matters, but when he has given this advice and the experimenter has decided the questions mentioned above, he can only devise experiments which explore the space the experimenter has defined as efficiently as possible. The ultimate success of the *experiment as a whole* (in contrast to the statistical exercise) must necessarily depend on the skill of the experimenter. No amount of artistry in statistical design can compensate for the omission of the most important factor.

It should be borne in mind that the major indeterminancies mentioned above are major indeterminancies of experimentation as a whole and are not associated with the use of any particular method. We can take comfort in the thought that in spite of them scientific research often manages to achieve useful results.

METHODS FOR EXPLORING THE RESPONSE SURFACE

4. *General Considerations.*

In order to determine and hence exploit multi-factor dependence we must at some stage perform groups of experiments capable of determining how the factors *jointly* influence the response. Any method which attempts to avoid the multi-factor problem by varying only one factor at a time will be almost valueless when the response surface contains a ridge, except possibly as a preliminary procedure.

As an illustration, consider the behavior of the 'one factor at a time' method when the surface is like one of those in Figures 3, 4, and 5.

(i) If the experimenter starts at the point A in Figure 5 a series of experiments in which x_1 is varied keeping x_2 constant performed along AB will lead to the conclusion that E is the best level of x_1. Experiments in which x_2 is now varied, keeping x_1 at its previous best level, along CD will lead to a point almost indistinguishable from E, and in the presence of experimental error will lead the experimenter to the conclusion that E is a maximum.

(ii) A similar difficulty occurs in the situation of Figure 4. Although the experimenter would in this case have reached a point of maximum yield on the ridge he would not know of the existence of the ridge and hence of the possibility of using alternative and often more convenient processes.

(iii) In the example of Figure 3 the experimenter would at best follow a tortuous path to the maximum advancing by small increments along a zig-zag route. At worst if the error was not very small he would often mistake a lower point on the "ridgy" surface for a true maximum.

The problems mentioned above in the two-factor case are even more acute when there are more than two factors and multi-factor dependence has to be dealt with.

The writer's attention has been repeatedly drawn to the behavior of the one factor at a time method in studies that have been carried out to improve processes already operating. Usually the current process will have been arrived at by the "one factor at a time method", which in various guises has been the normal procedure used by the chemist. When (as is often the case) further marked improvement is possible this is usually because the one factor at a time experiments have become "stuck on a ridge". Approximate determination of the local surface and exploitation of the local factor dependence is essential to further progress. The provision of methods for doing this is thus extremely important. The advantage to the experimenter in using such methods is not merely that fewer experiments are required to attain a *given* result which could ultimately have been reached by traditional methods, but that a result can be obtained that *could not have been got* by such methods.

We may explore the system by fitting some sufficiently elastic surface to the observations at a suitable set of pre-selected points and then examining the fitted surface. In the absence of specialized knowledge the most suitable model for this purpose seems to be the generalized polynomial equation which for two variables is

$$\eta = \beta_0 + (\beta_1 x_1 + \beta_2 x_2) + (\beta_{11} x_1^2 + \beta_{22} x_2^2 + \beta_{12} x_1 x_2) + (\beta_{111} x_1^3 + \beta_{222} x_2^3 + \beta_{112} x_1^2 x_2 + \beta_{122} x_1 x_2^2) + \text{etc.} \qquad (2)$$

and corresponds to representing the function by its Taylor's series. The brackets delimit the terms containing coefficients of first, second and third order. The equation containing all coefficients up to rth order is said to be of rth degree. The first degree equation defines a plane, the second degree equation a quadric surface. The surfaces defined by third and fourth degree equations we shall call cubic and quartic surfaces. Table 1 shows the number of terms (L) contained in equations of degree 1, 2, 3 and 4 when the number of factors is 2, 3, 4 or 5. The number of experiments N must be at least as great as L, the number of constants fitted. We see therefore that the number of experiments necessary to fit the equation will rise rapidly with its degree. However, the higher the order of the terms included, the smaller the effect they usually will have, and we shall include terms only to that order necessary to give an approximate representation of the surface in the region studied.

TABLE 1
NUMBER OF CONSTANTS (L) TO BE FITTED FOR EQUATIONS OF VARYING DEGREE

k Number of Factors	d = Degree of fitted equation			
	1(plane)	2(quadric)	3(cubic)	4(quartic)
2	3	6	10	15
3	4	10	20	35
4	5	15	35	70
5	6	21	56	126

The equation can readily be fitted to the observations by the method of least squares (which is in this context sometimes called the method of multiple regression). Using this technique estimates b_0, b_1, b_2, etc. of the coefficients β_0, β_1, β_2, etc. are determined which make the sum of squares of discrepancies between the observed values and those predicted by the fitted equation as small as possible. Estimates of this sort may be shown to have specially desirable properties. In practice (as we shall describe later) the N levels of the variables will be specially selected to give accurate estimates.

Essentially what is being done is to use the technique of "multiple regression" in the circumstance in which the values for the variables or factors are chosen in advance rather than merely observed and because of these specialized levels the calculations are greatly simplified.

For convenience we do not work directly with the "natural" variables like temperature, time, etc., but with the quantities x_1, x_2, etc., which will refer to "standardized variables" in which the origin is taken to be the center of the design and the units are fixed by the amounts the natural variables are changed in the design. We shall not necessarily limit the relation between natural and standardized variables to be a linear one. In some circumstances the use of logarithms or some other function may be appropriate.

An equation of fairly high degree would usually be needed to represent the response surface adequately over the whole *experimental region* (the whole region of possible variation of the factors). The fitting of such a surface could involve an excessive number of experiments. We can examine the surface in a smaller subregion however by a fairly simple equation.

If, for example, we were *close enough* to the ridge system or maximum we might represent its main features approximately by an equation of only second degree (see Figures 10 and 11). However the starting conditions would often not be close to the maximum or ridge system, particularly if the system were being investigated for the first time. The mathematical procedure of fitting selects from all possible surfaces of the degree fitted that which approximates the responses *at the experimental points* the best (in the sense of least squares). The features of the fitted surface at points remote from the region of the experiment would probably bear no resemblance to the features of the actual surface.

The experimenter needs some *preliminary procedure* therefore to bring him to a suitable point near the maximum or ridge where the second degree equation could most usefully be fitted. One such preliminary procedure is the one factor at a time method already discussed. An alternative which in the author's opinion is usually more effective and economical in experiments (at least in the field of application where it has been tried) is the "steepest ascent" method. This preliminary procedure followed by the fitting of an equation of second degree (or if necessary of higher degree) is the basis of the sequential method proposed in [1].

5. *A Sequential Method for Exploring the Response Surface.*

We have seen that the more elaborate the equation which is to be taken as a model the greater is the number of experiments which must be performed to fit it. If we can proceed sequentially therefore (that is to say if we can use the result of one set of experiments to decide the location of the next set), it would seem best to begin by fitting locally the simplest possible equation, abandoning it for a more elaborate one only when the circumstances showed this to be necessary.

(i) We should begin by fitting a first degree equation representing a plane. If this plane seemed to provide a reasonably close fit and was sloping in some particular direction then progress could be made by proceeding up it in the direction of greatest slope, or *steepest ascent.* Experiments performed in this direction would lead to a point where no further gain was obtained. Here a second plane could be fitted and the procedure repeated. Since we would only be fitting this very simple type of equation any advance which could be made by such a procedure would be progress attained very economically.

(ii) Sooner or later it would become clear that the slopes of the last fitted plane were not large compared with the second order effects and consequently that further progress by this method was not possible. This situation might be found in the original region in which experiments were conducted (i.e., it might be found after performing the initial set of exploratory experiments), or it might be found after one or two applications of the steepest ascent procedure had brought the experimenter to a region of higher yield than that at the starting conditions. Thus in one way or another would be attained a near-stationary region. That is a region in which little change occurred in the response when the factor levels were changed. In the situations illustrated in Figures 3, 4, and 5 the four points at the corners of a square indicate the four experiments arranged in a 2^2 factorial design which might be used to determine the first order effects. The arrow indicates the direction of steepest ascent. The steepest ascent procedure will in each case lead the experimenter to a near-stationary region on the crest of the ridge surface. Further progress by this method would not then be possible but in the neighborhood of the best point attained a suitable pattern of experiments would be performed and the second degree equation could now be fitted with maximum effect.

(iii) The analysis of this fitted second degree equation would indicate certain features of the surface requiring further study. For example the fitted surface might suggest the existence of a local maximum or some stationary or rising ridge system.

(iv) Further confirmatory points at positions indicated by the first fitted surface would therefore be added and the surface refitted including the information from these extra points. This may usually be done with least labor by employing a technique due to Plackett [4], and [1].

(v) If it appeared that the second degree equation was such a poor fit as to be an inadequate model then steps would be taken to fit a third degree equation. Extra constants may be added to the model without undue recalculation following a method given by Box and Hunter [5].

The basic ideas in these procedures are not of course new. In particular both the device of steepest ascent and that of fitting equations in the neighborhood of maxima have been employed in a problem closely analogous to ours, that of solving mathematical equations by approximate methods (see for example Booth [6] and Koshal [7]). Their application to the present problem, in the manner we propose, does however appear to be novel.

6. *Experimental Designs.*

Each combination of levels of the factors corresponds to a point in the factor space and the pattern of such points used to elucidate the surface is called the experimental design. For example, the sets of four points in Figures 3, 4, and 5 correspond to the 2^2 factorial design, and the eight points in Figure 7 to the 2^3 factorial.

It will be understood that arbitrary patterns of experimental points might fail not only to provide accurate estimates of the constants but might not even allow certain constants to be separately estimated at all. When attempting to obtain a best distribution of experimental points for the purpose of fitting a polynomial equation three considerations must be kept in mind.

(i) On the assumption that a polynomial of the degree assumed can adequately represent the surface in the region examined, the design should be such that the errors of estimate of points on the fitted surface should be small.

(ii) The biasses in the estimated coefficients, which might occur if the assumed equation were representationally inadequate, should be small.

(iii) If possible, provision should be made to estimate certain of the coefficients or combinations of them of higher order that those in the assumed form of equation to be fitted. Study of these sample members of estimated higher order effects will provide some indication of whether the assumption that these terms can be ignored is a reasonable one or not.

The number N of experimental points needed to determine L constants cannot be less than L and we should normally require that it should exceed this number. If N were equal to L the process of fitting would inevitably force the N values $Y_1, Y_2, \cdots, Y_N$ predicted by the fitted equation to agree exactly with the observed values $y_1, y_2, \cdots, y_N$. In this circumstance we could, of course, draw no conclusion at all as to whether the fitted surface were really representing the true surface or not. If N were greater than L, however, the N values $Y_1, Y_2, \cdots, Y_N$ predicted by the fitted equation at the N set of conditions would differ from the N observed values $y_1, y_2, \cdots, y_N$. The sum of squares of the discrepancies $S = \sum_{i=1}^{N} (y_i - Y_i)^2$ (which is of course the quantity minimized in the "least squares" procedure) is called the residual sum of squares. If this quantity is divided by $N - L$ (called the residual number of degrees of freedom) we have an estimate of σ^2 the experimental error variance *provided that the real surface can be represented by a function of the form assumed.*

If the real surface cannot be approximately represented by an equation of the form assumed $S/(N - L)$ tends on the average to exceed σ^2 by an amount which depends on the magnitude of the constants which have been ignored and on the sensitivity of the design chosen to the departures from assumption which have occurred. In general the design will be sensitive to the existence of those higher order terms which can be separately estimated in (iii).

If therefore some estimate of the experimental error variance is available (this might be quite a rough estimate from previous experience or some more precise estimate from replication of experiments), and if the design is suitably chosen, comparison of $S/(N - L)$ with the estimate enables the experimenter to obtain some conception of the *goodness of fit* of the postulated equation; that is to say, its representational adequacy in the particular circumstance of the experiment. A good fit does not necessarily imply that $S/(N - L)$ is small but only that it is of the magnitude anticipated for the experimental error variance. If $S/(N - L)$ is much larger than the experimental error variance this points to the necessity for including terms of higher order. It should be emphasized again that the sensitivity of this 'goodness of fit test' to particular types of departure from the assumed form of equation depends on the design used. A more extensive discussion of this question will be found in [5].

It is encouraging to find that experimental design is a field where virtue is rewarded, for designs which satisfy the criteria mentioned above usually result in extremely simple calculations.

The designs employed in [1] for determining the best fitting equation of *first* degree were the two-level factorials and fractional factorials. These designs provide efficient estimates of first order effects and also allow certain of the second order effects (some or all of the two-factor interactions) to be examined. This examination provides some indication of whether a model in which second order terms are ignored is adequate or not. A more fundamental consideration of the problem of first order design is given in [8].

Designs for determining the best fitting equation of *second* degree without undue expenditure of experiments were developed empirically in [1]. In these a two level factorial or fractional factorial which allowed the estimation of all linear and two factor interaction terms was augmented with further points which allowed the quadratic effects to be determined also. These designs were called *composite* designs.

When it appeared that the design was symmetrically placed with respect to the region it was desired to study, a "central" composite design was used (illustrated for three variables in Figure 8). When the region of interest lay towards one corner of the factorial design, it was augmented in that region with further points to form a non-central composite design (illustrated for three variables in Figure 9)

A more fundamental consideration of this design problem has

recently been undertaken [9]. This study indicates, that the composite designs which have, by now, been used in many practical examples are reasonably efficient but are capable of further improvement. The original composite designs are employed in the examples that follow in §§8, 9, and 10.

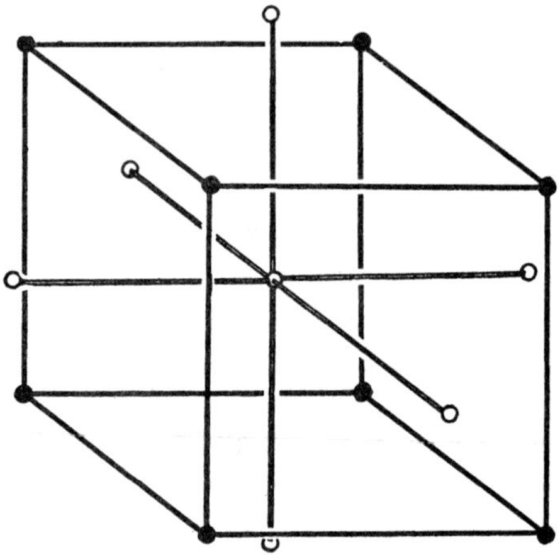

FIGURE 8
CENTRAL COMPOSITE DESIGN

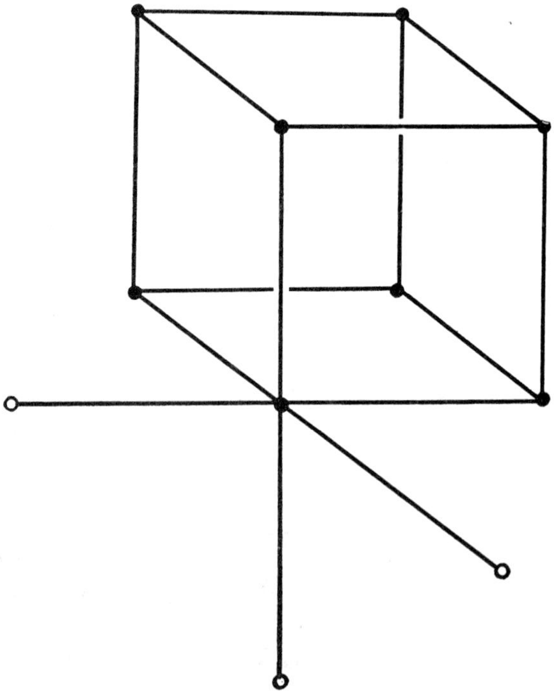

FIGURE 9
NON-CENTRAL COMPOSITE DESIGN

7. Canonical Analysis of the Fitted Second Degree Equation.

One of the most striking things to come out of the application of the technique described has been the power of elucidation afforded by the analysis of the second degree equation after a near stationary region was attained. This technique of the fitting and analysis of a second degree equation has been effective when marked factor dependence made the system difficult or impossible to study by other means. This we shall illustrate in the next section with a number of actual examples. Before we consider these, we explain briefly the basis of the method of analysis of the second degree equation. This is essentially the reduction of the fitted equation to canonical form. This is, of course, a standard process which is given in text books in coordinate geometry; however, it may be of value to explain it here essentially from the point of view of our problem.

Suppose the fitted second degree equation for the two variables x_1 and x_2 is

$$Y = b_0 + b_1x_1 + b_2x_2 + b_{11}x_1^2 + b_{22}x_2^2 + b_{12}x_1x_2 \tag{3}$$

The coefficients b_1 and b_2 are called the linear effects, b_{11} and b_{22} the quadratic effects, and b_{12} the interaction effect. By mere inspection of the coefficients of equation (3) we usually could not perceive the type of surface which it represented. Still less could we do this when more than two variables were involved. Our analysis consists essentially of restating the information contained in the coefficients of the original equation in another more readily comprehended form.

There are basically only two types of surface which this second degree equation can represent. The contours of these two basic surfaces are shown in Figures 10a and 10b and we shall later show that the surfaces of 10c and 10d may be regarded as special forms of these. For the surfaces of 10a and 10b, if we take the center of the system (S) as origin and the principal axes X_1 and X_2 of the system as axes of a new coordinate system, it will be seen by inspection that their equations in the new coordinate system will reduce to the form

$$Y - Y_S = B_{11}X_1^2 + B_{22}X_2^2 \tag{4}$$

where Y_S is the predicted response at S. By this transformation we have reduced the equation to one with only quadratic effects B_{11} and B_{22} (measured along the axes X_1 and X_2).

(i) If B_{11} and B_{22} are of the same sign we have a contour diagram like Figure 10a with elliptical contours. In this figure B_{11} and B_{22} are both negative and the center of the system S is a maximum. If the coefficients are both positive, the surface is like a trough instead of a mound and S is a minimum.

(ii) If B_{11} and B_{22} are of opposite signs we have a saddle point (sometimes called a col or minimax) illustrated in Figure 10b.

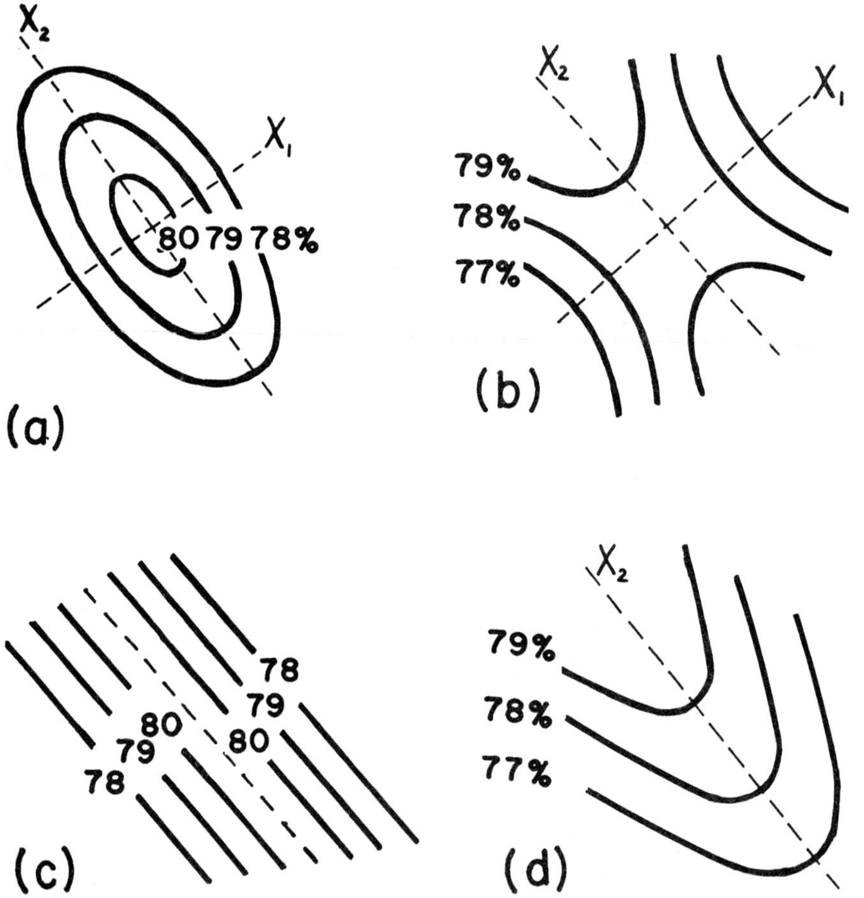

FIGURE 10. FUNDAMENTAL AND LIMITING SURFACES GENERATED BY A SECOND DEGREE EQUATION IN TWO DIMENSIONS

In either case if one of the coefficients is small in magnitude compared with the other then the surface is attenuated along the axis corresponding to the small coefficient. In Figure 10a, B_{22} is smaller than B_{11} .

(iii) If B_{22} were zero we would have the stationary ridge surface shown in Figure 10c (which might be regarded as a surface like that in Figure 10a or 10b infinitely attenuated along the X_2 axis).

(iv) If B_{22} were zero and the center of the system were at infinity we would have the rising ridge surface shown in Figure 10d in which the contours of the ridge were parabolas.

Denoting the coefficients (B_{11} , B_{22}) in the canonical equation by $-$, 0, or $+$ the systems 10a, b, c, and *d*, are typified by ($--$), ($-+$), (-0) and (-0, centre at infinity).

We see that the fitted surfaces shown in Figures 10a, 10c, and 10d could provide valuable approximations for the systems discussed in §1 and illustrated in Figures 3, 4, and 5. Figures 10c and 10d represent essentially limiting cases, for example Figure 10c represents the momentary situation between Figures 10a and 10b when the sign of B_{22} changes from negative to positive. These limiting cases would seldom if ever occur *exactly* in practice. When the underlying surface is like that in Figure 4 or in Figure 5 we obtain an attenuated form of one of the basic surfaces of Figures 10a and 10b which *approaches* one of the limiting cases.

In general if one coefficient is small compared with the other in the canonical equation a ridge of some kind is indicated. If the center is near the center of the design we have the central part of a system which is an attenuated form of 10a and 10b. In either case the system will approximate to the stationary ridge in Figure 10c.

If one coefficient is small compared with the other but the center is remote from the design then a sloping ridge of the kind illustrated in type 10d is usually indicated (although it could happen that the slope of the ridge was so slight that the situation was most closely approximated by 10c). We cannot, of course, draw any conclusions about the nature of the surface at the remote point corresponding to the center. We can however use the calculated center as a construction point which, together with the information concerning the directions of the axes will enable us to determine the approximate nature of the *local* surface in the region where it has actually been fitted. Suppose B_{22} is the small coefficient then, since we have already liquidated all large first order effects, the axis, X_2 corresponding to the coefficient B_{22} would normally be found to pass close to the design. We therefore take a new origin S' on X_2 close to the center of the design. Referred to this new origin the coordinates are X_1 and $X_2' = X_2 - a$. Substituting the latter relation in (4) we obtain the equation in the form

$$Y - Y_S' = B_{11}X_1^2 + B_{22}X_2'^2 + B_2'X_2' \tag{5}$$

where Y_S' is the value of Y at the new origin;

$$Y_S' = Y_S + a^2B_{22}$$

and

$$B_2' = 2aB_{22}$$

is the slope of the ridge X_2 at S'.

If B_2' is very small therefore, we would have an almost stationary ridge like Figure 10c. If (as is usually the case when S is remote) B_2' is not small we have a rising ridge, the gain in response being about equal to B_2' for unit movement up the ridge. We see that in this case we are again approximating to the surface by an attenuated form of either of the surfaces in Figures 10a or 10b, but the part of the surface used is not at the center but at some point along the axis of attenuation.

In general our analysis of k dimensional second degree fitted surfaces will follow the same lines.

From the $\frac{1}{2}(k + 1)(k + 2)$ coefficients of the original equation we calculate

(i) The k coordinates of the new center x_{1S} , x_{2S} , $\cdots$, x_{kS} and the value Y_S of the response at this point,

(ii) the canonical form of the equation

$$Y - Y_S = B_{11}X_1^2 + B_{22}X_2^2 + , \cdots , + B_{kk}X_k^2$$

which contains only k coefficients.

(iii) The k equations of the new coordinates (X's) in terms of the old coordinates (x's).

Figure 11 shows some of the possible three dimensional surfaces generated by second degree equations. Thus in figure 11a the ellipsoid illustrated is intended to represent one of a series fitting one inside the other. The centre thus corresponds to a point maximum if the response is decreasing on moving away from the centre, or a point minimum in the contrary case. Denoting as before by $-$, 0 or $+$ the values of the coefficients (B_{11} , B_{22} , B_{33}) in the canonical equation the system 11a is typified ($- - -$) for a minimum or ($+ + +$) for a maximum. We need consider only one of the alternative possibilities for this and the other examples. We then have 11a ($- - -$), 11b ($- - 0$), 11c ($- - +$), 11d (-00), 11e ($-0+$), 11f (-00 centre at infinity), 11g ($- - 0$, centre at infinity).

SOME EXAMPLES OF SURFACES MET IN PRACTICE

Rather more than half the examples so far studied have shown marked ridge systems of one form or another whilst most of the remaining examples have shown factor dependence of a less severe kind. The

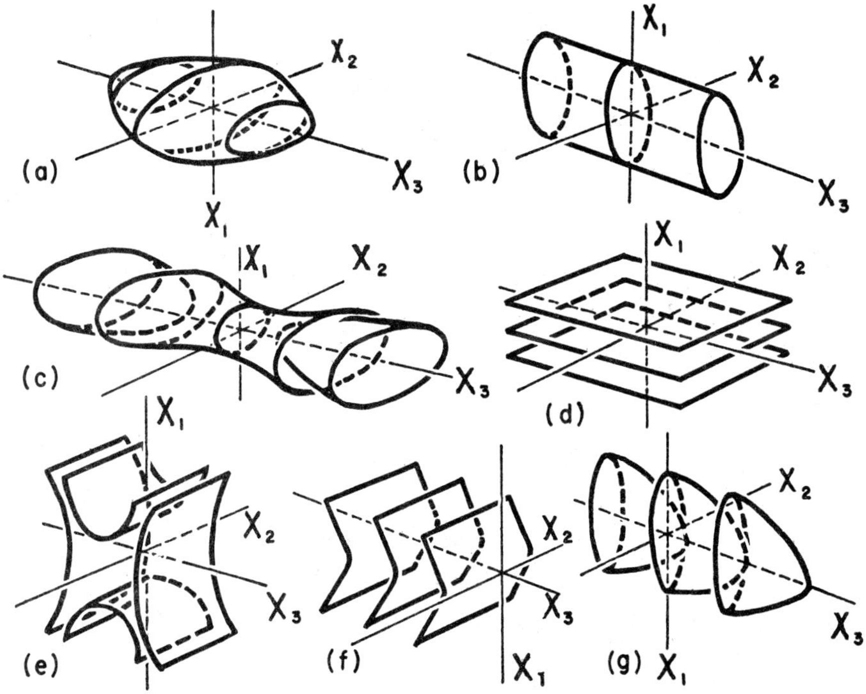

FIGURE 11. CONTOURS OF SOME POSSIBLE SURFACES GENERATED BY A SECOND DEGREE EQUATION IN THREE DIMENSIONS

following three examples of ridge systems were selected because of the interesting surfaces they reveal.

8. *An approximate stationary plane ridge in three variables.*

The detailed calculations for this example have been given in [2].* It concerns a reaction of the type

$$A + B \rightarrow C \qquad \text{followed by} \qquad A + C \rightarrow D$$

in which two reactants A and B formed a mixture of C and D. The object was to obtain the maximum for C subject to the condition that the yield of D should not exceed 20% (more than this amount would cause difficulty in purification). The quantity of B used was kept constant throughout, the factors varied being temperature (T), the concentration of $A(c)$, time of the reaction (t). Preliminary experimentation had led to the levels $T = 167°\text{C}$, $c = 27.5\%$, $t = 6.5$ hours. The levels of the factors used in the experiments to be described are

*The "natural units" given as an example in the book differ slightly from those given here.

listed below. The initial design consisted of a 2^3 factorial. It is convenient to regard the levels of x_1, x_2 and x_3 used in this design as ± 1 from which the relationships between standardized variables x_1, x_2, x_3, and natural variables T, c and t follow.

Factor levels in units of the design:		−2	−1	0	1	2
Factor levels in Natural Units.	T—Temperature (°C)	157	162	167	172	177
	c—Concentration of A (%)	22.5	25	27.5	30	32.5
	t—Time of Reaction (hours)	3.5	5	6.5	8	9.5

The standardized variables x_1, x_2, x_3 (that is the factor levels scaled in units of the factorial design) may be expressed in terms of the natural variables as follows;

$$x_1 = (T - 167)/5 \qquad x_2 = (c - 27.5)/2.5 \qquad x_3 = (t - 6.5)/1.5 \tag{6}$$

The first eight trials listed in Table 2 comprise the factorial design On the assumption that a second degree equation provided an adequate model, unbiased estimates of the linear and interaction coefficients were obtained as follows;

$$b_1 = 1.76 \qquad b_{12} = -3.09$$

$$b_2 = 1.19 \qquad b_{13} = -2.19$$

$$b_3 = -0.01 \qquad b_{23} = -1.21$$

Prior information suggested that the experimental error standard deviation was about 1% from which the standard errors for the b's would be about 0.4. The three first order effects were small compared with the two factor interactions whence it was concluded that a near-stationary region had been reached and seven further experiments (points 9–15 in Table 2) were added to complete a second order composite design. Estimates of all the coefficients in the second order equation could now be readily calculated. The fitted second degree equation thus obtained was

$$Y = 57.71 + 1.94x_1 + 0.91x_2 + 1.07x_3 - 1.54x_1^2 - 0.26x_2^2 - 0.68x_3^2 - 3.09x_1x_2 - 2.19x_1x_3 - 1.21x_2x_3 \tag{7}$$

The standard errors for linear and quadratic effects were about 0.3 and those for interaction effects about 0.4.

An estimate of the experimental error standard deviation calculated from the residual sum of squares and based on five degrees of

TABLE 2
LEVELS OF EXPERIMENTAL VARIABLES AND RESULTS OBTAINED

	Trial	x_1	x_2	x_3	Yield of C
2^3 Factorial	1	−1	−1	−1	45.9
	2	−1	−1	1	53.3
	3	−1	1	−1	57.5
	4	−1	1	1	58.8
	5	1	−1	−1	60.6
	6	1	−1	1	58.0
	7	1	1	−1	58.6
	8	1	1	1	52.6
Additional points to form a central composite design	9	0	0	0	56.9
	10	2	0	0	55.4
	11	−2	0	0	46.9
	12	0	2	0	57.5
	13	0	−2	0	55.0
	14	0	0	2	58.9
	15	0	0	−2	50.3
Confirmatory points.	16	2	−3	0	59.4
	17	2	−3	0	61.5
	18	−1.4	2.6	0.7	59.5
	19	−1.4	2.6	0.7	58.5

freedom was 1.5. This suggested that the fit of the equation was reasonably good.

The fitted surface was now reduced to canonical form.

(i) The coordinates x_{1S}, x_{2S}, x_{3S} of the center point S of the system and the predicted yield Y_S at this point were respectively

$$x_{1S} = 0.061, \quad x_{2S} = 0.215, \quad x_{3S} = 0.499, \quad \text{and} \quad Y_S = 58.14 \tag{8}$$

(ii) The canonical form of the fitted second equation (6) was

$$Y - 58.14 = -3.19X_1^2 - 0.07X_2^2 + 0.78X_3^2 \tag{9}$$

(iii) The new coordinates (X_1, X_2, X_3) for any point are given in terms of the old coordinates (x_1, x_2, x_3) by equations which may be written

	$(x_1 - 0.061)$	$(x_2 - 0.215)$	$(x_3 - 0.499)$	
X_1	0.7511	0.4884	0.4443	
X_2	0.3066	0.3383	−0.8897	(10)
X_3	0.5848	−0.8044	−0.1044	

The entries in the rows are the coefficients in the equation which express the X's in terms of the x's. Thus the first such equation is

$$X_1 = 0.7511(x_1 - 0.061) + 0.4884(x_2 - 0.215) + 0.4443(x_3 - 0.499) \qquad (11)$$

Because the transformation is orthogonal the entries in the *columns* of the same table are the coefficients in the equations which express the x's in terms of the X's. Thus x_1 is expressed in terms of X_1, X_2, and X_3 by the equation

$$(x_1 - 0.061) = 0.7511X_1 + 0.3066X_2 + 0.5848X_3 \qquad (12)$$

whose coefficients are taken from the first *column* of the table.

Inspection of the canonical form (9) of the equation shows that the first term is the predominant one. Any movement away from the center point S in the direction of the X_1 axis will lead to a rapid drop in yield, but changes in both X_2 and X_3 can be made with considerably smaller effects on yield. We note however that the coefficient of X_3^2 is positive so that there is a suggestion that further gains might be made by moving away from the center point S in either direction along this axis. Additional experiments (16) to (19) in these directions were conducted therefore and the new results added in to the solution the newly calculated canonical equation was then

$$Y - 59.23 = -3.51X_1^2 - 0.25X_2^2 + 0.24X_3^2$$

Bearing in mind that the standard errors of the coefficient in this equation are of roughly the same magnitude of those of the quadratic effects in the original equation (i.e. about 0.4) we see that the surface approximated to that in Figure 11d in which there is a plane of maximum passing through the axes X_2 and X_3. Subsequent experiments at various points on the plane confirmed that the surface was of this type. The plane stationary ridge system to which the surface approximates is shown in Figure 12. In this example we see that we have two dimensional stationary ridges of the type pictured in Figure 4 in all three pairs of variables at once. The yield is at its maximum value of nearly 60% on the second of the three planes, and on each side of it are shown planes on which the yield is about 50%.

The location of the maximum plane was little changed by the addition of the new observations. Points on the plane will be all those for which X_1 is zero so that the equation of the plane in terms of x_1, x_2, and x_3 will be obtained by putting equation (11) equal to zero. Equations (6) may now be used to get the equation of the plane in terms of the original variables T, c, and t. We find

$$0.1502T + 0.1954c + 0.2962t = 32.76 \qquad (13)$$

which defines the set of alternative conditions giving approximately the same maximum yield.

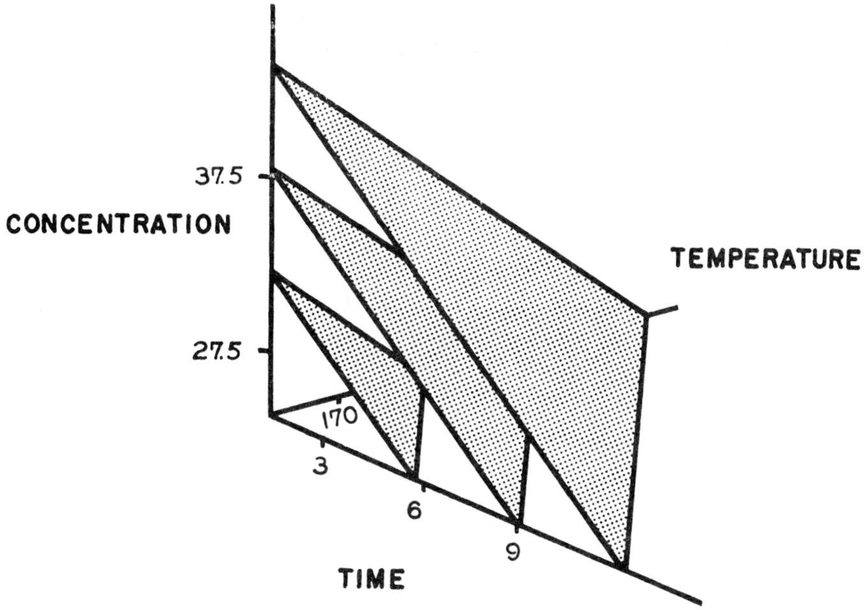

FIGURE 12. CONTOURS OF PLANE STATIONARY RIDGE SYSTEM TO WHICH SURFACE APPROXIMATES

The situation is seen to be that, over the range considered, the three variables temperature, concentration and time are approximately "compensating". For example if we had a process working at one point on the ridge and we wished to change one or possibly two of the factor levels so that the maximum yield was approximately maintained this could be done by changing the remaining variable, or variables so that equation (13) was satisfied. Thus if we wished to reduce the reaction time, the equation would indicate the higher values of temperature and/or concentration necessary to compensate the change.

In Figure 13 these approximate alternatives are shown on a tri-coordinate diagram. This is essentially the maximum plane of Figure 12 with the intersections of the coordinate planes drawn in.

The detection of ridge systems of this sort is extremely important in practice since when such systems exist they provide the experimenter with a number of alternative processes he can use, some of which may be very much cheaper or more convenient than others. Furthermore

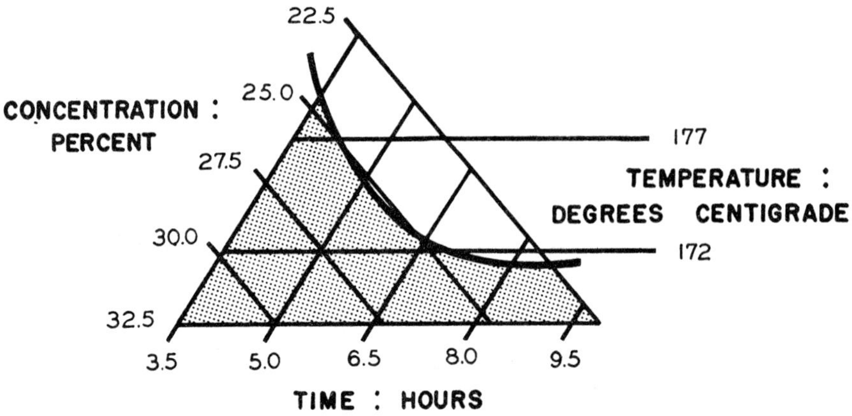

FIGURE 13. ALTERNATIVE PROCESSES GIVING MAXIMUM YIELD WITH SHADED REGION IN WHICH BI-PRODUCT GREATER THAN 20%

the existence of a ridge allows points to be chosen so that auxiliary responses, such as purity, are brought to their most satisfactory levels.

In the present investigation a second surface was fitted for the amount of by-product D which it was desired to maintain below the level of 20%. The 20% contour line is drawn on the figure, the shaded region is where amounts of by-product greater than 20% are produced. By choosing a point in the unshaded region the experimenter may obtain both maximum yield of C and a satisfactorily low yield of D.

9. *A two dimensional ridge system in five variables.*

Thus far in the paper we have discussed situations in which there were only one, two or three factors and the surface could be pictured geometrically. The technique of analysis of factor dependence by canonical analysis of the fitted equation becomes even more valuable when there are more than three factors for it enables the situation to be readily appreciated even though it cannot so easily be perceived geometrically. The following example is one in which five factors were studied and where a preliminary application of the steepest ascent procedure had (as it turned out) brought the experimenter close to a near-stationary region.

The process studied occurred in two stages and the factors considered were the temperatures and times of reaction at the two stages and the concentration of one of the reactants at the first stage. Evidence was available which suggested that the times of reaction were best varied on a logarithmic scale and it was known that the second reaction time needed to be varied over a wide range.

The table below shows the levels used in the experiments to be described.

Factor levels in Units of the Design				−3	−1	1	3
	Stage						
Factor levels in natural units	1	T_1—Temperature	(°C)		105	120	135
		t_1—Time of reaction	(hours)	1.25	2.5	5	
		c—Concentration	(%)	40	60	80	
	2	T_2—Temperature	(°C)		25	40	55
		t_2—Time of reaction	(hours)		0.25	1.25	6.25

From the above table it will be seen that the standardized variables to be used in the fitted equation expressed in terms of the natural variables were

$$x_1 = (T_1 - 112.5)/7.5, \qquad x_2 = \{2(\log t_1 - \log 5)/(\log 2)\} + 1,$$
$$x_3 = (c - 70)/10, \qquad x_4 = (T_2 - 32.5)/7.5, \tag{14}$$
$$x_5 = \{2(\log t_2 - \log 1.25)/(\log 5)\} + 1$$

A $\frac{1}{2}$ replicate of a 2^5 factorial experiment was first performed using the levels corresponding to -1 and $+1$ in the standardized variables. The defining contrast characterizing the fractional design was the single five factor interaction. The experiments in this design are the first 16 in Table 3.

On the assumption that the surface may be adequately represented by an equation of second degree, the five first order terms $b_1, b_2, \cdots, b_5$ and the ten two factor interactions terms $b_{12}, b_{13}, \cdots, b_{45}$ may now be estimated without bias.

The values obtained were as follows,

$b_1 = 3.194$	$b_{12} = -1.869$	$b_{24} = -0.169$
$b_2 = 1.456$	$b_{13} = 2.081$	$b_{25} = -0.769$
$b_3 = 0.981$	$b_{14} = -0.506$	$b_{34} = -4.019$
$b_4 = 3.419$	$b_{15} = -0.431$	$b_{35} = -0.744$
$b_5 = 1.419$	$b_{23} = 0.743$	$b_{45} = 0.144$

It will be seen that, compared with some of the two-factor interactions, the first order effects are neither so large as to suggest that we are so far removed from a near-stationary region that we must move to a new locale and start afresh there to build up the second order design,

TABLE 3

LEVEL OF EXPERIMENTAL VARIABLES AND RESULTS OBTAINED

	Trial	x_1	x_2	x_3	x_4	x_5	Yield
	1	−1	−1	−1	−1	1	49.8
	2	1	−1	−1	−1	−1	51.2
	3	−1	1	−1	−1	−1	50.4
	4	1	1	−1	−1	1	52.4
	5	−1	−1	1	−1	−1	49.2
	6	1	−1	1	−1	1	67.1
½ Replicate	7	−1	1	1	−1	1	59.6
of 2^5	8	1	1	1	−1	−1	67.9
factorial	9	−1	−1	−1	1	−1	59.3
	10	1	−1	−1	1	1	70.4
	11	−1	1	−1	1	1	69.6
	12	1	1	−1	1	−1	64.0
	13	−1	−1	1	1	1	53.1
	14	1	−1	1	1	−1	63.2
	15	−1	1	1	1	−1	58.4
	16	1	1	1	1	1	64.3
Additional	17	3	−1	−1	1	1	63.0
points to form	18	1	−3	−1	1	1	63.8
non-central	19	1	−1	−3	1	1	53.5
composite	20	1	−1	−1	3	1	66.8
design	21	1	−1	−1	1	3	67.4

Confirmatory points

Canonical Variables

X_1	X_2	X_3	X_4	X_5	Trial	x_1	x_2	x_3	x_4	x_5	Yield	(Predicted) Yield
0	0	0	0	0	22	1.23	−0.56	−0.03	0.69	0.70	72.3	(71.4)
2	0	0	0	0	23	0.77	−0.82	1.48	1.88	0.77	57.1	(52.6)
−2	0	0	0	0	24	1.69	−0.30	−1.55	−0.50	0.62	53.4	(52.6)
0	2	0	0	0	25	2.53	0.64	−0.10	1.51	1.12	62.3	(61.1)
0	−2	0	0	0	26	−0.08	−1.75	0.04	−0.13	0.27	61.3	(61.1)
0	0	2	0	0	27	0.78	−0.06	0.47	−0.12	2.32	64.8	(66.3)
0	0	−2	0	0	28	1.68	−1.06	−0.54	1.50	−0.93	63.4	(66.3)
0	0	0	2	0	29	2.08	−2.05	−0.32	1.00	1.63	72.5	(69.5)
0	0	0	−2	0	30	0.38	0.93	0.25	0.38	−0.24	72.0	(69.5)
0	0	0	0	2	31	0.15	−0.38	−1.20	1.76	1.24	70.4	(72.2)
0	0	0	0	−2	32	2.30	−0.74	1.13	−0.38	0.15	71.8	(72.2)

nor are they so small that we should expect the region covered by the design to be perfectly centered in the near stationary region which we wish to study further. In these circumstances it seemed appropriate to add the further points necessary to allow estimation of all effects up to second degree in a "non-central" composite design of the type discussed in §6 and in [1]. The idea is to add points on that "corner" of the factorial design which is in the direction of increasing yield. Although such a design is not so efficient as the centered composite design this type of augmentation will have the effect of shifting the center of gravity of the design closer to the center of the near-stationary region which we wish to study and thus reduce errors due to "lack of fit" of the postulated form of the equation.

In deciding where best to add the extra points it was noted that the highest single yield (70.4%) was recorded at the point (1, −1, −1, 1, 1) and this would therefore seem the best point at which to augment the design. If we had ignored the second order effects we would have been led to use the point (1, 1, 1, 1, 1) since all the first order effects are positive. We could not judge accurately the modifying effect of the two-factor interaction terms unless quadratic terms were available also; however in view of the (by no means negligible) size of the interaction terms b_{12}, b_{13}, and b_{34} and the nature of their signs the point at which the highest yield was obtained appeared a reasonable choice and was adopted. Five extra points having coordinates (3, −1, −1, 1, 1); (1, −3, −1, 1, 1); (1, −1, −3, 1, 1); (1, −1, −1, 3, 1) and (1, −1, −1, 1, 3) were therefore added; these are numbered 17-21 in Table 3. It will be noted that by using this design we were estimating 21 constants in 21 experiments. This would not normally be regarded as a good practice since no comparisons remained to test goodness of fit. However 11 confirmatory points were later added to the design so that some check on the fit was possible at a later stage. The least squares estimates of the linear and interaction effects are unaffected by the new points for this type of arrangement. Estimates of b_0, the predicted yield at the center of the design, and of the quadratic effects b_{11}, b_{22}, b_{33}, b_{44} and b_{55} could now be calculated as follows;

$$b_0 = 68.173, \quad b_{11} = -1.454, \quad b_{22} = -1.348,$$

$$b_{33} = -2.723, \quad b_{44} = -2.261, \quad b_{55} = -1.036.$$

We now have estimates of all the 21 constants in the second degree equation and we can carry out the canonical analysis as before. The coordinates of the center point S of the system and the predicted yield at this point are respectively

$$x_{1S} = 1.2263, \quad x_{2S} = -0.5611, \quad x_{3S} = -0.0334,$$
$$x_{4S} = 0.6916, \quad x_{5S} = 0.6977, \quad Y_S = 71.38 \tag{15}$$

whilst the canonical form of the equation becomes

$$Y - 71.38 = -4.70X_1^2 - 2.58X_2^2 - 1.25X_3^2 - 0.48X_4^2 + 0.20X_5^2 \tag{16}$$

The transformation which gives the new X coordinates in terms of the old x coordinate (or by transposition, the old coordinates in terms of the new) may be written

	$(x_1-1.2263)$	$(x_2+0.5611)$	$(x_3+0.0334)$	$(x_4-0.6916)$	$(x_5-0.6977)$	
X_1	−0.2300	−0.1289	0.7581	0.5953	0.0382	
X_2	0.6518	0.5994	−0.0343	0.4116	0.2128	
X_3	−0.2245	0.2489	0.2520	−0.4058	0.8120	(17)
X_4	0.4255	−0.7445	−0.1437	0.1561	0.4686	
X_5	−0.5392	0.0883	−0.5831	0.5359	0.2726	

We see from equation (16) that once again two of the canonical factors X_4 and X_5 are playing a comparatively minor role in describing the functions and are of a similar order of magnitude to their standard errors. At least locally most of the change can be described by the canonical variables X_1, X_2, X_3.

Further experiments were now performed to determine whether these general findings as to the nature of the surface could be confirmed. To do this eleven further experiments were carried out, one at S, the predicted center of the system, and the others in pairs about S along the axes X_1, X_2, X_3, X_4 and X_5. Each of these points was at a distance two units from the center. Thus in terms of *new* variables X_1, X_2, X_3, X_4, X_5 the coordinates of these points were (0, 0, 0, 0, 0); (±2, 0, 0, 0, 0); (0, ±2, 0, 0, 0), ···, (0, 0, 0, 0, ±2).

The levels of the original variables corresponding to these points were calculated from the equation expressing the x's in terms of the X's, and are the coordinates of the experimental points 22 to 32 in Table 3. This table also shows the yield predicted from the fitted second degree equation (in brackets) and the values actually found. Although the predicted yields are not always in very close agreement with those found, it is apparent that the additional experiments confirm the overall features of the surface in a rather striking manner.

It will be seen that the experimental points on either side of S along the X_1 axis are associated with the largest reductions in yield, approxi-

TABLE 4

SETS OF NEARLY ALTERNATIVE CONDITIONS ON PLANE OF X_4 AND X_5

T_1 °C	120	124	128	132	136	
t_1 hrs	1.9	1.8	1.7	1.7	1.6	
c %	55	61	66	72	78	2
T_2 °C	48	44	40	36	32	
t_2 hrs	3.2	2.6	2.1	1.7	1.3	
T_1 °C	117	121	125	129	133	
t_1 hrs	2.4	2.3	2.3	2.2	2.1	
c %	57	62	68	74	80	1
T_2 °C	47	43	39	35	31	
t_2 hrs	2.2	1.8	1.4	1.1	0.9	
T_1 °C	114	118	121	126	130	
t_1 hrs	3.1	3.0	2.9	2.8	2.7	
c %	58	64	70	75	81	0 X_4
T_2 °C	46	42	38	34	30	
t_2 hrs	1.5	1.2	1.0	0.8	0.6	
T_1 °C	110	114	119	123	127	
t_1 hrs	4.0	3.9	3.8	3.7	3.5	
c %	59	65	71	77	83	−1
T_2 °C	45	41	37	32	28	
t_2 hrs	1.0	0.8	0.7	0.5	0.4	
T_1 °C	107	111	115	119	123	
t_1 hrs	5.2	5.0	4.9	4.7	4.6	
c %	61	67	73	78	84	−2
T_2 °C	43	39	35	31	27	
t_2 hrs	0.7	0.6	0.5	0.4	0.3	
	−2	−1	0	1	2	
			X_5			

mately confirming the large coefficient of X_1^2 in the canonical equation. Somewhat smaller changes are associated with movement away from S along the X_2 and X_3 axes and these again are of the order expected. Finally movement along the X_4 and X_5 axes is associated with only minor changes in yield thus confirming that there is an approximately stationary plane and that we have wide local choice of almost equally satisfactory conditions from which to choose. After including the information from these 11 extra points the canonical form of the second degree equation becomes $Y - 73.38 = -4.46X_1^2 - 2.64X_2^2 - 1.80X_3^2 - 0.39X_4^2$

$- 0.004X_5^2$ which agrees closely with that found before. The transforming equations (17) are affected very little. The estimated experimental error variance based on the residual sum of squares having 11 degrees of freedom is 1.51; using this value it appears that the standard errors of the estimated coefficients are all less than 0.4. The sets of conditions which yield approximately the same results are in the plane defined by the equations $X_1 = 0, X_2 = 0, X_3 = 0$. Using (17) these equations are readily transformed to equations in the five unknowns x_1, x_2, x_3, x_4, and x_5 and using (14) to equations in the five natural variables. We have two "degrees of freedom" in the choice of conditions in the sense that within the region considered if we choose values for any two of the variables we obtain three equations in three unknowns which can be solved to give appropriate levels for the remaining three variables. To demonstrate the practical implications of these findings to the experimenter, a table of approximate alternatives, which covered the range of interest, was calculated as is shown in Table 4.

10. *A falling ridge in cost.*

This third example concerns the improvement of a process already in operation in which two solids A and B were fused at high temperature to give a third substance C. The object was to reduce, if possible the manufacturing cost of C. The amount of B was kept constant throughout the experiment and the three factors were studied at the levels indicated below.

Factor levels in units of the factorial design:		−3	−1	1
Factor levels in Natural Units	Fusion Temperature °C (T)	240	245	250
	Fusion Time (hours) (t)	32	24	16
	Molar Ratio of A to B (M)	3.5	4.5	5.5

The standardized variables x_1, x_2, x_3 are thus expressed in terms of the natural variables as follows,

$$x_1 = (T - 247.5)/2.5, \quad x_2 = (t - 20)/(-4), \quad x_3 = (M - 5.0)/0.5 \quad (18)$$

In the trials which are now described the yield corresponding to each set of factor combinations was determined experimentally and the expenditure involve in running the process with the factors at these levels was calculated. (High temperature and longer times both involved greater fuel consumption for example, and greater molar ratios involved the use of larger amounts of the expensive material A). Using the experimentally determined yield and the calculated expenditure it was now possible to calculate the cost of manufacturing one pound of product at each set of conditions tried. For convenience in calculation an arbitrary quantity has been subtracted from the cost per pound of product assessed in tenths of a penny. This is the response recorded below and we will refer to it subsequently as the "cost". The requirement was to find a minimum on the cost response surface.

TABLE 5

LEVELS OF EXPERIMENTAL VARIABLES AND RESULTS OBTAINED

	Trial	x_1	x_2	x_3	Cost
	1	1	1	1	37
	2	1	1	−1	70
	3	1	−1	1	70
2^2 Factorial	4	1	−1	−1	39
	5	−1	1	1	64
	6	−1	1	−1	74
	7	−1	−1	1	48
	8	−1	−1	−1	18
Additional points to form composite design	9	−3	−1	−1	90
	10	−1	−3	−1	52
	11	−1	−1	−3	16
Confirmatory points	12	0	0	0	38
	13	−3	−3	−3	48
	14	−1	−1	0	33

The experiments carried out are listed in Table 5. The first eight comprised a 2^3 factorial design with the standardized variables at the levels -1 and $+1$. Assuming that some second degree equation is adequate to express the cost function unbiased estimates of the linear terms and two-factor interaction terms may be obtained from the eight results as follows,

$$b_1 = 1.50, \qquad b_2 = 8.75, \qquad b_3 = 2.25,$$

$$b_{12} = -9.25, \qquad b_{13} = -2.75, \qquad b_{23} = -13.00,$$

The experimental error standard deviation was expected to be in the neighborhood of eight units so that an estimate of the standard errors of these coefficients would be about three units.

We see that this is again an example where the first order effects are neither dominant nor negligible compared with two factor interactions. The design was therefore extended to form a non-central composite design. The factor-combination associated with the smallest cost is at the point $(-1, -1, -1)$ and bearing in mind that a minimum is being sought, augmenting the design at this apex would seem to be supported by the signs of the estimated coefficients. Points were added therefore at $(-3, -1, -1)$, $(-1, -3, -1)$ and $(-1, -1, -3)$.

All first and second order coefficients in the second degree equation could now be estimated. Canonical analysis of the fitted equation indicated a falling ridge passing near the center of the factorial design in the direction of negative levels of the variables. Two further confirming experiments were conducted therefore, one at the center of the factorial design $(0, 0, 0)$ and the other in which the levels of all three variables were reduced by three units $(-3, -3, -3)$. A further experiment (performed in error) was at the point $(-1, -1, 0)$.

The information from these three additional experiments was now included using Plackett's technique and the coefficients in the equation fitted to all 14 points were now as follows,

$$b_0 = 28.19$$

$$\begin{array}{lll} b_1 = 1.53 & b_{12} = -7.28 & b_{11} = 11.23 \\ b_2 = 8.78 & b_{13} = -0.81 & b_{22} = 10.85 \\ b_3 = 2.31 & b_{23} = -11.06 & b_{33} = 3.11 \end{array}$$

The coordinates of the fitted equation and the predicted cost at the center were as follows,

$$x_{1S} = 3.87 \qquad x_{2S} = 10.13 \qquad x_{3S} = 18.12 \qquad y_S = 96.61 \tag{19}$$

The canonical form of the equation is

$$Y - 96.61 = -0.16X_1^2 + 9.49X_2^2 + 15.86X_3^2 \tag{20}$$

and the transforming equations

	$(x_1 - 3.87)$	$(x_2 - 10.13)$	$(x_3 - 18.12)$	
X_1	0.1871	0.4897	0.8516	
X_2	0.7995	0.4290	-0.4205	(21)
X_3	0.5695	-0.7599	0.3133	

We notice that although this example is similar to the two previous ones in that the canonical equation contains a coefficient small compared with the others, it differs from these in one very important respect. Whereas in the other examples the center of the system has been in the immediate vicinity of the center of the design here the center of the fitted system is remote from the design. (The design extends from -3 to 1 in each of the variables x_1 , x_2 , x_3 , but the center of the fitted contour surface is at the point (3.87, 10.13, 18.12).

We have then the situation in which,

i) no first order effects, which are very large compared with second order effects, occur,

ii) one of the coefficients in the canonical equation is small compared with the remainder,

iii) the center of the fitted system is remote from the center of the design

This, as we have seen, may indicate a sloping ridge.

The small coefficient -0.16* is associated with the axis X_1 , which (since b_1 , b_2 and b_3 are not large compared with b_{11} , b_{12} , etc.) is expected to pass close to the design to form the 'axis' of the ridge system. The point (which we will call S') on the X_1 axis closest to the center of the factorial design is at $x_1 = -0.08$, $x_2 = -0.21$, $x_3 = 0.14$, confirming that the axis does indeed pass close to the center of the design. These coordinates of S' are found as follows. Using the equations (21) it is readily found that the X_1 coordinate of the point $x_1 = 0$, $x_2 = 0$, $x_3 = 0$ is -21.116. Whence it follows that S' is at the point $X_1 = -21.116$, $X_2 = 0$, $X_3 = 0$. Its position in terms of the x coordinates given above may now be calculated using the transforming equations (21) once more. The predicted cost at S' is 27.32.

Taking S' as the new center (by writing $X_1 = X_1' - 21.116$ in equation (20)) we obtain the fitted equation referred to the local origin S' in the form

$$Y - 27.32 = +6.56X_1' - 0.16X_1'^2 + 9.49X_2^2 + 15.86X_3^2 \qquad (22)$$

and the surface approximates Figure (11g) but with the X_1 axis taking the place of the X_3 axis in that diagram.

The coefficient 6.56 measures the slope at S' down the ridge. Thus locally each unit which we move down the ridge will be accompanied by a reduction of about six units in cost.

Any point on the 'ridge axis' X_1 can readily be found from equations (21) and (18). In particular the points at which costs of 10, 0, and -10 were predicted are as follows,

*To avoid large rounding errors the more exact value -0.1554 is used in the subsequent calculations.

	Temperature	Time	Molar Ratio	Predicted Cost
(i)	246.1	25.7	4.01	10
(ii)	245.5	28.2	3.46	0 2.6 (found)
(iii)	244.9	30.7	2.91	−10

In practice the conditions (ii) above which involved a saving of about 20 units of cost represented the limit to which the ridge could be followed. Beyond this point mechanical difficulties arose in running the process. It is of interest to note that the process studied had already been investigated thoroughly (using the one factor at a time method). This had led to the ridge at which point it had been assumed that no further improvement was possible. Experiments which made possible the approximate elucidation of the nature of the multifactor dependence were able to lead to further marked improvement.

SOME CONSIDERATIONS ARISING FROM THESE INVESTIGATIONS

11. *Demonstration of Results.*

It is essential to the ultimate success of experimental work that those concerned with it, in particular those who must make decisions as a result of it, and who may have no special knowledge of statistics or mathematics, should clearly appreciate the conclusions that have been reached. In a chemical application, months of effort will often have been expended in the collection of the data and many days in computing the results. This labor will be wholly or partially wasted unless a further effort is expended to demonstrate what has been discovered in such a way as to be readily comprehensible.

Particular ways of demonstrating conclusions have been illustrated in the examples, thus the alternative processes available in the example of §8 were shown in Figure 13 using tri-linear coordinates. It was also possible to show on the same diagram the region in which an acceptably small amount of the by-product D was obtained. Again, even though in the example §9 we were dealing with five variables and consequently the geometrical surface could only be fully expressed in a space of five dimensions, yet the implications of the two dimensional ridge system were clearly brought out by listing alternative processes over the essential region in Table 4.

To allow ready appreciation of the possibilities for process improvement which exist in a given situation the geometrical method is most valuable. This is particularly true when more than one response is to be studied. Contour representation allows the interrelationships between as many as three variables and one or more responses to be easily comprehended, and the actual construction of three-dimension contour models to represent such surfaces as those shown in Figures 7 and 12, is often worth while. These models may be constructed in various ways. One useful method is to outline the contour surface by colored wires. The experimental points may be marked by plastic counters on which the observed responses may be written, thus allowing the experimental arrangement and the fitted contours to be seen simultaneously. A number of suitably placed wire grids produces a framework on which to build the model and these do not obstruct the view of the contours and experimental points. A photograph of one such model showing the fitted surfaces before the addition of confirmatory points for the example of §8 is shown below. In order that the experimental points would show more clearly on the photograph the plastic counters have here been replaced by marbles.

When there are two responses to consider, sets of contours for both responses may be shown on the same model or alternatively two models may be constructed and the results assessed by viewing them side by

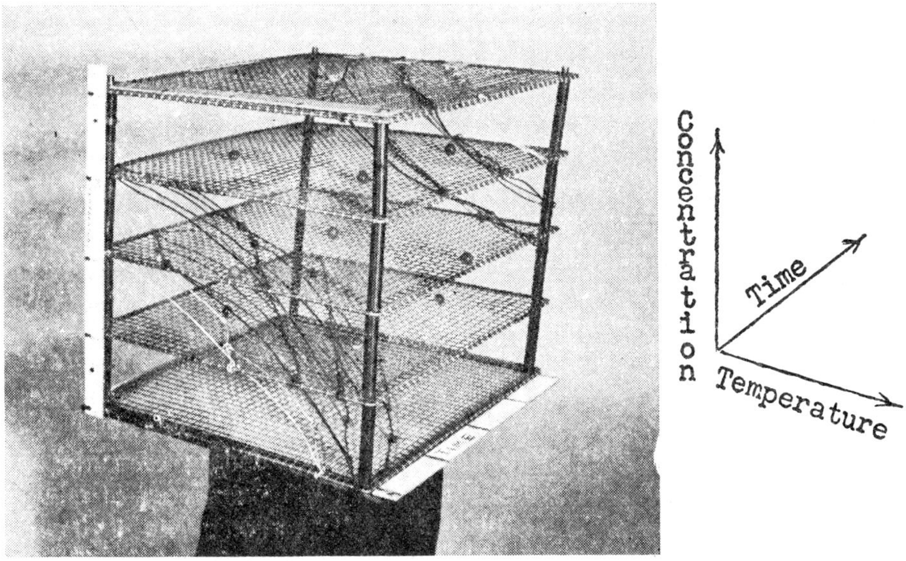

FIGURE 15. PHOTOGRAPH OF A THREE DIMENSIONAL MODEL SHOWING THE CONTOURS OF THE APPROXIMATE PLANE STATIONARY RIDGE SYSTEM CONSTRUCTED ON THE BASIS OF TRIALS RUN AT THE POINTS INDICATED BY THE MARBLES. A DESCRIPTION OF THIS PARTICULAR EXPERIMENT IS FOUND IN SECTION 8. THE TYPE OF EXPERIMENTAL DESIGN USED IS THAT ILLUSTRATED IN FIGURE 8.

side. With more than three variables, selected three dimensional sections or two dimensional sections of the k-dimensional contour surface may be constructed to illustrate important features of the total surface.

12. *The Analysis of Multifactor Dependence.*

It is well known of course that the existence of interactions between continuous variables implies factor dependence of some sort. In the past however interpretation of such dependence has not always been very satisfactory or helpful because,

i) the interactions were usually considered piece-meal and not as a whole,

ii) attempts were made to interpret interaction effects ignoring the influence of other terms of the same order (in particular two factor interactions estimated from two-level factorial experiments have been considered in the absence of the corresponding quadratic effects).

In an experiment (such as that in §9 above) in which there are five factors, there are no less than 10 two-factor interactions. Unless most of these were negligible it would be almost impossible by individual study to appreciate the joint effect of these constants.

To attempt to interpret two factor interactions without the corresponding quadratic effects in precisely analogous to considering covariances without the corresponding variances. If we were told that a certain covariance between two variables y_1 and y_2 was large (say equal to 1000) we would not know whether y_1 and y_2 were closely correlated or not. (For example if the variances were each equal to a million the dependence between y_1 and y_2 would be negligible, if they were each equal to 1001, y_1 and y_2 would be almost completely correlated.) In a similar way, knowledge of the interaction taken alone, without the corresponding quadratic effects will not enable us to appreciate the nature and importance of the factor dependence.

It is a deficiency of the three level factorial design so far as the present problem is concerned that the quadratic effects are estimated with only half the precision (twice the variance) of the interaction effects this mean that the second order derivative $\partial^2 y/(\partial x_1)^2$ at the centre of the design is estimated with only one eighth the precision of the mixed derivative $\partial^2 y/(\partial x_1 \partial x_2)$.

It is sometimes found in the analysis of variance of three level factorial designs that two-factor interactions are 'significant' whereas quadratic effects are not. This has led to a supposition that conditions frequently occur in which two-factor interactions are important but quadratic effects are unimportant, which apparently conflicts with the common sense view that for a smooth surface effects of the same order ought to be of equal importance. That this contradiction is apparent rather than real can be seen if we remember that the expected values of the mean squares in the analysis of variance are of the form

$$\sigma^2 + \frac{\beta^2}{V(b)/\sigma^2} \tag{23}$$

It will be noted that the second term in (23), which will cause the mean square to be inflated when real effects occur, is a function not only of the size of the real effect β but also of the variance $V(b)$ of the estimate of this quantity. Thus if real quadratic and interaction derivatives of equal magnitude occurred the inflation of the mean square for the interaction effect would be eight times as large on the average as the inflation of the mean squares for the quadratic effect.

It is of course extremely important that before attempting its interpretation we should ensure that the fitted equation is meaningful. This can be done by considering the magnitude of their standard errors of the coefficients and a convenient 'portmanteau test' is provided by the analysis of variance.

The whole sum of squares due to regression may be divided into components 'due to mean', 'due to all first order effects', 'due to all second order effects', and so on. Where there is some doubt as to whether a certain variable x_i has any influence at all, or any effect other than a linear one, mean squares for 'all effects involving the ith factor', or 'all second order effects involving the ith factor' can be computed.

There would however seem to be little merit and some danger in the practise sometimes adopted of testing coefficients individually and dropping those which were 'not significant' (that is could not be demonstrated to be different from zero). By so doing we would be replacing an unbiased estimate of smallest variance by an estimate (zero) which had neither of these qualities.

13. *Quantitative and Qualitative Factors*

The techniques discussed have been for the attainment of a maximum when the factors studied were *quantitative* like "temperature", "time", "concentration", not *qualitative* like "type of reactant" or "operator performing experiment".

The statistical designs chiefly used heretofore have been the complete and fractional factorials, the complete and incomplete block designs, latin squares and lattices. The analysis of results from experiments using these designs has usually been aimed only at estimating the effects, calculating confidence intervals, and testing significance by means of the analysis of variance technique. The formal application of such designs, and much of the analysis of the results, is identical whether the factors are qualitative or quantitative. It is perhaps this circumstance that has sometimes led to insufficient distinction being drawn between these two very different types of 'factors' and to a consequent idea that a method of experimentation ought to cope equally with both quantitative and qualitative factors.

When some of the factors are qualitative variables it is the writer's belief that there is no way of finding the absolute optimum other than carrying out separate investigations for each qualitative 'factor' combination unless some very specialized prior assumptions can be made about the qualitative factors.

If the experimental designs were so comprehensive that the whole experimental region for the quantitative variables were explored for each qualitative variable, this would amount to the same thing as carrying out a separate investigation for each variable. If less than the whole experimental region were included a design which involved qualitative factors with quantitative factors would seem only to provide a means for testing whether the response surfaces in the qualitative factors were similar or not. Suppose for example that the experimenter was attempting to find the optimum combination of three 'factors', two of which 'temperature' (T) and 'concentration' (c) were quantitative, and the remaining one 'type' of reactant used (A or B) was qualitative. We can imagine the temperature, concentration, yield-contour diagrams for reactants A and B in a particular temperature and concentration region being like those shown in Figure 14 so that, for example, a 2^3 factorial experiment performed for the three factors can be imagined to have its experimental points embedded in the contour diagrams in the manner illustrated.

i) If the mechanism of the reaction with A was different from that with B (as would usually happen if A and B were essentially different substances) then the two temperature-concentration yield surfaces would also be different (as are those illustrated). Large interactions between the temperature-concentration effects and reagent effects would now be found. The only conclusion the experimenter could reach would be that if both reactants were to be seriously considered the surface for each would have to be examined separately.

ii) If A and B were essentially the same substance but were perhaps from different batches of material it could well happen that the two response surfaces were identical apart perhaps from some differences in mean level. The interaction terms involving the type of reactant would then all be small and we would draw the conclusion that at least locally the yield surfaces were similar. This might lead to the tentative assumption that further experiments could be conducted with only

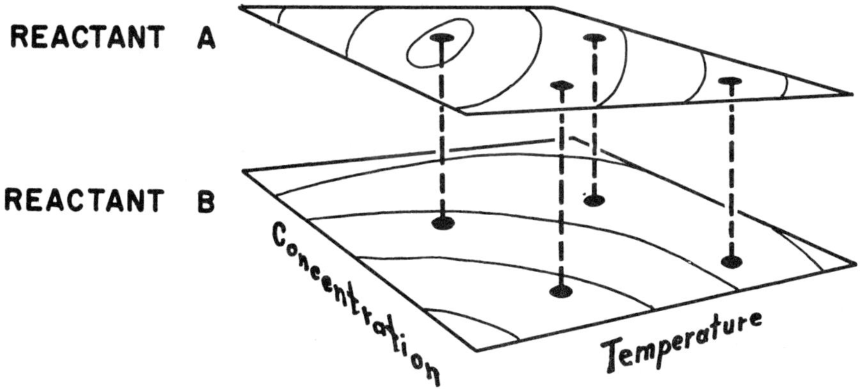

FIGURE 14. THREE FACTOR SYSTEM TWO QUANTITATIVE FACTORS ONE QUALITATIVE FACTOR

one reactant. It will be noted that the circumstances here are rather specialized.

iii) A further possibility might occasionally be worth considering. This is that the surface for reactant A is the same as that for reactant B but displaced. For example, if the design was such as to allow the fitting of a second degree temperature-concentration response surface at each level of the reactant then displacement would be indicated if the second order effects were nearly equal but the linear effects were different.

It will be noted in particular that in (i) if all the yields from trials using A were higher than those using B the experimenter should not conclude that in subsequent experiments only A need be considered for it might be that the experiments had been conducted near the optimum levels of temperature and concentration with A, but that a higher maximum was attainable using B in some other region.

In the writer's opinion, the type of experiment described above (limited as its purpose is to test the hypothesis that the yield surface at the two levels of the reactant are similar against the alternative hypothesis that they are different) would often not need to be performed. In particular, the experimenter would usually know in advance whether a qualitative factor was likely to behave like that in (i) or (ii).

I am indebted to Mr. J. S. Hunter and to Dr. R. J. Hader for assistance in the preparation of this paper.

REFERENCES

(1) Box, G. E. P., and Wilson, K. B. On the Experimental Attainment of Optimum Conditions. *J. R. Stat. Soc., B, 13:* 1, 1951.

(2) *The Design and Analysis of Industrial Experiments.* G. E. P. Box, L. R. Connor, W. R. Cousins, O. L. Davies, F. R. Himsworth and G. P. Sillitto, edited by O. L. Davies, Edinburgh: Oliver and Boyd, 1953.

(3) Thurstone, L. L. *Multiple Factor Analysis.* University of Chicago Press, Chicago. 1948.

(4) Plackett, R. L. Some Theorems on Least Squares. *Biometrika 36:* 458, 1949.

(5) Box, G. E. P. and Hunter, J. S. Technical Report No. 5 prepared under Ordnance Contract DA/36/034/ORD/1177.

(6) Booth, A. D. An Application of the Method of Steepest Descents to the Solution of Systems of Non-Linear Simultaneous Equations. *Quart. J. Appl. Math., 1:* 237, 1949.

(7) Koshal, R. S. Application of the Method of Maximum Likelihood to the Improvement of Curves Fitted by the Method of Moments. *J. R. Stat. Soc. 96:* 303–13, 1933.

(8) Box, G. E. P. Multi-Factor Designs on First Order. *Biometrika 39:* 49, 1952.

(9) Box, G. E. P. and Hunter, J. S. Study and Exploitation of Response Surfaces. Paper read before East North American Regional Meeting of Biometric Society, December 1953.

2.5

Multi-Factor Experimental Designs for Exploring Response Surfaces[1]

G. E. P. BOX[2] and J. S. HUNTER[3]

Summary. Suppose that a relationship $\eta = \varphi(\xi_1, \xi_2, \cdots, \xi_k)$ exists between a *response* η and the levels $\xi_1, \xi_2, \cdots, \xi_k$ of k quantitative *variables* or *factors*, and that nothing is assumed about the function φ except that, within a limited region of immediate interest in the space of the variables, it can be adequately represented by a polynomial of degree d.

A k-dimensional experimental design of order d is a set of N points in the k-dimensional space of the variables so chosen that, using the data generated by making one observation at each of the points, all the coefficients in the dth degree polynomial can be estimated.

The problem of selecting practically useful designs is discussed, and in this connection the concept of the *variance function* for an experimental design is introduced. Reasons are advanced for preferring designs having a "spherical" or nearly "spherical" variance function. Such designs insure that the estimated response has a constant variance at all points which are the same distance from the center of the design. Designs having this property are called *rotatable designs.* When such arrangements are submitted to rotation about the fixed center, the variances and covariances of the estimated coefficients in the fitted series remain constant.

Rotatable designs having satisfactory variance functions are given for $d = 1, 2$; and $k = 2, 3, \cdots, \infty$. Blocking arrangements are derived. The simplification in the form of the confidence region for a stationary point resulting from the use of a second order rotatable design is discussed.

1. Introduction. Suppose we have k *variables* or *factors* whose levels are denoted by $\xi_1, \xi_2, \cdots, \xi_k$ on which depend the level of some response η in accordance with an unknown relationship

$$\eta = \varphi(\xi_1, \xi_2, \cdots, \xi_k). \tag{1}$$

Suppose that in order to explore this relationship, N experiments are performed. The uth of these experiments consists in adjusting the factor levels to a certain set of k predecided values, $\xi_{1u}, \xi_{2u}, \cdots, \xi_{ku}$ and of observing a response y_u. The problem of experimental design discussed is that of choosing the N sets of levels at which observations are to be made. It is often convenient to view the problem geometrically and to regard Eq. (1) as defining a surface referred to as the *response surface.* The N sets of conditions at which the response is observed

Received June 21, 1955; revised November 26, 1956.

[1] Prepared under the Office of Ordnance Research, Contract No. DA-36-034-ORD-1177 at the Institute of Statistics, Raleigh N. C.

[2] Now with the Statistical Techniques Group, Princeton University.

[3] Now with the American Cyanamid Company, New York.

The Collected Works of George E. P. Box, 1984, Wadsworth, Inc., Belmont, CA 94002.
Originally published in *Ann. Math. Stat.,* vol. 28, no. 1 (1957), pp. 195–241.

will then correspond to N points in the space of the variables called *experimental points*.

1.1 *Notation*. Following the convention adopted in previous papers [1], [2] we shall define a set of standardized levels

$$(2) \qquad x_{iu} = \frac{(\xi_{iu} - \bar{\xi}_i)}{S_i}, \quad \text{where} \quad S_i = \left\{ \sum_{u=1}^{N} \frac{(\xi_{iu} - \bar{\xi}_i)^2}{N/c} \right\}^{1/2}$$

For these standardized levels therefore

$$(3) \qquad \sum_{u=1}^{N} x_{iu} = 0 \quad \text{and} \quad \sum_{u=1}^{N} x_{iu}^2 = \frac{N}{c}$$

and for the time being the convention is adopted that $c = 1$.

We shall denote by **D** the $N \times k$ *design matrix* which provides a program of the N experiments to be performed. The elements of the uth row of this matrix are the values of the standardized levels $x_{1u}, x_{2u}, \cdots, x_{ku}$ to be used in the uth experiment. These elements also define the uth experimental point in the k-dimensional space of the variables. Since the designs we consider may include many factors, they will be called *multi-factor* designs.

Using standardized factor levels in accordance with Eq. (3) we can prepare standard design matrices appropriate for various values of k, and for various types of assumptions concerning the function φ. In given circumstances the experimenter can select the appropriate design matrix and choose suitable average values $\bar{\xi}_1, \bar{\xi}_2, \cdots, \bar{\xi}_k$ and units $S_1, S_2, \cdots, S_k$ so that the design covers the region of immediate interest in the space of the variables. The level of the ith factor to be used in the uth trial is then $\xi_{iu} = \bar{\xi}_i + S_i x_{iu}$.

We shall assume in what follows that in the limited region of immediate interest φ can be represented by a polynomial of degree d so that the response at the uth point is assumed to be

$$(4) \qquad \begin{aligned} \eta_u = {} & \beta_0 x_{0u} + \beta_1 x_{1u} + \cdots + \beta_k x_{ku} + \beta_{11} x_{1u}^2 + \cdots + \beta_{kk} x_{ku}^2 + \cdots \\ & + \beta_{12} x_{1u} x_{2u} + \cdots + \beta_{k-1,k} x_{k-1,u} x_{ku} + \beta_{111} x_{1u}^3 + \text{etc.} \end{aligned}$$

Following convention, we call $x_0; x_1, \cdots, x_k; x_1^2, \cdots, x_k^2; x_1x_2, \cdots$ etc., the "independent" variables. When the polynomial is of degree higher than the first, the "independent" variables are not, of course, *functionally* independent.

We shall obtain least squares estimates b_0, b_1, etc., of the coefficients β_0, β_1, etc, by fitting Eq. (4) to the N observed values $y_1, y_2, \cdots, y_u, \cdots, y_N$. It is convenient to write down the constant term as $\beta_0 x_{0u}$ rather than as β_0 defining x_{0u} as unity for all values of u. We call β_i the ith *linear* coefficient, β_{ii} the ith *quadratic* coefficient, β_{ij} the ijth *linear* $\times$ *linear crossproduct* coefficient (or simply the ijth interaction coefficient where no ambiguity will arise) and so on. The independent variables $x_i, x_i^2, x_i x_j$ are similarly named.

A design which includes k variables and allows all constants up to order d to be determined will be called a k-dimensional design of order d. In a polynomial equation of degree d there are $\binom{k+d}{d}$ terms, so that for a k-dimensional design of order d, the number of experimental points must be at least $\binom{k+d}{d}$.

1.2 *Factorial designs.* A factorial design from which are to be determined all the polynomial coefficients of order d or less includes all combinations of $d + 1$ levels of each of the factors. The number $(d + 1)^k$ of experimental points so generated is often excessively large compared with the number $\binom{k+d}{d}$ of constants to be determined. In five variables for instance, the factorial design would require $3^5 = 243$ points to determine the 21 constants in the second order polynomial. The number of experimental points may sometimes be considerably reduced by fractional replication [3]. Unfortunately the device of fractional replication is not very effective in generating from the higher level factorials satisfactory designs of order greater than one. For the particular problem here considered of fitting multivariate polynomials to data there seems to be no reason for basing experimental arrangements on the factorials and a more fundamental approach will be attempted.

1.3 *Requirements.* The following are properties of an experimental design of order d which are desirable in the present context. The relative importance of these properties depends on the particular experimental situation. To be of value for specific purposes a design will not need to possess them all.

(a) The design should allow the approximating polynomial of degree d (tentatively assumed to be representationally adequate) to be estimated with satisfactory accuracy within the region of interest.

(b) It should allow a check to be made on the representational accuracy of the assumed polynomial.

(c) It should not contain an excessively large number of experimental points.

(d) It should lend itself to 'blocking'.

(e) It should form a nucleus from which a satisfactory design of order $d + 1$ can be built in case the assumed degree of polynomial proves inadequate.

In this paper we are concerned with interpreting (a) in such a way as, where possible, to satisfy the other properties also. In Section 2 some general results in Least Squares are stated. In Section 3 the criterion of orthogonality is discussed. In Section 4 the concept of the *variance function* for the designs is introduced. This indicates the desirability of designs for which the variance is constant at a constant distance from the origin of the design. Such designs are called *rotatable* designs and the conditions that such designs must satisfy are derived in Sections 5 and 6. In Section 7 second order rotatable designs are obtained. The arrangement into blocks of second order rotatable designs is discussed in Section 8. The details of the calculations required when using the designs is given in Section 9. Completely worked numerical examples will appear in [14]. Section 10 discusses the construction of a confidence region for a stationary point.

2. Least squares results. For any linear model, such as (4), in which there are L unknown coefficients, the N equations at the N experimental points may be written in an obvious matrix notation as

(4a) $$\boldsymbol{\eta} = \mathbf{X}\boldsymbol{\beta},$$

where the $N \times L$ matrix $\mathbf{X}$ is called the matrix of independent variables.

If the observed values found at the N experimental points are represented by a vector $\mathbf{Y}$ and

(5) $$\mathcal{E}(\mathbf{Y}) = \boldsymbol{\eta}; \qquad \mathcal{E}(\mathbf{Y} - \boldsymbol{\eta})(\mathbf{Y} - \boldsymbol{\eta})' = \mathbf{I}_N\sigma^2$$

then, on the supposition that the mathematical model (4a) exactly represents the true situation, the estimates $\mathbf{B}$ of $\boldsymbol{\beta}$ linear in the observations which are unbiased (i.e., $\mathcal{E}(\mathbf{B}) = \boldsymbol{\beta}$) and have severally the smallest possible variances, are those which reduce to a minimum the sums of squares of discrepancies $(\hat{\mathbf{Y}} - \mathbf{Y})'(\hat{\mathbf{Y}} - \mathbf{Y})$ between the observed values $\mathbf{Y}$ and the values $\hat{\mathbf{Y}} = \mathbf{XB}$ given by the fitted function. These are the "least squares" estimates

(6) $$\mathbf{B} = (\mathbf{X}'\mathbf{X})^{-1}\mathbf{X}'\mathbf{Y}.$$

Their variances and co-variances are the elements of the matrix

(7) $$\mathcal{E}(\mathbf{B} - \boldsymbol{\beta})(\mathbf{B} - \boldsymbol{\beta})' = (\mathbf{X}'\mathbf{X})^{-1}\sigma^2$$

and an unbiased estimate of $(N - L)\sigma^2$ is provided by the quantity

(8) $$(\hat{\mathbf{Y}} - \mathbf{Y})'(\hat{\mathbf{Y}} - \mathbf{Y}) = \mathbf{Y}'\mathbf{Y} - \mathbf{B}'\mathbf{X}'\mathbf{X}\mathbf{B}.$$

If, contrary to supposition, the mathematical model $\boldsymbol{\eta} = \mathbf{X}\boldsymbol{\beta}$ is inadequate and in fact L_1 further terms $\mathbf{X}_1\boldsymbol{\beta}_1$ are needed to ensure an adequate representation of the response so that

$$\boldsymbol{\eta} = \mathbf{X}\boldsymbol{\beta} + \mathbf{X}_1\boldsymbol{\beta}_1,$$

then the estimates given by (6) are biased for

(8a) $$\mathcal{E}(\mathbf{B}) = \boldsymbol{\beta} + \mathbf{A}\boldsymbol{\beta}_1,$$

where $\mathbf{A} = (\mathbf{X}'\mathbf{X})^{-1}\mathbf{X}'\mathbf{X}_1$ is an $L \times L_1$ matrix of bias coefficients which has been called, [1], the "alias" matrix. In this situation the residual sum of squares is also biased and we find

(8b) $$\begin{aligned}\mathcal{E}(\mathbf{Y}'\mathbf{Y} - \mathbf{B}'\mathbf{X}'\mathbf{X}\mathbf{B}) &= (N - L)\sigma^2 + \boldsymbol{\beta}_1'(\mathbf{X}_1 - \mathbf{X}\mathbf{A})'(\mathbf{X}_1 - \mathbf{X}\mathbf{A})\boldsymbol{\beta}_1 \\ &= (N - L)\sigma^2 + \boldsymbol{\beta}_1'\mathbf{X}_1'(\mathbf{I} - \mathbf{X}(\mathbf{X}'\mathbf{X})^{-1}\mathbf{X}')\mathbf{X}_1\boldsymbol{\beta}_1.\end{aligned}$$

2.1 *The moment matrix.* Equations (6), (7), (8), (8a), and (8b) contain the matrix $\mathbf{X}'\mathbf{X}$ of sums of squares and products of the independent variables. We notice that $N^{-1}\mathbf{X}'\mathbf{X}$ may be viewed as a matrix of *moments* of the design. For example, if there are $k = 2$ variables, and we are considering a design of order two, the equation to be fitted is

(9) $$\eta = \beta_0 x_0 + \beta_1 x_1 + \beta_2 x_2 + \beta_{11} x_1^2 + \beta_{22} x_2^2 + \beta_{12} x_1 x_2$$

and the matrix $N^{-1}\mathbf{X}'\mathbf{X}$ is

(10) $$\begin{array}{c} \\ 0 \\ 1 \\ 2 \\ 11 \\ 22 \\ 12 \end{array} \begin{array}{c} \begin{array}{cccccc} 0 & 1 & 2 & 11 & 22 & 12 \end{array} \\ \left[\begin{array}{cccccc} 1 & [1] & [2] & [11] & [22] & [12] \\ [1] & [11] & [12] & [111] & [122] & [112] \\ [2] & [12] & [22] & [112] & [222] & [122] \\ [11] & [111] & [112] & [1111] & [1122] & [1112] \\ [22] & [122] & [222] & [1122] & [2222] & [1222] \\ [12] & [112] & [122] & [1112] & [1222] & [1122] \end{array}\right] \end{array}.$$

The quantities in square brackets denote the moments of the design. For example, $N^{-1}\sum_{u=1}^{N} x_{1u} = [1]$, $N^{-1}\sum_{u=1}^{N} x_{1u}^2 x_{2u} = [112]$ and so on. We shall call $N^{-1}\mathbf{X}'\mathbf{X}$ the moment matrix and its inverse $N(\mathbf{X}'\mathbf{X})^{-1}$ the *precision* matrix. When $\sigma^2 = 1$ the elements of this latter matrix are the variances and covariances of the effects measured on a "per-observation" basis.

3. Orthogonal designs. The problem of choosing a "best" design for the fitting of a model $\boldsymbol{\eta} = \mathbf{X}\boldsymbol{\beta}$ has usually been interpreted as that of satisfying the requirement that $\mathbf{D}$ should be so chosen that the coefficients $\boldsymbol{\beta}$ are separately estimated with smallest variance. In references [4], [5], [2], and [6] a theorem is proved (for the case where the variables in the matrix $\mathbf{X}$ are functionally independent and the diagonal elements of $\mathbf{X}'\mathbf{X}$ are fixed by the definition of the problem) that the requirement of smallest variance is satisfied by so choosing $\mathbf{D}$ that the matrix $\mathbf{X}'\mathbf{X}$ is diagonal. Such an arrangement may be called an orthogonal design.

In the present context it is only in the case of designs of first order that the variables are functionally independent and that all the diagonal elements of $\mathbf{X}'\mathbf{X}$ are fixed by the definition of the problem. For this reason, as we see in more detail below, the above theorem is directly helpful only in the derivation of first order designs. For higher order designs an alternative approach is necessary.

3.1 *First order designs.* In this case the independent variables are $x_0, x_1, \cdots, x_k$. Since these variables are also functionally independent, and since $\sum x_{iu}^2 = N$ $(i = 0, 1, 2, \cdots, k)$ so that the diagonal elements of $\mathbf{X}'\mathbf{X}$ are fixed by definition of the problem, the smallest variance theorem referred to above leads at once to the conclusion that a best design matrix $\mathbf{D}$ is one for which $N^{-1}\mathbf{X}'\mathbf{X} = \mathbf{I}$. It will be noted that for this case we are led to a unique form of moment matrix. Such a moment matrix is realized in practice simply by choosing $\mathbf{D}$ to have orthogonal columns subject to Eqs. (3).

The construction and properties of such designs are discussed in [2]. Geometrically the designs consist of N points, at the vertices of an $N - 1$ dimensional regular simplex if $k = N - 1$, or the projections onto a space of k dimensions of the vertices of the $N - 1$ dimensional regular simplex if $k < (N - 1)$. The arbitrariness in the choice of **D** corresponds to the fact that the simplex may be taken in any orientation. This class of designs includes the factorials and fractional factorials. These latter designs are of special value because they are easy to carry out, they allow the adequacy of the first degree representation to be checked and the nature of departures from it to be readily identified, they form natural nuclei which can be augmented to form designs of higher order, and they are readily arranged in blocks.

3.2 *Second order "orthogonal" designs.* For designs of order higher than the first the quantities x_0 ; $x_1, \cdots, x_k$; $x_1^2, \cdots, x_k^2$; $x_1x_2, x_1x_3, \cdots, x_{k-1}x_k$; x_1^3, etc., are not all functionally independent and a diagonal moment matrix is impossible of attainment since, unless the x_{iu} are all zero, certain sums of products such as those between x_i^2 and x_0 and between x_i^2 and x_j^2 are necessarily positive.

Orthogonal second order designs of a sort can be obtained if we redefine the independent variables in terms of the orthogonal polynomials. We show below however that there is an infinite variety of such designs with widely different properties and that these designs do not provide a wholly satisfactory solution to our problem.

Let $x_i^{(m)}$ be the orthogonal polynomial of mth degree for the ith variable x_i. Thus

$$x_i^{(m)} = x_i^m + \alpha_{m-1,m}x_i^{m-1} + \cdots + \alpha_{1m}x_i + \alpha_{0m}, \tag{11}$$

where the α's are chosen so that

$$\sum_{u=1}^{N} x_{iu}^{(m)}x_{iu}^{(m-p)} = 0, \qquad p = 1, 2, \cdots, m. \tag{12}$$

Then we can express the original polynomial equations in terms of these orthogonal polynomials and their products in the form

$$\boldsymbol{\eta} = (\mathbf{XP})(\mathbf{P}^{-1}\boldsymbol{\beta}) = \dot{\mathbf{X}}\dot{\boldsymbol{\beta}},$$

where **P** is the matrix transforming the old independent variables to the new. For clarity we will discuss the particular case of a two dimensional design of order 2 but, as will be readily appreciated, the conclusions drawn will be quite general.

Using (3) with (11) and (12) we have

$$x_i^{(1)} = x_i, \quad x_i^{(2)} = x_i^2 - [iii]x_i - 1 \tag{13}$$

and the second degree equation (9) for $k = 2$ could be written as

$$\begin{aligned}\eta = {} & (\beta_0 + \beta_{11} + \beta_{22})\, x_0 + (\beta_1 + [111]\,\beta_{11})x_1 + (\beta_2 + [222]\beta_{22})x_2 \\ & + \beta_{11}(x_1^2 - [111]x_1 - 1) + \beta_{22}(x_2^2 - [222]x_2 - 1) + \beta_{12}x_1x_2 .\end{aligned} \tag{14}$$

The symmetric moment matrix $N^{-1}\,\dot{\mathbf{X}}'\dot{\mathbf{X}}$ is then that given below.

$$
(15)\qquad
\begin{array}{c|cccccc|}
 & 0 & 1 & 2 & 11 & 22 & 12 \\
0 & 1 & \cdot & \cdot & \cdot & \cdot & a \\
1 & \cdot & 1 & a & \cdot & b & c \\
2 & \cdot & a & 1 & d & \cdot & e \\
11 & \cdot & \cdot & d & f & g & h \\
22 & \cdot & b & \cdot & g & i & j \\
12 & a & c & e & h & j & k
\end{array}
$$

Here $a = [12]$, $b = [122] - [222][12]$, $c = [112]$, $d = [112] - [111][12]$,

$$e = [122],\ f = [1111] - [111]^2 - 1,$$

$$g = [1122] + [111][222][12] - [111][122] - [222][112] - 1,$$

$$h = [1112] - [111][112] - [12],\ i = [2222] - [222]^2 - 1,$$

$$j = [1222] - [222][122] - [12],\ k = [1122].$$

To obtain a diagonal matrix we must evidently choose the design so that

$$[12] = [112] = [122] = [1112] = [1222] = 0 \qquad \text{and} \qquad [1122] = 1,$$

which insures that all the elements of the matrix (15) vanish except those on the diagonal. Examples of such designs are the factorials with more than two levels and the orthogonal composite designs given by Box and Wilson [1].

There seems little justification for limiting consideration to only these arrangements. In particular nothing in our discussion has indicated that the choice $[1122] = 1$ is necessarily a good one, or that it would not be better to choose some other value and let the quadratic effects be correlated. Again, it is far from clear what constitutes a "good" choice of the diagonal elements $[iiii] - [iii]^2 - 1$ corresponding to the quadratic constants in the moment matrix.

Since the scaling of the design has been *standardized*, $[iii]^2$ and $[iiii]$ are measures of "skewness" and "kurtosis" for the ith variable. The choice of the moments $[iii]$ decides the question of whether the marginal distribution of the pattern of design points for ith variable is to be symmetric or skew. The choice of the moments $[iiii]$ decides whether there is to be a tendency to a uniform distribution of points or to a concentration of points at the center and at the extremes of the range. Since for all such designs in our conventional scaling the variances of linear, quadratic and interaction estimates corresponding to the ith variable are $\sigma^2 N^{-1}$, $\sigma^2 N^{-1}([iiii] - [iii]^2 - 1)^{-1}$ and $\sigma^2 N^{-1}$ respectively this choice also decides the relative precision with which linear quadratic and interaction coefficients are estimated.

It may be noted for example that, for the 3^k factorial design in conventional scaling, $[iii] = 0$ and $[iiii] = \frac{3}{2}$ $(i = 1, 2, \cdots, k)$. The variances of the estimates for the quadratic coefficients are thus twice as large as those for the interaction coefficients. In terms of estimated derivatives at the center of the design therefore, the estimated "quadratic" derivative $\partial^2\eta/(\partial x_i)^2$ has *eight times* the variance of the estimated "interaction" derivative $\partial^2\eta/\partial x_i\partial x_j$. This was pointed out in [1], where an intuitive attempt to reduce this apparent unbalance was made by introducing designs in which quadratic and interaction derivatives were determined with equal precision. In fact designs both orothogonal and non-orthogonal can be found for which the relative variances of estimated coefficients of different kinds can differ over a wide range. Up to this point the present discussion has provided no satisfactory basis on which a rational choice can be made.

In selecting from possible orthogonal designs it seems at first sight that the quantities $[iiii] - [iii]^2 - 1$ should be made as large as possible. This would seem to give the smallest possible variances for the quadratic effects without affecting the precision of the remaining constants. On closer inspection however the apparent advantage of such a choice turns out to be somewhat illusory because

(a) The apparent advantage of making the quantity $[iiii] - [iii]^2 - 1$ large arises only because of the particular scale convention adopted. If for example we scaled our designs on the basis of the size of the fourth moment instead of on the size of the second moment a contrary conclusion would be reached.

(b) The quantity $[iiii]$ enters not only into the precision matrix but also into the alias matrix. In fact for any orthogonal design of this type the expected value of the ith linear effect is

$$\mathcal{E}(b_i) = \beta_i + [iiii]\,\beta_{iii} + \sum_{j \neq i}^{k} \beta_{ijj}.$$

Thus the apparent reduction in the variance of the quadratic effects is gained only at the expense of an increase in possible bias in the linear effects.

(c) The quadratic effect β_{ii} measures curvature of the surface in the direction of the ith coordinate axis. It is shown in the next section that by attempting to measure the precision with which curvature is determined in the directions of the coordinate axes we may decrease the precision with which it is determined in some other direction which might be of equal importance to the experimenter.

3.3 *Effect of rotation on precision of the estimates.* If, as we shall assume, we wish to use the design to explore a surface about which little is known, we shall in particular not know how the design is oriented relative to the response surface. For example suppose the surface could be represented locally by an equation of second degree, then the response contours would be a set of conics which could be referred to their principal axes. The orientation of these axes relative to the axes of the variables would differ from one problem to another. It is of some interest therefore to consider how the variances and co-variances of the estimated coefficients are changed when the design is rotated.

As an example suppose that we were to use the symmetrical 3^2 factorial design to estimate the coefficients in the second degree Eq. (9). Then bearing in mind the conventions expressed by Eqs. (3) concerning the origin of the design and the size of the scale factor, we should use the nine combinations of the levels

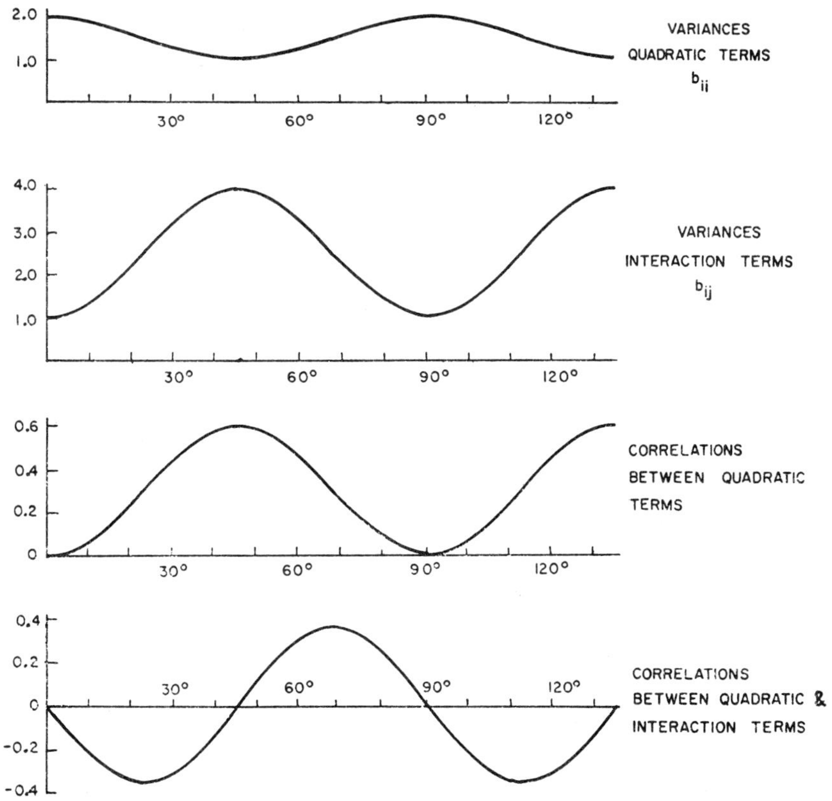

FIG. 1. Variances of, and correlations between, second order coefficients estimated from a 3^2 factorial design rotated through an angle θ

$x_1 = (-\alpha, 0, \alpha)$ and $x_2 = (-\alpha, 0, \alpha)$, where $\alpha = (\frac{3}{2})^{\frac{1}{2}}$. In the normal orientation of the design the variances of the linear effects, quadratic effects and interaction effects would be $\sigma^2/9$, $2\sigma^2/9$ and $\sigma^2/9$ respectively, and consequently the corresponding entries in the precision matrix (which measure the variance on a "per observation" basis for unit experimental error variance) would be 1, 2, and 1, respectively. All the covariances between these effects would be zero. If the design were rotated through some angle θ however then, as is illustrated in Fig. 1 the variances of the quadratic and interaction effects would undergo marked changes and the quadratic effects would become correlated with each other and with the interaction effect. Only the linear effects would have constant variance and would remain unchanged in all orientations.

We see that the variances of individual coefficients estimated using a design in a particular orientation may give a somewhat misleading impression of its efficiency. The condition of orthogonality refers to orthogonality in a particular orientation, and this property is in general lost on rotation of the design.

4. The variance function for the design. We have proceeded so far by considering the accuracy with which *individual* coefficients are estimated. This approach does not, for the case of designs of order higher than the first, seem to lead to any unique class of solutions, but points to the conclusion that we should in some way consider the joint accuracy of the coefficients. We are really interested in the individual coefficients only in so far as they supply information about the surface. To make further progress therefore we consider what we call the design "variance function".

We shall denote the k coordinates $x_1, \cdots, x_i, \cdots, x_k$ of a point in the space of the variables by the $k \times 1$ vector $\mathbf{x} = \{x_i\}$. Suppose that $\hat{y}_x$ is the response estimated at the point $\mathbf{x}$ using a polynomial fitted by least squares to N observations made in accordance with some experimental design $\mathbf{D}$. The variance $V(\hat{y}_x)$ of this estimated value is a function of $\mathbf{x}$ and σ^2 and we can reduce $V(\hat{y}_x)$ by increasing N (for example by replicating the points). The quantity $V(\mathbf{x}) = NV(\hat{y}_x)/\sigma^2$, or alternatively its reciprocal $W(\mathbf{x}) = \sigma^2/NV(\hat{y}_x)$, is thus a standardized measure of the accuracy with which the design $\mathbf{D}$ allows the response at the point $\mathbf{x}$ to be estimated. $NV(\hat{y}_x)/\sigma^2$ will be called the *variance function* of the design and $W(\mathbf{x}) = \{V(\mathbf{x})\}^{-1}$ the *weight function*. For any experimental design $V(\mathbf{x})$ provides a standardized measure of the precision of the estimated response at any point in the space of the variables. It is a function of $x_1, x_2, \cdots, x_k$ and the elements of the precision matrix alone and is uniquely defined for every k dimensional experimental design of order d.

For example suppose we used the nine points of the 3^2 symmetrical factorial as a second order two dimensional design. On the convention that the origin and scale are chosen so that $[1] = [2] = 0$ and $[11] = [22] = 1$, we have for the variances and covariances of the effects $V(b_0) = (\frac{5}{9})\sigma^2$, $V(b_1) = V(b_2) = (\frac{1}{9})\sigma^2$, $V(b_{11}) = V(b_{22}) = (\frac{2}{9})\sigma^2$, $V(b_{12}) = (\frac{1}{9})\sigma^2$ and $\text{Cov}(b_0b_{11}) = \text{Cov}(b_0b_{22}) = (-\frac{2}{9})\sigma^2$. The variance function for this design is therefore

$$V(\mathbf{x}) = \frac{N}{\sigma^2} V(\hat{y}_x) = 5 - 3x_1^2 - 3x_2^2 + 2x_1^4 + 2x_2^4 + x_1^2x_2^2 .$$

The variance contours for which are shown in Fig. 2(i).

In Figs. 2(ii) and 2(iii) are shown variance functions for other two-dimensional second order designs mentioned in [1]. The arrangement in figure 2(ii) is the "pentagonal design" and that in figure 2(iii) is an example of a class of designs, already referred to in Section 3.2, in which quadratic and interaction *derivatives* are determined with equal precision. It will be seen that the arrangement of points in the latter design is in fact almost the same as would be obtained by rotating the factorial through 45°.

If, as we shall suppose, nothing is known in advance about the orientation of the surface, it seems most appropriate to adopt designs which have variance functions like that of the pentagonal design. That is to say designs which generate information such that the response is estimated with constant variance at all points equidistant from the origin of the design. When we have no knowledge in

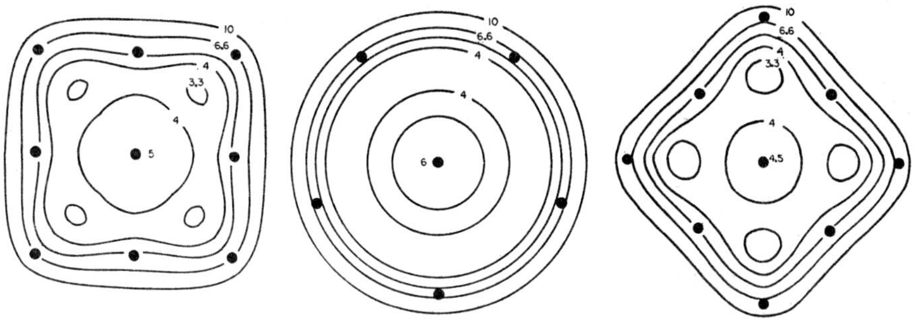

(i) 3^2 FACTORIAL (ii) PENTAGONAL (iii) COMPOSITE

FIG. 2. Variance contours for some 2 dimensional designs

advance of the orientation of the surface relative to the design this also seems instinctively to be a sensible requirement for designs other than those of second order.

In general, for any $k-$ dimensional design, if the variance of the response estimated by the fitted polynomial is a function only of

$$\rho^2 = \sum_{i=1}^{k} x_i^2,$$

so that the variance contours in the space of the variables are circles, spheres or hyperspheres centered at the origin the design will be said to have a *spherical variance function* $V(\rho)$. An arrangement of points giving such a variance function will be called a rotatable design.

The remainder of this paper is devoted to constructing rotatable or nearly rotatable designs, that is, arrangements of experimental points which symmetrically generate information in those coordinates regarded as most relevant by the experimenter. We shall interpret requirement (a) in Section 1.3 in this sense.

5. Condition for rotatability. In the developments which follow we need some properties of derived power and product vectors and Schläflian matrices [7], [8], and [9]. If $\mathbf{x}' = (x_1, x_2, \cdots, x_k)$ then we denote by $\mathbf{x}'^{[p]}$ the derived power vector of degree p. For example if $k = 2$

$$\mathbf{x}' = [x_1, x_2] \qquad \text{and} \qquad \mathbf{x}'^{[2]} = [x_1^2, x_2^2, 2^{1/2}x_1 x_2]$$

and in general $\mathbf{x}'^{[p]}$ will contain as elements all the powers and products of total degree p and less (duly ordered) of the elements in $\mathbf{x}'$ with suitable multipliers attached so that $\mathbf{x}'^{[p]}\mathbf{x}^{[p]} = [\mathbf{x}'\mathbf{x}]^p$. If a vector $\mathbf{x}$ is transformed to a vector $\mathbf{z}$ by $\mathbf{z} = \mathbf{H}\mathbf{x}$, the pth Schläflian matrix $\mathbf{H}^{[p]}$ is defined such that $\mathbf{z}^{[p]} = \mathbf{H}^{[p]}\mathbf{x}^{[p]}$. It is readily confirmed that $[\mathbf{HK}]^{[p]} = \mathbf{H}^{[p]}\mathbf{K}^{[p]}$ and also that if $\mathbf{H}$ is orthogonal then so also is $\mathbf{H}^{[p]}$.

We need some properties of spherical distribution functions discussed in references [2], [10]. These distribution functions are of some importance in basic statistical theory, and especially in randomization theory, but these aspects are not pursued here.

5.1 *Moments of a spherical distribution.* If we have a set of random variables, $z_1, z_2, \cdots, z_k$, which may be regarded as the elements of a random vector $\mathbf{z}$, and each of which has zero mean and unit variance and if their joint distribution can be written in the form

$$p(\mathbf{z}) = kf(\mathbf{z}'\mathbf{z}), \qquad 0 \leq \mathbf{z}'\mathbf{z} < W, \tag{16}$$

where W may be infinite and k is taken so that the integral over the whole space is unity, then since the density will be constant on hyper-spheres centered at the origin of the z's, we shall say that the variables have a spherical distribution.

Now if all the moments of a distribution exist, and the m.g.f. $\varphi(t)$ can be expanded in an infinite series, we can write this series

$$\varphi(\mathbf{t}) = 1 + \sum_{s=1}^{\infty} \frac{1}{s!} \mathbf{t}'^{[s]} \mathbf{m}_s , \tag{17}$$

where $\mathbf{m}_s$ is the vector of moments $\mathcal{E}\{\mathbf{z}^{[s]}\}$. But for a spherical distribution the moments are invariant under any orthogonal rotation of the coordinates whence the m.g.f. is

$$\varphi(\mathbf{t}) = \mathcal{E}(e^{\mathbf{t}'\mathbf{z}}) = \mathcal{E}(e^{\mathbf{t}'\mathbf{Hz}});$$

that is,

$$\varphi(\mathbf{t}) = \varphi(\mathbf{H}'\mathbf{t}) \tag{18}$$

for any orthogonal matrix $\mathbf{H}$. Regarding now the matrix $\mathbf{H}$ as transforming the matrix $\mathbf{t}'$, this implies that $\varphi(\mathbf{t})$ is unchanged by any transformation on $\mathbf{t}$ which leaves $\mathbf{t}'\mathbf{t}$ unchanged. The m.g.f. is therefore a function of $\mathbf{t}'\mathbf{t}$ and can be written in the form

$$\varphi(\mathbf{t}) = 1 + \sum_{p=1}^{\infty} \lambda_{2p} \frac{1}{p!2^p} (\mathbf{t}'\mathbf{t})^p, \tag{19}$$

where the λ's are real positive constants depending on the function f in (16). Equating terms in (17) and (19) and writing $[1^{\alpha_1}, 2^{\alpha_2}, \cdots, k^{\alpha_k}]$ for the moment $\mathcal{E}[x_1^{\alpha_1}, x_2^{\alpha_2}, \cdots, x^{\alpha_k}]$ we have

$$[1^{\alpha_1}, 2^{\alpha_2}, \cdots, k^{\alpha_k}] = \lambda_\alpha \frac{\prod_{i=1}^{k} \alpha_i!}{2^{\alpha/2} \prod_{i=1}^{k} (\frac{1}{2}\alpha_i)!}, \tag{20}$$

where $\alpha = \sum_{i=1}^{k} \alpha_i$ is called the order of the moment.

If the z's are independent so that

$$p(z) = \prod_{i=1}^{k} p(z_i), \tag{21}$$

then it has been shown, in references [11] and [12] that the only spherical distribution possible is the multi-variate normal with equal variances and zero covariances, which may be called the spherical multi-normal distribution. For this distribution the m.g.f. is

$$\varphi(t) = \exp[\tfrac{1}{2}(\mathbf{t}'\mathbf{t})] \tag{22}$$

and all the λ's are equal to unity. We see therefore that for any spherical distribution, the moments of the *same order* bear the same relationship to one another as do the moments for the spherical multi-normal. The moments of different orders will however depend on the λ's and hence on the function $f(\mathbf{z}'\mathbf{z})$.

5.2 *Variance of an estimated response.* Consider the response $\hat{y}_x$ estimated by a fitted polynomial of degree d at the point whose co-ordinates are given by the last k elements of the vector $\mathbf{x}'$ now defined as $\mathbf{x}' = (1, x_1, x_2, \cdots, x_k)$. The polynomial has $\binom{k+d}{d} = L$ terms and the estimated response at the point $x_1, x_2, \cdots, x_k$ is

$$\begin{aligned} \hat{y}_x = b_0 + b_1x_1 + b_2x_2 + \cdots + b_kx_k + b_{11}x_1^2 + b_{22}x_2^2 + \cdots \\ + b_{kk}x_k^2 + b_{12}x_1x_2 + \cdots + b_{k-1,k}x_{k-1}x_k + b_{111}x_1^3 + \text{etc.}, \end{aligned} \tag{23}$$

which may be written

$$\hat{y}_x = \mathbf{x}'^{[d]}\mathbf{b}, \tag{24}$$

where the $L \times 1$ vector $\mathbf{b}$ contains all the b's with suitable multipliers attached so that (24) is equivalent to (23). Suppose also that the true response at this point is given by

$$\eta_x = \mathbf{x}'^{[d]}\boldsymbol{\beta}. \tag{25}$$

Then for a given design matrix $\mathbf{D}$ for which there exists a matrix of independent variables $\mathbf{X}$ of full rank L the variance of $\hat{y}_x$ is

$$\begin{aligned} V(\hat{y}_x) = \mathcal{E}\{(\hat{y}_x - \eta_x)(\hat{y}_x - \eta_x)'\} = \mathbf{x}'^{[d]}\mathcal{E}\{(\mathbf{b} - \boldsymbol{\beta})(\mathbf{b} - \boldsymbol{\beta})'\}\mathbf{x}^{[d]} \\ = \mathbf{x}'^{[d]}[\mathbf{X}'\mathbf{X}]^{-1}\mathbf{x}^{[d]}\sigma^2. \end{aligned} \tag{26}$$

Consider now the variance of a second estimated value $\hat{y}_z$ which is the same distance ρ from the origin and whose co-ordinates are the last k elements of the vector $\mathbf{z} = \mathbf{R}\mathbf{x}$, where $\mathbf{R}$ is an orthogonal $(k+1) \times (k+1)$ matrix consisting of an arbitrary orthogonal matrix $\mathbf{H}$ bordered by a first row $\mathbf{u}' = (1, 0, 0, \cdots, 0)$ and a first column $\mathbf{u}$. Making the substitution in (26) we have

$$V(\hat{y}_z) = \mathbf{x}'^{[d]}\mathbf{R}'^{[d]}[\mathbf{X}'\mathbf{X}]^{-1}\mathbf{R}^{[d]}\mathbf{x}^{[d]}\sigma^2 \tag{27}$$

$$= \mathbf{x}'^{[d]}(\mathbf{R}'^{[d]}\mathbf{X}'\mathbf{X}\mathbf{R}^{[d]})^{-1}\mathbf{x}^{[d]}\sigma^2. \tag{28}$$

To satisfy the condition that the variance is constant on spheres centered at the origin of the design we require therefore that (28) and (26) are identically equal for every $\mathbf{x}$ and every $\mathbf{R}$. Whence

$$\text{(29)} \qquad \mathbf{X'X} = \mathbf{R}'^{[d]}\mathbf{X'XR}^{[d]}$$

for every orthogonal matrix $\mathbf{R}$. Now $N^{-1}\mathbf{R}'^{[d]}\mathbf{X'XR}^{[d]}$ is the moment matrix for the design matrix $\mathbf{HD}$, and consequently the variance is constant for every point a distant ρ from the origin if and only if the moment matrix is invariant under orthogonal transformation of the design matrix. This means that unlike the 3^2 factorial design whose behavior under rotation is illustrated in Fig. 1, every variance and covariance of the b's and all the moments and mixed moments of the design must remain constant under rotation. We now need to find the form of moment matrix $N^{-1}\mathbf{X'X}$ for which Eq. (29) is satisfied.

5.3 *Moments of a rotable design.* We redefine the vector $\mathbf{t}'$ to be $(1, t_1, t_2, \cdots, t_k)$ and consider expression

$$\text{(30)} \qquad Q = N^{-1}\mathbf{t}'^{[d]}\mathbf{X'Xt}^{[d]},$$

which is a generating function for the moments of order $2d$ and less of the design. More specifically, since if $\mathbf{x}'_u = (1, x_{1u}, x_{2u}, \cdots, x_{ku})$, $\mathbf{X'X} = \sum_{u=1}^{N} \mathbf{x}_u^{[d]}\mathbf{x}_u'^{[d]}$, we have

$$Q = N^{-1}\mathbf{t}'^{[d]}\left(\sum_{u=1}^{N} \mathbf{x}_u^{[d]}\mathbf{x}_u'^{[d]}\right)\mathbf{t}^{[d]}$$

$$\text{(31)} \qquad = N^{-1}\sum_{u=1}^{N} (\mathbf{t}'\mathbf{x}_u\mathbf{x}'_u\mathbf{t})^d$$

$$= N^{-1}\sum_{u=1}^{N} (1 + t_1x_{1u} + t_2x_{2u} + \cdots + t_kx_{ku})^{2d}.$$

Thus if we write $[1^{\alpha_1}, 2^{\alpha_2}, \cdots, k^{\alpha_k}]$ for the moment $N^{-1}\sum_{u=1}^{N} x_{1u}^{\alpha_1}\, x_{2u}^{\alpha_2} \cdots x_{ku}^{\alpha_k}$ then the coefficient of $t_1^{\alpha_1}, t_2^{\alpha_2}, \cdots, t_k^{\alpha_k}$ in Q is

$$\text{(32)} \qquad \frac{(2d)!}{\prod_{i=1}^{k} \alpha_i!(2d-\alpha)!}\,[1^{\alpha_1}, 2^{\alpha_2}, \cdots, k^{\alpha_k}].$$

Now from (29) the design is rotatable if and only if

$$\text{(33)} \qquad Q = N^{-1}\mathbf{t}'^{[d]}\mathbf{X'Xt}^{[d]} \equiv N^{-1}\mathbf{t}'^{[d]}\mathbf{R}'^{[d]}\mathbf{X'XR}^{[d]}\mathbf{t}^{[d]}$$

$$\text{(34)} \qquad = N^{-1}(\mathbf{t'R'})^{[d]}\mathbf{X'X}(\mathbf{Rt})^{[d]};$$

that is to say, if any transformation which leaves $\mathbf{t't}$ unchanged does not change Q. Hence the design is rotatable if and only if Q is some function of $\mathbf{t't}$ and since it is a polynomial in the t's it must be of the form

$$\text{(35)} \qquad Q = \sum_{s=0}^{d} a_{2s}\left(\sum_{i=1}^{k} t_i^2\right)^s.$$

The coefficient of $t_1^{\alpha_1}, t_2^{\alpha_2}, \cdots, t_k^{\alpha_k}$ in this expression is zero if any of the α_i are odd integers. If the α_i are even integers the coefficient is

$$\text{(36)} \qquad a_\alpha(\tfrac{1}{2}\alpha)! \Big/ \prod_{i=1}^{k} (\tfrac{1}{2}\alpha_i)!.$$

We may now equate coefficients to obtain specific values for the moments to order $\alpha = 2d$ as follows

$$[1^{\alpha_1}, 2^{\alpha_2}, \cdots, k^{\alpha_k}] = \frac{a_\alpha(\frac{1}{2}\alpha)!(2d - \alpha)!}{2d!} \cdot \frac{\prod_{i=1}^{k} \alpha_i!}{\prod_{i=1}^{k} (\frac{1}{2}\alpha_i)!}; \tag{37}$$

and if we write

$$\frac{a_\alpha 2^{\alpha/2}(\frac{1}{2}\alpha)!(2d - \alpha)!}{2d!} = \lambda_\alpha, \tag{38}$$

then finally the moments of a rotatable design of order d are

$$[1^{\alpha_1}, 2^{\alpha_2}, \cdots, k^{\alpha_k}] = 0 \qquad \text{if one or more of the } \alpha_i \text{ are odd,}$$

$$= \lambda_\alpha \frac{\prod_{i=1}^{k} \alpha_i!}{2^{\alpha/2} \prod_{i=1}^{k} (\frac{1}{2}\alpha_i)!} \qquad \text{if all of the } \alpha_i \text{ are even,} \tag{39}$$

which are the moments up to order $2d$ of a spherical distribution.

5.4 *Effect of transformations on zero moments.* It is readily seen that if the original variables $x_1, x_2, \cdots, x_k$ are transformed by any non-singular linear transformation to variables $X_1, X_2, \cdots, X_k$ then any moment of order α for the new variables will be a linear combination of moments all of order α for the old variables. It follows in particular that if an arrangement of points is such that *all* the moments of a given odd order are zero then they remain zero in every orientation of the arrangement.

6. Moment requirements for rotatable designs of first and second order. It has been shown that the moment matrix for a rotatable design of order d has a specific form which is invariant under orthogonal rotation of the axes of the variables. The moments which are the elements of this moment matrix are the same as those of a spherical distribution and are known apart from arbitrary constants $\lambda_0, \lambda_2, \cdots, \lambda_{2d}$. The variance function $V(\rho)$ for a rotatable design depends only on $\lambda_0, \lambda_2, \cdots, \lambda_{2d}$ and on $\rho = (\mathbf{x}'\mathbf{x})^{1/2}$. In selecting a design of order d we can proceed as follows:

(a) Using Eq. (39) we first obtain in terms of the λ's the form of the moment matrix for a rotatable design of the required order.

(b) Since for a rotatable design the variance function depends, apart from the λ's, only on ρ it is now comparatively easy to study the effect on the variance function of varying the λ's. Having selected the λ's to give a satisfactory variance function and alias matrix the required form of the moment matrix is completely defined.

(c) We have then to determine actual arrangements of points which, so far as possible, satisfy the requirements listed in Section 1.3 as well as these moment conditions.

From Eq. (39), $\lambda_0 = [0]$ and $\lambda_2 = [ii]$, $(i = 1, 2, \cdots, k)$. Since by convention $[0] = [ii] = 1$, λ_0 and λ_2 are always equal to unity, and no element of choice for the λ's arises with rotatable designs of first order, but only with designs of order 2, 3, etc., for which the values of λ_4, λ_6, etc., must be selected. In making this selection both the variance function and the alias matrix must be considered. It will be recalled that the alias matrix contains as elements the coefficients of biases which arise in the estimated coefficients when the assumed form of the model is inadequate. For a design of order d it is natural to first consider biases arising from terms of order $d + 1$ not allowed for in the assumed form of the model. If we suppose that the true form of the model is of order $d + 1$ it will be clear from the form of the alias matrix $\mathbf{A}$ in Eq. (8a) that for a rotatable design of order d the coefficients of the biases can be completely expressed in terms of $\lambda_0, \lambda_2, \cdots, \lambda_{2d}$ and the moments of order $2d + 1$. When possible, it is advantageous to use a rotatable design of order d for which all moments of order $2d + 1$ are zero. From Section 5.4 these moments are then zero in every orientation. The alias matrix $\mathbf{A}$ is a function only of the λ's and all biases arising from terms of order $d + 1$ are avoided except those which arise inevitably because the bias coefficients are functions of the λ's themselves.

The remainder of the present section is devoted to determining the necessary form of the moment matrix and to discussing the properties of designs of first and second order. Specific second order designs having the required form of moment matrix are derived in Section 7.

6.1 *Rotatable designs of order* 1. Suppose we have k variables $x_1, x_2, \cdots, x_k$ and we desire to fit a polynomial of degree $d = 1$, that is to say, a fitted equation representing a plane

$$\hat{y}_x = b_0 + b_1x_1 + b_2x_2 + \cdots + b_kx_k . \tag{40}$$

Then from Eq. (39), for a k-dimensional rotatable design of first order
all moments $[i]$ $(i = 1, 2, \cdots, k)$ of order 1 are zero,
mixed moments $[ij]$ $(i \neq j = 1, 2, \cdots, k)$ of order 2 are zero,
quadratic moments $[ii]$ $(i = 1, 2, \cdots, k)$ of order 2 $= \lambda_2 = 1$.
Thus the moment matrix $N^{-1}\mathbf{X}'\mathbf{X}$ is $\mathbf{I}_{k+1}$.

The condition that a first order design is rotatable is thus precisely the same as that it should have smallest variances, namely that its moment matrix should be the identity matrix.

The variance function for this type of design is given by

$$NV(\hat{y}_{\bar{x}})/\sigma^2 = V(\rho) = (1 + \rho^2), \tag{41}$$

where $\rho = \left\{ \sum_{i=1}^{k} x_i^2 \right\}^{1/2}$.

The standardized weight function $W(\rho) = [V(\rho)]^{-1}$, which shows the relative precision of the estimate $\hat{y}$ at a distance ρ from the center of the design, for any first order rotatable design is graphed in Fig. 3.

Setting aside our scaling convention we see that in general the variance of

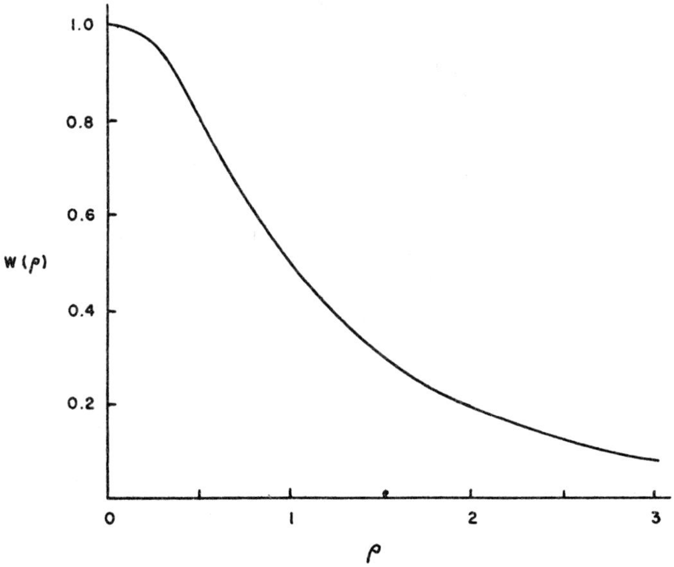

FIG. 3. Weight function for any first order rotatable design

the estimated response at a distance ρ from the center of any rotatable design of first order is given simply by

(41a) $$V(\hat{y}) = V(b_0) + V(b_i)\rho^2.$$

This is of the same form as the well-known formula for a single variable x with ρ replacing x.

6.11. *Biases due to second order coefficients.* If it happens, contrary to assumption, that terms of second order are not negligible, then using Eq. 8a, the expected values of the estimated coefficients for any first order rotatable design are as follows

(42) $$\begin{aligned} \mathcal{E}(b_0) &= \boldsymbol{\beta}_0 + \boldsymbol{\lambda}_2 \sum_{g=1}^{k} \boldsymbol{\beta}_{gg} \\ \mathcal{E}(b_i) &= \boldsymbol{\beta}_i + \sum_{g=1}^{k} \sum_{h=g}^{k} [ghi]\beta_{gh}. \qquad (i = 1, 2, \cdots, k) \end{aligned}$$

Those terms printed in boldface type inevitably arise but those in ordinary type may be eliminated by a suitable choice of the design. On our convention λ_2 is equal to unity and it follows that, if β_0 is estimated assuming a first order model, then when the quadratic coefficients β_{gg} are not zero, bias in this estimate is inevitable. In general the coefficients $[ghi]$ of the biases in the estimates of the linear coefficients β_i will vary as the design is rotated (see reference [2] for particular examples of this). By selecting a design for which all the third order moments are zero we may eliminate bias in the estimates of the linear coefficients in every orientation of the design (see Section 5.4). Specific designs of this sort were discussed in [1] under the name "first order designs of type B." They can be obtained by duplicating with reversed signs any orthogonal first order design, in particular any of the "simplex" designs of [2], [5]. The two-level factorial designs and many of the fractional factorials are also examples of particular orientations of designs of this sort.

6.2. *Rotatable designs of order two.* Suppose we have k variables, and desire to fit a polynomial of degree two. From (39) the moments of a k-dimensional second order rotatable design suitable for this situation are such that all odd moments are zero, and the remaining moments are $[ii] = \lambda_2 = 1$, $[iijj] = \lambda_4$, $[iiii] = 3\lambda_4$.

Thus for a second order rotatable design the moment matrix is of the form:

$$
(43) \qquad N^{-1}\mathbf{X}'\mathbf{X} =
\begin{array}{c}
 \\ 0 \\ 1 \\ 2 \\ \cdot \\ \cdot \\ \cdot \\ k \\ 11 \\ 22 \\ \cdot \\ \cdot \\ \cdot \\ kk \\ 12 \\ 13 \\ \cdot \\ \cdot \\ \cdot \\ k-1,k
\end{array}
\begin{array}{c}
\begin{array}{c|cccc|cccc|c}
0 & 1 & 2 & \ldots & k & 11 & 22 & \ldots & kk & 12\ 13 \ldots k-1,k
\end{array}\\
\left[\begin{array}{c|cccc|cccc|c}
1 & & * & & & 1 & 1 & \ldots & 1 & * \\ \hline
 & 1 & & & & & & & & \\
 & & 1 & & & & & & & \\
* & & & \ddots & & & * & & & * \\
 & & & & & & & & & \\
 & & & & & & & & & \\
 & & & & 1 & & & & & \\ \hline
1 & & & & & 3\lambda_4 & \lambda_4 & \ldots & \lambda_4 & \\
1 & & & & & \lambda_4 & 3\lambda_4 & \ldots & \lambda_4 & \\
\cdot & & * & & & \cdot & \cdot & & \cdot & * \\
\cdot & & & & & \cdot & \cdot & & \cdot & \\
\cdot & & & & & \cdot & \cdot & & \cdot & \\
1 & & & & & \lambda_4 & \lambda_4 & \ldots & 3\lambda_4 & \\ \hline
 & & & & & & & & & \lambda_4 \\
 & & & & & & & & & \quad\lambda_4 \\
* & & * & & & & * & & & \quad\ddots \\
 & & & & & & & & & \\
 & & & & & & & & & \\
 & & & & & & & & & \qquad\lambda_4
\end{array}\right]
\end{array}
$$

where the asterisks indicate null submatrices.

The inverse matrix (i.e. the precision matrix) is readily shown to be

$$
(44) \qquad N(X'X)^{-1} =
\begin{array}{c}
 \\ 0 \\ 1 \\ 2 \\ \cdot \\ \cdot \\ k \\ 11 \\ 22 \\ \cdot \\ \cdot \\ \cdot \\ kk \\ 12 \\ 13 \\ \cdot \\ \cdot \\ \cdot \\ k-1,k
\end{array}
\begin{array}{c}
\begin{array}{c|c|cccc|c}
0 & 1\ 2 \ldots k & 11 & 22 & \ldots & kk & 12\ 13 \ldots k-1,k
\end{array}\\
\left[\begin{array}{c|cccc|cccc|c}
2\lambda_4^2(k+2)A & & * & & & -2\lambda_4 A & -2\lambda_4 A & \ldots & -2\lambda_4 A & * \\ \hline
 & 1 & & & & & & & & \\
 & & 1 & & & & & & & \\
* & & & \ddots & & & * & & & * \\
 & & & & & & & & & \\
 & & & & 1 & & & & & \\ \hline
-2\lambda_4 A & & & & & [(k+1)\lambda_4-(k-1)]A & (1-\lambda_4)A & \ldots & (1-\lambda_4)A & \\
-2\lambda_4 A & & & & & (1-\lambda_4)A & [(k+1)\lambda_4-(k-1)]A & \ldots & (1-\lambda_4)A & \\
\cdot & & * & & & \cdot & \cdot & & \cdot & * \\
\cdot & & & & & \cdot & \cdot & & \cdot & \\
\cdot & & & & & \cdot & \cdot & & \cdot & \\
-2\lambda_4 A & & & & & (1-\lambda_4)A & (1-\lambda_4)A & \ldots & [(k+1)\lambda_4-(k-1)]A & \\ \hline
 & & & & & & & & & \lambda_4^{-1} \\
 & & & & & & & & & \quad\lambda_4^{-1} \\
* & & * & & & & & & & \quad\ddots \\
 & & & & & & & & & \\
 & & & & & & & & & \\
 & & & & & & & & & \qquad\lambda_4^{-1}
\end{array}\right]
\end{array}
$$

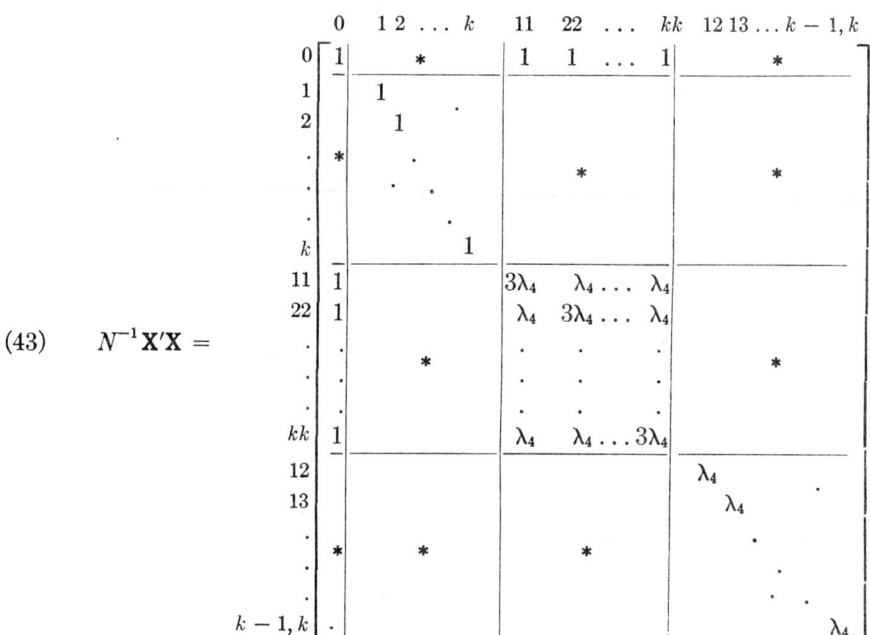

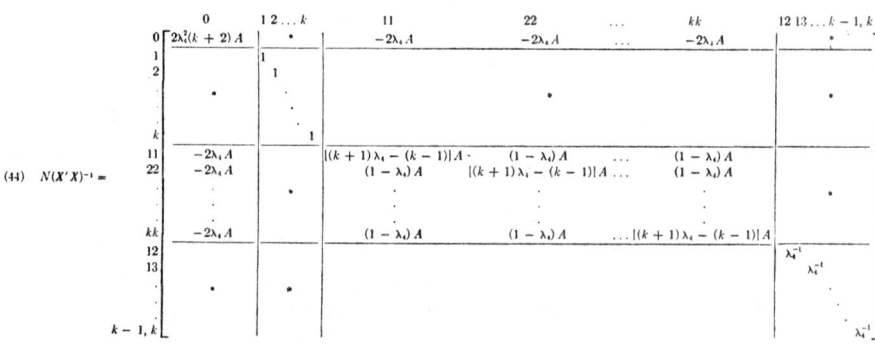

where

(45) $$A = [2\lambda_4\{(k+2)\lambda_4 - k\}]^{-1}.$$

The variances and covariances of the estimated coefficients using any second order rotatable design are therefore given by

(46) $$\frac{NV(b_0)}{\sigma^2} = 2\lambda_4^2(k+2)A; \quad \frac{NV(b_i)}{\sigma^2} = 1;$$
$$\frac{NV(b_{ii})}{\sigma^2} = [(k+1)\lambda_4 - (k-1)]A; \quad \frac{NV(b_{ij})}{\sigma^2} = \lambda_4^{-1};$$
$$\frac{N\,\text{Cov}(b_0\,, b_{ii})}{\sigma^2} = -2\lambda_4 A; \quad \frac{N\,\text{Cov}\,(b_{ii} b_{jj})}{\sigma^2} = (1-\lambda_4)A$$

and all the remaining covariances are zero. Thus all first and second degree coefficients are uncorrelated except the quadratic coefficients. These have a coefficient of correlation $\{[2/(1-\lambda_4)] - (k+1)\}^{-1}$.

If we put $\lambda_4 = 1$ the correlations between the quadratic coefficients are all zero and the design is orthogonal (in the sense of Section 3) as well as rotatable. We then have

(47) $$\frac{NV(b_0)}{\sigma^2} = \tfrac{1}{2}(k+2); \quad \frac{NV(b_i)}{\sigma^2} = 1; \quad \frac{NV(b_{ii})}{\sigma^2} = \tfrac{1}{2};$$
$$\frac{NV(b_{ij})}{\sigma^2} = 1; \quad \frac{N\,\text{Cov}\,(b_0 b_{ij})}{\sigma} = -\tfrac{1}{2}.$$

The conditions of rotatability and orthogonality together fix the relative variances for effects of different orders. In particular for designs of this sort the variances of the quadratic coefficients b_{ii} are one half those of the two-factor interaction coefficients b_{ij}, in contrast with three-level factorial designs for which the variances of the quadratic coefficients are twice those for the interaction coefficients. Compared with the factorial in standard orientation, the orthogonal rotatable design thus places four times as much emphasis on the quadratic coefficient relative to the interaction coefficients. From (44) the variance function for any general second order rotatable design is given by

(48) $$V(\rho) = A\{2(k+2)\lambda_4^2 + 2\lambda_4(\lambda_4 - 1)(k+2)\rho^2 + [(k+1)\lambda_4 - (k-1)]\rho^4\},$$

and for the particular case of orthogonal second order rotatable design by

(48a) $$V(\rho) = \tfrac{1}{2}(k + 2 + \rho^4).$$

Setting aside for the moment our scaling convention we see that in general the variances of the estimated response at a distance ρ from the center of any rotatable second order design is simply given by

$$V(\hat{y}) = V(b_0) + 2\,\text{Cov}\,(b_0 b_{ii})\rho^2 + V(b_i)\rho^2 + V(b_{ii})\rho^4.$$

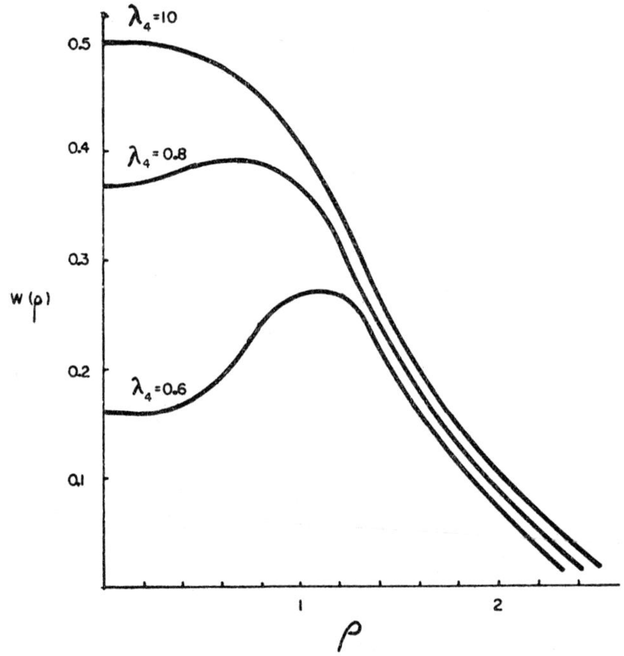

FIG. 4. Weight functions for second order rotatable designs having various values of λ when $k = 2$

This, with ρ replacing x, is of the form obtained for a single variable x when the design employed is any symmetric arrangement.

In Fig. 4 the standardized weight function $W(\rho) = \{V(\rho)\}^{-1}$ is graphed for various values of λ_4 for the case of a two dimensional second order design. Similar graphs are obtained for other values of k.

We notice that whatever value of λ_4 is chosen the precision falls off rapidly when ρ exceeds unity. If we chose $\lambda_4 = 1$ the design is orthogonal in the sense discussed previously. When λ_4 approaches or exceeds unity, the precision, particularly at the center of the design, is high but the bias coefficients are high also. It is well to remember at this point that we are comparing designs for which the "spread" of points, as measured by the marginal second moments $S_i^2 = N^{-1}\sum_{u=1}^{N}(\xi_{iu}^2 - \bar{\xi}_i)^2$, is constant. Such a convention is bound to favor designs with a high value of λ_4, so that although the general shape of the design weight function will be meaningful, the absolute height of the curve at any point will be to some extent an outcome of this convention. It seems reasonable to seek a relatively uniform distribution of precision in the immediate vicinity of the design. This is attained with a value of λ_4 somewhat less than unity and such a choice will give satisfactory values for the third order bias coefficient. In particular, if the variance at $\rho = 1$ is to equal the variance at $\rho = 0$, the values of λ_4 shown in Table 1 will be needed.

TABLE 1

Values of λ_4 required to make the variance at $\rho = 1$ equal to that at $\rho = 0$

k	2	3	4	5	6	7	8
λ_4	0.7844	0.8385	0.8704	0.8918	0.9070	0.9184	0.9274

So far we have supposed that our interest centers on the absolute value of η as estimated by $\hat{y}$. In some circumstances we would be more interested in the relative value of η rather than in its absolute magnitude. For example, the estimation of the slope of a first degree surface, or of the position of the stationary point on a fitted second degree surface, does not require knowledge of the absolute value of η but only of its value at one point in the space of the variables relative to its value at some other point.

A natural reference value is β_0, the response at the origin of the design, and the appropriate estimate of

$$\Delta = \eta - \beta_0 = \beta_1 x_1 + \cdots + \beta_k x_k + \beta_{11} x_1^2 + \text{etc.}$$

is

$$\hat{d} = b_1 x_1 + \cdots + b_k x_k + b_{11} x_1^2 + \text{etc.}$$

For the rotatable designs already discussed, $V(\hat{d})$ as well as $V(\hat{y})$ is constant at a constant distance ρ from the center of the design. At a point distance ρ from the center, for any first order rotatable design, we have simply

$$V(\hat{d}) = V(b_i)\rho^2 \tag{49}$$

and for any second order rotatable design

$$V(\hat{d}) = V(b_i)\rho^2 + V(b_{ii})\rho^4. \tag{49a}$$

It is not our intention at this time to claim optimal properties for the designs which are here derived. The only justification presented is that these arrangements symmetrically generate information in the coordinates regarded as most relevant by the experimenter. An approach which makes it possible to measure the practical efficiency of these and other arrangements has however been attempted and it is hoped to publish this elsewhere. In this more complete appraisal of the situation, a citerion on the effectiveness of the design is taken as the integrated mean square error of $\hat{y}$ over the region of interest to the experimenter. The scale of the design, and the constants such as λ_4, which reflect the distribution of the experimental points, are so chosen as to minimize this criterion. The integrated mean square error contains two terms: one which measures the variance of $\hat{y}$ and the other the bias in $\hat{y}$ due to possible inadequacies of the assumed model. As λ_4 is increased the variance term becomes smaller and the bias term becomes larger. The problem is to strike some compromise which will be satisfactory for practical situations. It appears that the values of λ_4 given in Table 1, although perfectly satisfactory, may be a little high in the light of this more recent work.

In the present paper we consider the nature of the bias in the estimates of the individual coefficients rather than the nature of the bias in $\hat{y}$.

6.21. *Biases due to third order constants.* If it happens, contrary to assumption, that terms of third order are not negligible, then using equation 8(a), the expected values of the estimated constants for any second order rotatable design are as follows:

$$\mathcal{E}(b_0) = \boldsymbol{\beta}_0 - 2\lambda_4 A \sum_{f=1}^{k} \sum_{g=f}^{k} \sum_{h=g}^{k} \sum_{i=1}^{k} [fghii]\beta_{fgh}, \tag{50}$$

$$\mathcal{E}(b_i) = \boldsymbol{\beta}_i + 3\boldsymbol{\lambda}_4 \boldsymbol{\beta}_{iii} + \boldsymbol{\lambda}_4 \sum_{h \neq i}^{k} \boldsymbol{\beta}_{hhi}, \tag{51}$$

$$\begin{aligned} \mathcal{E}(b_{ii}) = \boldsymbol{\beta}_{ii} &+ \{(k+2)\lambda_4 - k\} A [iiiii]\beta_{iii} \\ &+ (1 - \lambda_4) A \sum_{f=1}^{k} \sum_{g=f}^{k} \sum_{h=g}^{k} \sum_{i=1}^{k} [fghii]\beta_{fgh}, \end{aligned} \tag{52}$$

$$\mathcal{E}(b_{ij}) = \boldsymbol{\beta}_{ij} + \lambda_4^{-1} \sum_{f=1}^{k} \sum_{g=f}^{k} \sum_{h=g}^{k} [fghij]\, \beta_{fgh}. \tag{53}$$

Again terms printed in boldface type arise inevitably but those in ordinary type may be eliminated in every orientation by selecting a design for which all the fifth order moments are zero.

The requirements and properties of rotatable designs of third and higher order may be studied in the same way as for designs of order 1 and 2. We shall not pursue this topic here but will now consider how we may obtain actual arrangements of points which satisfy the requirements for rotatable designs of second order.

7. Examples of second order rotatable designs. The above discussion has been directed to deciding what type of design we should be seeking. We have shown the moment conditions which a design of any given order must satisfy to obtain constant precision on spheres centered at the origin of the design. We now consider the problem of finding arrangements of experimental points which satisfy these conditions. The classes of rotatable designs we discuss are by no means exhaustive but rather are intended to illustrate some of the possibilities.

We mention in passing that for any design of order d which is both orthogonal in the sense already discussed, and rotatable, has moments up to order $2d$ which are the same as those of the spherical multinormal distribution. This is of interest since it shows that in the usual multiple regression problem where the values of the independent variables $x_1, x_2, \cdots, x_k$ are not held at predetermined levels but are allowed to vary at random we should obtain a good arrangement, if it happened the x's followed a multivariate normal distribution with zero covariances—a conclusion which is intuitively very acceptable. Rotatable designs could be approximated simply by taking independent random samples from a normal distribution, but in fact it is possible to satisfy the criterion of rotatability exactly.

We have seen that for a rotatable design of order d the moments must be the same up to order $2d$ as those of a spherical density function. This suggests that we might construct rotatable designs by equally spacing the available finite number of experimental points on one or more spheres.

We find in fact that it is convenient to regard designs as built up from a number of component sets of points each set having its points all equidistant from the origin. This we call an equiradial set and ρ the distance of each point from the origin the radius of the set. If the moments to order $2d$ of such a set are unchanged by rotation we call this an equiradial rotatable set of order d.

An equiradial rotatable set of order d does not necessarily, or even usually, of itself provide a design. For example, n points at the origin provide an equiradial set of infinite order. Furthermore, no single equiradial set can provide a design of order greater than one, for if ρ is the common distance from the origin then

$$\sum_{i=1}^{k} x_{iu}^2 = \rho^2 x_{0u}, \qquad u = 1, 2, \cdots, N.$$

It follows that if $d > 1$ the matrix of independent variables for such an arrangement is singular and the quadratic coefficients $b_{11}, b_{22}, \cdots, b_{kk}$ and the constant term b_0 cannot be separately estimated.

As is shown later we can obtain equiradial sets of points which satisfy the moment conditions (39) for rotatable design of order 2 but only for such values of λ_4 as lead to a singular moment matrix. For such a set of points $\rho^2 = \sum_{i=1}^{k} x_{iu}^2$ for $u = 1, 2, \cdots, N$, whence

$$\rho^2 = N^{-1} \sum_{u=1}^{N} \sum_{i=1}^{k} x_{iu}^2 = \sum_{i=1}^{k} [ii] = k; \tag{54}$$

also,

$$\rho^4 = N^{-1} \sum_{u=1}^{N} \left\{ \sum_{i=1}^{k} x_{iu}^2 \right\}^2 = \sum_{i=1}^{k} [iiii] + \sum_{i=1}^{k} \sum_{j \neq i}^{k} [iijj], \tag{55}$$

whence

$$3k\lambda_4 + k(k-1)\lambda_4 = k^2 \tag{56}$$

and

$$\lambda_4 = k/(k+2). \tag{57}$$

When this value of λ_4 is substituted in (45) and (46) the quantity A becomes infinite and the quadratic terms and the constant term are not separately estimable. We shall refer to this value λ_4 as the "singular" value.

Now although no single equiradial set can supply a usable second order design, two or more sets can do so. Suppose we have s equiradial rotatable sets of points having the same origin such that in the wth set there are n_w points each at a distance ρ_w from the origin then the value of λ_4 for the whole aggregate of the $N = \sum_{w=1}^{s} n_w$ points is

$$(58) \qquad \lambda_4 = \frac{Nk \sum_{w=1}^{s} n_w \rho_w^4}{(k+2)\left(\sum_{w=1}^{s} n_w \rho_w^2\right)^2},$$

which will not in general have a singular value. By combining equiradial sets we can thus obtain rotatable designs. Since a set of points at the origin will affect only the value of N in (58) the formula may be applied without modification to designs for which one set of points is at the center. In practice, we shall find that the placing of one or more points at the center of an equiradial rotatable set provides one useful means of modifying the value of λ_4. If there are n_1 points at the origin and n_2 points in the equiradial rotatable set we see that for the aggregate of $n_1 + n_2$ points

$$(59) \qquad \lambda_4 = \frac{k(n_1 + n_2)}{(k+2)n_2}.$$

We now show that sets of points which are equally spaced on a circle, a sphere, or a hypersphere and which thus form the vertices of a regular polygon, polyhedron, or polytope, can provide rotatable sets which may be combined to form rotatable designs. Our study begins with the two dimensional figures.

7.1 *Two dimensional designs*. We first show that for the vertices of a regular n-gon, all moments up to order $n - 1$ are invarient under rotation.

Suppose the coordinates of the uth point are $\rho \cos(\varphi + 2\pi u/n)$ and $\rho \sin(\varphi + 2\pi u/n)$ and that $a = e^{i\varphi}$ and $\omega = e^{i2\pi/n}$. We have for the moment $[pq]$ which is of order $p + q$

$$(60) \qquad [pq]_\varphi = (\tfrac{1}{2}\rho)^{p+q} i^{-q} \sum_{u=0}^{n-1} (a\omega^u + a^{-1}\omega^{-u})^p (a\omega^u - a^{-1}\omega^{-u})^q.$$

After expanding the bracketed expressions and collecting terms

$$(61) \qquad [pq]_\varphi = (\tfrac{1}{2}\rho)^{p+q} \sum_{r=0}^{p} \sum_{t=0}^{q} i^{2t-q} \binom{p}{r}\binom{q}{t} a^{p+q-2(r+t)} \sum_{u=0}^{n-1} \omega^{u\{p+q-2(r+t)\}}.$$

By putting $a = 1$ in the expression and substituting the result from (61) we obtain the change in the value of the moment after rotation through an angle φ:

$$(62) \quad [pq]_\varphi - [pq]_0 = (\tfrac{1}{2}\rho)^{p+q} \sum_{r=0}^{p} \sum_{t=0}^{q} i^{2t-q} \binom{p}{r}\binom{q}{t} [a^{p+q-2(r+t)} - 1] \sum_{u=0}^{n-1} \omega^{u\{p+q-2(r+t)\}}.$$

Now

$$(63) \qquad \sum_{u=0}^{n-1} \omega^{u\{p+q-2(r+t)\}} \begin{array}{ll} = n & \text{if } p + q - 2(r+t) = 0 \text{ or } mn \text{ where } m \text{ is an integer,} \\ = 0 & \text{otherwise,} \end{array}$$

and $-(p + q) \leqq p + q - 2(r + t) \leqq p + q$. Thus if $(p + q) < n$ the expression on the left of (63) is zero unless $p + q = 2(r + t)$, but in this case a $a^{\{p+q-2(r+t)\}} - 1 = 0$. Hence if $p + q < n$ then (62) is zero whatever the value of φ and our assertion is proved.

A class of two-dimensional second order rotatable designs may be constructed therefore from two or more concentric rings of equispaced points with unequal radii. Points at the origin constitute a ring of zero radius and each ring which is not of zero radius must contain at least five points. The number of the points in each set and the radial distances will determine the value of λ_4 in accordance with equation (58).

Of this class of designs the simplest are those having one ring of $n_2 \geqq 5$ equally spaced points with additional n_1 points at the origin.

Pentagonal designs with center points. Putting $n_2 = 5$ we obtain the following values of λ_4 for specimen numbers of center points

Number of points at center of pentagon............	1	3	5
Value of λ_4....................................	0.6	0.8	1.0

By using three points at the center a value of $\lambda_4 = 0.8$ is obtained which is close to that given in Table 1. For orthogonality five center points are required.

Hexagonal designs with center points. If we put n_2 equal to six so that the external points are at the vertices of a hexagon, we obtain the added advantage that all the moments of order 5 are zero thus insuring that in every orientation of the design the estimate b_0 and the estimates b_{ii}, b_{ij} of the second order coefficients are not biased by any third order terms. A value of λ_4 close to that given in Table 1 is obtained by placing $n_1 = 3$ points at the center. "Orthogonality" is obtained when $n_1 = 6$.

Designs containing two rings of points. A variety of designs, which however use more than ten points, can be obtained by combining two or more concentric circles of equispaced points. Table 2, below, shows some of the alternatives. Values of the ratio of radii are shown (i) which give $\lambda_4 = 0.7844$, the value given in Table 1, and (ii) which give $\lambda_4 = 1.0$, the value required for orthogonality.

TABLE 2

Radii for equispaced points on concentric circles

n_1	5	5	5	6	6	7
n_2	6	7	8	7	8	8
ρ_2/ρ_1 for $\lambda_4 = 0.7844$	0.414	0.438	0.454	0.407	0.430	0.404
ρ_2/ρ_1 for $\lambda_4 = 1.0$	0.204	0.267	0.304	0.189	0.250	0.176

The arrangements so far discussed by no means exhaust the possible second order designs in two dimensions. A further class of designs is obtained by combining sets of equiradial points which are not individual rotatable sets of order 2. For example, sets of three points each of which form the vertices of an equilateral triangle with center coincident with the origin, may be combined to form such arrangements. Suppose that a line, assumed to be of length $\rho\sqrt{2}$ and connecting one of the vertices of an equilateral triangle to the center makes an angle φ with the x_2 axis. Then it is easily shown that the second order moment matrix for this arrangement is

$$N^{-1}\mathbf{X}'\mathbf{X} = \begin{array}{c} \\ \end{array}\begin{array}{cccccc} 0 & 1 & 2 & 11 & 22 & 12 \end{array}$$

$$N^{-1}\mathbf{X}'\mathbf{X} = \begin{bmatrix} 1 & \cdot & \cdot & \rho^2 & \rho^2 & \cdot \\ \cdot & \rho^2 & \cdot & \rho^3 a & -\rho^3 a & -\rho^3 b \\ \cdot & \cdot & \rho^2 & -\rho^3 b & \rho^3 b & -\rho^3 a \\ \rho^2 & \rho^3 a & -\rho^3 b & \frac{3}{2}\rho^4 & \frac{1}{2}\rho^4 & \cdot \\ \rho^2 & -\rho^3 a & \rho^3 b & \frac{1}{2}\rho^4 & \frac{3}{2}\rho^4 & \cdot \\ \cdot & -\rho^3 b & -\rho^3 a & \cdot & \cdot & \frac{1}{2}\rho^4 \end{bmatrix},$$

where $a = 2^{-\frac{1}{2}} \sin 3\varphi$ and $b = 2^{-\frac{1}{2}} \cos 3\varphi$.

This is of the form required for rotatability except for the elements containing a and b. Suppose s arrangements of this sort are combined, the wth such arrangement having $\rho = \rho_w$ and $\varphi = \varphi_w$. Then the moment matrix will be of the exact form required for rotatability if

$$(64) \qquad \sum_{w=1}^{s} \rho_w^3 \sin 3\varphi_w = 0 \quad \text{and} \quad \sum_{w=1}^{s} \rho_w^3 \cos 3\varphi_w = 0.$$

This implies simply that the sum of the s vectors

$$(\rho_1^3 \cos 3\varphi_1, \rho_1^3 \sin 3\varphi_1), \cdots, (\rho_s^3 \cos 3\varphi_s, \rho_s^3 \sin 3\varphi_s)$$

shall be zero (i.e., they form the sides of some polygon) and an infinity of solutions is at once obtainable.

The value of λ_4 for such a design is

$$(65) \qquad \lambda_4 = \tfrac{1}{2}s \sum_{w=1}^{s} \rho_w^4 \Big/ \left(\sum_{w=1}^{s} \rho_w^2\right)^2.$$

If $s = 2$, then to satisfy (64) the two vectors must be of the form

$$(\rho^3 \cos 3\varphi, \rho^3 \sin 3\varphi) \quad \text{and} \quad (-\rho^3 \cos 3\varphi, -\rho^3 \sin 3\varphi),$$

where φ is arbitrary. The design then consists of the vertices of a hexagon with λ_4 equal to its singular value.

If $s > 2$ an infinite variety of these designs, in which the sets of points have different radii and appropriate relative orientations, can be derived. The largest value of λ_4 is obtained when all but two of the ρ_w are zero. The two non-zero vectors are equal in magnitude and opposite in sign and $\lambda_4 = \frac{1}{4}s$. The resulting design then consists of the vertices of a hexagon in any orientation together with the remaining $3(s - 2)$ points at the center.

Designs may similarly be built up from combinations of regular figures having

2 and 4 points. As before the maximum value of λ_4 is obtained by having a single ring of equispaced points plus points at the center.

7.2 *Rotation of Set of Points in k dimensional space.* In order to investigate the possibilities in more than two dimensions we first consider a method for studying the effect on the moment matrix of rotating any k-dimensional arrangement of experimental points.

Consider some set of N points in k dimensional space and as before denote by **D** the $N \times k$ design matrix, the elements of whose rows are the coordinates of the points. To correspond to our definition of derived power vectors suppose a second degree model written so that product terms such as $x_i x_j$ have the coefficient $\sqrt{2}$ attached. For example, if $k = 3$

$$\eta = \beta_0 + \beta_1 x_1 + \beta_2 x_2 + \beta_3 x_3 + \beta_{11} x_1^2 + \beta_{22} x_2^2 + \beta_{33} x_3^2 \tag{66}$$
$$+ \frac{\beta_{12}}{\sqrt{2}}(\sqrt{2}\, x_1 x_2) + \frac{\beta_{13}}{\sqrt{2}}(\sqrt{2}\, x_1 x_3) + \frac{\beta_{23}}{\sqrt{2}}(\sqrt{2}\, x_2 x_3).$$

As before we denote by **X** the $N \times \frac{1}{2}(k+1)(k+2)$ matrix of independent variables corresponding to **D**. If the set of points is submitted to some rotation the new design matrix will be **DH** where **H** is $k \times k$ orthogonal matrix. The matrix of independent variables will be transformed to $\dot{\mathbf{X}} = \mathbf{XG}$ where **G** is a matrix derived from **H** which when partitioned after the 1st and $(k+1)$th row and column has the form

$$\mathbf{G} = \left[\begin{array}{c|c|c} 1 & & \\ \hline & \mathbf{H} & \\ \hline & & \mathbf{H}^{[2]} \end{array}\right] \tag{67}$$

The partitioning will be seen to correspond to the separation of the constant term, first order coefficients and second order coefficients. The matrix $\mathbf{H}^{[2]}$ is the second Schläflian matrix derived from **H** which may itself be conveniently partitioned after its kth row and column and is of the form

$$\mathbf{H}^{[2]} = \left[\begin{array}{c|c} \boldsymbol{\alpha} & \boldsymbol{\beta} \\ \hline \boldsymbol{\gamma} & \boldsymbol{\delta} \end{array}\right]. \tag{68}$$

This partitioning corresponds to the separation of quadratic and interaction effects. For example, for $k = 3$

$$\mathbf{H} = \begin{bmatrix} h_{11} & h_{12} & h_{13} \\ h_{21} & h_{22} & h_{23} \\ h_{31} & h_{32} & h_{33} \end{bmatrix}; \tag{69}$$

then

(70) and (71) $$\boldsymbol{\alpha} = \begin{bmatrix} h_{11}^2 & h_{12}^2 & h_{13}^2 \\ h_{21}^2 & h_{22}^2 & h_{23}^2 \\ h_{31}^2 & h_{32}^2 & h_{33}^2 \end{bmatrix}, \qquad \boldsymbol{\beta} = \begin{bmatrix} 2^{\frac{1}{2}}h_{11}h_{12} & 2^{\frac{1}{2}}h_{11}h_{13} & 2^{\frac{1}{2}}h_{12}h_{13} \\ 2^{\frac{1}{2}}h_{21}h_{22} & 2^{\frac{1}{2}}h_{21}h_{23} & 2^{\frac{1}{2}}h_{22}h_{23} \\ 2^{\frac{1}{2}}h_{31}h_{32} & 2^{\frac{1}{2}}h_{31}h_{33} & 2^{\frac{1}{2}}h_{32}h_{33} \end{bmatrix},$$

(72) $$\boldsymbol{\gamma} = \begin{bmatrix} 2^{\frac{1}{2}}h_{11}h_{21} & 2^{\frac{1}{2}}h_{12}h_{22} & 2^{\frac{1}{2}}h_{13}h_{23} \\ 2^{\frac{1}{2}}h_{11}h_{31} & 2^{\frac{1}{2}}h_{12}h_{32} & 2^{\frac{1}{2}}h_{13}h_{33} \\ 2^{\frac{1}{2}}h_{21}h_{31} & 2^{\frac{1}{2}}h_{22}h_{32} & 2^{\frac{1}{2}}h_{23}h_{33} \end{bmatrix},$$

(73) $$\boldsymbol{\delta} = \begin{bmatrix} h_{11}h_{22} + h_{12}h_{21} & h_{11}h_{23} + h_{13}h_{21} & h_{12}h_{23} + h_{13}h_{22} \\ h_{11}h_{32} + h_{12}h_{31} & h_{11}h_{33} + h_{13}h_{31} & h_{12}h_{33} + h_{13}h_{32} \\ h_{21}h_{32} + h_{22}h_{31} & h_{21}h_{33} + h_{23}h_{31} & h_{22}h_{33} + h_{23}h_{32} \end{bmatrix},$$

where the terms are arranged in the order shown in Eq. (66).

The original moment matrix is $N^{-1}\mathbf{X}'\mathbf{X}$. After rotation the moment matrix is $N^{-1}\dot{\mathbf{X}}'\dot{\mathbf{X}} = N^{-1}\mathbf{G}'\mathbf{X}'\mathbf{X}\mathbf{G}$. If the design is rotatable then the moment matrix before and after rotation are equal whatever the orthogonal matrix $\mathbf{H}$ from which $\mathbf{G}$ is derived. As we have seen, this implies that the matrix $N^{-1}\mathbf{X}'\mathbf{X}$ must be of a particular form. With the present definition of the interaction variables this form is

(74) $$N^{-1}(\mathbf{X}'\mathbf{X}) = \begin{bmatrix} 1 & \cdot & \mathbf{1}' & \cdot \\ \cdot & \mathbf{I} & \cdot & \cdot \\ \mathbf{1} & \cdot & \lambda_4(2\mathbf{I} + \mathbf{1}\mathbf{1}') & \cdot \\ \cdot & \cdot & \cdot & 2\lambda_4\mathbf{I} \end{bmatrix},$$

where the dots indicate null submatrices, $\mathbf{I}$ is the identity matrix and $\mathbf{1}$ is a column vector with elements all unity. The partitioning in this and all moment matrices that follow is after the 1st, $(k + 1)$th, and $(2k + 1)$th row and column. This partitioning separates the elements corresponding to the constant term, the first order terms, the quadratic terms and the interaction terms respectively.

7.3 *Three dimensional designs.* In three dimensions sets of n points equally spaced on a sphere are provided by the vertices of the five regular figures. These are the tetrahedron ($n = 4$), the octahedron ($n = 6$), the cube ($n = 8$), the icosahedron ($n = 12$), and the dodecahedron ($n = 20$). Using the method of the previous section we can study the moment matrices for these sets of points subject to an arbitrary rotation.

For example, in the case of the tetrahedron we may suppose that initially the coordinates of its four vertices, i.e., the rows of $\mathbf{D}$, are $(-1, -1, 1)$, $(1, -1, -1)$, $(-1, 1, -1)$, and $(1, 1, 1)$. We can then write down the matrix of independent variables $\mathbf{X}$ for a second order model, the moment matrix $\frac{1}{4}\cdot\mathbf{X}'\mathbf{X}$ and finally the corresponding matrix $\frac{1}{4}\cdot\dot{\mathbf{X}}'\dot{\mathbf{X}} = \frac{1}{4}\cdot\mathbf{G}'\mathbf{X}'\mathbf{X}\mathbf{G}$ after an arbitrary rotation $\mathbf{H}$ has been applied to $\mathbf{D}$. We thus obtain

$$\frac{1}{4}\cdot\dot{\mathbf{X}}'\dot{\mathbf{X}} = \left[\begin{array}{c|c|c|c} 1 & \cdot & \mathbf{1}' & \cdot \\ \hline \cdot & \mathbf{I} & \sqrt{2}\mathbf{H}'\mathbf{J}\boldsymbol{\gamma} & \sqrt{2}\mathbf{H}'\mathbf{J}\boldsymbol{\delta} \\ \hline \mathbf{1} & \sqrt{2}\boldsymbol{\gamma}'\mathbf{J}\mathbf{H} & \mathbf{1}\mathbf{1}' + 2\boldsymbol{\gamma}'\boldsymbol{\gamma} & 2\boldsymbol{\gamma}'\boldsymbol{\delta} \\ \hline \cdot & \sqrt{2}\boldsymbol{\delta}'\mathbf{J}\mathbf{H} & 2\boldsymbol{\delta}'\boldsymbol{\gamma} & 2\boldsymbol{\delta}'\boldsymbol{\delta} \end{array}\right],$$

Where $\mathbf{J}$ is a square matrix with unit elements along the diagonal running from the bottom left hand corner to the top right-hand corner and zero elements elsewhere.

In a similar way for the other regular figures with scales adjusted so that $\rho^2 = k = 3$ we consider equiradial points formed by the

6 vertices of the octahedron	$(\pm\sqrt{3}, 0, 0), (0, \pm\sqrt{3}, 0), (0, 0, \pm\sqrt{3})$,
8 vertices of a cube	$(\pm 1, \pm 1, \pm 1)$,
12 vertices of the icosahedron	$(0, \pm a, \pm b), (\pm b, 0, \pm a), (\pm a, 0, \pm b)$,
20 vertices of the dodecahedron	$(0, \pm c^{-1}, \pm c), (\pm c, 0, \pm c^{-1}), (\pm c^{-1}, \pm c, 0)$,
	$(\pm 1, \pm 1, \pm 1)$,

where $a = 1.473$, $b = 0.911$, and $c = 1.618$. For the octahedron and cube the moment matrices after applying the general rotation are obtained by setting $k = 3$ in the expressions:

(76)
$$\frac{1}{2k}\mathbf{X}'\mathbf{X} = \left[\begin{array}{c|c|c|c} 1 & \cdot & \mathbf{1}' & \cdot \\ \hline \cdot & \mathbf{I} & \cdot & \cdot \\ \hline \mathbf{1} & \cdot & k\boldsymbol{\alpha}'\boldsymbol{\alpha} & k\boldsymbol{\alpha}'\boldsymbol{\beta} \\ \hline \cdot & \cdot & k\boldsymbol{\beta}'\boldsymbol{\alpha} & k\boldsymbol{\beta}'\boldsymbol{\beta} \end{array}\right],$$

Octahedron

(77)
$$\frac{1}{2^k}\mathbf{X}'\mathbf{X} = \left[\begin{array}{c|c|c|c} 1 & \cdot & \mathbf{1}' & \cdot \\ \hline \cdot & \mathbf{I} & \cdot & \cdot \\ \hline \mathbf{1} & \cdot & \mathbf{1}\mathbf{1}' + 2\boldsymbol{\gamma}'\boldsymbol{\gamma} & 2\boldsymbol{\gamma}'\boldsymbol{\delta} \\ \hline \cdot & \cdot & 2\boldsymbol{\delta}'\boldsymbol{\gamma} & 2\boldsymbol{\delta}'\boldsymbol{\delta} \end{array}\right]$$

Cube

and for the icosahedron and dodecahedron by setting $n = 12$ and 20 respectively in the expression:

$$(78)\qquad \frac{1}{n}\,\mathbf{X}'\mathbf{X} = \left[\begin{array}{c|c|c|c} 1 & \cdot & \mathbf{1}' & \cdot \\ \hline \cdot & \mathbf{I} & \cdot & \cdot \\ \hline \mathbf{1} & \cdot & \frac{3}{5}(2\mathbf{I} + \mathbf{1}\mathbf{1}') & \cdot \\ \hline \cdot & \cdot & \cdot & \frac{3}{5}(2\mathbf{I}) \end{array}\right].$$

As is to be expected, the vertices of the tetrahedron, octahedron, and cube do not individually supply rotatable sets of order 2 although they all provide rotatable sets of order 1. The larger number of points provided by the icosahedron and the dodecahedron however provide rotatable sets of order 2. These have of course the singular value of $\lambda_4 = \frac{3}{5}$.

As before therefore we may combine icosahedral and dodecahedral sets either with points in the center or with one another to form second order rotatable designs. These designs are built up in any one of the following ways:

(1) the 12 vertices of an icosahedron in any orientation with $n_1 \geqq 1$ points at the center,

(2) the 20 vertices of a dodecahedron in any orientation with $n_1 \geqq 1$ points at the center,

(3) the vertices of two concentric icosahedra of differing radii ρ_1 and ρ_2 each in any relative and absolute orientation,

(4) the vertices of two concentric dodecahedra of differing radii ρ_1 and ρ_2 each in any relative and absolute orientation,

(5) the vertices of an icosahedron of radius ρ_1 together with the vertices of a dodecahedron having the same center and radius ρ_2 each in any relative and absolute orientation.

The choice of n_1 in designs (1) and (2) and of ρ_1/ρ_2 in designs (3), (4), and (5) determines the value of λ_4 and hence via Eq. (48) the manner in which the variance of $\hat{y}$ changes with ρ. In particular, the value given in Table 1 for $k = 3$ is $\lambda_4 = 0.84$. This value is most nearly obtained (Eq. 59) with $n_1 = 5$ for design (1), $n_1 = 8$ for design (2). Also from Eq. (58), ρ_1/ρ_2 should be 2.11 for designs (3) and (4), and 2.85 or 0.530 for design (5).

As with the two dimensional designs, sets of points not themselves rotatable arrangements of order two may be combined to give second order rotatable designs. Of particular importance because of the existence of parallel designs in k dimensions are the designs obtained by combining the vertices of a concentric cube and octahedron. The relative orientation of the figures is such that each line joining the origin to a vertex of the cube pierces the center of a face of the octahedron and vice versa. These designs are special cases of the central composite designs described in references [1] and [13].

Since $\mathbf{H}^{[2]}$ is orthogonal

$$(79)\qquad \boldsymbol{\alpha}'\boldsymbol{\beta} = -\boldsymbol{\gamma}'\boldsymbol{\delta}; \qquad \boldsymbol{\gamma}'\boldsymbol{\gamma} = \mathbf{I} - \boldsymbol{\alpha}'\boldsymbol{\alpha}; \qquad \boldsymbol{\beta}'\boldsymbol{\beta} = \mathbf{I} - \boldsymbol{\delta}'\boldsymbol{\delta}.$$

Using these identities, it will be seen from the nature of the moment matrices of the octahedron and cube (76) and (77) that by suitably choosing the relative sizes of the figures a combined arrangement can be obtained whose moment matrix is of the form (74) required for a rotatable design of second order. If ρ_a and ρ_c are the radii of circumscribing spheres of the octahedron and cube then $\rho_a/\rho_c = 2^{3/4}/3^{1/2} = 0.9710$ and $\lambda_4 = 0.6005$.

This value of λ_4 is very close to the singular value of $k/(k + 2) = 0.6$. By adding points at the center however satisfactory values of λ_4 may be obtained. The value $\lambda_4 = 0.84$ given in Table 1 which gives the same precision at $\rho = 0$ as at $\rho = 1$ is most nearly obtained with 6 points at the center, and the value of $\lambda_4 = 1$ required for "orthogonality" is most nearly attained with 9 points at the center.

From the nature of the moment matrices in Eqs. (76), (77), and (78) it is seen that in general an infinity of three-dimensional second order rotatable designs can be generated by combining in various ways the vertices of octahedra, cubes, icosahedra and dodecahedra with or without added center points. One such design of considerable practical interest used by De Baun employs a cube with *two* added octahedra but no center points.

7.4 *Designs in more than* 3 *dimensions*. In five or more dimensions there exist only three regular figures. These are the regular simplex (k-dimensional analogue of the tetrahedron having $k + 1$ vertices), the cross-polytope (k-dimensional analogue of the octahedron having $2k$ vertices) and the measure polytope or hypercube (k-dimensional analogue of the cube having 2^k vertices). In four dimensions other regular figures occur. These have 24 vertices, 120 vertices, and 600 vertices. The figures with 120 and 600 vertices are of little interest from the point of view of constructing usable experimental designs and the figure with 24 vertices may be obtained by combining the cross polytope with the hypercube. Our discussion will therefore be confined to designs constructed from the vertices of the regular simplex, the cross polytope and the hypercube.

The regular simplex always supplies a first order rotatable design and we shall show that the cross polytope and hypercube can always be combined to give an arrangement from which a second order rotatable design may be obtained.

If we suppose the cross polytope and the hypercube each to have radius $k^{1/2}$ and to be in 'standard orientation' so that the $2k$ vertices of the cross polytope have coordinates

$$(\pm k^{1/2}, 0, 0, \cdots, 0)\ (0, \pm k^{1/2}, 0, \cdots, 0) \cdots, (0, 0, 0, \cdots, \pm k^{1/2})$$

and the 2^k vertices of the cube have coordinates $(\pm 1, \pm 1, \cdots, \pm 1)$ then the moment matrices, after applying the same rotation $\mathbf{H}$, are those given in Eqs. (76) and (77).

By combining the vertices of the cross polytope of radius ρ_a with those of the measure polytope of radius ρ_c in the relative orientation indicated so that $2k^2\rho_a^4 = 2^{k+1}\rho_c^4$, that is so that $\rho_a/\rho_c = 2^{k/4}/k^{1/2}$; an arrangement with the desired moment matrix (74) results.

These rotatable arrangements when combined together or augmented with suitable numbers of points at the center provide second order rotatable designs of great practical value. In their "standard orientation" the resulting designs are particular examples of the composite designs discussed in references [1] and [13] and consequently lend themselves very conveniently to sequential experimentation. As is discussed more fully in [1] they may be built up in parts each of which supplies valuable interim information. They are particularly simple to use. In standard orientation the part of the design corresponding to the measure polytope or hypercube defines a set of experimental points which follow the familiar 2^k factorial design. To form the second order rotatable designs these are augmented with points at the center and with points corresponding to the cross polytope in which all the variables except one are held in turn at the 'center' levels, the remaining variable being maintained first at a level above its center value and then at a level below its center value. Because in standard orientation the latter points lie along the coordinate axes they may be referred to as "axial points".

When k is sufficiently large a suitable fractional replicate can replace the full hypercube. Since a second order moment matrix identical with that of the full factorial will be obtained with any fractional replicate of the 2^k design in which no effects of second order or lower order are confounded. The only result of this substitution will be to effect the alias matrix of the design.

For a k dimensional design with $n_a = 2k$ axial points, $n_c = 2^{(k-p)}$ points in the $(\frac{1}{2})^p$ replicate of the 2^k factorial, and n_0 points at the center, $\rho_a/\rho_c = n_c^{1/4}/k^{1/2}$ and

$$\lambda_4 = N/\{n_c + 4(1 + n_c^{1/2})\}, \tag{81}$$

where $N = n_a + n_c + n_0$.

When using these designs in their standard orientation it is simplest to regard them as scaled so that the hypercube or fractional hypercube has its coordinates equal to plus or minus unity. The N coordinates of the complete design are then

n_c points (2^k factorial or suitable fraction): $(\pm 1, \pm 1, \cdots, \pm 1)$,
n_a points (axial points): $(\pm\alpha, 0, 0, \cdots, 0)$, $(0, \pm\alpha, 0, \cdots, 0)$, $\cdots$, $(0, 0, \cdots, \pm\alpha)$,
n_0 points (center conditions): $(0, 0, 0, \cdots, 0)$.

Then $\alpha = n_c^{1/4}$ and the scale factor c of Eq. (3) is $N/(n_c + 2n_c^{1/2})$. The values of n_c, n_a, n_0, α, ρ_a/ρ_c and λ_4 for second order rotatable designs, which give the values of λ_4 set out in Table 1 and for orthogonal designs, are given in the Table 3 for $k = 2, 3, 4, \cdots, 8$ dimensions.

8. Arrangement of the designs in blocks. To avoid bias due to systematic disturbances the complete set of experimental trials corresponding to the points which form the design could be performed in random order. Frequently however it is possible to carry out limited groups of trials under more homogeneous conditions than can be attained for the complete set. It may then be possible to achieve greater accuracy by performing the designs in blocks, carrying out the individual trials within each block in random order. A block may for example refer to a group of experiments performed on the same day, or a group of experiments for which it was possible to use the same batch of starting material.

TABLE 3

Central composite rotatable second order designs

k	2		3		4		5		5(½ rep)		6		6(½ rep)		7		7(½ rep)		8		8(½ rep)		8(¼ rep)	
n_c	4		8		16		32		16		64		32		128		64		256		128		64	
n_a	4		6		8		10		10		12		12		14		14		16		16		16	
n_0 (Table 1)	5		6		7		10		6		15		9		21		14		28		20		13	
n_0 (Orthogonal)		8		9		12		17		10		24		15		35		22		52		33		20
N total	13	16	20	23	31	36	52	59	32	36	91	100	53	59	163	177	92	100	300	324	164	177	93	100
$\alpha = n_c^{\frac{1}{4}}$:	1.414		1.682		2.000		2.378		2.000		2.828		2.378		3.364		2.828		4.000		3.364		2.828	
λ_4 Table 1	0.81		0.86		0.86		0.89		0.89		0.91		0.90		0.92		0.92		0.93		0.93		0.93	
λ_4 Orthogonal		1		0.99		1		1.01		1		1		1.01		1.00		1		1		1.00		1
ρ_a/ρ_c	1.000		0.971		1.000		1.064		0.894		1.155		0.971		1.271		1.069		1.414		1.189		1.000	

We shall show how the designs we have discussed may be performed in orthogonal blocking arrangements. On the usual assumption, that the effect of carrying out a particular trial in one block rather than another is merely to change the expected value of the response by a fixed amount which depends only on the particular blocks involved, such arrangements insure that the estimated coefficients of the polynomial are completely independent of the block differences and their standard errors depend on the within-block variance only.

For N experimental points assigned to m blocks with n_w' points in the wth block we suppose that

$$\eta_u = \sum_{w=1}^{m} \beta_{0w} z_{wu} + \sum_{i=1}^{k} \beta_i x_{iu} + \sum_{i=1}^{k} \sum_{j=i}^{k} \beta_{ij} x_{iu} x_{ju} ,$$

where β_{0w} is the expected value of the response in the wth block at the experimental conditions corresponding to the origin of the design, and z_{wu} is a "dummy" variable taking the value unity for those experimental points which fall in the wth block and zero for all other experimental points. We shall call x_i , $x_i x_j$, etc., the polynomial variables and z_v , z_w , etc., the block variables.

In whatever manner the experimental points are assigned to the blocks we can except in the "pathological" cases detailed below estimate the coefficients of the polynomial equation allowing for the block effects, by the method of least squares. However the manner in which the experimental points are assigned to the blocks profoundly affects the efficiency of estimation and the ease with which the estimates are calculated. In particular, if it is possible so to allocate the experimental points to the blocks that the block variables and polynomial variables are orthogonal, then the inclusion of blocks does not at all influence the estimation of the polynomial coefficients and the only effect of blocking is the wholly desirable one of limiting the experimental error to that occurring within blocks. The analysis of the experiment proceeds exactly as if there were no blocking, except that in the estimation of the residual error the contribution due to blocks is subtracted from the residual sum of squares.

In order to find the conditions that must be satisfied to allow orthogonal blocking it is simplest to rewrite the model in the equivalent form

$$\eta_u = \beta_0 + \sum_{i=1}^{k} \beta_i x_{iu} + \sum_{i=1}^{k} \sum_{j=i}^{k} \beta_{ij} x_{iu} x_{ju} + \sum_{w=1}^{m} \delta_w (z_{wu} - \bar{z}_w),$$

(82)

$$\beta_0 = \sum_{w=1}^{m} \frac{n_w'}{N} \beta_{0w} , \qquad \delta_w = \beta_{0w} - \beta_0 , \qquad \bar{z}_w = n_w' / N,$$

We note that $z_{wu} - \bar{z}_w$ is equal to $1 - n_w'/N$ when the uth set of conditions is in the wth block, and to $-n_w'/N$ otherwise.

The conditions for orthogonal blocks, that is to say the conditions that the block variables $z_{wu} - \bar{z}_w$ shall be orthogonal to the variables

$$x_0 , x_1 , x_2 , \cdots x_k , x_1^2 , x_2^2 , \cdots x_k^2 , x_1 x_2 , \cdots , x_{k-1} x_k$$

in the second degree polynomial, can be written

(83) $$\sum_{u=1}^{N} x_{iu} x_{ju}(z_{wu} - \bar{z}_w) = 0, \qquad (i, j = 0, 1, \cdots, k)$$

that is

(84) $$\sum_{u=1}^{N} x_{iu} x_{ju} z_{wu} = \bar{z}_w \sum_{u=1}^{N} x_{iu} x_{ju}.$$

Now for any second order rotatable design, if $i \neq j$, $\sum_{u=1}^{N} x_{iu}x_{ju} = 0$, whence for orthogonal blocking we require

(85) $$\sum_{u}^{n_w'} x_{iu} x_{ju} = 0, \qquad i \neq j, w = 1, 2, \cdots m,$$

where the summation includes only those values of u in the wth block. *Thus (1) all the sums of products between* x_0, x_1, $\cdots$, x_k *must be zero for each block.*

A second condition arises from putting $i = j$ in (84) whence

(86) $$\frac{\sum_{u}^{n_w'} x_{iu}^2}{\sum_{u=1}^{N} x_{iu}^2} = \frac{n_w'}{N},$$

where the summation in the numerator again is for those values of u in the wth block. *Hence* (2) *the fraction of the total sum of squares for each variable contributed by each block must be proportional to the number of observations in each block.*

8.1 *Examples of orthogonal blocking with designs based on equiradial sets of points.* Where the rotatable design consists of an equiradial set of points with added points at the center, we can satisfy both conditions (85) and (86) if we can divide the equiradial set into subsets which are themselves first order rotatable (i.e., first order orthogonal) designs. When this is possible the sum of squares for each variable in each subset will be proportional to the number of points in the subset, and we have only to add a number of center points to each subset proportional to the number of points it already possesses to obtain a orthogonal block.

For example, consider a two dimensional "hexagonal" design consisting of 6 points at the vertices of a hexagon together with $2p$ center points. We can perform this design in two orthogonal blocks each consisting of three points at the vertices of an equilateral triangle and the remaining p points at the center. Similarly an octagonal design (the two dimensional rotatable composite design of Table 3) can be divided into two sets of four points at the vertices of a square with equal number of center points added to each to form two blocks. A "nonagonal" design may be divided into three equilateral triangles, which form the basis of three blocks and so on.

In three dimensions the vertices of the dodecahedron may be divided into five sub-sets each of which comprise the vertices of a tetrahedron. Thus for example a design consisting of twenty points at the vertices of the dodecahedron plus $5p$ points at the center is divisable into five orthogonal blocks of $4 + p$ points. Each block consists of a complete tetrahedron plus p center points.

8.2 *Blocking with composite rotatable designs.* The important composite rotatable designs lend themselves very conveniently to blocking and some valuable work on this topic has been carried out independently by De Baun [15]. Because the number of center points in any block must be an integer, exact rotatability and exact orthogonality between quadratic variables and block variables is not always attainable. We can however insure that either one of these desiderata is exactly satisfied and the other one nearly so. Although the extra labor involved in the calculations due to the slight non-orthogonality is not very great and the loss of efficiency is negligible, it is simplest in practice to use designs in which the block effects are exactly orthogonal but the condition of rotatability is slightly relaxed.

The central composite design in standard orientation consists of n_c points at the vertices of a cube corresponding to a 2^k factorial arrangement or some suitable fraction of it with coordinates $(\pm 1, \pm 1, \cdots, \pm 1)$, together with $n_a = 2^k$ "axial" points with coordinates $(\pm\alpha, 0, \cdots, 0)$, $(0, \pm\alpha, \cdots, 0)$, $\cdots$, $(0, 0, \cdots, \pm\alpha)$, and n_0 points at the center with coordinates $(0, 0, \cdots, 0)$. If $\alpha = n_c^{1/4}$ the design is rotatable, but let us for the moment not assume that this is so.

The sets of points at the vertices of the cube and the set of the axial points are each first order rotatable designs. These two parts of the design thus provide a basis for a first division of the composite design into two blocks. The blocking will be orthogonal if it is possible to allocate the center points to the two parts so that the total number of points in each part is proportional to the sum of squares for each variable contributed by that part. If n_{c0} and n_{a0} are the numbers of center points in the cubic part and the axial part respectively then we require

$$\frac{2\alpha^2}{n_c} = \frac{n_a + n_{a0}}{n_c + n_{c0}}; \tag{87}$$

thus for any composite design with

$$\alpha = \left\{\frac{n_c(n_a + n_{a0})}{2(n_c + n_{c0})}\right\}^{1/2} \tag{88}$$

we obtain orthogonal blocking. Now for rotatability $\alpha = n_c^{1/4}$ and hence to achieve both orthogonal blocking and rotatability we require that

$$\frac{n_c^{1/2}}{2} = \frac{n_c + n_{c0}}{n_a + n_{a0}}. \tag{89}$$

The set of axial points cannot be further sub-divided into sets which are first order rotatable designs. Such sub-division is possible however for the set of points at the vertices of the cube provided a system of confounding for the two-level factorial or fractional factorial design exists such that all the comparisons confounded correspond to interactions between three or more variables. If this is so the comparisons confounded will be unassociated with the comparisons used to estimate the coefficients of the polynomial. Also, since the comparisons confounded are the defining contrasts of the sub-sets regarded as fractional factorials and correspond to interactions between three or more factors, it follows that the sub-sets are individually first order rotatable designs. If the cube is divided up into sub-sets each containing the same number of points then in accordance with Eq. (86) we must add an equal number of center points to each sub-set to maintain orthogonality.

8.3 *Examples.*

(i) We first consider the four-dimensional design to illustrate the situation where both rotatability and orthogonality of blocking may be attained. From Table 4 it is seen that in standard orientation this design consists of the 2^4 factorial, with coordinates $(\pm 1, \pm 1, \pm 1, \pm 1)$, 8 axial points at distance $\alpha = 16^{1/4} = 2$ units from the center, and to approximately satisfy the requirement of Table 1 seven points added at the center.

We can achieve orthogonal blocking and rotatability if we can satisfy Eq. (89) which gives

$$2 = \frac{16 + n_{c0}}{8 + n_{a0}}. \tag{90}$$

Now the total number of points at the center $n_0 = n_{c0} + n_{a0}$ is not critical and if, for example we use six points at the center instead of seven the only effect will be to slightly change the variance function and in particular to decrease slightly the precision near the origin. If we now allocate $n_{c0} = 4$ points to the cube and $n_{a0} = 2$ points to the axial part, we satisfy (90). In this way the complete set of $16 + 8 + 6 = 30$ points is divided into two orthogonal blocks, one containing $16 + 4 = 20$ points and one with $8 + 2 = 10$ points. Now the 2^4 factorial designs corresponding to the cube may be further divided into two halves each of which is a rotatable design of order 1. This may be done without affecting the estimation of the polynomial coefficients by arranging that the contrast between the two halves corresponds to a three or four-factor interaction. The most suitable contrast is the four-factor interaction and to effect the division we allocate those points in the design for which the product $x_1x_2x_3x_4$ is 1 to one part and those for which it is -1 to the other. We may in accordance with Eq. (86) maintain orthogonal blocking by dividing the four center points assigned to the cube equally between the parts, two to each half. Finally therefore the design of 30 points is carried out in 3 blocks each of ten points consisting of the axial points together with two center points and the two half-replicates of the cube each with two center points. The blocking is completely orthogonal and the design exactly rotatable. It should be noted that since the separate blocks are themselves first order rotatable designs, this scheme ensures orthogonal blocking not only in the standard orientation of the design which we have specifically discussed but also in every other orientation.

(ii) To illustrate the situation where orthogonality of blocking and rotatability are not exactly attainable simultaneously, consider the three-dimensional composite design. From Table 3 we see that, in standard orientation, the design consists of the 2^3 factorial with coordinates $(\pm 1, \pm 1, \pm 1)$, the six axial points each at a distance $8^{1/4} = 1.6818$ and about six center points are needed to satisfy approximately the requirements of Table 1. From (89), to achieve orthogonal blocking and rotatability we require

$$\frac{8^{1/2}}{2} = \frac{8 + n_{c0}}{6 + n_{a0}}, \tag{91}$$

where $n_{c0} + n_{a0}$ is about 6. We come nearest to satisfying this requirement by putting $n_{c0} = 4$ and $n_{a0} = 2$; however, since the equation cannot be exactly satisfied with integral values, some slight nonorthogonality must occur. This non-orthogonality would be the same in every orientation of the design and the estimates of the coefficients corrected for block effects would in fact be very easily obtained without much extra labor. However since *exact* rotatability is not required in practice we choose therefore to adjust the value of α to obtain orthogonality at the expense of rotatability. From Eq. (88) it is seen that for $n_{c0} = 4$, $n_{a0} = 2$ and for orthogonal blocking we require

$$\alpha = \left\{\frac{8(8)}{2(12)}\right\}^{1/2} = 1.6330 \tag{92}$$

instead of $\alpha = 1.6818$ required for rotatability. For this value of α the variance contours will not be exact spheres, the difference from sphericity will however be so slight as to be of no practical importance. Since the sub-groups are first order rotatable designs the blocking will remain orthogonal in every orientation. The elements corresponding to the constant terms and second order terms in both the moment matrix and the precision matrix will change slightly as the design is rotated however.

As before the cube part may be divided into two portions. These are the two fractional replicates whose defining contrast is the three-factor interaction. Since four center points are allocated to the cube we can divide these equally between the fractions and so maintain orthogonality.

Finally therefore the 20 points of the design with $\alpha = 1.6330$ may be divided into three orthogonal blocks. One block consists of eight points containing the six axial points and two of the center points, and other two blocks each contain a half replicate of the cube together with two center points. A list of orthogonal blocking arrangements for rotatable and near-rotatable composite designs is given in Table 4 below.

An aspect which makes these blocking arrangements particularly useful arises out of the nature of the situation in which these designs are often used. In the exploration of response surfaces [1], [13] trials are usually performed sequentially and often have as their object the increase or maximization of some response. It has been shown that each block of the second order design is itself a first order rotatable design with added points at the center. Such a design allows estimates to be obtained not only of first order effects but also (assuming a second order equation is adequate to represent the surface) of the *sum* of the quadratic effects. For if $\bar{y}_0$ is the mean of the observations of the center and $\bar{y}_d$ the mean of the observations in the first order design then using (8a) it will be found that for any

TABLE 4

Blocking arrangements for rotatable and near-rotatable central composite design

k	2*	3	4	5	5(½ rep)	6	6(½ rep)	7	7(½ rep)
Blocks within cube									
n_c : Number of points in cube	4	8	16	32	16	64	32	128	64
Number of blocks in cube	1	2	2	4	1	8	2	16	8
Number of points in block from cube	4	4	8	8	16	8	16	8	8
Number of added center points	3	2	2	2	6	1	4	1	1
Total number of points in block	7	6	10	10	22	9	20	9	9
Axial block									
n_a Number of axial points	4	6	8	10	10	12	12	14	14
Number of added points	3	2	2	4	1	6	2	11	4
Total number of points in block	7	8	10	14	11	18	14	25	18
Grand total of points in the design	14	20	30	54	33	90	54	169	80
Value of α for orthogonal blocking	1.4142	1.6330	2.0000	2.3664	2.0000	2.8284	2.3664	3.3636	2.8284
Value of α for rotatability	1.4142	1.6818	2.0000	2.3784	2.0000	2.8284	2.3784	3.3333	2.8284

* A more economical design for $k = 2$ (which is not however a composite design as here defined) is that mentioned in Section 7.1, in which 6 points at the vertices of a hexagon are divided into two sets of three points. To attain the value of λ_4 given in Table 1 two points are added to each set to form two blocks.

first order rotatable design in any orientation (that is, for any orthogonal first order design),

$$\mathcal{E}(\bar{y}_d - \bar{y}_0) = \sum_{i=1}^{k} \beta_{ii} .$$

In the neighborhood of a true maximum $\bar{y}_d - \bar{y}_0$ is not small relative to the linear coefficients b_i and this provides an additional indication of the inadequacy of a linear model.

For example suppose that four variables were studied with the intention of finding an optimum set of conditions using the methods of Box and Wilson [1]. A half replicate of the 2^4 design with two center points could first be run in random order. This design would supply estimates of the four first order terms β_1, β_2, β_3, and β_4 and combinations of the interaction terms namely $(\beta_{12} + \beta_{34})$, $(\beta_{13} + \beta_{24})$, $(\beta_{14} + \beta_{23})$ and of the sum of the four quadratic terms $\sum \beta_{ii}$. If first order terms were dominant, progress could probably be made without a more elaborate design and moves in the indicated direction of increasing yield would be made until improvement in that direction was exhausted. At this point a new first order design would be begun at the improved set of conditions.

If the first order terms were not dominant, or if more precision were needed, or a tentative move had not met with success, the second half of the factorial design together with two further center points could be carried out, and the situation again reviewed in the light of the set of 20 trials now available. Finally, if the evidence indicated that further progress could be attained only by fitting the second degree equation, the third block of ten experiments consisting of the axial points and two center points would be added. The complete set of 30 trials could then be used to provide efficient estimates of the best fitting second degree equation and further progress would follow.

9. Details of calculations using the designs. From the observations generated by the second-order designs least squares estimates of the coefficients of the fitted polynomial, together with their variances and the appropriate analysis of variance table, are readily computed.

9.1 *Estimates of the coefficients and of their standard errors.* To estimate the coefficients we require only the sums of products of the observations with the independent variables. We shall use the notation

$$\sum_{u=1}^{N} x_{0u}\, y_u = \{0\ y\}, \qquad \sum_{u=1}^{N} x_{iu}\, y_u = \{i\ y\}, \qquad \sum_{u=1}^{N} x_{iu}\, x_{ju}\, y_u = \{i\ j\ y\},$$

where $i, j = 1, 2, \cdots, k$.

(i) *Rotatable designs.* The form of the inverse matrix for any rotatable design is given by (44) whence we have for any k-dimensional second order rotatable design with parameter λ_4

$$(93)\quad b_0 = AN^{-1}[2\lambda_4^2(k+2)\{0\ y\} - 2\lambda_4 c \textstyle\sum_{i=1}^{k} \{i\ i\ y\}],$$

$$(94)\quad b_i = cN^{-1}\{i\ y\},$$

$$(95)\quad b_{ii} = AcN^{-1}[c[(k+2)\lambda_4 - k]\cdot\{i\ i\ y\} + c(1-\lambda_4)\textstyle\sum_{j=1}^{k}\{j\ j\ y\} - 2\lambda_4\{0\ y\}],$$

$$(96)\quad b_{ij} = c^2N^{-1}\lambda_4^{-1}\{i\ j\ y\},$$

where

$$(97)\qquad A = [2\lambda_4\{(k+2)\lambda_4 - k\}]^{-1}$$

and the scale factor

$$(98)\qquad c = N/\textstyle\sum_{u=1}^{N} x_{iu}^2\,.$$

Again using (44) the variances of the estimates are

$$(99)\qquad V(b_0) = 2A\lambda_4^2(k+2)\sigma^2/N,$$

(100) $$V(b_i) = c\sigma^2/N,$$

(101) $$V(b_{ii}) = A\{(k+1)\lambda_4 - (k-1)\}c^2\sigma^2/N,$$

(102) $$V(b_{ij}) = c^2\sigma^2/N\lambda_4\,.$$

These formulae apply to any rotatable design. For the particular case of the rotatable central composite designs scaled so that in standard orientation the coordinates of the n_c points forming the two-level factorial or fractorial factorial part are $(\pm 1, \pm 1, \cdots, \pm 1)$ the value of the scale factor c is $N/(n_c + 2n_c^{1/2})$.

(ii) *Central composite designs not necessarily rotatable.* In order to obtain exactly orthogonal blocking for the central composite designs, we have proposed certain arrangements which are not exactly rotatable. For any central composite design the estimates of the coefficients and their variances are readily computed and these are of course unaffected by orthogonal blocking.

The non-diagonal elements of the moment matrix for any central composite design in standard orientation are zero apart from terms arising from cross products between the "constant term" variable x_0 and the quadratic variables $x_i^2 (i = 1, 2, \cdots, k)$. The sub-matrix of the moment matrix corresponding to these variables is of the form

(103) $$\begin{bmatrix} d & e & e & e & \cdots & e \\ e & f & g & g & \cdots & g \\ e & g & f & g & \cdots & g \\ \cdot & \cdot & & & & \cdot \\ \cdot & \cdot & & & & \cdot \\ \cdot & \cdot & & & & \cdot \\ e & g & g & g & \cdots & f \end{bmatrix}$$

The reciprocal of this matrix is of the same form. Denoting its elements in corresponding positions by capital letters we find

(104) $$D = H^{-1}\{f + (k-1)g\}(f-g),$$

(105) $$E = -H^{-1}e(f-g),$$

(106) $$F = H^{-1}\{df + (k-2)\,dg - (k-1)e^2\},$$

(107) $$G = H^{-1}(e^2 - dg),$$

(108) $$H = (f-g)\{df + (k-1)\,dg - ke^2\}.$$

For the composite designs in conventional scaling

(109) $$d = N, \qquad e = n_c + 2\alpha^2, \qquad f = n_c + 2\alpha^4, \qquad g = n_c\,.$$

Using (109) with (104), (105), (106), (107), and (108) the estimated coefficients are readily calculated from the formulae

(110) $$b_0 = D\{0\,y\} + E\textstyle\sum_{i=1}^{k}\{i\,i\,y\},$$

(111) $$b_i = (n_c + 2\alpha^2)^{-1}\{i\,y\},$$

(112) $$b_{ii} = E\{0\,y\} + F\{i\,i\,y\} + G\textstyle\sum_{i \neq j}^{k} \{j\,j\,y\},$$

(113) $$b_{ij} = n_c^{-1}\{i\,j\,y\}.$$

The standard errors of these estimates are

(114) $$V(b_0) = D\sigma^2,$$

(115) $$V(b_i) = (n_c + 2\alpha^2)^{-1}\sigma^2,$$

(116) $$V(b_{ii}) = F\sigma^2,$$

(117) $$V(b_{ij}) = n_c^{-1}\sigma^2.$$

In practice we normally estimate σ from the data in the manner described below.

9.2 *Analysis of variance.* The total sum of squares $\sum_{u=1}^{N} y_u^2 = S$ having N degrees of freedom may be split into two parts: (i) The sum of squares S_{012} attributable to the fitted second degree equation having $(\frac{1}{2})(k + 2)(k + 1)$ degrees of freedom. This is given by

(118) $$S_{012} = b_0\{0\,y\} + \sum_{i=1}^{k} b_i\{i\,y\} + \sum_{i=1}^{k}\sum_{j=i}^{k} b_{ij}\{i\,j\,y\}.$$

(ii) the "overall" residual sum of squares R having $N - (\frac{1}{2})(k + 2)(k + 1)$ degrees of freedom which is obtained by difference

(119) $$R = S - S_{012}.$$

Each of these sums of squares may be further subdivided.

9.21 *Analysis of sum of squares due to regression.* The sum S_{012} may be divided into three parts:

(i) The sum S_0, having one degree of freedom attributable to the fitting of a polynomial of zero-th degree (the so-called "correction due to the mean")

(120) $$S_0 = \{0\,y\}^2/N$$

(ii) The sum $S_{1.0}$ having k degrees of freedom. This is the "extra" sum of squares associated with the fitting of a first degree polynomial. Since, for the designs we have considered, b_0 is uncorrelated with any of the b_i

(121) $$S_{1.0} = \sum_{i=1}^{k} b_i\{i\,y\}.$$

(iii) The sum $S_{2.10}$ having $\frac{1}{2}k(k + 1)$ degrees of freedom, the "extra" sum of squares associated with the fitting of a second degree polynomial. Since for the designs we consider the b_i's are uncorrelated with b_0 and with the b_{ii}'s

(122) $$S_{2.10} = b_0\{0\,y\} + \sum_{i=1}^{k}\sum_{j=i}^{k} b_{ij}\{i\,j\,y\} - \{0\,y\}^2/N.$$

In specific examples other sub-divisions may be relevant. For instance, it might be suspected that a particular variable x_i had no effect at all on the result. In this case it would be appropriate to isolate a sum of squares S_i associated with the variable x_i. This could be done most conveniently by carrying out an analysis omitting x_i and $x_i x_j$ $(j = 1, 2, \cdots, k)$ from the model. The sum of squares associated with the reduced model would then be subtracted from the sum of squares associated with the full model to give S_i.

9.22 *Analysis of residual sum of squares.* Where blocking has not been used the overall residual sum of squares R may be analysed into two parts:

(i) The sum

$$S_E = \sum_{u=1}^{n_0} (y_{u0} - \bar{y}_0)^2, \tag{123}$$

where y_{u0} is the uth repeated observation at the center of the design and $\bar{y}_0$ is the mean of the observations at the center. This sum of squares has $n_0 - 1$ degrees of freedom and $S_E/(n_0 - 1)$ provides an estimate of the experimental error variance σ^2 on the assumption that this variance is independent of the levels of the variables x_i.

(ii) $R - S_E$ having $N - (\frac{1}{2})(k + 2)(k + 1) - n_0 + 1$ degrees of freedom, the residual sum of squares which measures experimental error together with lack of fit of the assumed form of equation. When the corresponding mean square is large compared with the estimate of pure error obtained in (i) above, this implies that the assumed form of the equation is inadequate. A full discussion will be found in [16] and will be published elsewhere.

9.23 *Analysis when blocking is employed.* When blocking is employed a further sum of squares S_B due to blocks is extracted so that the residual sum of squares R is now divided into three parts:

(i) S_B having $B - 1$ degrees of freedom

$$S_B = \sum_{b=1}^{B} n_b(\bar{y}_b - \bar{y})^2, \tag{124}$$

where B is the number of blocks, n_b is the number of observations in the bth block and $\bar{y}_b$ is the mean of the observations in the bth block.

(ii) The sum of squares S_E corresponding to pure error and having $n_0 - B$ degrees of freedom. This is the sum of the individual sums of squares for repeated observations at the center of each block

(iii) $R - S_B - S_E$ the residual which measures experimental error plus lack of fit.

The designs discussed are extremely convenient to use. The points at the center of the design allow a check to be made at a standard set of conditions while the experiment is being carried out and so helps to show up gross errors. Furthermore the center points provide an estimate of pure error from which it is possible to check the adequacy of the assumed form of equation without replicating the whole design.

10. Simplification of confidence region for maximum. On the assumption that a second degree equation can adequately represent a response surface in the region of interest, a confidence region for the stationary point of this surface has been derived in [17]. Unfortunately the boundary of the confidence region is, in general, not easy to compute but Wallace [20] has devised valuable approximations which are easy to compute and to appreciate. A very considerable simplification in the expression for the exact confidence interval occurs when a rotatable design is used to generate the experimental data.

Suppose the second degree equation in k variables $x_1, \cdots, x_i, \cdots, x_j, \cdots, x_k$ which has been fitted by least squares is written in the form

$$y - a_{00} = \mathbf{x}'\mathbf{a}_0 + (\tfrac{1}{2})\mathbf{x}'\mathbf{A}\mathbf{x}, \tag{125}$$

where $\mathbf{a}_0$ is the $k \times 1$ vector $\{a_{i0}\}$ and $\mathbf{A}$ is the $k \times k$ matrix $\{a_{ij}\}$ and in terms of the notation previously adopted $a_{00} = b_0$, $a_{i0} = b_i$, $\frac{1}{2}a_{ii} = b_{ii}$, $a_{ij} = b_{ij}$.

The position of the center of the fitted system is obtained by equating to zero the elements of the $k \times 1$ vector $\boldsymbol{\delta}$ defined by

$$\boldsymbol{\delta} = \mathbf{a}_0 + \mathbf{A}\mathbf{x}. \tag{126}$$

Thus if $\mathbf{x}_0$ is the vector of the k coordinates of this center then

$$\mathbf{a}_0 = -\mathbf{A}\mathbf{x}_0, \tag{127}$$

at which point the value of y is given by the equation

$$y_0 = a_{00} + (\tfrac{1}{2})\mathbf{x}_0'\mathbf{a}_0. \tag{128}$$

The confidence region given in [17] is defined by the inequality:

$$\boldsymbol{\delta}'\mathbf{V}^{-1}\boldsymbol{\delta} \leqq s^2kF_\alpha(k, \varphi), \tag{129}$$

where $\mathbf{V}\sigma^2 = \mathcal{E}(\boldsymbol{\delta}'\boldsymbol{\delta})$, s^2 is an estimate of variance having φ degrees of freedom and distributed as $\chi^2\sigma^2/\varphi$ independently of $\boldsymbol{\delta}$, and $F_\alpha(k, \varphi)$ is the α probability point of the F distribution having k and φ degrees of freedom.

For a rotatable design using (100, 101, and 102), the variances and covariances of the δ's are given by

$$V(\delta_i) = N^{-1}\sigma^2(c + \lambda^{-1}c^2\rho^2 + \ell c^2x_i^2), \qquad \text{Cov}\,(\delta_i\delta_j) = N^{-1}\sigma^2\ell c^2x_ix_j, \tag{130}$$

where

$$\rho^2 = \sum_{i=1}^{k} x_i^2 \qquad \ell = \frac{k(\lambda - 1) + 2}{\lambda\{(k+2)\lambda - k\}}, \tag{131}$$

c is the scale factor for the design defined in (3), and $\lambda = \lambda_4$. Thus

$$N\mathbf{V} = c(1 + \lambda^{-1}c\rho^2)\left\{\mathbf{I} + \frac{\ell c}{1 + \lambda^{-1}c\rho^2}\,\mathbf{x}\mathbf{x}'\right\}. \tag{132}$$

The matrix $\mathbf{V}$ is readily inverted using a theorem due to K. D. Tocher [18] to give

$$\mathbf{V}^{-1} = \frac{N}{c(1 + \lambda^{-1}c\rho^2)}\left\{\mathbf{I} - \frac{c\ell}{1 + (\lambda^{-1} + \ell)c\rho^2}\,\mathbf{x}\mathbf{x}'\right\}. \tag{133}$$

Thus

$$\boldsymbol{\delta}'\mathbf{V}^{-1}\boldsymbol{\delta} = \frac{N}{c(1+\lambda^{-1}c\rho^2)}\left\{\boldsymbol{\delta}'\boldsymbol{\delta} - \frac{c\ell}{1+(\lambda^{-1}+\ell)c\rho^2}(\boldsymbol{\delta}'\mathbf{x})^2\right\}. \tag{134}$$

Now using (126) with (127)

$$\boldsymbol{\delta} = \mathbf{A}(\mathbf{x} - \mathbf{x}_0). \tag{135}$$

It follows that when the data have been generated by a rotatable design the confidence region for the stationary point is given by the expression

$$\frac{N}{(1+\lambda^{-1}c\rho^2)}\left[\frac{1}{c}(\mathbf{x}-\mathbf{x}_0)'\mathbf{A}^2(\mathbf{x}-\mathbf{x}_0) - \frac{\ell}{1+(\lambda^{-1}+\ell)c\rho^2}\{(\mathbf{x}-\mathbf{x}_0)'\mathbf{A}\mathbf{x}\}^2\right] \leqq ks^2F_\alpha(k,\varphi). \tag{136}$$

Now a fitted second degree equation can be interpreted most readily by writing it in the canonical form

$$y - y_0 = \sum_{i=1}^{k} B_{ii}X_i^2. \tag{137}$$

The k elements B_{ii} are the latent roots of the matrix $(\frac{1}{2})\mathbf{A}$. If a diagonal matrix $\mathbf{B}$ is formed with the B_{ii} for its elements then the latent vectors of $\mathbf{A}$ form the k rows of an orthogonal matrix $\mathbf{M}$ such that

$$(\tfrac{1}{2})\mathbf{MA} = \mathbf{BM} \tag{138}$$

and the matrix $\mathbf{X}$ of "canonical variables" whose elements are $X_1, X_2, \cdots, X_k$ is defined by

$$\mathbf{X} = \mathbf{M}(\mathbf{x} - \mathbf{x}_0). \tag{139}$$

The coordinates of the initial origin $\mathbf{x} = \mathbf{0}$ in terms of the canonical variables is given by

$$\mathbf{X}_0 = -\mathbf{M}\mathbf{x}_0, \tag{140}$$

whence $\mathbf{x} = \mathbf{M}'(\mathbf{X} - \mathbf{X}_0)$. Thus the expression defining the confidence region reduces to the relatively simple expression

$$\frac{4N}{1+\lambda^{-1}c\rho^2}\left[\frac{1}{c}\sum_{i=1}^{k} B_{ii}^2X_i^2 - \frac{\ell}{1+(\lambda^{-1}+\ell)c\rho^2}\left\{\sum_{i=1}^{k} B_{ii}X_i(X_i - X_{i0})\right\}^2\right] \leqq ks^2F_\alpha(k,\varphi), \tag{141}$$

where

$$\rho^2 = \mathbf{x}'\mathbf{x} = (\mathbf{X}-\mathbf{X}_0)'\mathbf{MM}'(\mathbf{X}'-\mathbf{X}_0) = \sum_{i=1}^{k}(X_i - X_{i0})^2.$$

A rough delineation of the confidence region in a readily appreciated form can now be obtained by enumerating the points at which it cuts the canonical axes.

By direct substitution in (141) the point $(0, \cdots, 0, X_i, 0 \cdots 0)$ will be included in the confidence region provided that

$$B_{ii}^2 \leqq cks^2F_\alpha(k, \varphi)\left\{1 + \lambda^{-1}c \sum_{j \neq i}^{k} X_{j0}^2 + \lambda^{-1} c(X_i - X_{i0})^2\right\}$$

$$(142) \qquad \times \frac{\left\{1 + (\lambda^{-1} + \ell)c \sum_{j \neq i}^{k} X_{j0}^2 + (\lambda^{-1} + \ell)c(X_i - X_{i0})^2\right\}}{4NX_i^2\left\{1 + (\lambda^{-1} + \ell)c \sum_{j \neq i}^{k} X_{j0}^2 + \lambda^{-1}c(X_i - X_{i0})^2\right\}}.$$

If the quantities $X_{10}, X_{20}, \cdots, X_{k0}$ are finite then as X_j tends to infinity this becomes

$$(143) \qquad B_{ii}^2 \leqq (4N)^{-1}kF_\alpha(k, \varphi)\{\lambda^{-1} + \ell\}c^2s^2,$$

which is the condition that the confidence interval includes points at $\pm\infty$ on the canonical axis X_i

10.1 *Redundancy of canonical variables.* This result can be viewed from a somewhat different angle. In real problems one would expect (see for example reference [3]) that the underlying system would often be approximated by surfaces containing a stationary line, plane or hyperplane rather than a stationary point. When this occurred one or more of the B_{ii} calculated from the fitted equation would be close to zero and the corresponding canonical axes would delineate the stationary line, plane or space. Analysis of this sort is of considerable practical importance since it may help in the deduction of the underlying mechanism of the system [19].

An important question that normally arises is how small the canonical units B_{ii} must be before we may conclude that zero values are not inconsistent with the data. Now for any rotatable design the variances of the coefficients are the same in every orientation and since the B_{ii} are simply the "quadratic effects" in the directions of the canonical variables they have the same standard errors as have the quadratic effects b_{ii} before transformation. Were it not that the small values of the B_{ii} will be selected *because* they are small we might "test the significance" of the B_{ii} by dividing by their standard errors and referring the quotient to tables of the t distribution. Now the estimated variance of the quadratic coefficients selected from a rotatable design with scale factor c is $(4N)^{-1}$ $(\lambda^{-1} + \ell)c^2s^2$ and consequently the inequality (143) can be written

$$\left|\frac{B_{ii}}{\text{s.e. } (B_{ii})}\right| \leqq \sqrt{kF_\alpha(k, \varphi)}$$

implying that we can "test the significance" of the B_{ii} computed from a rotatable design by referring not to the t distribution but to the distribution of $kF_\alpha(k, \varphi)$.

This is a special case of a more general result derived by Wallace [20].

REFERENCES

[1] G. E. P. Box and K. B. Wilson, "On the experimental attainment of optimum conditions," *J. Roy. Stat. Soc., Ser. B*, Vol. 13 (1951), pp. 1–45.

[2] G. E. P. Box, "Multifactor designs of first order," *Biometrika*, Vol. 39 (1952), pp. 49–57.

[3] D. J. Finney, "The fractional replication of factorial experiments," *Ann. Eugenics*, Vol. 12 (1945), pp. 291–301.

[4] H. Hotelling, "Some improvements in weighing and other experimental techniques," *Ann. Math. Stat.*, Vol. 15 (1943), pp. 297–306.

[5] R. L. Plackett and J. P. Burman, "The design of multifactorial experiments," *Biometrika*, Vol. 33 (1946), pp. 305–325.

[6] K. D. Tocher, "A note on the design problem," *Biometrika*, Vol. 39 (1952), pp. 109–117.

[7] A. C. Aitken, *Determinants and Matrices*, Oliver & Boyd, London, 1948.

[8] A. C. Aitken, "On the Wishart distribution in statistics," *Biometrika*, Vol. 36 (1949), pp. 59–62.

[9] J. H. M. Wedderburn, *Lectures on Matrices*, Amer. Math. Soc., Coloquium Publication, Vol. 17 (1934).

[10] G. E. P. Box, "Spherical distributions (preliminary report)," *Ann. Math. Stat.*, Vol. 24 (1953), pp. 687–688.

[11] J. Clerk Maxwell, "Illustrations of the dynamical theory of gases. Part 1: On the motion and collisions of perfectly elastic spheres," *Philos. Mag.*, Vol. 19 (1860), pp. 19–32.

[12] M. S. Bartlett, "The vector representation of a sample," *Proc. Cambridge Philos. Soc.*, Vol. 30 (1934), pp. 327–340.

[13] G. E. P. Box, "The exploration and exploitation of response surfaces: Some general considerations and examples" *Biometrics*, Vol. 10 (1954), pp. 16–60.

[14] G. E. P. Box and J. S. Hunter, "The exploration and exploitation of response surfaces: Some useful designs," in preparation; to be submitted to *Biometrics*.

[15] R. M. DeBaun, "Block effects in the determination of optimum conditions," *Biometrics*, Vol. 12 (1956), pp. 20–22.

[16] G. E. P. Box, J. S. Hunter and R. J. Hader, "The effect of inadequate models in surface fitting," *Technical Report #5*, Institute of Statistics, Raleigh, N. C.

[17] G. E. P. Box and J. S. Hunter, "A confidence region for the solution of a set of simultaneous equations with an application to experimental design," *Biometrika*, Vol. 41 (1954), pp. 190–199.

[18] K. D. Tocher, Discussion to the paper "On the experimental attainment of optimum conditions," by Box and Wilson, *J. Roy. Stat. Soc., Ser. B*, Vol. 13 (1951), pp. 39–42.

[19] G. E. P. Box and P. V. Youle, "The exploration and exploitation of response surfaces: An example of the link between the fitted surface and the basic mechanism of the system," *Biometrics*, Vol. 11 (1955), pp. 287–323.

[20] D. Wallace, "Intersection region confidence procedures with an application to the location of the maximum in quadratic regression," submitted for publication.

2.6
Design of Experiments in Non-Linear Situations

G. E. P. BOX and H. L. LUCAS
Statistical Techniques Research Group, Princeton University, and Institute of Statistics, North Carolina State College

1. Outline of the problem

We suppose that some response η is a known function

$$\eta = f(\xi_1, \xi_2, \ldots, \xi_k; \theta_1, \theta_2, \ldots, \theta_p) = f(\boldsymbol{\xi}; \boldsymbol{\theta}) \tag{1}$$

of k variables whose levels are denoted by the elements $\xi_1, \ldots, \xi_i, \ldots, \xi_k$ of the vector $\boldsymbol{\xi}$ and of p parameters $\theta_1, \ldots, \theta_r, \ldots, \theta_p$ elements of the vector $\boldsymbol{\theta}$, and that this function is not necessarily linear in either the variables or the parameters. We further suppose that it is possible to perform a series of experiments or trials in each one of which the response is observed at a particular chosen set of levels of the variables decided in advance.

The problem here considered is that of selecting a programme of trials such that they may be expected to provide results from which the p parameters can be estimated with high accuracy. In general the experimental programme may be defined by an $N \times k$ matrix $\mathbf{D} = \{\xi_{iu}\}$ called the design matrix. The uth row $\boldsymbol{\xi}'_u$ of this matrix with elements $\xi_{1u}, \ldots, \xi_{iu}, \ldots, \xi_{ku}$ provides the levels of the k variables at which the response is to be observed in the uth trial.

1·1. *Example*

The following example illustrates the situation we have in mind. Suppose a consecutive chemical reaction is under study in which a substance A decomposes to form substance B which then in turn decomposes to form substance C. With the assumptions that the reactions are first order and irreversible it may be shown that, after time ξ_1 has elapsed, the yield η of intermediate product B is given by

$$\eta = \frac{\theta_1}{\theta_1 - \theta_2}\{\exp(-\theta_2 \xi_1) - \exp(-\theta_1 \xi_1)\}, \tag{2}$$

where θ_1 and θ_2 are constants measuring the specific rates of the first and second decompositions, respectively.

The problem is to choose a set of times ξ_{1u}, $u = 1, 2, \ldots, N$, at which to observe the yield so that from these observations θ_1 and θ_2 can be estimated as accurately as possible, allowing for experimental error in the observations. In this example the design matrix $\mathbf{D}$ would simply consist of a single column whose N elements were the times $\xi_{11}, \ldots, \xi_{1u}, \ldots, \xi_{1N}$ at which the yield was to be observed and $\boldsymbol{\theta}$ would contain the two elements θ_1 and θ_2. More generally there will be k variables (for example, ξ_1 = time, ξ_2 = temperature, ξ_3 = concentration, etc.) and the design matrix will contain k columns.

1·2. *Practical limits to levels of variables*

In practice, levels at which variables such as time, temperature, concentration and dosage can be set are usually restricted. For example, in the study of the chemical reaction above, the feasible range for the variable time (ξ_1) will be from zero to some value ξ_1(max) which will be the longest time which could be contemplated in an actual experiment. In general, there

The Collected Works of George E. P. Box, 1984, Wadsworth, Inc., Belmont, CA 94002.
Originally published in *Biometrika,* vol. 46, parts 1 and 2 (1959), pp. 77–90.

will exist some experimental region R in the $\boldsymbol{\xi}$ space which delimits the area within which trials can actually be conducted. In a chemical reaction, for example, such limitation may occur from explosive hazards associated with certain combinations of temperature and pressure. In other cases it may be simply that it is experimentally inconvenient to conduct trials at certain combinations of conditions. In some cases the experimental region can be defined by a series of inequalities in the ξ's of the type

$$\xi_i(\min) \leqslant \xi_i \leqslant \xi_i(\max) \quad (i = 1, 2, \ldots, k).$$

In general, however, the definition of R may be more complicated.

1·3. *Criterion for selection of design*

We denote the response observed at the uth set of experimental conditions by y_u and we suppose that

$$E(y_u) = \eta_u = f(\boldsymbol{\xi}_u, \boldsymbol{\theta}), \tag{3}$$

and

$$E(y_u - \eta_u)(y_v - \eta_v) = \begin{cases} \sigma^2 & u = v \\ 0 & u \neq v \end{cases} \quad (u, v = 1, 2, \ldots, N), \tag{4}$$

where, in general, σ^2 is unknown. The true values of the parameters are denoted by $\theta_{10}, \theta_{20}, \ldots, \theta_{p0}$, the elements of the vector $\boldsymbol{\theta}_0$. The partial derivatives of the response function with respect to the rth parameter θ_r for the uth set of experimental conditions, taken at the point $\boldsymbol{\theta}_0$, is denoted by

$$f_{ru} = \left[\frac{\partial f(\boldsymbol{\xi}, \boldsymbol{\theta})}{\partial \theta_r}\right]_{\boldsymbol{\theta} = \boldsymbol{\theta}_0}. \tag{5}$$

Finally, the $N \times p$ matrix of these derivatives is

$$\mathbf{F} = \{f_{ru}\}. \tag{6}$$

It is well known that the least squares estimates $\hat{\boldsymbol{\theta}}$ obtained by minimizing the sum of squares

$$\sum_{u=1}^{N} \{y_u - f(\boldsymbol{\xi}_u, \boldsymbol{\theta})\}^2 \tag{7}$$

and given by solving the normal equations

$$\sum_{u=1}^{N} \{y_u - f(\boldsymbol{\xi}_u, \hat{\boldsymbol{\theta}})\}\{\partial f(\boldsymbol{\xi}_u, \boldsymbol{\theta})/\partial \theta_r\}_{\boldsymbol{\theta} = \hat{\boldsymbol{\theta}}} = 0 \quad (r = 1, 2, \ldots, p) \tag{8}$$

have a variance covariance matrix which is approximated by $(\mathbf{F}'\,\mathbf{F})^{-1}\sigma^2$.

We shall proceed by attempting to choose $\mathbf{D}$ so that the determinant $|(\mathbf{F}'\mathbf{F})^{-1}|$ is made as small as possible. If we assume that each $f(\boldsymbol{\xi}_u, \boldsymbol{\theta})$ is approximately linear in the neighbourhood of $\boldsymbol{\theta}_0$ we are thus minimizing Wilk's generalized variance for the estimates. The logic of this choice, also made by Wald (1943), can be seen by noting that if the estimates are approximately normally distributed, the determinant $|(\mathbf{F}'\,\mathbf{F})^{-1}|$ is proportional to the volume contained within any specific ellipsoidal probability contour for $\boldsymbol{\theta}$ about $\boldsymbol{\theta}_0$ in the space of the parameters. The criterion chosen ensures that any such probability contour includes the smallest volume.

To understand the significance of this criterion in general non-linear situations it seems desirable to consider the relationship between sample space, the 'solution locus' within sample space, and parameter space. The point $\boldsymbol{\theta}$ in parameter space can be mapped on to the point $P(\boldsymbol{\theta})$ in sample space with co-ordinates $\eta_u = f(\boldsymbol{\xi}_u, \boldsymbol{\theta})$ $(u = 1, \ldots, N)$. We define the part of sample space covered by this mapping as the 'solution locus'.

Now, given any point $P(\boldsymbol{\theta})$ on the solution locus, we can make an orthogonal transformation $\boldsymbol{\eta} = \mathbf{H}\boldsymbol{\zeta}$ of co-ordinates in sample space such that the solution locus in the immediate neighbourhood of $P(\boldsymbol{\theta})$ coincides with part of the p-dimensional hyperplane $\zeta_i = \text{constant}$ $(i = p+1, \ldots, N)$. We can then interpret $|(\mathbf{F}'\mathbf{F})^{-1}|$ as the square of the Jacobian of the transformation from co-ordinates $(\zeta_1, \ldots, \zeta_p)$ in sample space to $(\theta_1, \ldots, \theta_p)$ in parameter space, when $P(\boldsymbol{\theta}) = P(\boldsymbol{\theta}_0)$. So the criterion that $|(\mathbf{F}'\mathbf{F})^{-1}|$ is to be made as small as possible means geometrically that the p-dimensional volume of that part of parameter space associated with a region of prescribed small volume round $P(\boldsymbol{\theta}_0)$ on the solution locus shall be as small as possible. Since we do not know $\boldsymbol{\theta}_0$ in advance, we must assume that we can find a design that makes this Jacobian about as small as possible for all points $P(\boldsymbol{\theta})$ near $P(\boldsymbol{\theta}_0)$.

Let us now assume that the part of the solution locus near $P(\boldsymbol{\theta}_0)$ can be approximated by a p-dimensional linear subspace of sample space. If we also assume that the experimental error distributions are normally distributed as well as satisfying (3) and (4), then the confidence region for $\boldsymbol{\theta}$ based on the likelihood ratio criterion consists of all points associated with points on the solution locus within a p-dimensional sphere centred on the least squares estimate $P(\hat{\boldsymbol{\theta}})$, and with radius independent of the experimental design. Any design that gives an adequate approximation to the smallest possible value of the Jacobian for all points round $P(\boldsymbol{\theta}_0)$ therefore approximately minimizes the p-dimensional volume of the confidence region in parameter space.

1·4. *Necessity for preliminary estimates*

We see that the efficiency of different possible designs depends upon the matrix $\mathbf{F}$ whose elements are the values of the derivatives of the response function with respect to the θ's at $\boldsymbol{\theta} = \boldsymbol{\theta}_0$. Now the derivatives $\partial f/\partial\theta_r$ can only be independent of the values actually taken by the parameter θ_r if the response function f is linear in θ_r. For non-linear response functions the values of the derivatives, and hence the efficiency of any particular design, depend upon the actual values of the p-parameters. If we are to suppose that effective design is possible at all, we must also assume therefore that *something* is known about the values of the parameters in advance.

In practical problems it will almost invariably be the case that some such information is available, and this will then provide the basis of a first design. As is inevitably the case here and elsewhere, the real effectiveness of the design will depend upon the reliability of the information on which it is based. In this paper, therefore, we assume that preliminary values $\boldsymbol{\theta}^*$ of the parameters are available and will proceed so that the design would be optimal if these quantities were the true values. Some further discussion of the general situation for which this theory is appropriate is given by Box (1957).

1·5. *Possible extensions of the problem*

It is clear that in some examples normality of distribution and constancy of variance would not be possible at all points on the response function if only because the variation might be limited either upwards or downwards. Two obvious extensions, therefore, of this work would be to generalize to a maximum likelihood solution in which any distribution is considered or to consider weighted least squares. Valuable work in this direction has already been carried out by Elfving (1952) and Chernoff (1953). Again, in some cases we might wish to make a transformation of the model before the present theory was applied.

Apart from questions of appropriate distribution assumptions it should be possible to extend this work in at least two other directions. In the first place in different problems the reliability of the preliminary values $\boldsymbol{\theta}^*$ will be different, and we might take account of this by using some rough estimate of the prior probability distribution. Again, some observations might be cheaper than others. One could then attempt to find the design giving the most information for a given cost. Finally, it would be desirable in many circumstances to use a sequential procedure in which estimates from a preliminary experiment would be used to plan a second experiment, the results of which would in turn be used to plan a third experiment, and so on. These possibilities will not be further discussed here but we hope it may be possible to pursue them in later work.

2. Solution

Suppose that we are given the functional form

$$\eta_u = f(\boldsymbol{\xi}_u, \boldsymbol{\theta}) \tag{9}$$

and a set of preliminary values $\boldsymbol{\theta}^*$. Then we shall choose the design $\mathbf{D}$ such that $|\mathbf{F}^{*\prime}\mathbf{F}^*|$ is a maximum within the experimental region R in the $\boldsymbol{\xi}$-space, where

$$\mathbf{F}^* = \{f^*_{ru}\} \tag{10}$$

and

$$f^*_{ru} = \left[\frac{\partial f(\boldsymbol{\xi}_u, \boldsymbol{\theta})}{\partial \theta_r}\right]_{\boldsymbol{\theta}=\boldsymbol{\theta}^*}. \tag{11}$$

To limit to manageable proportions the possibilities we need to consider, we will in this paper assume that the number of trials (that is the number of combinations of levels of the variables) is equal to the number p of parameters to be estimated. Each of the p trials could of course be replicated to attain any desired degree of precision, but it should be noted that an r-fold replication of a design that minimizes the generalized variance using p observations, does not necessarily minimize the generalized variance for all possible experiments involving rp observations. Experiments in which the number of parameters estimated is equal to the number of trials are not perhaps intuitively appealing. This is (i) because one is usually not completely assured that the model is correct; so that one ordinarily has in fact a twofold objective: to check the model and to estimate the parameters in it if it proves to be representationally adequate, and (ii) because designs in which there are more than the minimum number of points may provide greater protection against gross errors in guessing $\boldsymbol{\theta}^*$.

In order for a design to be efficient in 'checking' the model, we must be specific about what sorts of departure from the model we suspect. In so far as this can be done, we should be able to write an 'alternative' wider model which makes allowance for these possible departures *in terms of extra parameters to be estimated.* In this way, within the framework discussed we can achieve both objectives mentioned above. As a simple illustration consider the chemical reaction discussed earlier. Here A was decomposed to form B which in turn decomposed to C. In a particular problem the second decomposition might be unlikely, in which case the primary model in mind for the production of B would be

$$\eta = 1 - e^{-\theta_1 \xi}. \tag{12}$$

However, in the light of possible decomposition of B to C it would be appropriate to plan the experiment so as to be effective in estimating both θ_1 and θ_2 in the model of equation (2).

When nothing is known about the type of departures which may occur we are confronted once more with the fact already mentioned that the effectiveness of a design must depend on the reliability of information on which it is based. If this information is non-existent then we have no basis on which to plan. In practice, however, the situation may arise where some information is available which is difficult to express in terms of extra parameters in the model, but is manifested in feelings about the way the response surface may be distorted. In this case further points might be added to the design in an intuitive way, so as to tie down doubtful areas of the surface, while at the same time bearing in mind the locations already determined which ensure maximum reliability for the principal parameters when the model is adequate.

When gross errors may occur in guessing $\boldsymbol{\theta}^*$ the present theory can be used to explore the changes that would occur in the position of the optimal p design points with changes in $\boldsymbol{\theta}^*$ over the region of uncertainty and the losses in efficiency which occurred with given errors in $\boldsymbol{\theta}^*$. If more than p-points were available these might then be chosen so as to minimize the loss of efficiency as far as possible.

2·2. *A geometrical formulation*

For that class of designs for which the number of trials is equal to the number of parameters, $\mathbf{F}^*$ is a square $(p \times p)$ matrix. Consequently,

$$|\mathbf{F}^{*\prime}\mathbf{F}^*| = |\mathbf{F}^*|^2 \tag{13}$$

and we should choose the design $\mathbf{D}$ so that the modulus of $|\mathbf{F}^*|$ is a maximum in the experimental region R.

In understanding the problem to be solved and in showing the properties of solutions in simple cases, the following geometrical formulation has been found of help. Consider a p-dimensional space in which the co-ordinates are the derivatives

$$\left(\frac{\partial f(\boldsymbol{\xi}, \boldsymbol{\theta})}{\partial \theta_r}\right)_{\boldsymbol{\theta}=\boldsymbol{\theta}^*} = \phi_r(\boldsymbol{\xi}) \quad (r = 1, 2, \ldots, p), \tag{14}$$

where, it will be noted, $\phi_r(\boldsymbol{\xi})$ is a function of $\boldsymbol{\xi}$ only. The surface in the space of the derivatives generated by all permissible values of $\boldsymbol{\xi}$ we call the 'design locus'. Thus with any particular choice of p 'design points', $\boldsymbol{\xi}_1, \boldsymbol{\xi}_2, \ldots, \boldsymbol{\xi}_u, \ldots, \boldsymbol{\xi}_p$, in the ξ-space for a particular design matrix $\mathbf{D}$, there will be corresponding points on the design locus in the space of the derivatives $\phi_r(\boldsymbol{\xi})$.

Now the volume of a simplex in p-space formed by p-points

$$(z_{11}, z_{12}, \ldots, z_{1p});\ \ldots;\ (z_{s1}, z_{s2}, \ldots, z_{sp});\ \ldots;\ (z_{p1}, z_{p2}, \ldots, z_{pp})$$

and the origin $(0, 0, \ldots, 0)$ is proportional to the determinant of the $p \times p$ matrix $\{z_{sr}\}$. Consequently, we must find p design points $\boldsymbol{\xi}_1, \ldots, \boldsymbol{\xi}_u, \ldots, \boldsymbol{\xi}_p$ for which the corresponding points in the space of the derivatives, with co-ordinates defined by equation (14), form with the origin the simplex of greatest volume.

2·3. *Example*

As a simple illustration, consider the example given in § 1·1 of a consecutive first-order chemical reaction. Suppose that we are given preliminary guesses θ_1^*, θ_2^* for the values of the parameters. We wish to choose two values ξ_{11}, ξ_{12} of time so that the best estimates will be available for θ_1, θ_2. We have

$$\eta = f(\xi_1; \theta_1, \theta_2) = \frac{\theta_1}{\theta_1 - \theta_2}\{\exp(-\theta_2 \xi_1) - \exp(-\theta_1 \xi_1)\}. \tag{15}$$

The region R is defined by $$0 \leqslant \xi_{11} \leqslant \xi_{12} \leqslant \xi_1(\max). \tag{16}$$
but for the moment no upper limit $\xi_1(\max)$ will be imposed. We have to choose ξ_{11}, ξ_{12} so as to maximize the modulus of the determinant

$$\Delta = |\mathbf{F}^*| = \begin{vmatrix} \phi_1(\xi_{11}) & \phi_2(\xi_{11}) \\ \phi_1(\xi_{12}) & \phi_2(\xi_{12}) \end{vmatrix}. \tag{17}$$

If the preliminary guesses are

$$\theta_1^* = 0\cdot7, \quad \theta_2^* = 0\cdot2, \tag{18}$$

then $$\phi_1(\xi_1) = (0\cdot8 + 1\cdot4\xi_1)\exp(-0\cdot7\xi_1) - 0\cdot8\exp(-0\cdot2\xi_1), \tag{19}$$
$$\phi_2(\xi_1) = (2\cdot8 - 1\cdot4\xi_1)\exp(-0\cdot2\xi_1) - 2\cdot8\exp(-0\cdot7\xi_1). \tag{20}$$

We can of course substitute equations (19) and (20) in (17) and calculate the values ξ_{11}, ξ_{12} which maximize the modulus of Δ. It is instructive, however, to study this solution in some detail using the geometrical construction. The dotted line in Fig. 1 is a graph of the function

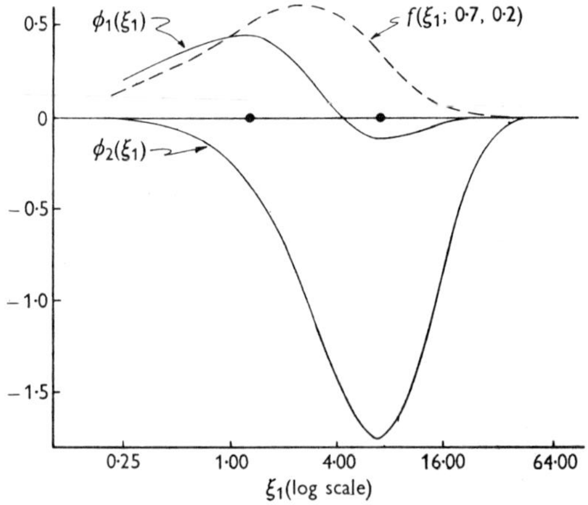

Fig. 1. The function $f(\xi_1; 0\cdot7, 0\cdot2)$ and its derivatives. The optimal solution for the design is indicated by the heavy dots.

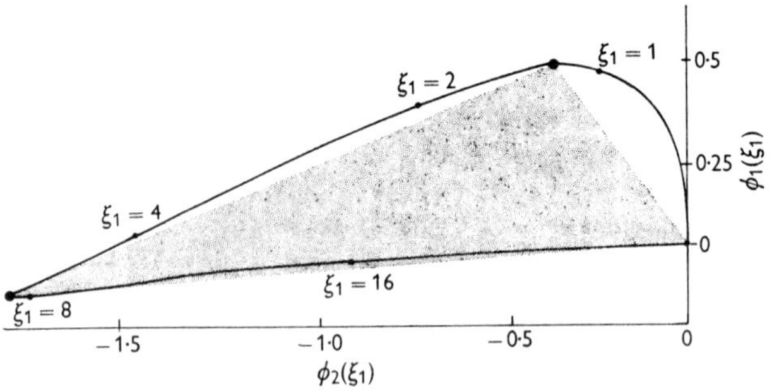

Fig. 2. The design locus for the function $f(\xi_1; 0\cdot7, 0\cdot2)$ with the optimal solution indicated by heavy dots.

$f(\xi_1; 0{\cdot}7, 0{\cdot}2)$. On the same diagram are shown the values of the derivatives $\phi_1(\xi_1)$, $\phi_2(\xi_1)$. Fig. 2 shows $\phi_1(\xi_1)$ plotted against $\phi_2(\xi_1)$ for values of ξ_1 covering the range of interest. This is the curve we call the design locus.

In terms of Fig. 2 the problem is to find those two points ξ_{11}, ξ_{12} on this locus, which together with the origin form the vertices of the triangle of greatest area. The exact solution obtained by numerical methods described later is $\xi_{11} = 1{\cdot}23$, $\xi_{12} = 6{\cdot}86$, corresponding to the triangle denoted by the shaded area in Fig. 2. This solution is indicated in Fig. 1 by the points at $\xi_{11} = 1{\cdot}23$ and $\xi_{12} = 6{\cdot}86$. From the graph of the function $f(\xi_1; 0{\cdot}7, 0{\cdot}2)$ it will be seen that for this particular example the values are chosen so as to lie on either side of the maximum at points of moderately high yield. The graphs of the derivatives illustrate how the criterion seeks points at which the derivatives are large in absolute value, but also, so far as possible, uncorrelated.

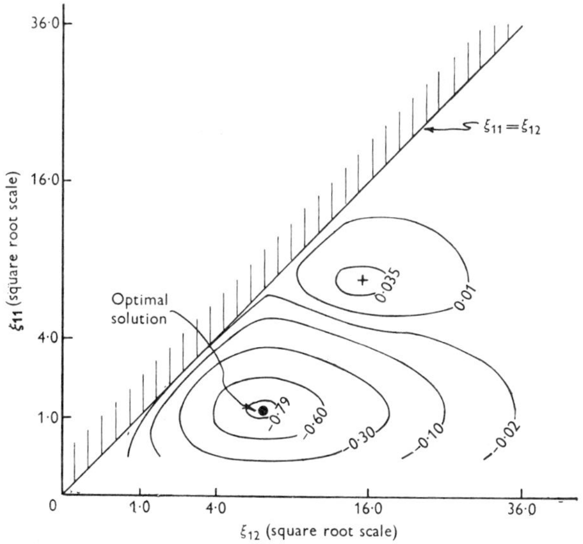

Fig. 3. Contours of Δ for the function $f(\xi_1; 0{\cdot}7, 0{\cdot}2)$ in the space of ξ_{11}, ξ_{12} for $0 \leqslant \xi_{11} \leqslant \xi_{12}$.

2·4. *Some problems in determining the solution*

This example may be used to illustrate some of the problems in determining the solution. Fig. 3 shows contours of Δ in the space ξ_{11}, ξ_{12} as defined by equation (17) within the region $0 \leqslant \xi_{11} \leqslant \xi_{12}$. It is seen that this surface contains *two* local extrema, both corresponding to local maxima in the *modulus* of the determinant. In terms of the actual value of the determinant, there is a minimum of $\Delta = -0{\cdot}810$ at the values $\xi_{11} = 1{\cdot}23$, $\xi_{12} = 6{\cdot}86$ and a maximum of $\Delta = 0{\cdot}039$ at about $\xi_{11} = 7{\cdot}5$, $\xi_{12} = 15{\cdot}0$. The latter value is proportional to the area of a small triangle that can be fitted into the concavity on the lower side of the design locus in Fig. 2. This solution is of no practical interest, since the absolute value of the determinant is only about one-twentieth of the value which corresponds to the other solution. However, it is of some importance to note that surfaces with multiple extrema can occur, and may complicate the problem of finding best solutions in more complex cases.

2·5. *Numerical solution*

Although in particular examples there is an analytic solution to the problem of maximizing the modulus of the determinant $|\mathbf{F}^*|$, often numerical methods must be employed. Such methods are attractive because they allow us to obtain solutions even when the function $f(\boldsymbol{\xi}, \boldsymbol{\theta})$ is not given explicitly. In chemical examples, for instance, a kinetic mechanism can often be postulated in terms of a series of differential equations which may have no explicit solution. The equations may be integrated numerically, however, on a computer and the quantities $\phi_r(\boldsymbol{\xi})$ for any trial value $\boldsymbol{\xi}$ obtained from difference formulae. Iterative methods may then be employed to find the maximum value of the modulus of the determinant.

In the particular example studied it is seen from Fig. 3 that in the neighbourhood of the optimum solution the surface is well conditioned, containing no oblique ridges, and that consequently a numerical method in which the variables were adjusted one at a time would be successful in determining the maximum. It is fairly easy to see that the satisfactory state of conditioning indicated by the 'symmetrical' maxima found in this problem cannot be expected to be general. Often the Δ surface would contain attenuated ridges which in the later stages of iteration would give rise to difficulties in numerical solution when the method of adjusting 'one variable at a time' or 'steepest ascent' procedures were used. A method which can be employed when the final solution is approached is the second-order procedure similar to that employed by Koshal (1933) in solving maximum likelihood equations. This consists of fitting a second-degree polynomial to values of the determinant calculated at suitably arranged series of points.

Calculations for improved values of ξ_{11} and ξ_{12}, starting from the values $\xi_{11} = 1{\cdot}10$, $\xi_{12} = 6{\cdot}50$ (obtained from Fig. 2) are shown for the present example in Table 1.

Table 1. *Calculation of improved values of maximum*

ξ_{11}	ξ_{12}	$-\Delta$
1·10	6·50	0·803 336
1·20	6·50	0·807 992
1·30	6·50	0·806 709
1·10	7·00	0·804 758
1·10	7·50	0·798 920
1·20	7·00	0·809 632

Writing $x_1 = 5(\xi_{11} - 1{\cdot}10)$, $x_2 = (\xi_{12} - 6{\cdot}50)$, the second-degree interpolation polynomial is

$$-\Delta = 0{\cdot}803336 + 0{\cdot}015251x_1 + 0{\cdot}010104x_2 - 0{\cdot}011878x_1^2 - 0{\cdot}014250x_2^2 + 0{\cdot}000872x_1x_2$$

which has a maximum at $\quad x_1 = 0{\cdot}66, \quad x_2 = 0{\cdot}37,$

that is $\quad \xi_{11} = 1{\cdot}23, \quad \xi_{12} = 6{\cdot}87.$

A second iteration provided values in substantial agreement with these, namely, $\xi_{11} = 1{\cdot}23$ and $\xi_{12} = 6{\cdot}86$. It is clear that because of difficulties of possible multiple maxima and the complexities of the Δ surface, considerable caution is necessary in applying numerical methods to problems in which there are a large number of parameters. It is hoped to discuss such cases in a later paper. The examples which follow contain only a few parameters but nevertheless are thought to be of some inherent interest.

3. Examples of non-linear designs

3·1. *Simple decay and growth law*

Consider the model

$$\eta = \theta_1 \exp(\theta_2 \xi_1), \quad 0 \leqslant \xi_1(\min) \leqslant \xi_{11} < \xi_{12} \leqslant \xi_1(\max). \tag{21}$$

With ξ representing time and with $\theta_1 > 0$, $\theta_2 < 0$, this is the well-known first-order decay law. With $\theta_1 > 0, \theta_2 > 0$, the function can represent growth phenomena. In application both lower and upper limits $\xi_1(\min)$ and $\xi_1(\max)$ may be imposed corresponding to the earliest and latest points in time at which observations can be made. The value of the determinant is

$$\Delta = \theta_1 \exp\{\theta_2(\xi_{11}+\xi_{12})\}(\xi_{12}-\xi_{11}). \tag{22}$$

The modulus of which for $\theta_2^* < 0$ attains its maximum value when

$$\left.\begin{aligned} \xi_{11} &= \xi_1(\min), \\ \xi_{12} &= \xi_1(\min) - 1/\theta_2^*. \end{aligned}\right\} \tag{23}$$

Substituting these values into equation (21) we see that the best design will be that in which a proportion $e^{-1} = 36{\cdot}8\,\%$ of the amount of substance present at the first observation remains at the second observation.

For $\theta_2^* > 0$ the modulus of Δ is a maximum when

$$\left.\begin{aligned} \xi_{11} &= \xi_1(\max) - 1/\theta_2^*, \\ \xi_{12} &= \xi_1(\max). \end{aligned}\right\} \tag{24}$$

The first observation should thus be made at a time when $36{\cdot}8\,\%$ of the growth expected at the final observation has occurred.

If instead of the amount of substance decayed we consider the amount of product formed from a simple decay law, we have

$$\eta = \theta_1\{1-\exp(\theta_2 \xi_1)\}$$

for

$$0 \leqslant \xi_1(\min) \leqslant \xi_{11} < \xi_{12} \leqslant \xi_1(\max). \tag{25}$$

The value of the determinant is

$$\Delta = \theta_1[\{1-\exp(\theta_2 \xi_{12})\}\xi_{11}\exp(\theta_2 \xi_{11}) - \{1-\exp(\theta_2 \xi_{11})\}\xi_{12}\exp(\theta_2 \xi_{12})], \tag{26}$$

the modulus of which has a maximum when

$$\xi_{12} = \xi_1(\max), \tag{27}$$

$$\xi_{11} = -\frac{1}{\theta_2^*} - \frac{\xi_{12}\exp(\theta_2^* \xi_{12})}{1-\exp(\theta_2^* \xi_{12})}. \tag{28}$$

Substituting these values into equation (25) and letting γ_2 be the fraction by which the yield at time $\xi_{12} = \xi_1(\max)$ falls short of the maximum obtainable, we find that the optimum value of ξ_{11} corresponds to a yield falling short of the maximum by a fraction

$$\gamma_1 = \exp\{-[1+\gamma_2 \log \gamma_2/(1-\gamma_2)]\}. \tag{29}$$

In particular if the second observation is taken after a sufficiently long time so that the asymptotic value is almost achieved, then γ_2 is close to zero and $\gamma_1 = 1/e$, so that the first observation should be made when the process is 63·2 % complete. As a second example, if the second observation is taken at a time when the process is 50 % complete, then $\gamma_1 = 0{\cdot}73$ and the first observation should be made at a time when the process is 27 % complete.

3·2. *Mitscherlich equation*

Consider the model

$$\eta = \theta_1 + \theta_2 \exp(\theta_3 \xi_1) \tag{30}$$

for

$$0 \leqslant \xi_1(\min) \leqslant \xi_{11} < \xi_{12} < \xi_{13} \leqslant \xi_1(\max).$$

With $\theta_3 < 0$ and $\theta_2 < 0$ this expression represents the well-known Mitcherlich law of diminishing returns. For example, in crop experiments with ξ_1 the amount of added fertilizer and η the amount of growth, $\theta_1 + \theta_2$ represents the growth when no fertilizer is added and θ_1 the hypothetical growth obtained from an infinite amount of fertilizer; θ_3 measures the rate at which the response to each additional increment of fertilizer decreases. The same model has been employed to relate the weight η of chickens to their cumulative feed consumption ξ_1 (Hendricks *et al.* 1932). With $\theta_3 > 0$ this model has been used to represent early growth (Brant, 1951).

The value of the determinant is

$$\Delta = \theta_2[\xi_{11}\exp(\theta_3\xi_{11})\{\exp(\theta_3\xi_{13}) - \exp(\theta_3\xi_{12})\} + \xi_{12}\exp(\theta_3\xi_{12})$$
$$\times\{\exp(\theta_3\xi_{11}) - \exp(\theta_3\xi_{13})\} + \xi_{13}\exp(\theta_3\xi_{13})\{\exp(\theta_3\xi_{12}) - \exp(\theta_3\xi_{11})\}]. \tag{31}$$

It is readily shown that irrespective of the signs of $\theta_1, \theta_2, \theta_3$, the optimum design is obtained with

$$\xi_{11} = \xi_1(\min), \tag{32}$$

$$\xi_{12} = -\frac{1}{\theta_3^*} + \frac{\xi_{11}\exp(\theta_3^*\xi_{11}) - \xi_{13}\exp(\theta_3^*\xi_{13})}{\exp(\theta_3^*\xi_{11}) - \exp(\theta_3^*\xi_{13})}, \tag{33}$$

$$\xi_{13} = \xi_1(\max). \tag{34}$$

In particular in a fertilizer experiment if $\xi_{11} = \xi_1(\min)$ corresponded to no dressing of fertilizer and $\xi_{13} = \xi_1(\max)$ corresponded to a dressing for which the response fell short of the maximum by a certain fraction γ_2, then ξ_{12} would be a dressing for which the response fell short of the maximum possible by a fraction γ_1 where again γ_1 would be given by (29). When $\xi_1(\max)$ is sufficiently large to give an almost maximal response then the intermediate value of the design should be that expected to yield a response of 63·2 % of the maximum possible.

3·3. *Simple first-order decay with rate a function of temperature*

Consider the model

$$\eta = \exp(-\theta_1 \xi_1 e^{-\theta_2 \xi_2}) \tag{35}$$

for

$$0 \leqslant \xi_1(\min) \leqslant \xi_{11} < \xi_{12} \leqslant \xi_1(\max),$$
$$\xi_2(\min) \leqslant \xi_{21} < \xi_{22} \leqslant \xi_2(\max).$$

This would apply for example where η represented the fractional amount remaining at time $t = \xi_1$ of a material decaying according to the first-order law, when the rate was temperature-sensitive. Assuming that the temperature effect follows the Arrhenius rule, θ_1 is the specific rate at some absolute temperature T_0, θ_2 is proportional to the 'activation energy' and $\xi_2 = T^{-1} - T_0^{-1}$ is the deviation of the reciprocal of the temperature, T, from the reciprocal of the base temperature T_0.

We find for preliminary values θ_1^*, θ_2^* that

$$\phi_1(\xi_1, \xi_2) = -\xi_1 \exp\{-(\theta_2^* \xi_2 + \theta_1^* \xi_1 e^{-\theta_2^* \xi_2})\} = \frac{\eta^* \ln \eta^*}{\theta_1^*},$$

$$\phi_2(\xi_1, \xi_2) = -\theta_1^* \xi_2 \phi_1(\xi_1, \xi_2) = -\xi_2 \eta^* \ln \eta^*, \tag{36}$$

where $\eta_u^* = \exp(-\theta_1^* \xi_{1u} e^{-\theta_2^* \xi_{2u}})$, the expected fraction of decaying substance remaining in the uth experiment. Thus

$$\operatorname{mod} \Delta = \operatorname{mod} \begin{vmatrix} \frac{\eta_1^*}{\theta_1^*} \ln \eta_1^* & -\xi_{21} \eta_1^* \ln \eta_1^* \\ \frac{\eta_2^*}{\theta_1^*} \ln \eta_2^* & -\xi_{22} \eta_2^* \ln \eta_2^* \end{vmatrix} = \frac{\eta_1^* \ln \eta_1^* \eta_2^* \ln \eta_2^*}{\theta_1^*} (\xi_{22} - \xi_{21}). \tag{37}$$

Now $-\eta \log \eta$ has its maximum value when $\eta = 1/e$. Also, if the range of permissible values for the times, ξ_1, is sufficiently wide then the expression (37) is maximized by ensuring that $\xi_{22} - \xi_{21}$ shall achieve the maximum possible range $\xi_2(\max) - \xi_2(\min)$ and adjusting the times of reaction so that the anticipated fractions remaining, η_1^*, η_2^*, are each equal to $1/e$. The actual values of ξ_{11} and ξ_{12} required in this case are obtained by substituting $\eta = e^{-1}$ in equation (35). The appropriate times ξ_{11}, ξ_{12}, associated with $\xi_2(\min)$ and $\xi_2(\max)$ are then given by

$$\xi_{11} = \frac{1}{\theta_1^*} \exp\{\theta_2^* \xi_2(\min)\},$$

$$\xi_{12} = \frac{1}{\theta_1^*} \exp\{\theta_2^* \xi_2(\max)\}. \tag{38}$$

It is instructive to consider a numerical example. Suppose the absolute temperature is restricted to the range 380–420° A. It is then convenient to take as an arbitrary base temperature the value $T_0 = 400°$ A. Suppose with time measured in minutes a guessed value for the rate θ_1 at this arbitrary base temperature is $\theta_1^* = 1{\cdot}0$ and a guessed value for $\theta_2^* = 16{,}000$, then we have

$$\theta_1^* = 1{\cdot}0, \quad \theta_2^* = 16{,}000$$

$$380° \text{ A.} \leqslant T = (\xi_2 + 0{\cdot}0025)^{-1} \leqslant 420° \text{ A.}$$

The optimum design will consist of two experiments, the first of which is performed at the minimum temperature of 380° A for time $\xi_{11} = 8{\cdot}21$ min. and the second at the maximum temperature of 420° A for 0·15 min. These times are found by substituting in equation (38) and are simply those values at which, if the guessed values are correct and there were no experimental error, the fraction of product remaining would be $1/e$. The design is illustrated geometrically in Fig. 4. In this figure the design locus in the plane of $\phi_1(\xi_1, \xi_2)$, $\phi_2(\xi_1, \xi_2)$ is bounded by the values $\phi_1(\xi_1, \xi_2) = 0$ and $\phi_1(\xi_1, \xi_2) = -1/\theta_1^* e$, the largest and the smallest values, respectively, that this derivative can take. The straight lines in Fig. 4 are for constant ξ_2 (i.e. for constant temperature) and the curved lines for constant ξ_1 (i.e. for constant time). Where, as we have assumed above, the permissible range of ξ_2 dictates the design points selected, the design locus is restricted to that portion of the plane within the triangle bounded by the three straight lines

$$\phi_1(\xi_1, \xi_2) = -1/\theta_1^* e,$$
$$\phi_2(\xi_1, \xi_2) = -\theta_1^* \phi_1(\xi_1, \xi_2)\, \xi_2(\text{min}),$$
$$\phi_2(\xi_1, \xi_2) = -\theta_1^* \phi_1(\xi_1, \xi_2)\, \xi_2(\text{max}).$$

The triangle so delineated is itself the triangle of largest area which can be formed with two points on the design locus and the origin with only temperature restricted.

The appropriate design for the case considered is indicated by the two heavy dots in the diagram. It will be seen from the figure that, if ξ_1 were the restricting variable, then a somewhat different result would be obtained, since now the triangle of maximum area would be achieved when the value of $\phi_1(\xi_1, \xi_2)$ fell somewhat short of $-1/\theta_1^* e$.

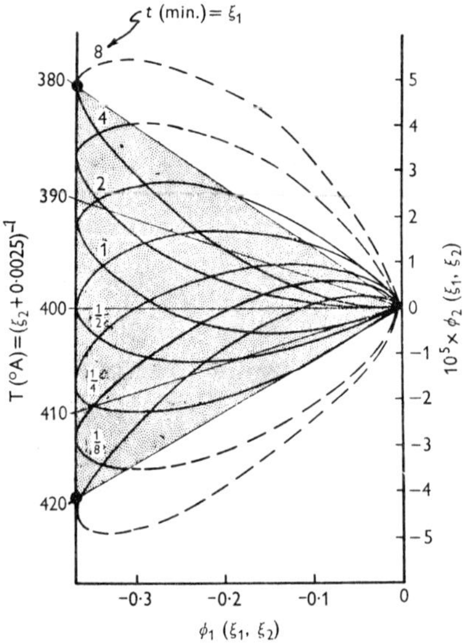

Fig. 4. The design locus for $\eta = \exp\{-1{\cdot}0te^{-16,000(T^{-1}-0{\cdot}0025)}\}$. The optimal solution for $T(\text{min}) = 380°$A., $T(\text{max}) = 420°$ A. and $0 \leqslant \xi_1 < \infty$ is indicated by heavy dots.

4. Calculation of the adjustments

Suppose a design of this type has been obtained for assumed values of the parameters $\boldsymbol{\theta}^*$, and suppose that the N calculated values at these sets of conditions for $\boldsymbol{\theta} = \boldsymbol{\theta}^*$ are given by the column vector $\boldsymbol{\eta}^*$. Suppose finally that observed responses from the N experiments performed are provided by the column vector $\mathbf{y}$. Then proceeding in the usual way a $p \times 1$ vector of adjustments $\boldsymbol{\delta}$ to be applied to the preliminary values $\boldsymbol{\theta}^*$ are given to a first approximation by

$$\delta = (\mathbf{F}^{*\prime}\mathbf{F}^*)^{-1}\mathbf{F}^{*\prime}(\mathbf{y}-\boldsymbol{\eta}^*)$$

or in the special case where the number of observations N is equal to the number of parameters p by

$$\delta = \mathbf{F}^{*-1}(\mathbf{y}-\boldsymbol{\eta}^*).$$

In the particular example of § 3·3,

$$\mathbf{F}^* = \frac{1}{e}\begin{bmatrix} -1/\theta_1^* & \xi_{21} \\ -1/\theta_1^* & \xi_{22} \end{bmatrix}$$

and

$$\mathbf{y}-\boldsymbol{\eta}^* = \begin{bmatrix} y_2 - e^{-1} \\ y_1 - e^{-1} \end{bmatrix},$$

so that the adjusted values will be

$$\theta_1' = \theta_1^* + \delta_1,$$
$$\theta_2' = \theta_2^* + \delta_2,$$

where δ_1 and δ_2 are the 'regression coefficients' of the elements in $\mathbf{y}-\boldsymbol{\eta}^*$ on the columns of the matrix $\mathbf{F}^*$.

To see the meaning of this suppose that the value of θ_1^* were taken at a base temperature such that $\xi_{21} = -\xi_{22}$, then the adjustments in θ_1^* and θ_2^* are simply

$$\delta_1 = -(\bar{y}-e^{-1})\, e\theta_1^*,$$
$$\delta_2 = (y_1 - y_2)\, e/2\xi_{21}.$$

As would be expected, therefore, when the observed values of y were equal but different from the value e^{-1}, this would imply that adjustment was needed only in the basic rate constant θ_1^*. Unequal values for y_1 and y_2 on the other hand would imply the necessity for adjustment in θ_2^*.

5. Discussion

When parameters are jointly estimated from observations made at particular sets of conditions, it frequently happens that the estimates are highly correlated and consequently have very large variances. This may be simply the result of poor design in which case the first group of estimates may be used as starting values to design a more informative set of trials from which greatly improved estimates can be obtained. Not infrequently, however, it turns out that the unsatisfactory nature of the estimates is a property of the model itself, and that even the best design still yields highly correlated estimates. This point is well illustrated by models such as

$$\eta = \frac{1+\theta_1\xi_1}{\theta_2+\theta_3\xi_1}.$$

In accordance with the method already outlined, the optimal design will be found by choosing three points on the design locus which together with the origin form the tetrahedron of greatest volume. But the design locus for such a model often turns out to be a nearly straight line, so that for any reasonable range of the variable, ξ_1, the volume of the tetrahedron and hence the value of the determinant is small. This arises from the nature of the model itself. Whatever values of the ξ's are chosen in such examples, when one has found one set of values of the θ's that most closely agrees with a given set of data, one can, by making suitable compensating changes in the parameters, obtain other quite different values of the θ's that agree almost as well.

In this connexion it is important to distinguish between the problem we are considering here, which is that of finding designs most efficient for estimating the parameters $\boldsymbol{\theta}$, and a somewhat different problem which is *not* discussed here, that of finding designs such that the response $\eta = f(\boldsymbol{\xi}; \boldsymbol{\theta})$ is determined most precisely in a given region. A particular design may be such that the estimates, $\boldsymbol{\theta}$, of the parameters are highly correlated with large variances and yet the response, η, is estimated reasonably accurately. The reason for this is not difficult to see. Instability in the $\boldsymbol{\theta}$'s which is *compensating* is by its very nature not such as will cause instability in the estimate of η.

The authors wish to acknowledge their indebtedness to a referee and to E. M. L. Beale for valuable suggestions for improving the presentation.

REFERENCES

Box, G. E. P. (1957). Use of statistical methods in the elucidation of basic mechanisms. *Bull. Inter. Statist. Inst.* **36**, part 3, 215–25.

Brant, J. W. A. (1951). Rate of early growth in domestic fowl. *Poult. Sci.* **30**, 343–61.

Chernoff, H. (1953). Locally optimal designs for estimating parameters. *Ann. Math. Statist.* **24**, 586–602.

Elfving, G. (1952). Optimum allocation in linear regression theory. *Ann. Math. Statist.* **23**, 255–62.

Hendricks, W. A., Jull, M. A. & Titus, H. W. (1932). The utilization of feed by chickens. I. The requirements for growth. *Poult. Sci.* **2**, 74–7.

Koshal, R. S. (1933). Application of the method of maximum likelihood to the improvement of curves fitted by the method of least squares. *J. Roy. Statist. Soc.* **96**, 303–13.

Wald, A. (1943). On the efficient design of statistical investigations. *Ann. Math. Statist.* **14**, 134–40.

2.7
A Basis for the Selection of a Response Surface Design*

G. E. P. BOX
Princeton University

NORMAN R. DRAPER
University of North Carolina and Imperial Chemical Industries Ltd.

The general problem is considered of choosing a design such that

(a) the polynomial $f(\boldsymbol{\xi}) = f(\xi_1, \xi_2, \cdots, \xi_k)$ in the k continuous variables $\boldsymbol{\xi}' = (\xi_1, \xi_2, \cdots, \xi_k)$ fitted by the method of least squares most closely represents the true function $g(\xi_1, \xi_2, \cdots, \xi_k)$ over some "region of interest" R in the $\boldsymbol{\xi}$ space, no restrictions being introduced that the experimental points should necessarily lie inside R; and

(b) subject to satisfaction of (a), there is a high chance that inadequacy of $f(\boldsymbol{\xi})$ to represent $g(\boldsymbol{\xi})$ will be detected.

When the observations are subject to error, discrepancies between the fitted polynomial and the true function occur:

(i) due to sampling error (called here "variance error"), and

(ii) due to the inadequacy of the polynomial $f(\boldsymbol{\xi})$ exactly to represent $g(\boldsymbol{\xi})$ (called here "bias error").

To meet requirement (a) the design is selected so as to minimize J, the expected mean square error averaged over the region R. J contains two components, one associated entirely with variance error and the other associated entirely with bias error.

There is a class of designs which satisfy requirement (a). To meet requirement (b) we select from this class a subclass for which the "non-centrality term" in the expectation of the residual sum of squares in the analysis of variance is large. This leads to a sensitive test of goodness of fit.

In this paper the theory is applied to the particular case where $f(\boldsymbol{\xi})$ is a polynomial of first degree and $g(\boldsymbol{\xi})$ a polynomial of second degree; that is, the experimenter is hopefully fitting a first degree equation over the region R in the circumstances where the true function is really quadratic. The somewhat unexpected conclusion is reached that, at least in the cases considered, the optimal design in typical situations in which both variance and bias occur is very nearly the same as would be obtained if *variance were ignored completely* and the experiment designed so as to *minimize bias alone*.

Particular examples of the class of optimal designs derived are fractional by replicated two-level factorial designs (in which no two-factor interaction is confounded with the main effect) with added center points.

It is proved (Appendix 1) that if a polynomial of *any* degree d_1 is fitted by the method of least squares over *any* region of interest R in the k variables, when the true function is a polynomial of any degree $d_2 > d_1$, then the bias averaged over R is minimized for all values of the coefficients of neglected terms, by making the moments of order $d_1 + d_2$ and less of the design points equal to the corresponding moments of a uniform distribution over R.

---

* Paper read at the 31st Session of the International Statistical Institute in Brussels 6th Sept. 1958. Material originally presented at the I.M.S. Summer Statistical Institute, July 1957, and prepared as *Technical Report* No. 23, August 1958 in connection with research sponsored by the Office of Ordnance Research, U. S. Army; Statistical Techniques Research Group, Princeton University, Contract No. DA 36-034-ORD 2297.

The Collected Works of George E. P. Box, 1984, Wadsworth, Inc., Belmont, CA 94002. Originally published in *J. Amer. Stat. Assoc.*, vol. 54 (1959), pp. 622–654.

1. INTRODUCTION

SUPPOSE a functional relationship

$$\eta = g(\xi_1, \xi_2, \cdots, \xi_k; \theta_1, \theta_2, \cdots, \theta_p) = g(\xi, \theta) \tag{1}$$

exists between a response η and k continuous variables $\xi_1, \xi_2, \cdots, \xi_k$. Suppose further that in order to elucidate certain aspects of this relationship, measurements of η are to be made for each of N combinations of levels of the variables

$$\xi_u' = (\xi_{1u}, \xi_{2u}, \cdots, \xi_{ku}), \quad (u = 1, 2, \cdots, N) \tag{2}$$

1.1 *The Problem.* The problem of experimental design here considered is the choice of the *design matrix* $\mathbf{D}$ of N rows and k columns. This matrix specifies the levels of all the factors for all the measurements to be made and its u-th row is given by the elements of ξ_u' in (2).

The considerations that influence the choice of $\mathbf{D}$ are different in different circumstances. Two particular cases that arise are:

(i) when, the form of the true functional relationship $\eta = g(\boldsymbol{\xi}, \boldsymbol{\theta})$ (which is of course not necessarily linear in the parameters $\boldsymbol{\theta}$ or in the variables $\boldsymbol{\xi}$) being assumed known, the object is to estimate the *parameters* $\theta_1, \cdots, \theta_p$; and

(ii) when, the form of the true functional relationship being unknown, the object is to approximate within a given region R of the k-dimensional space of the variables, the function $g(\boldsymbol{\xi}, \boldsymbol{\theta})$ by some graduating function $f(\xi_1, \xi_2, \cdots, \xi_k; \beta_1, \beta_2, \cdots, \beta_l)$. The function $f(\boldsymbol{\xi}, \boldsymbol{\beta})$ would often be a polynomial in which case $\boldsymbol{\beta}$ would be the $l \times 1$ vector of polynomial coefficients.

The two different objectives lead to different types of design. Designs appropriate for (i) may be called *designs for estimating parameters;* designs appropriate for (ii) may be called *designs for exploring a response surface* (the surface involved being that defined by the function (1) in the $(k+1)$-dimensional space of the response η, and the variables $\xi_1, \xi_2, \cdots, \xi_k$. The first problem has been discussed [11, 9, and 7]. This paper, however, is specifically concerned with the second problem and not with the first and it is supposed that the functional form $g(\boldsymbol{\xi}, \theta)$ is unknown and is to be graduated locally by a polynomial $f(\boldsymbol{\xi}, \boldsymbol{\beta})$ of degree d_1 in $\boldsymbol{\xi}$. In accordance with an already established notation [8, 6] a design suitable for fitting a polynomial of degree d_1 is called a design of *order* d_1.

Now, in practice, the nature of the variables whose levels are represented by ξ_{iu} will change from one application of a design to another. In one case, for example, ξ_i may refer to a temperature reading and in another to the dosage of a drug. Therefore, it is useful to define the general design in terms of "standardized" variables x_i which in any particular application are related linearly to the ξ_i by a transformation

$$x_{iu} = \frac{\xi_{iu} - \xi_i(0)}{S_i} \tag{3}$$

In what follows we shall choose $\xi_i(0)$ to be the mean of the ξ_{iu} so that

$$\sum_{u=1}^{N} x_{iu} = 0. \tag{4}$$

The scale factor S_i is a convenient constant relating the standardized levels x_i in the general design to the levels of the actual variables ξ_i to be used in a particular application.

1.2. *The Requirements.* In a recent paper [6], it was suggested that, when the function was to be graduated by a polynomial, suitable requirements for a response surface design were as follows:

(a) The design should allow the graduating polynomial of chosen degree d_1 to represent the true function as well as possible within the region of interest;
(b) It should allow a check to be made on the representational adequacy of the polynomial;
(c) It should not contain an excessively large number of experimental points;
(d) It should lend itself to "blocking";
(e) It should form a nucleus from which a satisfactory design of higher order could be built in case the polynomial of degree d_1 proved representationally inadequate.

1.3 *The Necessity for Considering Bias as Well as Variance.* Most approaches to the theory of experimental design have been concerned only with the errors arising from "sampling" variation. The graduating function is assumed to be capable of providing a perfect representation and the expectation of a fitted value $\hat{y}(\mathbf{x})$ is supposed equal to $\eta(\mathbf{x})$ the true value. In practice, of course, this specification is inadequate because a graduating function such as a polynomial will always fail, to some extent, to represent the true function. In reality therefore there are not one, but two, possible sources of discrepancy between the true function $\eta(\mathbf{x})$ and the fitted graduating function $\hat{y}(\mathbf{x})$. The first occurs because of sampling error and the second because of the inadequacy of the graduating function. We refer to the first as the "variance error" and the second as the "bias error."

Derivation of statistical designs as if "variance error" were the only cause of discrepancy has often led to conclusions which are at odds with the experimenter's natural intuition. Consideration of why this happens helps us to see how we ought to proceed. Usually there will be a large "operability region," O, of unknown, or vaguely known, extent within which it is *possible* to carry out experiments and within this, at a given stage of experimentation, a smaller "region of immediate interest," R. In this paper it is assumed that R is entirely within O and the boundary of O is never reached by any experimental point. Within the region of immediate interest R, the experimenter may feel it is reasonable to represent the response function by, for example, a polynomial of first or second degree, although he may know that such a representation would be quite inadequate over the whole operability region O.

Assuming a particular model, the variance of $\hat{y}(\mathbf{x})$ at some point $\mathbf{x}$ will normally decrease as the size of the design is increased, so that if variance error is treated as the only kind of discrepancy we are led to the conclusion that in order to obtain a good representation over R we ought to take as large a design as possible covering the *whole operability region O*. This result is at odds with common sense and is only reached because we ignore the decreased ability of the simple graduating function to represent the real relationship as wider and wider regions of the space of the variables are considered. It may be argued that to avoid this difficulty one could simply choose the design points to "cover" the limited region R. Against this however is the consideration that it might turn out that, even where bias was taken into account, to obtain a close fit within R it was best to locate some points outside R. Alternatively it might be best to confine all the design points to a region much smaller than R.

2. METHOD

Clearly what we require is some way in which the apparent added precision obtainable by making the design larger may be balanced against the loss of representational accuracy. So far we have mentioned only the choice of size of the design. In practice, the values of other quantities (such as λ_4 for a second order rotatable design [6]), which determine the distribution of the design points in the space of the variables, have to be decided upon and again similar questions arise.

The object of the present paper is to consider a general theory which will meet the first two requirements (a) and (b) listed above when both "variance error" and "bias error" are taken into account. To develop this theory we use a principle which might with profit be adopted more often in statistical investigations. This is that, rather than suppose, as is usually done, that the assumed model is always correct, we suppose that it is always to some extent incorrect. The general theory which is developed is applied to the particular problem of choosing a first order design, that is a design which is appropriate when the polynomial fitted is of first degree in $x_1, \cdots, x_k$.

2.1 *Interpretation of Requirement (a)*. We shall interpret requirement (a), that the design should allow a graduating polynomial of specific degree d_1 to represent the true function as well as possible within the region of interest R, in the following way.

We suppose that, although it is assumed that within R a polynomial of degree d_1 will be adequate, in fact the true function is a polynomial of higher degree $d_2 > d_1$ and that this polynomial of higher degree provides an *exact* representation within the whole operability region O. Let $\hat{y}(\mathbf{x})$ be the value of the estimated response at the point $\mathbf{x}$ where this value is obtained by the least squares fitting of the polynomial of degree d_1 using a design containing N points. Let $\eta(\mathbf{x})$ be the true response at this point given exactly by the polynomial of higher degree d_2 and let σ^2 be the experimental error variance. Then we should like to choose the design so as to minimize

$$J = \frac{N}{\sigma^2}\int_R E[\hat{y}(\mathbf{x})] - \eta(\mathbf{x})]^2 dx \Big/ \int_R dx \tag{5}$$

where $dx = dx_1, dx_2 \cdots dx_k$. Thus J is the mean squared deviation from the true response, averaged over the region R and normalized with respect to the number of observations and the variance. To illustrate the division into "variance error" and "bias error," (5) can be written

$$J = V + B \tag{6}$$

$$= \frac{N\Omega}{\sigma^2}\int_R V[\hat{y}(\mathbf{x})]dx + \frac{N\Omega}{\sigma^2}\int_R [E\hat{y}(\mathbf{x}) - \eta(\mathbf{x})]^2 dx \tag{7}$$

where

$$\Omega^{-1} = \int_R dx.$$

It will be noted that the expression V is the *variance function* of the reference [6] averaged over the region R. A similar criterion has been employed in recent work by David and Arens [10], and by Folks [12].

2.2 *Interpretation of Requirement* (*b*). We shall interpret requirement (b), that the design should allow a check to be made on the representational accuracy of the assumed polynomial in the following way.

We suppose that a test for lack of fit is to be made by the use of an analysis of variance in which the residual sum of squares

$$S_R = \sum_{u=1}^{N} (\hat{y}_u - y_u)^2 \tag{8}$$

is compared with the experimental error variance. This test may involve the comparison of S_R either with a prior value σ^2 of the experimental error variance, supposed to be known exactly, or with some independent estimate s^2. In either case, a parameter which determines the power of the test for goodness of fit will be the quantity

$$\sum_{u=1}^{N} [E(\hat{y}_u) - \eta_u]^2 = E(S_R) - \nu\sigma^2 \tag{9}$$

where ν is the number of degrees of freedom on which the residual sum of squares is based. While our ultimate object should be to make the power of the test as large as possible, in any particular instance in which ν is assumed fixed, an intermediate objective will be to make the expectation of S_R large.

We shall interpret requirement (b) therefore as implying that the design should be chosen so as to make $E(S_R)$ large. It seems reasonable to regard requirement (a) as being of major importance so that in practice we shall proceed by first attempting to find the class of designs which minimizes J in (5) and then attempting to satisfy criterion (b) by selecting from this class, a sub-class which makes large the expected value of S_R in (8).

3. APPLICATION: CHOOSING A DESIGN FOR FITTING A STRAIGHT LINE

We first illustrate the theory in the simple case $d_1 = 1$, $d_2 = 2$, $k = 1$. Here the situation is that a relationship

$$\eta = g(\xi) \tag{10}$$

is supposed to exist between the response η and a variable ξ such as temperature. The exact nature of the function $g(\xi)$ is unknown but the experimenter feels that he can usefully graduate this relationship by a linear equation in ξ over a limited region R within O extending from ξ' to ξ''. We assume that over some wider operability region O it can in fact be represented *exactly* by a quadratic equation in ξ. For simplicity and without loss of generality we make the transformation

$$x = \frac{2\xi - \xi' - \xi''}{\xi'' - \xi'} \tag{11}$$

so that in terms of x the region R extends from -1 to $+1$. Thus the fitted linear expression intended to approximate the true function over the region R is

$$\hat{y}(x) = b_0 + b_1 x \tag{12}$$

while the true relationship over the whole region O is assumed to be

$$\eta(x) = \beta_0 + \beta_1 x + \beta_{11} x^2 \tag{13}$$

We shall assume in what follows that

$$\sum_{u=1}^{N} x_u = 0,$$

which implies that the experimental design is centrally located in the region of interest R. We then apply our theory to decide what is the best distribution of the experimental points subject only to this restriction.

3.1 *Requirement* (*a*); *true model quadratic.* Denote the second and third moments of the design points respectively by

$$c = [11] = N^{-1} \sum_{u=1}^{N} x_u^2 \tag{14}$$

$$[111] = N^{-1} \sum_{u=1}^{N} x_u^3 \tag{15}$$

Then

$$V = \frac{N}{\sigma^2} \int_{-1}^{1} V\{\hat{y}(x)\} dx \Big/ \int_{-1}^{1} dx \tag{16}$$

$$= \frac{1}{2} \int_{-1}^{1} \left(1 + \frac{x^2}{c}\right) dx = 1 + \frac{1}{3c} \tag{17}$$

Now with $\eta(x)$ given by the quadratic model (13) the expected values of b_0 and b_1 in the fitted linear equation (12) are

$$E(b_0) = \beta_0 + c\beta_{11} \tag{18}$$

$$E(b_1) = \beta_1 + \frac{[111]}{c}\beta_{11}. \tag{19}$$

Therefore

$$\begin{aligned} E\hat{y}(x) &= E(b_0) + E(b_1)x \\ &= \beta_0 + \beta_1 x + \beta_{11}\left[c + \frac{[111]}{c}x\right], \end{aligned} \tag{20}$$

whence, with $\alpha_{11}^2 = N\sigma^{-2}\beta_{11}^2$,

$$B = \frac{1}{2}\alpha_{11}^2 \int_{-1}^{1} \{E\hat{y}(x) - \eta(x)\}^2 dx \tag{21}$$

$$= \frac{1}{2}\alpha_{11}^2 \int_{-1}^{1}\left[c - x^2 + \frac{[111]}{c}x\right]^2 dx \tag{22}$$

$$= \alpha_{11}^2\left[c^2 - \frac{2c}{3} + \frac{1}{5} + \frac{[111]^2}{3c^2}\right]. \tag{23}$$

Combining (17) and (23) we obtain

$$\begin{aligned} J &= V + B \\ &= \left[1 + \frac{1}{3c}\right] + \alpha_{11}^2\left[c^2 - \frac{2c}{3} + \frac{1}{5} + \frac{[111]^2}{3c^2}\right] \end{aligned} \tag{24}$$

The quantity V in the first bracket is the contribution from sampling error and the remaining quantity B is the contribution from bias. From (24) it is seen immediately that whatever the values of c and α_{11} the expression is minimized for changes in [111] when this third moment is zero. In the subsequent discussion therefore we suppose that $[111]=0$.

With [111] set equal to zero

$$J = V + B = \left[1 + \frac{1}{3c}\right] + \alpha_{11}^2\left[\left(c - \frac{1}{3}\right)^2 + \frac{4}{45}\right] \tag{25}$$

and the only design characteristic contained in this expression is c. It will be noted that while the part V does not contain any of the parameters $\boldsymbol{\beta}$, the part B, which expresses the integrated bias, contains $\alpha_{11}^2 = N\beta_{11}^2\sigma^{-2}$ a "standardized" measure of the quadratic curvature. The value of c which minimizes the whole expression J depends, therefore, on the size of α_{11}^2.

Now the sampling error in $\hat{y}$ is proportional to $\sigma/\sqrt{N}$ so that the quantity

$$\alpha_{11} = \beta_{11}/(\sigma/\sqrt{N}) \tag{26}$$

is a measure of the ratio of the quadratic curvature to the sampling error. When α_{11} is very small, the curvature is very small compared with the sampling error, and the optimum c tends to infinity. On the other hand, when α_{11} tends to infinity, the curvature is very large compared with the sampling error and c tends to the value $\frac{1}{3}$.

The quantity $\sqrt{c}$ is the root mean square deviation of the experimental points from the origin. It is a convenient measure of the spread of the design points which, if the x's were random variables, would correspond to the standard deviation σ_x of x.

In the case where there is no experimental error and all the discrepancy arises from bias, that is from the inadequacy of the linear model, we are instructed therefore to choose the x_u so that the root mean square deviation

$$\sqrt{c} = \frac{1}{\sqrt{3}} = 0.58 \tag{27}$$

To summarize: if we believed implicitly in the assumption of linearity over an indefinitely wide region then we should minimize V alone, which would lead us to allow the bounds of the design to extend as far as possible. At the other extreme, if with R extending from $-\theta$ to $+\theta$ we knew that errors which occurred in the observations were negligible but we were doubtful about the assumption of linearity, then we should minimize B alone, which would lead us to limit the range of the observations so that the root mean square deviation $\sqrt{c}=0.58\theta$.

It will be seen that the two extreme cases lead to widely different conclusions about the optimal choice of the spread of the design. To make further progress we must obtain some idea of what might be a "typical value" for α_{11} in the usual situation where variance and bias *both* occur. Now it might be expected that if the experimenter were graduating a function by a polynomial over a particular limited region R, he would try to choose the degree of polynomial so that, over the region considered, the average size of error arising from the bias part was at least no larger than the average size of error arising from the variance part and, in a typical case, things might perhaps be arranged so that these contributions were about equal in magnitude. To see what this would imply for the present example we write (25) in the form

$$J = V(c) + \alpha_{11}^2 B(c). \tag{28}$$

For every value of α_{11} there will exist a corresponding minimizing value of c and corresponding values for $V=V(c)$ and $B=\alpha_{11}^2B(c)$. We shall proceed by choosing c from these minimizing values so that the value of α_{11} makes $V=B$.

Now in general we can choose c from the minimizing values so that the value of α_{11} makes $V=gB$ where g is any desired positive constant by minimizing the product $\{V(c)\}^g B(c)$.

For, let c^* be the value of c that satisfies

$$\frac{\partial}{\partial c}\left[\{V(c)\}^g B(c)\right] = 0, \tag{29}$$

that is, at $c=c^*$,

$$g\{V(c)\}^{g-1}V'(c)B(c) + \{V(c)\}^g B'(c) = 0 \tag{30}$$

Provided the product $g\{V(c)\}^{g-1}B(c)B'(c)\neq 0$, we have

$$\frac{1}{g}\,\frac{V(c^*)}{B(c^*)} = -\left[\frac{V'(c)}{B'(c)}\right]_{c=c^*} = \alpha_{11}^{*2}, \quad \text{say.} \tag{31}$$

Thus equation (29) provides the values c^* of c, and α^* of α such that

$$\text{(a)} \quad V'(c) + \alpha_{11}{}^{*2}B'(c) = \frac{\partial}{\partial c}(V + B) = 0 \tag{32}$$

$$\text{(b)} \quad V(c) - g\alpha_{11}{}^{*2}B(c) = V - gB = 0 \tag{33}$$

Putting $g=1$ we find that the optimal values of $\sqrt{c}$ when $V=B$ is 0.62, when the value of α_{11} is 4.49. This compares with the value $\sqrt{c}=0.58$ when variance is completely ignored.

We are led therefore to a somewhat remarkable conclusion. This is that the optimal design in a "typical" situation in which the influence of bias and variance are equal is very nearly the same as that obtained *when variance is ignored completely* and the experiment is designed to minimize bias alone.

It is admitted that the concept, that the experimenter would typically arrange matters so that the average errors in the predicted response due to sampling variation and to bias were about the same, is not very precise and in practice considerable deviations from it would be expected to occur. It is seen from Table 630 however, that even in the case where the variance contribution is *four times* the bias contribution the resulting root mean square value $\sqrt{c}$ is very close to the value obtained when variance is ignored completely and entirely different from the value obtained when bias is ignored. It will be seen in the following sections that similar results are obtainable for the case of k variables. Almost all previous investigators (notable exceptions being Hotelling [14], and more recently David and Arens [10], and Folks [12]), have been principally concerned with the minimization of variance. The recent work raises the possibility that this preoccupation with variance is mistaken and, if in some design problems a simplification of the situation had to be made, it might be better to ignore sampling variation rather than to ignore bias.

TABLE 630. OPTIMAL VALUES FOR $\sqrt{c}$ WITH ASSOCIATED VALUES OF α_{11}

	$\sqrt{c}$	α_{11}	
	∞	0	(variance contribution completely dominant)
	0.72	1.82	(variance contribution four times that of bias)
	0.62	4.49	(variance and bias contributions equal)
	0.58	∞	(bias contribution completely dominant).

3.2 *The Effect of Bias When the True Model Is a General Polynomial.* We have seen that the situation where all the discrepancy arises from bias may be of considerable importance since, in the case studied above, it approximates to the situation in which *both* variance and bias occur. It is of some interest therefore to consider the case where a linear function is fitted and all the discrepancy arises from bias but the true function is *not* necessarily quadratic but is a polynomial of any degree d_2 whatever. It appears as a special case of a theorem proved in Appendix 1 that the integral B which measures the average error arising from bias alone is minimized when all the moments of the design up to that of order (d_2+1) are equal to the moments of a uniform distribution over the region R. For example with the true model quadratic, all moments up to order 3 must be the same as those of a uniform distribution over the range $[-1, 1]$. This means that $\sqrt{c}$ the root mean square deviation of the design points from the center of the region must be equal to 0.58 and the third moment [111] to zero: a result obtained already. In general, when the true model is a polynomial of degree d the i-th moment

$$N^{-1} \sum_{u=1}^{N} x_u{}^i = \begin{cases} 0 & i \text{ odd} \\ (i+1)^{-1} & i \text{ even} \end{cases} \quad i \leqq d_2 + 1. \tag{34}$$

This is intuitively a very reasonable result; in particular it implies that if bias only had to be considered and *nothing whatever* were known of the nature of the function $\eta(x)$ except that it could be represented by some polynomial having an indefinitely large number of terms, we would do best by spreading the design points *evenly* over the region R. It is thought that a completely general polynomial in which high order terms are given the same weight as lower order terms is not a particularly realistic model. Usually it could be safely assumed that the function would have smoothness properties which would place greater emphasis on lower order terms so that, except in the case of complete ignorance as to the nature of the function, even spacing of the design points is not necessarily indicated as the most desirable solution, even for the case where bias is completely dominant.

3.3 *Requirement* (*b*); *True Model Quadratic.* Having selected the best value of c for the design by use of criterion (a), we introduce criterion (b).

To satisfy this criterion we must make large

$$\sigma^{-2}E(S_R) - \nu = \sigma^{-2} \sum_{u=1}^{N} [E(\hat{y}_u) - \eta_u]^2, \tag{35}$$

For this example of the fitting of a straight line with the true function quadratic we obtain, by use of (12), (13) and (35),

$$\sigma^{-2}E(S_R) - \nu = \alpha_{11}{}^2 c^2 \left[\frac{[1111]}{c^2} - \frac{[111]^2}{c^3} - 1 \right].$$

Now criterion (a) requires that $[111]=0$. It is interesting to note that criterion (b) requires this independently; for in order that (36) be maximized [111] must be equal to zero. We see then from (36) that, with $\sqrt{c}$ set equal to some suitable value decided by criterion (a) and [111] set equal to zero, criterion (b) aditionally requires that the fourth moment coefficient should be made large.

It can be shown that the design of this kind that maximizes the fourth moment is obtained by placing $(N-2)$ of the N points at the center of the interval and the remaining two points symmetrically about this center and equidistant from it. Intuitively such an arrangement is not particularly acceptable. The reason can be seen if we remember that in our example it has been assumed that the true function is *exactly* represented by a quadratic polynomial. If this were not the case, and in fact a cubic term had to be included, then from Section 3.2, minimization of bias alone would require that the fourth moment

[1111] should be $\frac{1}{5}$. This implies that the measure of "kurtosis" $[1111]/c^2$ for the design points should be 1.8 —a small value. The practical implication is that to get good detectability of quadratic departure from the assumed model, the design should be chosen so that the fourth moment was not small but the value should not be taken so large as to cause serious cubic bias in B. The precise value chosen will depend on one's relative anxiety to get good graduation on the one hand and good detectability of quadratic lack of fit on the other.

The conclusions to be drawn from the application of criteria (a) and (b) to this simple case therefore are as follows. The design should be selected so that:

(i) the third moment of the distribution of the design points is zero;

(ii) the size of the design is such that the root mean square distance of the design points from the origin is approximately $.6\theta$, with 2θ the width of the region over which a linear approximation is required. (This is very close to the value which would be appropriate if there were *no experimental error* and is completely contrary to the conclusion obtained when bias is ignored.)

(iii) In order that a quadratic tendency in the true model can be readily detected, the fourth moment of the design should not be small.

4. APPLICATION: CHOOSING A DESIGN FOR FITTING A PLANE

Suppose now that there are k variables $\xi_1, \xi_2, \cdots, \xi_k$, that the nature of the true relationship $\eta = g(\boldsymbol{\xi}, \boldsymbol{\theta})$ is unknown but the experimenter believes it can usefully be graduated by a linear relationship over some region R of immediate interest.

The fitted function is then

$$\hat{y} = b_0 + b_1x_1 + b_2x_2 + \cdots + b_kx_k \tag{36}$$

or, in matrix notation,

$$\hat{y} = \mathbf{x}_1'\boldsymbol{b}_1 \tag{37}$$

where

$$\boldsymbol{b}_1' = (b_0, b_1, \cdots, b_k)$$
$$\mathbf{x}_1' = (1, x_1, \cdots, x_k)$$

The true relationship which applies over the whole operability region O is assumed to be some polynomial function which, for the moment, we will suppose is of some *unspecified* degree d_2 in the k variables.

$$\begin{aligned}\eta = \beta_0 + \beta_1x_1 + \cdots + \beta_kx_k + \beta_{11}x_1^2 + \cdots + \beta_{kk}x_k^2 \\ + \beta_{12}x_1x_2 + \cdots + \beta_{(k-1)k}x_{k-1}x_k + \beta_{111}x_1^3 + \cdots\end{aligned} \tag{38}$$

or, in matrix notation,

$$\eta = \mathbf{x}_1'\boldsymbol{\beta}_1 + \mathbf{x}_2'\boldsymbol{\beta}_2 \tag{39}$$

where

$$\boldsymbol{\beta}_1' = (\beta_0, \beta_1, \cdots, \beta_k)$$
$$\mathbf{x}_1' = (1, x_1, \cdots, x_k)$$

$$\boldsymbol{\beta}_2' = (\beta_{11}, \beta_{22}, \cdots, \beta_{kk}; \beta_{12}, \beta_{13}, \cdots, \beta_{(k-1)k}; \beta_{111}, \cdots)$$
$$\mathbf{x}_2' = (x_1^2, x_2^2, \cdots, x_k^2; x_1x_2, x_1x_3, \cdots, x_{k-1}x_k; x_1^3, \cdots)$$

and each of the vectors $\boldsymbol{\beta}_2'$ and $\mathbf{x}_2'$ contains p terms, where

$$p = \binom{d_2 + k}{d_2} - (k + 1)$$

We have from (7)

$$J = \Omega \frac{N}{\sigma^2} \int_R E[\hat{y}(\mathbf{x}) - \eta(\mathbf{x})]^2 d\mathbf{x} = V + B \tag{40}$$

with

$$V = N\Omega \int_R \mathbf{x}_1'(\boldsymbol{X}_1'\boldsymbol{X}_1)^{-1}\mathbf{x}_1 d\mathbf{x} \tag{41}$$

and

$$B = \frac{N}{\sigma^2} \Omega \int_R \boldsymbol{\beta}_2'[\boldsymbol{A}'\mathbf{x}_1 - \mathbf{x}_2][\mathbf{x}_1'\boldsymbol{A} - \mathbf{x}_2']\boldsymbol{\beta}_2 d\mathbf{x} \tag{42}$$

where

$$\boldsymbol{X}_1' = [\mathbf{x}_{11}, \cdots, \mathbf{x}_{1u}, \cdots, \mathbf{x}_{1N}] \tag{43}$$

is a $(k+1)\times N$ matrix with $\mathbf{x}_{1u}' = (1, x_{1u}, x_{2u}, \cdots, x_{ku})$,

$$\boldsymbol{X}_2' = [\mathbf{x}_{21}, \cdots, \mathbf{x}_{2u}, \cdots, \mathbf{x}_{2N}] \tag{44}$$

is a $p\times N$ matrix with

$$\mathbf{x}_{2u}' = (x_{1u}^2, x_{2u}^2, \cdots, x_{ku}^2; x_{1u}x_{2u}, \cdots, x_{(k-1)u}x_{ku}; x_{1u}^3, \cdots);$$

and

$$\boldsymbol{A} = (\boldsymbol{X}_1'\boldsymbol{X}_1)^{-1}\boldsymbol{X}_1'\boldsymbol{X}_2 \tag{45}$$

is the $(k+1)\times p$ "alias matrix." This last matrix (for example, [8]) has, for its elements, quantities which measure the extent to which thee stimates $\boldsymbol{b}_1$ are biased by higher order coefficients in accordance with the equation

$$E(\boldsymbol{b}_1) = \boldsymbol{\beta}_1 + \boldsymbol{A}\boldsymbol{\beta}_2. \tag{46}$$

Making the necessary substitutions we obtain

$$\Omega^{-1}V = \sum_{i=0}^{k}\sum_{j=0}^{k} c^{ij} \int_R x_i x_j d\mathbf{x} \tag{47}$$

$$\Omega^{-1}B = \boldsymbol{\alpha}_2'\boldsymbol{A}'\left(\int_R \mathbf{x}_1\mathbf{x}_1' d\mathbf{x}\right)\boldsymbol{A}\boldsymbol{\alpha}_2 - 2\boldsymbol{\alpha}_2'\left(\int_R \mathbf{x}_2\mathbf{x}_1' d\mathbf{x}\right)\boldsymbol{A}\boldsymbol{\alpha}_2 + \boldsymbol{\alpha}_2'\left(\int_R \mathbf{x}_2\mathbf{x}_2' d\mathbf{x}\right)\boldsymbol{\alpha}_2 \tag{48}$$

where

$$\alpha_2 = \frac{1}{\sigma/\sqrt{N}} \beta_2$$

and c^{ij} are the elements of the matrix $\{c^{ij}\} = \{c_{ij}\}^{-1} = C^{-1} = N(X_1'X_1)^{-1}$.

We note in passing that, if we were concerned only with minimizing the average squared bias B, then as a special case of the result of Appendix 1 *whatever* the nature of the region R and *whatever* the degree d_2 of the polynomial describing the true response, for all values of β_2 the bias term B is minimized when all the moments of the design up to order d_2+1 are made equal to the moments of a uniform distribution over R. This result has been still further generalized to include weighting functions, in unpublished work by C. L. Mallows. In particular this implies, as for a single variable, that, if only bias had to be considered and if the rather unrealistic assumption were made that nothing whatever was known of the nature of the function over the region R, then we should do best by spreading the points evenly over R.

4.1 *Choice of the Region R.* The specific results we obtain, whether from minimizing V, B or the whole integral J, will of course depend on how we define the region R. In the present paper we shall suppose that considerations of strategy (for example [1]) dictate the sequential exploration of subregions R entirely contained within O. Such a strategy is often appropriate, for example, in experiments designed to find and to explore a region in which some response or responses have optimal values (for example [8]). To give satisfactory expression to the experimenter's desire to use designs which symmetrically generate information, it seems reasonable to choose R to be a symmetric region in the coordinate system *currently* believed by the experimenter to be most appropriate. Of course the experimenter's ideas as to what *is* the best coordinate system in which to work will almost certainly change as the investigation proceeds. For example suppose he began his investigation with a simple 2^2 factorial in variables x_1 and x_2. This would seem to imply that he currently expected that the response surface could be conveniently represented in an orthogonal coordinate system (x_1, x_2) with the scales of measurement for the two variables proportional to the step size in the factorial design. In fact (and unknown at this time to him) when so represented the response surface might be highly unsymmetrical; for example it might have a ridge-like appearance. In such a case, as he proceeded iteratively from one group of experiments to another, the information built up about the response surface would probably lead him by a process such as is already described [4] to employ, in later stages, transformations and changes of scales and metrics in the variables in terms of which the response surface could be more symmetrically and simply described. At any given stage he would work with that coordinate system which his experience had so far led him to believe would provide the simplest and most symmetric representation of the response surface. Later experience would usually show that his ideas were capable of improvement and would lead to modifications.

We shall try to select designs therefore which generate information symmetrically in that coordinate system currently thought to be best and we will interpret "symmetric" to mean that the region R measured in this coordinate system is a sphere. Adopting this convention the particular case which we study further in this paper is that where a *linear* function is fitted over a *spherical* region R and the true model involving $p=\frac{1}{2}k(k+1)$ extra constants is in fact *quadratic* over the whole of the operability region O.

4.2 *Requirement* (*a*); *True Model Quadratic and Region R Spherical.* We assume that the "center of gravity" of the design points is at the center of the region R, defined by

$$\sum_{i=1}^{k} x_i^2 \leqq 1$$

where

$$x_i = \frac{\xi_i - \bar{\xi}_i}{S_i}, \tag{49}$$

Now for such a region

$$\int_R x_1^{\delta_1} x_2^{\delta_2} \cdots x_k^{\delta_k} dx_1 \cdots dx_k$$

$$= \frac{\Gamma\left(\frac{\delta_1+1}{2}\right)\Gamma\left(\frac{\delta_2+1}{2}\right)\cdots\Gamma\left(\frac{\delta_k+1}{2}\right)}{\Gamma\left\{\frac{\sum_{i=1}^{k}(\delta_i+1)}{2}+1\right\}} \tag{50}$$

unless any δ_i is odd, when the value of the integral is zero.

Hence

$$\Omega \int_R \mathbf{x}_1\mathbf{x}_1' d\mathbf{x} = \left[\begin{array}{c|c} 1 & 0 \\ \hline 0 & (k+2)^{-1}\boldsymbol{I}_k \end{array}\right] \tag{51}$$

and with $\mathbf{x}_2'$ now the vector of quadratic variables

$$(x_1^2, \cdots, x_k^2;\ x_1x_2, \cdots, x_{k-1}x_k),$$

$$\Omega \int_R \mathbf{x}_2\mathbf{x}_1' d\mathbf{x} = (k+2)^{-1}\left[\begin{array}{c|c} j_k & 0 \\ \hline 0 & 0 \end{array}\right] \tag{52}$$

$$\Omega \int_R \mathbf{x}_2\mathbf{x}' d\mathbf{x} = (k+2)^{-1}(k+4)^{-1}\left[\begin{array}{c|c} 2\boldsymbol{I}_k + j_k j_k' & 0 \\ \hline 0 & \boldsymbol{i}_{p-k} \end{array}\right] \tag{53}$$

where $j_k' = [1\ 1 \cdots 1]$ and $\boldsymbol{I}_k$ is the unit matrix of order k.

4.2a *Vanishing of the Third Order Moments.* Now with $\boldsymbol{X}_2$ a matrix of quadratic variables, the alias matrix $\boldsymbol{A}$ appearing in (48) can be partitioned after the first row as follows:

$$A = \left[\frac{A_1}{A_2}\right] \tag{54}$$

where $A_1 = (c_{11}, \cdots, c_{kk}; c_{12}, \cdots, c_{(k-1)k})$ is a $1 \times p$ row vector of second moments, with

$$c_{ij} = N^{-1} \sum_{u=1}^{N} x_{iu} x_{ju}$$

and A_2 is a $k \times p$ matrix of linear combinations of third moments only. Using (50) to obtain the values of the integrals in the equation (47) for V and substituting (51), (52), (53) and (54) in the equation (48) for B, the elements of the integral $J = V + B$ for this case of a plane fitted to a truly quadratic surface are found to be

$$V = 1 + \sum_{i=1}^{k} c^{ii}/(k+2) \tag{55}$$

$$B = \frac{1}{k+2} \sum_{g=1}^{k} \left\{ \sum_{i=1}^{k} \sum_{j=i}^{k} \alpha_{ij} \sum_{h=1}^{k} c^{gh}[hij] \right\}^2 + \left\{ \sum_{i=1}^{k} \sum_{j=i}^{k} \alpha_{ij}(c_{ij} - \delta_{ij}/(k+2)) \right\}^2$$
$$+ \frac{2(k+2) \sum_{i=1}^{k} \alpha_{ii}^2 + (k+2) \sum_{i=1}^{k} \sum_{j=i+1}^{k} \alpha_{ij}^2 - 2\left(\sum_{i=1}^{k} \alpha_{ii} \right)^2}{(k+2)^2(k+4)} \tag{56}$$

where

$$\delta_{ij} = \begin{cases} 0 & i \neq j \\ 1 & i = j \end{cases} \quad \text{and the} \quad \alpha_{ij} = \frac{\beta_{ij}}{(\sigma/\sqrt{N})}$$

measure the size of the second order constants β_{ij} relative to the sampling error. Now third moments $[hij]$ enter J only in the first term of B and from the form of this term we see that, if $\{c_{gh}\}$ is non-singular, J is minimized with respect to the $[hij]$ for general α_{ij} and whatever the c^{gh}, if all the $[hij]$ vanish. We proceed therefore in what follows by assuming that all third moments have these optimal zero values.

With third moments all zero, the expression J contains only the elements of $\boldsymbol{\alpha}_2$ and the second moments c_{ij} of the design. If the x's were random variables these c_{ij} would be the variances and covariances of the k-variate distribution of the x's and they will be so called. Now, whereas V does not contain $\boldsymbol{\alpha}_2$, the integrated bias B is a quadratic form in the elements of $\boldsymbol{\alpha}_2$. In general then, if contributions from bias and variance are considered simultaneously, values of c_{ij} which minimize J will depend on the elements of $\boldsymbol{\alpha}_2$, that is they will depend on the size of the quadratic parameters β_{ij} relative to the sampling error. Before studying this general case further we first consider the extreme cases of completely dominant variance and completely dominant bias.

4.2b *Case of Completely Dominant Variance.* In the case of completely dominant variance, the model is assumed to fit perfectly and there is no contribution from the component B. The integral J is minimized by minimizing V. In general, designs chosen to minimize V will extend to the limits of the operability region O.

For example suppose the size of the design is limited by limiting the variances of the k-variate distribution of points so that

$$c_{ii} \leqq \gamma_i \quad i = 1, \cdots, k \tag{57}$$

with γ_1 suitably chosen constants. It is shown [13, 16, 19, 2] that V is minimized when

$$\left.\begin{array}{ll} c_{ii} = \gamma_i & i = 1, \cdots, k \\ c_{ij} = 0 & \text{all } i, j \end{array}\right\} \tag{58}$$

Thus in these circumstances we should use a first order *orthogonal* design, that is a design which is such that the column vectors of the design matrix have zero inner products one with another and with a vector of ones. This orthogonal design would be chosen to have largest possible values for the "variances" of the x's.

4.2c *Case of Completely Dominant Bias.* In the other extreme case of completely dominant bias, where the model is not assumed to fit perfectly and the variance contribution approaches zero, J is minimized by minimizing B. From inspection of equation (56) with the $[hij]=0$, or from the general result of Appendix 1, the minimizing values for the c_{ij} are seen to be

$$\left.\begin{array}{ll} c_{ii} = c = (k+2)^{-1} & i = 1, \cdots, k \\ c_{ij} = 0 & \text{all } i, j \end{array}\right\} \tag{59}$$

Thus to minimize bias alone over the spherical region R, the design must be orthogonal with third moments all zero and the "variances" of the x's all equal to $(k+2)^{-1}$. What is implied concerning the *size* of the design can be seen as follows. If

$$r_u = \left(\sum_{i=1}^{k} x_{iu}^2\right)^{1/2}$$

is the distance of the u-th design point from the center of the region R, then for the optimal design

$$kc = \frac{k}{k+2} = \sum_{i=1}^{k} c_{ii} = N^{-1} \sum_{u=1}^{N} \sum_{i=1}^{k} x_{iu}^2 = N^{-1} \sum_{u=1}^{N} r_u^2 = \overline{r^2} = \dot{r}^2 \tag{60}$$

where $\dot{r}$ is the root mean square distance of the experimental points from the origin and so equals $\sqrt{c}$ if $k=1$. Thus for minimal bias $\dot{r}$ the root mean square distance of the experimental points from the center of R must be $\sqrt{k/(k+2)}$ times the radius of R.

It is interesting to see that the orthogonality condition arises here as a result of minimizing *bias alone* over a spherical region. In previous investigations the conclusion that an orthogonal design was optimal has usually been arrived at by minimizing only variance. The result concerning minimization of bias alone is of some practical interest in certain problems of numerical analysis [4, 3] where there are no sampling errors but planarity assumptions may not be completely justified.

4.2d *Intermediate Case of Contributions from Both Variance and Bias.* In practice usually both variance and bias occur simultaneously. Now for *known* α_{ij} the optimum values of the c_{ij} could readily be found; usually however, the α_{ij} are unknown. In these circumstances we can make some progress by proceeding in the following way. In practice we usually do not know the nature of the response surface with which we are dealing; in particular we do not know the orientation of the response surface with respect to the design. We proceed therefore by taking the average value of J over all orthogonal rotations of the response surface denoting the average thus obtained by $\tilde{J}$. Since V does not contain any of the constants of the response function, such a process leaves the value of (55) unchanged. The quantity B however is a quadratic form in the elements of the vector $\boldsymbol{\alpha}_2$ or equivalently of the vector $\boldsymbol{\beta}_2$. The average value of B over all rotations of the surface is obtained by substituting the averaged values over all rotations of the products $\alpha_{gh}\alpha_{ij}$ which occur in (56). The details of the averaging are given in Appendix 2. Using these averaged values, and with the optimal value of zero substituted for all the third moments we obtain

$$\begin{aligned}\tilde{J} &= V + \tilde{B}\\ &= \left[1 + (k+2)^{-1}\sum_{i=1}^{k} c^{ii}\right] + a\sum_{i=1}^{k}\left(c_{ii} - \frac{1}{k+2}\right)^2\\ &\quad + b\sum_{i=1}^{k}\sum_{j=1}^{k}\left(c_{ii} - \frac{1}{k+2}\right)\left(c_{jj} - \frac{1}{k+2}\right) + 2(a-b)\sum_{i=1}^{k}\sum_{j=1}^{k} c_{ij}{}^2\\ &\quad + k(k+2)^{-2}(k+4)^{-1}[k(k+3)a - (k-1)(k+4)b], \end{aligned} \tag{61}$$

where

$$a = \widetilde{\alpha_{ii}{}^2} = N\sigma^{-2}\widetilde{\beta_{ii}{}^2}, \qquad b = \widetilde{\alpha_{ii}\alpha_{jj}} = N\sigma^{-2}\widetilde{\beta_{ii}\beta_{jj}}$$

Now V is minimized with respect to the c_{ij} $(i \neq j)$ when $\mathbf{C} = N^{-1}(\boldsymbol{X}'\boldsymbol{X})$ is diagonal, i.e. when

$$c_{ij} = 0.$$

But these are also the values of c_{ij} $(i \neq j)$ which minimize $\tilde{B}$; hence minimization of J for variations in c_{ij} $(i \neq j)$, irrespective of the values of the c_{ii}, is achieved when

$$c_{ij} = 0. \tag{62}$$

This implies that

$$c^{ii} = c_{ii}{}^{-1}. \tag{63}$$

After substituting (62) and (63) in (61), differentiating with respect to the c_{ii} $(i = 1, 2, \cdots, k)$ and equating the result to zero, we obtain k equations the i-th one of which is

$$-(k+2)^{-1}c_{ii}^{-2} + 2a\left(c_{ii} - \frac{1}{k+2}\right) + 2b\sum_{j=1}^{k}\left(c_{jj} - \frac{1}{k+2}\right) = 0. \qquad (64)$$

Subtracting the g-th such equation from the i-th then gives

$$(k+2)^{-1}c_{ii}^{-2}c_{gg}^{-2}(c_{ii}^2 - c_{gg}^2) + 2a(c_{ii} - c_{gg}) = 0 \qquad (65)$$

which, since the c_{ii} are positive, implies that

$$c_{ii} = c_{gg} = c \quad (i, g = 1, 2, \cdots, k) \qquad (66)$$

We are thus led to the conclusion that the criterion J averaged for all orientations of the response surface is minimized when the design is a first order orthogonal design with all third moments zero and with the variances of the x_i all equal.

4.2c *Summary of Results.* We have shown then that if the surface is truly quadratic over the operability region O and a linear model is fitted over a smaller spherical region within O defined by $\sum x_i^2 \leq 1$:

(i) $V+B$ is minimized, for all β_{ij} and c_{ij}, when the third moments $[ijk]$ of the design are chosen to be zero.

(ii) V alone is minimized, for O defined by $c_{ii} \leq \gamma_i$, when the design is chosen to be first order orthogonal with $c_{ii} = \gamma_i$, that is with the design of maximum possible size.

(iii) B alone is minimized when the design is chosen to be first order orthogonal with all third order moments zero and $c_{ii} = 1/(k+2)$.

(iv) $V+B$ is minimized if values of β_{ij} averaged over all rotations are substituted, when the design has all third order moments zero and is first order orthogonal with the c_{ii} all equal.

These conclusions suggest that we should accept that the best design to use in the circumstances implied by our assumptions is first order orthogonal with the c_{ii} equal and with all third order moments zero, that is such that

$$\begin{aligned} c_{ii} &= c_{jj} = c \quad &&(\text{all } i \text{ and } j); \\ c_{ij} &= 0 &&(i \neq j); \qquad (67) \\ [ijk] &= 0. \end{aligned}$$

Adopting the terminology used before [8] we shall call designs of this class first order orthogonal designs *of type B*.

4.3 *Choice of Optimal Size for First Order Orthogonal Design of Type B.* The question remains of the optimal size of the first order orthogonal type B design. This involves the choice of c or equivalently of $\dot{r} = \sqrt{kc}$, the root mean square distance of the experimental points from the centre of the design. We have seen that for minimization of V alone the size of the design should be as large as possible whereas for minimization of B alone $\dot{r}$ the root mean square distance of the experimental points from the center of R should be $\sqrt{k/(k+2)}$, that is smaller than the radius of R. We need to reach some compromise in the practical situation where neither the contribution from variance nor that from bias can be ignored.

The variance and bias functions for any first order orthogonal design of type B in the general case where no assumptions are made concerning the nature of the β_{ij} may be found by setting $c_{ii}=c_{jj}=c$ (all i and j); $c_{ij}=0$ $(i\neq j)$ and $[ijk]=0$ (all i and j) in (55) and (56). We find

$$V = 1 + \frac{k}{k+2}\,\frac{1}{c} \tag{68}$$

$$B = \left(\sum_i \alpha_{ii}\right)^2\left(c - \frac{1}{k+2}\right)^2 + \frac{2(k+2)\sum_i \alpha_{ii}^2 + (k+2)\sum\sum_{i<j} \alpha_{ij}^2 - 2\left(\sum_i \alpha_{ii}\right)^2}{(k+2)^2(k+4)}. \tag{69}$$

Now consider a symmetric $k\times k$ matrix $\boldsymbol{\alpha}$ for which the i-th diagonal element is α_{ii} and for which the element occupying the intersection of the i-th row and j-th column is $\frac{1}{2}\alpha_{ij}$ $(i\neq j)$ and $\alpha_{ij}=\alpha_{ji}$. We have

$$\operatorname{tr}\boldsymbol{\alpha} = \sum_i \alpha_{ii}, \qquad \operatorname{tr}\{\boldsymbol{\alpha}^2\} = \sum_i \alpha_{ii}^2 + \frac{1}{4}\sum\sum_{i\neq j}\alpha_{ij}^2.$$

Writing

$$\theta = \operatorname{tr}\{\boldsymbol{\alpha}^2\} \qquad \varphi = (\operatorname{tr}\boldsymbol{\alpha})^2/\operatorname{tr}\{\boldsymbol{\alpha}^2\} \tag{70}$$

we have finally for any first order orthogonal design of type B

$$\begin{aligned} J &= V + B \\ &= \left\{1 + \frac{k}{(k+2)c}\right\} + \theta\left\{\varphi\left(c - \frac{1}{k+2}\right)^2 + \frac{2(k+2-\varphi)}{(k+2)^2(k+4)}\right\} \end{aligned} \tag{71}$$

Remembering that $\alpha_{ij}=\beta_{ij}\sqrt{N}/\sigma$, the matrix $\boldsymbol{\alpha}$ is seen to be a simple multiple of a symmetric matrix whose diagonal elements are the quadratic coefficients β_{ii} and whose off-diagonal elements are one-half the interaction coefficients β_{ij} of the true surface. Suppose the k latent roots of this latter matrix are $\lambda_1, \lambda_2, \cdots, \lambda_k$. Then

$$\theta = \operatorname{tr}\{\boldsymbol{\alpha}^2\} = \frac{N}{\sigma^2}\left\{\sum_{i=1}^{k}\beta_{ii}^2 + \frac{1}{2}\sum_{i=1}^{k}\sum_{j=i+1}^{k}\beta_{ij}^2\right\} = \frac{N}{\sigma^2}\sum_{i=1}^{k}\lambda_i^2 \tag{72}$$

$$\begin{aligned} \varphi &= (\operatorname{tr}\boldsymbol{\alpha})^2/\operatorname{tr}\{\boldsymbol{\alpha}^2\} = \left(\sum\beta_{ii}\right)^2 \Big/ \left\{\sum_{i=1}^{k}\beta_{ii}^2 + \frac{1}{2}\sum_{i=1}^{k}\sum_{j=i+1}^{k}\beta_{ij}^2\right\} \\ &= \left(\sum\lambda_i\right)^2/\sum\lambda_i^2 \end{aligned} \tag{73}$$

Now the λ_i's in the expression above are simply the coefficients in

$$\sum_{i=1}^{k}\lambda_i X_i^2$$

the canonical form containing no product terms, into which the quadratic part of the true model

$$\sum_{i=1}^{k} \sum_{j=i}^{k} \beta_{ij} x_i x_j$$

can be transformed by orthogonal rotation. These λ_i thus represent the quadratic effects of the canonical variable X_i in the new coordinate system and their signs and magnitudes determine the characteristics of the true response surface. For example, if all the λ_i are negative, the surface has a maximum with ellipsoidal contours, the length of the i-th principal axis being proportional to λ_i^{-1}. If one or more of the λ_i approaches zero, a line, plane or hyperplane of maxima results, while if certain of the λ_i are positive various minimax situations occur.

The quantity θ is therefore seen to be an overall measure of the magnitude of the quadratic tendency of the surface relative to the sampling error. The quantity φ on the other hand is a homogeneous function of degree zero in the λ's, being independent both of the sampling error and of the absolute magnitude of the λ's. It measures the "state of conditioning" (for example [5]) of the quadratic surface as evidenced by the variation among the λ's. In fact

$$\varphi = k/(1 + C_\lambda^2) \tag{74}$$

where

$$C_\lambda = \sqrt{\frac{\sum_i (\lambda_i - \bar{\lambda})^2}{k}} \bigg/ \bar{\lambda} \tag{75}$$

is the coefficient of variation of the λ's.

If all the λ's were equal and of the same sign corresponding to the best state of conditioning of the surface (which would then have spherical contours) C_λ would be zero and φ would take its maximum value of k. At the other extreme, if the λ's were of mixed signs and $\sum \lambda_i = 0$, C_λ would be infinite and φ would take its minimum value of zero. If it so happened that p of the λ's were equal and of the same sign and the remainder were zero the value of φ would be equal to p. Thus in this case p would be a measure of the number of non-redundant canonical variables. The latter situation approximates to that found in problems where the eventual objective is the location and exploration of maxima. Here, the most common situation is that the true response surface is approximated by a system in which a point, line, plane, or space of maxima occur. For such examples p of the λ's would be negative and the remainder zero.

The optimal value of c and hence of $\dot{r} = (kc)^{1/2}$, the root mean square distance of the experimental points from the center of the design, can of course be calculated for any given values of θ and φ by finding that value which minimizes J in (71). As we have seen however, the optimal value of $\dot{r}$ can take any value between ∞ and $\sqrt{k/(k+2)}$ depending on the values chosen for θ and φ. To make further progress we need to determine what might be typical values for these constants and to do this we may proceed in a manner similar to that adopted in the one-dimensional case. We have seen that the constant φ measures the state of conditioning of the response surface and values of particular interest lie between the limits $1 \leq \varphi \leq k$. Now for any fixed value of φ the integral J is of the form:

$$J = V(c) + \theta B(c)$$

As before, by minimizing the product $\{V(c)\}^{\varphi} B(c)$ we can choose c and hence $\dot{r}$ from these minimizing values so that $V = gB$, that is so that the value of θ makes the contribution $V = V(c)$ from variance any desired multiple g of the contribution $B = \theta B(c)$ from bias. The case $g = 1$ is of particular interest since it is plausible that an experimenter attempting to graduate a function over a particular limited region might make his choices of the size of region and degree of polynomial so that the average size of error arising from bias was about the same as that which he thought might arise from variance.

Table 642 shows the value of $\dot{r}$ for $\varphi = k, k-1, \cdots, 2, 1$ and 0.2; for $k = 2, 3$ and 5, for the four cases.

V Variance completely dominant
$V = 4B$ Variance contribution four times that of bias
$V = B$ Variance and bias contributions equal
B Bias completely dominant.

Considering first those entries for which the variance and bias contributions are equal, it is seen once more that over the important range $1 \leq \varphi \leq k$ the values for $\dot{r}$ are very little larger than those obtained if the variance contribution is ignored entirely, and that furthermore, even if the variance contribution is made as large as *four times* the bias contribution, the value of $\dot{r}$ still remains comparatively small.

As ϕ approaches the value 0 the optimal value of $\dot{r}$ becomes larger and eventually goes to infinity. The reason for this apparent anomaly is easily seen. If θ is not infinite, a zero value for φ implies that $\sum \lambda_i = 0$. Now $\sum \lambda_i = \sum \beta_{ii}$ is proportional to the bias in the estimate b_0 of the constant term β_0. With the particular class of designs considered b_0 is the only biased estimate, all the other estimates $b_1, b_2, \cdots, b_k$ being completely unbiased by second order terms. It follows that for the particular case in which $\sum \lambda_i = 0$ and hence $\varphi = 0$, *no second*

TABLE 642. OPTIMAL VALUE FOR $\dot{r}$ WITH ASSOCIATED VALUE OF $\sqrt{\theta}$

k \ ϕ		0.2		1		2		3		4		5	
		$\dot{r}$	$\sqrt{\theta}$	$\dot{r}$	$\sqrt{\theta}$	$\dot{r}$	$\sqrt{\theta}$	$\dot{r}$	$\sqrt{\theta}$	$\dot{r}$	$\sqrt{\theta}$	$\dot{r}$	$\sqrt{\theta}$
2	V	∞	0	∞	0	∞	0						
	$V=4B$	1.34	1.54	1.00	2.00	0.84	3.07						
	$V=B$	0.99	4.73	0.79	6.23	0.74	8.10						
	B	0.71	∞	0.71	∞	0.71	∞						
3	V	∞	0	∞	0	∞	0	∞	0				
	$V=4B$	1.36	3.11	0.99	4.80	0.87	6.56	0.82	8.75				
	$V=B$	1.01	9.58	0.84	12.55	0.80	15.18	0.79	18.97				
	B	0.77	∞	0.77	∞	0.77	∞	0.77	∞				
5	V	∞	0	∞	0	∞	0	∞	0	∞	0	∞	0
	$V=4B$	2.05	1.89	1.61	1.87	1.33	2.59	1.09	4.70	0.97	7.32	0.91	10.65
	$V=B$	1.30	9.02	1.01	12.15	0.93	14.57	0.89	17.04	0.87	20.20	0.86	25.13
	B	0.85	∞	0.85	∞	0.85	∞	0.85	∞	0.85	∞	0.85	∞

order bias exists in the estimate $\hat{y}$ and consequently the optimal design (which is now that which minimizes variance alone) is of *infinite* size. This case $\sum\lambda_i = 0$ is of course very atypical and we are left with the conclusion that in cases likely to be met in practice the optimal orthogonal design of type B should usually be chosen so that $\dot{r}$ is somewhat greater than $\sqrt{k/(k+2)}$ but rather less than unity.

4.4 *Requirement (b); True Model Quadratic.* In accordance with the plan outlined at the beginning of the paper, we will choose from among that class of designs which satisfy requirement (a) that sub-class which best meets requirement (b). It will be recalled that this latter requirement is designed to ensure that the experimental arrangement should allow a check to be made on the representational accuracy of the assumed class of polynomials. The requirement is met by choosing the expected value of the residual sum of squares S_R to be large. Before going ahead with this plan we make a small digression to show that if, following the indications of requirement (a), we choose the first order design to be orthogonal, then requirement (b) *independently* implies that all third order moments should be zero.

4.4a *Independent Vanishing of Third Order Moments.* Writing

$$\mathbf{x}_{1u}' = (1, x_{1u}, x_{2u}, \cdots, x_{ku}) \tag{77}$$

$$\mathbf{x}_{2u}' = (x_{1u}^2, x_{2u}^2, \cdots, x_{ku}^2, x_1x_2, x_1x_3, \cdots, x_{k-1}x_k) \tag{78}$$

we have from (9)

$$E(S_R) - \nu\sigma^2 = \sum_u \boldsymbol{\beta}_2'(A'\mathbf{x}_{1u} - \mathbf{x}_{2u})(\mathbf{x}_{1u}'A - \mathbf{x}_{2u}')\boldsymbol{\beta}_2$$

$$= \boldsymbol{\beta}_2'A'\sum_u \mathbf{x}_{1u}\mathbf{x}_{1u}'A\boldsymbol{\beta}_2 - 2\boldsymbol{\beta}_2'\sum_u \mathbf{x}_{2u}\mathbf{x}_{1u}'A\boldsymbol{\beta}_2 + \boldsymbol{\beta}_2\sum_u \mathbf{x}_{2u}\mathbf{x}_{2u}'\boldsymbol{\beta}_2 \tag{79}$$

Now if the design is orthogonal $N^{-1}\sum_u \mathbf{x}_{1u}\mathbf{x}_1' = N^{-1}(X_i'X_1)$ is a $(k+1)\times(k+1)$ diagonal matrix with first diagonal element unity and remaining diagonal elements equal to c. The matrix

$$N^{-1}\sum_u \mathbf{x}_{2u}\mathbf{x}_{1u}' = N^{-1}\boldsymbol{X}_2'\boldsymbol{X}_1 = [\boldsymbol{M}_2 \mid \boldsymbol{M}_3]$$

is $p\times(k+1)$ and may be partitioned after the first column into two submatrices M_2 and M_3, with M_2 a column vector containing p elements the first k of which are equal to c and the remainder to zero, and M_3 having each element a third order moment. Finally the elements of the matrix

$$N^{-1}\sum_u \mathbf{x}_{2u}\mathbf{x}_{2u}' = N^{-1}\boldsymbol{X}_2'\boldsymbol{X}_2 = \boldsymbol{M}_4$$

are all fourth order moments. Substitution in equation (79) and division by N defines a quantity

$$F = N^{-1}\{E(S_R) - \nu\sigma^2\} = -\boldsymbol{\beta}_2'\boldsymbol{M}_2\boldsymbol{M}_2'\boldsymbol{\beta}_2 - c^{-1}\boldsymbol{\beta}_2'\boldsymbol{M}_3\boldsymbol{M}_3'\boldsymbol{\beta}_2 + \boldsymbol{\beta}_2'\boldsymbol{M}^4\boldsymbol{\beta}_2 \tag{80}$$

$$= -c^2\left\{\sum_{g=1}^{k}\beta_{gg}\right\}^2 - c^{-1}\sum_{g=1}^{k}\left\{\sum_{h=1}^{k}\sum_{i=h}^{k}\beta_{hi}[ghi]\right\}^2$$

$$+\sum_{g=1}^{k}\sum_{h=g}^{k}\sum_{i=1}^{k}\sum_{j=i}^{k}\beta_{gh}\beta_{ij}[ghij]. \tag{81}$$

Third order moments are involved only in the second term whence it is seen that for any first order orthogonal design, F is maximized with respect to third order moments for all values of β_{hi} when all these third order moments are zero.

4.4b *Size of Fourth Moments.* We now return to the main theme. From the infinite class of first order orthogonal designs which have all third order moments zero and for which $\dot{r}$ is fixed at some specific value chosen on criterion (a), we have to choose a sub-class which makes large

$$F = -c^2\left\{\sum_{i=1}^{k}\beta_{ii}\right\}^2 + \sum_{g=1}^{k}\sum_{h=g}^{k}\sum_{i=1}^{k}\sum_{j=i}^{k}\beta_{gh}\beta_{ij}[ghij] \tag{82}$$

We notice that the only quantities which determine the design and which are at our choice are the fourth moments $[ghij]$. In general the optimal values of these moments depend upon the values of the elements of $\boldsymbol{\beta}_2$ and usually these elements are unknown. We can however obtain a design which is "good on the average" by arguing as before that, in a case where all orientations of the response surface with respect to the chosen coordinate system are equally likely, it is reasonable to average F over all rotations. Using equations (2.5), (2.6), (2.7), (2.18) and (2.19) from Appendix 2, we obtain

$$\frac{N}{\sigma^2}\bar{F} = \frac{1}{k(k+2)}\theta(\varphi+2)\left\{\sum_{i=1}^{k}[iiii] + \sum_{i\neq j}^{k}\sum^{k}[iijj]\right\} - \theta\varphi c^2 \tag{83}$$

$$= \frac{\theta}{k}\left\{\frac{\varphi+2}{k+2}\,\frac{\sum_u r_u^4}{N} - \frac{\varphi\dot{r}^4}{k}\right\} \tag{84}$$

where

$$r_u = \left\{\sum_{i=1}^{k} x_{iu}^2\right\}^{1/2}$$

is the distance of the u-th experimental point from the center of the region R and

$$\left\{N^{-1}\sum_{u=1}^{N} r_u^2\right\}^{1/2} = \sqrt{kc} = \dot{r}$$

is the root mean square distance.

The above expression can be written

$$\frac{N}{\sigma^2}\bar{F} = \dot{r}^4\frac{\theta}{k}\left\{\frac{\varphi+2}{k+2}G(r) - \frac{\varphi}{k}\right\} \tag{85}$$

where $G(r)$ is a fourth moment coefficient

$$\frac{\sum r_u^4}{N}\Big/\dot{r}^4$$

of the distances r_u.

Also we can write

$$G = 1 + \{C_{r^2}\}^2 \tag{86}$$

where C_{r^2} is the coefficient of variation of the r_{u^2}. C_{r^2} takes its minimal value of zero and $G(r)$ its minimal value of unity when all the experimental points are equidistant from the center of R. We can see therefore from (85) that the detectability of quadratic discrepancy from the assumed planar model is increased as the distances of the experimental points from the center of the region R have greater and greater "percentage variation."

With this in mind we can rewrite the expression in the form

$$\frac{N}{\sigma^2}\tilde{F} = \dot{r}^4 \frac{\theta}{k}\left\{\frac{2(k-\varphi)}{k(k+2)} + \frac{\varphi+2}{k+2}\{C_{r^2}\}^2\right\} \tag{87}$$

Now, since $0 \leq \varphi \leq k$, both terms within the bracket are necessarily non-negative. For designs in which the points are all equally spaced from the center the second term vanishes. For given k and with $\dot{r}$ fixed by criterion (a) the size of $\sigma^{-2}\tilde{F}$ then depends only on the value of $\theta(k-\varphi)$ and in particular takes the value zero when $\varphi = k$. To see the reason for this we note that when $\varphi = k$ the contours of the response surface are spheres and the response function is represented by a second degree expression containing quadratic terms only and no interaction terms. Now if all points of the design are equally spaced from the center it is easily shown that F in equation (82) and hence the expected value of the residual sum of squares contains only interaction terms. Second order effects of a purely quadratic nature are therefore undetectable.

As φ becomes smaller we encounter surfaces which cannot (except in particular orientations) be represented by quadratic terms alone and consequently the power to detect interaction terms becomes on the average more and more valuable to us. When the points are not all equally spread from the center, $\tilde{F}$ is non-zero even if $\varphi = k$, because unequal values for the r_u render pure quadratic effects detectable.

5. SOME DESIGNS WHICH MEET THE REQUIREMENTS

We now consider one particular way of generating first order designs which satisfy requirements (a) and (b) and which, in accordance with requirement (c), do not contain an excessively large number of points.

5.1 *Requirement (a)*. Designs satisfying requirement (a) must be first order orthogonal with third order moments zero. One simply class of designs of this sort can be generated as follows. Suppose k factors are to be investigated in N trials where N is even and $k \leq \frac{1}{2}N$. We first write down any $(\frac{1}{2}N \times k)$ matrix $\boldsymbol{Z}_1 = \{x_{ij}\}$ having orthogonal columns so that

$$\sum_{u=1}^{N/2} x_{iu}x_{ju} = 0, \quad (i \neq j = 1, 2, \cdots, k) \tag{88}$$

$$\sum_{u=1}^{N/2} x_{iu}^2 = \tfrac{1}{2}Nc,$$

the value of c being chosen on criterion (a). We now take for our design the $N \times k$ matrix

$$\boldsymbol{D} = \begin{bmatrix} \boldsymbol{Z}_1 \\ \hdashline -\boldsymbol{Z}_1 \end{bmatrix} \tag{89}$$

This resulting arrangement satisfies the requirements for a first order orthogonal design and (compare [8, Appendix 2]) the third moments of the design are all zero because the $N\times\frac{1}{2}k(k+1)$ matrix $\boldsymbol{Z}_2$ of independent variables $x_1^2, x_2^2, \cdots, x_k^2, x_1x_2, \cdots, x_{k-1}x_k$ is of the form

$$\begin{bmatrix} \boldsymbol{Z}_2 \\ \hline \boldsymbol{Z}_2 \end{bmatrix}$$

and consequently the matrix of third order moments is null since

$$\begin{bmatrix} \boldsymbol{Z}_1' & \vline & -\boldsymbol{Z}_1' \end{bmatrix} \cdot \begin{bmatrix} \boldsymbol{Z}_2 \\ \hline \boldsymbol{Z}_2 \end{bmatrix} = \boldsymbol{Z}_1'\boldsymbol{Z}_2 - \boldsymbol{Z}_1'\boldsymbol{Z}_2. \tag{90}$$

Now [2], if the levels of the factors are completely unrestricted and at our choice and $k<\frac{1}{2}N$, the original k vectors in the matrix $\boldsymbol{Z}_1$ may be chosen to be orthogonal to a set of up to $\frac{1}{2}N-k$ further vectors which correspond to extraneous systematic effects which it is desired to eliminate. These further vectors may for example represent block contrasts or they may be orthogonal polynomials corresponding to possible time trends. Finally, "spherical randomisation" [2] may be employed to ensure that normal theory may be validly applied in subsequent statistical analysis. The design $\boldsymbol{D}$ obtained by replicating $\boldsymbol{Z}_1$ with reversed signs will preserve all these properties and will possess the additional one (arising from the property that all third order moments are zero) that no estimate of a first order effect is biased by a term of second order, that is by a quadratic or interaction term.

With k *equal* to $\frac{1}{2}N$ we obtain the designs of this kind which allow the investigation of the maximum number of factors. These particular designs are such that the distances of the experimental points from the origin are all equal, for since $\boldsymbol{Z}_1$ is a square orthogonal matrix

$$\sum_{i=1}^{k} x_{iu}^2 = r_u^2 = kc \quad (u = 1, 2, \cdots, N) \tag{91}$$

Of particular interest are the designs of this kind which employ only two levels. It was shown by Plackett and Burman [16] that a square $m\times m$ orthogonal matrix whose first column consisted of $+1$'s, and whose remaining columns contained an equal number of $+1$'s and -1's, could be formed for m any multiple of 4. Using such a design for the matrix $\boldsymbol{Z}_1$, first order designs of type B suitable for testing up to four factors in eight trials, eight factors in 16 trials, twelve factors in 24 trials, etc., can be generated by replicating the basic design with reversed signs.

When m is a power of two, the original Plackett and Burman design and the design of type B derived by replicating it with reversed signs are each fractionally replicated 2^k factorials. The requirement that all third order moments are zero is equivalent to choosing the fractional replicates so that no three-factor interaction is included in the alias sub-group. That is to say, so that no two factor interaction is confounded with a main effect. Two level fractional factorials of this kind, thus provide a particular class of designs which satisfy requirement (a). These designs were derived some time ago [17, 18, and 8] and have been extensively used (for example [15]). They can readily be generaated in the manner illustrated in the example which follows.

Suppose we require a design to test eight factors in 16 experiments. We first generate an 8×8 orthogonal matrix whose first column, which we associate with the variable x_1, consists of +1's and whose remaining seven columns, associated with the variables $x_2, x_3, \cdots, x_8$, consist of equal numbers of −1's and +1's. These last seven columns can be obtained by writing out as column vectors the contrasts associated with three factors x_2, x_3, x_4 run in a 2^3 factorial design together with their interactions. These interaction contrasts are then associated with the remaining factors, for example $x_5=x_2x_3$, $x_6=x_2x_4$, $x_7=x_3x_4$, $x_8=x_2x_3x_4$. We thus obtain for the matrix $\boldsymbol{Z}_1$

$$\boldsymbol{Z}_1 = \begin{bmatrix} x_1 & x_2 & x_3 & x_4 & x_5 & x_6 & x_7 & x_8 \\ 1 & -1 & -1 & -1 & 1 & 1 & 1 & -1 \\ 1 & 1 & -1 & -1 & -1 & -1 & 1 & 1 \\ 1 & -1 & 1 & -1 & -1 & 1 & -1 & 1 \\ 1 & 1 & 1 & -1 & 1 & -1 & -1 & -1 \\ 1 & -1 & -1 & 1 & 1 & -1 & -1 & 1 \\ 1 & 1 & -1 & 1 & -1 & 1 & -1 & -1 \\ 1 & -1 & 1 & 1 & -1 & -1 & 1 & -1 \\ 1 & 1 & 1 & 1 & 1 & 1 & 1 & 1 \end{bmatrix}$$

The first order design of type B is then obtained by replicating this matrix $\boldsymbol{Z}_1$ with all signs reversed.

5.2 *Requirement* (*b*). We now need to try to satisfy the requirement that departures from the assumed model, which occur because the true function is quadratic rather than linear, should be readily detectable.

The experimental designs we have discussed, in which k factors are tested in $2k$ trials by replicating a square $k\times k$ orthogonal matrix with reversed signs, necessarily have points all equispaced from the origin. It should be noted however that even if the points of the design are all equispaced from the origin the expected value of the residual sum of squares although *minimal* is not necessarily *small* except for special types of response surfaces. The arrangements considered do in fact provide quite sensitive tests for all but the rather exceptional kinds of departure from linearity in which the true response functions are described in the original coordinate system by quadratic constants β_{ii} alone. These particular kinds of departures from linearity only become detectable when C_{r^2}, the coefficient of variation of the squared radial distances of experimental points, is non-zero. If the basic design is to be modified by the addition of extra points, the greatest increase in C_{r^2} can be achieved by adding n_0 extra points at the origin. As an example, consider the $n=16$ point factorial design of type B already discussed in Section 5.1. With n_0 additional points at the center we obtain a design containing $N=n+n_0$ points in all.

Using this design the least squares estimates of the coefficients in the linear model have expected values which, on the assumption that the model is truly quadratic, are of the form

$$E(b_0) = \beta_0 + c \sum_{i=1}^{8} \beta_{ii}$$

$$E(b_i) = \beta_i \qquad (i = 1, 2, \cdots, 8)$$

The expected value of the residual sum of squares which has $7+n_0$ degrees of freedom is

$$E(S_R) = c^2 n_0 \frac{N}{n} (\beta_{11} + \beta_{22} + \cdots + \beta_{88})^2 + c^2 \frac{N^2}{n} \{(\beta_{12} + \beta_{35} + \beta_{46} + \beta_{78})^2$$
$$+ (\beta_{13} + \beta_{25} + \beta_{47} + \beta_{68})^2 + (\beta_{14} + \beta_{26} + \beta_{37} + \beta_{58})^2$$
$$+ (\beta_{15} + \beta_{23} + \beta_{67} + \beta_{48})^2 + (\beta_{16} + \beta_{24} + \beta_{38} + \beta_{57})^2$$
$$+ (\beta_{17} + \beta_{28} + \beta_{34} + \beta_{56})^2 + (\beta_{18} + \beta_{27} + \beta_{36} + \beta_{45})^2\} + (7 + n_0)\sigma^2.$$

The residual sum of squares S_R based on $(7+n_0)$ degrees of freedom can be divided into two parts, the first part containing $7+1=8$ degrees of freedom associated exclusively with terms measuring components due to possible lack of fit and the second part S_e containing n_0-1 degrees of freedom associated exclusively with a measure of pure error obtained from the variation of the observations recorded at the center point. The sum of squares for lack of fit having 8 degrees of freedom can be further sub-divided into 8 separate components $S_1, S_2, S_3, \cdots, S_8$ each associated with a single degree of freedom. If we denote the n_0 observations at the center conditions by $y_{01}, \cdots, y_{0u}, \cdots, y_{0n_0}$, their average by $\bar{y}_0$, and the average of the $n=16$ observations in the fractional factorial design by $\bar{y}_d$, we can write a detailed analysis of this residual sum of squares as follows.

The seven sums of squares $S_2, S_3, \cdots, S_8$ correspond to the seven contrasts within each of which four interaction terms are confounded (for example [8, p. 14]). Even if no center points were added, so that S_1 did not appear, a fairly powerful test of the assumption of linearity would usually be possible, provided

(i) an independent estimate of σ^2 was available;
(ii) the contours of the true response surface in the eight-dimensional space of the variables were not too nearly spherical;
(iii) the "canonical axes" of the true system were not oriented too closely parallel to the coordinate axes of the variables;
(iv) the interaction terms were not such that the estimated linear functions of them were all zero.

Conditions (ii) and (iii) are essentially concerned with the possibility that the true surface can be described entirely by the quadratic terms β_{ii}, none of which appear in the expected value of the residual sum of squares when $n_0=0$. With $n_0>0$ an additional term containing

$$\sum_{i=1}^{8} \beta_{ii}$$

appears. As seen from the detailed analysis in Table 649, this is associated with the isolation of the contrast $\bar{y}_d-\bar{y}_0$, the difference between the average of the

TABLE 649. ANALYSIS OF RESIDUAL SUMS OF SQUARES

Sums of Squares	D/F	Expected Values of Mean Squares
$S_1 = n_0\left(\frac{n}{N}\right)(\bar{y}_d - \bar{y}_0)^2$	1	$c^2 n_0 \frac{N}{n}\left\{\sum_{i=1}^{8} \beta_{ii}\right\}^2 + \sigma^2$
$S_2 = \frac{n}{c^2N^2}\left\{\sum_{u=1}^{N} y_u x_{1u} x_{2u}\right\}^2$	1	$c^2 \frac{N^2}{n}(\beta_{12} + \beta_{35} + \beta_{46} + \beta_{78})^2 + \sigma^2$
$S_3 = \frac{n}{c^2N^2}\left\{\sum_{u=1}^{N} y_u x_{1u} x_{3u}\right\}^2$	1	$c^2 \frac{N^2}{n}(\beta_{13} + \beta_{25} + \beta_{47} + \beta_{68})^2 + \sigma^2$
$\vdots$		
$S_8 = \frac{n}{c^2N^2}\left\{\sum_{u=1}^{N} y_u x_{1u} x_{8u}\right\}^2$	1	$c^2 \frac{N^2}{n}(\beta_{18} + \beta_{27} + \beta_{36} + \beta_{45})^2 + \sigma^2$
$S_e = \sum_{u=1}^{n_0} (y_{0u} - \bar{y}_0)^2$	$n_0 - 1$	σ^2

center points and the average of the points in the fractional factorial design. The expected value of this difference is

$$c \frac{N}{n} \sum_{i=1}^{8} \beta_{ii}.$$

As n_0 is increased above the value unity, the experimental arrangement becomes more and more sensitive to discrepancies associated with the overall criterion of curvature

$$\sum_{i=1}^{8} \beta_{ii}.$$

In addition it becomes possible to isolate a sum of squares based on $n_0 - 1$ degrees of freedom which, on the basic assumption of homogeneity of the error variance, measures pure error. This makes possible tests of departures from the linear model based solely on the internal evidence supplied by the design.

6. DISCUSSION

The class of designs to which we have been led, of which the fractional factorial discussed above is a member, seems to be excellently suited to the task at hand, indeed designs of exactly this type have for some time actually been applied in response surface studies. For the case of first order designs studied here the property of orthogonality which has repeatedly arisen is equivalent to that of "rotatability," a concept introduced previously [6]. This agreement between the present theory and what has seemed desirable on an intuitive basis suggests that the present formulation of the problem is reasonably well conceived. By happy circumstance designs of the type discussed not only achieve the first three requirements of Section 1.2 (graduate the function as accurately as possible, allow check of representational accuracy of assumed form of polynomial, do not contain excessively large numbers of points) but also these arrangements ([6] for details) can form a nucleus upon which a satisfactory design of order 2 can be built in case the assumed degree of polynomial proves inadequate, and they are ideally suited to become part of blocking arrangements in these larger designs.

Much remains to be done and work is in progress on such topics as the effect of cubic bias in first order designs, the extension of the present ideas to second order designs, the effect of changing the criterion to the minimization of maximum mean square error instead of minimization of average mean square error. The modifications necessary when a measure of the absolute value of the response is not important, but only the change in response from one point to another in the space of the variables, are also under consideration.

APPENDIX 1

A General Result Concerning Averaged Squared Bias. Suppose a polynomial of degree d_1 is fitted by the method of least squares over any region of interest R in the space of the variables when the true function is a polynomial of degree d_2. Then B the squared bias averaged over the region R is minimized for all values of the coefficient of neglected terms, by making the moments of the design up to order d_1+d_2 equal to the corresponding moments of a uniform distribution over the region R.

We suppose that the fitted model is

$$\hat{y}(\mathbf{x}) = \mathbf{x}_1'\boldsymbol{b}_1 \tag{1.1}$$

with $\boldsymbol{b}_1$ the vector of least squares estimates, while the true model is

$$\eta(\mathbf{x}) = \mathbf{x}_1'\boldsymbol{\beta}_1 + \mathbf{x}_2'\boldsymbol{\beta}_2 \tag{1.2}$$

where

$$\mathbf{x}_1' = (1; x_1, \cdots, x_k; x_1^2, \cdots, x_k^2; x_1x_2, \cdots; x_1^3, \cdots) \tag{1.3}$$

$$\boldsymbol{\beta}_1' = (\beta_0; \beta_1, \cdots, \beta_k; \beta_{11}, \cdots, \beta_{kk}; \beta_1\beta_2, \cdots; \beta_{111}, \cdots) \tag{1.4}$$

contain all terms up to order d_1, and $\mathbf{x}_2'$ and $\boldsymbol{\beta}_2'$ are similar vectors containing all terms from order d_1+1 up to order $d_2(d_2>d_1)$. With observations at N points we can define the matrices $\boldsymbol{X}_1$ and $\boldsymbol{X}_2$ by

$$\boldsymbol{X}_1 = \begin{bmatrix} \mathbf{x}_{11}' \\ \vdots \\ \mathbf{x}_{1u}' \\ \vdots \\ \mathbf{x}_{1N}' \end{bmatrix} \quad \boldsymbol{X}_2 = \begin{bmatrix} \mathbf{x}_{21}' \\ \vdots \\ \mathbf{x}_{2u}' \\ \vdots \\ \mathbf{x}_{2N}' \end{bmatrix} \tag{1.5}$$

where the column vectors which make up $\boldsymbol{X}_1$ are assumed to be linearly independent.

Now the component of the integral J arising from bias alone is B, where

$$\frac{\sigma^2}{N}B = \Omega\int_R [E\hat{y}(\mathbf{x}) - \eta(\mathbf{x})]^2 dx = \Omega\int_R \boldsymbol{\beta}_2'(\mathbf{x}_1'A - \mathbf{x}_2')'(\mathbf{x}_1'A - \mathbf{x}_2')\boldsymbol{\beta}_2 dx \tag{1.6}$$

$$= \boldsymbol{\beta}_2'\boldsymbol{\Delta}\boldsymbol{\beta}_2 \tag{1.7}$$

where

$$A = (X_1'X_1)^{-1}X_1'X_2 \tag{1.8}$$

and

$$\boldsymbol{\Delta} = A'\boldsymbol{\mu}_1 A - \boldsymbol{\mu}_2' A - A'\boldsymbol{\mu}_2 + \boldsymbol{\mu}_3 \tag{1.9}$$

with

$$\boldsymbol{\mu}_1 = \Omega \int_R \mathbf{x}_1\mathbf{x}_1' dx \tag{1.10}$$

$$\boldsymbol{\mu}_2 = \Omega \int_R \mathbf{x}_1\mathbf{x}_2' dx \tag{1.11}$$

$$\boldsymbol{\mu}_3 = \Omega \int_R \mathbf{x}_2\mathbf{x}_2' dx. \tag{1.12}$$

Now write $M_1 = X_1'X_1$ and $M_2 = X_1'X_2$ so that

$$A = M_1^{-1}M_2. \tag{1.13}$$

Then

$$\begin{aligned}\boldsymbol{\Delta} &= (\boldsymbol{\mu}_3 - \boldsymbol{\mu}_2'\boldsymbol{\mu}_1^{-1}\boldsymbol{\mu}_2) + M_2'M_1^{-1}\boldsymbol{\mu}_1 M_1^{-1}M_2 - \boldsymbol{\mu}_2' M_1^{-1}M_2 - M_2'M_1^{-1}\boldsymbol{\mu}_2 \\ &\quad + \boldsymbol{\mu}_2'\boldsymbol{\mu}_1^{-1}\boldsymbol{\mu}_2 & (1.14)\\ &= (\boldsymbol{\mu}_3 - \boldsymbol{\mu}_2'\boldsymbol{\mu}_1^{-1}\boldsymbol{\mu}_2) + (M_1^{-1}M_2 - \boldsymbol{\mu}_1^{-1}\boldsymbol{\mu}_2)'\boldsymbol{\mu}_1(M_1^{-1}M_2 - \boldsymbol{\mu}_1^{-1}\boldsymbol{\mu}_2) & (1.15)\\ &= \boldsymbol{\Delta}_1 + \boldsymbol{\Delta}_2 \quad \text{(say)} & (1.16)\end{aligned}$$

Now it is shown below that $\boldsymbol{\Delta}_1$ is positive semi-definite as is $\boldsymbol{\mu}_1$. It follows that whatever the value of $\boldsymbol{\beta}_2$, $\boldsymbol{\beta}_2'\boldsymbol{\Delta}\boldsymbol{\beta}_2$ is minimized when $M_1^{-1}M_2 = \boldsymbol{\mu}_1^{-1}\boldsymbol{\mu}_2$ and in particular when $M_1 = \boldsymbol{\mu}_1$ and $M_2 = \boldsymbol{\mu}_2$. Now the elements of M_1 and M_2 include all the moments of the design up to order $d_1 + d_2$ while the elements of $\boldsymbol{\mu}_1$ and $\boldsymbol{\mu}_2$ include, in corresponding positions, all the moments of the region R up to order $d_1 + d_2$. The stated result follows.

To show that $\boldsymbol{\mu}_1$ and $\boldsymbol{\Delta}_1$ are positive semi-definite

$$\begin{aligned}0 &\leq \int_R [(\mathbf{x}_1' \mid \mathbf{x}_2')\boldsymbol{\beta}]^2 dx & (1.17)\\ &= \boldsymbol{\beta}' \left\{ \int_R \begin{pmatrix} \mathbf{x}_1 \\ \hline \mathbf{x}_2 \end{pmatrix} (\mathbf{x}_1' \mid \mathbf{x}_2') dx \right\} \boldsymbol{\beta} & (1.18)\\ &= \Omega^{-1}\boldsymbol{\beta}' \begin{pmatrix} \boldsymbol{\mu}_1 & \boldsymbol{\mu}_2 \\ \boldsymbol{\mu}_2' & \boldsymbol{\mu}_3 \end{pmatrix} \boldsymbol{\beta}; & (1.19)\end{aligned}$$

whence $\boldsymbol{\mu}_1$ and

$$\begin{pmatrix} \boldsymbol{\mu}_1 & \boldsymbol{\mu}_2 \\ \boldsymbol{\mu}_2' & \boldsymbol{\mu}_3 \end{pmatrix}$$

are both positive semi-definite; but

$$\begin{pmatrix} \boldsymbol{\mu}_1 & 0 \\ 0 & \boldsymbol{\Delta}_1 \end{pmatrix} = \begin{pmatrix} \boldsymbol{\mu}_1 & 0 \\ 0 & \boldsymbol{\mu}_3 - \boldsymbol{\mu}_2'\boldsymbol{\mu}_1^{-1}\boldsymbol{\mu}_2 \end{pmatrix} = T' \begin{pmatrix} \boldsymbol{\mu}_1 & \boldsymbol{\mu}_2 \\ \boldsymbol{\mu}_2' & \boldsymbol{\mu}_3 \end{pmatrix} \boldsymbol{T} \tag{1.20}$$

where

$$T = \begin{pmatrix} I_1 & -\mathbf{y}_1^{-1}\mathbf{y}_2 \\ 0 & I_3 \end{pmatrix} \tag{1.21}$$

whence $\boldsymbol{\Delta}_1$ is positive semi-definite also.

APPENDIX 2

Rotational Average of the Constants in the Second Order Response Surface. To appreciate the character of the rotational average which is being taken, consider first the special case where $k=2$. We have for the quadratic part of the model

$$Q = \beta_{11}x_1^2 + \beta_{22}x_2^2 + \beta_{12}x_1x_2. \tag{2.1}$$

Now if we make an orthogonal rotation to new coordinates X_1 and X_2

$$\begin{aligned} x_1 &= X_1 \sin\theta + X_2 \cos\theta \\ x_2 &= X_1 \cos\theta - X_2 \sin\theta \end{aligned} \tag{2.2}$$

then

$$\begin{aligned} Q = &\left\{\beta_{11}\sin^2\theta + \beta_{22}\cos^2\theta + \beta_{12}\sin\theta\cos\theta\right\}X_1^2 \\ &+ \left\{\beta_{11}\cos^2\theta + \beta_{22}\sin^2\theta - \beta_{12}\sin\theta\cos\theta\right\}X_2^2 \\ &+ \left\{2(\beta_{11} - \beta_{22})\sin\theta\cos\theta + \beta_{12}(\cos^2\theta - \sin^2\theta)\right\}X_1X_2 \end{aligned} \tag{2.3}$$

The quantities we are interested in are the values of the coefficients of X_1^2, X_2^2 and X_1X_2 averaged over values of θ from 0 to 2π.

When k is greater than 2 we can deduce the form of the averages as follows. With $\beta_2' = (\beta_{11}, \beta_{22}, \cdots, \beta_{kk}, \beta_{12}, \cdots, \beta_{k,k-1})$ the elements we wish to average are those of the matrix $\boldsymbol{\beta}_2\boldsymbol{\beta}_2'$. Now if the rotational averages of the elements are substituted in this matrix to give $\widetilde{\boldsymbol{\beta}_2\boldsymbol{\beta}_2'}$ then $\mathbf{x}_2'\widetilde{\boldsymbol{\beta}_2\boldsymbol{\beta}_2'}\mathbf{x}_2$ must be a function of

$$\sum_{i=1}^{k} x_i^2$$

only.

This implies that

$$\widetilde{\boldsymbol{\alpha}_2\boldsymbol{\alpha}_2'} = N\sigma^{-2}\widetilde{\boldsymbol{\beta}_2\boldsymbol{\beta}_2'} = \left[\begin{array}{cccc|ccc} a & b & \cdots & b & & & \\ b & a & & & & 0 & \\ \vdots & & \ddots & & & & \\ b & & & a & & & \\ \hline & & & & 2(a-b) & & \\ & 0 & & & & \ddots & \\ & & & & & & 2(a-b) \end{array}\right]. \tag{2.4}$$

That is

$$N\sigma^{-2}\widetilde{\beta_{ii}^2} = \widetilde{\alpha_{ii}^2} = a \quad (i = 1, 2, \cdots, k) \tag{2.5}$$

$$N\sigma^{-2}\widetilde{\beta_{ii}\beta_{jj}} = \widetilde{\alpha_{ii}\alpha_{jj}} = b \quad (i \neq j = 1, 2, \cdots, k) \tag{2.6}$$

$$N\sigma^{-2}\widetilde{\beta_{ij}^2} = \widetilde{\alpha_{ij}^2} = 2(a-b) \quad (i \neq j = 1, 2, \cdots, k) \tag{2.7}$$

with the remaining rotational averages zero. The quantities a and b are those substituted in equation (61).

To determine the nature of these quantities consider again the symmetric matrix referred to in section 4.3 with diagonal elements β_{ii}, off-diagonal elements $\frac{1}{2}\beta_{ij}$ and latent roots $\lambda_1, \lambda_2, \cdots, \lambda_k$. We can express the a's and b's of equation (2.4) in terms of the λ's as follows.

In all orthogonal rotations the trace of any power of the matrix of section 4.3 remains constant. It follows in particular that

$$\sum_i \lambda_i = \sum_i \beta_{ii}$$

that is

$$\left(\sum_i \lambda_i\right)^2 = \sum_i \beta_{ii}^2 + \sum_{i \neq j}\sum \beta_{ii}\beta_{jj}. \tag{2.8}$$

Also

$$\sum_i \lambda_i^2 = \sum_i \beta_{ii}^2 + \frac{1}{4}\sum_{i \neq j}\sum \beta_{ij}^2 \tag{2.9}$$

Now since (2.8) and (2.9) are true in every rotation, these equations are also true for the rotational averages. That is

$$(\textstyle\sum \lambda_i)^2 = k\widetilde{\beta_{ii}^2} + k(k-1)\widetilde{\beta_{ii}\beta_{jj}} \tag{2.10}$$

$$\textstyle\sum \lambda_i^2 = k\widetilde{\beta_{ii}^2} + \frac{1}{4}k(k-1)\widetilde{\beta_{ij}^2} \tag{2.11}$$

and, using (2.5), (2.6) and (2.7)

$$N\sigma^{-2}\left(\sum_i \lambda_i\right)^2 = ka + k(k-1)b \tag{2.12}$$

$$N\sigma^{-2}\sum_i \lambda_i^2 = ka + \tfrac{1}{2}k(k-1)(a-b) \tag{2.13}$$

$$= \tfrac{1}{2}k\{(k+1)a - (k-1)b\}. \tag{2.14}$$

Whence

$$\frac{\sigma^2}{N}a = \{(\textstyle\sum \lambda_i)^2 + 2\sum \lambda_i^2\}/k(k+2) \tag{2.15}$$

$$\frac{\sigma^2}{N}b = \{(k+1)(\textstyle\sum \lambda_i)^2 - 2\sum \lambda_i^2\}/(k-1)k(k+2). \tag{2.16}$$

Whence, with

$$\theta = \frac{N}{\sigma^2}\sum_i \lambda_i^2 \quad \text{and} \quad \varphi = \left(\sum_i \lambda_i\right)^2 \Big/ \sum_i \lambda_i^2, \tag{2.17}$$

$$a = \frac{\theta(\varphi+2)}{k(k+2)} \tag{2.18}$$

$$b = \frac{\theta[(k+1)\varphi - 2]}{(k-1)k(k+2)} \tag{2.19}$$

REFERENCES

[1] Box, G. E. P., "Integration of Techniques in Process Development," *Transactions of the 11th Annual Convention of the American Society for Quality Control*, 1957.

[2] Box, G. E. P., "Multifactor Designs of First Order," *Biometrika*, 39 (1952), 49–57.

[3] Box, G. E. P., "Use of Statistical Methods in the Elucidation of Basic Mechanisms," *Proceedings of the 30th Session of the International Statistical Institute*, Stockholm, 1957.

[4] Box, G. E. P. and Coutie, G. A., "Application of Digital Computers in the Exploration of Functional Relationships," Proceedings of the Institution of Electrical Engineers, 103, Part B, Supplement No. 1 (1956), 100–7.

[5] Box, G. E. P. and Hunter, J. S., "A Confidence Region for the Solution of a Set of Simultaneous Equations with an Application to Experimental Design," *Biometrika*, 41 (1954), 190–9.

[6] Box, G. E. P. and Hunter, J. S., "Multi-factor Experimental Designs for Exploring Response Surfaces," *Annals of Mathematical Statistics*, 28 (1957), 195–241.

[7] Box, G. E. P. and Lucas, H. L., "Design of Experiments in Non-Linear Situations," Statistical Techniques Research Group (Princeton University), Technical Report No. 15, submitted for publication in *Biometrika.*

[8] Box, G. E. P. and Wilson, K. B., "On the Experimental Attainment of Optimum Conditions," *Journal of the Royal Statistical Society* (Series B), 13 (1951), 1–45.

[9] Chernoff, H., "Locally Optimal Designs for Estimating Parameters," *Annals of Mathematical Statistics*, 24 (1953), 586–602.

[10] David, H. A. and Arens, Beverly E., "Optimal Spacing in Regression Analysis," Technical Report No. 38, Department of Statistics and Statistical Laboratory, Virginia Polytechnic Institute, 1958.

[11] Elfving, G., "Optimum Allocation in Linear Regression Theory," *Annals of Mathematical Statistics*, 23 (1952), 255–62.

[12] Folks, J. L., "Comparison of Designs for Exploration of Response Relationships," paper read at Chicago at the 18th Annual Meeting, American Statistical Association, December, 1958.

[13] Hotelling, H., "Some Improvements in Weighting and Other Experimental Techniques," *Annals of Mathematical Statistics*, 15 (1943), 297–306.

[14] Hotelling, H., "The Experimental Determination of the Maximum of a Function," *Annals of Mathematical Statistics*, 12 (1941), 20–45.

[15] Hromi, John D., "Application of Fractional Replications in the Food Industry," presented at the Biometric Society Joint Meeting with the American Statistical Association, Atlantic City, 1957.

[16] Plackett, R. L. and Burman, J. P., "The Design of Multifactorial Experiments," *Biometrika*, 33 (1946), 305–25.

[17] Rao, C. R., "Factorial Experiments Derivable from Factorial Arrangements of Arrays," *Journal of the Royal Statistical Society*, Supplement, 9 (1947), 128–30.

[18] Rao, C. R., "On Hypercubes of strength *d:* a system of confounding any factorial experiments," *Bulletin of the Calcutta Mathematical Society*, 38 (1946), 67–78.

[19] Tocher, K. D., "A Note on the Design Problem," *Biometrika*, 39 (1952), 109–17.

p. 622: 9 lines up: "fractionally," not "fractional by"
p. 635: left-hand side of equation 50 should read:

$$\int_R x_1^{\delta_1} x_2^{\delta_2} \ldots x_k^{\delta_k}\, dx_1 \ldots dx_k$$

2.8

Simplex-Sum Designs: A Class of Second Order Rotatable Designs Derivable from Those of First Order[1]

G. E. P. BOX and D. W. BEHNKEN
University of Wisconsin and American Cyanamid Company

1.0 Introduction. A functional relationship $\eta = g(\xi_1, \xi_2, \cdots, \xi_k) = g(\boldsymbol{\xi})$ is assumed to exist between a response η and k continuous variables $\xi_1, \xi_2, \cdots, \xi_k$. To elucidate certain aspects of this relationship measurements of η are to be made for each of N combinations of the levels of the variables

$$\boldsymbol{\xi}_u' = (\xi_{1u}, \xi_{2u}, \cdots, \xi_{ku}) \qquad u = 1, 2, \cdots, N.$$

The problem of experimental design considered is the choice of the *design matrix* **D** of N rows and k columns whose uth row is $\boldsymbol{\xi}_u'$ which specifies the levels of the variables to be used in each of the N trials. The design matrix can be regarded as specifying the coordinates of N *experimental points* in the k dimensional space of the variables. As mentioned for example in [1], [2], [3], and [4] a number of distinct problems can arise. Here we suppose as in [5] that the nature of the functional relationship $g(\boldsymbol{\xi})$ is unknown but that over a specific region R in the space of the variables $\xi_1, \xi_2, \cdots, \xi_k$ a polynomial of degree d, $f_d(\boldsymbol{\xi})$, adequately graduates the function $g(\boldsymbol{\xi})$ and the objective is to use the polynomial to estimate η within the region R. A design of order d is such that it allows the estimation of the polynomial $f_d(\boldsymbol{\xi})$. In this paper we shall be particularly concerned with the case of $d = 2$, that is with the fitting of a polynomial of second degree. Using specifically what is called a rotatable design, we shall develop a method of obtaining rotatable designs of second order from those of first order. In defining rotatable designs it may be appropriate here to discuss briefly why they are thought to be useful.

A general design may be expressed in terms of standardized variables, for which

$$\sum_{u=1}^{N} x_{iu} = 0, \qquad i = 1, 2, \cdots, k$$

and

$$N^{-1} \sum x_{iu}^2 = \lambda_2,$$

where λ_2 is a convenient constant. In actual application therefore the levels of the experimental variables ξ_i are given by $\xi_{iu} = S_i x_{iu} + \xi_{i0}$ where ξ_{i0} and S_i were suitably chosen so as to give appropriate location and spread to the design in the

Received November 7, 1958; revised April 20, 1960.

[1] Prepared in part at the Statistical Techniques Research Group, Princeton University, Princeton, under the Office of Ordnance Research, Contract No. DA-36-034-ORD 2297 and in part at the Department of Experimental Statistics, North Carolina State College, Raleigh.

The Collected Works of George E. P. Box, 1984, Wadsworth, Inc., Belmont, CA 94002. Originally published in *Ann. Math. Stat.,* vol. 31, no. 4 (1960), pp. 838–864.

particular application. We shall suppose that the functional relationship is to be estimated by standard least squares.

A polynomial of degree one in the x's may be written

$$\eta_u = \sum_{i=0}^{k} \beta_i x_{iu}$$

where $x_{0u} = 1$, $u = 1, 2, \cdots, N$ or in matrix notation $\eta = \mathbf{X}\boldsymbol{\beta}$ where $\mathbf{X}$ is a $N \times (k + 1)$ matrix

$$\mathbf{X} = [\mathbf{1} \vdots \mathbf{D}]$$

and $\mathbf{1}$ is an $N \times 1$ column vector with all its elements unity. Whether one's objective is to obtain minimum variance for the estimated linear coefficients, minimum volume of the confidence region for the coefficients, or minimum volume of the confidence cone for the direction of steepest ascent, one is led to the simple conclusion that the most desirable design is orthogonal, that is, it is such that $\mathbf{X}'\mathbf{X} = N\boldsymbol{\Delta}$ where Δ is a diagonal matrix with its first diagonal element equal to unity and its remaining diagonal elements equal to λ_2.

Often we are not particularly interested in estimating the individual coefficients β_i but in estimating the polynomial itself. Suppose a design has been carried out which allows us to fit the polynomial by least squares. Using the fitted polynomial the estimated response at the conditions $\mathbf{x}' = [x_1, x_2, \cdots, x_k]$ is denoted by $\hat{y}_{\mathbf{x}}$. If a polynomial of the degree assumed can exactly represent $g(\mathbf{x})$ then

$$E(\hat{y}_{\mathbf{x}}) = \eta_{\mathbf{x}}$$

and a measure of the accuracy of our estimation over the region of interest R is provided by $V(\hat{y}_{\mathbf{x}})$.

It is easy to show that a first order orthogonal design has the property that $V(\hat{y}_{\mathbf{x}})$ is a function of $\mathbf{x}'\mathbf{x} = \sum x_i^2$ and λ_2 alone.

$$V(\hat{y}_{\mathbf{x}}) = \varphi(\mathbf{x}'\mathbf{x}, \lambda_2)$$

For such a design therefore, this variance (and hence the reciprocal of the variance which can be regarded as a measure of the information supplied by the design about the response surface) is constant on circles, spheres or hyperspheres in the factor space, i.e., in the space of the variables $x_1, x_2, \cdots, x_k$. Designs which have the property of generating spherical variance contours are called *rotatable designs*. It is easily shown for first order designs that the converse proposition is true, that is in order to insure rotatability the design must be orthogonal. As is pointed out in [5] the criterion of orthogonality, which has a central place for the first order design, is not readily extendable to designs of higher order. We can, however, readily extend the property of rotatability to designs of higher order and it is found that in general for a design of order d it is possible to choose a design such that

$$V(\hat{y}) = \varphi(\mathbf{x}'\mathbf{x}, \lambda_2, \lambda_4, \cdots \lambda_{2d})$$

where λ_i are constants at our choice.

To ensure the design is of this form it is only necessary to arrange that the moments of the design up to order $2d$ shall have certain values. For the case of second order designs with which we are specifically concerned

$$V(\hat{y}) = \varphi(\mathbf{x}'\mathbf{x}, \lambda_2, \lambda_4)$$

where λ_2 and λ_4 are at our choice. λ_2 is merely a scaling factor while λ_4 is chosen to give a satisfactory variance profile along a radius vector.

The problems for which the designs we are discussing have particular application are those where we are gaining knowledge of certain features of an unknown functional relationship by a sequential process in which any one "design" is only a single step. The results of each such step are used to more effectively plan the next group of observations.

At a particular stage we are interested in the behavior of the response function "in the neighborhood" R of some particular point P. We have in mind that the operability region O, that is the region in the space of the variables in which experiments could be conducted, is fairly extensive and that P is not close to the boundary of O. We suppose that the neighborhood of interest about P is a region R which nowhere reaches the boundary of O and that scales, metrics and transformations are chosen either implicitly or explicitly such that R is very approximately spherical and is centered at P.

The science of designing experiments is principally a convenient way of giving expression to prior information about the experimental situation which is currently in the experimenters mind and utilizing this information so as to generate further information most likely to be of value. The prior information is expressed in the choice of metrics, scales and transformations employed and is based on the experimenter's current feelings concerning the nature of the function under study. To the extent that the choices are poor, the extra information obtained about the nature of the function after the next set of observations have been completed, will be less than might otherwise have been obtained. This would mean that a sequence of such experiments, in which the information gained at each stage is utilized to design further more effective experiments, would be somewhat longer when prior information was less. This of course is to be expected and is a reflection of the fact that the apparent indeterminacy is a property of the experimental problem of exploring unknown functions itself, rather than of a particular technique for solving it. To demonstrate that some such rationale as the above is necessary one should remember that *any* set of experimental points distributed through the factor space such that $\mathbf{X}$ is of rank $k + 1$ provides a first order orthogonal design in some set of transformed x's.

The discussion so far has been based on the nature of the variance function $V(\hat{y}_{\mathbf{x}}) = E[\hat{y}_{\mathbf{x}} - E(\hat{y}_{\mathbf{x}})]^2$. In practice it would seldom if ever be true that the polynomial would provide an exact representation of the unknown function and in a more recent paper [2] this assumption has been dropped. Designs which minimize the mean square error $E(\hat{y}_{\mathbf{x}} - \eta_{\mathbf{x}})^2$ are considered instead. Now

$$E(\hat{y}_{\mathbf{x}} - \eta_{\mathbf{x}})^2 = V(\hat{y}_{\mathbf{x}}) + [E(\hat{y}_{\mathbf{x}}) - \eta_{\mathbf{x}}]^2$$

where the additional term on the right hand side may be called the squared bias. A general theorem in the above paper shows that if we are fitting a polynomial of degree d_1 over a region R when a polynomial of higher degree d_2 is necessary to give an exact representation, then the value of the squared bias averaged over the region, is minimized when the moments of the design points are the same as those of a uniform distribution over the region R. If it seems plausible in accordance with the previous discussion that the region of interest should be regarded as spherical then the optimum design to minimize average bias is also a particular rotatable design.

2.0. Outline. If we accept then that rotatable designs are of interest it becomes necessary to discover how they may be obtained in practice. First order rotatable designs are readily obtained (they are simply the orthogonal designs) but useful second order designs are less easy to derive. The method used here for obtaining second order rotatable designs from those of first order will now be outlined. In what follows we shall use n for the number of points in a first order design, and N for the size of a general or higher order derived design.

In the fitted first degree equation there are $k + 1$ constants, consequently at

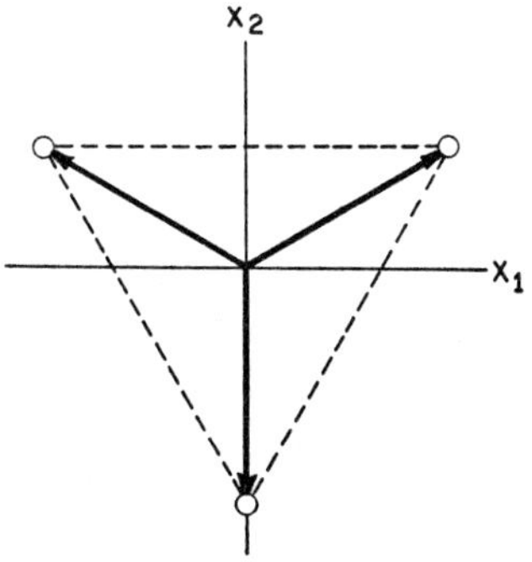

FIG. 1a. Two dimensional regular simplex

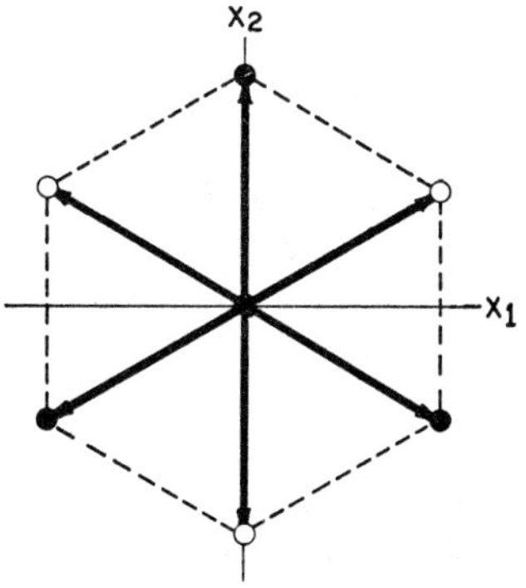

FIG. 1b. Generated second order rotatable design for two factors

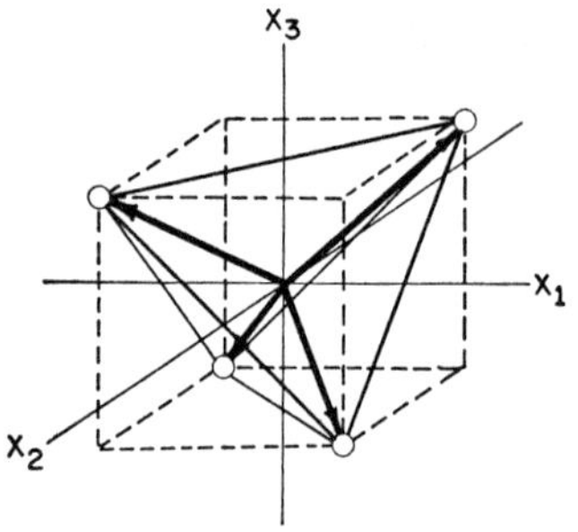

FIG. 2a. Three dimensional regular simplex

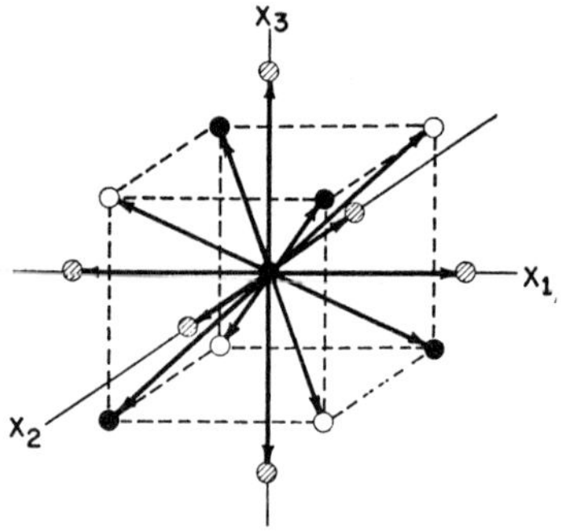

FIG. 2b. Generated second order rotatable design for three factors

least $k + 1$ observations must be made if the constants are to be separately estimable. Suppose the first order orthogonal (rotatable) design is used with the minimum number ($n = k + 1$) of experimental points. Then it is easily shown [6] that in the space of the x's these points lie at the vertices of a regular simplex. For example, if $k = 2$ the points are at the vertices of an equilateral triangle, if $k = 3$ at the corners of a regular tetrahedron. They can thus be called *first order simplex designs*. Now it can be observed that certain of the useful second order designs which have been found bear an interesting relation (illustrated for $k = 2$ and $k = 3$ in Figures 1 and 2) to the first order simplex designs. The three points at the vertices of an equilateral triangle in Figure 1a when joined to the origin at the center of the triangle, define three vectors. By adding these vectors two at a time we obtain a second equilateral triangle; by adding the vectors three at a time we obtain a center point. The original set of points plus the derived points generate the design shown in Figure 1b. This is the so-called hexagonal design which is known to be a second order rotatable design, [5]. The corresponding four vectors from the origin to the vertices of a tetrahedron shown in Figure 2a (a first order orthogonal or rotatable design) when added in all possible ways two at a time generate six further vectors passing through the midpoints of the edges of the tetrahedron, when added in all possible ways three at a time generate four vectors passing through the mid points of the faces of the original tetrahedron and when added four at a time generate a center point. If the lengths of the derived vectors are suitably chosen the resulting design coincides precisely with a previously derived second order rotatable design, namely the central composite rotatable design [5], [7]. These derived designs will be called *simplex-sum designs.*

In this paper we first demonstrate that the method suggested by these two examples for generating simplex-sum second order rotatable designs containing $2^n - 1$ points, from the first order rotatable simplex design containing $k + 1$ points, is a general one. For $k \geqq 5$ the number of points required by this method becomes large compared to the number of constants to be determined. A method is given for generating "fractions" and "replicated fractions" of the derived designs which have all the required properties and hence overcome this difficulty. Finally it is shown how the designs may be arranged in blocks so that they may be utilized in circumstances where insufficient homogeneous experimental material is available to complete the full quota of experimental runs.

To illustrate the method a second order rotatable design for seven variables is obtained, requiring only 66 experimental runs and only using three levels of each variable.

3.0. General Theory.

3.1. *Conditions for Rotatability.* We now define the design matrix **D** for the k standardized factors $x_1, x_2, \cdots, x_k$ as an $N \times k$ matrix whose uth row

$$\mathbf{x}_u' = (x_{1u}\, x_{2u} \cdots x_{ku})$$

defines the coded factor levels to be used in the uth of N experiments called for by the design. The general moment of the design will be denoted by

$$[1^{\alpha_1}2^{\alpha_2} \cdots k^{\alpha_k}] = N^{-1} \sum_{u=1}^{N} x_{1u}^{\alpha_1} x_{2u}^{\alpha_2} \cdots x_{ku}^{\alpha_k}.$$

and $\alpha = \sum \alpha_i$ will be called the order of the moment. The problem of finding rotatable designs is in essence one of finding configurations of points possessing the proper moments. It is in fact shown in [5] that when fitting the model

$$\eta = \beta_0 + \sum_{i=1}^{k} \beta_i x_i + \sum_{i=1}^{k} \sum_{j=i}^{k} \beta_{ij} x_i x_j + \sum_{i=1}^{k} \sum_{j=i}^{k} \sum_{l=j}^{k} \beta_{ijl} x_i x_j x_l + \cdots$$

including all terms through degree d, a rotatable design will be obtained when the moments through order $2d$ are of the form

$$(3.1) \qquad [1^{\alpha_1}2^{\alpha_2} \cdots k^{\alpha_k}] = \begin{cases} \lambda_\alpha \dfrac{\prod_{i=1}^{k} (\alpha_i)!}{2^{\alpha/2} \prod_{i=1}^{k} (\alpha_i/2)!}, & \text{all } \alpha_i \text{ even} \\ 0, & \text{any } \alpha_i \text{ odd,} \end{cases}$$

where λ_α is a constant for any design and α.

3.2. *Notation and Definition.* Consider a first order orthogonal (simplex) design in $k = n - 1$ variables with design matrix $\mathbf{D}_1$ and with $\sum_{u=1}^{n} x_{iu}^2$ set equal to n so that

$$\begin{bmatrix} \mathbf{1}' \\ \mathbf{D}_1' \end{bmatrix} [\mathbf{1}\ \mathbf{D}_1] = n\,\mathbf{I}_n .$$

It is conjectured that a second order rotatable design may be generated by using as design points the vectors obtained by taking all possible sums of the n rows of

$$\mathbf{D}_1 = \begin{bmatrix} \mathbf{x}_1' \\ \mathbf{x}_2' \\ \vdots \\ \mathbf{x}_u' \\ \vdots \\ \mathbf{x}_n' \end{bmatrix}$$

taken s at a time where $s = 1, 2, \cdots, k$. The problem thus reduces to one of finding the moments of a design matrix $\mathbf{D}$ derived in this way. We allow the vectors obtained by taking sums of s rows to be multiplied by a constant $a_s \geq 0$. The constants $a_1, a_2, \cdots, a_k$ will be called *radius multipliers*. Then the N by k matrix $\mathbf{D}$ for the derived design is

$$\mathbf{D} = \begin{bmatrix} a_1\,\mathbf{D}_1 \\ a_2\,\mathbf{D}_2 \\ \vdots \\ a_s\,\mathbf{D}_s \\ \vdots \\ a_k\,\mathbf{D}_k \end{bmatrix},$$

where $N = 2^n - 2$. Each $\mathbf{D}_s$ is an $\binom{n}{s}$ by k matrix whose rows consist of all possible sums of the rows of $\mathbf{D}_1$ taken s at a time. We omit for the moment the center point corresponding to $\mathbf{D}_{k+1}$ obtained when all n vectors are added together simultaneously. Since the columns of $\mathbf{D}_1$ are orthogonal to a vector of ones it follows that each vector obtained by summing rows s at a time is the negative of one obtained by summing rows $n - s$ at a time. The points in the factor space represented by $\mathbf{D}_s$ are, therefore, reflections through the origin of those represented by $\mathbf{D}_{n-s}$. (Of course, when n is even, $n - n/2 = n/2$ and half the rows of $\mathbf{D}_{n/2}$ are reflections of the other half.)

Let us define the *moment component* $[1^{\alpha_1}, 2^{\alpha_2}, \cdots, k^{\alpha_k}]_s$ as $\binom{n}{s} N^{-1}$ times the specified moment of $\mathbf{D}_s$, i.e.,

$$[1^{\alpha_1} 2^{\alpha_2} \cdots k^{\alpha_k}]_s = \frac{1}{N} \sum_{1 \leq u_1 < u_2} \sum \cdots \sum_{< u_s \leq n} (x_{1u_1} + x_{1u_2} + \cdots + x_{1u_s})^{\alpha_1}$$
$$\cdot (x_{2u_1} + x_{2u_2} + \cdots + x_{2u_s})^{\alpha_2} \cdots (x_{ku_1} + x_{ku_2} + \cdots + x_{ku_s})^{\alpha_k}.$$

Then the corresponding moment for the entire design can be written

$$[1^{\alpha_1}2^{\alpha_2} \cdots k^{\alpha_k}] = \sum_{s=1}^{k} (a_s)^{\alpha}[1^{\alpha_1}2^{\alpha_2} \cdots k^{\alpha_k}]_s .$$

3.3. *Analogy to Sampling from a Finite Population.* The problem of finding expressions for $[1^{\alpha_1}2^{\alpha_2} \cdots k^{\alpha_k}]_s$ in terms of either the moments of D_1 or of $[1^{\alpha_1}2^{\alpha_2} \cdots k^{\alpha_k}]_1$ corresponds to that of finding the sampling moments of means (or totals) of samples of s drawn from a k-variate finite population of n elements. These moments can be derived by a method due to Tukey [8] and elaborated by Wishart [9], Hooke [10] and Robson [11]. The necessary derivations are given in [12], [13] and the results utilized for our purposes here.

The sampling analogy is readily seen if we consider a k-variate vector of means obtained by averaging a random sample of s k-variate vectors chosen from a population of n such vectors,

$$(\bar{x}_1, \bar{x}_2, \cdots, \bar{x}_k) = \frac{1}{s}\sum_{u=1}^{s} (x_{1u}, x_{2u}, \cdots, x_{ku}).$$

Then the joint sampling moments of these multivariate means are

$$\text{Ave}\,\{\bar{x}_1^{\alpha_1}\bar{x}_2^{\alpha_2} \cdots \bar{x}_k^{\alpha_k}\},$$

Ave denoting the average value of the indicated power product over all combinations of samples of s. These expressions can be used to obtain the moment components of any submatrix $\mathbf{D}_s$ since

$$[1^{\alpha_1}2^{\alpha_2} \cdots k^{\alpha_k}]_s = N^{-1}s^{\alpha}\binom{n}{s}\text{Ave}\,\{\bar{x}_1^{\alpha_1}\bar{x}_2^{\alpha_2} \cdots \bar{x}_k^{\alpha_k}\}.$$

Using this equality and the results in [12] (recalling $\sum_{u=1}^{n} x_{iu}^2 = n$, $i = 1, 2, \cdots, k$ here) the expressions for the required moment components are readily obtained and are shown in Table 1a for $n \geqq \alpha$. Table 1b gives the one case required here for $n < \alpha$ not covered by Table 1a.

3.4. *Form of Moment Components.* Tables 1a and 1b show the moment components in terms of a notation designed to simplify their use and to make clear their general pattern. Certain of the coefficients $C(s)$ have single subscripts while the remainder have double subscripts. The former are not multiplied by unrestricted moment components of $\mathbf{D}_1$ and hence are constant terms in the moment component equations for a given n and s. The latter however, are multiplied by $\mathbf{D}_1$ moment components such as $[i\,j^2]_1$, or combinations of $\mathbf{D}_s$ moment components and are therefore coefficients of quantities which will not in general be constant for different choices of i, j, k, l, m.

The values taken on by any coefficient function $C(s)$, when n is held constant, possess a symmetry with respect to s as a result of the reflection relationship between vectors in $\mathbf{D}_s$ and $\mathbf{D}_{n-s}$. Since one matrix is the negative of the other their respective components must differ only by the factor $(-1)^{\alpha}$ or

$$[1^{\alpha_1}2^{\alpha_2} \cdots k^{\alpha_k}]_s = (-1)^{\alpha}[1^{\alpha_1}2^{\alpha_2} \cdots k^{\alpha_k}]_{n-s} .$$

TABLE 1a

Summary of general moment components of $\mathbf{D}_s$, $(n \geqq \alpha)$

	General Formulas	Abbreviations
$[i]_s$	0	
$[ij]_s$	0	
$[i^2]_s$	$C_2(s)$	$C_2(s) = \binom{n-2}{s-1}\frac{n}{N}$
$[ijk]_s$	$C_{31}(s)[ijk]_1$	$C_{31}(s) = \frac{(n-2s)}{(n-2)}\binom{n-2}{s-1}$
$[ij^2]_s$	$C_{31}(s)[ij^2]_1$	
$[i^3]_s$	$C_{31}(s)[i^3]_1$	
$[ijkl]_s$	$C_{41}(s)[ijkl]_1$	$C_{41}(s) = \left[\frac{(n-2s)(n-3s)-n(s-1)}{(n-2)(n-3)}\right]\binom{n-2}{s-1}$
$[ijk^2]_s$	$C_{41}(s)[ijk^2]_1$	
$[ij^3]_s$	$C_{41}(s)[ij^3]_1$	
$[i^2j^2]_s$	$C_4(s) + C_{41}(s)[i^2j^2]_1$	$C_4(s) = \binom{n-4}{s-2}\frac{n^2}{N}$
$[i^4]_s$	$3C_4(s) + C_{41}(s)[i^4]_1$	
$[ijklm]_s$	$C_{51}(s)[ijklm]_1$	$C_{51}(s) = (n-2s)\left[\frac{(n-3s)(n-4s)-5n(s-1)}{(n-2)(n-3)(n-4)}\right]\binom{n-2}{s-1}$
$[ijkl^2]_s$	$C_{51}(s)[ijkl^2]_1 + C_{52}(s)[ijkl^2 \mid 2]_1$	$C_{52}(s) = \frac{(n-2s)}{(n-4)}\binom{n-4}{s-2}n;$ $[ijkl^2 \mid 2]_1 = [ijk]_1$
$[i^5]_s$	$C_{51}(s)[i^5]_1 + 10C_{52}(s)[i^5 \mid 2]_1$	$[i^5 \mid 2]_1 = [i^3]_1$
$[i^2j^2k^2]_s$	$C_6(s) + C_{61}(s)[i^2j^2k^2]_1 + C_{62}(s)[i^2j^2k^2 \mid 2]_1 + 2C_{63}(s)[i^2j^2k^2 \mid 3]_1$	$C_6(s) = \binom{n-6}{s-3}\frac{n^3}{N};$ $C_{62}(s) = \left[\frac{n^2+3n-6sn+6s^2-4}{(n-4)(n-5)}\right]\binom{n-4}{s-2}n$

TABLE 1a—*Continued*

	General Formulas	Abbreviations
$[i^6]_s$	$15C_6(s) + C_{61}(s)[i^6]_1 + 15C_{62}(s)[i^6 \mid 2]_1 + 10C_{63}(s)[i^6 \mid 3]_1$	$C_{61}(s) = \binom{n-2}{s-1}\left[\frac{(n-2s)(n-3s)(n-4s)(n-5s)}{(n-2)(n-3)(n-4)(n-5)} - \frac{n(s-1)(16n^2 - 79sn + 11n + 86s^2 - 4s - 4)}{(n-2)(n-3)(n-4)(n-5)}\right]$ $C_{63}(s) = \left[\frac{n^2 - n - 4sn + 4s^2 + 4}{(n-4)(n-5)}\right]\binom{n-4}{s-2}N$ $[i^2j^2k^2 \mid 2]_1 = [i^2j^2]_1 + [i^2k^2]_1 + [j^2k^2]_1$ $[i^2j^2k^2 \mid 3]_1 = [ij^2]_1[ik^2]_1 + [i^2j]_1[jk^2]_1 + [i^2k]_1[j^2k]_1 + 2[ijk]_1^2$ $[i^6 \mid 2]_1 = [i^4]_1$ $[i^6 \mid 3]_1 = [i^3]_1^2$

TABLE 1b

Fourth order moment components of $\mathbf{D}_s$ *for* $n = 3$, $(n < \alpha)$

	General Formula	Abbreviation
$[ij^3]_s$	$C'_{41}(s)[ij^3]_1$	$C'_{41}(s) = \frac{s}{3!}\binom{3}{s}[2 - 7(s-1)]$
$[i^2j^2]_s$	$C'_4(s) + C'_{41}(s)[i^2j^2]_1$	$C'_4(s) = \frac{s}{3!}\binom{3}{s}(s-1)\frac{9}{N}$
$[i^4]_s$	$3C'_4(s) + C'_{41}(s)[i^4]_1$	

From Tables 1a and 1b we see that in general

$$(3.4) \qquad \begin{aligned} [1^{\alpha_1}2^{\alpha_2} \cdots k^{\alpha_k}]_s &= b_\alpha C_\alpha(s) + C_{\alpha 1}(s)[1^{\alpha_1}2^{\alpha_2} \cdots k^{\alpha_k}]_1 \\ &+ b_{\alpha 2}C_{\alpha 2}(s)[1^{\alpha_1}2^{\alpha_2} \cdots k^{\alpha_k} \mid 2]_1 + \cdots + b_{\alpha p}C_{\alpha p}(s)[1^{\alpha_1}2^{\alpha_2} \cdots k^{\alpha_k} \mid p]_1 , \end{aligned}$$

where the b_α values are zero or positive constants varying with the particular partition of $\boldsymbol{\alpha} = (\alpha_1\alpha_2 \cdots \alpha_k)$.

It is readily shown in general [13] and can be confirmed by direct substitution that $C_{\alpha i}(s) = (-1)^\alpha C_{\alpha i}(n - s)$.

4.0. Radius Multipliers and Rotatability. Having general formulas for the moment components contributed by each submatrix $\mathbf{D}_s$ of a derived design matrix $\mathbf{D}$, we now seek a suitable set of radius multipliers such that the moments of $\mathbf{D}$

$$(4.1) \qquad [1^{\alpha_1}2^{\alpha_2} \cdots k^{\alpha_k}] = \sum_{s=1}^{k} a_s^\alpha [1^{\alpha_1}2^{\alpha_2} \cdots k^{\alpha_k}]_s$$

will fulfill the requirements for rotatability listed under (3.1).

By *even-order moments* we mean those for which $\alpha = \sum \alpha_i$ is even and by *odd-order moments* we mean those for which α is odd. In addition we call those moments for which *any* α_i is odd, *odd moments* and those for which all α_i are even, *even moments*. For rotatability all odd moments must be zero and all even moments of the same order must be specified multiples of each other.

From Table 1a we see that the moments $[i]$, $[ij]$ and $[i^2]$ of $\mathbf{D}$, will satisfy the rotatability requirements for any choice of radius multipliers since the corresponding odd moment components $[i]_s$ and $[ij]_s$ are identically zero and $[i^2]_s$ is constant for all i. The other moments however all involve "variable terms" $C_{\alpha i}(s)[\ \ ]_1$ and only in the case of the even moments is the constant term $b_\alpha C_\alpha(s)$ added to this variable function. The moment requirements will be generally satisfied only if the radius multipliers are so chosen that each "variable term" sums to zero in the expression for all $[1^{\alpha_1}2^{\alpha_2} \cdots k^{\alpha_k}]$. For odd moments this is obviously required. For the even moments it would otherwise be impossible to attain the required constant ratio between moments of the same order since the quantities $[\ \ ]_1$ in general change in their relationships, from one moment to another. The only further requirement for rotatability is that the constant terms, $b_\alpha C_\alpha(s)$, are in the required ratios.

Using the general form of $[1^{\alpha_1}2^{\alpha_2} \cdots k^{\alpha_k}]_s$ from (3.4) in (4.1) we have

$$[1^{\alpha_1}2^{\alpha_2} \cdots k^{\alpha_k}] = \sum_{s=1}^{k} (a_s)^{\alpha}[1^{\alpha_1}2^{\alpha_2} \cdots k^{\alpha_k}]_s$$
$$= b_\alpha \sum_{s=1}^{k} (a_s)^{\alpha} C_\alpha(s) + [1^{\alpha_1}2^{\alpha_2} \cdots k^{\alpha_k}]_1 \cdot \sum_{s=1}^{k} (a_s)^{\alpha} C_{\alpha 1}(s)$$
$$+ b_{\alpha 2}[1^{\alpha_1}2^{\alpha_2} \cdots k^{\alpha_k} \mid 2]_1 \cdot \sum_{s=1}^{k} (a_s)^{\alpha} C_{\alpha 2}(s)$$
$$+ \cdots + b_{\alpha p}[1^{\alpha_1}2^{\alpha_2} \cdots k^{\alpha_k} \mid p]_1 \cdot \sum_{s=1}^{k} (a_s)^{\alpha} C_{\alpha p}(s),$$

where $b_\alpha C_\alpha(s)$ does not appear unless the moment is even and we require

$$\sum_{s=1}^{k} (a_s)^{\alpha} C_{\alpha i}(s) = 0, \qquad i = 1, 2 \cdots p.$$

Since we have seen previously that $C(s) = (-1)^{\alpha} C(n - s)$, then for all odd-order moments $C_{\alpha i}(s) = -C_{\alpha i}(n - s)$. We can say further, because of the factor $(n - 2s)$ in all such odd order moment coefficients, (Table 1a) that when α and k are both odd, $C_{\alpha i}(n/2) = C_{\alpha i}(k + 1/2) = 0$. Therefore as long as radius multipliers are selected such that $a_s = a_{n-s}$ *all* the odd-order moments will sum to zero for any value of a_s. Setting $m = k/2$ when k is even and $m = (k - 1)/2$when k is odd it then follows, for such a choice of radius multipliers, that

$$\sum_{s=1}^{k} (a_s)^{\alpha} C_{\alpha i}(s) = \sum_{s=1}^{m} (a_s)^{\alpha}[C_{\alpha i}(s) + C_{\alpha i}(n - s)] = 0 \text{ for all } i, \quad \alpha \text{ odd}.$$

We will call this type of solution for the radius multipliers, where $a_s = a_{n-s}$, a *symmetric solution.*

Having satisfied the odd-order moment requirements for rotatable designs of *any* order we must now find which symmetric solutions will also satisfy the requirements for even-order moments.

5.0. Second Order Requirements for Rotatability. For a design to be second order rotatable the even moments must have the following general form $[i^2] = \lambda_2$, $[i^2j^2] = \lambda_4$, $[i^4] = 3\lambda_4$ where λ_2 and λ_4 are constants at choice and the odd moments of order less than or equal to four must vanish.

It may be noted here that the addition of center points to a design matrix **D** does not change the general form of the moments since their only effect is to increase the denominator N.

5.1. *Application of Moment Requirements.* As noted previously in 4.0, the general second order moment $[i^2]$ places no restrictions on the choices of radius multipliers since

$$[i^2] = \sum_{s=1}^{k} (a_s)^2 C_2(s) = \frac{n}{N} \sum_{s=1}^{k} (a_s)^2 \binom{n-2}{s-1} = \lambda_2,$$

a constant for all values of i.

From Tables 1a and 1b it can be seen that the generalized moment component is obtained by letting the coefficient b_4 vanish for odd moments, and assume the values 3 and 1 for the even partitions of α, viz., (4) and (2,2). Hence

$$[i^{\alpha_1}j^{\alpha_2}k^{\alpha_3}l^{\alpha_4}]_s = b_4C_4(s) + C_{41}(s)[i^{\alpha_1}j^{\alpha_2}k^{\alpha_3}l^{\alpha_4}]_1$$

so that

$$[i^{\alpha_1}j^{\alpha_2}k^{\alpha_3}l^{\alpha_4}] = b_4\sum_{s=1}^{k} (a_s)^4C_4(s) + [i^{\alpha_1}j^{\alpha_2}k^{\alpha_3}l^{\alpha_4}]_1\sum_{s=1}^{k} (a_s)^4C_{41}(s).$$

In the previous section we showed that, for second order rotatability, we must have $\sum_{s=1}^{k} (a_s)^4C_{41}(s) = 0$ making

$$[i^{\alpha_1}j^{\alpha_2}k^{\alpha_3}l^{\alpha_4}] = b_4\sum_{s=1}^{k} (a_s)^4C_4(s).$$

This accomplished, all odd moments of order four would vanish with b_4 and

$$[i^2j^2] = \sum_{s=1}^{k} (a_s)^4C_4(s) = \lambda_4$$

$$[i^4] = 3\sum_{s=1}^{k} (a_s)^4C_4(s) = 3\lambda_4\,.$$

Clearly any symmetric solution for the radius multipliers such that

$$\sum_{s=1}^{k} (a_s)^4C_{41}(s) = 0$$

will provide a rotatable design of the simplex-sum type.

5.2. *Standard Solution for Radius Multipliers.* We will now demonstrate that a solution holding for any k is obtained by letting

$$a_s = \binom{n-2}{s-1}^{-\frac{1}{4}}, \qquad s = 1, 2, \cdots k.$$

This solution for the radius multipliers involving the binomial coefficients will be denoted by $B_s^{-\frac{1}{4}}$ and referred to as the *standard solution*.

It is immediately evident that all odd order moments will be zero since the choice for the a_s provides a symmetric solution as defined earlier. Further

$$[i^2] = \sum_{s=1}^{k} \binom{n-2}{s-1}^{-\frac{1}{2}} \cdot \binom{n-2}{s-1}\frac{n}{N} = \frac{n}{N}\sum_{s=1}^{k} \binom{n-2}{s-1}^{\frac{1}{2}} = \lambda_2$$

and, for each i, $[i^2]$ equals n/N times the sum of the square roots of the binomial coefficients of order $n-2$.

$$[i\,j\,k\,l] = \sum_{s=1}^{n-1} \binom{n-2}{s-1}^{-1} \frac{(n-2s)(n-3s) - n(s-1)}{(n-2)(n-3)} \binom{n-2}{s-1}[i\,j\,k\,l]_1$$

$$= \sum_{s=1}^{n-1} \frac{n^2 - 6sn + 6s^2 + n}{(n-2)(n-3)}[i\,j\,k\,l]_1 = \frac{[i\,j\,k\,l]_1}{(n-2)(n-3)}\,(0) = 0.$$

Since the zero quantity in brackets is the expression $\sum a_s^4 C_{41}(s)$, common to all fourth order moments, we have

$$[i\,j\,k^2] = [i\,j^3] = 0,$$

$$[i^2 j^2] = \sum_{s=1}^{n-1} \binom{n-2}{s-1}^{-1} \binom{n-4}{s-2} \frac{n^2}{N} = \frac{n^2(n-1)}{6N} = \lambda_4 ,$$

$$[i^4] = \sum_{s=1}^{n-1} \binom{n-2}{s-1}^{-1} \binom{n-4}{s-2} \frac{3n^2}{N} = \frac{n^2(n-1)}{2N} = 3\,\lambda_4 .$$

5.3. *Second Order Rotatability for the Case* $n = 3$. For $k = 2$ $(n = 3)$ the above demonstration does not apply since the fourth order moment formulas hold only for $n \geqq 4$ as noted previously. However by using the formulas in Table 1b, we can show that the above solution also applies here.

$$[i\,j^3] = \sum_{s=1}^{2} \binom{1}{s-1}^{-1} \frac{s}{3!} \binom{3}{s} [2 - 7\,(s-1)][i\,j^3]_1 = [i\,j^3]_1 - 5[i\,j^3]_1 .$$

Although apparently inconsistent with previous results this expression is zero because of a property of 3×3 matrices of the type $[\mathbf{1}\ \mathbf{x}_1\ \mathbf{x}_2]$ with orthogonal columns of equal vector length. Since we have already shown that a matrix of all rows taken s at a time is the negative of the matrix of sums taken $n - s$ at a time we have $\mathbf{D}_1 = -\mathbf{D}_2$ and $[i\,j^3]_1 = [i\,j^3]_2$. From the general moment formula for $n = 3$ we have $[i\,j^3]_2 = -5[i\,j^3]_1$ and hence $[i\,j^3]_1 = -5[i\,j^3]_1$ must vanish.

The moment

$$[i^2 j^2] = \sum_{s=1}^{2} \binom{1}{s-1}^{-1} \left\{ \frac{s}{3!} \binom{3}{s} [2 - 7\,(s-1)][i^2 j^2]_1 + (s-1)\, \frac{9}{N} \right\}$$

$$= [i^2 j^2]_1 - 5\,[i^2 j^2]_1 + \frac{9}{N} .$$

However since $[i^2 j^2]_1 = [i^2 j^2]_2$ and $[i^2 j^2]_2 = -5[i^2 j^2]_1 + 9/N$ we have $[i^2 j^2]_1 = 3/2N$, a constant for any matrix of this type. It then follows that $[i^2 j^2] = 2[i^2 j^2]_1 = 3/N = \lambda_4$ and similarly $[i^4] = 3\lambda_4$. Thus the moments are those of a rotatable design.

We have thus demonstrated that for $k \geqq 2$ a second order rotatable design can always be derived from the first order simplex design. It is possible to show however [13] that this method in its present form does not generate third order rotatable designs.

5.4. *Radius of Experimental Points.* As is illustrated in figures 1a and 1b for the case $k = 2$ and $k = 3$ the simplex-sum designs consist of subsets of vectors of experimental points corresponding to the rows of the submatrices $a_1\mathbf{D}_1$, $a_2\mathbf{D}_2$, $\cdots$, $a_k\mathbf{D}_k$. Geometrically these subsets are symmetrically oriented one to another in that the vectors for $a_2\mathbf{D}_2$ bisect the edges of the simplex defined by $a_1\mathbf{D}_1$, the vectors of $a_3\mathbf{D}_3$ pass symmetrically through the faces of the simplex defined by $a_1\mathbf{D}_1$ and so on. We can readily obtain an expression for r_s, the radius of the points in the sth subset. Denoting the uth row of $\mathbf{D}_s$ by $\mathbf{x}'_{su}$, $s = 2, 3, \cdots, k$

we have $\mathbf{x}'_{su} = \sum_{i=1}^{s} \mathbf{x}'_{u_i}$, where $\mathbf{x}'_{u_1}, \mathbf{x}'_{u_2}, \cdots, \mathbf{x}'_{u_s}$ is the uth set of s rows of the first order design matrix $\mathbf{D}_1$. Now since

$$[\mathbf{1}\ \mathbf{D}_1] = \begin{bmatrix} 1\ \mathbf{x}'_1 \\ 1\ \mathbf{x}'_2 \\ \vdots \\ 1\ \mathbf{x}'_n \end{bmatrix}$$

and $[\mathbf{1}\ \mathbf{D}_1][\mathbf{1}\ \mathbf{D}_1]' = n\ \mathbf{I}_n$ we have

$$\mathbf{x}'_i\ \mathbf{x}_j = \begin{cases} n - 1 = k, & i = j \\ -1, & i \neq j. \end{cases}$$

The square of the length of the row vector $\mathbf{x}'_{su}$ is therefore

$$\mathbf{x}'_{su}\mathbf{x}_{su} = (\mathbf{x}'_{u_1} + \mathbf{x}'_{u_2} + \cdots + \mathbf{x}'_{u_s})(\mathbf{x}_{u_1} + \mathbf{x}_{u_2} + \cdots + \mathbf{x}_{u_s})$$

$$= s(n - 1) + 2\binom{s}{2}(-1) = s(n - s).$$

Thus the radius of the experimental points in any submatrix $a_s\mathbf{D}_s$ is given by $r_s = a_s[s(n - s)]^{\frac{1}{2}}$, and since in a symmetric solution $a_s = a_{n-s}$, $r_s = r_{n-s}$.

For the particular set of radius multipliers of the standard solution $r_s = \binom{n-2}{s-1}^{-\frac{1}{2}}[s(n - s)]^{\frac{1}{2}}$. A summary of the radii for $k = 2$ through 8 of the standard solution rotatable designs is given in Table 2.

TABLE 2

Radii of experimental points for standard solution rotatable designs

k	r_1	r_2	r_3	r_4	r_5	r_6	r_7	r_8
2	1.41	1.41						
3	1.73	1.68	1.73					
4	2.00	1.86	1.86	2.00				
5	2.24	2.00	1.92	2.00	2.24			
6	2.45	2.11	1.95	1.95	2.11	2.45		
7	2.65	2.21	1.97	1.89	1.97	2.21	2.65	
8	2.83	2.30	1.98	1.84	1.84	1.98	2.30	2.83

5.5. *Singularity and Near Singularity of Moment Matrices.* A set of points can have the moments of a rotatable design but be impractical as a design since it leads to a singular moment matrix. The singularity arises from a dependency between the columns in the $\mathbf{X}$ matrix for the b_0 and quadratic terms, b_{11}, b_{22}, $\cdots$, b_{kk}. The situation is easily remedied, however, by the addition of center points to the design matrix. The moment matrix is singular [5] when the standardized fourth moment constant λ'_4 achieves the value $\lambda'_4 = \lambda_4/(\lambda_2)^2 = k/(k + 2)$, implying that the design points all lie on the same hypersphere [14]. For the

TABLE 3

Comparison of λ_4' to its singular value for standard solution designs

k	$\frac{k}{(k+2)}$	λ_4'
2	.500	.500
3	.600	.601
4	.667	.670
5	.714	.724
6	.750	.769
7	.778	.811
8	.800	.850

designs arising from the standard solution for a_s we have

$$\lambda_4' = \frac{n^2(n-1)}{6N}\left[\frac{n\sum_{s=1}^{n-1}\binom{n-2}{s-1}^{\frac{1}{2}}}{N}\right]^{-2} = \frac{(n-1)(2^n-2)}{6\left[\sum_{s=1}^{n-1}\binom{n-2}{s-1}^{\frac{1}{2}}\right]^2}$$

where we have used $N = 2^n - 2$, i.e., no center points having been added. The value for λ_4' is equal to the singular value $k/(k+2)$ when $k = 2$ and remains close to the singular value as k increases, as is shown in Table 3.

Since the addition of center points has no effect on the moments except to change N we see that the addition of N_0 center points will change λ_4' by a factor of $(2^n - 2 + N_0)/(2^n - 2)$. In practice sufficient center points were added to provide a satisfactory profile for the variance function $V(\hat{y}_x)$ taken along a radius vector. Denoting the distance from the center of the design by $\rho = (\mathbf{x}'\mathbf{x})^{\frac{1}{2}}$ it is suggested in general in [5] that sufficient points be added so that $V(\hat{y}_x)$ at $\rho = 0$ is equal to that at $\rho = (\lambda_2)^{\frac{1}{2}}$. Such an arrangement causes the variance to be approximately uniform over the important range $\rho = 0$ to $\rho = (\lambda_2)^{\frac{1}{2}}$. These designs will be said to attain "uniform variance".

6.0. Additional Second Order Rotatable Simplex-Sum Designs. The standard solution for a_s affords a set of rotatable designs for all $k \geqq 2$. When $k \geqq 5$ however, the number of experiments required by the standard solution becomes excessive. Fortunately, for such values of k smaller *reduced designs* are possible.

6.1. *Solution Space of Radius Multipliers.* We have shown in Section 4 that for second order rotatability we must find values for a_s, $s = 1, 2, \cdots, k$, such that $\sum a_s^3 C_{31}(s) = 0$ and $\sum a_s^4 C_{41}(s) = 0$ where $C_{31}(s)$ and $C_{41}(s)$ are the coefficients of the moment components of $\mathbf{D}_1$ (Tables 1a and 1b). When those values are found it was shown that the other moment requirements were automatically satisfied.

To state these requirements in a more convenient form for our present problem let us define the vectors

$$\begin{aligned}
\mathbf{a}' &= (a_1\, a_2 \cdots a_s \cdots a_k),\\
\mathbf{a}_3' &= (a_1^3\, a_2^3 \cdots a_s^3 \cdots a_k^3),\\
\mathbf{a}_4' &= (a_1^4\, a_2^4 \cdots a_s^4 \cdots a_k^4),\\
\mathbf{C}_{31}' &= (C_{31}(1)\; C_{31}(2) \cdots C_{31}(s) \cdots C_{31}(k)),\\
\mathbf{C}_{41}' &= (C_{41}(1)\; C_{41}(2) \cdots C_{41}(s) \cdots C_{41}(k)).
\end{aligned}$$

The requirements for second order rotatability are therefore $\mathbf{C}_{31}'\mathbf{a}_3 = 0$ and $\mathbf{C}_{41}'\mathbf{a}_4 = 0$. If we choose values of $\mathbf{a}_s$, such that $a_s = a_{n-s}$, we have shown previously that $\mathbf{C}_{31}'\mathbf{a}_3 = 0$. Therefore, calling any vectors $\mathbf{a}_3$ and $\mathbf{a}_4$ which are derived from symmetric solutions, *symmetric vectors*, we may further simplify our problem to that of finding all symmetric vectors $\mathbf{a}_4$ such that $\mathbf{C}_{41}'\mathbf{a}_4 = 0$. We must also add the restrictions of course, that all the elements of $\mathbf{a}_4$ are greater than zero.

The restriction of symmetry on the vector $\mathbf{a}_4$ has the effect of confining its values to an m dimensional subspace for which $m = k/2$, if k is even and $m = (k + 1)/2$, if k is odd. This is evident since $\mathbf{a}_4$ has exactly m elements which can be varied independently, the remaining $k - m$ elements then being determined by the relationship $a_s = a_{n-s}$. The elements of $\mathbf{C}_{41}$ are symmetric in a corresponding way as was shown earlier. Hence for convenience we might consider $\mathbf{a}_4$ and $\mathbf{C}_{41}$ as two m-dimensional vectors and use the fact that $m - 1$ independent vectors can be found orthogonal to any vector in m-space. Thus if we find $m - 1$ independent solutions to the equation $\mathbf{C}_{41}'\mathbf{a}_4 = 0$ they will form a basis for the solution space of all possible vectors satisfying this equation, that is of all vectors in the $m - 1$ space orthogonal to $\mathbf{C}_{41}$. Since the elements of $\mathbf{C}_{41}$ are of mixed sign it is clear that solution vectors can be found which fall in the positive 2^k-drant.

6.2. *Specific Solutions.* We will now obtain the $m - 1$ basis vectors $\boldsymbol{\Upsilon}_1, \boldsymbol{\Upsilon}_2, \cdots \boldsymbol{\Upsilon}_{m-1}$ for $k = 3, 4, \cdots 8$, selecting them to contain the maximum number of zero elements possible. Where zero's can be introduced, the equivalent designs will involve fewer points than the standard solution since any submatrix with a zero radius multiplier, may be eliminated from $\mathbf{D}$ without altering the moments. All other designs, resulting from the orthogonality relationship, can be derived from these basis vectors by taking linear combinations

$$\mathbf{a}_4 = d_1\boldsymbol{\Upsilon}_1 + d_2\boldsymbol{\Upsilon}_2 + \cdots + d_{m-1}\boldsymbol{\Upsilon}_{m-1},$$

where the d_i's are any constants such that $\mathbf{a}_4 \geqq \mathbf{0}$.

It will be recalled from the discussion of the standard solution that the two factor design is an anomaly in that its rotatability does not result from the orthogonality relationship. For $k = 2$, $\mathbf{C}_{41}'\mathbf{a}_4 \neq 0$ and hence a specific solution does not follow in the usual way. When $k = 3$, $m = 2$ and hence only one solution, the standard solution, is available, $(\boldsymbol{\Upsilon}_1 = \mathbf{a}_4)$. Similarly when $k = 4$, $m =$

2 so that for $k \leqq 4$ the standard solution is unique. When $k = 5$ however then $m = 3$ and two independent solutions $\boldsymbol{\Upsilon}_1$ and $\boldsymbol{\Upsilon}_2$ are possible. Specifically

$$\mathbf{C}'_{41}\boldsymbol{\Upsilon}_i = (1 \quad -2 \quad -6 \quad -2 \quad 1)\boldsymbol{\Upsilon}_i = 0,$$

and here for the first time we can obtain reduced designs. Two suitable basis vectors are

$$\boldsymbol{\Upsilon}'_1 = (1 \quad 0 \quad \tfrac{1}{3} \quad 0 \quad 1),$$
$$\boldsymbol{\Upsilon}'_2 = (1 \quad \tfrac{1}{2} \quad 0 \quad \tfrac{1}{2} \quad 1),$$

whence

$$\mathbf{a}' = (1 \quad 0 \quad 3^{-\frac{1}{4}} \quad 0 \quad 1),$$
$$\mathbf{a}' = (1 \quad 2^{-\frac{1}{4}} \quad 0 \quad 2^{-\frac{1}{4}} \quad 1).$$

The arrangement employing $\boldsymbol{\Upsilon}_1$ omits $a_2\mathbf{D}_2$ and $a_4\mathbf{D}_4$ while that employing $\boldsymbol{\Upsilon}_2$ omits $a_3\mathbf{D}_3$ from the design.

When $k = 6$, then $m = 3$ and the relationship

$$\mathbf{C}'_{41}\,\boldsymbol{\Upsilon}_i = (1 \quad -1 \quad -8 \quad -8 \quad -1 \quad 1)\,\boldsymbol{\Upsilon}_i = 0,$$

is satisfied by

$$\boldsymbol{\Upsilon}'_1 = (1 \quad 1 \quad 0 \quad 0 \quad 1 \quad 1),$$
$$\boldsymbol{\Upsilon}'_2 = (1 \quad 0 \quad \tfrac{1}{8} \quad \tfrac{1}{8} \quad 0 \quad 1).$$

When $k = 7$, then $m = 4$ and

$$\mathbf{C}'_{41}\,\boldsymbol{\Upsilon}_i = (1 \quad 0 \quad -9 \quad -16 \quad -9 \quad 0 \quad 1)\boldsymbol{\Upsilon}_i = 0$$

is satisfied by

$$\boldsymbol{\Upsilon}'_1 = (1 \quad 0 \quad \tfrac{1}{9} \quad 0 \quad \tfrac{1}{9} \quad 0 \quad 1),$$
$$\boldsymbol{\Upsilon}'_2 = (1 \quad 0 \quad 0 \quad \tfrac{1}{8} \quad 0 \quad 0 \quad 1),$$
$$\boldsymbol{\Upsilon}'_3 = (0 \quad 1 \quad 0 \quad 0 \quad 0 \quad 1 \quad 0).$$

When $k = 8$, then $m = 4$ and

$$\mathbf{C}'_{41}\,\boldsymbol{\Upsilon}_i = (1 \quad 1 \quad -9 \quad -25 \quad -25 \quad -9 \quad 1 \quad 1)\boldsymbol{\Upsilon}_i = 0,$$

which is satisfied by

$$\boldsymbol{\Upsilon}'_1 = (1 \quad 0 \quad \tfrac{1}{9} \quad 0 \quad 0 \quad \tfrac{1}{9} \quad 0 \quad 1),$$
$$\boldsymbol{\Upsilon}'_2 = (1 \quad 0 \quad 0 \quad \tfrac{1}{25} \quad \tfrac{1}{25} \quad 0 \quad 0 \quad 1),$$
$$\boldsymbol{\Upsilon}'_3 = (0 \quad 1 \quad \tfrac{1}{9} \quad 0 \quad 0 \quad \tfrac{1}{9} \quad 1 \quad 0).$$

A fourth reduced design can be derived from the vector

$$\mathbf{a}_4 = \boldsymbol{\Upsilon}_2 - \boldsymbol{\Upsilon}_1 + \boldsymbol{\Upsilon}_3 = (0 \quad 1 \quad 0 \quad \tfrac{1}{25} \quad \tfrac{1}{25} \quad 0 \quad 1 \quad 0).$$

TABLE 4

Radius multipliers for some second order rotatable designs

k	Design	Radius Multipliers								No. of Experimental Points[a]			
										Simplex-Sum Designs		Composite Designs	
		a_1	a_2	a_3	a_4	a_5	a_6	a_7	a_8	Radial Points	Center Points[b]	Radial Points	Center Points[b]
2	Std.	1	1							6	3	8	5
3	Std.	1	.8409	1						14	6	14	6
4	Std.	1	.7598	.7598	1					30	14	24	7
5	Std.	1	.7071	.6389	.7071	1				62	24		
	Υ_2	1	.8409	0	.8409	1				42	10		
	Υ_1	1	0	.7598	0	1				32	8	26	6
6	Std.	1	.6687	.5623	.5623	.6687	1			126	38		
	Υ_2	1	0	.5946	.5946	0	1			84	16		
	Υ_1	1	1	0	0	1	1			56	13	44	9
7	Std.	1	.6389	.5081	.4729	.5081	.6389	1		254	59		
	Υ_1	1	0	.5774	0	.5774	0	1		128	21		
	Υ_2	1	0	0	.5946	0	0	1		86	15		
	Υ_3	0	1	0	0	0	1	0		56	10	78	14
8	Std.	1	.6150	.4671	.4111	.4111	.4671	.6150	1	510	90		
	Υ_2	1	0	0	.4472	.4472	0	0	1	270	26		
	Υ_3	0	1	.5774	0	0	.5774	1	0	240	0		
	Υ_1	1	0	.5774	0	0	.5774	0	1	186	28	80	13

[a] The "Composite Design" values refer to the composite second order rotatable designs derived in [5] and are included for comparative purposes. Half replicates of the cube portion are used for $k = 5$, 6 and 7 and one quarter replicate for $k = 8$.

[b] Number of centerpoints required for "uniform variance" within $\rho = (\lambda_2)^{\frac{1}{2}}$.

A summary of the radius multipliers used to obtain the standard solution designs (B_s^{-1}) and the specific solution designs derived from the basis vectors, is given in Table 4. It can be seen that only the reduced designs will be practical in most instances when $k > 4$ since N increases rapidly. Also included in the table are the number of center points required to attain "uniform variance".

In order to produce a design using Table 4, it is only necessary to select a suitable matrix $\mathbf{D}_1$ and by taking all sums of rows s at a time, for each s of the non-zero a_s values, generate the required $\mathbf{D}_s$ matrices. Multiplication of $\mathbf{D}_s$ by a_s will then give the coordinates of the design points. An example is given in Section 9.

7.0. Replication. If it should be desired to replicate certain subsets of the derived matrices this can easily be done by making suitable adjustments to the radius multipliers. We will only consider the case where symmetric replication is used (i.e., $\mathbf{D}_s$ and $\mathbf{D}_{n-s}$ are replicated equally), thus ensuring that a symmetric solution for the radius multipliers can be found.

If we replicate a particular pair of submatrices $\mathbf{D}_s$ and $\mathbf{D}_{n-s}$ ν_s times, the elements $C_{31}(s)$, $C_{41}(s)$, $C_{31}(n - s)$ and $C_{41}(n - s)$ will be multiplied by ν_s and the moment equations will become

$$\sum_{s=1}^{k} \nu_s(a_s)^3 C_{31}(s) = 0,$$

$$\sum_{s=1}^{k} \nu_s(a_s)^4 C_{41}(s) = 0.$$

The first equation will still be negatively symmetric and will therefore be satisfied by any symmetric vector. The second equation will be satisfied if the new $\nu_s(a_s)^4$ equal the old $(a_s)^4$. Thus

$$a_s(\mathbf{D}_s \text{ replicated } \nu_s \text{ times}) = a_s(\text{unreplicated})/(\nu_s)^{\frac{1}{4}},$$

and a similar relation holds for radii.

For example, consider the standard solution for $k = 3$, and various patterns of replication. (We will always have $\nu_1 a_1^4 = 1$, $\nu_2 a_2^4 = \frac{1}{2}$, $\nu_3 a_3^4 = 1$.) Table 5 shows some results.

TABLE 5

The standard solution with $k = 3$ and various replication patterns

Pattern	Replications			Radius Multipliers			Radii		
	ν_1	ν_2	ν_3	a_1	a_2	a_3	r_1	r_2	r_3
1	1	1	1	1	$2^{-\frac{1}{4}}$	1	1.73	1.68	1.73
2	2	1	2	$2^{-\frac{1}{4}}$	$2^{-\frac{1}{4}}$	$2^{-\frac{1}{4}}$	1.45	1.68	1.45
3	1	8	1	1	2^{-1}	1	1.73	1.00	1.73

8.0. Blocking. When an experiment cannot be run under homogeneous conditions it is usually desirable to block the trials in such a way that the coefficients can be estimated efficiently while the error is confined to the magnitude of variation within blocks. We will assume that under the experimental conditions peculiar to any block the relationship of the response to the factors remains unchanged with the exception of a shift in level. Following the development in [5] then we assume the expected value of the uth experimental observation is represented by the model

$$\eta_u = \beta_0 + \sum_{i=1}^{k} \beta_i x_{iu} + \sum_{i=1}^{k} \sum_{j=i}^{k} \beta_{ij} x_{iu} x_{ju} + \sum_{w=1}^{m} \delta_w (z_{wu} - \bar{z}_w),$$

where

$$\beta_0 = \sum_{w=1}^{m} \frac{n_w}{N} \beta_{0w}, \qquad \delta_w = \beta_{0w} - \beta_0, \qquad \bar{z}_w = \frac{n_w}{N}$$

and β_{0w} is the level parameter for the wth block, z_{wu} is a dummy variable assuming the value unity when the uth experiment falls in block w and zero otherwise, n_w is the number of observations in the wth block (including center points) and $N = \sum_{w=1}^{m} n_w$.

8.1. *Orthogonal Blocking–Rotatable Designs.* It is shown in [5] that orthogonal blocking is obtained when the within block moment components of the design (denoted by $[i^{\alpha_1} j^{\alpha_2}]_{bw}$) have the following properties:

1. $$[i]_{bw} = \frac{1}{N} \sum_{u}^{n_w} x_{iu} = 0,$$

2. $$[i\,j]_{bw} = \frac{1}{N} \sum_{u}^{n_w} x_{iu}\, x_{ju} = 0, \qquad i \neq j$$

3. $$[i^2]_{bw} = \frac{1}{N} \sum_{u}^{n_w} x_{iu}^2 = \frac{n_w}{N} \lambda_2, \qquad w = 1, 2 \cdots m,$$

where $\sum_{u}^{n_w}$ indicates summation over the n_w design points within the wth block.

The blocking arrangements we consider here will be called submatrix blocking schemes since they utilize the submatrices $a_1\mathbf{D}_1$, $a_2\mathbf{D}_2$, $\cdots$, $a_k\mathbf{D}_k$, or combinations of them, as blocks. From the general formulas for the moment components of these submatrices it is clear that they individually satisfy the first two conditions above. To individually satisfy the third condition however it is necessary that the quantities a_s^2 be such that their ratios are rational numbers. Instead of using the submatrices themselves as the basis for blocking, combinations of these submatrices can be employed. If the a_s are such that they allow blocks to be formed which yield a ratio of $[i^2]_{bw}/\lambda_2$ which is equal to a rational number then orthogonal blocks can be obtained. Table 6 shows some blocking arrangements which are derived in this way for the designs in Table 4.

In general the *individual* submatrices can not be employed as blocks without sacrificing either orthogonal blocking or rotatability. It is naturally most reasonable to sacrifice rotatability since clearly we only require an approximately "symmetric distribution" of information. Unfortunately when the conditions for rotatability are relaxed in this way the general inverse of the resulting matrix is not easily written down. When an electronic computer is used in the analysis of data however this presents little difficulty. The radius multipliers required for orthogonal blocking differ little from those required for rotatability and the resulting designs are thus nearly rotatable. Table 7 provides these values of a_s together with the "uniform variance" number of center points for each sub-matrix block.

8.2. *Non-orthogonal Blocking of the Rotatable Designs.* An alternative would be to retain rotatability but to accept slightly non-orthogonal blocking. From the

TABLE 6

Summary of orthogonal blocking schemes for rotatable designs of Table 4

k	Design	Block	Number of points in Block from Submatrix								Total No. of Points in Block		
			a_1D_1	a_2D_2	a_3D_3	a_4D_4	a_5D_5	a_6D_6	a_7D_7	a_8D_8	Sans Center Points	Center Points Added[a]	Grand Total (n_w)
2	Std.	1	3								3	2	5
		2		3							3	2	5
3	none												
4	Std.	1	5	10							15	7	22
		2			10	5					15	7	22
5	Υ_2	1	6	15							21	5	26
		2				15	6				21	5	26
6	Std.	1	7	21	35						63	19	82
		2				35	21	7			63	19	82
	Υ_2	1	7		35						42	8	50
		2				35		7			42	8	50
	Υ_1	1	7	21							28	6	34
		2					21	7			28	6	34
	Υ_1	1	7								7	(0)	7
		2		21							21	(14)	35
		3					21				21	(14)	35
		4						7			7	(0)	7
7	Υ_1	1	8		56						64	10	74
		2					56		8		64	10	74
	Υ_3	1		28							28	5	33
		2						28			28	5	33
	Υ_1	1	8								8	(4)	12
		2			56						56	(4)	60
		3					56				56	(4)	60
		4							8		8	(4)	12
8	Std.	1	9	36	84	126					255	45	300
		2					126	84	36	9	255	45	300
	Υ_2	1	9			126					135	13	148
		2					126			9	135	13	148
	Υ_3	1		36	84						120	0	120
		2						84	36		120	0	120

TABLE 6—*Continued*

k	Design	Block	Number of points in Block from Submatrix								Total No. of Points in Block		
			a_1D_1	a_2D_2	a_3D_3	a_4D_4	a_5D_5	a_6D_6	a_7D_7	a_8D_8	Sans Center Points	Center Points Added[a]	Grand Total (n_w)
	Υ_1	1	9		84						93	14	107
		2						84		9	93	14	107
	Υ_2	1	9								9	(9) (10)	18 19
		2				126					126	(0) (7)	126 133
		3					126				126	(0) (7)	126 133
		4								9	9	(9) (10)	18 19
	Υ_3	1		36							36	(48)	84
		2			84						84	(0)	84
		3						84			84	(0)	84
		4							36		36	(48)	84
	Υ_1	1	9								9	(4)	13
		2			84						84	(7)	91
		3						84			84	(7)	91
		4								9	9	(4)	13

[a] Those values not in brackets are the number of centerpoints required for "uniform variance" and can be replaced by any other number evenly distributed between blocks. The values in brackets also provide uniform variance but can not be changed freely without loss of orthogonality.

point of view of computational difficulty this approach turns out to be much the simpler, while the loss of information due to the slight non-orthogonality in blocking is small. In reference [15] the moment conditions are given which the points within the individual blocks must satisfy in order to retain rotatability. In particular it is shown that these conditions are met by any blocks which satisfy conditions 1 and 2 in Section 8.1 and hence by the submatrices $a_1\mathbf{D}_1$, $a_2\mathbf{D}_2, \cdots, a_k\mathbf{D}_k$ whether or not they are augmented with center points. Thus when only condition 3 is violated in blocking a rotatable design the variance-covariance matrix of the response surface coefficients (adjusted for the block effects) retains the form necessary to give "spherical" variance contours. The form also readily lends itself to providing a general explicit solution for the normal equations. The estimates of the regression coefficients for any such arrangement are given below where we let $\bar{y}_w$ denote the average of the observations in block w and use the notation

$$\{iy\} = \sum_{u=1}^{n} y_u , \qquad \{iy\} = \sum_{u=1}^{n} x_{iu} y_u , \qquad \{ijy\} = \sum_{u=1}^{n} x_{iu} x_{ju} y_u ,$$

$$A_{\alpha}^{-1} = 2\lambda_4 \left[(k+2)\lambda_4 - kN \sum_{w}^{m} [i^2]_{bw}^2 / n_w \right]$$

to give

$$b_0 = N^{-1}\left[\{0y\} - 2A_\alpha\lambda_4\lambda_2\left(\sum_i^k \{iiy\} - kN\sum_w^m [i^2]_{bw}\bar{y}_w\right)\right],$$

$$b_{ii} = N^{-1}A_\alpha\left[\{iiy\}A_\alpha^{-1} + \left(N\sum_w^m \frac{[i^2]_{bw}^2}{n_w} - \lambda_4\right)\sum_i^k \{iiy\} - 2\lambda_4 N\sum_w^m [i^2]_{bw}\bar{y}_w\right],$$

$$b_i = (\lambda_2 N)^{-1}\{iy\},$$

$$b_{ij} = (\lambda_4 N)^{-1}\{ijy\}.$$

The variances and covariances are

$$V(b_0) = 2\sigma^2\lambda_4 N^{-1}A_\alpha\left[(k+2)\lambda_4 - k\left(N\sum_w^m \frac{[i^2]_{bw}^2}{n_w} - \lambda_2^2\right)\right],$$

$$V(b_i) = \sigma^2(N\lambda_2)^{-1}, \qquad V(b_{ij}) = \sigma^2(N\lambda_4)^{-1},$$

$$V(b_{ii}) = \sigma^2 N^{-1}A_\alpha\left[(k+1)\lambda_4 - (k-1)N\sum_w^m \frac{[i^2]_{bw}^2}{n_w}\right],$$

$$\text{Cov}\,(b_0 b_{ii}) = -2\sigma^2\lambda_4\lambda_2 N^{-1}A_\alpha, \qquad \text{Cov}\,(b_{ii}b_{jj}) = \sigma^2 N^{-1}A_\alpha\left[N\sum_w^m \frac{[i^2]_{bw}^2}{n_w} - \lambda_2^2\right].$$

It will be noted that the variances of b_i and b_{ij} are not affected by non-orthogonal blocking but the variance of the constant term b_0 and the quadratic terms b_{ii} are affected. In [15] it is shown that the loss of information introduced by the small degree of non-orthogonality is small.

The variance function from which the variance of an estimated value $\hat{y}$ can

TABLE 7

Radius multipliers and center points for orthogonal nearly rotatable submatrix blocking

k	Original Design	D_1		D_2		D_3		D_4		D_5		D_6		D_7		D_8	
		a_1	n_{10}	a_2	n_{20}	a_3	n_{30}	a_4	n_{40}	a_5	n_{50}	a_6	n_{60}	a_7	n_{70}	a_8	n_{80}
3	Standard	1	2	.8165	2	1	2										
4	Standard	1	3	.7638	4	.7638	4	1	3								
5	Standard	1	4	.7071	5	.6583	6	.7071	5	1	4						
	Υ_2	1	1	.8238	4	0	0	.8238	4	1	1						
	Υ_1	1	1	0	0	.7868	6	0	0	1	1						
6	Standard	1	6	.6679	8	.5547	5	.5547	5	.6679	8	1	6				
	Υ_2	1	4	0	0	.5954	4	.5954	4	0	0	1	4				
7	Standard	1	8	.6455	12	.5164	8	.4776	3	.5164	8	.6455	12	1	8		
	Υ_2	1	3	0	0	0	0	.5992	9	0	0	0	0	1	3		
8	Standard	1	12	.6172	20	.4690	13	.4140	0	.4140	0	.4690	13	.6172	20	1	12

readily be calculated is

$$V(\hat{y}) = \sigma^2 N^{-1} A_\alpha \left\{ 2(k+2)\lambda_4^2 - 2k\lambda_4 \left(N \sum_w^m \frac{[i^2]_{bw}^2}{n_w} - \lambda_2^2 \right) + 2\lambda_4 \lambda_2^{-1} \left[(k+2)\lambda_4 - \left(kN \sum_w^m \frac{[i^2]_{bw}^2}{n_w} + 2\lambda_2^2 \right) \right] \rho^2 + \left[(k+1)\lambda_4 - (k-1)N \sum_w^m \frac{[i^2]_{bw}^2}{n_w} \right] \rho^4 \right\}.$$

9.0. A Convenient Reduced Design for $k = 7$. The design derived from the basis vector, Υ_3 for the seven factor design in Section 6.2, has several interesting features which will be discussed here. Since it requires but 56 points (plus center points) to estimate the 36 coefficients of a seven factor second degree polynomial, it is extremely efficient. The comparable central composite design [5]

TABLE 8

Seven factor second order rotatable design in three levels

$\frac{1}{2}D_2$							$\frac{1}{2}D_6$						
1	1	0	1	0	0	0	−1	−1	0	−1	0	0	0
1	0	1	0	1	0	0	−1	0	−1	0	−1	0	0
1	0	0	0	0	1	1	−1	0	0	0	0	−1	−1
0	1	1	0	0	1	0	0	−1	−1	0	0	−1	0
0	1	0	0	1	0	1	0	−1	0	0	−1	0	−1
0	0	1	1	0	0	1	0	0	−1	−1	0	0	−1
0	0	0	1	1	1	0	0	0	0	−1	−1	−1	0
1	0	0	0	0	−1	−1	−1	0	0	0	0	1	1
1	0	−1	0	−1	0	0	−1	0	1	0	1	0	0
0	1	0	0	−1	0	−1	0	−1	0	0	1	0	1
0	1	−1	0	0	−1	0	0	−1	1	0	0	1	0
0	0	0	1	−1	−1	0	0	0	0	−1	1	1	0
0	0	−1	1	0	0	−1	0	0	1	−1	0	0	1
1	−1	0	−1	0	0	0	−1	1	0	1	0	0	0
0	0	1	−1	0	0	−1	0	0	−1	1	0	0	1
0	0	0	−1	1	−1	0	0	0	0	1	−1	1	0
0	−1	1	0	0	−1	0	0	1	−1	0	0	1	0
0	−1	0	0	1	0	−1	0	1	0	0	−1	0	1
0	0	0	−1	−1	1	0	0	0	0	1	1	−1	0
0	0	−1	−1	0	0	1	0	0	1	1	0	0	−1
0	−1	0	0	−1	0	1	0	1	0	0	1	0	−1
0	−1	−1	0	0	1	0	0	1	1	0	0	−1	0
−1	1	0	−1	0	0	0	1	−1	0	1	0	0	0
−1	0	1	0	−1	0	0	1	0	−1	0	1	0	0
−1	0	0	0	0	1	−1	1	0	0	0	0	−1	1
−1	0	0	0	0	−1	1	1	0	0	0	0	1	−1
−1	0	−1	0	1	0	0	1	0	1	0	−1	0	0
−1	−1	0	1	0	0	0	1	1	0	−1	0	0	0

requires 78 points (plus center points). The vector of radius multipliers that defines this design is $\mathbf{a}' = (0 \quad 1 \quad 0 \quad 0 \quad 0 \quad 1 \quad 0)$ and thus utilizes the points specified by the matrices $\mathbf{D}_2$ and $\mathbf{D}_6$ only.

In seven dimensions it is possible to find a matrix $\mathbf{D}_1$, giving the coordinates of a regular simplex, which involves only the two levels -1 and $+1$, for each factor. Consequently $\mathbf{D}_2$ and $\mathbf{D}_6$ need only involve three factor levels. Furthermore $\mathbf{D}_2$ and $\mathbf{D}_6$ provide orthogonal blocks.

The 8×8 matrix $[\mathbf{1}\ \mathbf{D}_1]$ which can be used to generate this design is

$$\begin{bmatrix} 1 & 1 & 1 & 1 & 1 & 1 & 1 & 1 \\ 1 & 1 & 1 & -1 & 1 & -1 & -1 & -1 \\ 1 & 1 & -1 & 1 & -1 & 1 & -1 & -1 \\ 1 & 1 & -1 & -1 & -1 & -1 & 1 & 1 \\ 1 & -1 & 1 & 1 & -1 & -1 & 1 & -1 \\ 1 & -1 & 1 & -1 & -1 & 1 & -1 & 1 \\ 1 & -1 & -1 & 1 & 1 & -1 & -1 & 1 \\ 1 & -1 & -1 & -1 & 1 & 1 & 1 & -1 \end{bmatrix}$$

Its squared vector length is eight, as required, and all rows and columns are orthogonal.

The derived matrices $\frac{1}{2}\mathbf{D}_2$ and $\frac{1}{2}\mathbf{D}_6$ are shown in Table 8. Since multiplication by a constant is permissible, we will define our derived design matrix $\mathbf{D}$ therefore as

$$\mathbf{D} = \begin{bmatrix} \frac{1}{2} & \mathbf{D}_2 \\ \frac{1}{2} & \mathbf{D}_6 \end{bmatrix}$$

The singularity of the moment matrix of this design is readily detectable by noting that all the points lie on a hypersphere of radius $(3)^{\frac{1}{2}}$ and hence center points must be added to make all coefficients separately estimable. The addition of ten such points will produce a design having the "uniform variance" property.

For this design (and whenever nonorthogonal blocking does not complicate the normal equations) the regression coefficients and their variances are easily obtained from the general solutions for rotatable designs given in [5].

REFERENCES

[1] Box, G. E. P., Integration of techniques in process development. Statistical Techniques Research Group, Technical Report No. 2, Princeton University, Princeton, N. J., 1957.

[2] Box, G. E. P. and Draper, N. R., A basis for the selection of a response surface design, Statistical Techniques Group, Technical Report No. 23, Princeton University, Princeton, N. J., 1958.

[3] Box, G. E. P., Use of statistical methods in the elucidation of basic mechanisms, Paper presented at International Institute of Statistics, Stockholm, 1957.

[4] Box, G. E. P. and Lucas, H. L., "Design of experiments in non-linear situations," *Biometrika*, Vol. 46 (1959), pp. 77–90.

[5] Box, G. E. P. and Hunter, J. S., "Multifactor experimental designs for exploring response surfaces," *Ann. Math. Stat.*, Vol. 28 (1957), pp. 195–241.

[6] Box, G. E. P., "Multifactor designs of first order," *Biometrika*, Vol. 39 (1958), pp. 49–57

[7] Box, G. E. P. and Wilson, K. B., "On the experimental attainment of optimum conditions," *J. Roy. Stat. Soc.*, Ser. B, Vol. 13 (1951), pp. 1–45.

[8] Tukey, John W., "Keeping moment-like sampling computations simple," *Ann. Math. Stat.*, Vol. 27 (1956), pp. 37–54.

[9] Wishart, J., "Moment coefficients of the k-statistics in samples from a finite population," *Biometrika*, Vol. 39 (1952), pp. 1–13.

[10] Hooke, Robert, "Symmetric functions of a two-way array," *Ann. Math. Stat.*, Vol. 27, (1956), pp. 55–79.

[11] Robson, D. S., "Applications of multivariate polykays to the theory of unbiased ratio-type estimation," *J. Amer. Stat. Assn.*, Vol. 52 (1957), pp. 511–522.

[12] Behnken, D.W., "Sampling moments of means from finite multivariate populations," submitted to *Ann. Math. Stat.*

[13] Box, G. E. P. and Behnken, D. W., "Derivation of second order rotatable designs from those of first order," Statistical Techniques Research Group, Technical Report No. 17, Princeton University, Princeton, New Jersey, 1958.

[14] Bose, R. C. and Draper, N. R., "Rotatable designs of second and third order in three or more dimensions," Inst. of Stat. Mimeo Series No. 197, University of North Carolina, Chapel Hill, 1958.

[15] Box, G. E. P. and Behnken, D. W., Simplex-sum designs, a class of second order rotatable designs derivable from those of first order, Inst. of Stat. Mimeo Series No. 232, North Carolina State College, Raleigh, North Carolina, 1959.

2.9

Some New Three Level Designs for the Study of Quantitative Variables

G. E. P. BOX and D. W. BEHNKEN
University of Wisconsin and American Cyanamid Company

A class of incomplete three level factorial designs useful for estimating the coefficients in a second degree graduating polynomial are described. The designs either meet, or approximately meet, the criterion of rotatability and for the most part can be orthogonally blocked. A fully worked example is included.

1.0. Introduction

A symmetrical factorial design is an experimental arrangement in which a small integral number p of levels is chosen for each of k factors (i.e. variables) and all p^k combinations of these levels are run. Classes of these designs which have proved to be of particular interest are those in which two levels or three levels are used for each of the k variables. These are called respectively 2^k and 3^k factorials. If not all the factorial combinations are employed but merely a selected subset, we call the design an incomplete factorial. Any factorial or incomplete factorial we call a factorial-type design.

A class of incomplete factorials of considerable interest are the fractional factorials of D. J. Finney [1] [2]. In these arrangements certain finite group properties are employed to select a $(1/p)^f$ fraction of the complete design which then requires only p^{k-f} combinations of levels and may be called a p^{k-f} factorial. A useful and different class of incomplete factorials in which the selected subset is not restricted to be a $(1/p)^f$ fraction is due to Plackett and Burman [3].

An infinite choice exists for the levels of quantitative variables such as temperature. In developing designs specifically for quantitative variables, there is therefore no essential need to restrict experimental conditions to combinations of a few basic levels of the component factors. Many useful designs have indeed been devised for the study of quantitative variables which do not employ the factorial principle [4] [5]. In spite of this, cases are not uncommon where even though the factors are all quantitative, convenience requires the use of only a few levels for each.

In this paper we discuss a particular class of three-level incomplete factorials specifically selected for the study of quantitative variables. The class of designs is not included among the types of incomplete factorials already discussed but nevertheless appears to be of considerable practical importance.

2.0. Incomplete Factorials for Quantitative Variables

When a design involving N runs is employed to separately estimate L constants we may define the ratio $R = N/L$ as the *redundancy factor* for the design. This factor is necessarily not less than unity.

The Collected Works of George E. P. Box, 1984, Wadsworth, Inc., Belmont, CA 94002.
Originally published in *Technometrics,* vol. 2, no. 4 (1960), pp. 455–475.

Suppose in what follows that the functional relationship between the response of interest and the levels of the k quantitative experimental variables may be graduated by a general polynomial of degree d in the levels of the variables. A design suitable for separately estimating the $(k + d)!/k!\, d!$ constants of such a polynomial is called a *design of order d*. The highest degree of polynomial that may be fitted to the observations from a p-level factorial is $p - 1$. Consequently when regarded as a design for the fitting of a general polynomial the p^k factorial is a design of order $p - 1$. The redundancy factor for such a design is therefore $p^k k!(p - 1)!/(k + p - 1)!$. When calculated in this way the redundancy factors for the complete factorials are usually large. For example, regarded as a first order design, the two-level factorial in five factors requires $2^5 = 32$ runs to estimate the 6 constants of the first degree polynomial. It therefore has a redundancy factor of $32/6 = 5.3$. Similarly, regarded as a second order design the three-level factorial in five factors requires $3^5 = 243$ runs to estimate the 21 constants of the second degree polynomial. It therefore has a redundancy factor of $243/21 = 11.6$.

In situations in which the experimental error variance is not so large as to require large numbers of observations to obtain necessary precision, designs having small redundacy factors are desirable. Small redundacy factors may sometimes be obtained by using incomplete rather than complete factorial designs. For example, if $k = 3, 7, 11, 15, \cdots, 4i - 1$ the two-level arrangements of Plackett and Burman provide first order designs requiring respectively only $4, 8, 12, 16, \cdots, 4i$ runs, where i is a positive integer. They are thus first order two-level designs of redundancy unity. Designs having this minimal redundancy are seldom employed in practice because they provide no residual degrees of freedom and so do not allow the possibility of partially checking [6] [7] the adequacy of the assumed form of model. Other incomplete two-level factorial designs are available however having low redundancy factors of two or less which do not suffer from this deficiency.

For the presently available three-level factorials the situation is less satisfactory than for the two-level designs. For example, the various one-ninth replicates of the 3^5 factorials all seem to lead to undesirable correlation or confounding of estimates of the coefficients and although a one-third replicate of the 3^5 factorial may be employed as a second order design it has a redundancy factor of $81/21 = 3.9$ which is somewhat high.

In developing the present class of designs we do not use the group properties exploited by Finney; rather we set out directly to select part of the 3^k factorial which allows efficient estimation of a second degree graduating polynomial. Specifically, we have where possible set out to generate second order rotatable designs. Arguments in favor of such a choice have been presented elsewhere [5]. Suppose we code the levels in standardized units so that the 3 values taken by each of the variables $x_1, x_2, \cdots x_k$ are -1, 0, and 1 and suppose also that the second degree graduating polynomial fitted by the method of least squares is

$$\hat{y} = b_0 + \sum_{i=1}^{k} b_i x_i + \sum_{i=1}^{k} \sum_{j=i}^{k} b_{ij} x_i x_j \,.$$

A second order rotatable design is such that the variance of $\hat{y}$ is constant for all points equidistant from the center of the design—that is, for all points for which $\rho = (\sum_i x_i^2)^{\frac{1}{2}}$ is constant. Among the class of rotatable designs we select those for which the variance of $\hat{y}$, regarded as a function of ρ, is reasonably constant in the region of the k-space covered by the design. The requirement of rotatability is introduced to ensure a symmetric generation of information in the space of the variables defined and scaled in a manner currently thought most appropriate by the experimenter. For a design to be useful it need not have the property of rotatability exactly. For certain values of k, it turns out that within the class of designs we consider, rotatability can be achieved exactly; in other cases, exact rotatability is not possible and here, as described more fully in Appendix A, we relax the requirement to some extent. All the designs we discuss possess a high degree of orthogonality; in fact, only the constant term b_0 and the quadratic estimates b_{ii} are correlated* one with another.

The requirement of rotatability or near-rotatability imposes certain restrictions [5] on the moments of the design. In Appendix A it is shown that when these restrictions are applied to variables which can take only the values -1, 0, and 1 certain simple combinatorial requirements emerge and that these requirements can be satisfied by combining two-level factorial designs and incomplete block designs in a particular manner exemplified in the next section.

The existence of the class of designs discussed here was suggested by the discovery in another connection [8] of a three-level rotatable design in seven variables which required only 56 points plus added points at the origin thus providing highly efficient estimates of the 36 constants in the polynomial of second degree. Further investigation led to the development of the present class of three-level designs utilizing the properties of incomplete blocks.

3.0. Method for Generating the Designs

The designs are formed by combining two-level factorial designs with incomplete block designs in a particular manner. This is best illustrated by an example. In Table 1 is shown a balanced incomplete block design for testing $k = 4$ varieties in $b = 6$ blocks of size $s = 2$.

Table 1

A balanced incomplete block design for four varieties in six blocks.

$k = 4$ varieties

		x_1	x_2	x_3	x_4
	1	*	*		
	2			*	*
$b = 6$	3	*			*
blocks	4		*	*	
	5		*		*
	6	*		*	

* Designs for which there is no correlation between either all or a subset of the quadratic coefficients can be obtained but they do not seem to possess any particular advantage [5] so far as estimating the response is concerned.

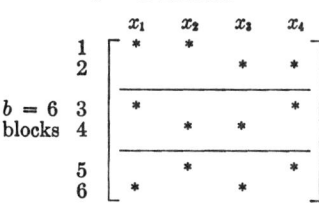

TABLE 2

A 2^2 factorial design.

x_i	x_j
−1	−1
1	−1
−1	1
1	1

If this design were being used in the usual way, varieties 1 and 2 denoted by x_1 and x_2 would be tested in the first block, varieties 3 and 4 in the second, and so on.

A basis for a three-level design in four variables is obtained by combining this incomplete block design with the 2^2 factorial of Table 2. The two asterisks in every row of the incomplete block design are replaced by the $s = 2$ columns of the two-level 2^2 design. Wherever an asterisk does *not* appear a column of zeros is inserted. The design is completed by the addition of a number of center points (0, 0, 0, 0), about three being desirable with this arrangement. The resulting design is shown in Table 3. As explained later, this design can in fact be run in three orthogonal blocks. These are indicated by dotted lines in the table.

The design obtained is a rotatable second order design suitable for studying four variables in 27 trials and is capable of being blocked in three sets of nine trials. It is shown in Appendix B that this particular design is in fact a rotation of the corresponding central composite rotatable design [5] in four variables. It is however not generally true that the present class of designs can be generated from the central composite designs by rotation.

TABLE 3

An incomplete 3^4 factorial in three blocks of nine experimental runs.

x_1	x_2	x_3	x_4	
−1	−1	0	0	
1	−1	0	0	
−1	1	0	0	
1	1	0	0	
0	0	−1	−1	Block 1
0	0	1	−1	
0	0	−1	1	
0	0	1	1	
0	0	0	0	
−1	0	0	−1	
1	0	0	−1	
−1	0	0	1	
1	0	0	1	
0	−1	−1	0	Block 2
0	1	−1	0	
0	−1	1	0	
0	1	1	0	
0	0	0	0	
0	−1	0	−1	
0	1	0	−1	
0	−1	0	1	
0	1	0	1	
−1	0	−1	0	Block 3
1	0	−1	0	
−1	0	1	0	
1	0	1	0	
0	0	0	0	

In Table 4 a number of designs of the class under study are given suitable for investigating 3, 4, 5, 6, 7, 9, 10, 11, 12, and 16 variables. In this table unless otherwise indicated the symbol ($\pm 1, \pm 1, \cdots, \pm 1$) means that all combinations of plus and minus levels are to be run. Whenever a fractional factorial is available which does not confound main effects and two factor interactions one with another, it may be used instead of the full factorial. For example, in design No. 8, s is equal to five and as indicated in the table rather than using a full 2^5 factorial we can achieve the desired result with a half-replicate.

Three members of the class of designs have been generated by other methods and have appeared elsewhere. Design No. 1 was first described by DeBaun [9], [10] and design No. 2 by Gardiner, Grandage and Hader [11]. The general method of Bose and Draper rederived design No. 2 in [12] and produced the points in designs No. 1 and No. 3 as identifiable subsets of rotatable designs in [13] and [12] respectively.

4.0. Blocking the Designs

Where insufficient homogeneous experimental material is available for all the experimental runs it becomes desirable to run them in blocks. Where possible it is desirable to achieve *orthogonal* blocking, that is to arrange that the block constrasts are uncorrelated with all the estimates of the coefficients in the polynomial. When this can be achieved the analysis may be carried out almost as if block differences did not exist. The only modification necessary is that in the analysis of variance table the sum of squares associated with block differences must be substracted from the residual sum of squares. On the assumption that the model is adequate, the residual sum of squares so adjusted may then be used to estimate the within-block variance and hence the standard errors of the coefficients.

The requirements for orthogonal blocking of second order designs have been given elsewhere [5]. Applying these results to the present problem, it is easy to see that:

(1) Where "replicate sets" can be found in the generating incomplete block design these provide a basis for orthogonal blocking. These replicate sets are subgroups within which each variety is tested the same number of times.
(2) Where the component factorial designs can be divided into blocks which only confound interactions of more than two factors these can provide a basis for orthogonal blocking.

An illustration of the first method of blocking has already been given in the example of Section 3.0. In Table 4 dotted lines indicate the appropriate divisions into replicate sets. Using these divisions design No. 2 can be split into three blocks, design No. 3 into two blocks, design No. 6 into five blocks and design No. 10 into six blocks. In these and other blocking schemes discussed below, the center points *must* be distributed equally among blocks to retain orthogonality.

The second method may be illustrated with design No. 4 for which the first method cannot be employed. The basis for the design consists of 48 trials gen-

Table 4

Some useful three-level designs

Design Number	Number of Factors(k)	Design Matrix	No. of Points	Blocking and Association Schemes
1	3	$\begin{bmatrix} \pm1 & \pm1 & 0 \\ \pm1 & 0 & \pm1 \\ 0 & \pm1 & \pm1 \\ 0 & 0 & 0 \end{bmatrix}$	12; 3; $N = 15$	No orthogonal blocking BIB (one associate class)
2	4	$\left[\begin{array}{cccc} \pm1 & \pm1 & 0 & 0 \\ 0 & 0 & \pm1 & \pm1 \\ 0 & 0 & 0 & 0 \\ \hdashline \pm1 & 0 & 0 & \pm1 \\ 0 & \pm1 & \pm1 & 0 \\ 0 & 0 & 0 & 0 \\ \hdashline \pm1 & 0 & \pm1 & 0 \\ 0 & \pm1 & 0 & \pm1 \\ 0 & 0 & 0 & 0 \end{array}\right]$	8; 1; 8; 1; 8; 1; $N = 27$	3 blocks of 9 BIB (one associate class)
3	5	$\left[\begin{array}{ccccc} \pm1 & \pm1 & 0 & 0 & 0 \\ 0 & 0 & \pm1 & \pm1 & 0 \\ 0 & \pm1 & 0 & 0 & \pm1 \\ \pm1 & 0 & \pm1 & 0 & 0 \\ 0 & 0 & 0 & \pm1 & \pm1 \\ 0 & 0 & 0 & 0 & 0 \\ \hdashline 0 & \pm1 & \pm1 & 0 & 0 \\ \pm1 & 0 & 0 & \pm1 & 0 \\ 0 & 0 & \pm1 & 0 & \pm1 \\ \pm1 & 0 & 0 & 0 & \pm1 \\ 0 & \pm1 & 0 & \pm1 & 0 \\ 0 & 0 & 0 & 0 & 0 \end{array}\right]$	20; 3; 20; 3; $N = 46$	2 blocks of 23 BIB (one associate class)

4	6	$\left[\begin{array}{cccccc} \pm1 & \pm1 & 0 & \pm1 & 0 & 0 \\ 0 & \pm1 & \pm1 & 0 & \pm1 & 0 \\ 0 & 0 & \pm1 & \pm1 & 0 & \pm1 \\ \pm1 & 0 & 0 & \pm1 & \pm1 & 0 \\ 0 & \pm1 & 0 & 0 & \pm1 & \pm1 \\ \pm1 & 0 & \pm1 & 0 & 0 & \pm1 \\ 0 & 0 & 0 & 0 & 0 & 0 \end{array}\right]$	48 6 $N = 54$	2 blocks of 27. First Associates: (1, 4); (2, 5); (3, 6).
5	7	$\left[\begin{array}{ccccccc} 0 & 0 & 0 & \pm1 & \pm1 & \pm1 & 0 \\ \pm1 & 0 & 0 & 0 & 0 & \pm1 & \pm1 \\ 0 & \pm1 & 0 & 0 & \pm1 & 0 & \pm1 \\ \pm1 & \pm1 & 0 & \pm1 & 0 & 0 & 0 \\ 0 & 0 & \pm1 & \pm1 & 0 & 0 & \pm1 \\ \pm1 & 0 & \pm1 & 0 & \pm1 & 0 & 0 \\ 0 & \pm1 & \pm1 & 0 & 0 & \pm1 & 0 \\ 0 & 0 & 0 & 0 & 0 & 0 & 0 \end{array}\right]$	56 6 $N = 62$	2 blocks of 31. BIB (one associate class).
6	9	$\left[\begin{array}{ccccccccc} \pm1 & 0 & 0 & \pm1 & 0 & 0 & \pm1 & 0 & 0 \\ 0 & \pm1 & 0 & 0 & \pm1 & 0 & 0 & \pm1 & 0 \\ 0 & 0 & \pm1 & 0 & 0 & \pm1 & 0 & 0 & \pm1 \\ 0 & 0 & 0 & 0 & 0 & 0 & 0 & 0 & 0 \\ \hline \pm1 & \pm1 & \pm1 & 0 & 0 & 0 & 0 & 0 & 0 \\ 0 & 0 & 0 & \pm1 & \pm1 & \pm1 & 0 & 0 & 0 \\ 0 & 0 & 0 & 0 & 0 & 0 & \pm1 & \pm1 & \pm1 \\ 0 & 0 & 0 & 0 & 0 & 0 & 0 & 0 & 0 \\ \hline \pm1 & 0 & 0 & 0 & \pm1 & 0 & 0 & 0 & \pm1 \\ 0 & 0 & \pm1 & \pm1 & 0 & 0 & 0 & \pm1 & 0 \\ 0 & \pm1 & 0 & 0 & 0 & \pm1 & \pm1 & 0 & 0 \\ 0 & 0 & 0 & 0 & 0 & 0 & 0 & 0 & 0 \\ \hline \pm1 & 0 & 0 & 0 & 0 & \pm1 & 0 & \pm1 & 0 \\ 0 & \pm1 & 0 & \pm1 & 0 & 0 & 0 & 0 & \pm1 \\ 0 & 0 & \pm1 & 0 & \pm1 & 0 & \pm1 & 0 & 0 \\ 0 & 0 & 0 & 0 & 0 & 0 & 0 & 0 & 0 \\ \hline \pm1 & 0 & 0 & \pm1 & 0 & 0 & \pm1 & 0 & 0 \\ 0 & \pm1 & 0 & 0 & \pm1 & 0 & 0 & \pm1 & 0 \\ 0 & 0 & \pm1 & 0 & 0 & \pm1 & 0 & 0 & \pm1 \\ 0 & 0 & 0 & 0 & 0 & 0 & 0 & 0 & 0 \end{array}\right]$	24 2 24 2 24 2 24 2 24 2 $N = 130$	(a) 5 blocks of 26. (b) 10 blocks of 13. First Associates: (1, 4); (1, 7); (4, 7); (2, 5); (2, 8); (5, 8); (3, 6); (3, 9); (6, 9).

TABLE 4—*Continued*

Design Number	Number of Factors(k)	Design Matrix	No. of Points	Blocking and Association Schemes
7	10	$\begin{bmatrix} 0 & \pm1 & 0 & 0 & 0 & \pm1 & \pm1 & 0 & 0 & \pm1 \\ \pm1 & \pm1 & 0 & 0 & \pm1 & 0 & 0 & 0 & 0 & \pm1 \\ 0 & \pm1 & \pm1 & 0 & 0 & 0 & \pm1 & \pm1 & 0 & 0 \\ 0 & \pm1 & 0 & \pm1 & 0 & \pm1 & 0 & 0 & \pm1 & 0 \\ \pm1 & 0 & 0 & 0 & 0 & 0 & 0 & \pm1 & \pm1 & \pm1 \\ 0 & 0 & \pm1 & \pm1 & \pm1 & 0 & 0 & 0 & 0 & \pm1 \\ \pm1 & 0 & 0 & \pm1 & 0 & 0 & \pm1 & \pm1 & 0 & 0 \\ 0 & 0 & \pm1 & 0 & \pm1 & 0 & \pm1 & 0 & \pm1 & 0 \\ \pm1 & 0 & \pm1 & 0 & 0 & \pm1 & 0 & 0 & \pm1 & 0 \\ 0 & 0 & 0 & \pm1 & \pm1 & \pm1 & 0 & \pm1 & 0 & 0 \\ 0 & 0 & 0 & 0 & 0 & 0 & 0 & 0 & 0 & 0 \end{bmatrix}$	160 10 $N = 170$	2 blocks of 85. Second Associates: (1, 8); (1, 9); (1, 10); (2, 6); (2, 7); (2, 10); (3, 5); (3, 7); (3, 9); (4, 5); (4, 6); (4, 8); (5, 10); (6, 9); (7, 8).
8	11	$\begin{bmatrix} 0 & 0 & \pm1 & 0 & 0 & 0 & \pm1 & \pm1 & \pm1 & 0 & \pm1 \\ \pm1 & 0 & 0 & \pm1 & 0 & 0 & 0 & \pm1 & \pm1 & \pm1 & 0 \\ 0 & \pm1 & 0 & 0 & \pm1 & 0 & 0 & 0 & \pm1 & \pm1 & \pm1 \\ \pm1 & 0 & \pm1 & 0 & 0 & \pm1 & 0 & 0 & 0 & \pm1 & \pm1 \\ \pm1 & \pm1 & 0 & \pm1 & 0 & 0 & \pm1 & 0 & 0 & 0 & \pm1 \\ \pm1 & \pm1 & \pm1 & 0 & \pm1 & 0 & 0 & \pm1 & 0 & 0 & 0 \\ 0 & \pm1 & \pm1 & \pm1 & 0 & \pm1 & 0 & 0 & \pm1 & 0 & 0 \\ 0 & 0 & \pm1 & \pm1 & \pm1 & 0 & \pm1 & 0 & 0 & \pm1 & 0 \\ 0 & 0 & 0 & \pm1 & \pm1 & \pm1 & 0 & \pm1 & 0 & 0 & \pm1 \\ \pm1 & 0 & 0 & 0 & \pm1 & \pm1 & \pm1 & 0 & \pm1 & 0 & 0 \\ 0 & \pm1 & 0 & 0 & 0 & \pm1 & \pm1 & \pm1 & 0 & \pm1 & 0 \\ 0 & 0 & 0 & 0 & 0 & 0 & 0 & 0 & 0 & 0 & 0 \end{bmatrix}$	176 12 $N = 188$	Use 2^{5-1} fractionated on $x_1x_2x_3x_4x_5$. No orthogonal blocking. BIB (one associate class)
9	12	$\begin{bmatrix} \pm1 & \pm1 & 0 & 0 & \pm1 & 0 & \pm1 & 0 & 0 & 0 & 0 & 0 \\ 0 & \pm1 & \pm1 & 0 & 0 & \pm1 & 0 & \pm1 & 0 & 0 & 0 & 0 \\ 0 & 0 & \pm1 & \pm1 & 0 & 0 & \pm1 & 0 & \pm1 & 0 & 0 & 0 \\ 0 & 0 & 0 & \pm1 & \pm1 & 0 & 0 & \pm1 & 0 & \pm1 & 0 & 0 \\ 0 & 0 & 0 & 0 & \pm1 & \pm1 & 0 & 0 & \pm1 & 0 & \pm1 & 0 \\ 0 & 0 & 0 & 0 & 0 & \pm1 & \pm1 & 0 & 0 & \pm1 & 0 & \pm1 \\ \pm1 & 0 & 0 & 0 & 0 & 0 & \pm1 & \pm1 & 0 & 0 & \pm1 & 0 \\ 0 & \pm1 & 0 & 0 & 0 & 0 & 0 & \pm1 & \pm1 & 0 & 0 & \pm1 \\ \pm1 & 0 & \pm1 & 0 & 0 & 0 & 0 & 0 & \pm1 & \pm1 & 0 & 0 \\ 0 & \pm1 & 0 & \pm1 & 0 & 0 & 0 & 0 & 0 & \pm1 & \pm1 & 0 \\ 0 & 0 & \pm1 & 0 & \pm1 & 0 & 0 & 0 & 0 & 0 & \pm1 & \pm1 \\ \pm1 & 0 & 0 & \pm1 & 0 & \pm1 & 0 & 0 & 0 & 0 & 0 & \pm1 \\ 0 & 0 & 0 & 0 & 0 & 0 & 0 & 0 & 0 & 0 & 0 & 0 \end{bmatrix}$	192 12 $N = 204$	2 blocks of 102. First Associates: (1, 7); (2, 8); (3, 9); (4, 10); (5, 11); (6, 12).

10	16	±1	±1	0	0	0	±1	0	0	±1	0	0	0	0	0	0	0	64
		0	0	±1	±1	0	0	0	±1	0	0	±1	0	0	0	0	0	
		0	0	0	0	±1	0	0	0	0	±1	0	0	±1	±1	0	0	
		0	0	0	0	0	0	±1	0	0	0	0	±1	0	0	±1	±1	
		0	0	0	0	0	0	0	0	0	0	0	0	0	0	0	0	2
		0	±1	±1	0	0	0	±1	0	0	±1	0	0	0	0	0	0	64
		±1	0	0	±1	±1	0	0	0	0	0	0	±1	0	0	0	0	
		0	0	0	0	0	±1	0	0	0	0	±1	0	0	±1	±1	0	
		0	0	0	0	0	0	0	±1	±1	0	0	0	±1	0	0	±1	
		0	0	0	0	0	0	0	0	0	0	0	0	0	0	0	0	2
		0	±1	0	0	±1	±1	0	0	0	0	0	0	±1	0	0	0	64
		0	0	0	±1	0	0	±1	±1	0	0	0	0	0	0	±1	0	
		±1	0	0	0	0	0	0	0	±1	±1	0	0	0	±1	0	0	
		0	0	±1	0	0	0	0	0	0	0	±1	±1	0	0	0	±1	
		0	0	0	0	0	0	0	0	0	0	0	0	0	0	0	0	2
		0	0	±1	0	0	±1	±1	0	0	0	0	0	0	±1	0	0	64
		±1	0	0	0	±1	0	0	±1	0	0	0	0	0	0	0	±1	
		0	±1	0	0	0	0	0	0	0	±1	±1	0	0	0	±1	0	
		0	0	0	±1	0	0	0	0	±1	0	0	±1	±1	0	0	0	
		0	0	0	0	0	0	0	0	0	0	0	0	0	0	0	0	2
		±1	0	±1	0	0	0	0	0	0	0	0	0	±1	0	±1	0	64
		0	±1	0	±1	0	0	0	0	0	0	0	0	0	±1	0	±1	
		0	0	0	0	±1	0	±1	0	±1	0	±1	0	0	0	0	0	
		0	0	0	0	0	±1	0	±1	0	±1	0	±1	0	0	0	0	
		0	0	0	0	0	0	0	0	0	0	0	0	0	0	0	0	2
		0	0	0	±1	0	±1	0	0	0	±1	0	0	0	0	0	±1	64
		0	0	±1	0	±1	0	0	0	±1	0	0	0	0	0	±1	0	
		±1	0	0	0	0	0	±1	0	0	0	±1	0	±1	0	0	0	
		0	±1	0	0	0	0	0	±1	0	0	0	±1	0	±1	0	0	
		0	0	0	0	0	0	0	0	0	0	0	0	0	0	0	0	2
																		$N = 396$

(a) 6 blocks of 66.
(b) 12 blocks of 33.

First Associates:
(1, 5); (1, 9); (1, 13);
(5, 9); (5, 13); (9, 13);
(2, 6); (2, 10); (2, 14);
(6, 10); (6, 14); (10, 14);
(3, 7); (3, 11); (3, 15);
(7, 11); (7, 15); (11, 15);
(4, 8); (4, 12); (4, 16);
(8, 12); (8, 16); (12, 16).

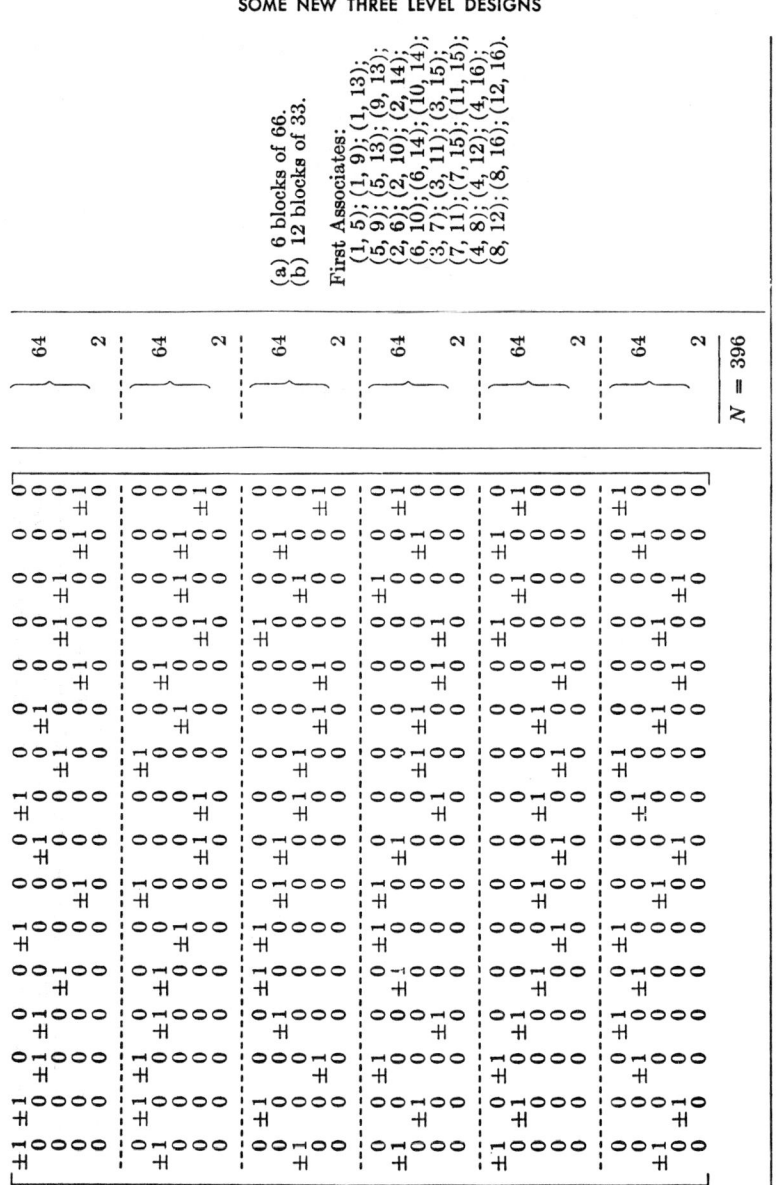

erated from six 2^3 factorial designs. If we were running a single 2^3 factorial design, it could be performed in two sets of four trials, confounding the three-factor interaction with blocks. Trials with levels (1, 1, 1), (1, −1, −1), (−1, −1, 1), (−1, 1, −1) would be included in one set (called the positive set) and trials with levels (−1, −1, −1), (−1, 1, 1), (1, −1, 1), (1, 1, −1) in the other (called the negative set). The complete group of 48 trials can be split into two orthogonal blocks of 24 by allocating one set (either positive or negative) from each of the 2^3 factorial designs to one block, and the remainder to the other.

This method is used where the block size $s > 2$ and employed for designs 4, 5, 6, 7, 9, and 10 in Table 4. In designs 7, 9, and 10 the basic factorial is a 2^4 design. This is split into two sets in such a way as to confound the four factor interaction, that is to say trials with levels whose product is positive are allocated to one group, and the remainder to the other.

In some cases, both methods may be used simultaneously. Thus in design 6 the basic incomplete block design contains five "replicates" indicated by the dotted lines in the table, providing a basis for generating five blocks of 24 runs. Each one of these blocks may now be split into two by allocating the positive sets of the component factorials to one block and the negative sets to the other. We obtain finally an arrangement for generating ten blocks of twelve runs. A similar procedure may be applied in blocking design No. 10.

While orthogonal blocking is desirable, since it minimizes the variance of the estimates of the regression coefficients, non-orthogonal blocking schemes may be employed without an excessive loss of precision when smaller block sizes than those given above are required. Such schemes will not be discussed in the present communication.

5.0. Inclusion of Center Points

In addition to the runs generated directly from the 2^s factorial design it is also necessary to include n_o center points in order to avoid singularity in the moment matrix. The number of center points affects the variance profile, that is, the variance of $\hat{y}$ regarded as a function of the distance $\rho = \sqrt{\Sigma x_i^2}$ from the center of the design. The exact number of center points is not critical. The numbers given in the table are chosen so that the variance profile will be reasonably uniform over the region of the experimental design and so that an even number of center points appear in each block. The variance profiles resulting from the designs here considered are shown in Figure 1 of Appendix A.

6.0. Analysis for the Designs

In Tables 5a, 5b, and 5c, formulae and constants are given which are needed for the analysis of the designs of Table 4. The notation is explained below.

6.1. *Calculation of the estimates.*

In order to calculate the estimates b_0, b_i, b_{ij}, b_{ii}, it is first necessary to write out the levels for each of the variables in the design and then to add further columns corresponding to x_1^2, x_2^2, $\cdots$, x_k^2, x_1x_2, x_1x_3, $\cdots$, $x_{k-1}x_k$. This is done in Table 6 for design No. 2 where a set of typical data is also shown

TABLE 5a

Estimates of the regression coefficients and their variances.

$$b_0 = \bar{y}_0$$
$$b_i = A\{iy\}$$
$$b_{ii} = B\{iiy\} + C_1 \sum_{j}^{n_1} \{jjy\} + C_2 \sum_{l}^{n_2} \{lly\} - (\bar{y}_0/s)$$

where $\sum_{j}^{n_1}$ and $\sum_{l}^{n_2}$ refer to summation over first and second associates of i.

$$b_{ij} = D_1\{ijy\} \qquad i, j \text{ first associates.}$$
$$b_{ij} = D_2\{ijy\} \qquad i, j, \text{ second associates.}$$
$$V(b_0) = \frac{1}{n_0}\sigma^2$$
$$V(b_i) = A\sigma^2$$
$$V(b_{ii}) = [B + 1/s^2n_0]\sigma^2$$
$$V(b_{ij}) = D_1\sigma^2 \qquad i, j \text{ first associates.}$$
$$= D_2\sigma^2 \qquad i, j \text{ second associates.}$$
$$\text{Cov}\,(b_0b_{ii}) = -\frac{1}{s^2n_0}\sigma^2$$
$$\text{Cov}\,(b_{ii}b_{jj}) = \left[C_1 + \frac{1}{s^2n_0}\right]\sigma^2, \qquad i, j \text{ first associates.}$$
$$= \left[C_2 + \frac{1}{s^2n_0}\right]\sigma^2, \qquad i, j \text{ second associates.}$$

NOTE: For BIB designs, all i, j are considered first associates and $C_2 = D_2 = 0$. The constants A, B, etc. for the various designs are given in Table 5c.

for illustration. The sum of products of the entries in the columns with the observations y are next calculated. In addition $\bar{y}_0$ the average value of the observations made at the center points is shown. The calculated quantities are next substituted in the formulae given in Table 5a to provide the required estimates using the constants of Table 5c.

The following notation is employed:

$$\{iy\} = \sum_{u=1}^{N} x_{iu}y_u\,, \qquad \{iiy\} = \sum_{u=1}^{N} x_{iu}^2y_u\,, \qquad \{ijy\} = \sum_{u=1}^{N} x_{iu}x_{ju}y_u\,.$$

The grand total can be regarded as the sum of products between y and a dummy variable x_0 which always takes the value 1 so that

$$\{0y\} = \sum_{u=1}^{N} y_u\,.$$

TABLE 5b

Formulae for the analysis of variance.

Correction due to the mean:	$\{0y\}^2/N$
Sum of squares due to linear terms:	$A \sum_{i=1}^{k} \{iy\}^2$
Sum of squares due to second degree terms:	
(a) Due to interaction terms:	$D_1 \sum_{i<j}^{n_1} \{ijy\}^2 + D_2 \sum_{i<j}^{n_2} \{ijy\}^2$
(b) Due to quadratic terms:	$b_0\{0y\} + \sum_{i=1}^{k} b_{ii}\{iiy\} - \{0y\}^2/N$
Total sum of squares after correction for the mean:	$\sum_{u=1}^{N} y_u^2 - \{0y\}^2/N$

In the present example the sums of products are:

$$\bar{y}_0 = 90.6; \quad \{0y\} = 2319.4;$$

$$\{1y\} = 23.2; \quad \{2y\} = -23.5; \quad \{3y\} = 13.6; \quad \{4y\} = -44.1;$$

$$\{11y\} = 1033.6; \quad \{22y\} = 1010.3; \quad \{33y\} = 1027.0; \quad \{44y\} = 1024.3;$$

$$\{12y\} = -6.7; \quad \{13y\} = -15.3; \quad \{14y\} = 3.8; \quad \{23y\} = -6.7;$$

$$\{24y\} = -10.5; \quad \{34y\} = -17.0;$$

TABLE 5c

Constants for the designs of Table 4.

Design	A	B	C_1	C_2	D_1	D_2	s	Center Points n_0	Redundancy Factor	Non-Sphericity Index I
1	1/8	1/4	−1/16	0	1/4	0	2	3	1.2	0.38
2	1/12	1/8	−1/48	0	1/4	0	2	3	1.6	0
3	1/16	1/12	−1/96	0	1/4	0	2	6	1.9	0.17
4	1/24	17/216	−10/216	−1/216	1/16	1/8	3	6	1.7	0.23
5	1/24	1/16	−1/144	0	1/8	0	3	6	1.6	0
6	1/40	1/30	−1/120	−1/720	1/16	1/8	3	10	2.2	0.25
7	1/64	17/512	1/512	−7/512	1/16	1/32	4	10	2.4	0.09
8	1/80	1/48	−1/600	0	1/32	0	5	12	2.3	0.06
9	1/64	23/1024	−9/1024	−1/1024	1/32	1/16	4	12	2.1	0.16
10	1/96	41/3072	−7/3072	−1/3072	1/32	1/16	4	12	2.5	0.18

TABLE 6

Sample calculation for the four-factor design (No. 2).

x_1	x_2	x_3	x_4	x_1^2	x_2^2	x_3^2	x_4^2	x_1x_2	x_1x_3	x_1x_4	x_2x_3	x_2x_4	x_3x_4	y
−1	−1	0	0	1	1	0	0	1	0	0	0	0	0	84.7
1	−1	0	0	1	1	0	0	−1	0	0	0	0	0	93.3
−1	1	0	0	1	1	0	0	−1	0	0	0	0	0	84.2
1	1	0	0	1	1	0	0	1	0	0	0	0	0	86.1
0	0	−1	−1	0	0	1	1	0	0	0	0	0	1	85.7
0	0	1	−1	0	0	1	1	0	0	0	0	0	−1	96.4
0	0	−1	1	0	0	1	1	0	0	0	0	0	−1	88.1
0	0	1	1	0	0	1	1	0	0	0	0	0	1	81.8
0	0	0	0	0	0	0	0	0	0	0	0	0	0	93.8
−1	0	0	−1	1	0	0	1	0	0	1	0	0	0	89.4
1	0	0	−1	1	0	0	1	0	0	−1	0	0	0	88.7
−1	0	0	1	1	0	0	1	0	0	−1	0	0	0	77.8
1	0	0	1	1	0	0	1	0	0	1	0	0	0	80.9
0	−1	−1	0	0	1	1	0	0	0	0	1	0	0	80.9
0	1	−1	0	0	1	1	0	0	0	0	−1	0	0	79.8
0	−1	1	0	0	1	1	0	0	0	0	−1	0	0	86.8
0	1	1	0	0	1	1	0	0	0	0	1	0	0	79.0
0	0	0	0	0	0	0	0	0	0	0	0	0	0	87.3
0	−1	0	−1	0	1	0	1	0	0	0	0	1	0	86.1
0	1	0	−1	0	1	0	1	0	0	0	0	−1	0	87.9
0	−1	0	1	0	1	0	1	0	0	0	0	−1	0	85.1
0	1	0	1	0	1	0	1	0	0	0	0	1	0	76.4
−1	0	−1	0	1	0	1	0	0	1	0	0	0	0	79.7
1	0	−1	0	1	0	1	0	0	−1	0	0	0	0	92.5
−1	0	1	0	1	0	1	0	0	−1	0	0	0	0	89.4
1	0	1	0	1	0	1	0	0	1	0	0	0	0	86.9
0	0	0	0	0	0	0	0	0	0	0	0	0	0	90.7

and from Table 5c for design No. 2 we have $A = 1/12$, $B = 1/8$, $C_1 = -1/48$, $D_1 = 1/4$, $s = 2$, $n_0 = 3$ whence, using the formulae of Table 5a,

$$b_0 = 90.6 \quad b_1 = 1.93 \quad b_{11} = -1.42 \quad b_{12} = -1.68$$

$$b_2 = -1.96 \quad b_{22} = -4.33 \quad b_{13} = -3.83$$

$$b_3 = 1.13 \quad b_{33} = -2.24 \quad b_{14} = 0.95$$

$$b_4 = -3.68 \quad b_{44} = -2.58 \quad b_{23} = -1.68$$

$$b_{24} = -2.63$$

$$b_{34} = -4.25.$$

For example,

$$b_1 = \frac{1}{12}(23.2) = 1.930$$

$$b_{11} = \frac{1}{8}(1033.6) - \frac{1}{48}(4095.2) - \frac{90.6}{2} = -1.416$$

$$b_{12} = \frac{1}{4}(-6.7) = -1.675$$

6.2. *The Analysis of Variance.*

The analysis of the variance table is readily calculated using the relations of Table 5b as follows.

Analysis of Variance Table

	s.s.	d.f.	m.s.
Due to linear terms	268.36	4	67.09
Due to second order terms	294.92	10	29.49
Residual	126.71	12	10.56
Total after eliminating the mean	689.99	26	

The observations recorded at the center point were 93.8, 87.3, and 90.7. Had there been no blocking of the design (that is if the runs had been made entirely in random order) these observations at the center point would have provided two degrees of freedom for estimating the error variance. Their sum of squares for deviations from their mean would have been 21.16 and the residual sum of squares could have been split into two parts, as follows

		s.s.	d.f.	m.s.
Residual	Replicated center points	21.16	2	10.58
	Remainder	105.57	10	10.56
		126.71	12	

to provide a basis for a possible test of goodness of fit for the model.

In this particular example, since the error sum of squares would have only two degrees of freedom, such a test would of course be very insensitive and provide no more than an indication that the remainder sum of squares was or was not of the right order of magnitude. Our main object here is to illustrate general principles.

6.3. *Elimination of Block Effects.*

The design illustrated was actually carried out in three blocks of nine observations. Since the blocking is orthogonal the elimination of blocks will only affect the residual sum of squares. The block means $\bar{y}_1$, $\bar{y}_2$ and $\bar{y}_3$ are respectively 749.1/9, 750.6/9, 774.7/9 and the sum of squares associated with blocks is

$$\frac{(794.1)^2 + (750.6)^2 + (774.7)^2}{9} - \frac{(2319.4)^2}{27} = 105.53.$$

We cannot now isolate the two degrees of freedom for the differences among the center points and the analysis of variance is as follows.

		s.s.	d.f.	m.s.
Due to linear terms		268.36	4	67.09
Due to second order terms		294.92	10	29.492
Residual	126.71 {	21.18	10	2.118
Blocks		105.53	2	52.765
Total after elimination of mean		689.99		

It is seen that in this example a large proportion of the residual variance is accounted for by the blocks. On the assumption that our model is adequate, the mean square of 2.118 provides an estimate $\hat{\sigma}^2$ of σ^2. This estimate will therefore be employed in calculating the standard errors of the variance coefficients. If extra runs at the center point could be made then an equal number of these should be allocated to each block. The pooled variances for replications at the center point within each block would then provide an estimate of error appropriate for testing the adequacy of the model.

6.4. *Variances, Covariances and Standard Errors.*

The variances and covariances of the various estimates are obtained from the formulae in Table 5a with an appropriate estimate $\hat{\sigma}^2$ of the experimental error variance replacing σ^2 in those formulae. In the present example we employ the estimate $\hat{\sigma}^2 = 2.118$. Taking square roots of the estimated variances we obtain the following values for the standard errors of the estimates:

$$\text{S.E.}(b_0) = \sqrt{\frac{2.118}{3}} = .84;$$

$$\text{S.E.}(b_i) = \sqrt{\frac{2.118}{12}} = .42;$$

$$\text{S.E.}(b_{ii}) = \sqrt{2.118 \cdot \frac{5}{24}} = .66;$$

$$\text{S.E.}(b_{ij}) = \sqrt{\frac{2.118}{4}} = .73.$$

6.5. *General Comments on the Analysis.*

The simple type of analysis illustrated above is appropriate for designs 1, 2, 3, 5, and 8. The analysis of designs 4, 6, 7, 9, and 10 is slightly more complicated. Estimates of b_0, the constant term, and the linear terms b_i are obtained exactly as before. The multiplier D for calculating the interaction effects however takes two values for these designs. The multiplier D_1 is appropriate for these combinations of variables listed as first associates in Table 4 and D_2 for those combinations listed as second associates. In Table 4 combinations belonging to only one of the associate classes are listed. All others belong to the other associate class. For example, in design No. 4 the interactions 1 4; 2 5; and 3 6 are between first associates and take the multiplier D_1 . For design No. 7 however it is more economical in space to list the second associates which take the multiplier D_2 . In calculating the estimate of b_{ii} (Table 5a), C_1 is the multiplier of $\sum_j^{n_1} \{jjy\}$ in which the j's are first associates of i while C_2 is the multiplier of $\sum_l^{n_2} \{lly\}$ in which the l's are second associates of i.

Appendix A

Derivation of the Class of Three Level Designs

The requirements which need to be satisfied in order that a design shall be second order rotatable are given elsewhere [5]. It is desirable [5] when possible to satisfy the additional condition that biases due to neglected third order terms are zero. The conditions which the design points must then satisfy are as follows:

$$(1) \quad \begin{cases} \sum_{u=1}^{N} x_{iu}^2 = \sum_{u=1}^{N} x_{ju}^2 > 0 & \text{all } i, j \\ \sum_{u=1}^{N} x_{iu}^4 = \sum_{u=1}^{N} x_{ju}^4 > 0 & \text{all } i, j \end{cases}$$

$$(2) \quad \sum_{u=1}^{N} x_{iu} = \sum_{u=1}^{N} x_{iu}^3 = \sum_{u=1}^{N} x_{iu}^5 = 0 \qquad \text{all } i$$

$$(3) \quad \sum_{u=1}^{N} x_{iu}^2 x_{ju}^2 = \sum_{u=1}^{N} x_{ku}^2 x_{lu}^2 > 0 \qquad i \neq j,\ k \neq l$$

$$(4) \quad 3 \sum_{u=1}^{N} x_{iu}^2 x_{ju}^2 = \sum_{u=1}^{N} x_{iu}^4 \qquad i \neq j$$

$$(5) \quad \sum_{u=1}^{N} x_{iu}^2 x_{ju} = \sum_{u=1}^{N} x_{iu}^4 x_{ju} = \sum_{u=1}^{N} x_{iu}^3 x_{ju}^2 = 0 \qquad i \neq j$$

$$(6) \quad \sum_{u=1}^{N} x_{iu} x_{ju} = \sum_{u=1}^{N} x_{iu}^3 x_{ju} = 0 \qquad i \neq j$$

$$(7) \quad \sum_{u=1}^{N} x_{iu}^2 x_{ju}^2 x_{ku} = 0 \qquad i \neq j \neq k$$

$$(8) \quad \sum_{u=1}^{N} x_{iu}^2 x_{ju} x_{ku} = 0 \qquad i \neq j \neq k$$

$$(9) \quad \sum_{u=1}^{N} x_{iu} x_{ju} x_{ku} = \sum_{u=1}^{N} x_{iu}^3 x_{ju} x_{ku} = 0 \qquad i \neq j \neq k$$

$$(10) \quad \sum_{u=1}^{N} x_{iu} x_{ju} x_{ku} x_{lu} = 0 \qquad i \neq j \neq k \neq l$$

$$(11) \quad \sum_{u=1}^{N} x_{iu}^2 x_{ju} x_{ku} x_{lu} = 0 \qquad i \neq j \neq k \neq l$$

$$(12) \quad \sum_{u=1}^{N} x_{iu} x_{ju} x_{ku} x_{lu} x_{mu} = 0 \qquad i \neq j \neq k \neq l \neq m$$

Bearing in mind that for our present purpose each x can take only the values -1, 0 or 1, we consider what is implied, first for single columns of the design, then for pairs of columns and so on. In what follows a coincidence means the occurrence of 1's (plus or minus) in the same row of the design matrix. In general where we refer to the occurrence of a "1" we mean a $+1$ or a -1. The equation numbers refer to the appropriate relations above.

(a) Single columns. The same number of 1's occur in each column. Half of these are $+1$ and half -1 (Equations 1 and 2).

(b) Two columns. The number of coincident 1's is greater than zero and the same for all sets of two columns. For these coincident 1's

$$\sum x_{iu} = 0 \quad \text{and} \quad \sum x_{iu}x_{ju} = 0$$

where, here and subsequently, the summation is taken over the relevant coincidences (Equations 3, 5 and 6).

(c) Three columns. For the coincident 1's occurring in any three columns

$$\sum x_{iu} = 0; \qquad \sum x_{iu}x_{ju} = 0; \qquad \sum x_{iu}x_{ju}x_{ku} = 0.$$

(Equations 7, 8 and 9)

(d) Four columns. For the coincident 1's occurring in any four columns

$$\sum x_{iu}x_{ju}x_{ku} = 0; \qquad \sum x_{iu}x_{ju}x_{ku}x_{lu} = 0.$$

(Equations 10 and 11)

(e) Five columns. For the coincident 1's occurring in any five columns

$$\sum x_{iu}x_{ju}x_{ku}x_{lu}x_{mu} = 0.$$

(Equation 12)

Considering the possible designs we see from (b) that we cannot use any arrangement for which no coincidences occur. It is on the other hand possible, in principle, to generate designs in which 1's are coincident only in pairs of columns. In this case requirements (c), (d), and (e) are automatically satisfied. To satisfy requirement (b) consider the coincidence of 1's in the ith and jth column. For these ones we require $\sum x_{iu} = 0$; $\sum x_{ju} = 0$ and $\sum x_{iu}x_{ju} = 0$. The fewest number of coincidences for which this can be satisfied is four. The actual values of the coincident 1's must then be some permutation of the rows of the 2^2 arrangement:

$$\begin{bmatrix} -1 & -1 \\ 1 & -1 \\ -1 & 1 \\ 1 & 1 \end{bmatrix}$$

We now need to include these component arrangements so that Equation (4) is also satisfied. This requires that the number of coincidences in each pair of columns is one third the number of 1's occurring in each column. The combinatorial properties required of the coincidences are seen to be exactly those of a balanced incomplete block design with $r = 3\mu$ (where, in the incomplete block design, r is the number of times each treatment is replicated and μ is the number of times each pair of treatments appear together in the same block). Precisely this method of construction is employed in design No. 2.

Designs may also be obtained in which 1's are coincident only in sets of three columns. Requirements (d) and (e) are automatically satisfied and requirement (c) can be met by arranging that the actual values of the coincident 1's form the elements of a 2^3 factorial. By arranging once more that the coincidences follow those of a balanced incomplete block design with $r = 3\mu$ all conditions are satisfied. Design No. 5 is an example of this type of arrangement. As has been shown [14], exactly similar arguments may be employed for designs with higher numbers of coincidences. Where coincidences of more than five columns are involved we could satisfy all the requirements with fractional factorials instead of full factorials for the basic units provided that the generators of the fractional factorials contain not less than six elements.

Among the designs listed in Table 4, the above method of generation accounts for arrangement No. 2 for four variables in twenty-four runs and arrangement No. 5 for seven variables in fifty-six runs. Other arrangements of this kind are available, but only those giving low redundancy factors are listed here. Balanced incomplete block designs for which $r = 3\mu$ and for which the redundancy factors are satisfactory are unfortunately not available for all k. To obtain useful designs for other values of k some relaxation in our requirements must be made. A natural modification is to employ balanced incomplete block designs for which $r \neq 3\mu$. It is easily seen that for such designs all the equations (1) through (12), excepting (4), will be satisfied. Instead the design will satisfy

$$\frac{r}{\mu} \sum_{u=1}^{N} x_{iu}^2 x_{ju}^2 = \sum_{u=1}^{N} x_{iu}^4 .$$

The ratio r/μ may be chosen to be as close to 3 as possible. Designs of this class in Table 4 are No. 1 ($k = 3$, $r/\mu = 2$), No. 3 ($k = 5$, $r/\mu = 4$), No. 8 ($k = 11$, $r/\mu = 2.5$). The resulting designs are not quite rotatable but, as has been pointed out already, the property of rotatability is desirable rather than critical and for the designs discussed the variance of $\hat{y}$ at points equidistant from the origin changes little. This is shown quantitatively in the last column of Table 5c which shows the non-sphericity factor "I" for the designs considered [14]. This non-sphericity factor measures the range of variance of $\hat{y}$ divided by its midrange on the unit sphere

$$\sum_{i=1}^{k} x_i^2 = 1.$$

For rotatable designs the factor is zero.

A further relaxation of the same kind is to allow the use of partially balanced incomplete block designs. Again all the conditions will be satisfied except those of equations (3) and (4). Instead of this relationship, we will have for these designs

$$\frac{r}{\mu_1} \sum_{u=1}^{N} x_{iu}^2 x_{ju}^2 = \sum_{u=1}^{N} x_{iu}^4 , \qquad \text{for } i, j \text{ first associates}$$

$$\frac{r}{\mu_2} \sum_{u=1}^{N} x_{iu}^2 x_{ju}^2 = \sum_{u=1}^{N} x_{iu}^4 , \qquad \text{for } i, j \text{ second associates}$$

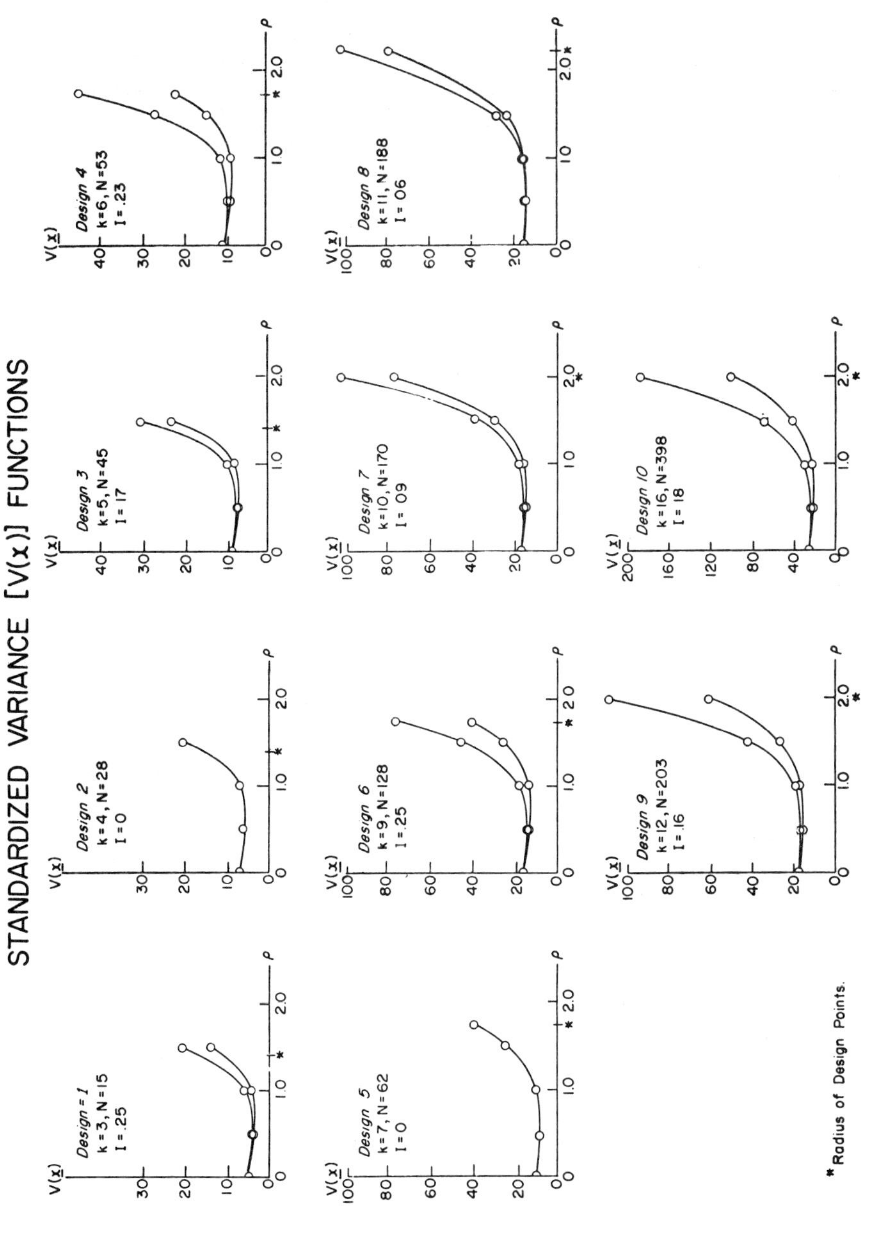
STANDARDIZED VARIANCE [V(x)] FUNCTIONS
Design = 1
k = 3, N = 15
I = .25
Design 2
k = 4, N = 28
I = 0
Design 3
k = 5, N = 45
I = 17
Design 4
k = 6, N = 53
I = .23
Design 5
k = 7, N = 62
I = 0
Design 6
k = 9, N = 128
I = .25
Design 7
k = 10, N = 170
I = 09
Design 8
k = 11, N = 188
I = 06
Design 9
k = 12, N = 203
I = .16
Design 10
k = 16, N = 398
I = 18
* Radius of Design Points.

where μ_1 is the number of times first associate treatments appear together in the same block and μ_2 is the corresponding parameter for second associates. Once more these designs are nearly rotatable and have low redundancy factors. The values of I and R for these designs also are shown in Table 5c. Characteristic of this classification of designs is that the variances of interaction coefficients (b_{ij}) are different depending upon whether i and j are first or second associates. In practice, as can be determined from the formulae and constants in Table 5a these differences in variance are not serious and the resulting designs are perfectly satisfactory. In Table 4 designs No. 4 (for $k = 6$), No. 6 (for $k = 9$), No. 7 (for $k = 10$), No. 9 (for $k = 12$) and No. 10 (for $k = 16$) are of this type.

Equation (12) of the moment conditions for three-level designs arises from the requirement that biases due to third order terms be made zero. The relaxation of this condition would preserve all the properties of the design except that if, contrary to assumption, three-factor interaction coefficients were not zero, these would cause the two-factor interaction coefficients to be biased. Condition (12) is relaxed in design No. 8 in which a half-replicate of the basis 2^5 design is employed.

Figure 1 gives the variance profiles for the designs of Table 4. These graphs show the standardized variance function

$$V(\mathbf{x}) = \frac{N}{\sigma^2} V(\hat{y})$$

plotted as a function of $\rho = (\sum_i x_i^2)^{\frac{1}{2}}$ the distance from the center of the design. A number of center points have been added to make the variance at $\rho = 0$ equal to the midrange variance at $\rho = 1$ and is close to the number recommended in Table 4. The small adjustment to n_0 required to distribute the center points equally among blocks has a negligible effect on these graphs. For non-rotatable designs the two curves indicate the maximum and minimum variance obtained [14] on a sphere of radius ρ. They thus represent the envelope of all possible variance functions that might be obtained by proceeding from the origin out along any arbitrary radius.

Our object here is merely to present a set of designs whose properties are sufficiently desirable to justify immediate application, it is by no means implied that the designs we have listed are exhaustive. In particular, as will be reported elsewhere, the method of generation here used can provide designs in which the number of ones occurring in each row is not constant. Even within the particular class of designs which we have considered (in which the number of ones in each row is constant), the designs presented are far from exhaustive. A wider but by no means complete selection of such designs is given in [14].

Appendix B

In Section 3.0 the three-level 24-point arrangement is described which forms the basis, with added center points, for a second order rotatable design. As is mentioned in the text, this design is in fact a rotation of the four-variable central composite rotatable arrangement. This may be readily confirmed in the following way.

Upon post-multiplying the matrix (excluding center points) for design No. 2

given in Table 3 by the orthogonal matrix

$$T = \frac{1}{\sqrt{2}} \begin{bmatrix} 1 & 1 & 0 & 0 \\ 1 & -1 & 0 & 0 \\ 0 & 0 & 1 & 1 \\ 0 & 0 & 1 & -1 \end{bmatrix}$$

we obtain, except for the scale factor $1/\sqrt{2}$, the design matrix of the rotatable central composite arrangement [5] which may in an obvious shorthand notation be denoted by

$$\begin{bmatrix} \pm 1 & \pm 1 & \pm 1 & \pm 1 \\ \pm 2 & 0 & 0 & 0 \\ 0 & \pm 2 & 0 & 0 \\ 0 & 0 & \pm 2 & 0 \\ 0 & 0 & 0 & \pm 2 \end{bmatrix}.$$

References

[1] Finney, D. J. 1945. The fractional replication of experiments. *Ann. Eugenics* 12:291–301.
[2] Kishen, K. 1948. On fractional replication of the general symmetrical factorial design. *J. Ind. Soc. Ag. Stat.* 1:91–106.
[3] Plackett, R. L. and Burman, J. P. 1946. The design of optimum multifactor experiments. *Biometrika* 33:305–325.
[4] Box, G. E. P. 1952. Multifactor designs of first order. *Biometrika* 39:49–57.
[5] Box, G. E. P. and Hunter, J. S. 1957. Multi-factor experimental designs for exploring response surfaces. *Ann. Math. Stat.* 28:195–241.
[6] Box, G. E. P. and Wilson, K. B. 1951. On the experimental attainment of optimum conditions. *Jour. Roy. Stat. Soc. B* 13:1–45.
[7] Box, G. E. P. and Draper, N. R. 1959. A basis for the selection of a response surface design. *Jour. Amer. Stat. Assoc.* 54:622–654.
[8] Box, G. E. P. and Behnken, D. W. 1958. Derivation of second order rotatable designs from those of first order. Stat. Techniques Research Group Technical Report No. 17. Princeton University. (Submitted for publication to the *Annals of Mathematical Statistics.*)
[9] DeBaun, R. M. 1956. An experimental design for three factors at three levels. *Nature* 181:209–210.
[10] DeBaun, R. M. 1959. Response surface designs for three factors at three levels. *Technometrics* 1:1–8.
[11] Gardiner, D. A., Grandage, A. H. E. and Hader, R. J. 1956. Some third order rotatable designs. Institute of Statistics Mimeo Series No. 149, Raleigh, North Carolina.
[12] Draper, N. R. 1960. Second order rotatable designs in four or more dimensions. *Ann. Math. Stat.* 31:23–33.
[13] Bose, R. C. and Draper, N. R. 1959. Second order rotatable designs in three dimensions. *Ann. Math. Stat.* 30:1097–1112.
[14] Box, G. E. P. and Behnken, D. W. 1958. A class of three level second order designs for surface fitting. Stat. Techniques Research Group Technical Report No. 26. Princeton University.

2.10

The 2^{k-p} Fractional Factorial Designs,* Part I

G. E. P. BOX and J. S. HUNTER
University of Wisconsin

"Cats is dogs and dogs is dogs and rabbits is dogs, and squirrels in cages is parrots . . . ."

1: The Two-Version Factorials and Fractionals

A full 2^k factorial design requires all combinations of two versions of each of k variables. If a variable is continuous, the two versions become the high and low level of that variable. If a variable is qualitative the two versions correspond to two types, sometimes the presence and absence of the variable.

The runs comprising the experimental design are conveniently set out in either of two notations as illustrated for the eight runs comprising a 2^3 factorial in Table 1.

TABLE 1
Alternative Notations for the 2^3 Factorial Design

Run Number	*Notation 1* Variables *A B C*	*Notation 2* Variables **1 2 3**
1	1	− − −
2	*a*	+ − −
3	*b*	− + −
4	*ab*	+ + −
5	*c*	− − +
6	*ac*	+ − +
7	*bc*	− + +
8	*abc*	+ + +

In the first notation the variables are identified by capital letters, and their two versions by the presence or absence of the corresponding lower case letter. When all the variables are at their "low" level or version a "1" is used. In the second notation the variables are identified by numbers and the two versions of each variable by either a minus and plus sign, or by minus and plus one. The experimental design can then be viewed geometrically. A run is represented by a point whose coordinates are the ± 1 versions for that run. For example,

* Sponsored by the United States Army under Contract No. DA-11-022-ORD-2059.

The Collected Works of George E. P. Box, 1984, Wadsworth, Inc., Belmont, CA 94002.
Originally published in *Technometrics,* vol. 3, no. 3 (1961), pp. 311–351.

the 2^3 factorial will provide the eight vertices of a cube in a three-dimensional coordinate system. The notation using minus and plus signs is used in this paper. The list of experimental runs is called the *design matrix* and is denoted by **D**. For a 2^k factorial, the design matrix contains k columns and $N = 2^k$ rows. There is a column for each of the k variables, and each row gives the combination of versions for each run.

In Table 1 the runs are listed in *standard order*. The elements of the first column are alternate minus and plus signs. The elements of the second column are alternate pairs of minus and plus signs, the elements of the third column alternate groups of four minus and plus signs and so on. The last column consists of 2^{k-1} minus signs followed by 2^{k-1} plus signs.

The Estimates

On the assumptions that the observations are uncorrelated and have equal variance, then the 2^k factorial designs provide independent minimum variance estimates of the grand average and of the 2^{k-1} effects:

$$\begin{array}{ll} k & \text{main effects,} \\ \dfrac{k(k-1)}{2} & \text{two-factor interaction effects,} \\ \dfrac{k(k-1)(k-2)}{2 \cdot 3} & \text{three-factor interaction effects,} \\ \vdots & \\ \dfrac{k(k-1)(k-2)\cdots(k-h-1)}{h!} & h\text{-factor interaction effects,} \end{array} \tag{1}$$

and finally a single k-factor interaction effect.

Although the Yates' Algorithm (1) provides a quick method for calculating these estimates, a longer, but more basic calculation technique will now be described. In Table 2, where for convenience a 2^3 design is used, a *matrix of independent variables* **X** is generated from the design matrix **D**. In general the individual elements for an ij interaction column in **X** are obtained by multi-

TABLE 2

Design Matrix D			Matrix of Independent Variables X								Observations Y
1	2	3	I	1	2	3	1 2	1 3	2 3	1 2 3	
−	−	−	+	−	−	−	+	+	+	−	2
+	−	−	+	+	−	−	−	−	+	+	10
−	+	−	+	−	+	−	−	+	−	+	8
+	+	−	+	+	+	−	+	−	−	−	12
−	−	+	+	−	−	+	+	−	−	+	6
+	−	+	+	+	−	+	−	+	−	−	8
−	+	+	+	−	+	+	−	−	+	−	6
+	+	+	+	+	+	+	+	+	+	+	4

plying the corresponding elements of the separate i and j columns. Similarly, the elements of the ijk interaction column are given by the product of the elements of the columns labelled ij and k and so on. The first column of **X** consists entirely of plus signs and is used to provide an estimate of the mean. For a 2^k design the full matrix of independent variables **X** contains 2^k columns as well as 2^k rows. The elements of the column **Y** in Table 2 are the observations recorded at each of the 2^k experiments. The estimate of the effect **ij** $\cdots$ **k** is obtained by taking the sum of products between the elements of **Y** and the corresponding elements of the column $ij \cdots k$ and dividing this product by $N/2$ where $N = 2^k$, e.g.,

$$\mathbf{ij \cdots k}\text{ effect} = \frac{2}{N} \sum y\{i.j. \cdots . k\} \tag{2}$$

where $\{i.j. \cdots . k\}$ stands for the elements of the $ij \cdots k$ column and the summation is taken over all N products. Thus, using the data from Table 2 the **1 3** interaction effect is

$$\mathbf{1\ 3} = \tfrac{1}{4}(2 - 10 + 8 - 12 - 6 + 8 - 6 + 4) = -3.0.$$

where here and henceforth numerals appearing in bold type are used to identify the main effects and interactions. Solving for all the effects gives:

Main Effects	Two-factor Interactions	Three-factor Interaction
1 = 3.0	**1 2** = −2.0	**1 2 3** = zero
2 = 1.0	**1 3** = −3.0	
3 = −2.0	**2 3** = −3.0	

Each estimated effect has variance

$$\text{Variance (effect)} = 4\sigma^2/N \tag{3}$$

where σ^2 is the variance of the individual observations.

The average is obtained by taking the sum of products of the column **I** with the observation column **Y** and dividing the result by N, thus

$$\text{average} = \bar{y} = \sum y\{\mathbf{I}\}/N. \tag{4}$$

Thus $\bar{y} = 56/8 = 7.0$ with variance σ^2/N. By this process 2^k estimates can be obtained from 2^k runs and when k is large the wealth of such estimates becomes almost an embarrassment. However, in many practical situations, the higher order interaction effects can often be hopefully supposed to be negligible in size. For example, with continuous variables it is reasonable to expect the response to vary *smoothly.* When factorial designs are correctly used to study qualitative variables it is because certain aspects of *similarity* are expected in the responses at the different versions. Thus, two solvents and two differently shaped particles may with profit be studied in a factorial design when at least some aspect of similarity in behavior of these variables might be expected.

In the conditions of smoothness and similarity commonly encountered, the three-factor and multi-factor interaction effects are often negligible. When this is the case, fractional designs using a smaller number of runs may be employed for although in these fractional designs the effects of the major interest are confused with higher order effects, nevertheless, the latter are small enough to be ignored. In some situations the total number of variables k is large, but only a few (say $p = 2$ or 3) are expected to have any effect. In this situation designs which are fractional in the k variables may be chosen which have the property that they are complete factorials in any sub-group of p variables.

For illustration, we first discuss the one half fraction of the 2^4 design.

One-half Fraction of the 2^4 Factorial

Since the design is to contain $2^{4-1} = 8$ runs a 2^3 factorial design is first written down. The $-$ and $+$ elements associated with the **1 2 3** interaction column then are used to identify the $-$ and $+$ versions of variable **4**. The resulting eight combinations shown in Table 3 give a particular half replicate or "fractional" of the complete 2^4 design. A $(\frac{1}{2})^p$ fraction of a 2^k factorial design is called a 2^{k-p} *fractional*, or more exactly, a 2^{k-p} fractional factorial. The present design is therefore a 2^{4-1} fractional.

TABLE 3

Constructing the 2^{4-1} Fractional Factorial Design

Design Matrix				Observations
1	2	3	1 2 3 = 4	Y
$-$	$-$	$-$	$-$	8.7
$+$	$-$	$-$	$+$	15.1
$-$	$+$	$-$	$+$	9.7
$+$	$+$	$-$	$-$	11.3
$-$	$-$	$+$	$+$	14.7
$+$	$-$	$+$	$-$	22.3
$-$	$+$	$+$	$-$	16.1
$+$	$+$	$+$	$+$	22.1

With a full 2^4 design, sixteen effects can be estimated: the grand average, four main effects, six two-factor interactions, four three-factor interactions and a single four factor interaction. With only eight observations it is clearly impossible to obtain sixteen independent estimates. We note that the combination of observations used to estimate the main effect **4** is identical to that used to estimate the three-factor interaction effect **1 2 3**. The estimates of **4** and **1 2 3** are said to be *confounded*. The "**4**" effect really estimates the *sum* of the effects of **4** and **1 2 3**.

Study of Table 3 will show that other estimates such as **1 2** and **3 4** are also confounded. It is desirable to have a general method which enables one to determine which effects are confounded. This is accomplished for this design by inducing the equality **4** = **1 2 3** where the multiplication product **1 2 3** refers to the multiplication of the individual elements in the corresponding columns **1**, **2** and **3**. Now it is obvious that by multiplying the elements in any column by a column of identical elements, a column of plus signs will result. Since a column of plus signs corresponds to **I** we have $\mathbf{1} \times \mathbf{1} = \mathbf{1}^2 = \mathbf{I}$ and similarly that $\mathbf{2}^2 = \mathbf{I}$, $\mathbf{3}^2 = \mathbf{I}$

and $\mathbf{4}^2 = \mathbf{I}$. This identity supplies the key to the remaining relationships. On multiplying both sides of the equation $\mathbf{4} = \mathbf{1\ 2\ 3}$ by $\mathbf{4}$ we get

$$\mathbf{4}^2 = \mathbf{1\ 2\ 3\ 4} \qquad \text{that is} \qquad \mathbf{I} = \mathbf{1\ 2\ 3\ 4}. \tag{5}$$

This identity is readily confirmed for if the elements in columns **1**, **2**, **3** and **4** are multiplied together we obtain a column of plus signs, that is **I**.

The interaction **1 2 3 4** associated with **I** is said to be a *generator* of the design. In this particular instance there is only one generator so this provides the *defining relation* $\mathbf{I} = \mathbf{1\ 2\ 3\ 4}$ which is the key to all the relationships which exist between the effects.

Aliases and Linear Combinations of Effects

Suppose we wish to know which effect is confounded with the main effect **3**. Multiplying both sides of the defining relation by **3** gives $\mathbf{3} = \mathbf{1\ 2\ 3}^2\ \mathbf{4} = \mathbf{1\ 2\ I\ 4} = \mathbf{1\ 2\ 4}$ since multiplication by **I** (a column of plus signs) leaves the elements in any column unchanged. Thus, the main effect **3** is confounded with the three-factor interaction **1 2 4**. Similarly, we find that the two-factor interaction **3 4** is confounded with **1 2** and so on. The quantities so associated are called *aliases*. If we now proceed to estimate the main effect **3** we will in fact obtain the *sum* of the estimates of the main effect **3** and the three-factor interaction **1 2 4**. The estimate of **3** is really an estimate of the combination of the effects $\mathbf{3} + \mathbf{1\ 2\ 4}$. Eight linear combinations of effects $\ell_I, \ell_1, \ell_2, \cdots$ are available. Thus $\ell_1, = \frac{1}{4} \sum y\{\mathbf{1}\}$ or equally $\ell_1 = \frac{1}{4} \sum y\{\mathbf{2\ 3\ 4}\}$. Similarly $\ell_{12} = \frac{1}{4} \sum y\{\mathbf{1\ 2}\}$ or equally $\ell_{12} = \frac{1}{4} \sum y\{\mathbf{3\ 4}\}$. Using the defining relation we find that these linear combinations estimate the quantities given in Table 4, the subscript on the ℓ's identifying the first effect in the linear combination.

TABLE 4

ℓ_I = **average + 1 2 3 4**	ℓ_4 = **4 + 1 2 3**
ℓ_1 = **1 + 2 3 4**	ℓ_{12} = **1 2 + 3 4**
ℓ_2 = **2 + 1 3 4**	ℓ_{13} = **1 3 + 2 4**
ℓ_3 = **3 + 1 2 4**	ℓ_{14} = **1 4 + 2 3**

The variance of these estimates is $\sigma^2/2$. The average $\bar{y}$ has variance $\sigma^2/8$.

On studying Table 4 we see that the two-factor interactions are mutually

TABLE 5

The eight linear combinations of effects from a 2^{4-1} design with defining relation $\mathbf{I} = \mathbf{1\ 2\ 3\ 4}$

ℓ_I = **average + 1 2 3 4** = 15.0	ℓ_4 = **4 + 1 2 3** = 0.8
ℓ_1 = **1 + 2 3 4** = 5.4	ℓ_{12} = **1 2 + 3 4** = −1.6
ℓ_2 = **2 + 1 3 4** = −0.4	ℓ_{13} = **1 3 + 2 4** = 1.4
ℓ_3 = **3 + 1 2 4** = 7.6	ℓ_{14} = **1 4 + 2 3** = 1.0

confounded in pairs, but assuming that the three and four factor interactions are either non-existent or negligible the estimates ℓ_I, ℓ_1, ℓ_2, ℓ_3 and ℓ_4 can be taken to be estimates of the average and the main effects **1**, **2**, **3** and **4**. If, furthermore, prior knowledge is available that, for example, the **3 4** interaction effect was negligible, then the estimate ℓ_{12} could be taken to estimate the **1 2** interaction effect alone.

The Alternative Fraction

In the above example, in forming the 2^{4-1} design, the factor **4** was associated with the three-factor interaction **1 2 3**. In standard ordering, the elements of the three-factor interaction column, and hence of factor **4**, are

$$- \quad + \quad + \quad - \quad + \quad - \quad - \quad +.$$

The factor **4** can either use these elements as they stand, or it can be associated with the negative of the **1 2 3** effect, that is, with the elements

$$+ \quad - \quad - \quad + \quad - \quad + \quad + \quad -.$$

In the first case **4** = **1 2 3** that is **I** = **1 2 3 4**, and in the second case −**4** = **1 2 3** that is **I** = −**1 2 3 4**. The designs for these two 2^{4-1} fractional factorials are given in Table 6. The two parts together constitute a complete 2^4 factorial design.

TABLE 6

The Design Matrices for the two 2^{4-1} Fractional Factorials with Defining Relations **I** = **1 2 3 4** *and* **I** = −**1 2 3 4**.

Defining Relation I = 1 2 3 4				Observations	Defining Relation I = −1 2 3 4				Observations
1	**2**	**3**	**4**		**1**	**2**	**3**	**4**	
−1	−1	−1	−1	8.7	−1	−1	−1	1	11.8
1	−1	−1	1	15.1	1	−1	−1	−1	13.6
−1	1	−1	1	9.7	−1	1	−1	−1	9.2
1	1	−1	−1	11.3	1	1	−1	1	14.6
−1	−1	1	1	14.7	−1	−1	1	−1	15.8
1	−1	1	−1	22.3	1	−1	1	1	24.0
−1	1	1	−1	16.1	−1	1	1	1	16.4
1	1	1	1	22.1	1	1	1	−1	24.2

Table 6 shows a further set of observations associated with the second fraction. In Table 7 eight linear combinations of effects ℓ'_I, ℓ'_1, ℓ'_2, $\cdots$ associated with the fraction having defining relation **I** = −**1 2 3 4** are given. If both fractions are present, then simple addition and subtraction of the ℓ and ℓ' linear combinations will provide unconfounded estimates of all the effects. For example, the main effect **1**, unconfounded with the **2 3 4** interaction is given by $\frac{1}{2}(\ell_1 + \ell'_1) = 5.6$. Similarly, the **2 3 4** interaction unconfounded by the main effect **1** is obtained from $\frac{1}{2}(\ell_1 - \ell'_1) = -0.20$. The average response, when both fractions are present, is given by $\frac{1}{2}(\ell'_I + \ell_I) = 15.6$.

The estimates obtained by taking the sums and differences of the linear

TABLE 7

The eight linear combinations of effects from a 2^{4-1} design with defining relation I = −**1 2 3 4**

ℓ'_I = average − **1 2 3 4** = 16.2	ℓ'_4 = **4** − **1 2 3** = −1.0
ℓ'_1 = **1** − **2 3 4** = 5.8	ℓ'_{12} = **1 2** − **3 4** = 0.8
ℓ'_2 = **2** − **1 3 4** = −0.2	ℓ'_{13} = **1 3** − **2 4** = 2.2
ℓ'_3 = **3** − **1 2 4** = 7.8	ℓ'_{14} = **1 4** − **2 3** = 0.6

combinations computed from the individual fractional factorials are the same as would be obtained from an analysis of a full 2^4 design.

The $\frac{1}{2}$ Fractions of the 2^k Designs

Any interaction or main effect can be used to split a full 2^k factorial into two half fractions. However, given the assumptions that the higher the order of the interaction the less likely the effect is to occur, there is clearly an advantage in using the interaction of highest order to make the split. The generator is then **1 2 3** ··· **k** and the defining relation **I** = **1 2 3** ··· **k**.

The $\frac{1}{2}$ fractions of all the 2^k factorial designs are best obtained by first writing down the design matrix for a full 2^{k-1} factorial and then adding the kth variable by identifying its + and − versions with the + and − signs of the highest order interaction **1 2 3** ··· (**k** − **1**). Thus the 2^{3-1} factorial is constructed by writing down the design matrix for the 2^2 factorial and then equating variable **3** with the **1 2** interaction. Similarly, the 2^{5-1} factorial is given by writing down the sixteen runs of the 2^4 and then equating the signs of variable **5** with the signs of the **1 2 3 4** interaction. The defining relations for these $\frac{1}{2}$ replicate designs are thus

Design	Defining Relations
2^{3-1}	I = **1 2 3**
2^{4-1}	I = **1 2 3 4**
2^{5-1}	I = **1 2 3 4 5**

(6)

The extension to the half-replicate designs for $k > 5$ is obvious. However, for $k > 5$ these half-replicate designs permit the estimation of a plethora of linear combinations of effects, many of which are combinations of higher order interactions solely. We are therefore interested in still smaller fractions of the 2^k designs, that is, in the 2^{k-p} fractional factorials for $p > 1$. For such designs there is not one, but p generators which combine to provide the defining relation. Before discussing these designs, it is profitable first to discuss their areas of application.

2: AREAS OF APPLICATION

Fractional designs are of value in a number of different circumstances:

1) where certain interactions can be assumed non-existent from prior knowledge,

2) in "screening" situations where it is expected that the effects of all but a few of the variables studied will be negligible,
3) where groups of experiments are run in sequence and ambiguities remaining at a given stage of experimentation can be resolved by later groups of experiments,
4) where certain variables, which may interact, are to be studied simultaneously with other variables whose influence, if any, can be described by main effects only.

Some Interactions Non-Existent, A Priori

As already noted, when properties of smoothness and similarity exist, interactions between three or more variables are often negligible. In addition, the physical nature of a problem is sometimes such that certain interactions must be small or non-existent. In these circumstances we can then use arrangements in which the effects expected to be real are confounded only with interactions expected to be negligible. For example, in Table 3 the estimate of the **1 2 3** interaction effect is perfectly confounded with the main effect **4**. Under the assumption that the three-factor interaction is small, the estimate can be taken as the main effect of **4** alone.

In most practical situations, to say that we *assume*, a priori, that certain effects are negligible would be too strong. Frequently, limitations of time and money do not allow the luxury of the certainty obtainable from exploring an entirely comprehensive model which allows for every contingency. We *tentatively entertain* the possibility of negligible interactions and try to check assumptions as the evidence unfolds.

Screening Situations

Situations often occur where not very much is known about the variables that influence some response. Any subset of a large number of variables might be important, but *which* variables form this subset is unknown. Although usually the number of variables under study will be greater than four, the application of fractionals to this situation can be illustrated with the 2^{4-1} design given in Table 3. It will be seen that if any one variable out of the four produces a large effect, then no matter which variable it is, the design may be regarded as a 2^1 factorial replicated four times in the important variable. If any two variables are producing large effects, the design becomes a full 2^2 factorial replicated twice in these variables. If any three variables are producing large effects, again the design becomes a full 2^3 factorial in these variables. Fractionals for use in screening situations which are replicated factorials for any number up to three variables out of sixteen can be obtained using only thirty-two runs. For picking out the two or three important variables from among a large group of variables, these designs are very useful.

Sequential Groups of Experiments

Fractional factorials are of considerable value in the common situation where experiments are performed in sequence. Having performed one fraction, the results can be reviewed and where there is ambiguity due to the confounding of particular estimates, or experimental error, a further group of experiments can be selected to resolve the uncertainty.

Simultaneous Study of "Major" and "Minor" Variables

It sometimes happens that there exists a group of "major" variables whose study is the chief objective of the investigation. In addition there may be a number of "minor" variables which are expected to have negligible effects. Fractional designs are available in which both kinds of variables are included simultaneously, the main effects and interactions of the major variables estimated without bias, and the main effects of the minor variables checked. The assumption made is that interactions between the minor variables will be negligible.

3: Special Types of 2^k Factorials

Fractional factorial designs can, for convenience, be divided into types. In general the higher the degree of fractionation the more comprehensive the assumptions needed to make unequivocal interpretation possible. The following three types of designs are discussed:

(i) Designs of Resolution III in which no main effect is confounded with any other main effect, but main effects are confounded with two-factor interactions and two-factor interactions with one another. The 2^{3-1} design is of Resolution III.

(ii) Designs of Resolution IV in which no main effect is confounded with any other main effect *or* two-factor interaction, but where two-factor interactions are confounded with one another. The 2^{4-1} design is of Resolution IV.

(iii) Designs of Resolution V in which no main effect or two-factor interaction is confounded with any other main effect or two-factor interaction but two factor interactions are confounded with three factor interactions. The 2^{5-1} design is of Resolution V.

In general, a design of resolution R is one in which no p factor effect is confounded with any other effect containing less than $R - p$ factors.

To identify the resolution of a fractional factorial design, the appropriate Roman numeral subscript is used. Thus, rewriting Equation (6) along with the defining relations for both one-half functions we have

Design	Defining Relations	
2_{III}^{3-1}	$\text{I} = \pm\mathbf{1\ 2\ 3}$	
2_{IV}^{4-1}	$\text{I} = \pm\mathbf{1\ 2\ 3\ 4}$	(7)
2_{V}^{5-1}	$\text{I} = \pm\mathbf{1\ 2\ 3\ 4\ 5}$	

In the above a *word* refers to a combination of elements such as **1 2 3**, **1 2 3 4**. In general the *resolution* of a design is equal to the smallest number of characters in any word appearing in the defining relation.

4: Resolution III Designs

Designs of resolution III are available which require only N runs to study up to $N - 1$ variables, where N is a multiple of four. We first discuss the arrangements for which N is a power of two. Particularly important designs are those for testing three variables in four runs, seven variables in eight runs and fifteen in sixteen runs. Two level designs for studying eleven variables in twelve runs, nineteen variables in twenty runs, etc., are derived by a somewhat different method due to Plackett & Burman (6), and are described later.

Designs for studying $k = N - 1$ variables in N runs may be called *saturated designs*. We introduce these designs by first considering a fractional for testing $k = 7$ variables in $N = 8$ runs. The complete factorial would require $2^7 = 128$ runs. We are considering therefore a one-sixteenth (i.e., a 2^{-4}) fractional, that is, a 2^{7-4}_{III} design. Since the design uses $2^3 = 8$ runs, we start construction of the design matrix with the 2^3 factorial, and then associate four additional variables with the plus and minus signs of the four interaction columns. For example, we may set

$$4 = 1\,2, \qquad 5 = 1\,3, \qquad 6 = 2\,3, \qquad 7 = 1\,2\,3 \tag{8}$$

to obtain the following 2^{7-4}_{III} design

TABLE 8

The Design Matrix for a 2^{7-4}_{III} Design

1	2	3	4 = 1 2	5 = 1 3	6 = 2 3	7 = 1 2 3
−	−	−	+	+	+	−
+	−	−	−	−	+	+
−	+	−	−	+	−	+
+	+	−	+	−	−	−
−	−	+	+	−	−	+
+	−	+	−	+	−	−
−	+	+	−	−	+	−
+	+	+	+	+	+	+

The identifications in Equation (8) provide the *generating relations*

$$\mathbf{I} = 1\,2\,4, \qquad \mathbf{I} = 1\,3\,5, \qquad \mathbf{I} = 2\,3\,6, \qquad \mathbf{I} = 1\,2\,3\,7 \tag{9}$$

associated with the generators **1 2 4**, **1 3 5**, **2 3 6** and **1 2 3 7**. Now clearly, if $\mathbf{I} = 1\,2\,4$ *and* $\mathbf{I} = 1\,3\,5$ then also $\mathbf{I} = 1\,2\,4 \times 1\,3\,5 = 1^2\,2\,3\,4\,5 = 2\,3\,4\,5$. Whence it follows for example that **2 3** and **4 5** are confounded. Thus, when there is more than one generator, the defining relation must contain not only the relations provided by the generators themselves, but all those obtained from all their possible products. The complete defining relation for this 2^{7-4}_{III} design is obtained, for example, by taking the generators first one at a time and then multiplying them together in all possible ways. Taking them one at a time gives $\mathbf{I} = 1\,2\,4 = 1\,3\,5 = 2\,3\,6 = 1\,2\,3\,7$. Multiplying them together two at a time gives

$$\mathbf{I} = 2\,3\,4\,5 = 1\,3\,4\,6 = 3\,4\,7 = 1\,2\,5\,6 = 2\,5\,7 = 1\,6\,7,$$

three at a time gives:

$$\mathbf{I} = 4\,5\,6 = 1\,4\,5\,7 = 2\,4\,6\,7 = 3\,5\,6\,7,$$

and finally, four at a time gives:

$$\mathbf{I} = \mathbf{1\,2\,3\,4\,5\,6\,7}.$$

The complete *defining relation* for this 2^{7-4}_{III} design is therefore

$$\begin{aligned}\mathbf{I} &= \mathbf{1\,2\,4} = \mathbf{1\,3\,5} = \mathbf{2\,3\,6} = \mathbf{1\,2\,3\,7} = \mathbf{2\,3\,4\,5} = \mathbf{1\,3\,4\,6} = \mathbf{3\,4\,7}\\ &= \mathbf{1\,2\,5\,6} = \mathbf{2\,5\,7} = \mathbf{1\,6\,7} = \mathbf{4\,5\,6} = \mathbf{1\,4\,5\,7} = \mathbf{2\,4\,6\,7} = \mathbf{3\,5\,6\,7} \qquad (10)\\ &= \mathbf{1\,2\,3\,4\,5\,6\,7}.\end{aligned}$$

As before, the defining relation quickly provides the alias structure for any effect, that is, indicates which effects are confounded. For example, multiplying the defining relation through by **1** we obtain

$$\begin{aligned}\mathbf{1} &= \mathbf{2\,4} = \mathbf{3\,5} = \mathbf{1\,2\,3\,6} = \mathbf{2\,3\,7} = \mathbf{1\,2\,3\,4\,5} = \mathbf{3\,4\,6} = \mathbf{1\,3\,4\,7} = \mathbf{2\,5\,6}\\ &= \mathbf{1\,2\,5\,7} = \mathbf{6\,7} = \mathbf{1\,4\,5\,6} = \mathbf{4\,5\,7} = \mathbf{1\,2\,4\,6\,7} = \mathbf{1\,3\,5\,6\,7} = \mathbf{2\,3\,4\,5\,6\,7}.\end{aligned}$$

Thus the interactions **2 4, 3 5, 1 2 3 6** etc., are seen to be aliases of, or confounded with, the main effect **1**. Similarly, multiplying through by **1 2 3** we obtain

$$\begin{aligned}\mathbf{1\,2\,3} &= \mathbf{3\,4} = \mathbf{2\,5} = \mathbf{1\,6} = \mathbf{7} = \mathbf{1\,4\,5} = \mathbf{2\,4\,6} = \mathbf{1\,2\,4\,7}\\ &= \mathbf{3\,5\,6} = \mathbf{1\,3\,5\,7} = \mathbf{2\,3\,6\,7} = \mathbf{1\,2\,3\,4\,5\,6} = \mathbf{2\,3\,4\,5\,7}\\ &= \mathbf{1\,3\,4\,6\,7} = \mathbf{1\,2\,5\,6\,7} = \mathbf{4\,5\,6\,7}.\end{aligned}$$

Thus the three-factor interaction **1 2 3** is an alias of, or confounded with **3 4, 2 5, 1 6**, etc. Since the resolution is determined by the smallest number of symbols forming any word in the defining relation, the design is of resolution III, as we have already noted.

In this example, if we write $\ell_1 = \frac{1}{4} \sum y\{1\}$, $\ell_2 = \frac{1}{4} \sum y\{2\}$, etc, and if we assume that all interactions between three of more variables are negligible, then by repeated use of the defining relation we obtain:

$$\begin{aligned}\ell_{\text{I}} &= \text{average}\\ \ell_1 &= \mathbf{1} + \mathbf{2\,4} + \mathbf{3\,5} + \mathbf{6\,7}\\ \ell_2 &= \mathbf{2} + \mathbf{1\,4} + \mathbf{3\,6} + \mathbf{5\,7}\\ \ell_3 &= \mathbf{3} + \mathbf{1\,5} + \mathbf{2\,6} + \mathbf{4\,7}\\ \ell_4 &= \mathbf{4} + \mathbf{1\,2} + \mathbf{5\,6} + \mathbf{3\,7}\\ \ell_5 &= \mathbf{5} + \mathbf{1\,3} + \mathbf{4\,6} + \mathbf{2\,7}\\ \ell_6 &= \mathbf{6} + \mathbf{2\,3} + \mathbf{4\,5} + \mathbf{1\,7}\\ \ell_7 &= \mathbf{7} + \mathbf{3\,4} + \mathbf{2\,5} + \mathbf{1\,6}.\end{aligned} \qquad (11)$$

The Alternative Fractions

In writing down the design matrix for the 2^{7-4}_{III} fractional, the variables **4, 5, 6** and **7** were identified positively with the elements of the interactions **1 2, 1 3,**

2 3 and **1 2 3** respectively. However, each of these identifications could have been made with either a plus or minus sign. For example, instead of associating the variable **4** positively with the interaction **1 2**, that is taking

$$+ \quad - \quad - \quad + \quad + \quad - \quad - \quad +$$

for its elements, the variable **4** could be associated negatively with the elements of **1 2**, that is:

$$- \quad + \quad + \quad - \quad - \quad + \quad + \quad -.$$

The first association gives $\mathbf{4} = \mathbf{1\,2}$ or equivalently $\mathbf{I} = \mathbf{1\,2\,4}$. The second association yields $\mathbf{4} = -\mathbf{1\,2}$ or equivalently $\mathbf{I} = -\mathbf{1\,2\,4}$. We could, in fact, have used any one of the sixteen identifications corresponding to the sixteen possible choices of signs

$$\mathbf{I} = \pm\mathbf{1\,2\,4}, \qquad \mathbf{I} = \pm\mathbf{1\,3\,5}, \qquad \mathbf{I} = \pm\mathbf{2\,3\,6}, \qquad \mathbf{I} = \pm\mathbf{1\,2\,3\,7}. \tag{12}$$

The sixteen possible identifications give the sixteen individual fractions which together yield the complete 2^7 design. In composing the defining relation for any one of the sixteen designs the usual rules of algebraic multiplication determine the signs in the defining relation and hence in the alias pattern.

Another one of these sixteen fractions is, for example, that in which variables **5** and **6** are associated with the elements of the interaction vectors **1 3** and **2 3** taken *negatively*, The generators for this design are:

$$\mathbf{1\,2\,4}, \qquad -\mathbf{1\,3\,5}, \qquad -\mathbf{2\,3\,6}, \qquad \mathbf{1\,2\,3\,7}, \tag{13}$$

and the corresponding defining relation is:

$$\begin{aligned} \mathbf{I} &= \mathbf{1\,2\,4} = -\mathbf{1\,3\,5} = -\mathbf{2\,3\,6} = \mathbf{1\,2\,3\,7} = -\mathbf{2\,3\,4\,5} = -\mathbf{1\,3\,4\,6} = \mathbf{3\,4\,7} \\ &= \mathbf{1\,2\,5\,6} = -\mathbf{2\,5\,7} = -\mathbf{1\,6\,7} = \mathbf{4\,5\,6} = -\mathbf{1\,4\,5\,7} = -\mathbf{2\,4\,6\,7} \\ &= \mathbf{3\,5\,6\,7} = \mathbf{1\,2\,3\,4\,5\,6\,7}. \end{aligned}$$

Assuming as before that all interactions between three or more variables are negligible, we see that this fraction allows the estimation of eight somewhat different combinations of effects

$$\begin{aligned} \ell'_I &= \text{average} \\ \ell'_1 &= \mathbf{1} + \mathbf{2\,4} - \mathbf{3\,5} - \mathbf{6\,7} \\ \ell'_2 &= \mathbf{2} + \mathbf{1\,4} - \mathbf{3\,6} - \mathbf{5\,7} \\ \ell'_3 &= \mathbf{3} - \mathbf{1\,5} - \mathbf{2\,6} + \mathbf{4\,7} \\ \ell'_4 &= \mathbf{4} + \mathbf{1\,2} + \mathbf{5\,6} + \mathbf{3\,7} \\ \ell'_5 &= -\mathbf{5} + \mathbf{1\,3} - \mathbf{4\,6} + \mathbf{2\,7} \\ \ell'_6 &= -\mathbf{6} + \mathbf{2\,3} - \mathbf{4\,5} + \mathbf{1\,7} \\ \ell'_7 &= \mathbf{7} + \mathbf{3\,4} - \mathbf{2\,5} - \mathbf{1\,6} \end{aligned} \tag{14}$$

where the use of the prime notation on the ℓ's indicates only that some alternative function is under consideration. We see that Eq. (14) is identical to Eq. (11) with the numerals 5 and 6 having minus instead of plus signs.

Families of Fractionals

In the above example, there are $2^4 = 16$ different 2^{7-4}_{III} designs, each design corresponding to a particular choice of signs from among the generators $\pm\mathbf{1\ 2\ 4}$, $\pm\mathbf{1\ 3\ 5}$, $\pm\mathbf{2\ 3\ 6}$ and $\pm\mathbf{1\ 2\ 3\ 7}$. When the generators of a fractional factorial design associated with the identity $\mathbf{I}$ all have positive signs, they are called the *principal generators.* The defining relation obtained by multiplying out the generators is similarly called the *principal defining relation*, and the corresponding fractional factorial the *principal fraction.* Individual member fractions obtained from changes of sign in the generators are said to belong to the same *family.* In general, a 2^{k-p} fractional factorial design will have p generators, and the 2^p ways of allocating plus and minus signs to the generators will produce the 2^p different fractions belonging to the same family.

In general, a 2^{k-f} design will have f independent generators $\mathbf{G}_1, \mathbf{G}_2, \cdots, \mathbf{G}_f$. An *independent* generator is such that it cannot be obtained by multiplying together the other generators, and is identified by the original association adopted in writing down the design. A defining relation for a particular fraction will contain 2^f words obtained by multiplying out $(\mathbf{I} \pm \mathbf{G}_1)(\mathbf{I} \pm \mathbf{G}_2) \cdots (\mathbf{I} \pm \mathbf{G}_f)$. The 2^f different fractions have defining relations given by the 2^f different ways of allocating plus and minus signs in this product. The defining relation for the *principal* fraction is given when all signs are plus. The alias pattern for any of the non-principal fractions is simply obtained by making the appropriate changes of sign in the alias pattern for the principal fraction.

Resolution III *Designs Containing* 16 *and* 32 *runs.*

The principal fraction of the 2^{15-11}_{III} design is obtained by first writing down the sixteen runs of the complete 2^4 design and then associating an additional eleven variables with the interactions **1 2, 1 3, 1 4, 2 3, 2 4, 3 4, 1 2 3, 1 2 4, 1 3 4, 2 3 4,** and **1 2 3 4**. Similarly, the thirty-two runs comprising the 2^{31-26}_{III} factorial are obtained by writing down the complete factorial for five variables and then equating the additional twenty-six new variables with their interactions between the original five variables.

Effect of Dropping Variables

For intermediate values of k resolution III designs may be obtained by omitting variables from the resolution III design of next higher order. For example, to test six factors in eight runs we can use the 2^{7-4}_{III} design dropping out any one column in its design matrix. The alias relationships remain the same except that all words containing the characters associated with the dropped variables are omitted from the alias structure, and from any estimates of linear combinations. For example, dropping the columns **3** and **5** from the design matrix for the 2^{7-4}_{III} fractional given in Table 8 yields the 2^{5-2}_{III} design shown in Table 9. We can select the variables to be dropped out so that the most satisfactory alias arrangements exist among those remaining.

TABLE 9

Design Matrix 2^{5-2}_{III}

Defining Relation **I** = **1 2 4** = **1 6 7** = **2 4 6 7**.

1	2	4	6	7
−	−	+	+	−
+	−	−	+	+
−	+	−	−	+
+	+	+	−	−
−	−	+	−	+
+	−	−	−	−
−	+	−	+	−
+	+	+	+	+

Although it is true that a fractional of resolution R in a reduced number of $k - d$ variables can always be obtained by omitting d variables from a k variable fractional of resolution R, nevertheless a particular design obtained in this manner does not necessarily provide the best arrangement possible. For instance, if we drop variables **3**, **5**, **6** and **7** from the principal fraction 2^{7-4}_{III} design with generators **1 2 4**, **1 3 5**, **2 3 6** and **1 2 3 7** we are left with a design in the three variables **1**, **2** and **4** along with the unresolved generator **1 2 4** and hence the defining relation appropriate to a design having only *four* runs. On inspection we find that our eight factor combinations in the three remaining variables consist of *two replications* of the four run half-replicate design defined by **I** = **1 2 4**. This design is of resolution III, of course, but in many cases we would prefer to use the eight runs to perform a full factorial in the variables **1**, **2** and **4**. A full factorial would have been obtained had we, for example, dropped variables **1**, **2**, **3** and **7**.

The defining relation for the design obtained after dropping d variables will contain all those words in the original defining relation which do not contain any of the dropped numerals. Suppose among the f generators of the original design there are d generators that contain dropped variables, and $f - d$ generators that do not. A set of generators for the derived design will contain all the $f - d$ generators *not* containing dropped variables together with the largest set of independent products not containing dropped variables which can be found by multiplying the remaining generators.

For example consider again the resolution III design with generators $\mathbf{G_1} = \mathbf{1\ 2\ 4}$, $\mathbf{G_2} = \mathbf{1\ 3\ 5}$, $\mathbf{G_3} = \mathbf{2\ 3\ 6}$ and $\mathbf{G_4} = \mathbf{1\ 2\ 3\ 7}$. Suppose variable **1** is dropped. Since $\mathbf{G_3} = \mathbf{2\ 3\ 6}$ does not contain **1** this generator will be included in the generators for the derived design. From the remaining generators we can obtain the products $\mathbf{G_1G_2}$, $\mathbf{G_1G_4}$ and $\mathbf{G_2G_4}$ none of which contain the dropped variable **1**. Only two of the three products may be used since, having taken two of them, the third may be obtained by multiplication. For example, $\mathbf{G_1G_2 \cdot G_1G_4} = \mathbf{G_2G_4}$. In general, a group of p words (such as the products we are considering here) are said to be independent if no one of them can be obtained by multiplying together some subset of the remaining $p - 1$. In this example then, a set of generators for the design derived after dropping **1** are **2 3 6**, **2 3 4 5** and **3 4 7** (that is, G_3, G_1G_2 and G_1G_4).

At best, the effect of dropping d variables is to produce a design having d fewer generators. However, this represents the maximum reduction in generators possible, and particular choices of dropped variables may produce a smaller reduction in the number of generators. Of course, the greater the number of generators, the more words there will be in the defining relation and correspondingly, the more aliases for the remaining effects.

Effect of Combining Fractions from the Same Family

If we take the original fraction of the 2^{7-4}_{III} together with the second fraction in which the signs of **5** and **6** are switched, and take one-half the sums and differences of the respective linear combinations of effects we can estimate the following quantities (assuming all interactions with more than two factors to be nil).

From $\frac{1}{2}$ the Sums	From $\frac{1}{2}$ the Differences	
$\frac{1}{2}(\ell_I + \ell'_I)$ = **Grand average**	$\frac{1}{2}(\ell_I - \ell'_I)$ = **Block effect**	
$\frac{1}{2}(\ell_1 + \ell'_1)$ = **1 + 2 4**	$\frac{1}{2}(\ell_1 - \ell'_1)$ = **3 5 + 6 7**	
$\frac{1}{2}(\ell_2 + \ell'_2)$ = **2 + 1 4**	$\frac{1}{2}(\ell_2 - \ell'_2)$ = **3 6 + 5 7**	
$\frac{1}{2}(\ell_3 + \ell'_3)$ = **3 + 4 7**	$\frac{1}{2}(\ell_3 - \ell'_3)$ = **1 5 + 2 6**	(15)
$\frac{1}{2}(\ell_4 + \ell'_4)$ = **4 + 1 2 + 5 6 + 3 7**	$\frac{1}{2}(\ell_4 - \ell'_4)$ = higher order interactions	
$\frac{1}{2}(\ell_5 + \ell'_5)$ = **1 3 + 2 7**	$\frac{1}{2}(\ell_5 - \ell'_5)$ = **5 + 4 6**	
$\frac{1}{2}(\ell_6 + \ell'_6)$ = **2 3 + 1 7**	$\frac{1}{2}(\ell_6 - \ell'_6)$ = **6 + 4 5**	
$\frac{1}{2}(\ell_7 + \ell'_7)$ = **7 + 3 4**	$\frac{1}{2}(\ell_7 - \ell'_7)$ = **2 5 + 1 6**	

In general when two fractions from the same family are combined, the sums and differences of the corresponding linear combinations of the effects determine the effects which can be estimated from the combined design. The "block effect" referred to in Eq. (15) is the difference in average level between the first and second groups of eight runs.

Combining Fractional Factorials to Separate Effects

The procedure of adding fractions in sequence with suitably switched signs provides a useful method for the systematic isolation and confirmation of important effects in multi-variable systems. The method is very flexible and can be used in different ways as different situations unfold.

Mention will be made of two particular uses of this device: (1) the addition of a second fraction in which the signs in a *single column* are switched and, (2) the addition of a second fraction in which the signs in *all the columns* are switched.

Switching Signs for a Single Variable

Suppose a fractional factorial is generated by switching the signs associated with only the variable **1** in the 2^{7-4}_{III} factorial given in Table 8. Then the linear combinations that can be estimated from this fraction (given that the three-factor and higher order interactions are negligible) are the following

$$
\begin{aligned}
\ell_I &= \text{Average} \\
\ell_1 &= -1 + 24 + 35 + 67 \\
\ell_2 &= 2 - 14 + 36 + 57 \\
\ell_3 &= 3 - 15 + 26 + 47 \\
\ell_4 &= 4 - 12 + 56 + 37 \\
\ell_5 &= 5 - 13 + 46 + 27 \\
\ell_6 &= 6 + 23 + 45 - 17 \\
\ell_7 &= 7 + 34 + 25 - 16
\end{aligned} \tag{16}
$$

Combining this fraction with the principal fraction, the following linear combination of effects are obtained from the combined design

From $\frac{1}{2}(\ell + \ell')$	From $\frac{1}{2}(\ell - \ell')$
Average	Block effect
24 + 35 + 67	1
2 + 36 + 57	14
3 + 26 + 47	15
4 + 56 + 37	12
5 + 46 + 27	13
6 + 23 + 45	17
7 + 34 + 25	16

(17)

We see that by adding to a fraction a further fraction with the signs for a single variable reversed, we isolate the main effect of that variable together with all of its two-factor interactions. Given any fractional of resolution III or higher and a second fractional identical to the first except that the signs of a single variable are switched, then the combined design will provide estimates of the main effect of the switched variable and all its associated two-factor interactions unbiased by any other main effect or two-factor interaction.

Switching Signs for All Variables

By switching signs for all seven variables given in the principal fraction we can estimate the following linear combinations

$$
\begin{aligned}
\ell'_I &= \text{Average} \\
\ell'_1 &= -1 + 24 + 35 + 67 \\
\ell'_2 &= -2 + 14 + 36 + 57 \\
\ell'_3 &= -3 + 15 + 26 + 47 \\
\ell'_4 &= -4 + 12 + 56 + 37 \\
\ell'_5 &= -5 + 13 + 46 + 27 \\
\ell'_6 &= -6 + 23 + 45 + 17 \\
\ell'_7 &= -7 + 34 + 25 + 16
\end{aligned} \tag{18}
$$

By combining this fraction with the principal fraction all the main effects can be estimated clear of all the two-factor interactions. The two-factor interactions in turn will associate themselves in groups of three in accordance with the following scheme

From $\frac{1}{2}(\ell + \ell')$	From $\frac{1}{2}(\ell - \ell')$	
Average	**Block effect**	
2 4 + 3 5 + 6 7	1	
1 4 + 3 6 + 5 7	2	
1 5 + 2 6 + 4 7	3	
1 2 + 5 6 + 3 7	4	(19)
1 3 + 4 6 + 2 7	5	
2 3 + 4 5 + 1 7	6	
3 4 + 2 5 + 1 6	7	

This is a special example of a general principle, [14], which states that if any fractional is replicated with reversed signs, then all alias links between main effects and two-factor interactions are broken.

It should be noticed that although there are $2^7 = 128$ ways of switching signs, there are only $2^4 = 16$ of these switches that result in different designs. This must be so since there are only 2^4 different 2^{7-4} fractions belonging to the same family. It is easily confirmed by actual trial that the same design can be produced by a number of alternative sign switching arrangements, although the order in which the experimental runs appear may be different. The situation is made clear by considering only the generating relations for the principal fraction of the 2^{7-4}_{III}, that is:

$$\mathbf{I} = \mathbf{1\,2\,4}, \qquad \mathbf{I} = \mathbf{1\,3\,5}, \qquad \mathbf{I} = \mathbf{2\,3\,6}, \qquad \mathbf{I} = \mathbf{1\,2\,3\,7}.$$

It will be obvious for example that switching the signs of variables **4**, **5**, **7**, or of variable **1**, produces exactly the same effect. In each case the generating relations are:

$$\mathbf{I} = -\mathbf{1\,2\,4}, \qquad \mathbf{I} = -\mathbf{1\,3\,5}, \qquad \mathbf{I} = \mathbf{2\,3\,6}, \qquad \mathbf{I} = -\mathbf{1\,2\,3\,7}$$

Generators for Aggregate Designs

Suppose the principal fraction of the 2^{7-4}_{III} given in Table 8 is run. The generating relations for this design are

$$\mathbf{I}_8 = \mathbf{1\,2\,4}, \qquad \mathbf{I}_8 = \mathbf{1\,3\,5}, \qquad \mathbf{I}_8 = \mathbf{2\,3\,6}, \qquad \mathbf{I}_8 = \mathbf{1\,2\,3\,7}$$

where the notation $\mathbf{I}_8$ refers to a column of eight plus signs. Now suppose we perform a further series using a second 2^{7-4}_{III} from the same family as, for example, the fraction in which the variable **1** is run with reversed signs. The combined design formed from the two pieces is now a 2^{7-3} factorial. Since it is a one-eight replicate, it will have three generators, not four. How can these generators be identified? We note now that the generators for the second fraction are

$$\mathbf{I}_8 = -\mathbf{1\,2\,4}, \qquad \mathbf{I}_8 = -\mathbf{1\,3\,5}, \qquad \mathbf{I}_8 = \mathbf{2\,3\,6}, \qquad \mathbf{I}_8 = -\mathbf{1\,2\,3\,7}.$$

It is clear that the generator **2 3 6** must be one of the generators for the combined design for in both pieces of the design $\mathbf{I}_8 = \mathbf{2\,3\,6}$. Consequently if $\mathbf{I}_{16}$ represents the column of sixteen plus signs associated with the complete design, then also $\mathbf{I}_{16} = \mathbf{2\,3\,6}$.

In asking what are the generating relations for the complete design we must first ask the question, "For which combinations are the products of the elements everywhere equal to $\mathbf{I}_{16}$?" Now we observe that **1 2 3 7** has the value $\mathbf{I}_8$ in the first set of eight runs, and $-\mathbf{I}_8$ in the second set. Thus, **1 2 3 7** is not equal to $\mathbf{I}_{16}$ and is therefore not a generator of the combined design. Similarly, **1 2 4** and **1 3 5** also are not generators for the complete design.

Now clearly for the first part of the design

$$\mathbf{I}_8 = (\mathbf{1\,2\,4})(\mathbf{1\,3\,5}) = \mathbf{2\,3\,4\,5},$$

and also for the second part

$$\mathbf{I}_8 = (-\mathbf{1\,2\,4})(-\mathbf{1\,3\,5}) = \mathbf{2\,3\,4\,5}.$$

Thus it is true for the complete design that

$$\mathbf{I}_{16} = \mathbf{2\,3\,4\,5}.$$

Similarly, multiplying **1 2 4** by **1 2 3 7** it is true for the complete design that

$$\mathbf{I}_{16} = \mathbf{3\,4\,7}.$$

A third product is possible, obtained by multiplying **1 3 5** by **1 2 3 7** to give

$$\mathbf{I}_{16} = \mathbf{2\,5\,7}.$$

Now $(\mathbf{2\,3\,4\,5})(\mathbf{3\,4\,7}) = \mathbf{2\,5\,7}$ and since it is a property of generators that no individual generator can be obtained from the others, we include in the new set of generators any two of the three derived above. Thus, the generating relations for this 2^{7-3} design are

$$\mathbf{I}_{16} = \mathbf{2\,3\,6}, \qquad \mathbf{I}_{16} = \mathbf{2\,3\,4\,5}, \qquad \mathbf{I}_{16} = \mathbf{3\,4\,7}$$

and the corresponding defining relation is

$$\mathbf{I}_{16} = \mathbf{2\,3\,6} = \mathbf{2\,3\,4\,5} = \mathbf{3\,4\,7} = \mathbf{4\,5\,6} = \mathbf{2\,4\,6\,7} = \mathbf{2\,5\,7} = \mathbf{3\,5\,6\,7}$$

From the above it will be seen that a general rule for finding generators for a design derived from two fractions from the same family each defined by generating relations of the kind

$$\text{Fraction 1: } I = \pm A = \pm B = \pm C = \cdots$$

$$\text{Fraction 2: } I = \pm A = \pm B = \pm C = \cdots$$

is as follows:

> Suppose there are U words of unlike sign and L words of like sign in the two identities. Then $U + L - 1$ words which are generators of the

new design will contain the L words of like sign together with $U - 1$ words obtained as independent even products of the U words of unlike sign.

In the above an *even* product is a product between an even number of words (usually two). This rule can be applied quite generally not only for combining designs of resolution III, but for combining any pair of fractionals belonging to the same family. As a further example, suppose two 2_{III}^{7-4} fractions were combined with generating relations:

$$I = -1\,2\,4, \qquad I = -1\,3\,5, \qquad I = 2\,3\,6, \qquad I = -1\,2\,3\,7$$

and

$$I = -1\,2\,4, \qquad I = 1\,3\,5, \qquad I = -2\,3\,6, \qquad I = 1\,2\,3\,7$$

(The first fraction can be obtained from the principal fraction of the 2_{III}^{7-4} by reversing the sign of variable **1**, the second fraction by reversing the signs of variables **1** and **3**.) Then the generators for the complete design are **−1 2 4** and any two of the three words obtained from the even products of **−1 3 5**, **2 3 6** and **−1 2 3 7** to give the generating relations:

$$I = -1\,2\,4, \qquad I = -1\,2\,5\,6, \qquad I = 2\,5\,7.$$

The reader will notice that switching to an alternative set of permissable generators leaves the design unchanged for it produces the same defining relation. Thus, in the above, if we had used the generators **−1 2 4**, **−1 2 5 6** and **−1 6 7** the defining relation obtained by multiplying out these generators would have been identical to that obtained before.

Alternative Choice of Generators

A particular fractional has an *unique* defining relation for a given design. There are however a number of different but equivalent choices of generators all of which lead to the same defining relation and the same design. Therefore, although we may speak of *the* defining relation for a design, we should properly refer to *a* choice of generators. In general, suppose $G_1, G_2, \cdots, G_f$ are a set of generators, necessarily independent, for a particular design. Then any other set of f independent generators derived by multiplication will be equivalent and will produce the same defining relation and be associated with the same design. The generators satisfy the same rules of multiplication as before, that is, $G_1^2 = G_2^2 = \cdots = G_f^2 = I$. To see that this is so suppose that $G_1 = 1\,2\,3$, then $G_1^2 = 1^2 2^2 3^2 = I$. If we have four generators G_1, G_2, G_3 and G_4 for a particular design, then G_1G_2, G_1G_3, G_1G_4 and $G_1G_2G_3$ will be an alternative set of generators, but G_1G_2, G_1G_3, G_1G_4, and $G_1G_2G_3G_4$ will not since $G_1G_2 \cdot G_1G_3 \cdot G_1G_4 = G_1G_2G_3G_4$. In particular, suppose we are interested in the fully saturated 2_{III}^{7-3} design with generators $G_1 = 1\,2\,4$, $G_2 = 1\,3\,5$, $G_3 = 2\,3\,6$ and $G_4 = 1\,2\,3\,7$, then the first legimate alternative set of generators will give **2 3 4 5**, **1 3 4 6**, **3 4 7** and **4 5 6** whereas the second "illigimate" choice gives **2 3 4 5**, **1 3 4 6**, **3 4 7** and **1 2 3 4 5 6 7**, for it is readily confirmed that the last generator is the product of the first three.

Combining Fractionals Not of the Same Family

We have seen how by switching signs, fractional factorials may be combined together to isolate particular effects of interest, and that when fractional designs have the same generators except for their signs they are classified as being from the same family. Another method for isolating effects that is often of value is to combine fractions which are not of the same family. In one interesting species the numbers are switched in the generators as well as the signs. Possibilities arising from designs of this sort are presently being investigated.

Blocking Designs of Resolution III

Frequently an experimenter may fear that his results may be upset by shifts in average performance that occur from day to day, or with different batches of raw material. Such systematic sources of variation can often be successfully eliminated without biasing the estimates of the effects, or inflating the error variance by grouping the runs into "blocks".

The resolution III designs can be broken into two blocks of equal size by identifying the two blocks with the + and − versions of a single variable. For example, using the principal fraction of the 2^{7-4}_{III} design with generators **1 2 4**, **1 3 5**, **2 3 6** and **1 2 3 7** and using variable **7** for blocking we have the design given in Table 10. This design is a 2^{6-3}_{III} in blocks of four runs each. The generators

TABLE 10

1	2	3	4	5	6	7 = B	
+	−	−	−	−	+	+	Block 1
−	+	−	−	+	−	+	
−	−	+	+	−	−	+	
+	+	+	+	+	+	+	
−	−	−	+	+	+	−	Block 2
+	+	−	+	−	−	−	
+	−	+	−	+	−	−	
−	+	+	−	−	+	−	

for the design can now be written

$$\mathbf{1\,2\,4}, \quad \mathbf{1\,3\,5}, \quad \mathbf{2\,3\,6} \text{ and } \mathbf{1\,2\,3\,B} \tag{20}$$

where the letter **B** replaces the numeral **7** in the last generator to indicate the blocking variable. Assuming that three factor and higher order interactions are negligible, the defining relation for this design shows that the six main effects **1**, **2**, **3**, ⋯ , **6** are each confounded with three two-factor interactions, one of which is a two-factor interaction with the blocks. The block effect itself is confounded with three two-factor interactions among the variables. In general, any resolution III design can be broken into two blocks of equal size by selecting the + and − signs of any one of the variables in the design matrix to identify the two blocks.

Resolution III designs can be broken into four blocks of equal size by identifying two block variables $\mathbf{B}_1$ and $\mathbf{B}_2$ with the + and − versions of two of the

TABLE 11

Run Number	Variables $1 = B_1 B_2$	2	3	4	5	Blocking Variables $6 = B_1$	$7 = B_2$		Block Variables B_1	B_2	$B_1 B_2$
1	+	+	−	+	−	−	−	Block 1	−	−	+
2	+	−	+	−	+	−	−				
3	−	−	−	+	+	+	−	Block 2	+	−	−
4	−	+	+	−	−	+	−				
5	−	+	−	−	+	−	+	Block 3	−	+	−
6	−	−	+	+	−	−	+				
7	+	−	−	−	−	+	+	Block 4	+	+	+
8	+	+	+	+	+	+	+				

variables. For example, starting with the principal fraction of the 2_{III}^{7-4} design and using variables **6** and **7** for blocking we obtain the four blocks of two runs each as illustrated in Table 11. Among these four blocks there are three degrees of freedom associated with the main effects and the two-factor interaction of pseudo-block variables $\mathbf{B}_1$, $\mathbf{B}_2$ identified with the two-way table

		$\mathbf{B}_1$ −	$\mathbf{B}_1$ +
$\mathbf{B}_2$	−	1, 2	3, 4
$\mathbf{B}_2$	+	5, 6	7, 8

The pairs of numbers in the cells of the table denote the runs comprising the four individual blocks. The "interaction variable" $\mathbf{B}_1 \mathbf{B}_2$ has precisely the same importance as the main effects $\mathbf{B}_1$ and $\mathbf{B}_2$. We see that on associating $\mathbf{B}_1$ with variable **6** and $\mathbf{B}_2$ with variable **7** we automatically associate a comparison between blocks, that is, the interaction $\mathbf{B}_1 \mathbf{B}_2$ with the interaction **6 7**. In this particular example **1** = **6 7** and hence the plus and minus signs of column **1** are now no longer available to accommodate an experimental variable. The variable **1**, therefore, is dropped from the experimental design. Thus, using variables **6**, **7** and **6 7** to identify the four blocks we obtain the design in the variables **2**, **3**, **4** and **5** in four blocks of two, as shown in Table 12.

It should be noted here that the two runs comprising each block are "mirror images" of one another, that is, within a block the versions of one run are exactly reversed in the second run. We will later see that this attribute of blocks of size two has important consequences.

It is usually assumed that block variables corresponding to such characteristics as the time of day, batches of raw material, operators, etc., do not interact with

TABLE 12

2	3	4	5	
+	−	+	−	
−	+	−	+	The 2_{IV}^{4-1} in four
−	−	+	+	
+	+	−	−	blocks of two runs
+	−	−	+	each
−	+	+	−	
−	−	−	−	
+	+	+	+	

the experimental variables. As always it is wise to regard this as a supposition to be tentatively entertained. (Unchecked assumptions are never safe in an applied subject.) We can be reminded of our supposition by setting out the analysis as if the supposition were *not* true, that is, by taking $\mathbf{B}_1$, $\mathbf{B}_2$ and $\mathbf{B}_1\mathbf{B}_2$ as if they were capable of interacting with the variables. If we treat $\mathbf{B}_1$, $\mathbf{B}_2$ and $\mathbf{B}_1\mathbf{B}_2$ on the same basis as the experimental variables, we have for the generators of the design given in Table 12:

$$\mathbf{B}_1\mathbf{B}_2\ \mathbf{2}\ \mathbf{4}, \qquad \mathbf{B}_1\mathbf{B}_2\ \mathbf{3}\ \mathbf{5}, \qquad \mathbf{B}_1\ \mathbf{2}\ \mathbf{3} \tag{21}$$

As mentioned above, the "interaction" $\mathbf{B}_1\mathbf{B}_2$ between the psuedo block factors $\mathbf{B}_1$ and $\mathbf{B}_2$ represents a contrast between the blocks which is on exactly the same footing as $\mathbf{B}_1$ or $\mathbf{B}_2$. (A mere relabeling of the blocks could change the "interaction" contrast to a main effect contrast $\mathbf{B}_1$.) Consequently, the combination $\mathbf{B}_1\mathbf{B}_2$ must be treated as a group having the same status as a single variable. In particular, a word such as $\mathbf{B}_1\mathbf{B}_2\ \mathbf{1}$ must count as a two-factor interaction (between variable **1** and one of the block contrasts) and *not* as a three-factor interaction.

The generators given in Equation 21 can be used to construct the defining relation for the design. Assuming all three-factor and higher order interactions negligible we obtain the linear combinations of effects given in Table 13.

The bracketed values in Table 13 indicate the two-factor interactions which could, if they existed, bias the various effects. Usually of course, all interactions with blocks can be safely supposed to be negligible. On this assumption the

TABLE 13

$$\begin{aligned}
\ell_I &= \text{Average} \\
\ell_1 &= \mathbf{B}_1\mathbf{B}_2 + (\mathbf{2\,4} + \mathbf{3\,5}) \\
\ell_2 &= \mathbf{2} + (\mathbf{B}_1\mathbf{B}_2\mathbf{4} + \mathbf{B}_1\mathbf{3} + \mathbf{B}_2\mathbf{5}) \\
\ell_3 &= \mathbf{3} + (\mathbf{B}_1\mathbf{B}_2\mathbf{5} + \mathbf{B}_1\mathbf{2} + \mathbf{B}_2\mathbf{4}) \\
\ell_4 &= \mathbf{4} + (\mathbf{B}_1\mathbf{B}_2\mathbf{2} + \mathbf{B}_1\mathbf{5} + \mathbf{B}_2\mathbf{3}) \\
\ell_5 &= \mathbf{5} + (\mathbf{B}_1\mathbf{B}_2\mathbf{3} + \mathbf{B}_1\mathbf{4} + \mathbf{B}_2\mathbf{2}) \\
\ell_6 &= \mathbf{B}_1 + (\mathbf{2\,3} + \mathbf{4\,5}) \\
\ell_7 &= \mathbf{B}_2 + (\mathbf{3\,4} + \mathbf{2\,5})
\end{aligned}$$

Linear Combinations of Effects Provided by 2_{IV}^{4-1} in Four Blocks of Two Runs Each.

main effects of the variables **2**, **3**, **4** and **5** in this design are clear of two-factor interactions and the design given in Table 12 is in fact of resolution IV, that is, a 2_{IV}^{4-1} fractional in four blocks of two runs each.

The Plackett and Burman Designs

The methods given here allow us to construct resolution III designs suitable for exploring $k = 3$ variables in $N = 4$ runs, $k = 7$ in $N = 8$, $k = 15$ in $N = 16$ and $k = 31$ in $N = 32$ runs. It was pointed out by Plackett and Burman in 1946 [6] that two version designs which gave uncorrelated estimates of first order effects were available for exploring $k = N - 1$ variables in N runs where N was any multiple of four, and they presented the design matrices for these designs for 4 up to 100 (except for the isolated case of $N = 92$). When N is a power of two, the designs provided by Plackett and Burman are identical with one or the other of the families of resolution III designs derived by the methods given above. For the cases $N = 12$, 20, 24, 28 and 36 however, the Plackett and Burman designs allow useful gaps to be filled and are presented below.

The rows of plus and minus signs given in Table 14A are used to construct the design matrices for $N = 12$, 20, 24 and 36 while the design matrix for $N = 28$ is constructed from the nine rows shows in Table 14B.

TABLE 14A

$k = 11$	$N = 12$	++−+++−−−+−
$k = 19$	$N = 20$	++−−++−+−+−+−−−−++−
$k = 23$	$N = 24$	+++++−+−++−−++−−+−+−−−−
$k = 35$	$N = 36$	−+−+++−−−+++++−+++−−+−−−−+−+−++−−+−

TABLE 14B

$k = 27 \quad N = 28$

A	B	C
+ − + + + + − − −	− + − − − + − − +	+ + − + − + + − +
+ + − + + + − − −	− − + + − − + − −	− + + + + − + + −
− + + + + + − − −	+ − − − + − − + −	+ − + − + + − + +
− − − + − + + + +	− − + − + − − − +	+ − + + + − + − +
− − − + + − + + +	+ − − − − + + − −	+ + − − + + + + −
− − − − + + + + +	− + − + − − − + −	− + + + − + − + +
+ + + − − − + − +	− − + − − + − + −	+ − + + − + + + −
+ + + − − − + + −	+ − − + − − − − +	+ + − + + − − + +
+ + + − − − − + +	− + − − + − + − −	− + + − + + + − −

To construct the designs for $N = 12$, 20, 24 and 36 the plus and minus signs appearing in the appropriate row of Table 14A are first written down as a column. A second column is obtained from the first by moving down the elements of the first column once, and placing the last element in first position. This procedure is then repeated, moving down the second column one element to produce the third, and so on until all k columns are obtained. Finally a *row* of minus signs is added to complete the design. Thus, for the case of $k = 11$ variables

TABLE 14C

1	2	3	4	5	6	7	8	9	10	11
+	−	+	−	−	−	+	+	+	−	+
+	+	−	+	−	−	−	+	+	+	−
−	+	+	−	+	−	−	−	+	+	+
+	−	+	+	−	+	−	−	−	+	+
+	+	−	+	+	−	+	−	−	−	+
+	+	+	−	+	+	−	+	−	−	−
−	+	+	+	−	+	+	−	+	−	−
−	−	+	+	+	−	+	+	−	+	−
−	−	−	+	+	+	−	+	+	−	+
+	−	−	−	+	+	+	−	+	+	−
−	+	−	−	−	+	+	+	−	+	+
−	−	−	−	−	−	−	−	−	−	−

in $N = 12$ runs, the design of Table 14C is obtained. To construct the design for $k = 27, N = 28$ the three blocks, A, B and C illustrated in Table 14B are written down cyclically

$$\begin{matrix} A & B & C \\ C & A & B \\ B & C & A \end{matrix}$$

and these twenty-seven rows followed by a row of minus signs.

An Example

In the start up of a new manufacturing unit considerable difficulty was experienced at the filtration stage. Other similar units operated satisfactorily at other sites but this particular new unit, although apparently similar in most major respects to the other units, gave a crude product which required very much longer filtration times. A meeting was called to discuss possible explanations and to consider ways of curing the trouble. The following variables were proposed as being possibly responsible.

(1) *The water supply*: The new plant used piped water from the local municipal reservoir. An alternative but somewhat limited supply of water was available from a local well. It was proposed that the effect of changing to the weli water should be tried since it was argued that the well water corresponded more closely to the water used at other sites.

(2) *Raw Material*: The raw material used was manufactured on the site and it was suggested that this might be in some way deficient. It was proposed that raw material which had been satisfactorily used in the manufacturing of the product at another site should be shipped in and tested locally.

(3) *Temperature of Filtration*: This was not thought to be a critical factor over the range involved and no special attempt to control this temperature

had been made. However, the physical arrangement of the new process was such that filtration was accomplished at a somewhat lower temperature than had been experienced at other plants. By temporarily covering pipes and equipment, provision could be made to raise the temperature to the level experienced elsewhere.

(4) *Hold up Time*: Prior to filtration the product was held in a stirred tank. The average period of hold-up in the new plant was somewhat less than that used in the other plants but it could be easily increased.

(5) *Recycle*: The only major difference between production facilities at the other plants and the present one lay in the introduction of a recycle stage which slightly increased conversion of the reagents prior to precipitation and filtration. Arguments were advanced which accounted for the longer filtration time in terms of this recycle stage. Arrangements could be made to temporarily eliminate the recycle stage.

(6) *Rate of Addition of Caustic Soda*: Immediately prior to filtration a quantity of caustic soda liquor was added resulting in precipitation of the product. The addition rate was somewhat faster with the new plant but it was possible to produce slower rates of addition.

(7) *Type of Filter Cloth*: The filter cloths employed in this plant were very similar to those used at the other sites. However, they did come from more recently supplied batch and it was suggested that their performance should be compared with cloths from previously supplied batches which were still available.

In the following design the minus version corresponds to the usual operation for the new plant and the plus version to the change. Thus we have

	−	+
(1) water	town	well
(2) raw material	on site	other
(3) temperature of filtration	low	high
(4) hold up time	low	high
(5) recycle	included	omitted
(6) rate of addition NaOH	fast	slow
(7) filter cloth	new	old

The 2^{7-4}_{III} design with generators

$$\mathbf{I} = \mathbf{1\,2\,5}, \qquad \mathbf{I} = \mathbf{1\,3\,6}, \qquad \mathbf{I} = \mathbf{2\,3\,7}, \qquad \mathbf{I} = \mathbf{1\,2\,3\,4}$$

was chosen. This design is equivalent to the 2^{7-4}_{III} design considered earlier, but is obtained by a different association of variables, that is, $\mathbf{5} = \mathbf{12}$, $\mathbf{6} = \mathbf{13}$, $\mathbf{7} = \mathbf{23}$ and $\mathbf{4} = \mathbf{123}$. Eight experiments run in random order gave the filtration times listed below.

	1	2	3	4	5	6	7	Filtration Time
1	−	−	−	−	+	+	+	68.4
2	+	−	−	+	−	−	+	77.7
3	−	+	−	+	−	+	−	66.4
4	+	+	−	−	+	−	−	81.0
5	−	−	+	+	+	−	−	78.6
6	+	−	+	−	−	+	−	41.2
7	−	+	+	−	−	−	+	68.7
8	+	+	+	+	+	+	+	38.7

The usual analysis gives the estimates

water	$\ell_1 = 1 + 25 + 36 + 47 = -10.9$
raw material	$\ell_2 = 2 + 15 + 37 + 46 = -\ 2.8$
temperature	$\ell_3 = 3 + 16 + 27 + 45 = -16.6$
hold up	$\ell_4 = 4 + 35 + 26 + 17 = 0.5$
recycle	$\ell_5 = 5 + 12 + 34 + 67 = 3.2$
rate of addition NaOH	$\ell_6 = 6 + 13 + 24 + 57 = -22.8$
filter cloth	$\ell_7 = 7 + 23 + 14 + 56 = -\ 3.4$.

The estimates -10.9, -16.6, and -22.8 are suspiciously large when compared to the others. The simplest interpretation of the results would be that the main effect of the factors **1**, **3** and **6** were important. However, many other interpretations are possible. Among these would be that the main effects of factor **3** and **6** and the interaction **3 6** (which is associated with **1**) were responsible for the observed results. Equivalently the main effects of **1** and **6** with **1 6**, or **1** and **3** with **1 3**, could be responsible. It was decided therefore to repeat the design with reverse signs, yielding the following results:

1	2	3	4	5	6	7	Filtration
+	+	+	+	−	−	−	66.7
−	+	+	−	+	+	−	65.0
+	−	+	−	+	−	+	86.4
−	−	+	+	−	+	+	61.9
+	+	−	−	−	+	+	47.8
−	+	−	+	+	−	+	59.0
+	−	−	+	+	+	−	42.6
−	−	−	−	−	−	−	67.6.

The estimates from this second design alone are

$$\ell_1' = -1 + 25 + 36 + 47 = 2.5$$
$$\ell_2' = -2 + 15 + 37 + 46 = 5.0$$
$$\ell_3' = -3 + 16 + 27 + 45 = -15.8$$
$$\ell_4' = -4 + 35 + 26 + 17 = 9.2$$
$$\ell_5' = -5 + 12 + 34 + 67 = -\ 2.3$$
$$\ell_6' = -6 + 13 + 24 + 57 = 15.6$$
$$\ell_7' = -7 + 23 + 14 + 56 = -\ 3.3$$

Whence by taking sums and differences of the linear combinations provided by the two component designs we obtain for the aggregate design

1 = − 6.7	**2 5** + **3 6** + **4 7** = − 4.2
2 = − 3.9	**1 5** + **3 7** + **4 6** = 1.1
3 = − 0.4	**1 6** + **2 7** + **4 5** = −16.2
4 = − 4.4	**3 5** + **2 6** + **1 7** = 4.9
5 = 2.8	**1 2** + **3 4** + **6 7** = 0.5
6 = −19.2	**1 3** + **2 4** + **5 7** = − 3.6
7 = 0.1	**2 3** + **1 4** + **5 6** = − 3.4

on review it seemed likely that the effect -19.2 associated with factor **6**, and the effect -16.2 associated with the linear combination (**1 6** + **2 7** + **4 5**) were probably real. It was also to be noted that the largest of the remaining effects, -6.7, was associated with factor **1**. The most likely explanation of the data

Rate of Addition of NaOH		Water Supply: Reservoir	Water Supply: Well
	Slow	65.4	42.6
	Fast	68.5	78.0

FIGURE 1
Two way table of average responses

therefore was that variables **1** and **6** both have effects and that they interacted. A two-way table of average values exemplifying these effects is shown below in Figure 1. It should be noted that the other explanations of the data are quite possible. For example the large effect attributed to the interaction between the factors **1** and **6** could be attributed equally well to the interaction **2 7** or **4 5**. The fact that none of the factors **2**, **4**, **5** and **7** have main effects does not, of course, preclude the possibility that their interactions exist. In fact in terms of the response surface if the center conditions of the experiment are located on the crest of a diagonally running ridge we should expect exactly this situation to occur. Of the possible explanations, however, that involving **1** and **6** and their two-factor interaction seemed by far the most likely. The crucial test was whether the trouble would be cured by using well water and the slow addition rate of caustic soda while leaving the other variables at their usual levels.

A number of additional trials were run on the plant in which the only modifications made were the use of well water with a slow rate of addition of caustic. These runs did give satisfactorily short filtration times in the neighborhood of forty minutes and the modification was adopted.

5: RESOLUTION IV DESIGNS

We have seen that a valuable design can be generated by switching the signs of all the variables in a 2^{7-4}_{III} fractional factorial and adding the resultant design to the original fraction. This aggregate design, which uses sixteen runs, makes it possible to estimate all seven main effects clear of the two-factor interactions. The design is thus of resolution IV. In fact, it is a 2^{7-3}_{IV} design. It is possible to do even slightly better than this. The signs of the elements corresponding to the identity column I can also be switched, and the resulting set of eight positive and eight negative signs can be associated with an eighth factor. The final design is shown in Table 15 on page 338.

We call such a design a "fold over" design.

We must now consider what the generators and hence the defining relations are for this design. Each component group of eight runs can be regarded as a 2^{8-5} design with generating relations

$$\mathbf{I} = \mathbf{8} = \mathbf{1\,2\,4} = \mathbf{1\,3\,5} = \mathbf{2\,3\,6} = \mathbf{1\,2\,3\,7}$$

and

$$\mathbf{I} = -\mathbf{8} = -\mathbf{1\,2\,4} = -\mathbf{1\,3\,5} = -\mathbf{2\,3\,6} = \mathbf{1\,2\,3\,7}$$

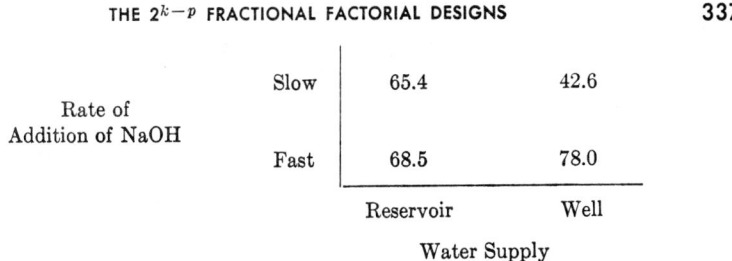

TABLE 15
A 2^{8-4}_{IV} fold over design

8				4	5	6	7	
I	1	2	3	1 2	1 3	2 3	1 2 3	
+	−	−	−	+	+	+	−	
+	+	−	−	−	−	+	+	
+	−	+	−	−	+	−	+	
+	+	+	−	+	−	−	−	Principal fraction 2^{7-4}_{III}
+	−	−	+	+	−	−	+	
+	+	−	+	−	+	−	−	
+	−	+	+	−	−	+	−	
+	+	+	+	+	+	+	+	
−	+	+	+	−	−	−	+	
−	−	+	+	+	+	−	−	
−	+	−	+	+	−	+	−	
−	−	−	+	−	+	+	+	Principal fraction with all signs reversed
−	+	+	−	−	+	+	−	
−	−	+	−	+	−	+	+	
−	+	−	−	+	+	−	+	
−	−	−	−	−	−	−	−	

respectively. Applying the rule for combining fractions we notice at once that **1 2 3 7** is a generator for the aggregate design, and the remaining three generators are independent even products of **8**, **1 2 4**, **1 3 5**, and **2 3 6**. In particular, we can use **1 2 4 8**, **1 3 5 8** and **2 3 6 8** so that finally the generating relations for the aggregate design is:

$$\mathbf{I} = \mathbf{1\,2\,4\,8} = \mathbf{1\,3\,5\,8} = \mathbf{2\,3\,6\,8} = \mathbf{1\,2\,3\,7} \tag{22}$$

The defining relation is therefore

$$\begin{aligned}\mathbf{I} &= \mathbf{1\,2\,4\,8} = \mathbf{1\,3\,5\,8} = \mathbf{2\,3\,6\,8} = \mathbf{1\,2\,3\,7} = \mathbf{2\,3\,4\,5} = \mathbf{1\,3\,4\,6} = \mathbf{3\,4\,7\,8}\\ &= \mathbf{1\,2\,5\,6} = \mathbf{2\,5\,7\,8} = \mathbf{1\,6\,7\,8} = \mathbf{4\,5\,6\,8} = \mathbf{2\,4\,6\,7} = \mathbf{1\,4\,5\,7} = \mathbf{3\,5\,6\,7}\\ &= \mathbf{1\,2\,3\,4\,5\,6\,7\,8}\end{aligned}$$

The generating relations for all sixteen of the 2^{8-4}_{IV} factionals are:

$$\mathbf{I} = \pm\mathbf{1\,2\,4\,8}, \qquad \mathbf{I} = \pm\mathbf{1\,3\,5\,8}, \qquad \mathbf{I} = \pm\mathbf{2\,3\,6\,8} \text{ and } \mathbf{I} = \pm\mathbf{1\,2\,3\,7}\,.$$

Ignoring interactions between three or more factors, and using the principal defining relation, the sixteen quantities which can now be estimated from the principal one-sixteenth fraction are given in Table 16.

As before further fractions can be performed in combination with the original fraction to isolate particular two-factor interactions or combinations of two-factor interactions. It will be seen now that when a design is formed containing 2^{k+1} runs from a design containing 2^k runs by replicating the 2^k design with reversed signs and associating some further factor **X** with the 2^k plus ones and 2^k minus ones, then a general rule for obtaining the generators and defining relation of the new design from the generators and defining relation of the old design is as follows: 1) All generators which contain an even number of

TABLE 16

Effects estimable using the 2_{IV}^{8-4} design

	Average
8 main effects	**1**
	2
	3
	4
	5
	6
	7
	8
7 sets of two-factor interactions confounded in groups of four	**1 2 + 3 7 + 4 8 + 5 6**
	1 3 + 2 7 + 5 8 + 4 6
	1 4 + 2 8 + 3 6 + 5 7
	1 5 + 3 8 + 2 6 + 4 7
	1 6 + 7 8 + 3 4 + 2 5
	1 7 + 2 3 + 6 8 + 4 5
	1 8 + 2 4 + 3 5 + 6 7

characters in the original design are retained as generators in the new design, 2) All generators which contain an odd number of characters in the original designs will be reproduced containing the extra character **X** as generators in the new design. For example, the generator **1 3 4** will become **1 3 4 X**.

An Alternative Method for Generating Designs of Resolution IV

An inspection of the generators for the 2_{IV}^{8-4} design just described will show that an alternative method for constructing this design would be to write down in standard order the sixteen combinations of variables for a complete 2^4 factorial, and then to associate further factors with the four three-factor interactions. To demonstrate, let the 2^4 factorial be written down in terms of the variables **1**, **2**, **3** and **8**. The four three-factor interactions are then **1 2 8**, **1 3 8**, **2 3 8** and **1 2 3**. These can now be associated with the four new variables **4**, **5**, **6** and **7** to give the set of four generators

$$\begin{array}{cccccc} 1 & 2 & & 8 & 4 \\ 1 & & 3 & 8 & 5 \\ & 2 & 3 & 8 & 6 \\ 1 & 2 & 3 & & 7 \end{array} \qquad (24)$$

The design thus constructed is identical to that given in Table 15. The only reason, of course, for starting off with variables **1**, **2**, **3** and **8** instead of **1**, **2**, **3** and **4** is to show the identity between this method of construction and the previous one.

As a further example of this second method for constructing resolution IV designs let us construct the 2_{IV}^{16-11} design. Since the design contains 32 runs we begin by writing down the full 2^5 factorial in the variables **1**, **2**, **3**, **4** and **5**. Eleven additional variables are now introduced by associating them with the ten three-factor interactions and the single five-factor interaction. We thus have for the set of eleven generators

1	2	3			6
1	2		4		7
1	2			5	8
1		3	4		9
1		3		5	10
1			4	5	11
	2	3	4		12
	2	3		5	13
	2		4	5	14
		3	4	5	15
1	2	3	4	5	16

(25)

If three-factor and higher order interaction terms are negligible, thirty-two independent estimates can be obtained. They include the grand average, the sixteen main effects **1**, **2**, **3**, $\cdots$, **16**; and the fifteen combinations of two-factor interactions displayed below

1 2 + 15 16 + 3 6 + 4 7 + 5 8 + 9 12 + 10 13 + 11 14
1 3 + 2 6 + 14 16 + 4 9 + 5 10 + 11 15 + 7 12 + 8 13
1 4 + 2 7 + 3 9 + 13 16 + 5 11 + 6 12 + 10 15 + 8 14
1 5 + 2 8 + 3 10 + 4 11 + 12 16 + 6 13 + 7 14 + 9 15
1 6 + 2 3 + 14 15 + 4 12 + 5 13 + 11 16 + 7 9 + 8 10
1 7 + 2 4 + 3 12 + 13 15 + 5 14 + 6 9 + 10 16 + 8 11
1 8 + 2 5 + 3 13 + 4 14 + 12 15 + 6 10 + 7 11 + 9 16
1 9 + 2 12 + 3 4 + 10 11 + 5 15 + 6 7 + 13 14 + 8 16
1 10 + 2 13 + 3 5 + 4 15 + 12 14 + 6 8 + 7 16 + 9 11
1 11 + 2 14 + 3 15 + 4 5 + 9 10 + 6 16 + 7 8 + 12 13
1 12 + 2 9 + 3 7 + 4 6 + 5 16 + 10 14 + 8 15 + 11 13
1 13 + 2 10 + 3 8 + 4 16 + 5 6 + 11 12 + 7 15 + 9 14
1 14 + 2 11 + 3 16 + 4 8 + 5 7 + 6 15 + 9 13 + 10 12
1 15 + 2 16 + 3 11 + 4 10 + 5 9 + 6 14 + 7 13 + 8 12
1 16 + 2 15 + 3 14 + 4 13 + 5 12 + 6 11 + 7 10 + 8 9

In general, a resolution IV design may always be constructed by first writing down the design matrix for a two-level factorial and then associating new variables with all those interaction columns having an *odd* number of numerals.

Of course, this 2_{IV}^{16-11} design could have been obtained by fold-over by first writing down the 2_{III}^{15-11} design, the saturated resolution III design for fifteen variables in sixteen runs. The eleven generators for this design is given in Table 17a. In Table 17a the variables are numbered from **2** to **16** to make the equivalence between the two methods of construction evident. The generators for the 2_{IV}^{16-11} obtained by fold-over is shown in Table 17b. These generators are obtained by attaching the variable **1** to every word in the generating relation of the 2_{III}^{15-11} having an odd number of symbols. The generators, and hence the design obtained by fold-over, are thus identical to those displayed earlier in Eq(25). The same principle of fold-over may be used with the Plackett and Burman designs. For example, using the Plackett and Burman design for $k = 11$

TABLE 17a

Generating Relation 2_{III}^{15-11}

2	3			6
2		4		7
2			5	8
	3	4		9
	3		5	10
		4	5	11
2	3	4		12
2	3		5	13
2		4	5	14
	3	4	5	15
2	3	4	5	16

TABLE 17b

Generation Relation for 2_{IV}^{16-11} *Obtained by Fold-over*

1	2	3			6
1	2		4		7
1	2			5	8
1		3	4		9
1		3		5	10
1			4	5	11
	2	3	4		12
	2	3		5	13
	2		4	5	14
		3	4	5	15
1	2	3	4	5	16

variables in twelve runs we may derive a design usable for studying twelve variables in twenty-four runs in which no two-factor interaction is aliased with any main effect.

Complete Factorials within Fractionals Applied to Screening

When little is known about the variables which effect a particular response we are in what may be called a screening situation. That is to say, that although it is necessary to test a rather large number of variables which might conceivably have important effects, it can be realistically postulated that only a few, perhaps one, two or three of the variables, will be of major importance. Whichever variables do turn out to be of major influence may of course interact with one another. To put this argument in another way, we may have a fairly large number, say eight variables, which are of possible importance, but we believe

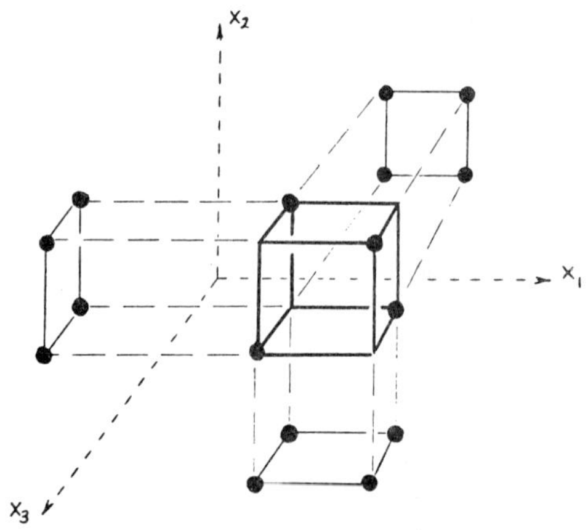

FIGURE 2—Projection of 2^{3-1}_{III} into three 2^2 factorials.

the effects of all but, say, three of these are likely to be negligible. Thus, we tentatively entertain the idea that at least five of the variables can be regarded essentially as dummies, but we don't know *which* five. In these circumstances we need a design in the complete set of eight variables which will produce a complete factorial in *any* three of the component variables. Thus, although we don't know which subset of the variables will turn out to be important, whichever subset does, provides a full factorial, or even a replicated factorial, in those variables.

The basic idea is illustrated in the very simpliest case for the one-half replicate of the 2^3 factorial shown in Figure 2. Suppose the total number of variables considered is three, but it can be reasonably postulated that not more than two have any real effects. Then we see from Figure 2 that the design supplies a complete factorial in any of the three pairs of variables since each projection of the 2^{3-1}_{III} design into a two dimensional plane produces a complete factorial design. This is also apparent from inspection of the design matrix since, if we drop any one column of the design matrix, the remaining two columns provide a full 2^2 factorial. This can be seen even more simply, for the generating relation for this design is $\mathbf{I} = \mathbf{1\ 2\ 3}$ and if any one of the variables is dropped the generator will vanish showing that the resulting design is not a fractional factorial.

In general, it is clear that a design of resolution R will provide a complete factorial in any sub-set of the $(R - 1)$ variables. This must be so since every word in the defining relation contains R or more characters. It follows that is all but $(R - 1)$ characters are treated as dummies, then every word in the defining relation will disappear.

2^3 *Factorials within Resolution* IV *Designs*

If a design of resolution IV contains $r \times 2^3$ runs then it can be regarded as providing r replicates of a full factorial in any three variables. As an example, consider the sixteen-run resolution IV design for eight variables i.e. the 2^{8-4}_{IV} design.

This design can be regarded as providing a twice replicated 2^3 factorial for every one of the fifty-six choices of three variables out of eight. Geometrically, this means that the sixteen points in eight dimensional space can be projected into any one of the fifty-six three dimensional coordinate sub-spaces to produce a replicated cube. The reader can readily confirm for himself that the omission of any five columns from Table 15 provides a twice replicated factorial in the remaining variables.

As always, evidence from experiments of this kind should only be regarded as suggestive and subject to confirmation rather than as supplying definite proof. Alternative explanations of the results obtained from such experiments involving higher order interactions could be easily produced. However, in selecting alternative explanations as worthy of further study we rest heavily upon our prior beliefs about the plausibility of these alternatives.

It is interesting to note an early use of designs of this kind by Tippett [15]. An adequate statement of the proper attitude towards the results is to be found in a discussion by R. A. Fisher [12] of Tippett's example.

General Rules for Designs Obtained by Projection

We have seen that a design of resolution R provides a complete factorial in any sub-set of $(R - 1)$ variables. In particular, designs of resolution III may be used for screening up to two variables out of $N - 1$ variables, designs of resolution IV may be used for screening up to three variables out of $N/2$ variables, and designs of resolution V, which we shall discuss later, may be used for screening up to four variables out of a larger number. If a design of resolution R is used to screen subsets of R variables, then full factorials will result for certain subsets, and fractional factorial for others. Those subsets of variables providing fractional factorials are simply subsets which appear as words in the final defining relation. For example, consider the 2_{IV}^{8-4} design discussed earlier. Its defining relation is

$$\begin{aligned} \mathbf{I} &= \mathbf{1\,2\,4\,8} = \mathbf{1\,3\,5\,8} = \mathbf{2\,3\,6\,8} = \mathbf{1\,2\,3\,7} = \mathbf{2\,3\,4\,5} = \mathbf{1\,3\,4\,6} = \mathbf{3\,4\,7\,8} \\ &= \mathbf{1\,2\,5\,6} = \mathbf{2\,5\,7\,8} = \mathbf{1\,6\,7\,8} = \mathbf{4\,5\,6\,8} = \mathbf{2\,4\,6\,7} = \mathbf{1\,4\,5\,7} = \mathbf{3\,5\,6\,7} \\ &= \mathbf{1\,2\,3\,4\,5\,6\,7\,8}. \end{aligned}$$

Regarded as a design to screen sub-sets of four variables, this design will provide replicated half-fractions for the fourteen combinations of variables **1**, **2**, **4** and **8**; **1**, **3**, **5** and **8**; **2**, **3**, **6** and **8**; etc., which appear as forming words in the defining relation, and complete 2^4 factorials designs for any one of the remaining fifty-six combinations of four variables. In the case of resolution V designs we can, in accordance to our general rule, obtain full factorials in any set of four variables. These designs would, for most purposes, also be adequate for screening five variables because even for those combinations of variables which appear as words in the defining relation, one-half replicates would be available, and these would permit all main effects and two factor interactions to be distinguished, on the assumption of course that higher order interaction effects are negligible.

Example

The problem of analyzing these designs can be thought of either as picking out the one, two or three variables whose main effects and interactions can account for all the effects found, or equivalently for looking for sets of replicates within the runs. As an example, consider the data given in Table 18 obtained from a screening experiment containing eight variables, using the generators **1 2 4 8**, **1 3 5 8**, **2 3 6 8** and **1 2 3 7**

TABLE 18

	Variables								
Run #	1	2	3	8	4	5	6	7	
1	−	−	−	−	−	−	−	−	60.4
2	+	−	−	−	+	+	−	+	66.0
3	−	+	−	−	+	−	+	+	62.1
4	+	+	−	−	−	+	+	−	63.3
5	−	−	+	−	−	+	+	+	82.9
6	+	−	+	−	+	−	+	−	75.4
7	−	+	+	−	+	+	−	−	82.4
8	+	+	+	−	−	−	−	+	73.0
9	−	−	−	+	+	+	+	−	68.1
10	+	−	−	+	−	−	+	+	61.2
11	−	+	−	+	−	+	−	+	71.3
12	+	+	−	+	+	−	−	−	59.6
13	−	−	+	+	+	−	−	+	67.3
14	+	−	+	+	−	+	−	−	75.3
15	−	+	+	+	−	−	+	−	66.7
16	+	+	+	+	+	+	+	+	77.1

The estimated effects are given in Table 19a.

TABLE 19a

Average	69.5
1	−1.3
2	−0.1
3	11.0
4	0.5
5	7.6
6	0.2
7	1.2
8	−2.4
12 + 37 + 48 + 56	−1.1
13 + 27 + 58 + 46	1.7
14 + 28 + 36 + 57	0.8
15 + 38 + 26 + 47	−4.5
16 + 78 + 34 + 25	0.6
17 + 23 + 68 + 45	−0.3
18 + 24 + 35 + 67	1.2

TABLE 19b

3	8	5	Responses	
−	−	−	60.4	62.1
+	−	−	75.4	73.0
−	+	−	61.2	59.6
+	+	−	67.3	66.7
−	−	+	66.0	63.3
+	−	+	82.9	82.4
−	+	+	68.1	71.3
+	+	+	75.3	77.1

It was not expected that more than a few of the eight variables would in fact have important effects upon the response, and it will be seen that the data is readily explained by supposing that the important variables are **3**, **5** and **8**. The main effects and two-factor interactions associated with these variables are underlined in Table 19a. On this explanation runs 1 and 3, 2 and 4, 5 and 7, 6 and 8, 9 and 11, and 10 and 12, 13 and 15 and finally 14 and 16 are essentially duplicates one of the other differing mainly because of experimental error and partly because of effects of the other variables of lesser importance. The data are rearranged as a duplicated 2^3 factorial in variables **3**, **8** and **5** in Table 19b.

In an experiment of this kind it would have been advantageous to have available some independent estimate of pure error obtained, for example, from duplication of certain of the runs selected in accordance with principles described elsewhere [10]. In such a case we could then compare the size of the error obtained from the "constructed" duplicates with that from known duplicates.

Blocking for Designs of Resolution IV

Assuming that interactions between three or more variables are negligible, the 2_{IV}^{8-4} design with generators **1 2 3 7**, **1 2 4 8**, **1 3 5 8** and **2 3 6 8** provides independent estimates of the eight main effects and of seven groups of two-factor interactions. By using the + and − signs associated with the interaction columns, this design can be broken into either two, four or eight equal sized blocks which are unconfounded with main effects.

For example, we may use the + and − signs associated with the two-factor interaction set **1 2 + 3 7 + 4 8 + 5 6**, to define two blocks, if we call the block contrast $\mathbf{B}_1$, and put $\mathbf{B}_1$ = **1 2 + 3 7 + 4 8 + 5 6**. To break the design into four equal blocks, two columns associated with the interaction sets may be used. For example, we might choose $\mathbf{B}_1$ = **1 2 + 3 7 + 4 8 + 5 6** and $\mathbf{B}_2$ = **1 3 + 2 7 + 5 8 + 4 6**. Each of the four blocks will contain the four runs identified by the pairs of versions (±, ±), that is, the four sets of versions (−, −), (+, −), (−, +), (+, +) provided by the two interaction columns. The block interaction effect $\mathbf{B}_1\mathbf{B}_2$ will then be found to be associated with another two-factor interaction set, that is, $\mathbf{B}_1\mathbf{B}_2$ = **1 7 + 2 3 + 6 8 + 4 5**. This can be confirmed by actually multiplying out the elements of the columns associated with $\mathbf{B}_1$ and $\mathbf{B}_2$, or simply by noting, for example, that products of the interaction elements in $\mathbf{B}_1$ and $\mathbf{B}_2$ are **1 2** × **1 3** = **2 3**, **1 2** × **2 7** = **1 7**; **5 6** × **4 8** = **4 5** etc.

To break the design into eight equal blocks, three of the interaction sets must be used. However, in choosing the third set we may not use the column associated with the interaction $\mathbf{B}_1\mathbf{B}_2$, although any of the remaining interaction columns may be used. Let us choose $\mathbf{B}_3$ = **1 8 + 2 4 + 3 5 + 6 7**. Each of the blocks will now contain the two runs identified by the eight sets of versions (±, ±, ±) provided by $\mathbf{B}_1$, $\mathbf{B}_2$ and $\mathbf{B}_3$. It can be readily confirmed by multiplying out the elements of the block columns that the complete set of seven two-factor interaction comparisons are now used up, that is,

TABLE 20a

Construction of 2_{IV}^{8-4} in Blocks of Two

1	2	3	8	7 1 2 3	4 1 2 8	5 1 3 8	6 2 3 8	B_1 1 2	B_2 1 3	B_3 1 8
−	−	−	−	−	−	−	−	+	+	+
+	−	−	−	+	+	+	−	−	−	−
−	+	−	−	+	+	−	+	−	+	+
+	+	−	−	−	−	+	+	+	−	−
−	−	+	−	+	−	+	+	+	−	+
+	−	+	−	−	+	−	+	−	+	−
−	+	+	−	−	+	+	−	−	−	+
+	+	+	−	+	−	−	−	+	+	−
−	−	−	+	−	+	+	+	+	+	−
+	−	−	+	+	−	−	+	−	−	+
−	+	−	+	+	−	+	−	−	+	−
+	+	−	+	−	+	−	−	+	−	+
−	−	+	+	+	+	−	−	+	−	−
+	−	+	+	−	−	+	−	−	+	+
−	+	+	+	−	−	−	+	−	−	−
+	+	+	+	+	+	+	+	+	+	+

TABLE 20b

Construction of 2_{IV}^{8-4} in Blocks of Two

1	2	3	8	7	4	5	6	B_1	B_2	B_3	
+	−	−	−	+	+	+	−	−	−	−	Block 1
−	+	+	+	−	−	−	+				
+	+	−	−	−	−	+	+	+	−	−	Block 2
−	−	+	+	+	+	−	−				
+	−	+	−	−	+	−	+	−	+	−	Block 3
−	+	−	+	+	−	+	−				
+	+	+	−	+	−	−	−	+	+	−	Block 4
−	−	−	+	−	+	+	+				
+	−	−	+	+	−	−	+	−	−	+	Block 5
−	+	+	−	−	+	+	−				
+	+	−	+	−	+	−	−	+	−	+	Block 6
−	−	+	−	+	−	+	+				
+	−	+	+	−	−	+	−	−	+	+	Block 7
−	+	−	−	+	+	−	+				
+	+	+	+	+	+	+	+	+	+	+	Block 8
−	−	−	−	−	−	−	−				

$$
\begin{aligned}
\mathbf{B}_1 &= \mathbf{1\,2} + \mathbf{3\,7} + \mathbf{4\,8} + \mathbf{5\,6} \\
\mathbf{B}_2 &= \mathbf{1\,3} + \mathbf{2\,7} + \mathbf{5\,8} + \mathbf{4\,6} \\
\mathbf{B}_3 &= \mathbf{1\,8} + \mathbf{2\,4} + \mathbf{3\,5} + \mathbf{6\,7} \\
\mathbf{B}_1\mathbf{B}_2 &= \mathbf{1\,7} + \mathbf{2\,3} + \mathbf{6\,8} + \mathbf{4\,5} \\
\mathbf{B}_1\mathbf{B}_3 &= \mathbf{1\,4} + \mathbf{2\,8} + \mathbf{3\,6} + \mathbf{5\,7} \\
\mathbf{B}_2\mathbf{B}_3 &= \mathbf{1\,5} + \mathbf{3\,8} + \mathbf{2\,6} + \mathbf{4\,7} \\
\mathbf{B}_1\mathbf{B}_2\mathbf{B}_3 &= \mathbf{1\,6} + \mathbf{7\,8} + \mathbf{3\,4} + \mathbf{2\,5}
\end{aligned}
\tag{26}
$$

As the reader can confirm for himself, subject only to the condition that the $\mathbf{B}_3$ must not be chosen so as to coincide with $\mathbf{B}_1\mathbf{B}_2$, the association between blocks and two-factor interactions can be made in any other way whatever.

Tables 20a and 20b show how we can write out a 2_{IV}^{8-4} design arranged in eight blocks of two runs each. Since the complete design contains sixteen runs, we begin by writing down a 2^4 factorial in four of the eight variables. (In order that the final design may be compared with designs obtained previously, we chose these variables to be **1**, **2**, **3** and **8** although they could just as easily been chosen to be **1**, **2**, **3** and **4**.) The variables **4**, **5**, **6** and **7** are then associated with the three-factor interactions. The generators are **1 2 3 7**, **1 2 4 8**, **1 3 5 8** and **2 3 6 8**. As illustrated in Table 20a we now write down the three columns corresponding to the interactions **1 2**, **1 3**, and **1 8** and associate these with the block factors $\mathbf{B}_1$, $\mathbf{B}_2$ and $\mathbf{B}_3$. The eight blocks are then obtained by putting those pairs of runs for which $\mathbf{B}_1$, $\mathbf{B}_2$ and $\mathbf{B}_3$ are $(-\ -\ -)$ into the first block, the pair of runs for which $\mathbf{B}_1$, $\mathbf{B}_2$ and $\mathbf{B}_3$ are $(+\ -\ -)$ in the second block and so on.

In Table 20b it will be noted that the second run in each block is the fold-over, or mirror image, of the first run, that is, the versions of one run are exactly reversed in the second. Suppose now that a single run is taken from each block such that one of the variables always appears with the same sign. Choosing, for example, those runs with the + version of variable 1 we obtain the array given in Table 21.

The reader will note that the result, omitting variable **1**, is the 2_{III}^{7-4} design in the variables **2**, **3**, **4**, $\cdots$, **8** with generating relation $\mathbf{I} = \mathbf{2\,3\,7} = \mathbf{2\,8\,4} = \mathbf{3\,8\,5} = \mathbf{2\,3\,8\,6}$. This design is identical, except for the number identification given the seven variables, to the 2_{III}^{7-4} design described earlier.

TABLE 21

1	2	3	8	7	4	5	6
+	−	−	−	+	+	+	−
+	+	−	−	−	−	+	+
+	−	+	−	−	+	−	+
+	+	+	−	+	−	−	−
+	−	−	+	+	−	−	+
+	+	−	+	−	+	−	−
+	−	+	+	−	−	+	−
+	+	+	+	+	+	+	+

We see now that the principle of fold-over can be modified slightly to provide resolution IV designs automatically broken into blocks of two runs such that the block effects are unconfounded with the main effects. We begin by writing down the design matrix for the appropriate resolution III design in k variables plus an additional column **I** consisting solely of plus signs. Each row of the design matrix is then folded-over, that is, repeated with all signs reversed. The pairs of rows form blocks of two of a resolution IV design in $k + 1$ variables. For example, the 2_{IV}^{4-2} in blocks of two is constructed by first writing down the 2_{III}^{3-1} along with the column **I**, and then folding over each row to provide four blocks of two runs each as illustrated in Table 22. This 2_{IV}^{4-1} design is identical

TABLE 22

Original 2^{3-1}, I = 1 2 3: 1	2	3	I	2_{IV}^{4-1} in Blocks of two, I = 1 2 3 4: 1	2	3	4	
−	−	+	+	−	−	+	+	Block 1
+	−	−	+	+	+	−	−	
−	+	−	+	+	−	−	+	Block 2
+	+	+	+	−	+	+	−	
				−	+	−	+	Block 3
				+	−	+	−	
				+	+	+	+	Block 4
				−	−	−	−	

to that obtained earlier, and illustrated in Table 12, in the discussion of blocking designs of resolution III. Similarly, the 2_{IV}^{8-4} design in blocks of two obtained in Table 20b could also have been formed by using the principle of fold-over starting out with the 2_{III}^{7-3} along with a column vector of plus signs and pairing each run with its fold-over. The same is true of the 2_{IV}^{16-11} obtained by fold-over described in Table 17b. The designs obtained by folding over the Plackett and Burman designs can also be broken into blocks of two runs each using precisely the same device.

In general, any resolution IV design provides an opportunity to obtain blocks of size two whose effects do not confound any of the main effects. In doing this, of course, we confound the two-factor interactions with blocks. Nevertheless, the resulting designs are of considerable interest. Often, the comparisons between blocks merely represent influences upon the response having a somewhat higher variation than that responsible for differences within blocks. In these circumstances it is reasonable to think of the strings of two-factor interactions simply as being estimated with a variance somewhat higher than that appropriate for the main effects. For instance, these designs may be used where it is suspected that a time trend may occur during the course of the trials. Provided proper randomization is applied both to the order of runs within the blocks and to the order of running the blocks themselves, the design is such that whereas main effects were determined with a variance appropriate to successive observations, the strings of two-factor interactions would be estimated with a variance appropriate to pairs of runs made in random order in the presence of a time trend.

"Major and "Minor" Variables in Resolution IV *Designs*

We have already seen that a 2^{p-q} design of resolution IV can be regarded from two points of view: (1) it is a design suitable for providing estimates of the p main effects even though two-factor interactions may occur, and (2) it is a design suitable for providing unbiased estimates of all main effects and interactions between *any* three of the factors if the others are of no importance. The designs can be considered from still another point of view. Considering the 2_{IV}^{16-11} design as an example, we have seen how this design may be run in sixteen blocks of two runs each, the blocks being obtained from four block generators associated with two-factor interactions. Alternatively we can choose the four block generators to represent actual variables. Suppose for example, we have four "major" variables for which we wish to estimate all the main effects and all the interactions and we have sixteen further variables which we believe exert at most main effects, and may be conveniently viewed as "minor" variables. Then this design may be employed associating the four major variables with the block generators and the remaining minor variables with the sixteen "main effect" factors. Of course, all the effects among the major variables will now be confounded with the sets of two-factor interactions of the minor variables. However, since minor variables are believed to exert at most main effects, the two-factor interactions between these variables are tentatively assumed to be nil.

In this connection, there is an opportunity to make use of any prior feeling which the experimenter may have concerning the possibility of interaction in the minor variables. Should he feel particularly anxious about a possible interaction between two minor variables, then he can usually arrange, by inspecting tables such as that given in Equation (26), that this interaction is associated with an unimportant interaction between the major variables. For instance, in this present example, the interactions **1 6**, **7 8**, **3 4** and **2 5** between the minor variables are all confounded with the three-factor interaction $\mathbf{B}_1\mathbf{B}_2\mathbf{B}_3$ between major variables. This three factor interaction might be expected to be unimportant *a priori*. It should be noticed that so long as $\mathbf{B}_1$, $\mathbf{B}_2$ and $\mathbf{B}_3$ are pseudo variables representing comparisons among blocks, then interactions such as $\mathbf{B}_1\mathbf{B}_2\mathbf{B}_3$ will represent comparisons of precisely the same potential as are represented by the main effects $\mathbf{B}_1$, $\mathbf{B}_2$, etc. When, however, $\mathbf{B}_1$, $\mathbf{B}_2$, etc., are used to represent real variables, main effects and interactions revert to their former relative status.

When thirty-two runs are to be made and where there are four major variables along with sixteen minor variables, the 2_{IV}^{16-11} design may be employed. With sixteen runs three variables and all their interactions may be investigated by associating the block generators with these major variables and the eight minor variables then introduced. For eight runs, two major variables and their interaction plus four main effect variables are possible. Of course, when the designs are used in this way no blocking is permissible. However, even here a certain degree of flexibility is possible. For instance, for the thirty-two run design we might wish to have only two principal factors in which case we could associate these with two of the block generators using the other two block generators to form blocks of eight. It will be clear to the reader that these arrangements provide a very versatile set of designs which may be used in a variety of circumstances.

From Resolution IV *Designs to Resolution* III *Designs*

When the 2_{IV}^{16-11} designs is used to study simultaneously four major variables along with sixteen minor variables, a convenient notation for the design is

$$2^4 \subset 2_{IV}^{16-11}$$

where the symbol $\subset$ is read "contained in" or "embedded in". Thus $2^3 \subset 2_{IV}^{8-4}$ and $2^2 \subset 2_{IV}^{4-1}$ identify the sixteen and eight run designs described above.

The construction of a resolution III design from one of resolution IV now becomes obvious. The $2^4 \subset 2_{IV}^{16-11}$ is clearly a design for studying twenty variables in thirty-two runs. Suppose now that one of the interactions between the major variables is used to bring in still another variable. In fact, we might be willing to assume that all the interactions between the four major variables are negligible and in this instance eleven new variables (one for each of the interactions) could be introduced. The result, of course, is the 2_{III}^{31-26} design, that is, the fully saturated resolution III design for studying thirty-one variables in thirty-two runs.

Other Embedded Designs

The principle of embedded fold-over pairs producing blocks of two runs has wide application. As an example less orthodox than those mentioned above, we note that the thirty-two run 2_{IV}^{16-11} design in sixteen blocks is one in which a central composite design in three variables can be embedded. The composite design [14] would employ factor combinations of three major variables consisting of a 2^3 factorial along with six axial points and two center points for a total of sixteen points. Each of these factor combinations would then be duplicated, one duplicate containing one combination of versions of sixteen minor variables with the second duplicate the mirror image of these same sixteen variables. The additional sixteen variables would have to be such that they were not expected to have any effect other than a linear one.

Part II of this paper will contain a discussion of Resolution V designs along with an appendix.

Bibliography

(1) Yates, F., The Design and Analysis of Factorial Experiments., *Imper. Bur. Soil Sci. Tech. Comm.* 35, 1937.

(2) Fisher, R. A., The Theory of Confounding in Factorial Experiments, in Relation to the Theory of Groups, *Ann. Eugen.* 11, 1942.

(3) Daniel, C., Fractional Replication in Industrial Research, *Proc. 3rd Berkeley Symp.* V., 1956.

(4) Finney, D. J., The Fractional Replication of Factorial Arrangements, *Ann. Eugen.* 12, 1945.

(5) Brownlee, K. A., B. K. Kelly and P. K. Loraine, Fractional Replication Arrangements for Factorial Experiments with Factors at Two Levels, *Biometrika*, 35, 1948.

(6) Plackett, R. L., and J. P. Burman, The Design of Optimum Multifactorial Experiments, *Biometrika*, 33, 1946.
(7) Clatworthy, W. H., W. S. Connor, Deming and M. Zelen, Fractional Factorial Experiment Designs for Factors at Two Levels, U. S. *Dept. Comm. Nat. Bur. of Stnds., Applied Math.* Ser. 48, 1957.
(8) Kempthorne, O., A Simple Approach to Confounding and Fractional Replication in Factorial Experiments, *Biometrika*, 34, 1947.
(9) Bose, R. C. and Kishen, K., On the Problem of Confounding in the General Symmetrical Factorial Design, *Sankhya* 5, 1940.
(10) Dykstra, O., Partial Duplication of Factorial Experiments, *Technometrics*, 1, 1959.
(11) Plackett, R. L., Some Generalizations in the Multifactorial Design, *Biometrika*, 33, 1946.
(12) Fisher, R. A., *The Design of Experiments*, Oliver & Boyd, London, 1935.
(13) Box, G. E. P., Multifactor Designs of First Order, *Biometrika*, 39, 1951.
(14) Box, G. E. P. & K. B. Wilson, On the Experimental Attainment of Optimum Condition, *J. Roy. Stat. Soc., B.*, 13, 1951.
(15) Tippett, L. H. C., "Applications of Statistical Methods to the Control of Quality in Industrial Production", Manchester Statistical Society, 1934.
(16) Yates, F., Complex Experiments, *J. Roy. Stat. Soc., Suppl.*, 2, 1935.
(17) Rao, C. R., Factorial Experiments Derivable from Combinatorial Arrangements of Arrays, *J. Roy. Stat. Soc., Suppl.*, 9, 1947.

2.11

The 2^{k-p} Fractional Factorial Designs,* Part II

G. E. P. BOX and J. S. HUNTER†
*University of Wisconsin**

"......... but that there turtle is an insect."

6. Resolution V Designs

In Part I of this paper the construction of two version factorials of resolution III and IV was discussed. In some situations we need experimental designs in which all main effects and two factor interactions are unconfounded with all other main effects and two factor interactions. Such designs are called designs of resolution V since each word in the defining relation has five or more characters. The one-half replicate of the 2^5 factorial with defining relation $\mathbf{I} = \mathbf{1\,2\,3\,4\,5}$ mentioned in Part I of this paper is a resolution V design, that is, a 2_{V}^{5-1} fractional.

Since there are $k(k + 1)/2$ main effects and two factor interactions to be estimated, these designs require considerably larger numbers of experimental runs than designs of resolution III or IV. No simple fractional of resolution V exist for two, three or four variables. In fact, the largest number of factors which can be included in a fractional of resolution V is as follows:

Resolution V Designs

Number of Runs	Number of Factors
16	5
32	6
64	8
128	11

Construction of Resolution V Designs

Table 23 lists the resolution V designs for $k = 5, 6, \cdots, 11$. The use of this table for constructing the appropriate design and arranging it in blocks may be illustrated in the case of the one-sixteenth replicate of the 2^{11} design, i.e., the 2_{V}^{11-4} design. In this design there are $2^{11-4} = 2^7 = 128$ runs so that we first write down in standard order the seven columns of the full 2^7 design. The first column

* Research Sponsored by the United States Army under Contract No. DA-11-022-ORD-2059.

† Now with the Chemical Engineering Department, Princeton University.

The Collected Works of George E. P. Box, 1984, Wadsworth, Inc., Belmont, CA 94002. Originally published in *Technometrics,* vol. 3, no. 4 (1961), pp. 449-458.

TABLE 23

Number of variables	Number of Runs	Degree of Replication	Type of Design	Method of Introducing "new" factors	Blocking	Method of Introducing Blocks
5	16	$\frac{1}{2}$	2_V^{5-1}	$\pm$ 5 = 1 2 3 4	Not Available	
6†	32	$\frac{1}{2}$	2_V^{6-1}	$\pm$ 6 = 1 2 3 4 5	Two blocks of sixteen runs	B_1 = 1 2 3
7†	64	$\frac{1}{2}$	2_V^{7-1}	$\pm$ 7 = 1 2 3 4 5 6	Eight blocks of eight runs	B_1 = 1 3 5 7 B_2 = 1 2 5 6 B_3 = 1 2 3 4
8	64	$\frac{1}{4}$	2_V^{8-2}	$\pm$ 7 = 1 2 3 4 $\pm$ 8 = 1 2 5 6	Four blocks of sixteen runs	B_1 = 1 3 5 B_2 = 3 4 8
9*†	128	$\frac{1}{4}$	2_V^{9-2}	$\pm$ 9 = 1 4 5 7 8 $\pm$ 10 = 2 4 6 7 8	Eight blocks of sixteen runs	B_1 = 1 4 9 B_2 = 1 2 10 B_3 = 8 9 10
10	128	$\frac{1}{8}$	2_V^{10-3}	$\pm$ 8 = 1 2 3 7 $\pm$ 9 = 2 3 4 5 $\pm$ 10 = 1 3 4 6	Eight blocks of sixteen runs	B_1 = 1 4 9 B_2 = 1 2 10 B_3 = 8 9 10
11	128	$\frac{1}{16}$	2_V^{11-4}	$\pm$ 8 = 1 2 3 7 $\pm$ 9 = 2 3 4 5 $\pm$ 10 = 1 3 4 6 $\pm$ 11 = 1 2 3 4 5 6 7	Eight blocks of sixteen runs	B_1 = 1 4 9 B_2 = 1 2 10 B_3 = 8 9 10

Summary of Resolution V designs and their blocks. (All main effects and two-factor interactions clear of one another.)

*The nine factors in this design are **1, 2, 4, 5, 6, 7, 8, 9, 10**. In the text this design is derived from the 2^{11-4} and, as is there explained, it is convenient to drop factors **3** and **11**.

† The designs for 6, 7 and 9 variables are of resolution VI, VII and VI respectively *before* blocking.

corresponding to variable **1** consists of alternating minus and plus signs, the second column alternating pairs of minus and plus signs and so on. We now add four further columns using the relations found in column five of Table 23. The new vector associated with the variable **8** is generated by multiplying together the minus and plus signs of columns **1**, **2**, **3** and **7** to obtain the elements for the interaction column **1 2 3 7**. These are then used to identify the minus and plus signs of variable **8**, that is, **8** = **1 2 3 7**. A second new column is written down in a similar way for variable **9** using **9** = **2 3 4 5**. Similarly **10** = **1 3 4** 6 and **11** = **1 2 3 4 5 6 7**. We now have a complete one-sixteenth replicate of the 2^{11} design with generators

$$\mathbf{1\,2\,3\,7\,8}, \qquad \mathbf{2\,3\,4\,5\,9}, \qquad \mathbf{1\,3\,4\,6\,10}, \qquad \mathbf{1\,2\,3\,4\,5\,6\,7\,11},$$

The full defining relation of this design is therefore made up of the words:

I	**1 2 5 6 9 10**
1 2 3 7 8	**1 6 7 9 11**
2 3 4 5 9	**2 5 7 10 11**
1 3 4 6 10	**3 5 6 7 8 9 10**
1 2 3 4 5 6 7 11	**1 3 5 8 10 11**
1 4 5 7 8 9	**2 3 6 8 9 11**
2 4 6 7 8 10	**3 4 7 9 10 11**
4 5 6 8 11	**1 2 4 8 9 10 11**

We observe that all of the words in this defining relation have five or more characters identifying the design as one of resolution V. The generators given above are for the principal fraction. The generators for all sixteen fractions are given by

$$\pm\mathbf{1\,2\,3\,7\,8}, \qquad \pm\mathbf{2\,3\,4\,5\,9}, \qquad \pm\mathbf{1\,3\,4\,6\,10}, \qquad \pm\mathbf{1\,2\,3\,4\,5\,6\,7\,11}$$

To block this design in eight blocks of 16 runs we write down three further columns for $\mathbf{B}_1$, $\mathbf{B}_2$ and $\mathbf{B}_3$ using the relations in column seven of Table 23. The two versions of $\mathbf{B}_1$ are obtained by multiplying together the minus and plus signs of columns **1**, **4** and **9**, the versions of $\mathbf{B}_2$ using columns **1 2** and **10**, and the versions of $\mathbf{B}_2$ using columns **8 9** and **10**. The eight blocks are then identified by those runs having the signs of $\mathbf{B}_1$, $\mathbf{B}_2$, and $\mathbf{B}_3$ equal to $(-,-,-)$, $(+,-,-)$, $(-,+,-)$, $(+,+,-)$, $(-,-,+)$, $(+,-,+)$, $(-,+,+)$, and $(+,+,+)$.

Designs with 16 Runs: The 2_{V}^{5-1} Design

With five variables, a 2_{V}^{5-1} design is obtained using the defining relation

$$\mathbf{I} = \pm\mathbf{1\,2\,3\,4\,5}.$$

The 16 version combinations for such a design are written down by first writing out the design matrix for the 2^4 complete factorial and then identifying the versions of the fifth variable with those of the interaction **1 2 3 4**. Clearly all main effects are here confounded with four-factor interactions since the alias relationships of main effects are of the type

$$\pm\mathbf{1} = \mathbf{2\,3\,4\,5}$$

while those for the two-factor interactions are of the type

$$\pm\mathbf{1\ 2} = \mathbf{3\ 4\ 5}.$$

It follows that if we can ignore three and four-factor interactions we can estimate with these sixteen runs the average, the five main effects and the ten two-factor interactions. This design is remarkable in that every degree of freedom is used to estimate effects of interest. However none are available for confounding so that it is not possible for this design to be run in blocks without associating one or more main effects or two-factor interactions with the block variables.

*Designs with 32 Runs: The 2_{V}^{6-1} Design**

For six variables there are 21 effects to estimate (the six main effects and the 15 two-factor interactions). It is clear that a quarter replicate of a 2^6 factorial involving 16 points would not be large enough to estimate all the quantities of interest. The half replicate has as its defining relation

$$\mathbf{I} = \pm\mathbf{1\ 2\ 3\ 4\ 5\ 6}.$$

Using this design main effects are associated with five-factor interactions, for example, $\pm\mathbf{1} = \mathbf{2\ 3\ 4\ 5\ 6}$ and two-factor interactions are associated with four-factor interactions, for example, $\pm\mathbf{1\ 2} = \mathbf{3\ 4\ 5\ 6}$.

32 Runs, The 2_{V}^{6-1} Design in Two Blocks of 16

In blocking this and other fractional factorial designs it is important to bear in mind that any factor associated with blocks will have its aliases also associated with blocks. For the 2_{V}^{6-1} design for instance, where the alias relation $\mathbf{I} = \mathbf{1\ 2\ 3\ 4\ 5\ 6}$ already exists, if we adopt for the block arrangement $\mathbf{B}_1 = \mathbf{1\ 2\ 3\ 4\ 5}$ then we should also have $\mathbf{B}_1 = \mathbf{6}$. Thus we would confound the main effect of variable **6** with blocks. In general, when choosing a suitable factor to be associated with the block variables we must be careful to ensure that not only the chosen factor but also its aliases correspond to at least three-factor interactions. The present design can be run in two blocks of 16 by using any three-factor interaction to define the block. For example, for block variable $\mathbf{B}_1$ we can set

$$\mathbf{B}_1 = \mathbf{1\ 2\ 3}$$

whence, because of the alias relationship $\mathbf{I} = \mathbf{1\ 2\ 3\ 4\ 5\ 6}$, we find that $\mathbf{B}_1 = \mathbf{1\ 2\ 3} = \mathbf{4\ 5\ 6}$, and no two-factor interaction is confounded.

Designs with 64 Runs: The 2_{V}^{7-1} Design

With seven variables there are 28 effects to be determined (the seven main effects and twenty-one two-factor interactions) so that in principle it might be possible to use a quarter replicate of the full 2^7 design which would involve 32 experimental runs. In practice however this is not possible for the type of fractional factorial designs discussed here. Other designs, such as the three quarter replicate designs discussed elsewhere in this issue, can however be em-

* The 2^{6-1} and 2^{7-1} designs are properly of resolution VI and VII respectively. The block arrangements however insure that first and second order effects will not be associated with blocks or with each other.

ployed. The best that can be done for this series of conventional fractional factorials is to use of half replicate and once again it is desirable to use for the defining contrast the highest possible order interaction, i.e.,

$$\mathbf{I} = \mathbf{1\ 2\ 3\ 4\ 5\ 6\ 7}.$$

This seven variable design can be run in eight blocks of eight experimental runs each using the following four-factor interactions to define the block variables: $\mathbf{B}_1 = \mathbf{1\ 3\ 5\ 7}$, $\mathbf{B}_2 = \mathbf{1\ 2\ 5\ 6}$ and $\mathbf{B}_3 = \mathbf{1\ 2\ 3\ 4}$. To write down the complete design, first write down all the minus and plus signs for the 2^6 factorial in standard order. Introduce the seventh variable using $\mathbf{7} = \pm\mathbf{1\ 2\ 3\ 4\ 5\ 6}$. The eight blocks can then be obtained by writing down the three additional vectors $\mathbf{B}_1 = \mathbf{1\ 3\ 5\ 7}$, $\mathbf{B}_2 = \mathbf{1\ 2\ 5\ 6}$ and $\mathbf{B}_3 = \mathbf{1\ 2\ 3\ 4}$ and allocating the experimental points to the first block, second block, $\cdots$, eighth block in accordance as the signs of $\mathbf{B}_1$, $\mathbf{B}_2$ and $\mathbf{B}_3$ are $(-, -, -)$, $(+, -, -)$, $(-, +, -)$, $(+, +, -)$, $(-, -, +)$, $(+, -, +)$, $(-, +, +)$ and $(+, +, +)$. The derivation of this design is given in the Appendix.

64 Runs: The 2_{V}^{8-2} Design

For eight variables there are 36 effects to be estimated, the eight main effects and twenty-eight two-factor interactions. These estimates can be obtained with a one-quarter replicate of the 2^8 design involving 64 experimental points, that is, a 2_{V}^{8-2} design. The defining relation will include three words in addition to the identity **I** and if all main effects and two-factor interactions are not to be confounded then all the words in the defining contrast must be at least five-factor interactions. Bearing in mind that the three words in the defining contrast will be such that any two of them multiply to give the third, the best arrangement will be of the type indicated below, in which the interactions involved are **1 2 3 4 7**, **1 2 5 6 8** and their product **3 4 5 6 7 8** giving:

1	2	3	4			7	
1	2			5	6		8
		3	4	5	6	7	8

From the above array it will be seen that we are trying to pick two interactions containing five factors or more which have "minimum overlap", so that their product will also be an interaction of five factors or more. We therefore take for the generating relations

$$\mathbf{I} = \pm\mathbf{1\ 2\ 3\ 4\ 7}; \qquad \mathbf{I} = \pm\mathbf{1\ 2\ 5\ 6\ 8}$$

and write down the design by first setting out the full 2^6 design for the variables **1**, **2**, **3**, **4**, **5** and **6**. The versions of the variables **7** and **8** then follow the minus and plus elements of the interactions **1 2 3 4** and **1 2 5 6**.

It is possible to arrange this 2_{V}^{8-2} design in four blocks of 16 runs each without confounding block variables with any main effect or two-factor interaction. To see how this can be done, we remember that although it would be possible to write down 63 columns associated with all the main effects and interactions of the basic 2^6 design only 36 of these will be associated with main effects and two-factor interactions for the eight variables of interest. There remain therefore $63 - 36 = 27$ columns containing + and − signs which can be utilized to accommodate block effects. If the experiment is run in *two* blocks of 32 *any* one of these 27 columns can be used to accommodate the blocks. In practice it is not necessary to write down all the 63 columns. In fact we select the columns for blocking by inspection of the generators. Some care is needed in this selection. For example, the **1 2 3 4** interaction contains four symbols and so on first sight appears to provide as a suitable column to be used in blocking. However, it cannot be used since **1 2 3 4** is an alias of **7** and therefore the main effect of **7** would be confounded with blocks. This is so since one of the generators for the quarter replicate design is **1 2 3 4 7** and consequently **1 2 3 4** = **7**.

Clearly it is necessary to check not only that the interaction employed involves three factors or more but also that all its aliases involve three factors or more. The simplest way to pick out a suitable interaction is to write out the defining relation for the fractional factorial

$$\mathbf{I} = \mathbf{1\,2\,3\,4\,7} = \mathbf{1\,2\,5\,6\,8} = \mathbf{3\,4\,5\,6\,7\,8}$$

and then by inspection select interactions whose multiples with the words of the defining relation contain three or more characters. For instance, suppose **1 3 5** is selected as a possible interaction to confound. Then

$$\mathbf{1\,3\,5} = \mathbf{2\,4\,5\,7} = \mathbf{2\,3\,6\,8} = \mathbf{1\,4\,6\,7\,8}.$$

Thus the three-factor interaction **1 3 5** is confounded with no interaction containing fewer than three factors, and hence, **1 3 5** is a suitable contrast to use for blocking.

If we wish to run the experiment in four blocks of 16 then another interaction of three factors or more having no aliases with less than three factors must be chosen with the additional requirement that its interaction with the first interaction used to identify blocks has aliases all of three factors or more. For example, take **1 3 5** and **3 4 8**. Using these, the three degrees of freedom among the four blocks will be associated with **1 3 5** and **3 4 8** and with their interaction **1 4 5 8**. That these give satisfactory aliases is seen by obtaining their multiples with the defining relation of the fractional factorial giving:

$$\begin{aligned}
\mathbf{1\,3\,5} &= \mathbf{2\,4\,5\,7} = \mathbf{2\,3\,6\,8} = \mathbf{1\,4\,6\,7\,8}\\
\mathbf{3\,4\,8} &= \mathbf{1\,2\,7\,8} = \mathbf{1\,2\,3\,4\,5\,6} = \mathbf{5\,6\,7}\\
\mathbf{1\,4\,5\,8} &= \mathbf{2\,3\,5\,7\,8} = \mathbf{2\,4\,6} = \mathbf{1\,3\,6\,7}
\end{aligned}$$

Thus as indicated in column six of Table 23, the + and − signs associated with the interaction vectors **1 3 5** and **3 4 8** can be used to identify four blocks of sixteen runs each.

Designs with 128 Runs: The 2_{V}^{9-2}, 2_{V}^{10-3} and 2_{V}^{11-4} Designs

The designs for nine and ten variables using 128 runs can be regarded as a special case of the 2_{V}^{11-4} design from which one or two variables have been dropped. In many cases therefore where nine or ten variables are to be studied and where there are one or two further variables of possible importance, it would be worthwhile to include these and in fact to run the full eleven variable design. We shall first discuss the 2_{V}^{11-4} design and then discuss the designs for nine and ten variables as special cases. To construct the 2_{V}^{11-4} design we begin by writing down the − and + signs comprising the seven columns for the complete 2^7 factorial. The sixteenth fraction of the 2^{11} design is then obtained by associating certain interactions between the seven variables with the new variables **8**, **9**, **10** and **11**. Let us call these interactions **W**, **X**, **Y** and **Z**, so that the column headings for our final design matrix will be as follows:

W X Y Z

1 2 3 4 5 6 7 8 9 10 11

It will be understood that each of the symbols **W**, **X**, **Y**, **Z**, corresponds to an interaction of the first set of seven variables. For example, if **W** is the interaction **1 2 3 4**, the generators produced by associating **W** with the variable **8** would be **1 2 3 4 8** or **W8**. In general the generating relation for the complete design are

$$\mathbf{I} = \mathbf{W\,8}, \qquad \mathbf{I} = \mathbf{X\,9}, \qquad \mathbf{I} = \mathbf{Y\,10}, \qquad \mathbf{I} = \mathbf{Z\,11}$$

and hence the defining relation is **I** = **W 8** = **X 9** = **Y 10** = **Z 11** = **W X 8 9** = **W Y 8 10** = **W Z 8 11** = **X Y 9 10** = **X Z 9 11** = **Y Z 10 11** = **W X Y 8 9 10** = **W X Z 8 9 11** = **W Y Z 8 10 11** = **X Y Z 9 10 11** = **W X Y Z 8 9 10 11**. Now if this design is to be of resolution V all the words in the defining relation must contain at least five symbols; it follows that:

(i) **W**, **X**, **Y** and **Z** must themselves each contain at least four symbols.
(ii) The products **W X**, **W Y**, **W Z** etc., must contain at least three symbols since all words in the defining relation containing these pairs will also contain two further symbols from the group **8**, **9**, **10**, **11**.
(iii) Similarly the products in threes **W X Y**, **W X Z**, etc., must contain two symbols and
(iv) the product **W X Y Z** must contain at least one symbol.

The problem is therefore to find four interactions each containing at least four symbols with all the above properties. A set of interactions having these properties has in fact already been derived. It was shown earlier that the defining relation for the 2_{III}^{7-3} design is:

$$\mathbf{I} = \mathbf{124} = \mathbf{135} = \mathbf{236} = \mathbf{1237} = \mathbf{2345} = \mathbf{1346} = \mathbf{347} = \mathbf{1256}$$
$$= \mathbf{257} = \mathbf{167} = \mathbf{456} = \mathbf{1457} = \mathbf{2467} = \mathbf{3567} = \mathbf{1234567}.$$

A set of generators for this defining relation are **1 2 3 7**, **2 3 4 5**, **1 3 4 6** and **1 2 3 4 5 6 7**. Suppose then we choose:

$$\mathbf{W} = \mathbf{1\,2\,3\,7}$$

$$\mathbf{X} = \mathbf{2\,3\,4\,5}$$

$$\mathbf{Y} = \mathbf{1\,3\,4\,6}$$

$$\mathbf{Z} = \mathbf{1\,2\,3\,4\,5\,6\,7}$$

Then clearly all conditions (i), (ii), (iii) and (iv) mentioned above will be satisfied. A one-sixteenth replicate of the 2^{11} for which no main effect or two-factor interaction is confounded with any other main effect or interaction, that is, the 2_{V}^{11-4} design can be obtained by setting

$$\pm\ 8 = 1\ 2\ 3\ 7$$

$$\pm\ 9 = 2\ 3\ 4\ 5$$

$$\pm 10 = 1\ 3\ 4\ 6$$

$$\pm 11 = 1\ 2\ 3\ 4\ 5\ 6\ 7$$

to give the generators

$$\pm 1\ 2\ 3\ 7\ 8;\ \pm 2\ 3\ 4\ 5\ 9,\ \pm 1\ 3\ 4\ 6\ 10,\ \pm 1\ 2\ 3\ 4\ 5\ 6\ 7\ 11$$

Given that three-factor and higher order interaction effects are negligible the resulting 2_{V}^{11-4} design provides separate estimates of the eleven main effects and fifty-five two-factor interactions.

The 2_{V}^{10-3} and 2_{V}^{9-2} designs may be obtained from the 2_{V}^{11-4} design by dropping out respectively one or two of the variables. The defining relations for these designs omit all words containing the dropped variables. All designs obtained by dropping variables will have the properties of the parent design. However, selections can be made which improve the reduced design. For example, if we drop variable **11** the generators for the resultant 2_{V}^{10-3} design are

$$\pm 1\ 2\ 3\ 7\ 8; \qquad \pm 2\ 3\ 4\ 5\ 9; \qquad \pm 1\ 3\ 4\ 6\ 10$$

The corresponding defining relation contains three words with five characters, three with six and one word with seven characters. However, if variable **10** is dropped the generators are

$$\pm 1\ 2\ 3\ 7\ 8; \qquad \pm 2\ 3\ 4\ 5\ 9; \qquad \pm 1\ 2\ 3\ 4\ 5\ 6\ 7\ 11$$

and the defining relation now contains four words with five characters two with six and one word with eight. The first arrangement is slightly preferable since, with fewer five character words, fewer two-factor interactions will beconfounded with three-factor interactions.

Similarly for the 2_{V}^{9-7} design it is best to drop variable **3** in addition to variable **11**. The generators for this design are

$$\pm 1\ 4\ 5\ 7\ 8\ 9; \qquad \pm 2\ 4\ 6\ 7\ 8\ 10$$

from which we see that all main effects will be confounded with only five-factor interactions and two-factor interactions only with four-factor interactions. (This design is, in fact, of resolution VI).

128 Runs: The 2_{V}^{11-4} in 8 Blocks of 16 Runs

It is a somewhat remarkable fact that the 2_{V}^{11-4} design can be blocked into eight groups of sixteen runs. Using the generators for the principal fraction of the 2_{V}^{11-4} given in the previous section, we obtain the defining relation:

$$
\begin{aligned}
\mathbf{I} &= \mathbf{1\,2\,3\,7\,8} = \mathbf{2\,3\,4\,5\,9} = \mathbf{1\,3\,4\,6\,10} = \mathbf{1\,2\,3\,4\,5\,6\,7\,11} = \mathbf{1\,4\,5\,7\,8\,9} \\
&= \mathbf{2\,4\,6\,7\,8\,10} = \mathbf{4\,5\,6\,8\,11} = \mathbf{1\,2\,5\,6\,9\,10} = \mathbf{1\,6\,7\,9\,11} = \mathbf{2\,5\,7\,10\,11} \\
&= \mathbf{3\,5\,6\,7\,8\,9\,10} = \mathbf{2\,3\,6\,8\,9\,11} = \mathbf{1\,3\,5\,8\,10\,11} \\
&= \mathbf{3\,4\,7\,9\,10\,11} = \mathbf{1\,2\,4\,8\,9\,10\,11}.
\end{aligned}
$$

To obtain eight blocks we require seven interactions produced by three block generators which, when multiplied together and by the words in the above defining relation produce words with three or more characters. It can be confirmed by actual multiplication that a suitable group of block generators is

$$\mathbf{B}_1 = \mathbf{1\,4\,9}, \qquad \mathbf{B}_2 = \mathbf{1\,2\,10}, \qquad \mathbf{B}_3 = \mathbf{8\,9\,10}.$$

We find that the following seven interactions will be confounded with block effects:

$$
\begin{aligned}
\mathbf{B}_1 &= \mathbf{1\,4\,9} & \mathbf{B}_1\mathbf{B}_2 &= \mathbf{2\,4\,9\,10} \\
\mathbf{B}_2 &= \mathbf{1\,2\,10} & \mathbf{B}_1\mathbf{B}_3 &= \mathbf{1\,4\,8\,10} \\
\mathbf{B}_3 &= \mathbf{8\,9\,10} & \mathbf{B}_2\mathbf{B}_3 &= \mathbf{1\,2\,8\,9} \\
& & \mathbf{B}_1\mathbf{B}_2\mathbf{B}_3 &= \mathbf{2\,4\,8}
\end{aligned}
$$

When variables **3** and **11** are dropped from the eleven variable design these same block generators may also be used to generate the blocks for the ten and nine factor resolution V designs.

Appendix

The Derivation of the 2_{V}^{7-1} in Blocks of Eight

The 2_{V}^{7-1} contains sixty-four experimental runs and is constructed by first writing down a full 2^6 factorial and then setting $\mathbf{7} = \mathbf{1\,2\,3\,4\,5\,6}$. To divide the design into eight blocks of eight runs such that no main effect or two-factor interaction is confounded with any block effect we must be able to find seven interactions to associate with the seven degrees of freedom between the eight blocks and each of these interactions and their aliases must be at least three-factor interactions. To see how this can be done we refer once again to the table of minus and plus signs appropriate to the 2^3 factorial. This table, shown below, has been used to produce a second table in which a number appears if there is a minus sign in the first table, and does not appear if there is a plus sign. If we associate the first three columns of the first table with B_1 , B_2 and B_3 we see, that the interactions B_1B_2 , B_1B_3 , B_2B_3 and $B_1B_2B_3$ will identify the remaining columns. We also note that all of the columns contain four minus and four plus signs. Referring now to the second table, when two columns are multiplied together only those numerals which appear once, or an odd number of times, will appear in the product, and those numerals appearing an even number of times will not appear in the product. This is exactly the same rule adopted earlier to assist us in identifying the alias terms. It is clear therefore that setting $\mathbf{B}_1$ equal to the **1 3 5 7** interaction effect, $\mathbf{B}_2 = \mathbf{1\,2\,5\,6}$ and $\mathbf{B}_3 =$

TABLE

	B_1	B_2	B_3	B_1B_2	B_1B_3	B_2B_3	$B_1B_2B_3$	
(1)	−1	−1	−1	1	1	1	−1	
(2)	1	−1	−1	−1	−1	1	1	
(3)	−1	1	−1	−1	1	−1	1	
(4)	1	1	−1	1	−1	−1	−1	± sign in
(5)	−1	−1	1	1	−1	−1	1	2^3 factorial
(6)	1	−1	1	−1	1	−1	−1	
(7)	−1	1	1	−1	−1	1	−1	
(8)	1	1	1	1	1	1	1	
	1	1	1				1	
		2	2	2	2			confounded inter-
	3		3	3		3		actions for the
			4		4	4	4	$\frac{1}{2}$th replicate of
	5	5			5	5		the 2^7 design
		6		6		6	6	
	7			7	7		7	

1 2 3 4 that all seven of the block effects will be confounded with four-factor interactions. Since the 2_{V}^{7-1} has the defining relation **I** = **1 2 3 4 5 6 7**, each of the block defining contrasts will therefore be aliased with those three of the seven numerals which do not appear in it. Thus, the 2_{V}^{7-1} is blocked such that all block effects are aliased with at least a three factor interaction and hence clear of both main effects and two factor interactions as required. Of course, *before blocking*, the design with defining relation **I** = **1 2 3 4 5 6 7** is of resolution VII.

2.12
The Choice of a Second Order Rotatable Design*

G. E. P. BOX and NORMAN R. DRAPER
University of Wisconsin

1. Introduction

1·1. *General remarks.* It frequently happens that an experimenter is interested in exploring a functional relationship

$$\eta = \eta(\xi_1, \xi_2, \ldots, \xi_k).$$

When the actual functional form is not known, useful information about the nature of the actual relationship in some particular region R of the $\boldsymbol{\xi}$ space can often be obtained by approximating the relationship by a graduating function $g(\boldsymbol{\xi}, \boldsymbol{\beta})$ where $\boldsymbol{\beta}$ is a vector of adjustable constants. The graduating functions usually employed have been polynomials in the variables $\boldsymbol{\xi}$. The problem of experimental design which arises in the fitting of a graduating function has been discussed by Box & Draper (1959) and will be briefly restated here.

We desire to choose a design matrix $\mathbf{D}$ of N rows and k columns which will specify the levels of the k variables to be run in N experiments. Denote the uth row of this matrix by $\boldsymbol{\xi}_u'$. This vector has as elements the levels $(\xi_{1u}, \xi_{2u}, \ldots, \xi_{ku})$ of the k factors to be employed in the uth experiment, $u = 1, 2, \ldots, N$. Our primary objective will be to choose these levels so that when the graduating function $g(\boldsymbol{\xi})$ is fitted by least squares, it will closely represent the true function $\eta(\boldsymbol{\xi})$, within the region of interest R. Subject to the satisfaction of our primary objective, we shall also require that the factor levels be such that there is a high chance that the inadequacy of the graduating function $g(\boldsymbol{\xi}, \boldsymbol{\beta})$ to represent $\eta(\boldsymbol{\xi})$ will be detected. As will be seen, a subclass of all possible designs can be selected which will satisfy our primary requirement. We can then make use of our secondary requirement to make a selection of a particular design from this subclass.

We now define 'region of interest', 'closely represent' and 'detection of inadequacy of model'.

1·2. *Interpretation of 'region of interest'.* Let us call the region in the $\boldsymbol{\xi}$ space, in which experiments can actually be performed, the operability region O. This region is usually bounded although its limits are often known only vaguely. For example, in chemical experiments there will often be conditions of temperature or pressure which will be too severe to use safely on the apparatus. In biological work certain combinations of drugs for therapeutic use will produce death. For some applications the experimenter may wish to explore the whole region O, but this is comparatively rare. Usually a particular group of experiments is used to explore a rather limited region of interest R entirely contained within the operability region O. Frequently, experiments are conducted sequentially and a group of experiments designed for the exploration of one current region of interest may lead to a further set exploring a different region. Often an alternative statement of the problem

* This research was supported by the United States Navy through the Office of Naval Research, and by the United States Army at the Mathematics Research Center, U.S. Army, Madison, Wisconsin.

The Collected Works of George E. P. Box, 1984, Wadsworth, Inc., Belmont, CA 94002.
Originally published in *Biometrika,* vol. 50, parts 3 and 4 (1963), pp. 335–352.

would be that it is desired to explore the nature of a functional relationship 'in the neighbourhood' of a point P. The latter statement is perhaps closer to the real desires of some experimenters with its implication that the situation is one of a falling off of interest at points more and more distant from P rather than of equal interest at all points within R and no interest outside R and within O.

As we shall indicate briefly below, by dealing with various types of weight functions in specified regions (for example, if R is a k-dimensional sphere centred at the origin we can use weights which are functions of distance of a point in space from the origin), these various possible desires can be combined into a unified treatment. For the immediate purposes of this paper, however, we shall soon revert to considering the 'interest within R, no interest outside R' formulation; while this may not suit all tastes, a great many experimental investigations are undertaken with this thought in mind and the formulation is thus not unrealistic.

1·3. *Interpretation of 'closeness'*. Let $\hat{y}(\boldsymbol{\xi})$ denote the response estimated by the graduating function at the point $\boldsymbol{\xi}$. Then we desire to choose $\mathbf{D}$ so that the difference $\hat{y}(\boldsymbol{\xi}) - \eta(\boldsymbol{\xi})$ will be small over the region of interest R. The measure of closeness which we shall use at a particular point $\boldsymbol{\xi}$ is

$$E[\hat{y}(\boldsymbol{\xi}) - \eta(\boldsymbol{\xi})]^2.$$

Over the whole region we may use the average

$$\Omega \int_R E[\hat{y}(\boldsymbol{\xi}) - \eta(\boldsymbol{\xi})]^2 \, d\boldsymbol{\xi} \tag{1·3·1}$$

where

$$\Omega^{-1} = \int_R d\boldsymbol{\xi}.$$

In certain circumstances we might wish more weight given to errors at one value of $\boldsymbol{\xi}$ than at another. We may therefore introduce a weight function $W(\boldsymbol{\xi})$ such that

$$\int_O W(\boldsymbol{\xi}) \, d\boldsymbol{\xi} = 1.$$

Our measure will then take the form

$$\int_O W(\boldsymbol{\xi}) \, E[\hat{y}(\boldsymbol{\xi}) - \eta(\boldsymbol{\xi})]^2 \, d\boldsymbol{\xi}.$$

The previous formulation is a special case of this as is easily seen by setting

$$W(\boldsymbol{\xi}) = \begin{cases} \Omega & \text{in } R, \\ 0 & \text{elsewhere.} \end{cases}$$

It is desirable that we should be able to compare designs which do not contain the same number of points and that our criterion of closeness should be independent of the variance σ^2 of the observations, which we assume to be constant. Thus we shall choose as our measure of closeness

$$J = \int_O w(\boldsymbol{\xi}) \, E[\hat{y}(\boldsymbol{\xi}) - \eta(\boldsymbol{\xi})]^2 \, d\boldsymbol{\xi} \tag{1·3·2}$$

where

$$w(\boldsymbol{\xi}) = NW(\boldsymbol{\xi})/\sigma^2.$$

Writing

$$\hat{y}(\boldsymbol{\xi}) - \eta(\boldsymbol{\xi}) = \{\hat{y}(\boldsymbol{\xi}) - E\hat{y}(\boldsymbol{\xi})\} + \{E\hat{y}(\boldsymbol{\xi}) - \eta(\boldsymbol{\xi})\}$$

we can split J into two parts $$J = V + B,$$

where V is the average weighted variance

$$V = \int_O w(\boldsymbol{\xi})\,[\hat{y}(\boldsymbol{\xi}) - E\hat{y}(\boldsymbol{\xi})]^2\,d\boldsymbol{\xi} \tag{1·3·3}$$

and B is the average squared bias

$$B = \int_O w(\boldsymbol{\xi})\,[E\hat{y}(\boldsymbol{\xi}) - \eta(\boldsymbol{\xi})]^2\,d\boldsymbol{\xi}. \tag{1·3·4}$$

In what follows we shall suppose that the graduating function is a polynomial of degree d_1 in $\boldsymbol{\xi}$

$$g(\boldsymbol{\xi}) = \boldsymbol{\xi}_1'\boldsymbol{\beta}_1,$$

where the vector $\boldsymbol{\xi}_1$ contains p_1 elements, all of which are powers and products of $\xi_i\,(i=1,2,\ldots,k)$, of order d_1 or less. The true functional form over the whole region O is assumed to be a polynomial of degree d_2 in $\boldsymbol{\xi}$

$$\eta(\boldsymbol{\xi}) = \boldsymbol{\xi}_1'\boldsymbol{\beta}_1 + \boldsymbol{\xi}_2'\boldsymbol{\beta}_2,$$

where $\boldsymbol{\xi}_2'$ contains p_2 elements, all of which are powers and products of $\xi_1, \xi_2, \ldots, \xi_k$ of order d_2 or less but greater than d_1.

Corresponding to any design matrix $\mathbf{D}$, there will exist an $N \times p_1$ matrix $\mathbf{X}_1$ with $\boldsymbol{\xi}_{1u}'$ as its uth row, whose elements are powers and products of order d_1 or less of the elements of the vector $\boldsymbol{\xi}_u$. There will also be a matrix $\mathbf{X}_2$ with uth row $\boldsymbol{\xi}_{2u}'$ whose elements are the power and products of orders $(d_1+1), \ldots, d_2$ of the elements of the vector $\boldsymbol{\xi}_u$.

Let
$$M_{11} = N^{-1}\mathbf{X}_1'\mathbf{X}_1, \quad M_{12} = N^{-1}\mathbf{X}_1'\mathbf{X}_2, \quad M_{22} = N^{-1}\mathbf{X}_2'\mathbf{X}_2. \tag{1·3·5}$$

It will be seen that the elements of these matrices are of the form

$$N^{-1}\sum_{u=1}^{N} \xi_{1u}^{\alpha_1}\xi_{2u}^{\alpha_2}\ldots\xi_{ku}^{\alpha_k}.$$

They will thus be referred to as the moments of the design points and will be said to be of order α if $\alpha = \alpha_1 + \alpha_2 + \ldots + \alpha_k$. Also write

$$\boldsymbol{\mu}_{11} = \int w(\boldsymbol{\xi})\,\boldsymbol{\xi}_1\boldsymbol{\xi}_1'\,d\boldsymbol{\xi}, \quad \boldsymbol{\mu}_{12} = \int w(\boldsymbol{\xi})\,\boldsymbol{\xi}_1\boldsymbol{\xi}_2'\,d\boldsymbol{\xi}, \quad \boldsymbol{\mu}_{22} = \int w(\boldsymbol{\xi})\,\boldsymbol{\xi}_2\boldsymbol{\xi}_2'\,d\boldsymbol{\xi}, \tag{1·3·6}$$

where all integrals are taken over the region O.

The elements of the matrices (1·3·6) are of the form

$$\int_O w(\boldsymbol{\xi})\,\xi_1^{\alpha_1}\xi_2^{\alpha_2}\ldots\xi_k^{\alpha_k}\,d\boldsymbol{\xi}$$

and this is a moment of the weight function of order $\alpha = \alpha_1 + \alpha_2 + \ldots + \alpha_k$. Proceeding exactly as in Box & Draper (1959), we obtain

$$\begin{aligned} J &= \operatorname{trace}\,[\boldsymbol{\mu}_{11}\mathbf{M}_{11}^{-1}] \\ &\quad + \boldsymbol{\beta}_2'[(\boldsymbol{\mu}_{22} - \boldsymbol{\mu}_{12}\boldsymbol{\mu}_{11}^{-1}\boldsymbol{\mu}_{12}) + (\mathbf{M}_{11}^{-1}\mathbf{M}_{12} - \boldsymbol{\mu}_{11}^{-1}\boldsymbol{\mu}_{12})'\boldsymbol{\mu}_{11}(\mathbf{M}_{11}^{-1}\mathbf{M}_{12} - \boldsymbol{\mu}_{11}^{-1}\boldsymbol{\mu}_{12})]\boldsymbol{\beta}_2 \\ &= V + B, \quad \text{say,} \end{aligned} \tag{1·3·7}$$

and, as our first objective, we shall choose the design matrix $\mathbf{D}$ in such a way that this quantity is a minimum. Our formulation will thus ensure that the graduating function will closely represent the true function $\eta(\boldsymbol{\xi})$, in the way we have described, after a suitable weight function has been chosen. We see from (1·3·7) that V does not contain $\boldsymbol{\beta}_2$ at all and depends only on $\mathbf{D}$, while B depends on both $\mathbf{D}$ and $\boldsymbol{\beta}_2$. Thus, minimization of J depends on what value we assign to $\boldsymbol{\beta}_2$. We shall return to this point later.

1·4. *Interpretation of 'detection of inadequacy of model'.* We shall suppose that a test for lack of fit is to be made by the use of an analysis of variance in which the residual sum of squares

$$S_R = \sum_{u=1}^{N} (\hat{y}_u - y_u)^2,$$

where y_u are the actual observations, is compared either with a prior value σ^2 of the experimental error variance, supposed to be known exactly, or with some independent estimate s^2. In either case, a parameter which determines the power of the test for goodness of fit will be the quantity

$$\sum_{u=1}^{N} [E(\hat{y}_u) - \eta_u]^2 = E(S_R) - \nu\sigma^2,$$

where ν is the number of degrees of freedom on which the residual sum of squares is based. While our ultimate object should be to make the power of the test as large as possible, in any particular instance in which ν is assumed fixed this will be equivalent to making the expectation of S_R large.

We shall interpret our secondary requirement, therefore, as implying that the design should be chosen so as to make $E(S_R)$ large. It seems reasonable to regard our primary requirement as being of major importance so that in practice we shall proceed by first attempting to find the class of designs which minimizes J and then attempting to satisfy the secondary criterion by selecting from this class a subclass which makes large the expected value of S_R.

1·5. *Choice of R as a spherical region.* Up to this point we have said nothing about the shape of the region of interest R. R can be of any shape one can imagine and, given any particular region, the theory can be applied in a way similar to the way shown below. However, it is impossible to forsee all conceivable choices of R and in order to develop results we must make a reasonable assumption. Two reasonable assumptions are

(i) R is spherical or ellipsoidal, that is, some deformation of a sphere attained because of change of scale, so that mathematically only a sphere need be considered (this case we shall treat).

(ii) R is cuboidal, or is some deformation of a k-dimensional cube attained because of change of scale, so that mathematically only a cube need be considered. We shall not treat this case but remark how it would affect succeeding paragraphs. Instead of being later led to rotatable designs where all odd moments are zero and even moments bear certain relationships to one another as given by Box & Hunter (1957), we should be led to 'rectangular' designs in which all odd moments are zero and even moments bear certain relationships to one another. As a possible example in certain circumstances: Instead of obtaining, as for case (i), a second-order rotatable design with ratio (pure fourth moment)/(mixed fourth moment) = 3, we should obtain a symmetrical design with all odd moments zero but with the ratio (pure fourth moment)/(mixed fourth moment) = 1·8.

From these two reasonable formulations we select the first for further development. It is, in our opinion, probably the one more frequently in an experimenter's mind.

1·6. *Reasons for the consideration of rotatable designs only.* We now intend to consider only rotatable designs and this choice is closely related to our choice of a spherical region R.

Box & Draper (1959) showed that, no matter what the shape of R, a sufficient condition for the bias B alone to be minimized is that the moments of the design should be equal to the moments of the region R up to and including order (d_1+d_2). It is clear from an inspection of equation (1·3·7) above that a necessary and sufficient condition for the minimization of B alone is simply

$$\mathbf{M}_{11}^{-1}\mathbf{M}_{12} = \boldsymbol{\mu}_{11}^{-1}\boldsymbol{\mu}_{12}$$

since the first term of B is always positive as was previously shown (1959). (This result is interesting in a numerical analysis context. The details are in Appendix 1.) This implies of course that a sufficient condition is $\{\mathbf{M}_{11} = \boldsymbol{\mu}_{11}, \mathbf{M}_{12} = \boldsymbol{\mu}_{12}\}$ which is just a statement that moments of the design equal moments of the region R up to and including order (d_1+d_2).

If the region R is spherical it follows that designs which will minimize bias B only are rotatable designs of certain orders which depend on d_1 and d_2. If $d_1+d_2 = 2m$, say, then the appropriate design is an mth order rotatable design. If $d_1+d_2 = 2m+1$, then the appropriate design is mth order rotatable with moments of order $(2m+1)$ all zero.

A spherical weight function has no effect whatsoever on this conclusion, since rotatability is entirely dependent on ratios between moments of the same order. These ratios, which are attained for spheres, must therefore be attained for shells, hence attained for any collection of shells and it follows that a spherical weight function, which merely attaches weights to various spherical shells, cannot affect these ratios.

This persuades us to consider only rotatable designs when both V and B enter into consideration. We must then decide on the values of the design parameters to be used for any given situation as well as the particular rotatable design.

1·7. *Choice of weight function.* We shall choose in what follows the weight function

$$W(\boldsymbol{\xi}) = \begin{cases} \Omega & \text{in } R \\ 0 & \text{elsewhere.} \end{cases}$$

It might be more appropriate in some applications to choose a weight function which decreased as we moved away from the centre of the region. In that case, questions which would naturally arise would be 'How quickly should $W(\boldsymbol{\xi})$ fall off?' 'At what point should $W(\boldsymbol{\xi})$ be made zero?' 'Should the rate of fall-off vary for successive "zones" as we move from the centre of R?' It is clear that the weight function could be chosen in numerous ways and for any particular weight function the problem could be treated as below. Clever choice of the weight function might also contribute to the ease with which J can be minimized but we shall not discuss this point further here. Since choice of the weight function as given corresponds to the choice made in Box & Draper (1959), the results below are logical extensions of those previously found.

1·8. *Recapitulation of previous work.* In Box & Draper (1959) it was assumed that $\eta(\boldsymbol{\xi})$ was a quadratic polynomial in k factor variables $\xi_1, \xi_2, \ldots, \xi_k$ and that the graduating function $g(\boldsymbol{\xi})$ was a linear function of these same variables. It emerged, somewhat surprisingly, that in typical experimental situations choice of the design depended far more on the effect of bias error than on variance error. Moreover, designs suitable when both variance and bias contributions contributed about equally to the total error were close to designs suitable for the 'all bias, no variance' situation and completely different from those suitable for the 'all variance' situation on which most previous conclusions have been based.

2. THE PROBLEM AND ITS SOLUTION

2·1. *Assumptions in the present paper.* We shall assume that $\eta(\boldsymbol{\xi})$ is a cubic polynomial and $g(\boldsymbol{\xi})$ is a quadratic polynomial in $\xi_1, \xi_2, \ldots, \xi_k$. In other words, $d_1 = 2$, $d_2 = 3$, $d_1+d_2 = 5$ and we shall consider designs which are second-order rotatable with fifth moments zero, for the reasons which were described earlier. We shall also assume that the variables $\boldsymbol{\xi}$ have been linearly transformed to variables $\mathbf{x}$ in such a way that the centre of the design is at the origin $(0, 0, \ldots, 0)$ and that the region R is the k-dimensional unit sphere. The graduating function $g(\mathbf{x})$ is

$$\hat{y} = b_0+b_1x_1+b_2x_2+\ldots+b_kx_k+b_{11}x_1^2+\ldots+b_{kk}x_k^2+b_{12}x_1x_2+\ldots+b_{k-1,k}x_{k-1}x_k,$$

or, in matrix notation,

$$\hat{y} = \mathbf{x}_1'\mathbf{b}_1$$

where

$$\mathbf{b}_1' = (b_0;\, b_1\ldots b_k;\, b_{11}, \ldots, b_{kk};\, b_{12}, \ldots, b_{k-1,k}),$$
$$\mathbf{x}_1' = (1;\, x_1\ldots x_k;\, x_1^2, \ldots, x_k^2;\, x_1x_2, \ldots, x_{k-1}x_k).$$

The true relationship which applies over the whole region O is assumed to be the cubic polynomial

$$\eta = \mathbf{x}_1'\boldsymbol{\beta}_1+\mathbf{x}_2'\boldsymbol{\beta}_2,$$

where $\mathbf{x}_1$ is as above, $\boldsymbol{\beta}_1$ is defined like $\mathbf{b}_1$ and

$$\boldsymbol{\beta}_2' = (\beta_{111}, \beta_{122}, \ldots, \beta_{1kk};\, \beta_{222}, \beta_{211}, \ldots, \beta_{2kk};\, \ldots\beta_{kk,k-1};\, \beta_{123}, \beta_{124}, \ldots, \beta_{k-2,k-1,k}),$$
$$\mathbf{x}_2' = (x_1^3, x_1x_2^2, \ldots, x_1x_k^2;\, x_2^3, x_2x_1^2, \ldots, x_2x_k^2;\, \ldots\, x_{k-1}x_k^2;\, x_1x_2x_3, x_1x_2x_4, \ldots, x_{k-2}x_{k-1}x_k).$$

Exactly as in Box & Draper (1959) we have $J = V+B$ where

$$V = N\Omega\int_R \mathbf{x}_1'(\mathbf{X}_1'\mathbf{X}_1)^{-1}\mathbf{x}_1\,d\mathbf{x}$$

and

$$B = N\sigma^{-2}\Omega\int_R \boldsymbol{\beta}_2'[\mathbf{A}'\mathbf{x}_1-\mathbf{x}_2][\mathbf{x}_1'\mathbf{A}-\mathbf{x}_2']\boldsymbol{\beta}_2\,d\mathbf{x},$$

where now

$$\mathbf{X}_1' = [\mathbf{x}_{11}, \ldots, \mathbf{x}_{1u}, \ldots, \mathbf{x}_{1N}]$$

is a $\frac{1}{2}(k+1)(k+2)$ by N matrix with

$$\mathbf{x}_{1u}' = (1, x_{1u}, x_{2u}, \ldots, x_{ku}, x_{1u}^2, \ldots, x_{ku}^2, x_{1u}x_{2u}, \ldots, x_{k-1,u}x_{ku}),$$
$$\mathbf{X}_2' = [\mathbf{x}_{21}, \ldots, \mathbf{x}_{2u}, \ldots, \mathbf{x}_{2N}]$$

is a $k(k+1)(k+2)/6$ by N matrix with

$$\mathbf{x}_{2u}' = (x_{1u}^3, x_{1u}x_{2u}^2, \ldots, x_{1u}x_{ku}^2;\, x_{2u}^3, \ldots;\, x_{1u}x_{2u}x_{3u}\ldots, x_{k-2,u}x_{k-1,u}x_{ku})$$

and $\mathbf{A} = (\mathbf{X}_1'\mathbf{X}_1)^{-1}\mathbf{X}_1'\mathbf{X}_2$ is the $\frac{1}{2}(k+1)(k+2)$ by $k(k+1)(k+2)/6$ 'alias matrix'. This last matrix has for its elements quantities which measure the extent to which the estimates $\mathbf{b}_1$ are biased by higher order coefficients in accordance with the equation

$$E(\mathbf{b}_1) = \boldsymbol{\beta}_1+\mathbf{A}\boldsymbol{\beta}_2.$$

By making the necessary substitutions we can evaluate V and B as follows.

2·2. *Evaluation of B.* Writing $\boldsymbol{\alpha}_2 = \boldsymbol{\beta}_2 N^{\frac{1}{2}}/\sigma$ we see that

$$\Omega^{-1}B = \boldsymbol{\alpha}_2'\mathbf{A}'\left(\int_R \mathbf{x}_1\mathbf{x}_2'\,d\mathbf{x}\right)\mathbf{A}\boldsymbol{\alpha}_2 - 2\boldsymbol{\alpha}_2'\left(\int_R \mathbf{x}_2\mathbf{x}_1'\,d\mathbf{x}\right)\mathbf{A}\boldsymbol{\alpha}_2 + \boldsymbol{\alpha}_2'\left(\int_R \mathbf{x}_2\mathbf{x}_2'\,d\mathbf{x}\right)\boldsymbol{\alpha}_2. \quad (2·2·1)$$

Since the design is second-order rotatable with fifth-order moments zero

$$\mathbf{A} = \frac{\lambda_4}{\lambda_2}\left[\begin{array}{c|c|c|c|c} 0 & 0 & & 0 & 0 \\ \hline 31\ldots1 & & & & \\ & 31\ldots1 & & & \\ & & \cdots & & 0 \\ & & & 31\ldots1 & \\ \hline 0 & 0 & & 0 & 0 \end{array}\right],$$

where $\quad 3\lambda_4 N = \sum_{u=1}^{N} x_{iu}^4 = 3\sum_{\substack{u=1\\ i\neq j}}^{N} x_{iu}^2 x_{ju}^2 \quad$ and $\quad \lambda_2 N = \sum_{u=1}^{N} x_{iu}^2$

are the parameters of the design. The columns of $\mathbf{A}$ correspond to the elements of $\mathbf{x}_2'$ and the rows of $\mathbf{A}$ correspond to the elements of $\mathbf{x}_1$. Let us denote this fact by saying that $\mathbf{A}$ is $(\mathbf{x}_1)(\mathbf{x}_2')$. Only k^2 elements of $\mathbf{A}$ are non-zero, and these are shown. They occupy the second, third, ..., $(k+1)$th rows. In the second row they are in the first k columns, ..., in the $(k+1)$th row they are in the columns numbered (k^2-k+1) to k^2. The divisions in both rows and columns of $\mathbf{A}$ correspond to the semicolons in the $\mathbf{x}$-vectors. Furthermore

$$\Omega\int_R \mathbf{x}_1\mathbf{x}_1'\,d\mathbf{x} = \left[\begin{array}{c|c|c|c} 1 & 0 & u\mathbf{j}_k' & 0 \\ \hline 0 & u\mathbf{I}_k & 0 & 0 \\ \hline u\mathbf{j}_k & 0 & v(2\mathbf{I}_k+\mathbf{j}_k\mathbf{j}_k') & 0 \\ \hline 0 & 0 & 0 & v\mathbf{I}_p \end{array}\right],$$

where $u(k+2) = v(k+2)(k+4) = 1$, $\mathbf{I}_k$ denotes the k by k unit matrix and $\mathbf{j}_k$ is a column vector of ones. This matrix is of shape $(\mathbf{x}_1)(\mathbf{x}_2')$ in our notation, the divisions again corresponding to the semicolons, and, since $\frac{1}{2}(k+1)(k+2)$ is the number of elements in $\mathbf{x}_1$, $p = \frac{1}{2}(k+1)(k+2)-2k-1 = \frac{1}{2}k(k-1)$. Similarly

$$\Omega\int_R \mathbf{x}_2\mathbf{x}_1'\,d\mathbf{x} = v\left[\begin{array}{c|c|c} & 3 & \\ & 1 & \\ & \vdots & \\ & 1 & \\ & \quad 3 & \\ & \quad 1 & \\ 0 & \quad \vdots & 0 \\ & \quad 1 & \\ & \quad\quad \vdots & \\ & \quad\quad\quad 3 & \\ & \quad\quad\quad 1 & \\ & \quad\quad\quad \vdots & \\ & \quad\quad\quad 1 & \\ \hline 0 & 0 & 0 \end{array}\right],$$

where $v(k+2)(k+4) = 1$. This matrix has the same dimensions and is similar element-wise to the transpose $\mathbf{A}'$ of $\mathbf{A}$. Again similarly

$$\Omega \int_R \mathbf{x}_2 \mathbf{x}_2' \, d\mathbf{x} = w \begin{bmatrix} \mathbf{G}_1 & & & & \\ & \mathbf{G}_2 & & & \\ & & \ddots & & \\ & & & \mathbf{G}_k & \\ & & & & \mathbf{I}_q \end{bmatrix},$$

where
$$G_i = \begin{bmatrix} 15 & 3 & 3 & \cdots & 3 \\ 3 & 3 & 1 & \cdots & 1 \\ 3 & 1 & 3 & & 1 \\ \vdots & & & \ddots & \vdots \\ 3 & 1 & 1 & \cdots & 3 \end{bmatrix}, \quad \text{all } i,$$

and $w(k+2)(k+4)(k+6) = 1$. The matrix divisions correspond to the semicolons in the vector $\mathbf{x}_2$. Hence, since each G_i is a k by k matrix, $q = k(k^2+3k+2)/6 - k^2 = k(k-1)(k-2)/6$.

Substituting in equation (2·2·1), carrying out the appropriate matrix multiplications and collecting the terms element-wise in the matrices we find that the bias contribution B is given by

$$B = \boldsymbol{\alpha}_2' \mathbf{Q} \boldsymbol{\alpha}_2,$$

where
$$\mathbf{Q} = \begin{bmatrix} \mathbf{Q}_1 & & & & \\ & \mathbf{Q}_1 & & & \\ & & \ddots & & \\ & & & \mathbf{Q}_1 & \\ & & & & \mathbf{Q}_2 \end{bmatrix},$$

$$\mathbf{Q}_1 = \begin{array}{c} \begin{matrix} \alpha_{i11} & \alpha_{i22} & \alpha_{i33} & \cdots & \alpha_{ikk} \end{matrix} \\ \begin{bmatrix} A & E & E & \cdots & E \\ E & C & D & \cdots & D \\ E & D & C & \cdots & \vdots \\ \vdots & \vdots & \vdots & C & D \\ E & D & \cdots & D & C \end{bmatrix} \end{array} \begin{matrix} \alpha_{i11} \\ \alpha_{i22} \\ \alpha_{i33} \\ \vdots \\ \alpha_{ikk} \end{matrix}$$

and
$$\mathbf{Q}_2 = \mathbf{I}_q/(k+2)(k+4)(k+6) = w\mathbf{I}_q.$$

The α_{ijj} indicate the positions of the elements of $\mathbf{Q}_1$ and show how the quadratic form will arise. The elements of $\mathbf{Q}_2$ will be multiplied by terms like α_{ijl} where i, j and l are all different.

If we define

$$\theta = 3\lambda_4/\lambda_2, \quad U = [\theta - 3/(k+4)]^2/9(k+2), \quad W = 1/(k+2)(k+4)^2(k+6);$$

then $A = 9U + 6(k+1)W, \quad E = 3U - 6W, \quad C = U + 2(k+3)W, \quad D = U - 2W.$

Evaluation of the quadratic form now gives $B = PU + [(k+4)Q - 2P]W$ where U and W are as defined above and where

$$\sigma^2 P/N \equiv (3\beta_{111} + \beta_{122} + \ldots + \beta_{1kk})^2 + \ldots + (3\beta_{kkk} + \beta_{k11} + \ldots + \beta_{k,k-1,k-1})^2,$$
$$\sigma^2 Q/N \equiv 2(3\beta_{111}^2 + \beta_{122}^2 + \ldots + \beta_{1kk}^2) + \ldots + 2(3\beta_{kkk}^2 + \beta_{k11}^2 + \ldots + \beta_{k,k-1,k-1}^2)$$
$$+ (\beta_{123}^2 + \ldots + \beta_{k-2,k-1,k}^2).$$

It is shown in Appendix 2 that P and Q are both invariant under rotation. This means that given a true relationship of any particular kind, the bias will be independent of the orientation of the contours.

2·3. *Evaluation of V.* The matrix $(\mathbf{X}_1'\mathbf{X}_1)^{-1}$ is found from the formulae given in Box & Hunter (1957) for the inverse of certain matrices which frequently arise in response surface work. If we pre-multiply this inverse by $\mathbf{x}_1'$, post-multiply by $\mathbf{x}_1$ and carry out the appropriate integration we find that

$$V = \frac{1}{\lambda_2} + \frac{3(k-1)}{2(k+4)\,\theta\lambda_2} + \frac{(k+2)\,(k+4)\,\theta\lambda_2 + 3 - 2(k+4)\,\theta}{(k+4)\,\lambda_2[(k+2)\,\theta - 3\lambda_2]}, \tag{2·3·1}$$

where $\theta = 3\lambda_4/\lambda_2$.

2·4. *Minimization of J.* Altogether, then, we have

$$J = V + P[\theta - 3/(k+4)]^2/9(k+2) + [(k+4)\,Q - 2P]/(k+2)\,(k+4)^2\,(k+6), \tag{2·4·1}$$

where V is given by equation (2·3·1), and we should like to choose λ_2 and θ, which is equivalent to choosing λ_2 and λ_4, in order to minimize J. Suppose now that we fix θ. Then B is fixed and V depends only on λ_2. Thus, it is possible to choose λ_2 as a function of θ so that, for each fixed θ, V (and thus J overall) takes on the lowest possible value. This gives us $J(\theta)$ in terms of θ alone, after we substitute the appropriate value for $\lambda_2 = \lambda_2(\theta)$ and we can then minimize J in terms of θ if we are given the values of P and Q which are functions of the β_{ijk}. Thus, for each pair of values (P, Q) we can choose θ so that the linked pair $[\lambda_2(\theta), \theta]$ gives rise to a minimum value of J. In fact, the minimizing design parameters $\lambda_2(\theta)$ and θ depend only on P since Q enters only in the constant portion of B. As a matter of practical calculation we shall not specify P and Q and then find the design which minimizes J. Computationally, it is far simpler to specify a value of θ, then determine λ_2 as a function of θ, $\lambda_2(\theta)$, so that V is minimized and, finally, see for what value of P this design would be best, i.e. which P would give these specified $\lambda_2(\theta)$ and θ as the ones which minimize J.

2·5. *Obtaining designs which will minimize J.* We refer back to equation (2·4·1) for

$$J = V(\lambda_2, \theta) + B(\theta, P, Q). \tag{2·5·1}$$

Fix θ and remember that P and Q, though their values are unknown, are constants. Then we must choose $\lambda_2 = \lambda_2(\theta)$ so that V (and hence J) are minimized for this θ; thus we set

$$\frac{\partial V}{\partial \lambda_2}(\lambda_2, \theta) = 0,$$

which leads to

$$\lambda_2(\theta) = \theta\left\{\frac{\{6[2(k+4)\theta + 3(k+1)]\,[(k+2)^2\,(k+4)\,\theta^2 - 6k(k+4)\,\theta + 9k]\}^{\frac{1}{2}} - 3k[2(k+4)\,\theta + 3(k+1)]}{3[2\theta^2(k+2)\,(k+4) - 6k(k+4)\,\theta - 9k(k-1)]}\right\} \tag{2·5·2}$$

and we can, in principle, substitute this value in equation (2·5·1) so that now

$$J = V(\theta) + B(\theta, P, Q).$$

Fortunately, as mentioned above, it is not necessary actually to make the substitution.

We now differentiate J with respect to θ. Differentiating equation (2·5·1) with respect to θ, remembering that λ_2 is a function of θ, and equating the result to zero we obtain

$$\frac{\partial V}{\partial \lambda_2}\frac{d\lambda_2}{d\theta}+\frac{\partial V}{d\theta}+\frac{dB}{d\theta}=0, \quad \text{or} \quad \frac{\partial V}{\partial \theta}+\frac{dB}{d\theta}=0, \quad \text{since}\ \frac{\partial V}{\partial \lambda_2}\equiv 0.$$

Now
$$\frac{\partial V}{\partial \theta}=-\frac{3}{(k+4)\lambda_2}\left\{\frac{k-1}{2\theta^2}+\frac{k(k+2)(k+4)\lambda_2^2-2k(k+4)\lambda_2+(k+2)}{[(k+2)\theta-3k\lambda_2]^2}\right\}$$

and
$$\frac{dB}{d\theta}=\frac{2}{9(k+2)}\left(\theta-\frac{3}{k+4}\right)P.$$

Thus $\partial V/\partial\theta+dB/d\theta=0$ implies that

$$P=\frac{27(k+2)}{2\lambda_2[(k+4)\theta-3]}\left\{\frac{k-1}{2\theta^2}+\frac{k(k+2)(k+4)\lambda_2^2-2k(k+4)\lambda_2+(k+2)}{[(k+2)\theta-3k\lambda_2]^2}\right\}, \qquad (2\cdot5\cdot3)$$

where $\lambda_2=\lambda_2(\theta)$ as given in equation (2·5·2). If we select a value for k and then a value for θ, we can use equations (2·5·2) and (2·5·3) to tabulate sets of values of (θ, λ_2, P). Although obtained in that order, they can be interpreted as follows. A design which minimizes J must necessarily have moments related by $\lambda_2=\lambda_2(\theta)$, as in equation (2·5·2), when there is a contribution from V. We shall choose the appropriate 'all-bias' design so that it, too, has moments related by the equation $\lambda_2=\lambda_2(\theta)$. Then, for a given P, we can find the appropriate value for θ from equation (2·5·3). We shall then use these calculations to arrive at some general conclusions about the correct design to use in practical situations.

2·6. *The calculations and their interpretation.* The calculations described above have been performed for $k=1, 2, 3, 4$, and 5 and for sufficiently many representative values of P so that the behaviour for all possible P can be predicted.

In the case $k=1$, P and Q are both multiples of β_{111} which is the only cubic coefficient and so Q is fixed if P is given and the total bias can be determined. Possible situations are thus $0\leqslant P\leqslant\infty$. When $k\geqslant 2$, the possible situations are $0\leqslant P\leqslant\infty$, $0\leqslant Q\leqslant\infty$. The best design in a given situation depends only on P, however, though the value of Q (when P is considered fixed) affects the relative values of V and B and hence affects the ratio V/B.

Fig. 1 shows eleven curves. Three of these provide values of $\lambda_2^{\frac{1}{2}}$, $\lambda=3\lambda_4/\lambda_2^2$ and V/B which correspond to the best design choice for a given P when $k=1$. These may be interpreted as follows. For the all-bias situation, $P=\infty$, we should choose $\lambda_2^{\frac{1}{2}}=0\cdot606$, $\lambda=1\cdot632$. As situations arise where the bias contribution to J becomes smaller and smaller we see that the best design moment values increase. For example, when $V=8B$ approximately, the best design is such that $\lambda_2^{\frac{1}{2}}=0\cdot7$, $\lambda=2\cdot0$ approximately, which is quite close to the appropriate design for the 'all-bias' situation and far from the appropriate 'all-variance' design which is, as always, the largest possible (denoted in the table by infinite moments, but in practice as large as possible until restricted by the operability region O). When $V=B$ approximately, a situation we can regard as 'typical' (as described in our earlier paper), the best design is such that, approximately, $\lambda_2^{\frac{1}{2}}=0\cdot621$, $\lambda=1\cdot669$, very close to the 'all-bias' figures.

The other eight curves of Fig. 1 provide values of $\lambda_2^{\frac{1}{2}}$ and λ which correspond to the best design choice for a given P when $k=2, 3, 4$ and 5. Over a very large range of possible values of P, the optimum design for each k changes only slightly. Only when quite small values of P are postulated do the moments of the best design increase appreciably compared with the

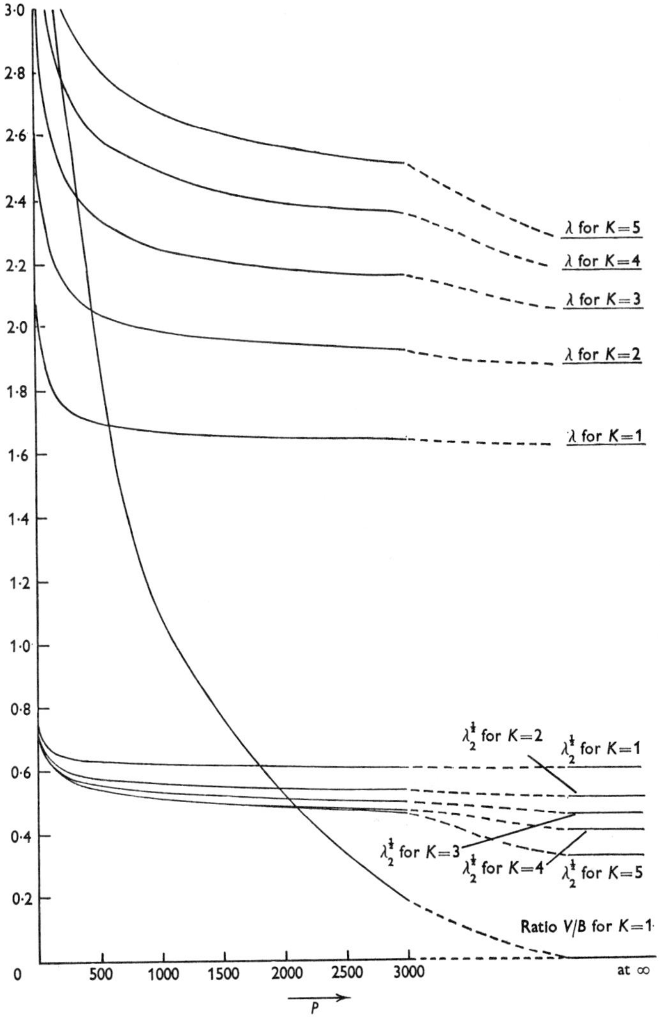

Fig. 1. Best design moments $\lambda_2^{\frac{1}{2}}$ and λ for varying P, $1 \leqslant K \leqslant 5$.

all-bias values. As in all cases, when there is no contribution from bias at all ($P = 0$) the best design is the largest possible. Note that a situation can arise where P is very small or zero and, at the same time, Q is large; thus, the total bias could be quite large (because of the size of Q) but the appropriate best design would be the largest possible because only the V part of J can be affected by altering the moments. In such a case, the lack of fit would be large but the coefficients would be either unbiased ($P = 0$) or not very biased (P small).

Thus, overall, we can conclude that in most situations the appropriate experimental designs to use to minimize J have moments slightly larger than the moments of the appropriate all-bias design. As a practical matter in situations where no information about the possible size of P exists, about 10 % greater is suggested as a rough rule of thumb.

2·7. *Use of the secondary consideration to select a design.* Even now we do not have a specific design, but only a certain subset of designs, since the requirements 'second-order rotatable, with zero fifth moments, and λ_2 and λ_4 of a given size' can be satisfied by a number of designs for every value of k. We select an individual design from all those with the correct sized moments by appealing to our second criterion which specifies that our chosen design should make large the quantity

$$\sum_{u=1}^{N} \{E(\hat{y}_u)-\eta_u\}^2 = E(S_R)-\nu\sigma^2 = NF, \quad \text{say.}$$

As explained in Box & Draper (1959), this quantity can be written

$$NF = -\boldsymbol{\beta}_2'\mathbf{X}_2'\mathbf{X}_1\mathbf{A}\boldsymbol{\beta}_2+\boldsymbol{\beta}_2\mathbf{X}_2'\mathbf{X}_2\boldsymbol{\beta}_2. \tag{2·7·1}$$

The matrix $N^{-1}(\mathbf{X}_2'\mathbf{X}_1\mathbf{A})$ is square and of dimension $k(k-1)(k-2)/6$. It consists of a number of submatrices down the main diagonal. The first k of these are of dimension k by k and have the form

$$\frac{\lambda_4^2}{\lambda_2}\begin{bmatrix} 9 & 3 & \dots & 3 \\ 3 & 1 & \dots & 1 \\ \vdots & \vdots & & \\ 3 & 1 & \dots & 1 \end{bmatrix}.$$

All other elements are zero. Thus,

$$\begin{aligned} N^{-1}\boldsymbol{\beta}_2'\mathbf{X}_2'\mathbf{X}_1\mathbf{A}\boldsymbol{\beta}_2 &= \lambda_4^2\lambda_2^{-1}\{9(\beta_{111}^2+\dots)+6(\beta_{111}\beta_{122}+\dots)+(\beta_{122}^2+\dots)+2(\beta_{122}\beta_{133}+\dots)\} \\ &= \lambda_4^2\lambda_2^{-1}\{(3\beta_{111}+\beta_{122}+\dots+\beta_{1kk})^2+\dots+(3\beta_{kkk}+\beta_{k11}+\dots+\beta_{k,k-1,k-1})^2\} \\ &= \lambda_4^2\lambda_2^{-1}P\sigma^2/N. \end{aligned}$$

The matrix $N^{-1}(\mathbf{X}_2'\mathbf{X}_2)$ is also square and of dimension $k(k-1)(k-2)/6$. It consists of $(k+1)$ submatrices down the main diagonal. The first of these is of the form

$$N^{-1}\begin{bmatrix} \Sigma x_{1u}^6 & \Sigma x_{1u}^4x_{2u}^2 & \dots & \Sigma x_{1u}^4x_{ku}^2 \\ \Sigma x_{1u}^4x_{2u}^2 & \Sigma x_{1u}^2x_{2u}^4 & \dots & \Sigma x_{1u}^2x_{2u}^2x_{ku}^2 \\ \vdots & & & \\ \Sigma x_{1u}^4x_{ku}^2 & \Sigma x_{1u}^2x_{2u}^2x_{ku}^2 & \dots & \Sigma x_{1u}^2x_{ku}^4 \end{bmatrix}$$

and the second, third, ..., down to the kth are similar but with the obvious variation in suffices. The $(k+1)$th matrix is diagonal with terms such as $N^{-1}\Sigma x_{1u}^2x_{2u}^2x_{3u}^2$. All summations are over u. The pattern of suffices can be seen by imagining the vector $\boldsymbol{\beta}_2$ to be written out along the top and side of the matrix $\mathbf{X}_2'\mathbf{X}_2$. Apart from the elements already mentioned, all elements of $\mathbf{X}_2'\mathbf{X}_2$ are zero.

We now make use of a concept introduced in Box & Draper (1959). Since the orientation of our true surface with respect to our design is unknown, we shall average the value of F over all orthogonal rotations, denoting the average value by $\bar{F}$. Only the second term on the right-hand side of equation (2·7·1) for F is affected by this rotation; as we have seen above, the first term, which is a multiple of P, is independent of the rotation. The details of the averaging process are given in Appendix 3. Our result is that

$$N\tilde{F}/\sigma^2 = (P+Q) \sum_{u=1}^{N} r_u^6/Nk(k+2)(k+4) - \lambda_4^2\lambda_2^{-1}P,$$

where $r_u^2 = x_{1u}^2 + x_{2u}^2 + \ldots + x_{ku}^2$. Since λ_2 and λ_4 are already determined by the work given above, for given (or suggested) P, only the quantity $\sum_u r_u^6$ is capable of allocation as far as the design is concerned. It follows that our requirement that the design should be such that $\tilde{F}$ is made large implies that $\sum_u r_u^6$ should be large. It should be noted that this conclusion is based on the assumption that the true model is of exactly third order.

3. Application of the solution to particular problems

3·1. *An application of the results to the case $k = 1$.* When $k = 1$ we have

$$\hat{y} = b_0 + b_1 x + b_{11} x^2,$$

$$\eta = \beta_0 + \beta_1 x + \beta_{11} x^2 + \beta_{111} x^3.$$

Suppose that we wished to have a design with $N = 10$ points and with not more than five distinct levels of x, one of the five being the origin and the others symmetrically placed. Thus we can denote the five levels by $(-b, -a, 0, a, b)$. How should we allocate the ten points to the five levels to get a design which would be judged best by the criteria used in this paper? The appropriate design for the all-bias situation is one for which $\lambda_2^{\frac{1}{2}} = 0{\cdot}606$, $\lambda = 1{\cdot}632$. However, if we felt that, probably, V and B would be of the same size, we could choose $\lambda_2^{\frac{1}{2}} = 0{\cdot}62$, $\lambda = 1{\cdot}67$; or if we felt that variance error would probably be about eight times as big as bias error we could choose $\lambda_2^{\frac{1}{2}} = 0{\cdot}70$, $\lambda = 2{\cdot}00$. For any particular choice of moment values, several ten point designs at five or less levels are possible as shown in Tables 1 and 2. Note that both levels a and b must be used in order to satisfy the specified values of $\lambda_2^{\frac{1}{2}}$ and λ.

Table 1. *Possible designs when $\lambda_2^{\frac{1}{2}} = 0{\cdot}62$, $\lambda = 1{\cdot}67$*

No. of points at					Value of		No. proportional to
$-b$	$-a$	0	a	b	a	b	$\sum_u r^6$
1	4	0	4	1	0·478	1·009	142
1	3	2	3	1	0·567	0·984	125
1	2	4	2	1	0·790	0·828	75
2	1	4	1	2	0·777	0·815	68
2	3	0	3	2	0·353	0·880	62
2	2	2	2	2	0·452	0·873	60

Table 2. *Possible designs when $\lambda_2^{\frac{1}{2}} = 0{\cdot}70$, $\lambda = 2{\cdot}00$*

No. of points at					Value of		No. proportional to
$-b$	$-a$	0	a	b	a	b	$\sum_u r^6$
1	4	0	4	1	0·500	1·225	45
1	3	2	3	1	0·587	1·210	42
1	2	4	2	1	0·755	1·166	39
2	3	0	3	2	0·303	1·054	37
2	2	2	2	2	0·376	1·053	37
2	1	4	1	2	0·553	1·047	36

We can now use our secondary consideration to choose a particular design from whichever group we decided to use. We recall that our theory told us to choose a design which makes large the quantity Σr_u^6 in order to obtain good detection of cubic bias. On this basis we should select the first design in each group, since it provides the maximum value of the sixth moment. However, since the second design in each table has experiments at all five levels and the sixth moment is still quite high, the second design in each table would also be a good choice. Note that the design with two experiments at each of the five levels is not a particularly good one in our assumed circumstances, despite the fact that it might seem the natural arrangement to choose.

3·2. *An application to certain rotatable design classes.* We now consider a different situation. Suppose an experimenter wishes to use a basic central composite rotatable design consisting of a cube plus octahedron plus n_0 centre points to investigate k factors. Altogether $N = 2^k + 2k + n_0$ experiments are required. Let $(\pm a, \pm a, \ldots, \pm a)$ be the points of the cube; let $(\pm b, 0, \ldots, 0), \ldots, (0, 0, \ldots, \pm b)$ be the points of the octahedron. Necessarily $b = 2^{\frac{1}{4}k} a$ for rotatability. The region of interest R is the unit sphere. The experimenter would like to know how many centre points to use and how to choose the size of the design in relation to the region. In other words, what values should he assign to n_0 and a? Our first criterion may be used to answer this question as follows. Allocating values of a and n_0 is equivalent to selecting values of $\lambda_2^{\frac{1}{2}}$ and λ, since

$$\lambda_2 = (2^k + 2^{\frac{1}{2}(k+2)})\, a^2/N \quad \text{and} \quad \lambda = 3\lambda_4/\lambda_2^2 = 3N/(2+2^{\frac{1}{2}k})^2.$$

By the first criterion of this paper the values of $\lambda_2^{\frac{1}{2}}$ and λ are linked for optimum designs. It is thus clear that, if we choose a value for n_0, λ becomes fixed, hence λ_2 is fixed through (2·5·2) and in turn a is fixed. We can examine possible variations of this type of design, therefore, by considering various numbers of centre points to be added. Table 3 shows the parameters of the designs which result when $0 \leqslant n_0 \leqslant 12$, for the case $k = 2$. We can see from Fig. 1 that pairs $(\lambda_2^{\frac{1}{2}}, \lambda)$ between approximately (0·52, 1·89) and (0·73, 2·80) are suitable design 'sizes' for possible values of P between infinity and 15, where

$$P = N\{(3\beta_{111}+\beta_{122})^2 + (3\beta_{222}+\beta_{112})^2\}/\sigma^2,$$

Table 3. *Parameter values for a central composite rotatable design series when* $k = 2$

n_0	$\lambda_2^{\frac{1}{2}}$	λ	a	b
0	0·628	1·500	0·628	0·888
1	0·578	1·688	0·613	0·867
2	0·505	1·875	0·565	0·799
3	0·583	2·063	0·684	0·967
4	0·627	2·250	0·768	1·086
5	0·663	2·438	0·846	1·196
6	0·696	2·625	0·921	1·303
7	0·727	2·813	0·996	1·408
8	0·757	3·000	1·070	1·514
9	0·785	3·188	1·145	1·619
10	0·813	3·375	1·220	1·725
11	0·840	3·563	1·295	1·832
12	0·867	3·750	1·371	1·939

(Note: The recommended 'all bias' design for $k = 2$ has $\lambda_2^{\frac{1}{2}} = 0·515$, $\lambda = 1·887$.)

and where $N = 8+n_0$. Thus the designs in Table 3 for which $n_0 = 0$, 1 and 2 cannot be optimal designs for any values of P (even though their parameters are related by 2·5·2) since their moments are smaller than the appropriate 'all bias' moments. All the other designs listed will be optimal for certain values of P. The extreme possible designs for $\infty \geqslant P \geqslant 15$ are those for which ($n_0 = 3, a = 0{\cdot}684, b = 0{\cdot}967$) and ($n_0 = 7$, $a = 0{\cdot}966, b = 1{\cdot}408$), while values of P less than 15 would require a design with even more centre points and even larger second and fourth moments. The region of interest, R, it should be remembered, is the unit circle. We can thus observe that, as we expect less and less effect from the biases of the coefficients we add more and more centre points to the composite design and place the points further and further from the origin even outside R. (Recall that for two factors, $E(b_1) = \beta_1+(3\beta_{111}+\beta_{122})\lambda_4/\lambda_2$ and $E(b_2) = \beta_2+(3\beta_{222}+\beta_{112})\lambda_4/\lambda_2$ are the expectations of the estimates of the linear coefficients in a fitted quadratic model when cubic terms exist and this particular type of design is used. Hence P is proportional to the sum of squares of biases in the estimates of the linear coefficients.) This is extremely reasonable. As we become surer of our model (i.e. bias is thought to be small) we spread out the design; but since variance error becomes a greater and greater part of the total discrepancy between fitted and actual model we add more centre points to provide a better estimate of the error variation. On the other hand, if we doubt our model (i.e. bias is thought to be large) and we believe variance error to be a small part of the total discrepancy between fitted and actual model we contract our design into the region of interest R and use only enough centre points (three or four say) to provide some estimate of σ^2.

Calculations of this type for the cases $k = 3$, 4 and 5 and similar calculations for the 'cube plus doubled octahedron plus centre points' design series for $k = 2$, 3, 4 and 5 are given in an earlier version of this paper (Box & Draper, 1962). Similar remarks apply to all cases.

APPENDIX 1

A problem in numerical analysis (see § 1·6)

Suppose we have a function $\eta(\mathbf{x})$, known exactly without error. Suppose we wish to approximate to this function over a region R by a polynomial of form $\hat{\eta}(\mathbf{x}) = \mathbf{x}_1'\boldsymbol{\gamma}_1$, say, and of order d_1.

We shall choose $\boldsymbol{\gamma}_1$ so that the integral defined by

$$\Sigma \equiv \Omega\int_R (\eta(\mathbf{x})-\hat{\eta}(\mathbf{x}))^2\,d\mathbf{x} \quad \text{is minimized.}$$

Now

$$\Sigma = \Omega\int_R (\eta(\mathbf{x})-\mathbf{x}_1'\boldsymbol{\gamma}_1)^2\,d\mathbf{x}.$$

Thus

$$\frac{\partial\Sigma}{\partial\boldsymbol{\gamma}_1} = -2\Omega\int_R \mathbf{x}_1\{\eta(\mathbf{x})-\mathbf{x}_1\boldsymbol{\gamma}_1\}\,d\mathbf{x} = 0$$

implies that

$$\Omega\int_R \mathbf{x}_1\eta(\mathbf{x})\,d\mathbf{x} = \left\{\Omega\int_R \mathbf{x}_1\mathbf{x}_1'\,d\mathbf{x}\right\}\boldsymbol{\gamma}_1.$$

Using earlier definitions on the right-hand side and calling the left-hand side $\boldsymbol{\mu}_{11\eta}$ by analogy, we can write this as

$$\boldsymbol{\mu}_{11\eta} = \boldsymbol{\mu}_{11}\boldsymbol{\gamma}_1.$$

Therefore,

$$\boldsymbol{\gamma}_1 = \boldsymbol{\mu}_{11}^{-1}\boldsymbol{\mu}_{11\eta}$$

no matter what η may be, assuming that $\boldsymbol{\mu}_{11}$ is non-singular. Applying this to the case where

$$\eta(\mathbf{x}) = \mathbf{x}_1'\boldsymbol{\beta}_1+\mathbf{x}_2'\boldsymbol{\beta}_2,$$

a polynomial of order d_2, with $\mathbf{x}_1$ as before, we find that

$$\begin{aligned}\boldsymbol{\gamma}_1 &= \boldsymbol{\mu}_{11}^{-1}\left\{\Omega \int_R \mathbf{x}_1(\mathbf{x}_1'\boldsymbol{\beta}_1+\mathbf{x}_2'\boldsymbol{\beta}_2)\,d\mathbf{x}\right\}\\ &= \boldsymbol{\mu}_{11}^{-1}\{\boldsymbol{\mu}_{11}\boldsymbol{\beta}_1+\boldsymbol{\mu}_{12}\boldsymbol{\beta}_2\}\\ &= \boldsymbol{\beta}_1+\boldsymbol{\mu}_{11}^{-1}\boldsymbol{\mu}_{12}\boldsymbol{\beta}_2\\ &= \boldsymbol{\beta}_1+\mathfrak{A}\boldsymbol{\beta}_2 \quad (\text{where } \mathfrak{A} = \boldsymbol{\mu}_{11}^{-1}\boldsymbol{\mu}_{12}).\end{aligned}$$

The value of the average integrated discrepancy between the fitted and true model is

$$\begin{aligned}\Sigma_{\text{min.}} &= \Omega \int_R \{\eta(\mathbf{x})-\hat{\eta}(\mathbf{x})\}^2\,d\mathbf{x} \quad (\text{where } \hat{\eta}(\mathbf{x}) = \mathbf{x}_1'\boldsymbol{\gamma}_1)\\ &= \Omega \int_R \{\mathbf{x}_1'\boldsymbol{\beta}_1+\mathbf{x}_2'\boldsymbol{\beta}_2-\mathbf{x}_1'\boldsymbol{\beta}_1-\mathbf{x}_1'\mathfrak{A}\boldsymbol{\beta}_2\}^2\,d\mathbf{x}\\ &= \Omega \int_R \boldsymbol{\beta}_2'(\mathbf{x}_2'-\mathbf{x}_1'\mathfrak{A})'(\mathbf{x}_2'-\mathbf{x}_1'\mathfrak{A})\boldsymbol{\beta}_2\,d\mathbf{x}\\ &= \Omega \int_R \boldsymbol{\beta}_2'\{\mathbf{x}_2\mathbf{x}_2'-\mathbf{x}_2\mathbf{x}_1'\mathfrak{A}-\mathfrak{A}'\mathbf{x}_1\mathbf{x}_2'+\mathfrak{A}'\mathbf{x}_1\mathbf{x}_1'\mathfrak{A}\}\boldsymbol{\beta}_2\,d\mathbf{x}\\ &= \boldsymbol{\beta}_2'(\boldsymbol{\mu}_{22}-\boldsymbol{\mu}_{12}'\boldsymbol{\mu}_{11}^{-1}\boldsymbol{\mu}_{12}-\boldsymbol{\mu}_{12}'\boldsymbol{\mu}_{11}^{-1}\boldsymbol{\mu}_{12}+\boldsymbol{\mu}_{12}'\boldsymbol{\mu}_{11}^{-1}\boldsymbol{\mu}_{11}\boldsymbol{\mu}_{11}^{-1}\boldsymbol{\mu}_{12})\boldsymbol{\beta}_2\\ &= \boldsymbol{\beta}_2'(\boldsymbol{\mu}_{22}-\boldsymbol{\mu}_{12}'\boldsymbol{\mu}_{11}^{-1}\boldsymbol{\mu}_{12})\boldsymbol{\beta}_2.\end{aligned}$$

Thus, if we knew the polynomial $\eta(\mathbf{x}) = \mathbf{x}_1'\boldsymbol{\beta}_1+\mathbf{x}_2'\boldsymbol{\beta}_2$ exactly and fitted $\hat{\eta}(\mathbf{x}) = \mathbf{x}_1'\boldsymbol{\gamma}_1$ to it to minimize Σ, then $\Sigma_{\text{min.}}$ is the smallest value of the average integrated squared discrepancy we can achieve.

But, if we fit, not to a known function, but to a function whose value is known only at N points, then we know that the bias which arises is

$$B_{\text{min.}} = \Sigma_{\text{min.}}+\boldsymbol{\beta}_2'(\mathbf{A}-\mathfrak{A})'\boldsymbol{\mu}_1(\mathbf{A}-\mathfrak{A})\boldsymbol{\beta}_2$$

as shown in Box & Draper (1959), where as above $\mathbf{A} = (\mathbf{X}_1'\mathbf{X}_1)^{-1}\mathbf{X}_1'\mathbf{X}_2$ and the rows of $\mathbf{X}_1$ and $\mathbf{X}_2$ are, respectively, the values of $\mathbf{x}_1$ and $\mathbf{x}_2$ which correspond to the N points.

Thus by choosing $\mathbf{A}$, which is at our disposal and involves design moments, in such a way that $\mathbf{A} = \mathfrak{A}$ we can make

$$B_{\text{min.}} = \Sigma_{\text{min.}}.$$

Thus even if we do not know our polynomial function $\eta(\mathbf{x})$ exactly, we can choose the points at which to evaluate $\eta(\mathbf{x})$ in such a way that the average integrated squared discrepancy incurred when we graduate by $\hat{\eta}(\mathbf{x})$ is exactly what it would be in the situation where we know the function exactly at every point.

APPENDIX 2

To show that P and Q are invariant under rotation (see § 2·2).

We know that

$$\begin{aligned}P &= (3\alpha_{111}+\alpha_{122}+\ldots+\alpha_{1kk})^2+\ldots+(3\alpha_{kkk}+\alpha_{k11}+\ldots+\alpha_{k,\,k-1,\,k-1})^2,\\ Q &= 2(3\alpha_{111}^2+\alpha_{122}^2+\ldots+\alpha_{1kk}^2)+\ldots+2(3\alpha_{kkk}^2+a_{k11}^2+\ldots+\alpha_{k,\,k-1,\,k-1}^2)+(\alpha_{123}^2+\ldots+\alpha_{k-2,\,k-1,\,k}^2),\end{aligned}$$

where $\alpha_{jik}^2 = N\beta_{jik}^2/\sigma^2$.

For the proof that these quantities P and Q are invariant under rotations of the surface about the origin we shall make use of the matrix direct product which has the following properties (Marcus, 1960):

$$\mathbf{A}*\mathbf{B} = \begin{bmatrix} b_{11}\mathbf{A} & b_{12}\mathbf{A} & \ldots & b_{1n}\mathbf{A}\\ b_{21}\mathbf{A} & b_{22}\mathbf{A} & \ldots & b_{2n}\mathbf{A}\\ \vdots & & & \\ b_{m1}\mathbf{A} & b_{m2}\mathbf{A} & \ldots & b_{mn}\mathbf{A}\end{bmatrix}$$

$$\begin{aligned}(A*B)' &= A'*B',\\ (A*B)*C &= A*C+B*C,\\ (A*B)(C*D) &= AC*BD.\end{aligned}$$

Let $\mathbf{H}$ be the matrix of an orthogonal transformation taking $\mathbf{x}$ into $\mathbf{X}$ by $\mathbf{x} = \mathbf{H}\mathbf{X}'$ where

$$\mathbf{x}' = (x_1, x_2, \ldots, x_k),$$
$$\mathbf{X}' = (X_1, X_2, \ldots, X_k).$$

Now

$$\mathbf{x}' * \mathbf{x}' = \{x_1(x_1, x_2, \ldots, x_k); \ldots; x_k(x_1, x_2, \ldots, x_k)\}$$

and

$$\mathbf{x}' * \mathbf{x}' * \mathbf{x}' = [x_1\{x_1(x_1, x_2, \ldots, x_k); \ldots; x_k(x_1, x_2, \ldots, x_k)\}, \ldots, x_k\{x_1(x_1, x_2, \ldots, x_k); \ldots; x_k(x_1, x_2, \ldots, x_k)\}].$$

Write

$$\boldsymbol{\beta}' = \{\beta_{111}, \tfrac{1}{3}\beta_{112}, \tfrac{1}{3}\beta_{113}, \ldots, \tfrac{1}{3}\beta_{11k}; \tfrac{1}{3}\beta_{112}, \tfrac{1}{3}\beta_{122}, \tfrac{1}{6}\beta_{123}, \ldots\}.$$

Then

$$\begin{aligned}\boldsymbol{\beta}'(\mathbf{x}' * \mathbf{x}' * \mathbf{x}') &= \beta_{111}x_1^3 + \beta_{112}x_1^2x_2 + \ldots \\ &= \mathbf{x}_2'\boldsymbol{\beta}_2,\end{aligned}$$

where the right-hand side consists of all the cubic terms, with appropriate coefficients, of the cubic response surface model. Employing the transformation $\mathbf{x}' = \mathbf{H}\mathbf{X}'$ we can write

$$\begin{aligned}\boldsymbol{\beta}'(\mathbf{x}' * \mathbf{x}' * \mathbf{x}') &= \boldsymbol{\beta}'(\mathbf{H}\mathbf{X}' * \mathbf{H}\mathbf{X}' * \mathbf{H}\mathbf{X}') \\ &= \boldsymbol{\beta}'(\mathbf{H} * \mathbf{H} * \mathbf{H})(\mathbf{X}' * \mathbf{X}' * \mathbf{X}') \\ &= \mathbf{B}'(\mathbf{X}' * \mathbf{X}' * \mathbf{X}'),\end{aligned}$$

say, where $\mathbf{B}' = \boldsymbol{\beta}'(\mathbf{H} * \mathbf{H} * \mathbf{H})$ is the new '$\boldsymbol{\beta}$ vector' for the transformed co-ordinates. Thus, for these transformed co-ordinates, Q would be given by

$$\begin{aligned}\sigma^2Q/6N &= \mathbf{B}'\mathbf{B} \\ &= \boldsymbol{\beta}'(\mathbf{H} * \mathbf{H} * \mathbf{H})(\mathbf{H}' * \mathbf{H}' * \mathbf{H}')\boldsymbol{\beta} \\ &= \boldsymbol{\beta}'(\mathbf{H}\mathbf{H}' * \mathbf{H}\mathbf{H}' * \mathbf{H}\mathbf{H}')\boldsymbol{\beta} \\ &= \boldsymbol{\beta}'(\mathbf{I} * \mathbf{I} * \mathbf{I})\boldsymbol{\beta} \\ &= \boldsymbol{\beta}'\boldsymbol{\beta},\end{aligned}$$

using the fact that, since $\mathbf{H}$ is orthogonal, $\mathbf{H}\mathbf{H}' = \mathbf{I}$, the unit matrix of appropriate dimension. Hence Q is invariant under rotation of axes (or equivalently under rotation of the response surface relative to the axes).

Now let $\mathbf{u}_i' = (0, 0, \ldots, 0, 1, 0, \ldots, 0)$ be a 1 by k vector with unity in the ith place and zeros elsewhere. Further, let $\mathbf{u}' = (\mathbf{u}_1', \mathbf{u}_2', \ldots, \mathbf{u}_k')$ so that $\mathbf{u}'$ is a 1 by k^2 vector with k unities in the appropriate positions and zeros elsewhere. Finally, let $\mathbf{U} = \mathbf{u} * \mathbf{I}$ where $\mathbf{I}$ is a k by k unit matrix. Then $\mathbf{U}$ is a k^3 by k matrix consisting of submatrices $\mathbf{u}$ in the diagonal and zeros elsewhere as follows:

$$\mathbf{U} = \begin{bmatrix}\mathbf{u} & & & \mathbf{0} \\ & \mathbf{u} & & \\ & & \ddots & \\ \mathbf{0} & & & \mathbf{u}\end{bmatrix}, \quad \text{where } \mathbf{u} = \begin{bmatrix}\mathbf{u}_1 \\ \mathbf{u}_2 \\ \vdots \\ \mathbf{u}_k\end{bmatrix}, \quad \mathbf{u}_i = \begin{bmatrix}0 \\ \vdots \\ 0 \\ 1 \\ 0 \\ \vdots \\ 0\end{bmatrix}.$$

In these terms, P is given by

$$\begin{aligned}\sigma^2P/9N &= \mathbf{B}'\mathbf{U}\mathbf{U}'\mathbf{B} \\ &= \boldsymbol{\beta}'(\mathbf{H} * \mathbf{H} * \mathbf{H})(\mathbf{u} * \mathbf{I})(\mathbf{u}' * \mathbf{I})(\mathbf{H}' * \mathbf{H}' * \mathbf{H}')\boldsymbol{\beta} \\ &= \boldsymbol{\beta}'\{(\mathbf{H} * \mathbf{H})\mathbf{u} * \mathbf{H}\}\{\mathbf{u}'(\mathbf{H}' * \mathbf{H}') * \mathbf{H}'\}\boldsymbol{\beta}.\end{aligned}$$

(We here use the fact that, since $\mathbf{H}$ is orthogonal $\{\mathbf{H} * \mathbf{H}\}\mathbf{u} = \mathbf{u}$.)

$$\begin{aligned}&= \boldsymbol{\beta}'(\mathbf{u} * \mathbf{H})(\mathbf{u}' * \mathbf{H}')\boldsymbol{\beta} \\ &= \boldsymbol{\beta}'(\mathbf{u}\mathbf{u}' * \mathbf{H}\mathbf{H}')\boldsymbol{\beta} \\ &= \boldsymbol{\beta}'(\mathbf{u}\mathbf{u}' * \mathbf{I})\boldsymbol{\beta} \\ &= \boldsymbol{\beta}'(\mathbf{u} * \mathbf{I})(\mathbf{u}' * \mathbf{I})\boldsymbol{\beta} \\ &= \boldsymbol{\beta}'\mathbf{U}\mathbf{U}'\boldsymbol{\beta}.\end{aligned}$$

Thus, P also is invariant under the transformation $\mathbf{H}$.

APPENDIX 3

The average of F over all orthogonal rotations of the response surface (see § 2·7)

We need consideronly $\boldsymbol{\beta}_2 \mathbf{X}_2' \mathbf{X}_2 \boldsymbol{\beta}_2$ since the other portion of NF is unaffected by the rotation. Write

A = average value of β_{111}^2 and similar terms,
B = average value of $\beta_{111}\beta_{122}$ and similar terms,
C = average value of β_{122}^2 and similar terms,
D = average value of $\beta_{122}\beta_{133}$ and similar terms,
E = average value of β_{123}^2 and similar terms.

Only three of these quantities are independent and two relations exist between them as will be shown. Then

$$\begin{aligned} NF' &= \text{Rotational average of } \boldsymbol{\beta}_2' \mathbf{X}_2 \mathbf{X}_2 \boldsymbol{\beta}_2 \\ &= A\sum_i\sum_u x_{iu}^6 + 2B\sum_i\sum_u x_{iu}^4(r_u^2 - x_{iu}^2) + C\sum_i\sum_u x_{iu}^2\Big(\sum_j x_{ju}^4 - x_{iu}^4\Big) + D\sum_i\sum_u x_{iu}^2\Big(\sum_{\substack{j\neq i\\ l\neq i\\ j\neq l}} x_{ju}^2 x_{lu}^2\Big) + (\tfrac{1}{6}E)\sum_{i\neq j\neq l}\sum_u x_{iu}^2 x_{ju}^2 x_{lu}^2 \\ &= A\sum_u r_u^6 + (2B + C - 3A)\Big\{\sum_i x_{iu}^4(r_u^2 - x_{iu}^2)\Big\} + \{D + (\tfrac{1}{6}E) - A\}\Big\{\sum_u \sum_{i\neq j\neq l} x_{iu}^2 x_{ju}^2 x_{lu}^2\Big\}, \end{aligned}$$

where $r_u^2 = x_{1u}^2 + \ldots + x_{ku}^2$. Since this is a rotational average it is necessarily a function of the r_u^2 only. This implies the relationships $2B + C - 3A = 0$, $D + (\frac{1}{6}E) - A = 0$. It follows that $NF' = A\sum_u r_u^6$. We now recall our earlier definition

$$\begin{aligned} \sigma^2 P/N &= (3\beta_{111} + \beta_{122} + \ldots + \beta_{1kk})^2 + (k-1)\text{ similar terms}, \\ \sigma^2 Q/N &= 2(3\beta_{111}^2 + \beta^2_{22} + \ldots + \beta_{1kk}^2) + (k-1)\text{ similar terms} + (\beta_{123}^2 + \ldots). \end{aligned}$$

Hence, averaging over all rotations, we see that

$$\begin{aligned} \sigma^2 P/N &= k\{9A + 6(k-1)B + (k-1)C + (k-1)(k-2)D\}, \\ \sigma^2 Q/N &= k\{6A + 2(k-1)C\} + k(k-1)(k-2)\tfrac{1}{6}E \end{aligned}$$

and

$$\begin{aligned} \sigma^2(P+Q)/N &= k\{15A + 6(k-1)B + 3(k-1)C + (k-1)(k-2)D + (k-1)(k-2)\tfrac{1}{6}E\} \\ &= k\{(k+2)(k+4)A + 3(k-1)(2B + C - 3A) + (k-1)(k-2)(D + [\tfrac{1}{6}E] - A)\} \\ &= k(k+2)(k+4)A, \end{aligned}$$

because of the relationships mentioned above. Substituting for A, we find

$$NF'/\sigma^2 = (P+Q)\Big(\sum_u r_u^6\Big)/Nk(k+2)(k+4)$$

and so

$$N\tilde{F}/\sigma^2 = (P+Q)\Big(\sum_u r_u^6\Big)/Nk(k+2)(k+4) - \lambda_4^2\lambda_2^{-1}P.$$

REFERENCES

BOX, G. E. P. & DRAPER, NORMAN R. (1959). A basis for the selection of a response surface design. *J. Amer. Statist. Ass.* **54**, 622–54.

BOX, G. E. P. & DRAPER, NORMAN R. (1962). The choice of a second order rotatable design. *Univ. Wisc. Stat. Dept. Tech. Rep.* **10**.

BOX, G. E. P. & HUNTER, J. S. (1957). Multifactor experimental designs for exploring response surfaces. *Ann. Math. Statist.* **28**, 195–241.

MARCUS, MARVIN (1960). Basic theorems in matrix theory. *Nat. Bur. Stand. Appl. Math. Ser.* **57**.

2.13
Sequential Design of Experiments for Nonlinear Models[1]

G. E. P. BOX and WILLIAM G. HUNTER
University of Wisconsin

OBJECTIVE OF THE EXPERIMENTER

Suppose that an experimenter is interested in studying a particular system for which there exists a mathematical model $\eta = f(\underline{\theta}, \underline{\xi})$, nonlinear in the parameters $\underline{\theta}$, which relates a measurable response η to the controllable variables $\underline{\xi}$. The objective of the experimenter may be: (1) to obtain an estimate of a response η over some particular region of interest in the space of the variables or (2) to determine the underlying physical mechanism of the phenomenon under investigation.

Mathematically, we could say that for problem (2) the whole object is to discover the nature of the function $f(\underline{\theta}, \underline{\xi})$. In practical situations, we can never know this completely. However, we shall say that we have an adequate theoretical model when we have derived from a consideration of the mechanism a function which closely predicts the results of actual experiments.

In problem (1), which has come to be called the *response surface problem,* it is useful but not essential to employ such a theoretical model (see Box, 1958 and 1960, Box and Coutie, 1956, and Box and Youle, 1955). In many circumstances, even though no theoretical model is available, perfectly good empirical approximations can be obtained by fitting a polynomial or some other flexible, graduating function over the region of interest (see Box, 1954, Box and Draper, 1959, Box and J. S. Hunter, 1957, and Davies, 1956).

Empirical models, however, are of limited value when the aim is to develop a suitable mechanistic theory. The search for underlying

[1]This research was supported by the Wisconsin Alumni Research Foundation and the United States Navy through the Office of Naval Research under Contract Nonr-1202, Project NR 042-222. Reproduction in whole or in part for any purpose of the United States government is permitted.

The Collected Works of George E. P. Box, 1984, Wadsworth, Inc., Belmont, CA 94002.
Originally published in *Proceedings of IBM Scientific Computing Symposium on Statistics* (1963), pp. 113–137.

physical mechanisms constitutes a major portion of effort in a number of scientific fields. To engineers, for example, basic-mechanism studies are of interest principally because a deeper understanding of the mechanism involved makes it possible to cope with engineering design problems in a more intelligent and useful manner than would be possible if the mechanism were entirely unknown.

In what follows we shall be concerned with a particular aspect of this second objective of trying to elucidate the mechanism. Such mechanism studies consist essentially of two steps: (1) establishing an adequate form for the theoretical model and (2) determining precisely the values of its parameters.

Step (1) is the model-building problem which has been discussed in Box and W. G. Hunter (1962), Cox (1961 and 1962) and W. G. Hunter and Mezaki (1964). Further facets of this important problem are currently being investigated. In this paper we shall suppose that step (1) has been accomplished and that the form of the theoretical model is therefore known. The problem which now confronts the experimenter is the evaluation of the physical parameters (for example, rate constants in chemical kinetics examples). The problem of statistical *analysis* of data in these situations has been discussed by Box (1960). The purpose of this paper is to consider the problem of *generation* of data, that is, the statistical design of experiments.

Bayes' Theorem

The essential machinery we shall use in drawing inferences from data is the well-known formula due to the Rev. Thomas Bayes

$$p_N(\underline{\theta}|\mathbf{y}, \mathbf{D}) = p_N(\underline{\theta}|\mathbf{y}) = \frac{p_0(\underline{\theta})\, L(\underline{\theta}|\mathbf{y})}{\int p_0(\underline{\theta})\, L(\underline{\theta}|\mathbf{y})}, \qquad (6.01)$$

where $p_N(\underline{\theta}|\mathbf{y}, \mathbf{D})$ is the posterior distribution of the parameters $\underline{\theta}$ after N observations $\mathbf{y}$ have been obtained, where $p_0(\underline{\theta})$ is the prior distribution that exists at stage $N = 0$, that is, before any observations are available from the experimental program, and where $L(\underline{\theta}|\mathbf{y})$ is the likelihood function.

The dependence of the posterior distribution on the design matrix $\mathbf{D}$ as well as on the experimental results $\mathbf{y}$ is made explicit when it is written as $p_N(\underline{\theta}|\mathbf{y}, \mathbf{D})$. For convenience, we shall write the posterior distribution as $p_N(\underline{\theta}|\mathbf{y})$, suppressing $\mathbf{D}$; however, the dependence of $p(\underline{\theta}|\mathbf{y})$ on $\mathbf{D}$ should always be borne in mind.

DESIGNS FOR PARAMETERS

If experiments are not carefully planned, the experimental points may be so situated in the space of the variables that the estimates which can be obtained for the parameters $\underline{\theta}$ are not only imprecise but also highly correlated. Once the data are collected, a statistical analysis, no matter how elaborate, can do nothing to remedy this unfortunate situation. However, by the selection of a suitable experimental design in advance, these shortcomings can often be overcome.

The problem of designing experiments in nonlinear situations has received comparatively little attention. Some possible approaches have been suggested by Box and Lucas (1959), Chernoff (1953), Elfving (1952), Kiefer (1959 and 1961), Stone (1959) and Wald (1943). In the next section of this paper, we shall present a Bayesian approach to the problem.

Box and Lucas (1959) proceed by attempting to choose **D** in such a manner that the volume of the approximate confidence region for $\underline{\theta}$ is minimized or, equivalently, under suitable assumptions, by trying to choose **D** to minimize the volume in the parameter space which contains a given percentage of the posterior distribution. If the experimental errors are approximately normally distributed, and if the response relationship is approximately linear in the vicinity of the least squares estimates $\hat{\underline{\theta}}$, then the volume of this region is proportional to the reciprocal of the determinant $\Delta = |X'X|$, where $X = \{x_{ru}\}$ and where

$$x_{ru} = \left[\frac{\partial f(\underline{\theta}, \underline{\xi}_u)}{\partial \theta_r}\right]_{\underline{\theta}=\hat{\underline{\theta}}}. \tag{6.02}$$

Unfortunately, since we do not know the values of $\hat{\underline{\theta}}$ in advance, we do not know the derivatives x_{ru} on which the design is to be based. In most cases, however, some knowledge of the size of the θ's will be available. It was suggested by Box and Lucas (1959) that preliminary guesses $\underline{\theta}^0$ should be made and that the derivatives should be determined at these values $\underline{\theta}^0$ instead of $\hat{\underline{\theta}}$. The resulting determinant $\Delta^0 = |(X^0)'X^0|$ is an explicit function of the settings of the experimental variables $\underline{\xi}$. It is therefore possible to find (perhaps analytically but, in any event, numerically) those values for $\underline{\xi}$ which maximize the determinant Δ^0.

At first sight, it may seem strange that in order to use this scheme, one must initially have estimates of the parameters; after all, it is the purpose of the experiment to obtain such estimates. Actually, however, this paradox is merely an example of the fact that any experimental design uses the experimenter's beliefs about the situation being studied. The scheme is thus efficient if the experimenter turns out to be nearly right.

In general, the more one initially knows, the better he can design experiments. As has been pointed out by Box (1960), if nothing is known about the experimental situation, then, strictly speaking, no experiment can be planned; or, as Daniel (1963) has stated: "All experimental plans reflect what you know, what you think you know but don't, what you don't know, and what you think you don't know but do."

A CRITERION FOR DESIGN

Usually, in the study of physical systems, experiments can be performed sequentially; that is, information from previous experimental results can be used in planning further experiments. If this procedure is adopted, all available information about the parameters $\underline{\theta}$ after N experiments have been performed is contained in the posterior distribution function $p_N(\underline{\theta}|\mathbf{y})$, and a careful analysis of the estimation situation involves a thorough study of this function.

To decide at what values the variables should be set in further experiments, we select those which will yield the most desirable posterior distribution or, equivalently, will produce the most desirable modification of the present posterior distribution. Ideally, barring purely technical difficulties, we would display the various possible posterior distributions which could result from different choices of the experimental conditions and then let the experimenter select the one which he thought was best.

This general approach to planning experiments involves no restrictions on the distribution of errors, the form of the response relationship, the nature of the prior distribution or the definition of the "best posterior distribution" that can be considered. In many common situations, however, the actual plotting of the posterior distribution for every combination of experimental conditions for a multiparameter system would be virtually impossible. Fortunately, this plotting is often unnecessary, for by making a set of plausible assumptions, we can completely describe the posterior distribution by a few summary statistics.

ASSUMPTIONS

We shall consider, specifically, the situation where information from the previous N experiments is available in planning the $(N + 1)$th experiment and where experiments are planned one at a time. Such a procedure will usually be most economical when it can be adopted.

Suppose that a posterior distribution $p_N(\underline{\theta})$ has been calculated after N observations have been obtained and that at this stage $\hat{\underline{\theta}}_N$ denotes the maximum likelihood estimates of $\underline{\theta}$. Suppose that we are about to take a further observation y_{N+1} whose true value is given by a known function of the settings for the k variables $\underline{\xi}_{N+1}$, that is,

$$\eta_{N+1} = f(\underline{\theta}, \underline{\xi}_{N+1}). \tag{6.03}$$

We now make two assumptions:

1. The y_u's are distributed normally and independently as

$$p(y_u) = \frac{1}{\sqrt{2\pi}\,\sigma} e^{-\frac{1}{2\sigma^2}(y_u - \eta_u)^2}, \quad u = 1, 2, \ldots, N+1 \qquad (6.04)$$

with mean η_u and common variance σ^2.

2. For a region in the $\underline{\theta}$ space sufficiently close to the maximum likelihood estimates $\hat{\underline{\theta}}_N$, we have approximately

$$f(\underline{\theta}, \underline{\xi}_{N+1}) = f(\hat{\underline{\theta}}_N, \underline{\xi}_{N+1}) + \sum_{i=1}^{p} (\theta_i - \hat{\theta}_{iN})\, x_{N+1}^{(i)}, \qquad (6.05)$$

where

$$x_{N+1}^{(i)} = \left[\frac{\partial f(\underline{\theta}, \underline{\xi}_{N+1})}{\partial \theta_i}\right]_{\underline{\theta} = \hat{\underline{\theta}}_N}. \qquad (6.06)$$

Then we have

$$\begin{aligned} y_{N+1} - \eta_{N+1} &= y_{N+1} - f(\underline{\theta}, \underline{\xi}_{N+1}) \\ &= r_{N+1} - \mathbf{x}'_{N+1}(\underline{\theta} - \hat{\underline{\theta}}_N), \end{aligned} \qquad (6.07)$$

where

$$r_{N+1} = y_{N+1} - f(\hat{\underline{\theta}}_N, \underline{\xi}_{N+1}). \qquad (6.08)$$

If the observation y_{N+1} were actually available, then Bayes' formula for the posterior distribution of $\underline{\theta}$ would give

$$p_{N+1}(\underline{\theta} \mid y_{N+1}) = \frac{p_N(\underline{\theta})\, L(\underline{\theta} \mid y_{N+1})}{\int p_N(\underline{\theta})\, L(\underline{\theta} \mid y_{N+1})\, d\underline{\theta}} \qquad (6.09)$$

$$= \frac{p_N(\underline{\theta})\, e^{-\frac{1}{2\sigma^2}\{y_{N+1} - f(\underline{\theta}, \underline{\xi}_{N+1})\}^2}}{\int p_N(\underline{\theta})\, e^{-\frac{1}{2\sigma^2}\{y_{N+1} - f(\underline{\theta}, \underline{\xi}_{N+1})\}^2}\, d\underline{\theta}} \qquad (6.10)$$

$$= \frac{p_N(\underline{\theta})\, e^{-\frac{1}{2\sigma^2}\{r_{N+1} - \mathbf{x}'_{N+1}(\underline{\theta} - \hat{\underline{\theta}}_N)\}^2}}{\int p_N(\underline{\theta})\, e^{-\frac{1}{2\sigma^2}\{r_{N+1} - \mathbf{x}'_{N+1}(\underline{\theta} - \hat{\underline{\theta}}_N)\}^2}\, d\underline{\theta}} \tag{6.11}$$

$$= \frac{p_N(\underline{\theta})\, e^{-\frac{1}{2\sigma^2}[(\underline{\theta} - \hat{\underline{\theta}}_N)'\mathbf{x}_{N+1}\mathbf{x}'_{N+1}(\underline{\theta} - \hat{\underline{\theta}}_N) - 2r_{N+1}\mathbf{x}'_{N+1}(\underline{\theta} - \hat{\underline{\theta}}_N)]}}{\int p_N(\underline{\theta})\, e^{-\frac{1}{2\sigma^2}[(\underline{\theta} - \hat{\underline{\theta}}_N)'\mathbf{x}_{N+1}\mathbf{x}'_{N+1}(\underline{\theta} - \hat{\underline{\theta}}_N) - 2r_{N+1}\mathbf{x}'_{N+1}(\underline{\theta} - \hat{\underline{\theta}}_{N+1})]}\, d\underline{\theta}}\,. \tag{6.12}$$

This expression is true for any prior distribution $p_N(\underline{\theta})$, but to make further progress, we need to be more specific.

The Principle of Precise Measurement

One difficulty associated with Bayes' theorem has been the question of what to take for the prior distribution. In most experimental situations, this is not as troublesome as it might seem. Consider two cases with regard to the parameters $\underline{\theta}$ of the response relationship itself:

1. The prior distribution is nearly constant over a region where the likelihood function has an appreciable value, and the prior distribution outside of this region does not become sufficiently great so that its contribution is appreciable when combined with the likelihood function; that is, the likelihood function dominates the prior distribution.

2. The case where the situation is reversed; that is, the prior distribution dominates the likelihood function.

In case (2), almost all of the information about $\underline{\theta}$ will come from the prior distribution, and very little, if any, will come from the data. In most instances of this kind, there will be little motivation for carrying out experiments since knowledge already available is so much more precise than any that could be expected from the data. Consequently, in experimental situations, case (1) is the one that usually occurs.

In case (1), since the prior distribution is virtually constant over the range where the likelihood is appreciable, it is spoken of as being *locally uniform*. In this instance, it is not necessary to know the exact mathematical form of the prior distribution since it cancels out from both the numerator and the denominator in Bayes' formula. The posterior distribution is then very nearly proportional to the likelihood. The *principle of precise measurement* (see Savage *et al.*, 1962) refers to this situation in which most of the information comes from the data and not from the prior distribution.

As Box and Tiao (1963) have pointed out, our assumption that the prior distribution is locally uniform is appropriate in those situations in which, if we were to compute a sensible confidence region for the parameters, we could honestly state that any point in this region *a priori* would have been as acceptable as any other point. This serves to indicate that the assumption of a locally uniform prior distribution is of rather general application since a statement similar to the one above could be made in most experimental situations.

Returning to equation (6.12) and appealing to the principle of precise measurement just described, we could regard the initial prior distribution as being locally uniform. This is a particularly innocuous assumption in this instance since, certainly after a few observations had been taken, the effect of any moderate nonuniformity *a priori* would have become negligible.

To obtain the posterior distribution $p_{N+1}(\underline{\theta}\,\mathbf{y})$ by the use of Bayes' formula, the posterior distribution at the Nth stage $p_N(\underline{\theta}|\mathbf{y})$ can be used as the prior distribution $p_N(\underline{\theta})$ for the $(N+1)$th stage.

OBTAINING $p_N(\underline{\theta})$

The probability density for the first N observations is

$$p(\mathbf{y}) = p(y_1, y_2, \ldots, y_N)$$
$$= \frac{1}{(\sqrt{2\pi}\,\sigma)^N} \exp\left[-\frac{1}{2\sigma^2}\sum_{u=1}^{N}(y_u - \eta_u)^2\right]. \tag{6.13}$$

Using the above assumption (2) and equation (6.08), we have

$$y_u - \eta_u = y_u - f(\underline{\theta}, \underline{\xi}_u) = r_u - \mathbf{x}'_u(\underline{\theta} - \underline{\hat{\theta}}_N). \tag{6.14}$$

The likelihood function $L(\underline{\theta}|\mathbf{y})$ for the parameters $\underline{\theta}$ can then be written as

$$L(\underline{\theta}|\mathbf{y}) = \frac{1}{(\sqrt{2\pi}\,\sigma)^N} e^{-\frac{1}{2\sigma^2}\{R_N - X_N(\underline{\theta} - \underline{\hat{\theta}}_N)\}'\{R_N - X_N(\underline{\theta} - \underline{\hat{\theta}}_N)\}}, \tag{6.15}$$

where

$$\begin{bmatrix} r_1 \\ r_2 \\ \vdots \\ r_N \end{bmatrix} = \begin{bmatrix} y_1 - f(\underline{\hat{\theta}}_N, \underline{\xi}_1) \\ y_2 - f(\underline{\hat{\theta}}_N, \underline{\xi}_2) \\ \vdots \\ y_N - f(\underline{\hat{\theta}}_N, \underline{\xi}_N) \end{bmatrix}, \tag{6.16}$$

$$\underline{\theta} - \underline{\hat{\theta}} = \begin{bmatrix} \theta_1 - \hat{\theta}_1 \\ \theta_2 - \hat{\theta}_2 \\ \vdots \\ \theta_p - \hat{\theta}_p \end{bmatrix}, \tag{6.17}$$

$$X_N = \begin{bmatrix} \mathbf{x}_1' \\ \mathbf{x}_2' \\ \vdots \\ \mathbf{x}_N' \end{bmatrix} = \begin{bmatrix} x_{11} & x_{21} & \dots & x_{p1} \\ x_{12} & x_{22} & \dots & x_{p2} \\ & & \vdots & \\ x_{1N} & x_{2N} & \dots & x_{pN} \end{bmatrix} \tag{6.18}$$

and

$$x_{ij} = \left[\frac{\partial f(\underline{\theta}, \underline{\xi}_j)}{\partial \theta_i} \right]_{\underline{\theta} = \underline{\hat{\theta}}_N} . \tag{6.19}$$

Now the likelihood (6.15) is maximized if and only if

$$R'_N X_N = \mathbf{0} . \tag{6.20}$$

Consequently, we have

$$L(\underline{\theta}|\mathbf{y}) = \frac{1}{(\sqrt{2\pi}\,\sigma)^N} e^{-\frac{1}{2\sigma^2}\{R'_N R_N + (\underline{\theta} - \underline{\hat{\theta}}_N)' C_N (\underline{\theta} - \underline{\hat{\theta}}_N)\}} , \tag{6.21}$$

where

$$C_N = X'_N X_N . \tag{6.22}$$

Using Bayes' formula, we obtain

$$p_N(\underline{\theta}|\mathbf{y}) = \frac{p_0(\underline{\theta})\, L(\underline{\theta}|\mathbf{y})}{\int p_0(\underline{\theta})\, L(\underline{\theta}|\mathbf{y})\, d\underline{\theta}}, \tag{6.23}$$

but if $p_0(\underline{\theta})$ is locally uniform, as we have assumed, then we have

$$p_N(\underline{\theta}|\mathbf{y}) = \frac{L(\underline{\theta}|\mathbf{y})}{\int L(\underline{\theta}|\mathbf{y})\, d\underline{\theta}} \tag{6.24}$$

and

$$\int L(\underline{\theta}|\mathbf{y})\, d\underline{\theta} = \frac{e^{-\frac{1}{2\sigma^2} R'_N R_N}}{(\sqrt{2\pi}\ \sigma)^N} \int e^{-\frac{1}{2\sigma^2}(\underline{\theta} - \hat{\underline{\theta}}_N)' C_N (\underline{\theta} - \hat{\underline{\theta}})}\, d\underline{\theta}. \tag{6.25}$$

Using the well-known integral

$$\int e^{-\frac{1}{2\sigma^2}\mathbf{x}' A \mathbf{x}}\, d\mathbf{x} = \frac{(\sqrt{2\pi}\ \sigma)^p}{|A|^{1/2}}, \tag{6.26}$$

we obtain

$$\int L(\underline{\theta}|\mathbf{y})\, d\underline{\theta} = \frac{e^{-\frac{1}{2\sigma^2} R'_N R_N}}{(\sqrt{2\pi}\ \sigma)^N} \frac{(\sqrt{2\pi}\ \sigma)^p}{|C_N|^{1/2}}. \tag{6.27}$$

Hence, by substituting equations (6.21) and (6.27) into (6.24), we obtain the posterior distribution of $\underline{\theta}$ after N observations, that is,

$$p_N(\underline{\theta}|\mathbf{y}) = \frac{|C_N|^{1/2}}{(\sqrt{2\pi}\ \sigma)^p} e^{-\frac{1}{2\sigma^2}(\underline{\theta} - \hat{\underline{\theta}}_N)' C_N (\underline{\theta} - \hat{\underline{\theta}}_N)}. \tag{6.28}$$

This expression can be used as the prior distribution $p_N(\underline{\theta})$ for the $(N + 1)$th stage.

THE POSTERIOR DISTRIBUTION $p_{N+1}(\underline{\theta}|\mathbf{y})$

The posterior distribution after $N + 1$ observations can, of course, be obtained by writing $N + 1$ for N in equation (6.28). For our purposes, however, we wish to express this distribution in terms of the information available to the experimenter at stage N, the contemplated levels $\underline{\xi}_{N+1}$ of the experimental variables, and the observation y_{N+1}. This can conveniently be done by deriving $p_{N+1}(\underline{\theta}|\mathbf{y})$ from $p_N(\underline{\theta})$ by a further application of Bayes' theorem, that is, substituting equation (6.28) into (6.12). Then we have

$$p_{N+1}(\underline{\theta}|\mathbf{y}_{N+1})$$

$$= \frac{e^{-\frac{1}{2\sigma^2}[(\underline{\theta}-\hat{\underline{\theta}}_N)'C_N(\underline{\theta}-\hat{\underline{\theta}}_N)+(\underline{\theta}-\hat{\underline{\theta}}_N)'\mathbf{x}_{N+1}\mathbf{x}'_{N+1}(\underline{\theta}-\hat{\underline{\theta}}_N)-2r_{N+1}\mathbf{x}'_{N+1}(\underline{\theta}-\hat{\underline{\theta}}_N)]}}{\int e^{-\frac{1}{2\sigma^2}[(\underline{\theta}-\hat{\underline{\theta}}_N)'C_N(\underline{\theta}-\hat{\underline{\theta}}_N)+(\underline{\theta}-\hat{\underline{\theta}}_N)'\mathbf{x}_{N+1}\mathbf{x}'_{N+1}(\underline{\theta}-\hat{\underline{\theta}}_N)-2r_{N+1}\mathbf{x}'_{N+1}(\underline{\theta}-\hat{\underline{\theta}}_N)]}\,d\underline{\theta}} \tag{6.29}$$

$$= \frac{e^{-\frac{1}{2\sigma^2}[(\underline{\theta}-\hat{\underline{\theta}}_N)'C_{N+1}(\underline{\theta}-\hat{\underline{\theta}}_N)-2r_{N+1}\mathbf{x}'_{N+1}(\underline{\theta}-\hat{\underline{\theta}}_N)]}}{\int e^{-\frac{1}{2\sigma^2}[(\underline{\theta}-\hat{\underline{\theta}}_N)'C_{N+1}(\underline{\theta}-\hat{\underline{\theta}}_N)-2r_{N+1}\mathbf{x}'_{N+1}(\underline{\theta}-\hat{\underline{\theta}}_N)]}\,d\underline{\theta}} \tag{6.30}$$

$$= \frac{|C_{N+1}|^{1/2}}{(\sqrt{2\pi}\,\sigma)^p}\exp\left\{-\frac{1}{2\sigma^2}[(\underline{\theta}-\hat{\underline{\theta}}_N)'C_{N+1}(\underline{\theta}-\hat{\underline{\theta}}_N) - 2r_{N+1}\mathbf{x}'_{N+1}(\underline{\theta}-\hat{\underline{\theta}}_N) + r_{N+1}\mathbf{x}'_{N+1}C_{N+1}{}^{-1}\mathbf{x}_{N+1}r_{N+1}]\right\} \tag{6.31}$$

$$= \frac{|C_{N+1}|^{1/2}}{(\sqrt{2\pi}\,\sigma)^p}\exp\left\{-\frac{1}{2\sigma^2}[(\underline{\theta}-\hat{\underline{\theta}}_N) - C_{N+1}{}^{-1}\mathbf{x}_{N+1}r_{N+1})' C_{N+1}(\underline{\theta}-\hat{\underline{\theta}} - C_{N+1}{}^{-1}\mathbf{x}_{N+1}r_{N+1})]\right\}, \tag{6.32}$$

but

$$C_{N+1}{}^{-1} = (X'_{N+1}X_{N+1})^{-1} = (X'_NX_N + \mathbf{x}_{N+1}\mathbf{x}'_{N+1})^{-1}$$
$$= (C_N + \mathbf{x}_{N+1}\mathbf{x}'_{N+1})^{-1} \tag{6.33}$$

$$= [C_N(I + C_N{}^{-1}\mathbf{x}_{N+1}\mathbf{x}'_{N+1})]^{-1} \tag{6.34}$$

$$= (I + C_N{}^{-1}\mathbf{x}_{N+1}\mathbf{x}'_{N+1})^{-1}C_N{}^{-1}. \tag{6.35}$$

We now employ a useful matrix result mentioned by Tocher (1951), that is,

$$(I_p + AB)^{-1} = I_p - A(I_q + AB)^{-1}B, \tag{6.36}$$

where A is a $p \times q$ matrix and where B is a $q \times p$ matrix. The advantage of (6.36) is that where q is less than p, the size of the matrix that needs to be inverted is smaller on the righthand side than it is on the lefthand side. In our particular case, $q = 1$ so that $I_q + BA$ is a scalar.

Letting

$$A = C_N^{-1}\mathbf{x}_{N+1} \tag{6.37}$$

and

$$B = \mathbf{x}'_{N+1}, \tag{6.38}$$

we obtain

$$C_{N+1}^{-1}$$

$$= I - C_N^{-1}\mathbf{x}_{N+1}[1 + \mathbf{x}'_{N+1}C_N^{-1}\mathbf{x}_{N+1}]^{-1}\mathbf{x}'_{N+1}C_N^{-1} \tag{6.39}$$

$$= \left\{ I - \frac{C_N^{-1}\mathbf{x}_{N+1}\mathbf{x}'_{N+1}}{1 + \mathbf{x}'_{N+1}C_N^{-1}\mathbf{x}_{N+1}} \right\} C_N^{-1} \tag{6.40}$$

$$= C_N^{-1} - \frac{C_N^{-1}\mathbf{x}_{N+1}\mathbf{x}'_{N+1}C_N^{-1}}{1 + \mathbf{x}'_{N+1}C_N^{-1}\mathbf{x}_{N+1}} \tag{6.41}$$

$$= C_N^{-1} - \frac{tt'}{1 + t'\mathbf{x}_{N+1}}, \tag{6.42}$$

where t is a $p \times 1$ vector $C_N^{-1}\mathbf{x}_{N+1}$. It follows that

$$\mathbf{x}'_{N+1}C_{N+1}^{-1}r_{N+1} = \mathbf{x}'_{N+1}C_N^{-1}r_{N+1}$$

$$- \frac{\mathbf{x}'_{N+1}C_N^{-1}\mathbf{x}_{N+1}\mathbf{x}'_{N+1}C_N^{-1}}{1 + \mathbf{x}'_{N+1}C_N^{-1}\mathbf{x}_{N+1}} r_{N+1}, \tag{6.43}$$

that is,

$$\mathbf{x}'_{N+1}C_{N+1}^{-1}r_{N+1} = \frac{\mathbf{x}'_{N+1}C_N^{-1}r_{N+1}}{1 + \mathbf{x}'_{N+1}C_N^{-1}\mathbf{x}_{N+1}}. \tag{6.44}$$

Thus, we may write the posterior distribution after $N + 1$ observations entirely in terms of quantities $\hat{\underline{\theta}}_N$ and C_N that are known before the last observation is taken and quantities r_{N+1} and $\mathbf{x}_{N+1}$ associated with the choice of the final experiment:

$$p_{N+1}(\underline{\theta}|\mathbf{y}) = \frac{|C_N + \mathbf{x}_{N+1}\mathbf{x}'_{N+1}|^{1/2}}{(\sqrt{2\pi}\,\sigma)^p}$$

$$\times \exp\left[-\frac{1}{2\sigma^2}\left\{\underline{\theta} - \hat{\underline{\theta}}_N - \frac{C_N^{-1}\mathbf{x}_{N+1}r_{N+1}}{1 + \mathbf{x}'_{N+1}C_N^{-1}\mathbf{x}_{N+1}}\right\}'\right.$$

$$\left.(C_N + \mathbf{x}_{N+1}\mathbf{x}'_{N+1})\left\{\underline{\theta} - \hat{\underline{\theta}}_N - \frac{C_N^{-1}\mathbf{x}_{N+1}r_{N+1}}{1 + \mathbf{x}'_{N+1}C_N^{-1}\mathbf{x}_{N+1}}\right\}\right]. \quad (6.45)$$

This posterior distribution $p_{N+1}(\underline{\theta}|\mathbf{y})$ is multinormal with mean

$$\hat{\underline{\theta}}_{N+1} = \hat{\underline{\theta}}_N + \frac{C_N^{-1}\mathbf{x}_{N+1}r_{N+1}}{1 + \mathbf{x}'_{N+1}C_N^{-1}\mathbf{x}_{N+1}}, \quad (6.46)$$

where $\hat{\underline{\theta}}_{N+1}$ is the maximum likelihood estimate at stage $N + 1$ and where the dispersion matrix is

$$C_{N+1}^{-1} = [C_N + \mathbf{x}_{N+1}\mathbf{x}'_{N+1}]^{-1}. \quad (6.47)$$

The expression for the dispersion matrix C_{N+1}^{-1} does not depend on y_{N+1} and can be calculated exactly at stage N for any given set of contemplated experimental conditions $\underline{\xi}_{N+1}$. Since the posterior distribution is multinormal, this dispersion matrix contains all of the information concerning the precision of the estimates that will result from running a particular set of experimental conditions $\underline{\xi}_{N+1}$. On the basis of C_{N+1}^{-1}, the experimenter can therefore choose the best experimental conditions for the next run.

A PORTMANTEAU CRITERION

In practice, even the calculation of the elements of the dispersion matrix for a number of different possible experimental conditions may prove to be too prodigious a task. If there were only three parameters, for example, there would be three variances and three covariances that would have to be calculated for each set of contemplated experimental conditions. The number n of quantities to be calculated increases rapidly as the number p of parameters increases, in fact, $n = \frac{1}{2}(p^2 + p)$.

It is desirable in many situations to have some kind of overall criterion that involves the calculation of only a single quantity; however, it is clear that as soon as one tries to economize by using such a portmanteau criterion, every possible need cannot be satisfied.

There will be investigations where there is particular interest in one parameter but less interest in the remaining parameters, and, in other circumstances, there may be a special reason for wanting to minimize the correlation between a certain pair of estimates. In such situations, one could proceed by calculating the variance and covariance terms which are of special interest in addition to using the overall criterion. At any rate, if an overall criterion is adopted for the sequential planning of experiments, a facility should be provided for looking at these other quantities if the experimenter so desires.

If one overall criterion must be chosen, in the absence of special needs, it is reasonable to take for the next experiment those conditions which give the maximum posterior density to the most probable values, that is, to maximize the posterior density with respect to both $\underline{\theta}$ and $\underline{\xi}_{N+1}$.

The posterior distribution after $N + 1$ observations will be

$$p_{N+1}(\underline{\theta}|\mathbf{y}) = \frac{|C_{N+1}|^{1/2}}{(\sqrt{2\pi}\,\sigma)^p} e^{-\frac{1}{2}(\underline{\theta} - \hat{\underline{\theta}}_{N+1})C_{N+1}(\underline{\theta} - \hat{\underline{\theta}}_{N+1})}. \quad (6.48)$$

The maximum probability density will be at the point $\underline{\theta} = \hat{\underline{\theta}}_{N+1}$, whatever the settings of $\underline{\xi}_{N+1}$, that is,

$$p_{\underline{\theta}} = \max_{\underline{\theta}} p_{N+1}(\underline{\theta}|\mathbf{y}) = \frac{|C_{N+1}|^{1/2}}{(\sqrt{2\pi}\,\sigma)^p}, \quad (6.49)$$

where σ is a positive constant. (The quantity $p_{\underline{\theta}}$ is necessarily positive if C_{N+1} is positive-definite.) Now we have

$$C_{N+1} = C_N + \mathbf{x}_{N+1}\mathbf{x}'_{N+1}. \quad (6.50)$$

If we are at stage N, then C_N is fixed, but $\mathbf{x}_{N+1}$ is a function of $\underline{\xi}_{N+1}$ and $\hat{\underline{\theta}}_N$, so settings can be chosen for $\underline{\xi}_{N+1}$ to maximize the determinant

$$\Delta = |C_{N+1}| = |C_N + \mathbf{x}_{N+1}\mathbf{x}'_{N+1}|. \quad (6.51)$$

This criterion of maximizing the determinant Δ has previously been suggested in other situations and on other grounds (see, for example, Box and Lucas, 1959, Kiefer, 1959 and 1961, and Wald, 1943). A number of alternatives for an overall criterion are discussed by Kiefer (1959). Of these, the most important competitor is perhaps the trace of the dispersion matrix which is to be minimized (see Elfving, 1952). This criterion suffers from the disadvantage that it is not independent of the manner in which the parameters $\underline{\theta}$ are scaled, and we shall not consider it further here.

Examination of the Portmanteau Criterion

The portmanteau criterion has some interesting implications which we shall now discuss. To maximize the determinant Δ, we can maximize

$$ln\Delta = ln|C_N + \mathbf{x}_{N+1}\mathbf{x}'_{N+1}| \tag{6.52}$$

$$= ln|C_N|\ |I + C_N^{-1}\mathbf{x}_{N+1}\mathbf{x}'_{N+1}| \tag{6.53}$$

$$= ln|C_N| + ln|I + C_N^{-1}\mathbf{x}_{N+1}\mathbf{x}'_{N+1}|. \tag{6.54}$$

But at stage N we are faced with the choice of $\underline{\xi}_{N+1}$ with C_N fixed, so we want to maximize $ln|I + C_N^{-1}\mathbf{x}_{N+1}\mathbf{x}'_{N+1}|$.

Now if λ is a latent root of a $p \times p$ matrix A so that

$$u'A = u'\lambda, \tag{6.55}$$

then $1 + \lambda$ is a latent root of $I + A$ since

$$u'(I + A) = u' + u'\lambda = u'(1 + \lambda). \tag{6.56}$$

Thus, we have

$$|I + A| = \prod_{i=1}^{p} (1 + \lambda_i) \tag{6.57}$$

and

$$ln|I + A| = \sum_{i=1}^{p} ln(1 + \lambda_i). \tag{6.58}$$

If the λ's are less than 1, then we have

$$ln|I + A| = \sum_{i=1}^{p} \lambda_i - \sum_{i=1}^{p} \frac{\lambda_i^2}{2} + \sum_{i=1}^{p} \frac{\lambda_i^3}{3} - \ldots \tag{6.59}$$

$$= \text{tr}\, A - \frac{\text{tr}\, A^2}{2} + \frac{\text{tr}\, A^3}{3} - \ldots, \tag{6.60}$$

where tr A is the trace of A. If the λ's are sufficiently small, then we have

$$ln|I + A| = \text{tr}\, A. \tag{6.61}$$

Since the latent roots of $C_N^{-1} \mathbf{x}_{N+1} \mathbf{x}'_{N+1}$ will be of order $1/N$, maximizing $ln|I + C_N^{-1} \mathbf{x}_{N+1} \mathbf{x}'_{N+1}|$ is equivalent to maximizing

$$\text{tr}\, C_N^{-1}\mathbf{x}_{N+1}\mathbf{x}'_{N+1} = \text{tr}\, \mathbf{x}'_{N+1} C_N^{-1}\mathbf{x}_{N+1}, \tag{6.62}$$

and $\mathbf{x}'_{N+1} C_N^{-1} \mathbf{x}_{N+1}$ is a 1×1 matrix or scalar. Thus, we have

$$\begin{aligned} \text{tr}\, \mathbf{x}'_{N+1} C_N^{-1} \mathbf{x}_{N+1} &= \mathbf{x}'_{N+1} C_N^{-1} \mathbf{x}_{N+1} \\ &= c^{11}x_1^2 + c^{22}x_2^2 + \ldots + c^{pp}x_p^2 \\ &+ c^{12}x_1x_2 + \ldots + c^{p-1,p}x_{p-1}x_p, \end{aligned} \tag{6.63}$$

where c^{ij} is the (i, j)th element of C_N^{-1} and $\{c_{ij}\} = C_N$, that is,

$$c_{ij} = \sum_{u=1}^{N} \left[\frac{\partial f(\underline{\theta}, \xi_u)}{\partial \theta_i} \cdot \frac{\partial f(\underline{\theta}, \xi_u)}{\partial \theta_j} \right]_{\underline{\theta} = \hat{\underline{\theta}}_N}, \tag{6.64}$$

and where x_i is the ith element of $\mathbf{x}_{N+1}$, that is,

$$\mathbf{x}'_{N+1} = (x_1, x_2, \ldots, x_i, \ldots, x_p). \tag{6.65}$$

Therefore, to the degree of approximation employed, maximizing the determinant Δ is equivalent to maximizing the quantity

$$c^{11}x_1^2 + c^{22}x_2^2 + \ldots + c^{pp}x_p^2 + c^{12}x_1x_2 + \ldots + c^{p-1,p}x_{p-1}x_p. \tag{6.66}$$

The terms c^{ii} are proportional to the variances of the estimated parameters $\hat{\theta}_i$ at stage N and c^{ij} $(i \neq j)$ proportional to the covariances.

If there is no correlation between the estimates, it is clearly desirable to make x_i^2 as large as possible. Using the criterion of maximum Δ, we are essentially weighting the x_i^2 terms with the corresponding variances $V(\hat{\theta}_i)$, and we are thus giving most weight to those terms x_i^2 which are associated with the estimates $\hat{\theta}_i$ that are known *least precisely*.

To shed some light on the role of weighting the cross-product terms, let us consider the case in which there are just two parameters. Let

$$C_N = \begin{bmatrix} c_{11} & c_{12} \\ c_{12} & c_{22} \end{bmatrix} \tag{6.67}$$

and

$$C_N^{-1} = \begin{bmatrix} c^{11} & c^{12} \\ c^{12} & c^{22} \end{bmatrix}. \tag{6.68}$$

If c^{12} is positive (negative), that is, if the correlation between $\hat{\theta}_1$ and $\hat{\theta}_2$ is positive (negative), then c_{12} is negative (positive). To maximize the cross-product term $c^{12}x_1x_2$ if c^{12} is negative, the quantities x_1 and x_2 will be chosen to be of different signs if possible. If c^{12} is positive, then x_1 and x_2 will be chosen to be of the same sign if possible. This means that if a correlation exists between $\hat{\theta}_1$ and $\hat{\theta}_2$ at stage N, the overall criterion which we have adopted will tend to pick out a set of conditions $\underline{\xi}_{N+1}$ which will cancel out this correlation.

CONCLUSIONS

Under the assumptions of normality and independence of homoscedastic errors [see equation (6.04)], approximate linearity near $\hat{\underline{\theta}}_N$ [see equation (6.05)], and locally uniform prior distribution $p_0(\underline{\theta})$, we reach the following conclusions:

1. The posterior distribution of $\underline{\theta}$ after $N + 1$ observations is multinormal and is therefore completely described by the vector of least squares estimates $\hat{\underline{\theta}}_{N+1}$ and by the dispersion matrix C_{N+1}^{-1}.

2. Information on how the precision of our estimates is improved is supplied by the change in the dispersion matrix

$$C_N^{-1} - C_{N+1}^{-1} = \frac{C_N^{-1}\mathbf{x}_{N+1}\mathbf{x}'_{N+1}C_N^{-1}}{1 + \mathbf{x}'_{N+1}C_N^{-1}\mathbf{x}_{N+1}}. \tag{6.69}$$

3. If we are going to use the general design criterion of trying to ensure the best posterior distribution, we can calculate all elements in the dispersion matrix or, alternatively, the changes in these quantities from equation (6.69) for any given values $\underline{\xi}_{N+1}$.

4. If we are going to use the special, overall design criterion of trying to choose those conditions which yield the maximum posterior density for the most probable values, we find those settings for $\underline{\xi}_{N+1}$ which maximize the determinant $|C_{N+1}|$.

5. Maximizing the determinant $|C_{N+1}|$ is approximately equivalent to maximizing the quadratic form

$$\sum_{i=1}^{p}\sum_{j=1}^{p} \operatorname{cov}(\hat{\theta}_i\hat{\theta}_j)\, x_i x_j, \tag{6.70}$$

where

$$x_i = \left[\frac{\partial f(\underline{\theta}, \underline{\xi}_{N+1})}{\partial \theta_i}\right]_{\underline{\theta} = \hat{\underline{\theta}}_N}. \tag{6.71}$$

This quadratic form can be regarded as a weighted sum of the squares and cross products of the $x_i x_j$ terms. A large value of x_i^2 is desirable for increasing the precision with which a particular parameter is estimated; under certain conditions, if the overall criterion is used, most weight is given to those quantities x_i associated with the parameters about which least is known.

EXAMPLE

To illustrate the method for sequential design discussed above, we apply it to a constructed example. A chemical reaction of the type

$$R \rightarrow P + P_1 \tag{6.72}$$

is being studied, and the true model is

$$\eta = f(\underline{\theta}, \underline{\xi}) = \frac{\theta_3\,\theta_1\,\xi_1}{1 + \theta_1\xi_1 + \theta_2\xi_2}, \tag{6.73}$$

where

η = true rate of the chemical reaction,
ξ_1 = partial pressure of reactant R,
ξ_2 = partial pressure of product P,
θ_1 = adsorption equilibrium constant for reactant R,
θ_2 = adsorption equilibrium constant for product P,
θ_3 = effective reaction rate constant.

This model has been reported by Laible (1959) to be applicable to a number of catalytic reactions of the type (6.72), where the reactant R is one of certain tertiary or long-chain primary alcohols, where the product P is an olefin and where the product P_1 is water.

For realistic values of the parameters $\underline{\theta}$, whose values, of course, are assumed to be unknown to the experimenter, the results found by Laible (1959) for the catalytic dehydration of n-hexyl alcohol at 550° F. are used, namely,

$$\theta_1 = 2.9, \qquad \theta_2 = 12.2, \qquad \theta_3 = 0.69. \tag{6.74}$$

We further suppose that the region of operability is defined by those values of ξ_1 and ξ_2 for which

$$0 \leq \xi_1 \leq 3,\ 0 \leq \xi_2 \leq 3. \tag{6.75}$$

If the true situation (that is, the situation with no experimental error) were depicted as a contour diagram (see Figure 1) for η with ξ_1 as the abscissa and ξ_2 as the ordinate, then it could be seen directly from equation (6.73) that all constant η contour lines are straight lines with the same intercept, that is,

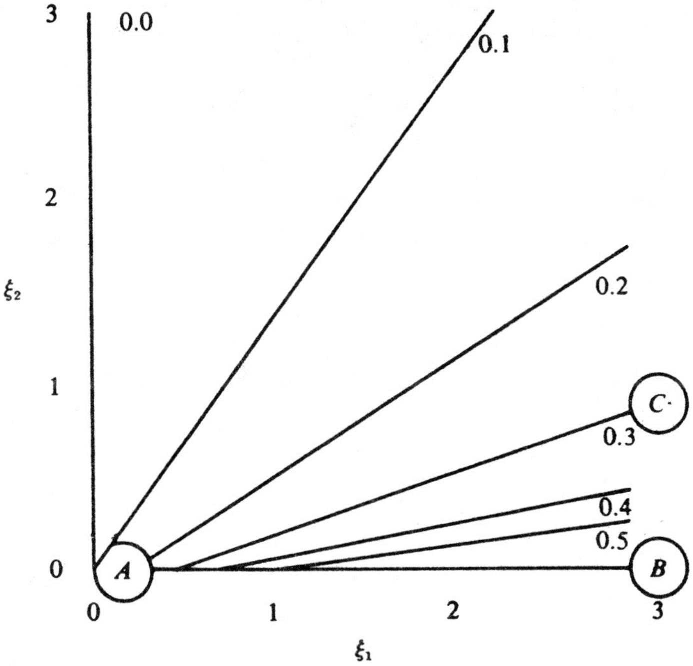

FIGURE 1. Contour diagram of η versus ξ_1 and ξ_2 for the model $\eta = f(\underline{\theta}, \underline{\xi}) = \theta_3 \theta_1 \xi_1/(1 + \theta_1 \xi_1 + \theta_2 \xi_2)$

$$a = -\frac{1}{\theta_2}, \tag{6.76}$$

but with different slopes b depending on the value of η, that is,

$$b = \frac{\theta_1}{\theta_2}\left(\frac{\theta_3}{\eta} - 1\right). \tag{6.77}$$

A contour line for a fixed value η_0 could, therefore, be represented by

$$\xi_2 = -\frac{1}{\theta_2} + \frac{\theta_1}{\theta_2}\left(\frac{\theta_3}{\eta_0} - 1\right)\xi_1 . \tag{6.78}$$

A contour line for a high value η_0 would have a small slope b.

To construct an observation y for a fixed pair of values ξ_1 and ξ_2, a random normal deviate with standard deviation = 0.01 was added to the true value η obtained from equation (6.73). It was supposed that initially a 2^2 factorial design was performed with the results shown in Table 1. Whence, applying the method of nonlinear least squares

TABLE 1
Results from 2^2 Factorial Design

Experiment No.	ξ_1	ξ_2	y
1	1	1	0.126
2	2	1	0.219
3	1	2	0.076
4	2	2	0.126

(see Meeter, 1961) to these data, we obtain

$$\hat{\theta}_1 = 10.39, \qquad \hat{\theta}_2 = 48.83, \qquad \hat{\theta}_3 = 0.74. \tag{6.79}$$

We are now in a position to select the settings of the levels of ξ_1 and ξ_2 for the fifth experiment. In accordance with the criterion we are using, we choose those levels which maximize the determinant

$$\Delta_5 = \begin{vmatrix} \sum x_{1u}^2 & \sum x_{1u}x_{2u} & \sum x_{1u}x_{3u} \\ & \sum x_{2u}^2 & \sum x_{2u}x_{3u} \\ \text{sym.} & & \sum x_{3u}^2 \end{vmatrix}$$

$$= \begin{vmatrix} c_{11} + x_{15}^2 & c_{12} + x_{15}x_{25} & c_{13} + x_{15}x_{25} \\ & c_{22} + x_{25}^2 & c_{23} + x_{25}x_{35} \\ \text{sym.} & & c_{33} + x_{35}^2 \end{vmatrix}, \tag{6.80}$$

where each summation goes from $u = 1$ to $u = 5$ and where

$$x_{i5} = \left[\frac{\partial f(\underline{\theta}, \underline{\xi}_5)}{\partial \theta_i}\right] \tag{6.81}$$

is evaluated at $\underline{\theta} = \hat{\underline{\theta}}_4$, the current least squares values given in equation (6.79). The quantities

$$c_{ij} = \sum_{u=1}^{4} x_{iu} x_{ju} \tag{6.82}$$

have fixed and known values.

We suppose that the experimenter wants values for the best settings of ξ_1 and ξ_2 to the nearest tenth. In this case, it is feasible to calculate the value of Δ_5 at each point of the 31×31 grid for ξ_1 and ξ_2, this computation being easily programmed for a digital computer. A printout of this kind, which can also be set up to produce the whole or selected parts of the dispersion matrix, would normally be presented to the experimenter for study after each run. In this case, we have assumed that the conditions which give the maximum value for the determinant are selected.

The maximum value for the determinant Δ_5 occurs at $\xi_1 = 0.1$ and $\xi_2 = 0.0$. The fifth experiment was "run" with these settings, and the result was $y_5 = 0.186$. Fitting the first five observations yields

$$\hat{\theta}_1 = 3.11, \qquad \hat{\theta}_2 = 15.19, \qquad \hat{\theta}_3 = 0.79. \tag{6.83}$$

(Upon comparing these estimates with those obtained after four observations, we notice that both $\hat{\theta}_1$ and $\hat{\theta}_2$ are much closer to their true values and that $\hat{\theta}_3$ is now slightly farther away from its true value.)

With these current values for $\hat{\underline{\theta}}$ and $N = 5$, we can maximize Δ_6 with respect to the settings of ξ_1 and ξ_2 for the sixth experiment. The computer printout after the fifth experiment is summarized in Figure 2. The maximum determinant Δ_6 now occurs at $\xi_1 = 3.0$ and $\xi_2 = 0.0$. The sixth experiment was "run" there, y_6 was obtained, new estimates

	3.0	4	4	5	5	5	5	5
	2.5	4	4	5	5	6	5	5
	2.0	4	5	5	5	5	5	6
ξ_2	1.5	4	5	5	5	5	6	8
	1.0	4	5	5	5	8	15	27
	0.5	4	5	7	22	57	112	182
	0	4	389	1155	1737	2154	2460	2693
		0	0.5	1.0	1.5	2.0	2.5	3.0
					ξ_1			

FIGURE 2. Condensation of computer printout of 31×31 grid of determinant $\Delta_6 \times 10^{10}$ as a function of ξ_1 and ξ_2

$\hat{\theta}_1$ and $\hat{\theta}_2$ were obtained, etc. The results from 13 cycles of this kind are summarized in Table 2.

TABLE 2
Results from Sequential Design Example

Experiment No.	ξ_1	ξ_2	y	θ_1	θ_2	θ_3
1	1.0	1.0	0.126			
2	2.0	1.0	0.219			
3	1.0	2.0	0.076			
4	2.0	2.0	0.126	10.39	48.83	0.74
5	0.1	0.0	0.186	3.11	15.19	0.79
6	3.0	0.0	0.606	3.96	15.32	0.66
7	0.2	0.0	0.268	3.61	14.00	0.66
8	3.0	0.0	0.614	3.56	13.96	0.67
9	0.3	0.0	0.318	3.32	13.04	0.67
10	3.0	0.8	0.298	3.33	13.48	0.67
11	3.0	0.0	0.509	3.74	13.71	0.63
12	0.2	0.0	0.247	3.58	13.15	0.63
13	3.0	0.8	0.319	3.57	12.77	0.63

Discussion on Example

The example has demonstrated the feasibility of the scheme for sequentially designing experiments. In summary, at stage N the experimenter would, in general, supply the computer with (1) the model, (2) the data and (3) the current least squares estimates for the parameters; the computer would produce (1) the new least squares estimates, (2) the best conditions for the next experiment, (3) information on the nature of the relationship between the determinant Δ and the conditions $\underline{\xi}_{N+1}$ in the neighborhood of the maximum and perhaps (4) additional information on elements in the dispersion matrix corresponding to particular variances and covariances that are of special interest.

We notice that the experimental points from the fifth experiment onward are all contained in three distinct regions of the factor space. Let us designate these regions as A, B and C, where A is the region in the neighborhood of $\xi_1 = 0.2$ and $\xi_2 = 0.0$, where B is the region in the neighborhood of $\xi_1 = 3.0$ and $\xi_2 = 0.0$ and where C is the region in the neighborhood of $\xi_1 = 3.0$ and $\xi_2 = 0.8$. These sites A, B and C are shown in Figure 1. The experimental points in turn fall into the regions $ABABACBAC$. Since there are three parameters, we would expect that there would be at least three sites for observation points and, furthermore, that an optimal design might require a different number of points at each site, as in the case above.

As is true in most situations in which a maximum is being sought (for example, in determining maximum yield conditions by the use of response surface methods or in finding the maximum likelihood estimates), it is useful to determine not only the point in the operability region for which the value of the determinant is maximized but also the nature of the dependence of the determinant on the variables $\underline{\xi}$ in the neighborhood of this maximum.

A printout of the type we are using (see Figure 2) indicates far more than the point at which the determinant is maximized and enables the experimenter to use informed judgment in selecting his actual runs; for instance, he might find that the determinant fell off very rapidly in the direction of the variable ξ_2, indicating that the control of this variable is of critical importance relative to the others. Information of this kind can be of considerable value to the experimenter.

The example further illustrates the fundamental difference between the problem in which the object is to estimate the response η and the problem in which the object is to estimate parameters $\underline{\theta}$. After four experiments, for example, although the parameter estimates $\hat{\theta}_1$ and $\hat{\theta}_2$ are widely discrepant from the true values, the estimates for η that are produced in the region of the experimental design are in close agreement with the true values.

The point that is illustrated here is the compensating nature of the errors in $\hat{\theta}_1$ and $\hat{\theta}_2$. Although both estimates are much greater than their corresponding true values, the errors are such that when these estimated values are substituted into equation (6.73) in place of θ_1 and θ_2, the results in terms of estimated values of η over the region in which the data have been taken are in close agreement with the true values. Such correlation among the estimates is common in models such as equation (6.73); in particular, this correlation is characteristic of estimates of parameters in catalytic reaction kinetics since the models are often of this form.

REFERENCES

Box, G. E. P. 1954. The exploration and exploitation of response surfaces: some general considerations and examples. Biometrics, 10:16-60.

———. 1958. Use of statistical methods in the elucidation of basic mechanisms. Bull. Inst. Internat. Statist., 36:215-25.

———. 1960. Fitting empirical data. Ann. New York Acad. Sci., 86:792-816.

Box, G. E. P., and G. A. Coutie. 1956. Application of digital computers in the exploration of functional relationships. Proc. Inst. Elec. Engrs. B, 103(suppl. 1):100-7.

Box, G. E. P., and N. R. Draper. 1959. A basis for the selection of a response surface design. J. Amer. Statist. Assoc., 54:622-54.

Box, G. E. P., and J. S. Hunter. 1957. Multifactor experimental designs for exploring response surfaces. Ann. Math. Statist., 28:195-241.

Box, G. E. P., and W. G. Hunter. 1962. A useful method for model-building. Technometrics, 4:301-18.

BOX, G. E. P., and H. L. LUCAS. 1959. Design of experiments in non-linear situations. Biometrika, 46:77-90.

BOX, G. E. P., and G. C. TIAO. 1963. Personal communication.

BOX, G. E. P., and P. V. YOULE. 1955. The exploration and exploitation of response surfaces: an example of the link between the fitted surface and the basic mechanism of the system. Biometrics, 11:287-323.

CHERNOFF, H. 1953. Locally optimal designs for estimating parameters. Ann. Math. Statist., 24:586-602.

COX, D. R. 1961. Tests of separate families of hypotheses *in* Proceedings of the fourth Berkeley symposium on mathematical statistics and probability, vol. I. Berkeley: Univ. of California Press, 105-23.

———. 1962. Further results on tests of separate families of hypotheses. J. Roy. Statist. Soc. Ser. B, 24:406-24.

DANIEL, C. 1963. Factor screening in process development. Indus. Eng. Chem., 55:45-48.

DAVIES, O. L. (ed.). 1956. Chapter 11 *in* The design and analysis of industrial experiments. 2d ed. New York: Hafner, 495-578.

ELFVING, G. 1952. Optimal allocation in linear regression theory. Ann. Math. Statist., 23:255-62.

HUNTER, W. G., and R. MEZAKI. 1964. A model-building technique for chemical engineering kinetics. Amer. Inst. Chem. Eng. J., 10:315-22.

KIEFER, J. C. 1959. Optimum experimental designs. J. Roy. Statist. Soc. Ser. B, 21:272-319.

———. 1961. Optimum experimental designs V, with applications to systematic and rotatable designs *in* Proceedings of the fourth Berkeley symposium on mathematical statistics and probability, vol. I. Berkeley: Univ. of California Press, 381-405.

LAIBLE, J. R. 1959. The kinetics of the catalytic dehydration of certain tertiary and long chain primary alcohols. Microfilmed Ph.D. dissertation. Madison, Wis.: Univ. of Wisconsin.

MEETER, D. A. 1961. Non-linear estimation program. Madison, Wis.: Univ. of Wisconsin, Numerical Analysis Laboratory.

SAVAGE, L. J., *et al.* 1962. The foundations of statistical inference. New York: Wiley.

STONE, M. 1959. Application of a measure of information to the design and comparison of regression experiments. Ann. Math. Statist., 30:55-70.

TOCHER, K. D. 1951. Discussion on Mr. Box and Dr. Wilson's paper. J. Roy. Statist. Soc. Ser. B, 13:39-42.

WALD, A. 1943. On the efficient design of statistical investigations. Ann. Math. Statist., 14:134-40.

Discussion

J. B. WILLIS: Have you made any quantitative investigation of how good the Taylor's approximation has to be?

W. G. HUNTER: No. However, I don't think the Taylor's approximation has to be too good because that is not the form we actually fit. The original nonlinear model is in fact used when the parameters are re-estimated.

J. B. WILLIS: Have you made any investigation of alternate models? You indicated that one might have a handful of models, and it has been our experience that sometimes the scientist has more than a handful. Have you done any investigation on testing alternate models with this procedure?

W. G. HUNTER: No, not with the procedure I just described. This procedure is appropriate when the experimenter knows the correct model. We like to think of testing more than one model as being a different kind of problem, namely, a model-building problem. That is a problem that we very much want to look at, and the first problem of this kind that we would consider looking at, I would think, would be the case where there are only two models. Now, if we want to discriminate between two models, critical experiments would have to be devised. We at Wisconsin—particularly in the Departments of Statistics and Chemical Engineering—are very much interested in the general model-building problem. We are really looking forward to working together in this area.

2.14
Some Aspects of Randomization

G. E. P. BOX and IRWIN GUTTMAN
University of Wisconsin

[Received March 1966]

SUMMARY

A key feature of Sir Ronald Fisher's work in designed experiments is randomization. The randomization schemes described by Fisher ensure that four important principles are met. Using these four principles as starting point, we investigate the requirements which these four conditions impose on the choice of design.

1. INTRODUCTION

IT WAS realized by Sir Ronald Fisher in the early 1920's that conclusions to be drawn from the analysis of routinely recorded data were of limited validity. To overcome the ambiguities and uncertainties connected with such analysis, he introduced the concept of designed experiments in which randomization, replication, blocking and orthogonality of design were central.

A key feature of Fisher's theory was randomization. Since that time much research has been done on various aspects of randomization, in particular by Welch (1937), Pitman (1937), Kempthorne (1955), Wilk (1955), Wilk and Kempthorne (1955).

In a Fisherian design such as a randomized block, not one but a number of requirements are met by the randomization scheme. The design ensures
(1) that unbiased estimates of the parameters of interest are obtained;
(2) that a relevant estimate of error variance is obtained;
(3) that this error variance and the variance of the estimators which supply information on the parameters of interest are small;
(4) that a specific distribution theory arising out of the physical conduct of the experiment itself can be applied to these results.

The randomization schemes described by Fisher (1935) provide an *equal* opportunity for a *finite* set of experimental arrangements to be applied to the experimental units. Most research has been conducted within this framework. However, specific designs in which not all arrangements had an equal opportunity of appearing have been suggested (Youden, 1956; Cox, 1956; Kempthorne, Zyskind and Sutter, 1963). Furthermore, designs in which the randomization set was infinite have also been proposed (Box, 1952). Thus it is clear that other arrangements are possible and could be of importance. To investigate such possibilities, in this paper, instead of starting from given randomized designs and investigating their properties, we start with the four principles enumerated above and find the requirements which must be satisfied as these four conditions are imposed one after the other.

The Collected Works of George E. P. Box, 1984, Wadsworth, Inc., Belmont, CA 94002.
Originally published in *J. Roy. Stat. Soc.*, Series B, vol. 28, no. 3 (1966), pp. 543–558.

2. FORMULATION

Suppose we conduct an experiment whose object is to determine the linear effects of p variables in n runs. Frequently, a model of the following kind would be postulated.

$$\mathbf{y} = \mathbf{S}\boldsymbol{\alpha} + \mathbf{Z}_1\boldsymbol{\beta}_1 + \mathbf{u}, \tag{2.1}$$

where $\boldsymbol{\alpha}$ and $\boldsymbol{\beta}_1$ are $(k \times 1)$ and $(p \times 1)$ column vectors of coefficients,

$\mathbf{S}$ is a fixed matrix of order $(n \times k)$ of variables to be "eliminated",

$\mathbf{Z}_1$ is a matrix of order $(n \times p)$ of the design variables,

$\mathbf{u}$ is a vector of random variables with zero mean vector, and variance–covariance matrix $\boldsymbol{\Sigma}_u = \sigma^2 \mathbf{I}_n$.

We think of the $\mathbf{S}\boldsymbol{\alpha}$ as that part of the model that takes out fixed effects which are not of direct interest and which are to be eliminated. For example, elimination of the overall mean would be achieved by letting $\mathbf{S}$ be an $(n \times 1)$ column vector of 1's and setting $\boldsymbol{\alpha}$ equal to the scalar μ.

Again in a randomized block design $\mathbf{S}$ could be an $(n \times k)$ matrix carrying k indicator variables which would achieve the elimination of the k block means $\alpha_1, \ldots, \alpha_k$.

We suppose that the design matrix $\mathbf{Z}_1$ satisfies the conditions

$$\mathbf{S}'\mathbf{Z}_1 = \mathbf{0} \quad \text{and} \quad \mathbf{Z}_1'\mathbf{Z}_1 = \mathbf{I}_p, \tag{2.2}$$

but that otherwise $\mathbf{Z}_1$ is at the experimenter's choice. Finally, $\mathbf{y}$ denotes the $(n \times 1)$ column vector of observations obtained when n runs are carried out according to the particular $\mathbf{Z}_1$ used.

In the present research it is convenient to isolate specifically a further $m-p$ "error degrees of freedom", so that valid estimates of σ^2 may be easily extracted. To do this, we add to the model a term $\mathbf{Z}_2\boldsymbol{\epsilon}$ where $\mathbf{Z}_2$ is an $n \times (m-p)$ matrix with orthogonal columns, which are also orthogonal to those of $\mathbf{Z}_1$ and of $\mathbf{S}$ and $\boldsymbol{\epsilon}$ is an $(m-p) \times 1$ vector of coefficients known to be null *a priori* so that:

$$\mathbf{y} = \mathbf{S}\boldsymbol{\alpha} + \mathbf{Z}_1\boldsymbol{\beta}_1 + \mathbf{Z}_2\boldsymbol{\epsilon} + \mathbf{u}. \tag{2.3}$$

We could then rewrite the model of (2.1) as

$$\mathbf{y} = \mathbf{S}\boldsymbol{\alpha} + \mathbf{W}\boldsymbol{\beta} + \mathbf{u}, \tag{2.3a}$$

where $\boldsymbol{\beta}' = (\boldsymbol{\beta}_1' \,|\, \boldsymbol{\epsilon}')$ is a partitioned vector of order $(1 \times m)$ containing both the coefficient parameters and the "error" parameters and $\mathbf{W} = (\mathbf{Z}_1 \,|\, \mathbf{Z}_2)$ is an $(n \times m)$ partitioned matrix. From previous assumptions

$$\mathbf{S}'\mathbf{W} = \mathbf{0} \quad \text{and} \quad \mathbf{W}'\mathbf{W} = \mathbf{I}_m. \tag{2.4}$$

If we perform the usual least squares analysis, we may obtain a vector of least squares estimators, say $\mathbf{b} = \hat{\boldsymbol{\beta}}$ of $\boldsymbol{\beta}$, which is, of course, such that $\hat{\boldsymbol{\beta}}' = \mathbf{b}' = (\hat{\boldsymbol{\beta}}_1' \,|\, \hat{\boldsymbol{\epsilon}}')$, and the estimates $\hat{\boldsymbol{\epsilon}}$ of $\boldsymbol{\epsilon}$ could be used to provide an estimate of the variance σ^2, viz.

$$s^2 = \sum_{j=1}^{m-p} \hat{\epsilon}_j^2/(m-p).$$

Using the linear model given by (2.3a) and (2.4), we now consider what assumptions we need on ordinary distribution theory (as distinct from randomization theory) to provide various desirable properties for the estimates.

2.1. *Unbiased Nature of Estimates and Error*

Whatever the distribution of the u_j's, provided that for each u_j, $E(u_j) = 0$, and the model (2.3a) was adequate, we have that

$$E(\mathbf{b}) = E\begin{pmatrix}\hat{\boldsymbol{\beta}}_1\\ \hat{\boldsymbol{\epsilon}}\end{pmatrix} = (\mathbf{W}'\mathbf{W})^{-1}\mathbf{W}'E(\mathbf{y})$$

$$= (\mathbf{W}'\mathbf{W})^{-1}\mathbf{W}'\{\mathbf{S}\boldsymbol{\alpha}+\mathbf{W}\boldsymbol{\beta}\}$$

$$= \boldsymbol{\beta} = \begin{pmatrix}\boldsymbol{\beta}_1\\ \mathbf{0}\end{pmatrix}. \tag{2.5}$$

That is to say, the mean values of the estimates $\hat{\boldsymbol{\beta}}_1$ would be the true parameters $\boldsymbol{\beta}_1$ and the mean values of the error estimates $\hat{\boldsymbol{\epsilon}}$ would be zero. We say that a vector **b** satisfying condition (2.5) has the property of unbiasedness.

2.2. *Variances and Covariances of Estimates*

We have assumed that $\boldsymbol{\Sigma}_u = E(\mathbf{uu}') = \sigma^2\mathbf{I}_n$, and this has the well-known consequence (on any distribution theory of the u's) that the variance–covariance matrix of the b's is

$$\boldsymbol{\Sigma}_b = E\{(\mathbf{b}-\boldsymbol{\beta})(\mathbf{b}-\boldsymbol{\beta})'\} = \sigma^2(\mathbf{W}'\mathbf{W})^{-1} = \sigma^2\mathbf{I}_m, \tag{2.6}$$

that is, the $\hat{\beta}_1$'s are *uncorrelated* with each other and with the $\hat{\epsilon}$'s (the error estimates) and each has the same variance σ^2. Hence, the s^2 mentioned above would supply an unbiased estimate of σ^2. In summary, if (2.3a) and (2.4) hold, then the b's are *uncorrelated*, with the further consequence that a *valid* estimate of σ^2 exists.

2.3. *Minimization of Variance*

One example of the use of the matrix **S** is the elimination of block variables as in the randomized block designs of R. A. Fisher. The use of the randomized block design ensures that the estimates of the effects of interest $\boldsymbol{\beta}_1$ and the relevant error are uncontaminated.

We note that if **W** were not orthogonal to **S** the elements of the variance–covariance matrix (2.6) would be increased. The selection of a design in which **W** were orthogonal to **S** thus ensures minimization of the variances of the b's. A general objective in the choice of **W** will be the minimization of these variances.

2.4. *Distribution of the Estimates*

With the assumption that the joint distribution $f(\mathbf{u})$ of the u's is spherical normal the m $(b-\beta)$'s, representing "error effects" as well as the effects of the variables, are jointly, normally and independently distributed with the same variance σ^2. Thus the standard distributions derived from the normal theory could be used in making inferences about $\boldsymbol{\beta}_1$ and σ^2.

When we do impose the condition that the u's are independent normal, we will say that we have imposed the *normality assumption.*

3. PROTECTION VIA RANDOMIZATION

It is often the case that an experimenter, postulating the model (2.3a), fears that the correct model might really be

$$\mathbf{y} = \mathbf{S}\boldsymbol{\alpha}+\mathbf{W}\boldsymbol{\beta}+\mathbf{X}\boldsymbol{\gamma}+\mathbf{u}, \tag{3.1}$$

where (1) the vector $\mathbf{X\gamma}$ represents the effect of a group of variables which are unmeasured, unknown, or both;

(2) the vector $\mathbf{u}$ has distribution which is not necessarily spherically normal and, in particular, $u_1, \ldots, u_n$ may not be independently distributed.

The experimenter would like to be able to analyse his results *as if* the "simple" assumptions were true with *adequate protection* in case they are not. Under suitable circumstances such protection can be gained by randomization.

Generalized randomization. In this paper we extend the word "randomization" to mean only that $\mathbf{W}$ is to be chosen as a realization from *some* known distribution $P(\mathbf{W})$, which the experimental procedure guarantees is independent of the distribution $f(\mathbf{u})$.

We would like to choose $P(\mathbf{W})$ first so that the condition of unbiasedness holds. We shall find a large number of randomization schemes that satisfy this requirement, so second, we proceed to find the set of randomization schemes which provide in addition a *valid estimate of error*. Third, we select from this remaining large set a subset of schemes $P(\mathbf{W})$ that *minimize* the variance of the estimates. Finally, we impose conditions which would ensure "desirable" properties for the distributions of the estimates. (It is, for example, possible to so choose the randomization scheme as to yield normally distributed estimates. Insistence on normality would in general be too severe a restriction, however.)

We shall proceed then on the assumption only that the very general model (3.1)

$$\mathbf{y} = \mathbf{S\alpha} + \mathbf{W\beta} + \mathbf{X\gamma} + \mathbf{u}$$

applies where $\mathbf{X}$ represents the levels of variables which are initially unknown and nothing is known of the distribution of $\mathbf{u}$. We shall seek to choose $P(\mathbf{W})$ so as to validate the usual least squares analysis and to meet the four requirements listed above.

3.1. *Unbiased Estimation*

If (3.1) holds, then the usual least squares estimates $\mathbf{b} = (\mathbf{W'W})^{-1}\mathbf{W'y} = \mathbf{W'y}$ are such that

$$\mathbf{b} = \mathbf{\beta} + \mathbf{W'X\gamma} + \mathbf{W'u}. \qquad (3.2)$$

For any randomization scheme, that is, for any $P(\mathbf{W})$,

$$E_w(\mathbf{b}) = \mathbf{\beta} + E_w(\mathbf{W'})\mathbf{X\gamma} + E_w(\mathbf{W'})\mathbf{u}, \qquad (3.3)$$

so that in general we have

$$E_w(\mathbf{b}) = \mathbf{\beta} \quad \text{if and only if} \quad E_w(\mathbf{W'}) = \mathbf{0}. \qquad (3.4)$$

We now accept this restriction and consider what further restrictions we must place on $P(\mathbf{W})$.

3.2. *Variance–Covariance Matrix of Effect Estimates and of Error Estimates to be of the Form* $\mathbf{I}_m \sigma_b^2$

The design and error vectors (the columns of $\mathbf{W}$) lie in an m-dimensional subspace and are orthogonal to $\mathbf{S}$. Suppose we denote this subspace by $\mathscr{W}$, and let $\mathbf{v}_1, \ldots, \mathbf{v}_m$ be any m fixed orthogonal $(n \times 1)$ vectors lying in the space $\mathscr{W}$. We can think of the $\mathbf{v}_j$ as columns of an $(n \times m)$ matrix $\mathbf{V}$ and providing a basis for the space $\mathscr{W}$, and we note that

$$\mathbf{V'V} = \mathbf{I}_m \quad \text{and} \quad \mathbf{V'S} = \mathbf{0}. \qquad (3.5)$$

Thus, there exists an $(n \times m)$ matrix $\mathbf{A}$, where

$$\mathbf{W}' = \mathbf{A}\mathbf{V}' \tag{3.6}$$

and such that $\mathbf{A}$ is orthogonal. Since $\mathbf{V}$ is fixed, $P(\mathbf{A})$ completely determines $P(\mathbf{W})$. We note that to satisfy condition (3.4), that is, $E_w(\mathbf{W}') = \mathbf{0}$, we require that

$$E_w(\mathbf{A}) = \mathbf{0}. \tag{3.7}$$

Now consulting (3.2) we have that

$$(\mathbf{b}-\boldsymbol{\beta}) = \mathbf{W}'(\mathbf{X}\boldsymbol{\gamma}+\mathbf{u}) = \mathbf{A}\mathbf{V}'(\mathbf{X}\boldsymbol{\gamma}+\mathbf{u}), \tag{3.8}$$

and we require that

$$\begin{aligned} E_w\{(\mathbf{b}-\boldsymbol{\beta})(\mathbf{b}-\boldsymbol{\beta})'\} &= E_w\{\mathbf{A}\mathbf{V}'(\mathbf{X}\boldsymbol{\gamma}+\mathbf{u})(\boldsymbol{\gamma}'\mathbf{X}'+\mathbf{u}')\mathbf{V}\mathbf{A}'\} \\ &= \mathbf{I}_m\,\sigma_b^2. \end{aligned} \tag{3.9}$$

If we now let the $(m \times 1)$ vector $\mathbf{c}$ be defined by

$$\mathbf{c} = \mathbf{V}'(\mathbf{X}\boldsymbol{\gamma}+\mathbf{u}), \tag{3.10}$$

we may write (3.9) as

$$E_w\{\mathbf{A}\mathbf{c}\mathbf{c}'\mathbf{A}'\} = \mathbf{I}_m\,\sigma_b^2. \tag{3.11}$$

Suppose we take the trace of both sides of (3.11), then the left-hand side gives

$$\begin{aligned} \text{trace}\, E_w\{\mathbf{A}\mathbf{c}\mathbf{c}'\mathbf{A}'\} &= E_w\{\text{trace}\,(\mathbf{A}\mathbf{c}\mathbf{c}'\mathbf{A}')\} \\ &= E_w\{\text{trace}\,(\mathbf{c}'\mathbf{A}'\mathbf{A}\mathbf{c})\} \\ &= E_w\{\mathbf{c}'\mathbf{c}\} \\ &= \mathbf{c}'\mathbf{c}, \end{aligned} \tag{3.12}$$

while the trace of the right-hand side of (3.11) gives

$$m\sigma_b^2 = m\sigma^2. \tag{3.13}$$

Thus we require that

$$\sigma^2 = \frac{\mathbf{c}'\mathbf{c}}{m}. \tag{3.14}$$

Now (3.14) together with (3.11) implies that

$$E_w(\mathbf{a}_i'\,\mathbf{c}\mathbf{c}'\mathbf{a}_j) = \begin{cases} \dfrac{\mathbf{c}'\mathbf{c}}{m} & \text{if } i=j, \\ 0 & \text{if } i \neq j, \end{cases} \tag{3.15}$$

for every fixed $\mathbf{c}$. Indeed, if $i=j$ then

$$\sum_{g=1}^{m}\sum_{h=1}^{m} E_w(a_{ig}\,a_{ih})\,c_g\,c_h = \frac{1}{m}\sum_{g=1}^{m} c_g^2 \tag{3.16}$$

for every $c_1, \ldots, c_m$. Therefore we easily deduce that

$$E_w(a_{ig}^2) = \frac{1}{m} \quad \text{and} \quad E_w(a_{ig}\,a_{ih}) = 0 \quad (g \neq h). \tag{3.17}$$

Further, if $i \neq j$, then we have that

$$\sum_{g=1}^{m} \sum_{h=1}^{m} E_w(a_{ig}\, a_{jh})\, c_g\, c_h = 0 \tag{3.18}$$

for every $c_1, \ldots, c_m$, so that we may easily deduce that

$$E_w(a_{ig}\, a_{jh}) = 0 \quad \text{for all } g \text{ and } h \quad (i \neq j). \tag{3.19}$$

That is, finally,

$$E_w\{(\mathbf{b}-\boldsymbol{\beta})(\mathbf{b}-\boldsymbol{\beta}')\} = \mathbf{I}_m\, \sigma_b^2 \quad \text{if and only if}$$

$$E_w(a_{ig}\, a_{jh}) = \begin{cases} \dfrac{1}{m} & \text{if } i = j, g = h, \\ 0 & \text{otherwise.} \end{cases} \tag{3.20}$$

It is easy to see that one way of obtaining a class of matrices satisfying the conditions (3.7) and (3.20) is as follows.

(1) Select an orthogonal matrix which has as one row

$$\left(\frac{1}{\sqrt{m}}, \ldots, \frac{1}{\sqrt{m}}\right). \tag{3.20a}$$

(2) Generate the class of matrices from this orthogonal matrix by permuting rows, permuting columns and switching signs in each row and each column.

We show in Appendix 1 that such a class of matrices does indeed satisfy conditions (3.7) and (3.20).

3.3. *Minimization of Variance*

We now suppose that we have a randomization scheme, summarized by $P(\mathbf{A})$, $\mathbf{A} \in \{\mathbf{A}\}$, which satisfies conditions (3.20) and (3.7). Recalling that $\mathbf{W}' = \mathbf{A}\mathbf{V}'$ or $\mathbf{W} = \mathbf{V}\mathbf{A}'$, then for a particular set of orthogonal base vectors which are the columns of $\mathbf{V}$, and subject only to the condition $\mathbf{V}'\mathbf{S} = \mathbf{0}$, we have from (3.14) that the effective error variance is given by $\mathbf{c}'\mathbf{c}/m$, where $\mathbf{c} = \mathbf{V}'(\mathbf{X}\boldsymbol{\gamma}+\mathbf{u})$. We would like to choose $\mathbf{V}$ so that the elements of $\mathbf{c}$ are small in absolute magnitude. If we confine our attention only to the randomization set, i.e. the set generated by $P(\mathbf{W})$ for given $\mathbf{y}$, then nothing can be done, because $\mathbf{X}\boldsymbol{\gamma}$ and $\mathbf{u}$ are then fixed but unknown. We consider, therefore, the "total variational set" generated by varying over $\mathbf{W}$ and over $\mathbf{u}$.

Suppose then that $\mathbf{u}$ is such that

$$E(\mathbf{u}) = 0 \quad \text{and} \quad \boldsymbol{\Sigma}_u = E(\mathbf{u}\mathbf{u}') = \boldsymbol{\Omega}\sigma^2, \tag{3.21}$$

where $\boldsymbol{\Omega}$ is an $n \times n$ positive definite matrix. Using (3.9), it is easy to see that

$$m\sigma^2 = \text{trace}\,[E_w\{\mathbf{V}'(\mathbf{X}\boldsymbol{\gamma}+\mathbf{u})(\boldsymbol{\gamma}'\mathbf{X}'+\mathbf{u}')\mathbf{V}\}], \tag{3.22}$$

and since $E(\mathbf{u}) = \mathbf{0}$, we have

$$m\sigma^2 = \text{trace}\,(\mathbf{V}'\mathbf{X}\boldsymbol{\gamma}\boldsymbol{\gamma}'\mathbf{X}'\mathbf{V} + \sigma^2\mathbf{V}'\boldsymbol{\Omega}\mathbf{V}), \tag{3.23}$$

that is, $m\sigma^2$ is the sum of two terms, the first we call the bias part, the second we call the variance part. Because $\mathbf{X}\boldsymbol{\gamma}$ is fixed and unknown, we concentrate on minimizing the second term.

Consider a set of variables $t_1, \ldots, t_m$, where $t_j = \mathbf{v}_j'\mathbf{y}$ and where the $\mathbf{v}_j$ are the m mutually orthogonal columns of $\mathbf{V}$ which were used to define the subspace $\mathscr{W}$ of Section 3.2. Referring to our model (2.3a), the least squares estimator $\mathbf{b}$ of $\boldsymbol{\beta}$ (with the design matrix $\mathbf{W} = \mathbf{VA}'$) is given by

$$\mathbf{b} = \mathbf{AV}'\mathbf{y} = \mathbf{At}. \tag{3.24}$$

Hence, the variance–covariance matrix of the b's is given by

$$\boldsymbol{\Sigma}_b = \sigma^2 \mathbf{AV}'\boldsymbol{\Omega}\mathbf{VA}' = \mathbf{A}\boldsymbol{\Sigma}_t\mathbf{A}'. \tag{3.25}$$

Using the fact that trace $\mathbf{AB}$ = trace $\mathbf{BA}$ we easily deduce that

$$\sum_{i=1}^{m} \sigma_{b_i}^2 = \text{trace}\, \sigma^2 \mathbf{V}'\boldsymbol{\Omega}\mathbf{V} = \sum_{i=1}^{m} \sigma_{t_i}^2. \tag{3.26}$$

Thus we should for preference choose $\mathbf{v}_1, \ldots, \mathbf{v}_m$ so that $t_1, \ldots, t_m$ will have small variances. This can be done if sufficient is known about $\boldsymbol{\Omega}$. There may be varying degrees of knowledge.

3.3.1. $\boldsymbol{\Omega}$ *known completely*

If $\boldsymbol{\Omega}$ were completely known, then we should select $\mathbf{v}_1, \ldots, \mathbf{v}_m$ to be the m latent vectors of $\boldsymbol{\Omega}$ corresponding to the m smallest latent roots $\lambda_1, \ldots, \lambda_m$. Then we would have

$$\sum_{i=1}^{m} \sigma_{b_i}^2 = \sigma^2 \sum_{i=1}^{m} \lambda_i. \tag{3.27}$$

3.3.2. *Blocks*

It is unlikely that $\boldsymbol{\Omega}$ would ever be completely known. However, we would often have sufficient knowledge of $\boldsymbol{\Omega}$ to identify k subsets or blocks of size s within the n experimental units, where $n = ks$. These blocks would be such that variation between experimental units in a given block would be less than variation between blocks. For experimental units distributed in space or time, it is often known that units which were close together can be expected to be more nearly alike than those far apart. In this situation, the vectors $\mathbf{v}_j$ that make up the columns of $\mathbf{V}$ could be such that comparisons were made only within blocks. The jth column of $\mathbf{V}$ would be such that

$$\sum_{t=1}^{s} v_{tj} = \sum_{s+1}^{2s} v_{tj} \ldots = \sum_{(k-1)s+1}^{ks} v_{tj} = 0. \tag{3.28}$$

3.3.3. *A serial correlation model*

The model of Section 3.3.1 assumes that we know everything about $\boldsymbol{\Omega}$, while the model of Section 3.3.2 assumes that we know very little. It is of interest that most of the advantage of complete knowledge of $\boldsymbol{\Omega}$ can be obtained if we can make an assumption that requires very little more knowledge than that needed for blocking.

Choosing design vectors having small error components. Suppose it can be assumed that the variance $V(y_s - y_t)$ of $(y_s - y_t)$ is a function only of $|s-t|$; that is, that the y's are subject to serial correlation. This model as it stands is awkward to deal with because of "end corrections" whose effects, although often unimportant, are troublesome mathematically.

Following Hotelling—see Anderson (1942)—we approximate this model then, by treating y_s and y_{n+s}. If, for example, only lag 1 serial correlation was important, then this means that in this approximation we should assume that y_n and y_1 have the same correlation as y_1 and y_2, y_2 and $y_3, \ldots, y_{n-1}$ and y_n. This is called a circular serial correlation model of lag 1.

Suppose the circular serial correlation is generated by a first-order autoregressive process,

$$u_i = \rho u_{i-1} + \delta_i, \tag{3.29}$$

where the δ_i are independent standard normal deviates. Then it may be verified that the latent vectors and latent roots of $\mathbf{\Omega}$ are of the form given in Table 1 below.

TABLE 1

Latent vectors and roots of the circular Markov process of lag 1

Vectors	*Corresponding latent roots* ($\alpha = -\rho/(1+\rho^2)$)	*Frequency*
One vector of form $(1, 1, \ldots, 1, 1)$	$\lambda_1 = 1/(1+2\alpha)$	0
Pairs of vectors of form $[1, \cos(2\pi j/n), \ldots, \cos\{2\pi j/n(n-1)\}]$ $[0, \sin(2\pi j/n), \ldots, \sin\{2\pi j/n(n-1)\}]$ where $j = 1, \ldots, \frac{1}{2}n-1$ if n even, $j = 1, \ldots, \frac{1}{2}(n-1)$ if n odd.	Twin roots $\lambda_j = 1/(1+2\alpha)(\cos 2\pi j/n)$	$2\pi j/n$
If n even, one vector of form $(1, -1, 1, -1, \ldots, 1, -1)$	$\lambda_n = 1/(1-2\alpha)$	π

In Table 1 we have put $\alpha = -\rho/(1+\rho^2)$, and it is easy to see that if ρ is positive, the vectors $\mathbf{v}_j$ corresponding to the smallest latent roots correspond to vectors with highest frequency $2\pi j/n$. Indeed if $t_j = \mathbf{v}_j'\mathbf{y}$, then the variates t_j corresponding to latent vectors of highest frequency have smallest variance, or, put another way, if ρ is positive, the variance of the t_j is a monotonic decreasing function of the frequency. On the other hand, if ρ is negative the variance would be a monotonic increasing function of the frequency.

Now it is true for *any* circulant that the latent vectors are of the form given in Table 1. Thus for any circular correlation scheme, these vectors would supply a set of orthogonal components. Although precise variances associated with each component would depend on the actual correlation scheme, therefore, this would always supply a means of selecting a subset of vectors. We would expect that, where positive serial correlation was induced by propinquity in space or time, high frequencies would carry low variances, so that the selection of vectors in accordance with the scheme mentioned above would be quite insensitive to the exact form of the correlation process. Furthermore, it can be readily verified that the latent vectors of the circular serial correlation matrices are, for moderate values of n, very close to those in which the circularity assumption is not made.

It follows that, in any case where a predominantly positive serial correlation was expected, the use of high-frequency vectors of Table 1 would provide the basis for a near optimal design. Once having selected these vectors, a suitable randomization process should be applied which would satisfy the requirements (3.7) and (3.20).

Example. Suppose we had an experiment which was to contain 16 observations and we decided to employ eight vectors for carrying errors and treatments, then in the common case where predominantly positive correlation is expected, we should choose the eight highest frequency vectors. In fact, by so doing we would reduce the variance by the factor D, where

$$D = \frac{P(\alpha)}{Q(\alpha)}, \tag{3.30}$$

$$\begin{aligned} P(\alpha) = \frac{1}{1+2\alpha} + \frac{2}{1+2\alpha\cos(\pi/8)} + \frac{2}{1+2\alpha\cos(2\pi/8)} \\ + \frac{2}{1+2\alpha\cos(3\pi/8)} + \frac{1}{1+2\alpha\cos(4\pi/8)}, \end{aligned} \tag{3.31}$$

$$\begin{aligned} Q(\alpha) = \frac{1}{1+2\alpha\cos(4\pi/8)} + \frac{2}{1+2\alpha\cos(5\pi/8)} + \frac{2}{1+2\alpha\cos(6\pi/8)} \\ + \frac{2}{1+2\alpha\cos(7\pi/8)} + \frac{1}{1-2\alpha}, \end{aligned} \tag{3.32}$$

and, of course, $\alpha = -\rho/(1+\rho^2)$.

As an example, if $\rho = 0{\cdot}5$, $D = 3{\cdot}84$, so we would reduce the variance by a factor of almost four. With higher degrees of correlation, even greater reductions of variance are possible. Thus with $\rho = 0{\cdot}9$, the variance is reduced by a factor of over 40. The values of D for various values of ρ are given in Table 2. (Similar gains are obtained by basing designs on the low-frequency vectors when ρ is negative.)

TABLE 2

ρ	D	ρ	D	ρ	D
−0·9	0·0239	−0·3	0·4627	+0·4	2·849
−0·8	0·0723	−0·2	0·6020	+0·5	3·836
−0·7	0·1258	−0·1	0·7773	+0·6	5·347
−0·6	0·1870	+0·1	1·287	+0·7	7·948
−0·5	0·2607	+0·2	1·661	+0·8	13·84
−0·4	0·3510	+0·3	2·161	+0·9	41·92

One way of selecting an appropriate randomization scheme would be to take any (8×8) orthogonal matrix, randomly permute rows and columns and randomly apply sign switching. The final orthogonal matrix could then be used to apply a rotation to the eight selected latent vectors. An alternative randomization set would be supplied by employing "spherical" randomization; Box and Hunter (1957). In this case an (8×8) orthogonal matrix representing a spherically random rotation would be applied to the basic latent vectors.

This particular application has a further interesting interpretation. It will be seen that the latent roots λ_j directly measure the power at the various frequencies, and the plot of the sum of the λ_k's against the frequencies $f_k = (2\pi/n)k$ (or simply against the k's if desired) would then reproduce a finite version of the power spectrum. In the above example, what is being done is to introduce the signal (and accompanying error components) into the region of the spectrum where the power is lowest. Table 3 tabulates the power P at the various frequencies for the above example.

TABLE 3

f	0	$\pi/8$	$2\pi/8$	$3\pi/8$	$4\pi/8$	$5\pi/8$	$6\pi/8$	$7\pi/8$	π
P	5·00	7·66	4·60	2·88	2·00	1·54	1·28	1·16	0·55

The idea of blocking can be seen to be an approximation of the same idea. For example, if we had an experiment taking place over a period of 16 days and one observation is made daily, we might very well block into four blocks of four days and the comparison which we could then make between blocks would correspond to the low frequencies; in fact, roughly to combinations of the frequencies $\pi/8$ and $2\pi/8$.

For instance, the three degrees of freedom could correspond to the linear contrasts of the observations having coefficients

$$\begin{array}{cccc} +1,+1,+1,+1 & -1,-1,-1,-1 & -1,-1,-1,-1 & +1,+1,+1,+1, \\ +1,+1,+1,+1 & +1,+1,+1,+1 & -1,-1,-1,-1 & -1,-1,-1,-1, \\ +1,+1,+1,+1 & -1,-1,-1,-1 & +1,+1,+1,+1 & -1,-1,-1,-1. \end{array}$$

The first and second sets of coefficients could be thought of as crude approximations to $\cos(\pi j/8)$ and $\sin(\pi j/8)$ $(0 \leqslant j \leqslant 15)$ and the last as an approximation to $\sin(\pi j/4)$.

It is important to notice also that, although a randomized block design will always ensure validity of comparisons, there are some correlation situations where designs such as those considered above do not ensure efficiency. For instance, in the chemical industry, negative serial correlation between batches can occur due to "carry-over". In these situations we should not try to compare treatments within short periods of time, but, conversely, try to run each treatment over a series of adjacent batches. In our spectral interpretation this corresponds to the fact that, with ρ negative, we do better to put the signal in at the low frequencies where the "noise", that is, power, is then lowest (since $\rho < 0$).

3.4. *Some Distribution Theory*

We have considered the case where the correct model may be, as in (3.1),

$$\mathbf{y} = \mathbf{S}\boldsymbol{\alpha} + \mathbf{W}\boldsymbol{\beta} + \mathbf{X}\boldsymbol{\gamma} + \mathbf{u} \tag{3.33}$$

or

$$\mathbf{y} = \mathbf{V}\mathbf{A}'\boldsymbol{\beta} + \boldsymbol{\varepsilon}, \tag{3.34}$$

where $\boldsymbol{\varepsilon} = \mathbf{S}\boldsymbol{\alpha} + \mathbf{X}\boldsymbol{\gamma} + \mathbf{u}$ are the "errors" in the y's.

Hence we have that

$$(\mathbf{V}'\mathbf{V})^{-1}\mathbf{V}'\mathbf{y} = \mathbf{V}'\mathbf{y} = \mathbf{A}^{-1}\hat{\boldsymbol{\beta}}, \tag{3.35}$$

so that

$$\mathbf{A}^{-1}\hat{\boldsymbol{\beta}} = \mathbf{A}'\boldsymbol{\beta}+\mathbf{V}'\boldsymbol{\varepsilon}, \tag{3.36}$$

which we write as

$$\mathbf{A}^{-1}\hat{\boldsymbol{\beta}} = \mathbf{A}'\boldsymbol{\beta}+\mathbf{d}, \tag{3.37}$$

where

$$\mathbf{d} = \mathbf{V}'\boldsymbol{\varepsilon}. \tag{3.38}$$

From (3.36), then we have

$$\hat{\boldsymbol{\beta}} = \boldsymbol{\beta}+\mathbf{A}\mathbf{d}, \tag{3.39}$$

since $\mathbf{A}$ is orthogonal.

Hence to find the nature of the randomized distribution of the quantities that give information about $\boldsymbol{\beta}$, it is sufficient to find the randomized distribution of the quantities

$$\mathbf{e} = \mathbf{A}\mathbf{d}. \tag{3.40}$$

Consider e_i and its randomization moments under the schemes of Section 3.3. We have that

$$e_i = a_{i1}\,d_1+\ldots+a_{im}\,d_m, \tag{3.41}$$

so that use of the schemes of Section 3.3 which have $E_w(\mathbf{A}) = \mathbf{0}$ gives immediately that the first (permutation) moment of e_i is

$$E_w(e_i) = 0 \quad \text{for all} \quad i. \tag{3.42}$$

Further, it can be shown that as

$$E_w(e_i^2) = E_w(a_{ij}^2) \sum_{i=1}^{m} d_i^2+2E_w(a_{ij}\,a_{ih}) \sum\sum_{j<h} d_j\,d_h \tag{3.43}$$

and that

$$E_w(e_i^2) = \frac{1}{m}\sum_{i=1}^{m} d_i^2 = d_2, \quad \text{that is,} \quad \text{var}_w(e_i) = d_2, \tag{3.44}$$

since we have from (3.20) that $E_w(a_{ij}\,a_{ih}) = 0\;(j\neq h)$ and $E_w(a_{ij}^2) = 1/m$. It is also easy to see that $E_w(e_i^3) = 0$.

We consider now the fourth moment about the origin of e_i (and in view of (3.42) moments about the origin are moments about the mean). We have

$$E_w(e_i^4) = E_w(a_{ij}^4) \sum_{i=1}^{m} d_i^4+6E_w(a_{ij}^2\,a_{ih}^2) \sum\sum_{j<h} d_j^2\,d_h^2, \tag{3.45}$$

since it can be shown that $E_w(a_{ij}^3\,a_{ih}) = E_w(a_{ij}\,a_{ih}\,a_{ir}\,a_{is}) = 0$ for the scheme of Section 3.2. It is easily verified that

$$E_w(a_{ij}^2\,a_{ih}^2) = \frac{1-ma_4}{m(m-1)}, \tag{3.46}$$

where

$$a_4 = \sum_i \sum_j a_{ij}^4/m^2. \tag{3.46a}$$

Further, denoting $\sum_{i=1}^m d_i^4/m$ and (as before) $\sum_{i=1}^m d_i^2/m$ by d_4 and d_2 respectively, we obtain

$$2\sum_{j<h}\sum d_j^2 d_h^2 = m^2 d_2^2 - md_4. \tag{3.46b}$$

Thus (3.45) may be written as

$$E_w(e_i^4) = ma_4 d_4 + 3\frac{1-ma_4}{m-1}(md_2^2 - d_4). \tag{3.47}$$

Now in some applications it is convenient to have an expression for the randomization variance of e_i^2. Using (3.47) and (3.44) we find

$$\operatorname{var}_w(e_i^2) = \frac{1}{m-1}\{(m^2 a_4 + 2ma_4 - 3)\, d_4 - (3m^2 a_4 - 2m - 1)\, d_2^2\}. \tag{3.48}$$

As an interesting illustration of the above, we turn to the "paired-comparison" situation. We discuss the special case where $n = 2^j$ $(j = 2, 3, \ldots)$. Then, $m = 2^{j-1}$, $m-1 = 2^{j-1}-1$ and it is easy to see that $\mathbf{W} = \mathbf{VA}'$ takes the form such that:

(i) $\mathbf{V}$ has columns made up of m pairs of a plus one and a minus one in any order, but such that the columns are orthogonal. For example, in the $n = 8$ case $(m = 4)$ we would have

$$\mathbf{V} = \begin{pmatrix} - & + & + & - \\ + & - & - & + \\ - & - & + & + \\ + & + & - & - \\ - & + & - & + \\ + & - & + & - \\ - & - & - & - \\ + & + & + & + \end{pmatrix}.$$

The $(n \times m)$ matrix $\mathbf{V}$ could be constructed by writing out a full 2^{2m} factorial design, say in Yates's standard order, and choosing the 2^m columns with the above structure.

(ii) The $\mathbf{A}$'s are generated by the set of matrices obtained by sign switching and permuting rows and columns of the $m \times m$ matrix

$$\begin{bmatrix} 1 & 0 & . & . & . & 0 \\ 0 & 1 & 0 & . & . & . \\ . & & 1 & & & . \\ . & & & 1 & & . \\ . & & & & . & . \\ 0 & . & . & . & . & 1 \end{bmatrix}.$$

The usual paired-comparison test uses the statistic

$$t^2 = \frac{e_i^2}{\sum_{j \neq i} e_j^2/(m-1)},$$

where $e_i = \mathbf{v}'_{.i}\boldsymbol{\varepsilon}$, with $\mathbf{v}_{.i}$ representing the ith column of $\mathbf{V}$. Now it is easy to see that the permutation test statistic W for this paired-comparison situation used by Welch (1937) and discussed in Box and Andersen (1955) is such that

$$W = \frac{\sum_{j \neq i} e_j^2}{\sum_{j=1}^{m} e_j^2} = \frac{m-1}{(m-1)+t^2}. \tag{3.49}$$

In any case, when permuting we have that $\sum_{j=1}^{m} e_j^2$ is constant. Hence we have

$$E_w(W) = \frac{1}{\mathbf{e}'\mathbf{e}} \sum_{j \neq i} E_w(e_j^2) \tag{3.50}$$

and using (3.40) and (3.44) and remembering that $\mathbf{A}$ is orthogonal we have

$$E_w(W) = \frac{1}{\sum d_i^2}(m-1)\frac{1}{m}\sum d_i^2 = \frac{m-1}{m}. \tag{3.50a}$$

To find the variance of W it is sufficient to find the variance of

$$1 - W = \frac{e_i^2}{\sum_{j=1}^{m} e_j^2},$$

where we bear in mind that $\mathbf{e} = \mathbf{Ad} = \mathbf{AV}'\boldsymbol{\varepsilon}$. We note that the elements of an $\mathbf{A}$ matrix are either ± 1 or zero and indeed that for the paired-comparison design of this section, $a_2 = a_4 = 1/m$. Using this with (3.48) we have

$$\operatorname{var}_w(W) = \operatorname{var}_w(1-W) = \frac{1}{m^2}\left(\frac{d_4}{d_2^2} - 1\right). \tag{3.51}$$

But $\mathbf{d} = \mathbf{V}'\boldsymbol{\varepsilon}$, and denoting $e_{2i} - e_{2i-1}$ by δ_i $(i = 1, \ldots, m)$—the "error" due to the treatments applied to the ith experimental unit—it is easy to see that

$$\sum_{i=1}^{m} d_i^2 = m\sum \delta_i^2 \quad \text{or} \quad d_2 = \sum \delta_i^2,$$

$$\sum_{i=1}^{m} d_i^4 = m\left(\sum \delta_i^4 + 6\sum_{i<j}\sum \delta_i^2 \delta_j^2\right) \quad \text{or} \quad d_4 = 3\left(\sum_{i=1}^{m} \delta_i^2\right)^2 - 2\sum_{i=1}^{m} \delta_i^4. \tag{3.52}$$

Hence

$$\operatorname{var}_w(W) = \frac{1}{m^2}\left\{\frac{3\left(\sum_{i=1}^{m}\delta_i^2\right)^2 - 2\sum_{i=1}^{m}\delta_i^4}{(\sum \delta_i^2)^2} - 1\right\},$$

and after some algebra this may be written as

$$\operatorname{var}_w(W) = \frac{2(m-1)}{m^2(m+2)}\left(1 - \frac{b_2 - 3}{m-1}\right), \tag{3.53}$$

where

$$b_2 = (m+2) \sum_{i=1}^{m} \delta_i^4 \Big/ \left(\sum_{i=1}^{m} \delta_i^2 \right)^2,$$

agreeing with the result of Welch (1937) and Box and Andersen (1955).

APPENDIX 1

In this appendix we show that the procedure (3.20a) generates a class of matrices satisfying the conditions (3.7) and (3.20).

Consider then the set of matrices generated by an orthogonal $(m \times m)$ matrix $\mathbf{A}$ which has one row (or column) with elements

$$\frac{1}{\sqrt{m}}, \frac{1}{\sqrt{m}}, \ldots, \frac{1}{\sqrt{m}}. \tag{A.1}$$

Let this row without loss of generality be row 1. Then we have

$$\sum_{g=1}^{m} a_{ig} = 0 \quad (i = 2, \ldots, m)$$

and

$$\sum_{g=1}^{m} a_{ig} a_{jg} = \begin{cases} 1 & \text{if} \quad i = j, \\ 0 & \text{if} \quad i \neq j, \end{cases} \tag{A.2}$$

since $\mathbf{A}$ is orthogonal. We also have, of course, that

$$\sum_{i=1}^{m} a_{ig} a_{ih} = \begin{cases} 1 & \text{if} \quad g = h, \\ 0 & \text{if} \quad g \neq h. \end{cases} \tag{A.3}$$

We note that $\sum_i \sum_j a_{ij}^2 = m$. Two examples of such matrices are provided by the Helmert matrix and the "factorial" matrix; for example, if $m = 4$, then $\mathbf{A}$ may be

$$\begin{pmatrix} \frac{1}{2} & \frac{1}{2} & \frac{1}{2} & \frac{1}{2} \\ \frac{1}{\sqrt{2}} & \frac{-1}{\sqrt{2}} & 0 & 0 \\ \frac{1}{\sqrt{6}} & \frac{1}{\sqrt{6}} & \frac{-2}{\sqrt{6}} & 0 \\ \frac{1}{\sqrt{12}} & \frac{1}{\sqrt{12}} & \frac{1}{\sqrt{12}} & \frac{-3}{\sqrt{12}} \end{pmatrix} \quad \text{or} \quad \begin{pmatrix} \frac{1}{2} & \frac{1}{2} & \frac{1}{2} & \frac{1}{2} \\ -\frac{1}{2} & -\frac{1}{2} & \frac{1}{2} & \frac{1}{2} \\ -\frac{1}{2} & \frac{1}{2} & \frac{1}{2} & -\frac{1}{2} \\ -\frac{1}{2} & \frac{1}{2} & -\frac{1}{2} & \frac{1}{2} \end{pmatrix}, \quad \text{etc.} \tag{A.4}$$

Our "randomization" set of matrices is generated from $\mathbf{A}$ by permuting rows, permuting columns and switching signs in each row and each column of any matrix. We denote this set of $2^{2m}(m!)^2$ matrices by $\{\mathbf{A}\}$, and the associated randomization scheme selects a particular $\mathbf{A} \in \{\mathbf{A}\}$ with probability equal to $1/2^{2m}(m!)^2$.

Because of the sign switching, it is easy to see that $E_w(\mathbf{A}) = \mathbf{0}$. We note that $E_w = E_R E_C E_S$, where E_S denotes expectation with respect to sign-switching, E_C denotes expectation with respect to columns and finally E_R denotes expectation taken with respect to rows. Further, we again note from conditions (A.2) that our matrices {**A**} are such that for any particular **A**, there is one row, say i, such that

$$\sum_{g=1}^{m} a_{ig} = \pm\sqrt{m} \quad \text{and} \quad \sum_{g=1}^{m} a_{i'g} = 0 \quad \text{for all} \quad i' \neq i. \tag{A.5}$$

Hence, the mean of all elements in any particular $\mathbf{A} \in \{\mathbf{A}\}$ has value $\bar{a}$, say, where

$$\bar{a} = \frac{1}{m^2} \sum_{i=1}^{m} \sum_{g=1}^{m} a_{ig} = \pm \frac{\sqrt{m}}{m^2} \tag{A.6}$$

or

$$m\bar{a} = \pm \frac{\sqrt{m}}{m}; \tag{A.6a}$$

that is, $m^2 \bar{a}^2 = 1/m$. Another way of interpreting (3.25) is to say that for any $\mathbf{A} \in \{\mathbf{A}\}$, there exists one row mean, say $\bar{a}_{i.}$ which is such that

$$\bar{a}_{i.} = \frac{1}{m} \sum_{g=1}^{m} a_{ig} = \pm \frac{1}{\sqrt{m}} \quad \text{and} \quad \bar{a}_{i'.} = 0 \quad \text{for all} \quad i' \neq i. \tag{A.7}$$

We now check the conditions (3.20). Consider first, the expectation

$$E_w(a_{ig}\, a_{jh}) = E_R\, E_C\, E_S(a_{ig}\, a_{jh}), \tag{A.8}$$

where $i \neq j$, $g \neq h$.

It is clear that we may evaluate (A.8) by first computing $E_S(a_{ig}\, a_{jh})$ $(i \neq j, g \neq h)$, and then averaging the result "with respect to rows and columns", that is, by taking $E_R\, E_C$ of the result. We have then that

$$\begin{aligned} E_S(a_{ig}\, a_{jh}) &= (a_{ig}\, a_{jh}) \frac{1}{2^{2m-4}} + (a_{ig})(-a_{jh}) \frac{1}{2^{2m-4}} \\ &\quad + (-a_{ig})(a_{jh}) \frac{1}{2^{2m-4}} + (-a_{ig})(-a_{jh}) \frac{1}{2^{2m-4}} \\ &= 0, \end{aligned} \tag{A.9}$$

and hence

$$E_R\, E_C\, E_S(a_{ig}\, a_{jh}) = 0. \tag{A.10}$$

Similarly, one can show that $E_w(a_{ig}\, a_{jg}) = 0$ if $i \neq j$, and $E_w(a_{ig}\, a_{ih}) = 0\; (g \neq h)$. Finally, we have that

$$\begin{aligned} E_w(a_{ig}^2) &= \frac{2^{2m-2}}{2^{2m-1}} \sum_{i=1}^{m} \sum_{g=1}^{m} a_{ig}^2 + \frac{2^{2m-2}}{2^{2m-1} m^2} \sum_{i=1}^{m} \sum_{g=1}^{m} (-a_{ig})^2 \\ &= \frac{1}{m^2} \left(\sum_{i=1}^{m} \sum_{g=1}^{m} a_{ig}^2 \right) \\ &= \frac{1}{m}. \end{aligned} \tag{A.11}$$

Hence, the procedure (3.20a) generates a class {**A**} satisfying (3.7) and (3.20).

ACKNOWLEDGEMENT

This research was supported by the National Science Foundation under contract GP-5459. Reproduction in whole or in part is permitted for any purpose of the United States Government.

REFERENCES

ANDERSON, R. L. (1942), "Distribution of the serial correlation coefficient", *Ann. math. Statist.*, **13**, 1–13.

BOX, G. E. P. (1952), "Multi-factor designs of first order", *Biometrika*, **39**, 49–57.

BOX, G. E. P. and ANDERSEN, S. L. (1955), "Permutation theory in the derivation of robust criteria and the study of departures from assumptions", *J. R. statist. Soc.* B, **17**, 1–34.

BOX, G. E. P. and HUNTER, J. S. (1957), "Multifactor experimental designs for exploring response surfaces", *Ann. math. Statist.*, **28**, 195–241.

COX, D. R. (1956), "A note on weighted randomization", *Ann. math. Statist.*, **27**, 1144–1150.

FISHER, R. A. (1935), *The Design and Analysis of Experiments.* Edinburgh: Oliver & Boyd.

KEMPTHORNE, O. (1955), "The randomization theory of experimental inference", *J. Amer. statist. Ass.*, **50**, 946–967.

KEMPTHORNE, O., ZYSKIND, G. and SUTTER, G. J. (1963), *Some Aspects of Constrained Randomization.* Aeronautical Research Laboratories, Wright-Patterson Air Force Base.

PITMAN, E. J. G. (1937), "Significance tests which may be applied to samples from any populations: III. The analysis of variance test", *Biometrika*, **29**, 322–335.

WELCH, B. L. (1937), "On the z-test in randomized blocks and Latin squares", *Biometrika*, **29**, 21–52.

WILK, M. B. (1955), "Linear models and randomized experiments", Ph.D. thesis, Iowa State University.

WILK, M. B. and KEMPTHORNE, O. (1955), "Fixed, mixed, and random models in the analysis of variance", *J. Amer. statist. Ass.*, **50**, 1144–1167.

YOUDEN, W. J. (1956), *Randomization and Experimentation.* National Bureau of Standards Technical Report.

Books and Articles Written by Box

Books

1. *Statistical Methods in Research and Production* (with W. R. Cousins, O. L. Davies, F. R. Himsworth, H. Kenney, M. Milbourn, W. Spendley, and W. L. Stevens). Edinburgh: Oliver and Boyd, 1963.
2. *Design and Analysis of Industrial Experiments* (with L. R. Connor, W. R. Cousins, O. L. Davies, F. R. Himsworth, and G. P. Sillitto). Edinburgh: Oliver and Boyd, 1963.
3. *Evolutionary Operation—A Statistical Method for Process Improvement* (with N. R. Draper). New York: John Wiley & Sons, 1969.
4. *Time Series Analysis Forecasting and Control,* 2nd ed. (with G. M. Jenkins). Oakland, Calif: Holden-Day, 1970.
5. *Bayesian Inference in Statistical Analysis* (with G. C. Tiao). Reading, Mass.: Addison-Wesley, 1973.
6. *Statistics for Experimenters* (with W. G. Hunter and J. S. Hunter). New York: John Wiley & Sons, 1977.

Articles

1. "The effect of exposure to sub-lethal doses of phosgene on the subsequent L(*ct*)50 for rats and mice" (with H. Cullumbine). *British Journal of Pharmacology and Chemotherapy,* vol. 2, no. 1 (1947), pp. 38–55.
2. "The relationship between survival time and dosage with certain toxic agents" (with H. Cullumbine). *British Journal of Pharmacology and Chemotherapy,* vol. 2, no. 1 (1947), pp. 27–37.
3. "A general distribution theory for a class of likelihood criteria." *Biometrika,* vol. XXXVI, parts IXX and IV (1949), pp. 317–346.
4. "Problems in the analysis of growth and wear curves." *Biometrics,* vol. 6, no. 4. (1950), pp. 362–389.
5. "On the experimental attainment of optimum conditions" (with K. B. Wilson). *J. Roy. Stat. Soc.,* Series B, vol. XIII, no. 1 (1951), pp. 1–45.
6. "Multifactorial designs of first order." *Biometrika,* vol. 39 (1952), pp. 49–57.
7. "Plan statistique dans l'etude des methodes de l'analyse chimique." *The Analyst,* vol. 77 (1952), pp. 879–891; Proceedings of the International Congress of Analytical Chemistry (1952), pp. 323–355.
8. "Non-normality and tests on variances." *Biometrika,* vol. 40 (1953), pp. 318–335.
9. "Pigment strength testing with the automatic muller" (with M. T. Hobbs and P. North). *Journal of the Oil and Colour Chemists' Assoc.,* vol. XXXVI, no. 396 (1953), pp. 283–299.

10. "A note on regions for tests of kurtosis." *Biometrika,* vol. 40, parts 3 and 4 (1953), pp. 465–466.
11. "A statistical design for the efficient removal of trends occurring in a comparative experiment with an application in biological assay" (with W. A. Hay). *Biometrics,* vol. 9, no. 3 (1953), pp. 304–319.
12. "The exploration and exploitation of response surfaces: Some general considerations and examples." *Biometrics,* vol. 10, no. 1 (1954), pp.16–60.
13. "A confidence region for the solutions of a set of simultaneous equations with an application to experimental design" (with J. S. Hunter). *Biometrika,* vol. 41, parts 1 and 2 (1954), pp. 190–198.
14. "Some theorems on quadratic forms applied in the study of analysis of variance problems: I. Effect on inequality of variance in the one way classification." *Ann. Math. Stat.,* vol. 25, no. 2 (1954), pp. 290–302.
15. "Some theorems on quadratic forms applied in the study of analysis of variance problems: II. Effects on inequality of variance and of correlation between errors in the two way classification." *Ann. Math. Stat.,* vol. 25, no. 3 (1954), pp. 484–498.
16. "Mathematical statistics and rubber technology." *Imperial Chemical Industries Technical Publication* (1955).
17. "Permutation theory in the derivation of robust criteria and the study of departures from assumptions" (with S. L. Andersen). *J. Roy. Stat. Soc.,* Series B, vol. XVII, part 1 (1955), pp. 1–34.
18. "The exploration and exploitation of response surfaces: An example of the link between the fitted surface and the basic mechanism of the system" (with P. V. Youle). *Biometrics,* vol. 11, no. 3 (1955), pp. 287–323.
19. "Application of digital computers in the exploration of functional relationships" (with G. A. Coutie). *Proc. Inst. Elec. Engrs.,* vol. 103, part B, supplement no. 1 (1956), pp. 100–107.
20. "Evolutionary operation: A method for increasing industrial productivity." *Applied Statistics,* vol. VI, no. 2 (1957), pp. 3–23.
21. "Multifactor experimental designs for exploring response surfaces" (with J. S. Hunter). *Ann. Math. Stat.,* vol. 28, no. 1 (1957), pp. 195–241.
22. "Integration of techniques in process development." *Trans. 11th Annual Convention of American Society for Quality Control* (1957), pp. 687–702.
23. "Use of statistical methods in the elucidation of basic mechanisms." *Bull. Int. Inst. of Stat.,* Stockholm (1957), pp. 215–225.
24. "A note on the generation of random normal deviates" (with M. E. Muller). *Ann. Math. Stat.,* vol. 29, no. 2 (1958), pp. 610–613.
25. "Experimental designs for the exploration and exploitation of response surfaces" (with J. S. Hunter). In *Experimental Designs in Industry,* V. Chew (ed.). New York: John Wiley & Sons (1958), pp. 610–613.
26. "Discussion of the papers of Messrs. Satterthwaite and Budne." *Technometrics,* vol. 1, no. 2 (1959), pp. 174–180.
27. "Design of experiments in non-linear situations" (with H. L. Lucas). *Biometrika,* vol. 46, parts 1 and 2 (1959), pp. 77–90.
28. "A basis for the selection of a response surface design" (with N. R. Draper). *J. Am. Stat. Assoc.,* vol. 54 (1959), pp. 622–654.
29. "Condensed calculations for evolutionary operation programs" (with J. S. Hunter). *Technometrics,* vol. 1, no. 1 (1959), pp. 77–95.
30. "Some general considerations in process optimization." *J. Basic Engineering* (1960), pp. 113–119.

31. "Simplex-sum designs: A class of second order rotatable designs derivable from those of first order" (with D. W. Behnken). *Ann. Math. Stat.*, vol. 31, no. 4 (1960), pp. 838–864.
32. "Some new three level designs for the study of quantitative variables" (with D. W. Behnken). *Technometrics,* vol. 2, no. 4 (1960), pp. 455–475.
33. "Fitting empirical data." *Ann. N. Y. Acad. of Sciences,* vol. 86, no. 3 (1960), pp. 792–816.
34. "The effects of errors in the factor levels and experimental design." *Bull. International Stat. Inst.*, vol. XXXVII, part III (1961). Reprinted in *Technometrics,* vol. 5, no. 2 (1963), pp. 247–262.
35. "The theory of errors." *Encyclopedia Britannica* (1961).
36. "The 2^{k-p} fractional factorial designs, part I" (with J. S. Hunter). *Technometrics,* vol. 3, no. 3 (1961), pp. 311–351.
37. "The 2^{k-p} fractional factorial designs, part II" (with J. S. Hunter). *Technometrics,* vol. 3, no. 4 (1961), pp. 449–458.
38. "Adaptive optimization of continuous processes" (with J. Chanmugam). *I and EC Fundamentals,* vol. 1 (1962), pp. 2–16.
39. "A useful method for model-building" (with W. G. Hunter). *Technometrics,* vol. 4, no. 3 (1962), pp. 301–318.
40. "Robustness to non-normality regression tests" (with G. S. Watson). *Biometrika,* vol. 49, parts 1 and 2 (1962), pp. 93–106.
41. "Some statistical aspects of adaptive optimization and control" (with G. M. Jenkins). *J. Roy. Stat. Soc.*, Series B, vol. 24, no. 2 (1962), pp. 297–343.
42. "A further look at robustness via Bayes' Theorem" (with G. C. Tiao). *Biometrika,* vol. 49, parts 3 and 4 (1962), pp. 419–432.
43. "Transformation of the independent variables" (with P. W. Tidwell) *Technometrics,* vol. 4, no. 4 (1962), pp. 531–550.
44. "Further contributions to adaptive quality control: Simultaneous estimation of dynamics: Non-zero costs" (with G. M. Jenkins). *Proceedings of the International Statistical Institute* (1963), pp. 943–974.
45. "The choice of a second order rotatable design" (with N. R. Draper). *Biometrika,* vol. 50, parts 3 and 4 (1963), pp. 335–352.
46. "An analysis of transformations" (with D. R. Cox). *J. Roy. Stat. Soc.*, Series B, vol. 26, no. 2 (1964), pp. 211–252.
47. "A Bayesian approach to the importance of assumptions applied to the comparison of variances" (with G. C. Tiao). *Biometrika,* vol. 51, parts 1 and 2 (1964), pp. 153–167.
48. "A note on criterion robustness and inference robustness" (with G. C. Tiao). *Biometrika,* vol. 51, parts 1 and 2 (1964), pp. 169–173.
49. "Sequential design of experiments for non-linear models" (with W. G. Hunter). *Proceedings of IBM Scientific Computing Symposium on Statistics* (1963), pp. 113–137.
50. "A simple system of evolutionary operation subject to empirical feedback." *Technometrics,* vol. 8, no. 1 (1966), pp. 19–26.
51. "The experimental study of physical mechanisms" (with W. G. Hunter). *Technometrics,* vol. 7, no. 1 (1965), pp. 23–42.
52. "A change in level of a non-stationary time series" (with G. C. Tiao). *Biometrika,* vol. 52, parts 1 and 2 (1965), pp. 181–192.
53. "Mathematical models for adaptive control and optimization" (with G. M. Jenkins). *A. I. Ch. E. Joint Meetings* (1965), pp. 4:61–4:68.

54. "The Bayesian estimation of common parameters from several responses" (with N. R. Draper). *Biometrika,* vol. 52, parts 3 and 4 (1965), pp. 355–365.
55. "Multi-parameter problems from a Bayesian point of view" (with G. C. Tiao). *Ann. Math. Stat.,* vol. 36, no. 5 (1965), pp. 1468–1482.
56. "A note on augmented designs." *Technometrics,* vol. 8, no. 1 (1966), pp. 184–188.
57. "A discrete predictor controller applied to sinusoidal perturbation adaptive optimization" (with R. J. Altpeter and K. D. Kotnour). *Instrument Society of America Transactions,* vol. 5, no. 3 (1966), pp. 255–262.
58. "Use and abuse of regression." *Technometrics,* vol. 8, no. 4 (1966), pp. 625–629.
59. "Some aspects of randomization" (with I. Guttman). *J. Roy. Stat. Soc.,* Series B, vol. 28, no. 3 (1966), pp. 543–558.
60. "Discrimination among mechanistic models" (with W. J. Hill). *Technometrics,* vol. 9, no. 1 (1967), pp. 57–71.
61. "Models for forecasting seasonal and non-seasonal time series" (with G. M. Jenkins and D. W. Bacon). In *Spectral Analysis of Time Series,* B. Harris (ed.). New York: John Wiley & Sons (1967), pp. 271–311.
62. "Bayesian analysis of a three-component hierarchical design model" (with G. C. Tiao). *Biometrika,* vol. 54, parts 1 and 2 (1967), pp. 109–125.
63. "Experimental strategy." *Proceedings of 6th International Biometric Conference,* Sydney, Australia (1967).
64. "A Bayesian approach to some outlier problems" (with G. C. Tiao). *Biometrika,* vol. 55, no. 1 (1968), pp. 119–130.
65. "Bayesian estimation of means for the random effect model" (with G. C. Tiao). *J. Amer. Stat. Assoc.,* vol. 63 (1968), pp. 174–181.
66. "Bayesian approaches to some bothersome problems in data analysis." *Proceedings of Seventh Annual Phi Delta Kappa Symposium on Education Research* (1968), pp. 61–101.
67. "Discrete models for forecasting and control." *Encyclopedia of Linguistics, Information, and Control.* Oxford: Pergamon Press (1968), pp. 1–6.
68. "Experimental design: Response surfaces." *International Encyclopedia of the Social Sciences.* New York: Macmillan (1968), pp. 254–259.
69. "The future of department of statistics." Panel discussion chaired by J. W. Tukey in *The Future of Statistics,* D. G. Watts (ed.). New York: Academic Press (1968), pp. 103–137.
70. "Discrete models for feedback and feedforward control" (with G. M. Jenkins). In *The Future Of Statistics,* D. G. Watts (ed.). New York: Academic Press (1968), pp. 201–240.
71. "Some recent advances in forecasting and control" (with G. M. Jenkins). *Applied Statistics,* vol. 17, no. 2 (1968), pp. 91–109.
72. "Isn't my process too variable for EVOP?" (with N. R. Draper). *Technometrics,* vol. 10 (1969), pp. 439–444.
73. "The challenge of statistical computation." In *Statistical Computation* (1969). New York: Academic Press, pp. 3–10.
74. "Distributions of residual autocorrelations in autoregressive-integrated moving average time series models" (with D. A. Pierce). *J. Amer. Stat. Assoc.,* vol. 65, no. 332 (1970), pp. 1509–1526.
75. "Some comments on a paper of Coen, Gomme, and Kendall" (with P. Newbold). *J. Roy. Stat. Soc.,* Series A, vol. 134, no. 2 (1971), pp. 229–240.

76. "Statistical techniques for mechanistic modelling" (with W. G. Hunter). *Proceedings of Second International Symposium on Chemical Reaction Engineering,* Amsterdam (1972), pp. B 4.9–4.19.
77. "Partial autocorrelations from a Bayesian viewpoint and orthogonal parameterization" (with G. M. Jenkins and I. Guttman). *METRON,* vol. XXX — N.1–4, 31-XII (1972), pp. 87–112.
78. "Some problems associated with the analysis of multiresponse models" (with W. G. Hunter, J. Erjavec, and J. F. MacGregor). *Technometrics,* vol. 15, no. 1 (1973), pp. 33–51.
79. "Some comments on Bayes' estimators" (with G. C. Tiao). *The American Statistician,* vol. 27, no. 1 (1973), pp. 12–14.
80. "Some comments on a paper by Chatfield and Prothero and on a review by Kendall" (with G. M. Jenkins). *J. Roy. Stat. Soc.,* Series A, vol. 136, part 3 (1973), pp. 337–352.
81. "Some recent advances in forecasting and control, Part II" (with G. M. Jenkins and J. F. MacGregor). *Applied Statistics,* vol. 23, no. 2 (1974), pp. 158–179.
82. "Statistics and the environment." *J. Wash. Acad. Sci.,* vol. 64, no. 2 (1974), pp. 52–59.
83. "Correcting inhomogeneity of variance with power transformation weighting" (with W. J. Hill). *Technometrics,* vol. 16, no. 3 (1974), pp. 385–389.
84. "The analysis of closed-loop dynamic-stochastic systems" (with J. F. MacGregor). *Technometrics,* vol. 16, no. 3 (1974), pp. 391–398.
85. "Analysis of Los Angeles photochemical smog data: A statistical overview" (with G. C. Tiao and W. J. Hamming). *APCA Journal,* vol. 25, no. 3 (1975), pp. 260–268.
86. "Intervention analysis with applications to economic and environmental problems" (with G. C. Tiao). *J. Amer. Stat. Assoc.,* vol. 70, no. 349 (1975), pp. 70–79.
87. "Parameter estimation for dynamic-stochastic models using closed-loop operating data" (with J. F. MacGregor). Invited address, *Proceedings of Sixth Triennial World Congress of International Federation of Automatic Control,* Boston, Mass., August (1975). Reprinted *Technometrics,* vol. 18, no. 4 (1976), pp. 371–380.
88. "Robust designs" (with N. R. Draper). *Biometrika,* vol. 62, no. 2 (1975), pp. 347–352.
89. "A statistical analysis of the Los Angeles ambient carbon monoxide data 1955–1972" (with G. C. Tiao and W. J. Hamming). *APCA Journal,* vol. 25, no. 11 (1975), pp. 1129–1136.
90. "Some empirical models for the Los Angeles photochemical smog data" (with G. C. Tiao and M. S. Phadke). *APCA Journal,* vol. 26, no. 5 (1976), pp. 485–490.
91. "Comparison of forecasts and actuality" (with G. C. Tiao). *Applied Statistics,* vol. 25, no. 3 (1975), pp. 195–200.
92. "Science and statistics." *J. Amer. Stat. Assoc.,* vol. 71, no. 356 (1976), pp. 791–799.
93. "Comment on 'Strong inconsistency from uniform priors' by M. Stone" (with G. C. Tiao). *J. Amer. Stat. Assoc.,* vol. 71, no. 353 (1976), p. 122.
94. "Identification of dynamic regression (distributed lag) models connecting two time series" (with L. D. Haugh). *J. Amer. Stat. Assoc.,* vol. 72, no. 357 (1977), pp. 121–130.

95. "Empirical-mechanistic modeling of air pollution" (with M. S. Phadke and G. C. Tiao). *Proceedings of 4th Symposium on Statistics and the Environment* (1976), pp. 91–101.

96. "Analysis and modeling of seasonal time series" (with S. C. Hillmer and G. C. Tiao). *Proceedings of Conference on Seasonal Analysis of Economic Time Series* (1976), pp. 309–344.

97. "A canonical analysis of multiple time series" (with G. C. Tiao). *Biometrika,* vol. 64, no. 2 (1977), pp. 355–365.

98. "On a measure of lack of fit in time series models" (with G. Ljung). *Biometrika,* vol. 65, no. 2 (1978), pp. 297–303.

99. "Deterministic and forecast-adaptive time-dependent models" (with B. Abraham). *Applied Statistics,* vol. 27, no. 3 (1978), pp. 120–130.

100. "Applications of time series analysis" (with G. C. Tiao). *Contributions to Survey Sampling and Applied Statistics,* papers in honor of H. O. Hartley, H. A. David (ed.). New York: Academic Press (1978), pp. 203–219.

101. "Conditions for the optimality of exponential smoothing forecasts procedures" (with J. Ledolter). *Metrika,* vol. 25 (1978), pp. 77–93.

102. "Linear models and spurious observations" (with B. Abraham). *Applied Statistics,* vol. 27, no. 2 (1978), pp. 131–138.

103. "Sampling interval and feedback control" (with B. Abraham). *Technometrics,* vol. 21, no. 1 (1979), pp. 1–8.

104. "The likelihood function of stationary autoregressive-moving average models" (with G. Ljung). *Biometrika,* vol. 66, no. 2 (1979), pp. 265–270.

105. "Some problems of statistics and everyday life." *J. Amer. Stat. Assoc.,* vol. 74, no. 365 (1979), pp. 1–4.

106. "Robustness in the strategy of scientific model building." In *Robustness in Statistics.* New York: Academic Press (1979), pp. 229–236.

107. "Bayesian analysis of some outlier problems in time series" (with B. Abraham). *Biometrika,* vol. 66, no. 2 (1979), pp. 229–236.

108. "The variance function of the difference between two estimated responses" (with N. Draper). *J. Roy. Stat. Soc.,* Series B, vol. 42, no. 1 (1980), pp. 79–82.

109. "Sampling and Bayes' inference in scientific modelling and robustness." *J. Roy. Stat. Soc.,* Series A, vol. 143, part 4 (1980), pp. 383–430.

110. "Analysis of variance and autocorrelated errors" (with G. Ljung). *Scandinavian Journal of Statistics,* vol. 7, part 4 (1980), pp. 172–180.

111. "Sampling inference, Bayes' inference, and robustness in the advancement of learning." In *Bayesian Statistics,* J. M. Bernardo, M. H. DeGroot, D. V. Lindley, and A. F. M. Smith (eds.) (1980), pp. 366–381.

112. "Modelling multiple time series with applications" (with G. C. Tiao). *J. Amer. Stat. Assoc.,* vol. 76, no. 376 (1981), pp. 802–816.

113. "Measures of lack of fit for response surface designs and predictor variable transformations" (with N. R. Draper). *Technometrics,* vol. 24, no. 1 (1982), pp. 1–8.

114. "An analysis of transformations revisited, rebutted" (with D. R. Cox). *J. Amer. Stat. Assoc.,* vol. 77, no. 377 (1982), pp. 209–210.

115. "Choice of response surface design and alphabetic optimality." *Utilitas Mathematica,* vol. 21B (1982), pp. 11–55.

116. "An apology for ecumenism in statistics." In *Scientific Inference, Data Analysis, and Robustness,* G. E. P. Box, Tom Leonard, Chien-Fu Wu (eds). New York: Academic Press (1983), pp. 51–84.

117. "The importance of practice in the development of statistics." To appear in *Technometrics,* Feb. 1984.

118. "Anatomy of time series models." To appear in *Statistics: An Appreciation,* H. A. David and H. T. David (eds.). Ames, Iowa: Iowa State University Press.

119. "Constrained nonlinear least squares" (with H. Kanemasu). To appear in *Contributions to Experimental Design Linear Models, and Genetic Statistics.* Essays in honor of Oscar Kempthorne. New York: Marcel Dekker (1984).

120. "Gwilym Jenkins, experimental design and time series." (1983). Invited address, *First Catalan Symposium on Statistics,* with special time series sessions dedicated to the memory of G. M. Jenkins (1983).